Aims and Scope

Optimization has continued to expand in all directions at an astonishing rate. New algorithmic and theoretical techniques are continually developing and the diffusion into other disciplines is proceeding at a rapid pace, with a spot light on machine learning, artificial intelligence, and quantum computing. Our knowledge of all aspects of the field has grown even more profound. At the same time, one of the most striking trends in optimization is the constantly increasing emphasis on the interdisciplinary nature of the field. Optimization has been a basic tool in areas not limited to applied mathematics, engineering, medicine, economics, computer science, operations research, and other sciences.

The series **Springer Optimization and Its Applications (SOIA)** aims to publish state-of-the-art expository works (monographs, contributed volumes, textbooks, handbooks) that focus on theory, methods, and applications of optimization. Topics covered include, but are not limited to, nonlinear optimization, combinatorial optimization, continuous optimization, stochastic optimization, Bayesian optimization, optimal control, discrete optimization, multi-objective optimization, and more. New to the series portfolio include Works at the intersection of optimization and machine learning, artificial intelligence, and quantum computing.

Volumes from this series are indexed by Web of Science, zbMATH, Mathematical Reviews, and SCOPUS.

More information about this series at http://www.springer.com/series/7393

Springer Optimization and Its Applications

Volume 187

Jean-Pierre Corriou

Numerical Methods and Optimization

Theory and Practice for Engineers

Springer

Jean-Pierre Corriou
LRGP-CNRS-ENSIC
University of Lorraine
Nancy, France

ISSN 1931-6828 ISSN 1931-6836 (electronic)
Springer Optimization and Its Applications
ISBN 978-3-030-89368-2 ISBN 978-3-030-89366-8 (eBook)
https://doi.org/10.1007/978-3-030-89366-8

This Springer imprint is published by the registered company Springer Nature Switzerland AG
The registered company address is: Gewerbestrasse 11, 6330 Cham, Switzerland

Preface

There are many ways to study and use numerical analysis and optimization, and several texts deal with either numerical methods or optimization. It seems useful to us to gather numerical methods and optimization in the same work because an engineer often has to solve optimization problems that involve numerical aspects or vice versa numerical problems resorting to optimization. The objective has been to cover a maximum of numerical and optimization methods that are useful to the engineer.

Some books deal with numerical methods in general or mathematical methods in engineering, while others study a specific point, such as matrix calculation, algebraic-differential systems, ordinary differential equations, partial differential equations of a certain type with given methods such as the finite difference method, spectral methods, finite volume method, finite element method, or boundary element method.

In the same way, for optimization, general works are frequently cited, while there are books dealing with particular points such as conjugated gradients, linear programming, convex optimization, global optimization, dynamic optimization, dynamic programming, optimal control, and nonlinear programming. Even if many authors of books are cited in the references, it is not possible to list them all and I apologize to the uncited authors.

An engineer solving a physical problem very often has to resort to the use of numerical methods. These techniques are based on mathematics, which have been developed extensively in the pure numerical analysis literature. The goal of this book is not to present the underlying mathematical theory but rather to unveil different numerical methods available to the user. Nevertheless, it is necessary to remain aware of the conditions of using a particular algorithm, and it will often be appropriate to use more detailed fundamental references. In particular, a programmer must be attentive to problems of convergence and error and may abandon a particular method to the benefit of another one that is better adapted to the case.

Choices have been made on the optimization methods to be presented. For example, integer and mixed-integer programming are not addressed. Dynamic optimization is discussed in detail with respect to the pure control aspect in the book "Process Control—Theory and Applications" (Springer, 2nd edition, 2018) by the author. However, a large number of optimization methods are presented, as the analytical aspects up to the numerical methods adapted to different types of problems.

While it is possible for everyone to develop their own programs, it is important to be aware of the high-quality programs produced by specialists (e.g., GAMS, NAG, IMSL) and available in numerical libraries; many of them are freely available on the Internet (e.g., LAPACK, BLAS, EISPACK, LINPACK, ODEPACK, MINPACK) while also being of excellent quality. For partial differential equations with 2D and 3D applications in heat transfer, fluid mechanics, material resistance, and commercial software (e.g., Comsol, Fluent) as well as excellent free software (e.g., FreeFEM, OpenFOAM) is available. In computer-aided design based on finite element analysis, CATIA, Solidworks, and Ansys are often used for very complex applications. The already cited software (i.e., GAMS, NAG, IMSL) also contains codes for optimization. NEOS uses advanced optimization solvers to solve different types of problems.

A set of programs made in Matlab™ is available at the author's personal website "http://jp.corriou.free.fr." They deal with both numerical and optimization methods, and the objective is to decompose the functioning of most of the methods presented. They can be useful in the cases of relatively small problems though many of them do not have the robustness of programs that have been extensively tested and designed to meet the needs of the most demanding cases. In almost all cases, free software, such as Octave™ or Scilab™ can be used instead of Matlab. Programs made in Matlab can be easily converted for Octave and Scilab. For 2D and 3D finite element calculations, the free mesh software Gmsh©, the free simulation software Elmer©, and the free postprocessing software ParaView© were used.

I would like to mention one of my favorite comments to my students: "When you have to choose a method, you have two possibilities. Either you choose a simple method and in general the quality of the results is limited, possibly unsatisfactory, or you choose a method that requires more effort of understanding and reflection, but you will be rewarded with the precision of the results obtained." During my career, I have often been able to verify the validity of this assertion.

Nancy, France
August 2021

Jean-Pierre Corriou

Acknowledgments

This book is the result of some 20 years of teaching of numerical and optimization methods, as well as statistical methods and process control, at the Ecole Nationale Supérieure des Industries Chimiques, Nancy, France. Discussions with other teachers and researchers and contacts with students have helped to refine and enrich the content of this book. The numerous research applications have contributed to highlight the importance of the right choice of numerical and optimization methods in the simulation of physical problems.

My colleague and friend, Xavier Joulia, a professor at the ENSIACET, Toulouse, and a process simulation specialist, has kindly examined in detail the contents of the first edition of this book. His comments were valuable and I thank him very much. My colleague, Abderrazak Latifi, a professor at the ENSIC, Nancy, shared a number of thoughts, particularly on optimization, and I thank him for that.

I would also like to thank my family, my wife, Cécile, and her great patience, my children, Alain and Gilles, to whom I took a part of their time during my many years of research that inevitably spilled over on the family setting, my grandchildren, Alexandre, Anne, and Lucie, who will 1 day ensure our succession, in one form or another.

Contents

Nomenclature

Mathematical Variables

$\boldsymbol{A}$	Matrix
$\mathbb{C}$	Space of complex numbers
f, g, h	Functions
i, j, k, m, n	Integer numbers
H	Hamiltonian
$\boldsymbol{I}$	Identity matrix
J	Jacobian
$\mathcal{L}$	Lagrangian
$\mathbb{N}$	Space of natural numbers
P	Polynomial
$\mathbb{R}$	Space of real numbers
t	Time
u	Control variable
$\boldsymbol{V}$	Vector
x, y, z	Variables

Mathematical Greek Variables

α	Root of an equation
δ	Central difference or Dirac delta function
Δ	Forward difference
Γ	Boundary for curvilinear integration
∇	Backward difference
ϵ	Small value, threshold
λ	Eigenvalue or Lagrange multiplier
μ	Karush–Kuhn–Tucker parameter
σ	Singular value
ψ	Adjoint variable
Ω	Domain for integration

Mathematical Symbols

$\overline{\boldsymbol{A}}$	Complex conjugate of matrix $\boldsymbol{A}$
$\boldsymbol{A}^*$	Transpose complex conjugate of matrix $\boldsymbol{A}$
$\boldsymbol{A}^{-1}$	Inverse of matrix $\boldsymbol{A}$
$\boldsymbol{A}^T$	Transpose of matrix $\boldsymbol{A}$
$\det(\boldsymbol{A})$	Determinant of matrix $\boldsymbol{A}$
$\mathrm{diag}(\boldsymbol{A})$	Diagonal matrix of matrix $\boldsymbol{A}$
$\mathrm{adj}(\boldsymbol{A})$	Adjoint of matrix $\boldsymbol{A}$
$\mathrm{cond}(\boldsymbol{A})$	Condition number of matrix $\boldsymbol{A}$
$\boldsymbol{A}^{(i)}$	Matrix $\boldsymbol{A}^{(i)}$ at iteration i
$\frac{df}{dx}$	Derivative of f with respect to x
$\frac{\partial f}{\partial x}$	Partial derivative of f with respect to x
$\int f(x)dx$	Integral of f with respect to x
$\oint f(s)ds$	Line or curvilinear integral of f
df	Total differential of f
$\prod_i x_i$	Product of all x_i
$\sum_i x_i$	Sum of all x_i
$\dot{f}$	First derivative of f
f'	First derivative of f
f''	Second derivative of f
$f^{(i)}$	ith derivative of f
f_x	Partial derivative of f with respect to x
$\mathrm{grad}(f)$	Gradient of f
$\mathrm{Im}(z)$	Imaginary part of complex number z
$\mathrm{Re}(z)$	Real part of complex number z
$\mathrm{lub}(\boldsymbol{A})$	Subordinate norm of $\boldsymbol{A}$
max	Maximum
min	Minimum
$\mathrm{sign}(x)$	Sign of x
x^n	nth power of x
$0(\Delta x^n)$	Error term of order n
$[a, b]$	Interval $[a, b]$
(u, v)	Scalar or dot product of u by v
$< u, v >$	Scalar or dot product of u by v
u_x	Partial derivative of u with respect to x
u_{xy}	Second partial derivative of u with respect to x and y
x^*	Solution of the extremum of a function with respect to x
$\bar{x}$	Mean of $\{x_1, \ldots, x_n\}$
$\tilde{y}$	Approximation or estimation of y
$\hat{y}$	Approximation or estimation of y
$x \rightarrow x_0$	x tends to x_0
$\lvert . \rvert$	Absolute value or modulus
$\lVert . \rVert$	Norm

Physical Variables (SI Units)

A	Surface area	$[m^2]$
Bi	Biot number	[-]
C	Concentration	$[mol\,m^{-3}]$
C or C_p	Heat capacity (at constant pressure)	$[J\,kg^{-1}\,K^{-1}]$
D	Diameter	[m]
$\mathcal{D}$	Diffusion coefficient	$[m^2\,s^{-1}]$
E	Young's modulus	[Pa]
F	Force	[N]
h	Heat transfer coefficient	$[W\,m^{-2}\,K^{-1}]$
L	Length	[m]
m	Mass	[kg]
$\dot{m}$	Mass flow rate	$[kg\,s^{-1}]$
Nu	Nusselt number	[-]
P	Pressure	[Pa]
q'	Thermal power (per length unit)	$[W\,m^{-1}]$
q''	Heat flux (thermal power per surface area)	$[W\,m^{-2}]$
q'''	Generated heat power (per volume unit)	$[W\,m^{-3}]$
r	Radius or distance	[m]
Re	Reynolds number	[-]
S	Surface area	$[m^2]$
T	Temperature	[K]
u, v	Velocity	$[m\,s^{-1}]$
V	Volume	$[m^3]$

Physical Greek Variables (SI Units)

α	Thermal diffusivity	$[m^2\,s^{-1}]$
ϵ	Strain	[-]
λ	Thermal conductivity	$[W\,m^{-1}\,K^{-1}]$
ϕ	Heat flux	$[W\,m^{-2}]$
ρ	Density	$[kg\,m^{-3}]$
σ	Tensile stress	[Pa]

Acronyms

BEM	Boundary element method
FDM	Finite difference method
FEM	Finite element method
FVM	Finite volume method
LQ	Linear quadratic
LU	Lower upper
MINLP	Mixed-integer nonlinear programming
NLP	Nonlinear programming
ODE	Ordinary differential equation
PDE	Partial differential equation

QP	Quadratic programming
SQP	Sequential quadratic programming
SVD	Singular value decomposition

Chapter 1
Interpolation and Approximation

1.1 Introduction

The application performed on a computer is based on a mathematical model, particularly in physical sciences. Supposing that the model is perfect (which is never the case), the main considered error is the calculation error, related to the use of a numerical method called *algorithm* allowing us to solve the problem. This error is due to the approximation of our mathematical model by a numerical method and is called *truncation error*.

The calculation being done on a computer, another type of error occurs, the *rounding error* related to the finite number of figures used for a number representation.

The issue of approximating a complex function or a group of numerical data by another function is called approximation and occupies an important place in many numerical and optimization studies.

1.2 Approximation of a Function by Another Function

Suppose that a function $f(x)$ has a form such that it is difficult to obtain. In this case, it is necessary to approximate it by another function $g(x)$ easier to get. Thus, the functions $\sin(x)$ or $\log(x)$ are approximated by an nth-order Taylor expansion, i.e. a Taylor polynomial of degree n. We thus get, for a sequence of points x_i, the corresponding values $f(x_i)$ in a table. The information on function $f(x)$ is fundamental. If the values of $f(x_i)$ are known with some precision, the function $g(x)$ will provide approximations of $f(x)$ with a lower precision

$$f(x) \doteq g(x) \tag{1.2.1}$$

J.-P. Corriou, *Numerical Methods and Optimization*, Springer Optimization and Its Applications 187, https://doi.org/10.1007/978-3-030-89366-8_1

1.2.1 Approximation Functions

The approximation functions are often present under the form of a linear combination of a class of given functions $g_i(x)$ for example as a Fourier series expansion

$$\begin{aligned} g(x) = a_0 + a_1 \cos x + a_2 \cos(2x) + \quad \dots \quad + a_n \cos(nx) + \\ b_1 \sin x + b_2 \sin(2x) + \quad \dots \quad + b_n \sin(nx) \end{aligned} \tag{1.2.2}$$

or an exponential series expansion

$$g(x) = a_0 \exp(b_0 x) + a_1 \exp(b_1 x) + \quad \dots \quad + a_n \exp(b_n x) \tag{1.2.3}$$

Among all expansions of this type, the simplest is the approximation polynomial which is a linear combination of monomials

$$g(x) = a_0 + a_1 x + \quad \dots \quad + a_n x^n \tag{1.2.4}$$

The advantage of the approximation polynomial is its easiness to be differentiated or integrated.

It is possible to approximate any continuous function on a given interval to any degree of precision by a polynomial of degree n according to *Weierstrass approximation theorem*:

$$\begin{aligned} &\forall\ f \text{ continuous on } [a, b],\ \forall \epsilon > 0, \exists \text{ a polynomial } P_n(x) \text{ of degree } n(\epsilon) \text{ such that} \\ &|f(x) - P_n(x)| < \epsilon\,, \quad a \le x \le b \end{aligned} \tag{1.2.5}$$

1.2.2 Polynomial Approximation

1.2.2.1 Interpolation Polynomial of Degree *n*

A criterion can be that, given $(n + 1)$ couples $(x_i, f(x_i))$, the *interpolation polynomial* passes through the $(n + 1)$ points

$$\forall i = 1, \dots, n + 1\,, \quad P_n(x_i) = f(x_i) \tag{1.2.6}$$

The approximation is not at all guaranteed for any value of x different from x_i. As the polynomial goes through all points, the term of *collocation* is also used.

1.2.2.2 Least Squares Polynomial

When the number n of values is large or when the function is known with inaccuracy in a reduced number of points, it may be interesting to do the *approximation* by a polynomial of degree m lower than n. The least squares approximation consists in searching the polynomial coefficients such that the sum of the squares of the errors is minimized

$$\min E = \sum_{i=0}^{n} [P_m(x_i) - f(x_i)]^2 \tag{1.2.7}$$

with

$$P_m(x) = \sum_{j=0}^{m} a_j x^j = a_0 + a_1 x + \cdots + a_m x^m \tag{1.2.8}$$

The coefficients a_j are found by minimizing the criterion E, hence the nullity of the gradient of E with respect to the coefficients

$$\frac{\partial E}{\partial a_0} = \cdots = \frac{\partial E}{\partial a_m} = 0 \tag{1.2.9}$$

which results in a system of $m + 1$ linear equations with respect to the parameters a_j.

These $(m + 1)$ equations can be written under the form

$$\begin{bmatrix} S_0 & S_1 & \ldots & S_m \\ S_1 & S_2 & \ldots & S_{m+1} \\ \vdots & \vdots & & \vdots \\ S_m & S_{m+1} & \ldots & S_{2m} \end{bmatrix} \begin{bmatrix} a_0 \\ a_1 \\ \vdots \\ a_m \end{bmatrix} = \begin{bmatrix} t_0 \\ t_1 \\ \vdots \\ t_m \end{bmatrix} \tag{1.2.10}$$

with

$$S_k = \sum_{i=0}^{n} x_i^k \,, \quad t_k = \sum_{i=0}^{n} x_i^k f(x_i) \tag{1.2.11}$$

that is

$$S\, a = t \quad \text{or} \quad a = S^{-1}\, t \tag{1.2.12}$$

When m is equal to 1, we thus get the least squares straight line.

1.2.2.3 Minimax Polynomial

The coefficients of the approximation polynomial $P_m(x)$ must be chosen so that the absolute value of the largest deviation $f(x_i) - P_m(x_i)$, be minimum

$$\min \left(\max_i |f(x_i) - P_m(x_i)| \right) \tag{1.2.13}$$

1.2.2.4 Series Expansion

If a function $f(x)$ is continuous and continuously differentiable on the interval $[x, x_0]$, its *Taylor series expansion* can be used

$$f(x) = f(x_0) + (x - x_0)f'(x_0) + \cdots + \frac{(x - x_0)^n}{n!} f^{(n)}(x_0) + \dots$$
$$= \sum_{i=0}^{\infty} \frac{(x - x_0)^i}{i!} f^{(i)}(x_0) \tag{1.2.14}$$

Note that we used the convention $0! = 1$.

In the case where $x_0 = 0$, this is called *Maclaurin series expansion.*

Very often, the *nth degree Taylor polynomial* or Taylor approximation of degree n is used. It is defined as the $(n + 1)$ first terms of the Taylor series expansion

$$\begin{aligned} f(x) &= f(x_0) + (x - x_0)f'(x_0) + \dots \\ &\quad + \frac{(x - x_0)^n}{n!} f^{(n)}(x_0) + \frac{(x - x_0)^{n+1}}{(n + 1)!} f^{(n+1)}(\xi) \quad \text{with } x_0 \le \xi \le x \\ &= f(x_0) + (x - x_0)f'(x_0) + \dots + \frac{(x - x_0)^n}{n!} f^{(n)}(x_0) + 0((x - x_0)^n) \end{aligned} \tag{1.2.15}$$

The *remainder* $(x - x_0)^{n+1}/(n + 1)!\, f^{(n+1)}(\xi)$ can be upper bounded as well as an evaluation of the error committed by truncating the Taylor series expansion at order n, hence the term of truncation error.

The Taylor series expansion is little used as such in the numerical practice, but the nth degree Taylor polynomial is often used as the reference when designing a new numerical method based on discretization.

1.2.2.5 Calculation of Polynomial $P_n(x)$

Horner's rule allows to calculate the polynomial $P_n(x)$ with a minimum of operations, n multiplications and n additions, according to

$$P_n(x) = a_0 + x(a_1 + x(a_2 + x(a_3) + x(a_4 + \cdots + x(a_{n-1} + xa_n))\dots)) \tag{1.2.16}$$

1.3 Determination of Interpolation Polynomials

1.3.1 Calculation of the Interpolation Polynomial

Given n points of coordinates $(x_i, f(x_i))$, there exists only one interpolation polynomial of degree lower than or equal to $n - 1$ passing through the n points

$$P_{n-1}(x_i) = f(x_i) \quad \forall i = 1, \dots, n \tag{1.3.1}$$

The coefficients a_i of the interpolation polynomial are solution of a linear system of n equations

$$\begin{aligned} a_0 + a_1x_1 + a_2x_1^2 + \cdots + a_{n-1}x_1^{n-1} &= f(x_1) \\ a_0 + a_1x_2 + a_2x_2^2 + \cdots + a_{n-1}x_2^{n-1} &= f(x_2) \\ \ldots &= \ldots \\ a_0 + a_1x_n + a_2x_n^2 + \cdots + a_{n-1}x_n^{n-1} &= f(x_n) \end{aligned} \tag{1.3.2}$$

The determinant of the matrix of coefficients of these equations is known as Vandermonde determinant

$$\begin{vmatrix} 1 & x_1 & x_1^2 & \ldots & x_1^{n-1} \\ 1 & x_2 & x_2^2 & \ldots & x_2^{n-1} \\ \vdots & \vdots & \vdots & \ddots & \vdots \\ 1 & x_n & x_n^2 & \ldots & x_n^{n-1} \end{vmatrix} \tag{1.3.3}$$

Recall that the approximation is not at all guaranteed for any value of x different from x_i.

1.3.2 Newton Interpolation Polynomial

From the definition of the derivative of a continuous function $f(x)$

$$\left[\frac{df(x)}{dx}\right]_{x_0} = f'(x_0) = \lim_{x \to x_0} \frac{f(x) - f(x_0)}{x - x_0} \tag{1.3.4}$$

we can define the first divided difference

$$f[x, x_0] = \frac{f(x) - f(x_0)}{x - x_0} \tag{1.3.5}$$

Indeed, this divided difference is the *ratio* or *quotient* of the *finite difference* $(f(x) - f(x_0))$ by *the finite difference* $(x - x_0)$. It is a rate of change of the function. Thus, it is a finite divided difference in general called *divided difference* that we maintain (Burden and Faires 2011; Gautschi 2012; Sauer 2012; Stoer and Bulirsch 1996). However, as soon as we deal with the discretization of ordinary differential equations and partial differential equations, with respect to the use of difference schemes in the *finite difference method*, frequently the term of *finite difference* is simply used even if it is a quotient (Allaire 2007; Sastry 2006; Epperson 2013).

From the mean value theorem (Figure 1.1), we know that:
$\forall f(x)$ continuous on $a \le x \le b$ and differentiable on $a \le x \le b, \exists\ \xi \in [a, b]$ such that

$$f'(\xi) = \frac{f(b) - f(a)}{b - a} \tag{1.3.6}$$

It results that the first divided difference is

$$f[x, x_0] = \frac{f(x) - f(x_0)}{x - x_0} = f'(\xi), \quad \xi \in [x, x_0] \tag{1.3.7}$$

The concept of divided difference can be generalized (Table 1.1).

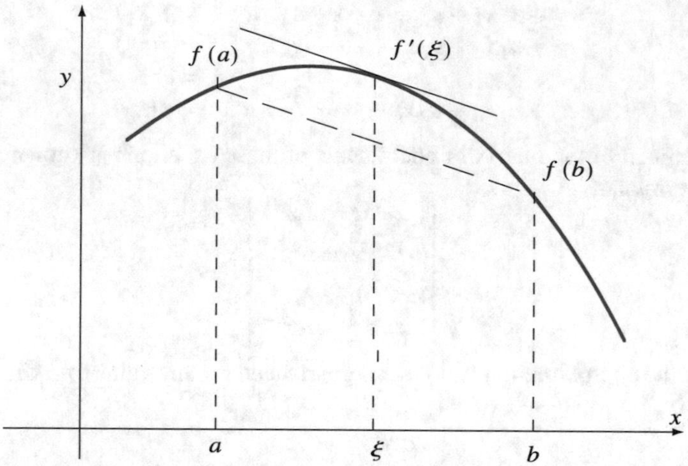

Fig. 1.1 Mean value theorem

Table 1.1 Divided differences

Order	Divided difference	Definition
0	$f[x_0]$	$f(x_0)$
1	$f[x_1, x_0]$	$\dfrac{f[x_1] - f[x_0]}{x_1 - x_0}$
2	$f[x_2, x_1, x_0]$	$\dfrac{f[x_2, x_1] - f[x_1, x_0]}{x_2 - x_0}$
n	$f[x_n, \ldots, x_1, x_0]$	$\dfrac{f[x_n, \ldots, x_2, x_1] - f[x_{n-1}, \ldots, x_1, x_0]}{x_n - x_0}$

It results that the nth divided difference is equal to

$$f[x_n, \ldots, x_1, x_0] = \sum_{i=0}^{n} \frac{f(x_i)}{\prod\limits_{\substack{j=0 \\ j \neq i}}^{n} (x_i - x_j)} \tag{1.3.8}$$

or

$$f[x_n, \ldots, x_1, x_0] = \frac{f(\xi)}{n!}, \quad \xi \in [x_0, x_1, \ldots, x_n] \tag{1.3.9}$$

The linear interpolation of function $f(x)$ at a point $x \in [x_0, x_1]$ will be done according to

$$f(x) \approx P_1(x) = f(x_0) + (x - x_0) f[x_1, x_0] = f[x_0] + (x - x_0) f[x_1, x_0] \tag{1.3.10}$$

Indeed, the interpolation thus defined is an approximation if $f(x)$ is not a linear function (Figure 1.2) and the error term $R_1(x)$ will be such that

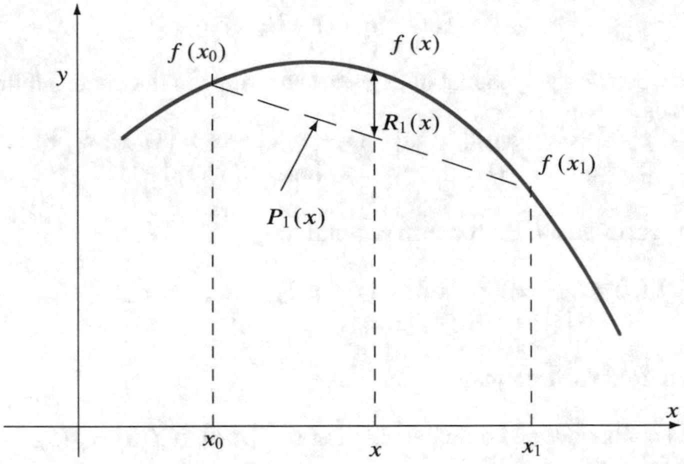

Fig. 1.2 Approximation of $f(x)$ by $P_1(x)$

$$f(x) = f[x_0] + (x - x_0)f[x_1, x_0] + R_1(x) = P_1(x) + R_1(x) \tag{1.3.11}$$

hence the error term which is zero at the limits x_0, x_1

$$R_1(x) = (x - x_0)(f[x, x_0] - f[x_1, x_0]) = (x - x_0)(x - x_1)f[x, x_1, x_0] \tag{1.3.12}$$

thus for the function $f(x)$

$$f(x) = f[x_0] + (x - x_0)f[x_1, x_0] + (x - x_0)(x - x_1)f[x, x_1, x_0] \tag{1.3.13}$$

The error term or *remainder* can be evaluated by introducing a known point $(x_2, f(x_2))$ if it is assumed that the function $f(x)$ does not change too rapidly on the interval containing x_0, x_1, x_2. In this case,

$$R_1(x) \approx (x - x_0)(x - x_1)f[x_2, x_1, x_0] \tag{1.3.14}$$

If the linear interpolation is insufficient, it is possible to add a curvature term with the second divided difference to evaluate the function $f(x)$ according to a second degree polynomial

$$f(x) \approx P_2(x) = f[x_0] + (x - x_0)f[x_1, x_0] + (x - x_0)(x - x_1)f[x_2, x_1, x_0] \tag{1.3.15}$$

while the exact expression of $f(x)$ is indeed

$$\begin{aligned} f(x) = {} & f[x_0] + (x - x_0)f[x_1, x_0] + (x - x_0)(x - x_1)f[x_2, x_1, x_0] + \\ & (x - x_0)(x - x_1)(x - x_2)f[x, x_2, x_1, x_0] \\ = {} & P_2(x) + R_2(x) \end{aligned} \tag{1.3.16}$$

The formula can be generalized at order n; the function $f(x)$ is the sum of an interpolation polynomial of degree n and a remainder

$$f(x) = P_n(x) + R_n(x) \tag{1.3.17}$$

with the interpolation polynomial of degree n according to the divided difference

$$\begin{aligned} P_n(x) = f[x_0] + (x - x_0)f[x_1, x_0] + (x - x_0)(x - x_1)f[x_2, x_1, x_0] + \quad \dots \quad + \\ (x - x_0)(x - x_1)\dots(x - x_{n-1})f[x_n, \dots, x_1, x_0] \end{aligned} \tag{1.3.18}$$

whereas the corresponding error term is equal to

$$\begin{aligned} R_n(x) &= (x - x_0)(x - x_1)\dots(x - x_n)f[x, x_n, \dots, x_1, x_0] \\ &= [\textstyle\prod_{i=0}^{n}(x - x_i)]f[x, x_n, \dots, x_1, x_0] \end{aligned} \tag{1.3.19}$$

According to Rolle's theorem:

$$\begin{aligned} &\forall f(x) \text{ continuous and differentiable for } x \in [a, b], \text{ if } f(a) = f(b), \\ &\exists\, \xi \in [a, b] \text{ such that } f'(\xi) = 0 \end{aligned} \tag{1.3.20}$$

Using iteratively this theorem and noting that the remainder $R_n(x)$ becomes zero at all points $x_0, x_1, \dots, x_n$, $R_n(x)$ is estimated

$$R_n(x) \doteq [\prod_{i=0}^{n}(x - x_i)]\frac{f^{(n+1)}(\xi)}{(n+1)!}, \quad \xi \in (x, x_n, \dots x_1, x_0) \tag{1.3.21}$$

which allows us to find an upper bound for the remainder if the function $f(x)$ is analytically known.

The interpolation polynomial $P_n(x)$ can also be seen under the form

$$P_n(x) = a_0 + a_1(x - x_0) + a_2(x - x_0)(x - x_1) + \cdots + a_n(x - x_0)\dots(x - x_{n-1}) \tag{1.3.22}$$

the coefficients a_i being evaluated according to the recursive scheme

$$\begin{aligned} f_0 &= P(x_0) = a_0 \\ f_1 &= P(x_1) = a_0 + a_1(x_1 - x_0) \\ f_2 &= P(x_2) = a_0 + a_1(x_2 - x_0) + a_2(x_2 - x_0)(x_2 - x_1) \\ &\vdots \end{aligned} \tag{1.3.23}$$

The divided differences can be calculated according to a triangular iterative scheme symbolized by Table 1.2. The column $k = 0$ corresponds to the ordinates of the function at points of abscissa x_i according to the relation $f[x_i] = f(x_i)$, the column $k = 1$ is simply deduced from column $k = 0$ by a formula of type

$$f[x_i, x_{i-1}] = \frac{f[x_i] - f[x_{i-1}]}{x_i - x_{i-1}} \tag{1.3.24}$$

the column $k = 2$ is simply deduced from column $k = 1$ by a formula of type

$$f[x_i, x_{i-1}, x_{i-2}] = \frac{f[x_i, x_{i-1}] - f[x_{i-1}, x_{i-2}]}{x_i - x_{i-2}} \tag{1.3.25}$$

and so on; the column i is deduced from column $i-1$ by a similar triangular relation.

Table 1.2 Divided differences, iterative triangular calculation

	$k=0$	$k=1$	$k=2$	$\dots$	$k=n$
x_0	$f[x_0]$				
		$f[x_1, x_0]$			
x_1	$f[x_1]$		$f[x_2, x_1, x_0]$		
				$\ddots$	
		$f[x_2, x_1]$			
x_2	$f[x_2]$		$\vdots$		$f[x_n, x_{n-1}, \dots, x_1, x_0]$
		$\vdots$		$\dots$	
$\vdots$	$\vdots$		$f[x_n, x_{n-1}, x_{n-2}]$		
		$f[x_n, x_{n-1}]$			
x_n	$f[x_n]$				

1.3.3 *Lagrange Interpolation Polynomial*

The Lagrange interpolation polynomial of degree 2 (see an example of Lagrange polynomial on Figure 1.3) is equal to

$$\begin{aligned} P_2(x) &= \frac{(x-x_1)(x-x_2)}{(x_0-x_1)(x_0-x_2)} f(x_0) + \frac{(x-x_0)(x-x_2)}{(x_1-x_0)(x_1-x_2)} f(x_1) \\ &+ \frac{(x-x_0)(x-x_1)}{(x_2-x_0)(x_2-x_1)} f(x_2) \\ &= \sum_{i=0}^{2} \left[\prod_{\substack{j=0 \\ j \neq i}}^{2} \frac{(x-x_j)}{(x_i-x_j)} \right] f(x_i) = \sum_{i=0}^{2} L_i(x) f(x_i) \end{aligned} \tag{1.3.26}$$

Generalizing at degree n, the Lagrange interpolation polynomial is written as

$$P_n(x) = \sum_{i=0}^{n} L_i(x) f(x_i) \tag{1.3.27}$$

where the factors $L_i(x)$ are equal to

$$L_i(x) = \prod_{\substack{j=0 \\ j \neq i}}^{n} \frac{(x-x_j)}{(x_i-x_j)} \tag{1.3.28}$$

As previously, the function $f(x)$ is equal to

$$f(x) = P_n(x) + R_n(x) \tag{1.3.29}$$

with the remainder $R_n(x)$ equal to

$$R_n(x) = [\prod_{i=0}^{n}(x - x_i)]f[x, x_n, \ldots, x_1, x_0] \tag{1.3.30}$$

$$R_n(x) = [\prod_{i=0}^{n}(x - x_i)]\frac{f^{(n+1)}(\xi)}{(n+1)!}, \quad \xi \in (x, x_n, \ldots x_1, x_0) \tag{1.3.31}$$

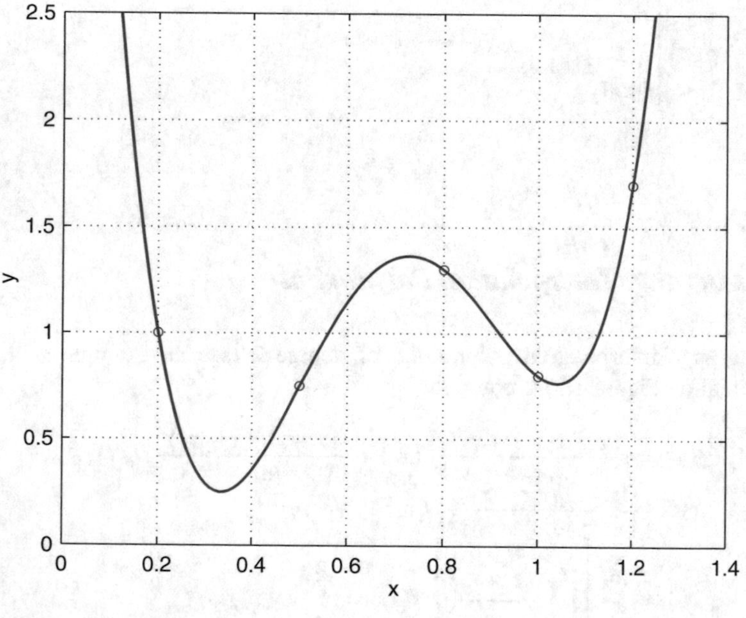

Fig. 1.3 Lagrange interpolation polynomial of degree 4 passing exactly through 5 given points

The Lagrange interpolation polynomial is determined by writing it under the form

$$\begin{aligned} P_n(x) = {} & a_0(x - x_1)(x - x_2)(x - x_3)\ldots(x - x_n) \\ & +a_1(x - x_0)(x - x_2)(x - x_3)\ldots(x - x_n) \\ & +a_2(x - x_0)(x - x_1)(x - x_3)\ldots(x - x_n) \\ & + \quad \ldots \\ & +a_i(x - x_0)(x - x_1)\ldots(x - x_{i-1})(x - x_{i+1})\ldots(x - x_n) \\ & + \quad \ldots \\ & +a_n(x - x_0)(x - x_1)(x - x_2)\ldots(x - x_{n-1}) \end{aligned} \tag{1.3.32}$$

where the coefficients a_i are equal to

$$a_i = \frac{f(x_i)}{(x_i - x_0)(x_i - x_1)\ldots(x_i - x_{i-1})(x_i - x_{i+1})\ldots(x_i - x_n)} \tag{1.3.33}$$

1.3.4 Polynomial Interpolation with Regularly Spaced Points

The spacing h is equal to

$$h = x_i - x_{i-1} \qquad \forall i = 1, \ldots, n \tag{1.3.34}$$

The interpolation polynomial can be expressed using a linear difference operator. According to the case, it will be the forward difference operator noted "Δ," backward difference operator noted "∇," central (or centered) difference operator noted "δ," as below:
Forward difference: $\Delta f(x) = f(x+h) - f(x)$
Backward difference: $\nabla f(x) = f(x) - f(x-h)$
Central difference: $\delta f(x) = f(x+\frac{h}{2}) - f(x-\frac{h}{2})$

1.3.4.1 Forward Differences

The forward difference operator is defined by

$$\Delta f(x) = f(x+h) - f(x) \tag{1.3.35}$$

$\Delta f(x)$ is called first forward difference (Figure 1.4).

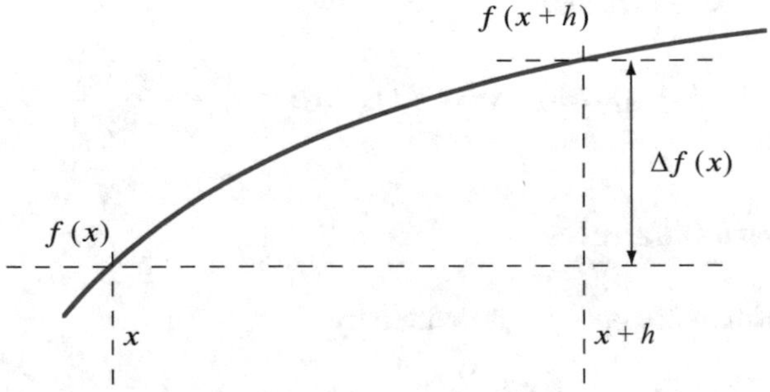

Fig. 1.4 Forward difference

It is possible to again use the operator, hence the second forward difference

$$\begin{aligned} \Delta^2 f(x) &= \Delta(\Delta f(x)) \\ &= \Delta(f(x+h) - f(x)) \\ &= \Delta f(x+h) - \Delta f(x) \\ &= f(x+2h) - 2f(x+h) + f(x) \end{aligned} \tag{1.3.36}$$

and so on until the nth forward difference

$$\Delta^n f(x) = \Delta^{n-1} f(x+h) - \Delta^{n-1} f(x) \tag{1.3.37}$$

The forward differences can be calculated and gathered in a table.

The correspondences between the forward differences and the divided differences are the following:

$$\begin{aligned} f[x_1, x_0] &= \frac{\Delta f(x_0)}{h} \\ &\vdots \\ f[x_n, \dots, x_1, x_0] &= \frac{\Delta^n f(x_0)}{n!\, h^n} \end{aligned} \tag{1.3.38}$$

Defining α such that

$$x = x_0 + \alpha h \qquad \text{with } x_0 \le x \le x_n \text{ and } 0 \le \alpha \le n \tag{1.3.39}$$

the fundamental Newton interpolation formula is

$$\begin{aligned} f(x_0 + \alpha h) = & f(x_0) + \alpha \Delta f(x_0) + \frac{\alpha(\alpha-1)}{2!}\Delta^2 f(x_0) + \quad \dots \quad + \\ & \frac{\alpha(\alpha-1)\dots(\alpha-n+1)}{n!}\Delta^n f(x_0) + R_n(x_0 + \alpha h) \\ & = P_n(x_0 + \alpha h) + R_n(x_0 + \alpha h) \end{aligned} \tag{1.3.40}$$

The residual R_n is equal to

$$R_n(x_0 + \alpha h) = h^{n+1}\alpha(\alpha-1)\dots(\alpha-n)\frac{f^{(n+1)}(\xi)}{(n+1)!} \quad \text{with } \xi \in (x, x_0, \dots, x_n) \tag{1.3.41}$$

an estimation of the remainder is

$$R_n(x_0 + \alpha h) \approx \alpha(\alpha-1)\dots(\alpha-n)\frac{\Delta^{n+1} f(x_0)}{(n+1)!} \tag{1.3.42}$$

1.3.4.2 Backward Differences

The backward difference operator is defined by

$$\nabla f(x) = f(x) - f(x-h) \tag{1.3.43}$$

The interpolation based on backward differences (Figure 1.5) is very similar to that based on forward differences. Thus, the function $f(x)$ is expressed as

$$\begin{aligned} f(x_0 + \alpha h) = & f(x_0) + \alpha \nabla f(x_0) + \frac{\alpha(\alpha+1)}{2!}\nabla^2 f(x_0) + \quad \dots \quad + \\ & \frac{\alpha(\alpha+1)\dots(\alpha+n-1)}{n!}\nabla^n f(x_0) + R_n(x_0 + \alpha h) \\ & = P_n(x_0 + \alpha h) + R_n(x_0 + \alpha h) \end{aligned} \tag{1.3.44}$$

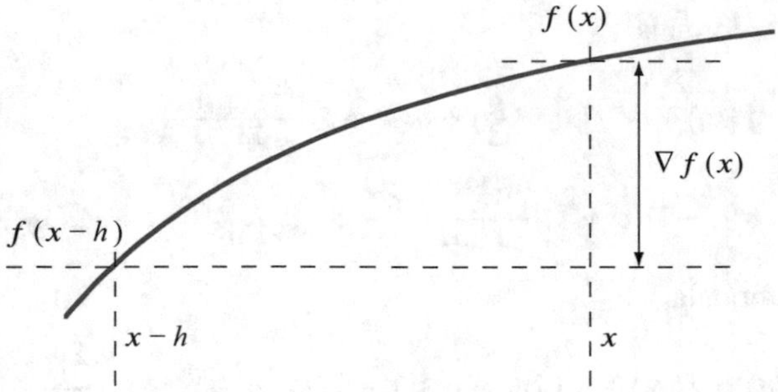

Fig. 1.5 Backward difference

1.3.4.3 Central Differences

The central difference operator is defined by

$$\delta f(x) = f(x + \frac{h}{2}) - f(x - \frac{h}{2}) \tag{1.3.45}$$

The interpolation error by using a first central difference (Figure 1.6) is of order 2 with respect to h, whereas it is of order 1 for forward or backward differences.

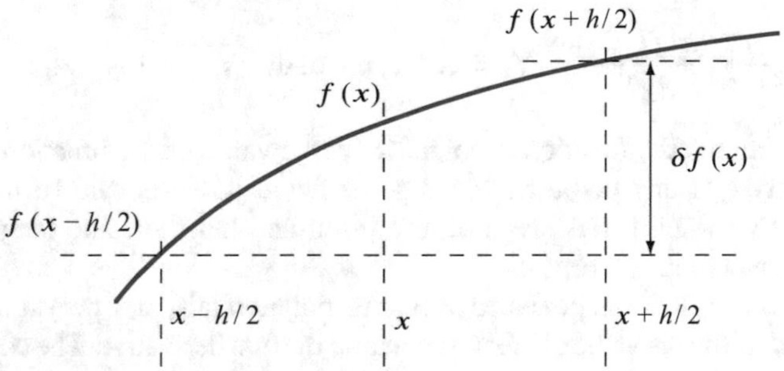

Fig. 1.6 Central difference

Thus, it will be possible to perform interpolations with polynomials of lower degree than for forward or backward differences without increasing the error.

The notations are more intricate than for forward or backward differences. Gauss formulas are
Forward Gauss formula

$$\begin{aligned} f(x_0 + \alpha h) = f(x_0) + \alpha \delta f(x_0 + \frac{h}{2}) + \alpha(\alpha - 1)\frac{\delta^2 f(x_0)}{2!} + \\ \alpha(\alpha - 1)(\alpha + 1)\frac{\delta^3 f(x_0 + \frac{h}{2})}{3!} + \quad \dots \quad + R_n(x_0 + \alpha h) \end{aligned} \tag{1.3.46}$$

Backward Gauss formula

$$f(x_0+\alpha h) = f(x_0) + \alpha\delta f(x_0 - \frac{h}{2}) + \alpha(\alpha+1)\frac{\delta^2 f(x_0)}{2!} + \alpha(\alpha-1)(\alpha+1)\frac{\delta^3 f(x_0 - \frac{h}{2})}{3!} + \quad \dots \quad + R_n(x_0+\alpha h) \tag{1.3.47}$$

Central Gauss formula

$$\begin{aligned} f(x_0+\alpha h) = {} & f(x_0) + \frac{\alpha}{2}[\delta f(x_0 - \frac{h}{2}) + \delta f(x_0 + \frac{h}{2})] + \alpha^2 \frac{\delta^2 f(x_0)}{2!} \\ & + \frac{\alpha(\alpha-1)(\alpha+1)}{2}\frac{\delta^3 f(x_0 - \frac{h}{2}) + \delta^3 f(x_0 + \frac{h}{2})}{3!} \\ & + \quad \dots \quad + R_n(x_0+\alpha h) \end{aligned} \tag{1.3.48}$$

1.3.5 Hermite Polynomials

Consider a set of $(n+1)$ points $(x_i, f(x_i))$ with $a \le x_0 \cdots \le x_n \le b$. f is a function of type C^m (function having continuous derivatives up to order m) with $m = \max(m_0, \dots, m_n)$ (each m_i is the value of m at x_i). The *osculatory* polynomial approximating the function f is the polynomial $P(x)$ of lower degree such that

$$\frac{d^k P(x_i)}{dx^k} = \frac{d^k f(x_i)}{dx^k} \qquad \forall i = 0, \dots, n \quad \text{and} \quad \forall k = 0, \dots, m_i \tag{1.3.49}$$

The polynomial and the function f coincide by the value of the function and its kth-order derivatives at any point x_i. More generally, a curve is said to be osculatory to another curve when it touches it at any point and has the same tangent line and curvature at this point.

The case $m_i = 1$ corresponds to Hermite polynomials; the polynomial and the function coincide by the value of the function and the first derivative. The corresponding Hermite polynomial of degree $(2n+1)$ is given by

$$H_{2n+1}(x) = \sum_{i=0}^{n} f(x_i) H_{n,i}(x) + \sum_{i=0}^{n} f'(x_i)\hat{H}_{n,i}(x) \tag{1.3.50}$$

with

$$H_{n,i}(x) = [1 - 2(x - x_i)L'_{n,i}(x_i)]L^2_{n,i}(x) \qquad \text{and} \quad \hat{H}_{n,i}(x) = (x - x_i)L^2_{n,i}(x) \tag{1.3.51}$$

where $L_{n,i}$ is the ith factor (L_i of Equation (1.3.28)) of the Lagrange polynomial of degree n.

It is possible to show that

$$f(x) = H_{2n+1}(x) + \frac{(x-x_0)^2 \dots (x-x_n)^2}{(2n+2)!} f^{(2n+2)}(\xi) \qquad \text{with } \xi \in [a,b] \tag{1.3.52}$$

if f is C^{2n+2} on $[a, b]$. For that purpose, the following properties are useful:

$$\begin{aligned} L_{n,i}(x_j) &= 0 \quad \text{if } i \neq j \\ L_{n,i}(x_i) &= 0 \end{aligned} \tag{1.3.53}$$

hence

$$\begin{aligned} H_{n,i}(x_j) &= 0 \quad \text{if } i \neq j \\ H_{n,i}(x_i) &= 1 \\ \hat{H}_{n,i}(x_j) &= 0 \quad \text{if } i \neq j \\ \hat{H}_{n,i}(x_i) &= 0 \end{aligned} \tag{1.3.54}$$

inducing the coincidence of the polynomial and the functions as well as their respective derivatives

$$H_{2n+1}(x_i) = f(x_i) \qquad \text{and } H'_{2n+1}(x_i) = f'(x_i) \tag{1.3.55}$$

resulting in Equation (1.3.52).

An interesting property of Hermite polynomials is that it is possible to determine them from the divided differences. First, we use Newton interpolation polynomial defined by

$$P_n(x) = f[x_0] + \sum_{i=1}^{n} f[x_0, \ldots, x_i](x - x_0) \ldots (x - x_{i-1}) \tag{1.3.56}$$

Defining

$$z_{2i} = z_{2i+1} = x_i \qquad \forall i = 0, \ldots, n \tag{1.3.57}$$

the divided difference results theoretically

$$f[z_{2i}, z_{2i+1}] = \frac{f[z_{2i}] - f[z_{2i+1}]}{z_{2i} - z_{2i+1}} \tag{1.3.58}$$

which is not suitable as $z_{2i} = z_{2i+1}$, but taking into account the definition of a derivative

$$f'(x_0) = \lim_{x \to x_0} \frac{f(x) - f(x_0)}{x - x_0} \tag{1.3.59}$$

we can write

$$f'(z_{2i}) = f'(z_{2i+1}) = f'(x_i) = f[z_{2i}, z_{2i+1}] \qquad \forall i = 0, \ldots, n \tag{1.3.60}$$

Thus, Hermite polynomial results with respect to divided differences

$$H_{2n+1}(x) = f[z_0] + \sum_{i=1}^{2n+1} f[z_0, \ldots, z_i](x - z_0) \ldots (x - z_{i-1}) \tag{1.3.61}$$

In particular, we will find an application of Hermite polynomials for the approximation of function in section 1.3.7 or differential equations in section 7.10.2.

1.3.6 Chebyshev Polynomials and Irregularly Spaced Points

Suppose that the variable x varies in $[-1, 1]$. In this case, the interpolation polynomial $P_n(x)$ of $f(x)$ defined as a sum of monomials produces a minimum error when x is close to 0 and maximum when $|x|$ is close to 1. Thus, the error is irregularly distributed on the interval $[0, 1]$. In the general case where $x \in [a, b]$, if we do a change of variable such that

$$t = \frac{2x - a - b}{b - a} \tag{1.3.62}$$

we are brought back to the previous case with $t \in [-1, 1]$.

It is interesting to find a series expansion of functions such that the error is better distributed and the maximum error is minimum.

The cosinus functions offer this type of advantage; their values are regularly spread on $[0, \pi]$. Moreover, the extreme values of two functions $\cos j\alpha$ and $\cos k\alpha$ with $j \neq k$ are in general reached for different values of α. The cosinus function is numerically evaluated by a series. Indeed, we can use the expression of $\cos n\alpha$ with respect to $x = \cos\alpha$. Thus Chebyshev polynomials $T_i(x) = \cos(i\alpha)$ are defined by

$$T_i(x) = \cos(i\alpha) = 2\cos(\alpha)\cos((i-1)\alpha) - \cos((i-2)\alpha)\,, \quad i = 0, 1, \ldots \tag{1.3.63}$$

giving the recurrence

$$T_i(x) = 2xT_{i-1}(x) - T_{i-2}(x) \tag{1.3.64}$$

In this way, we get

$$\begin{aligned} T_0(x) &= 1 \\ T_1(x) &= x \\ T_2(x) &= 2x^2 - 1 \\ T_3(x) &= 4x^3 - 3x \\ T_4(x) &= 8x^4 - 8x^2 + 1 \\ &\ldots \end{aligned} \tag{1.3.65}$$

and reciprocally the monomials $1, x, x^2, x^3, \ldots$ are simply expressed with respect to Chebyshev polynomials

$$\begin{aligned} 1 &= T_0 \\ x &= T_1 \\ x^2 &= \frac{1}{2}(T_0 + T_2) \\ x^3 &= \frac{1}{4}(3T_1 + T_3) \\ x^4 &= \frac{1}{8}(3T_0 + 4T_2 + T_4) \\ &\ldots \end{aligned} \tag{1.3.66}$$

The Chebyshev polynomials $T_n(x) = \cos(n\alpha)$ have $(n + 1)$ extrema of modulus 1, alternatively positive and negative on the interval $[-1, 1]$ (Figure 1.7). The n roots of $T_n(x)$ are real, in $[-1, 1]$ and given by

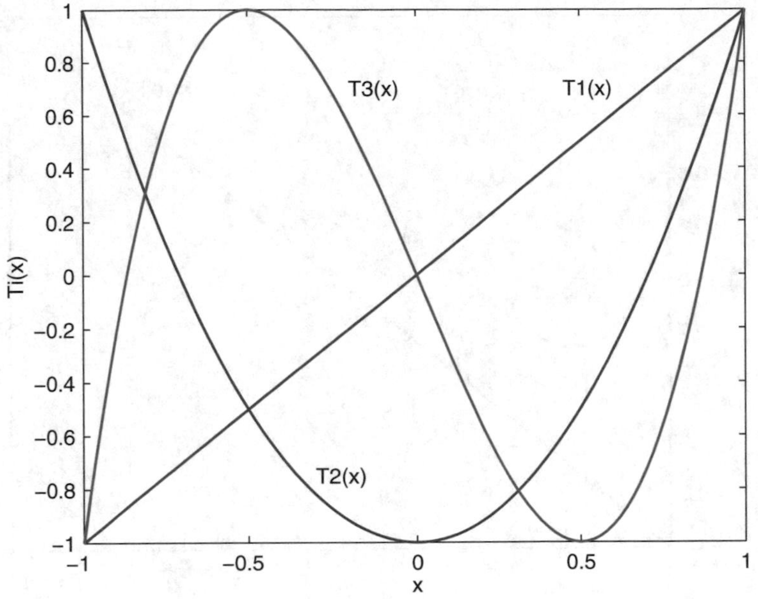

Fig. 1.7 Chebyshev polynomials

$$x_i = \cos\left[\frac{(2i-1)\pi}{2n}\right], \quad i = 1, 2, \ldots n \tag{1.3.67}$$

The coefficient of x^n in the polynomial $T_n(x)$ is equal to 2^{n-1}, and it can be shown that among all polynomials $P_n(x)$ having 1 as coefficient of x^n (such a polynomial is called *monic*), the monic Chebyshev polynomial

$$\phi_n(x) = \frac{T_n(x)}{2^{n-1}} \tag{1.3.68}$$

is the one which has, in absolute value, the lowest upper bound in the interval $[-1, 1]$. The monic Chebyshev polynomials are represented in Figure 1.8 where it can be noticed that the band surrounding each polynomial $\phi_n(x)$ has a decreasing amplitude as it is situated in $[-\frac{1}{2^{n-1}}, \frac{1}{2^{n-1}}]$.

Demonstration:

Suppose that a polynomial $P_n(x)$ exists which has the upper bound of its absolute value in the interval $[-1, 1]$ which would be lower than that of polynomial $\phi_n(x)$. We form the difference polynomial $D_n(x) = \phi_n(x) - P_n(x)$. As ϕ_n and P_n both have 1 as coefficient of x^n, $D_n(x)$ is a polynomial of degree $(n-1)$. The polynomial ϕ_n is zero and crosses the x axis n times in $[-1, 1]$; it possesses $(n+1)$ extrema $x_0, x_1, x_2, \ldots$. As at each extremum, P_n is lower in absolute value than ϕ_n, it implies that the difference $D_n(x)$ changes its sign for each extremum, sign of $D_n(x_{i-1}) = -$ sign of $D_n(x_i)$, which implies that $D_n(x)$ changes its sign n times in $[-1, 1]$. Thus, $D_n(x)$ would at least have

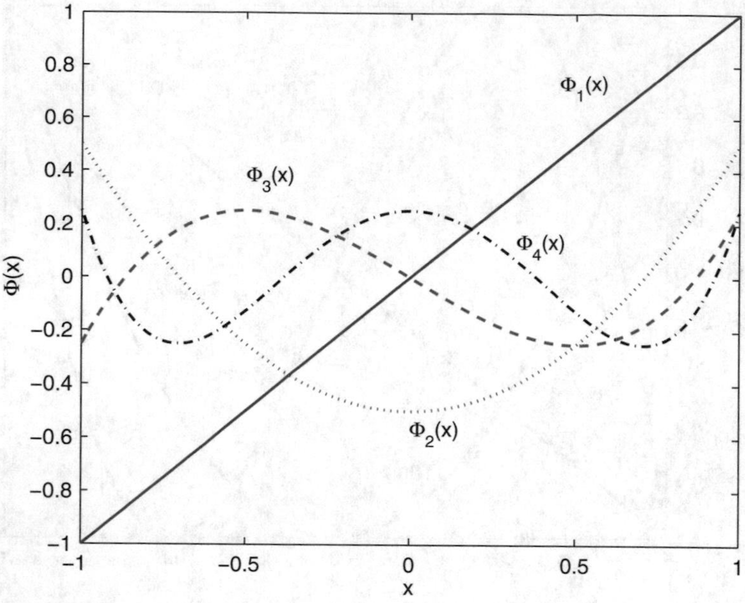

Fig. 1.8 Monic Chebyshev polynomials

n roots and would be of degree larger than or equal to n which contradicts the initial hypotheses.

1.3.6.1 Minimization of the Maximum Error

According to Chebyshev polynomial expansion, we know that ϕ_n presents the lowest maximum on the interval $[-1, 1]$. Thus, it will be possible to minimize the error expressed under the form of a polynomial of degree n on the interval $[-1, 1]$ by equaling it to $\phi_n(x)$.

For example, we can minimize the error term of the interpolation polynomial

$$R_n(x) = [\prod_{i=0}^{n}(x - x_i)]\frac{f^{(n+1)}(\xi)}{(n+1)!} \tag{1.3.69}$$

The term $f^{(n+1)}(\xi)$ can be considered as constant and we can simply minimize the polynomial of degree $(n + 1)$ which appears under the form of the product that we make equal to $\phi_{n+1}(x)$. Thus, the terms $(x - x_i)$ are simply the $(n + 1)$ factors of $\phi_{n+1}(x)$; it results that the roots of $\phi_{n+1}(x)$ are the roots of the corresponding Chebyshev polynomial, that is

$$x_i = \cos\left[\frac{(2i-1)\pi}{2n+2}\right], \quad i = 1, \ldots, n+1 \tag{1.3.70}$$

The previous reasoning was made on the interval $x \in [-1, 1]$; it can be generalized to any interval $z \in [a, b]$ by doing the change of variable

$$z = \frac{x(b-a) + (b+a)}{2} \tag{1.3.71}$$

The minimization thus performed of the maximum error constitutes the *minimax principle*.

Generalization:

In general, we desire to find a polynomial $P_n^*(x)$ of degree n which minimizes the maximum deviation with respect to a function $f(x)$ on the interval $[-1, 1]$. If the function $f(x)$ can be expanded with respect to Chebyshev polynomials (Fourier series expansion)

$$f(x) = \sum_{i=0}^{\infty} a_i T_i(x) \tag{1.3.72}$$

then the partial sum

$$P_n(x) = \sum_{i=0}^{n} a_i T_i(x) \tag{1.3.73}$$

constitutes a good approximation of $P_n^*(x)$, and $P_n(x)$ will nearly be minimax.

Example 1.1 :

Lagrange polynomials: interpolation on irregularly spaced points

Consider the function $y = \exp(-x^2)$ that we want to interpolate on the interval $[-2, 2]$. The function can be interpolated by a Lagrange interpolation polynomial of degree n by using $(n+1)$ points regularly spaced (left Figure 1.9).

It is possible to act differently by choosing as interpolation points the $(n+1)$ roots of Chebyshev polynomial $T_{n+1}(x)$ according to Equation (1.3.70) as

$$x_i = \frac{b+a}{2} + \frac{(b-a)}{2} \cos\left[\frac{(2i-1)\pi}{2n+2}\right], \quad i = 1, \ldots, n+1 \tag{1.3.74}$$

and perform the interpolation of the function by a Lagrange interpolation polynomial based on these irregularly spaced points (right Figure 1.9). The interpolation thus performed is better than the one based on the regularly spaced points, which appears clearly in Figure 1.9 at both extremities of the interval.

1.3.6.2 Chebyshev Economization

- The polynomial of degree n is written as

$$P_n(z) = \sum_{i=0}^{n} a_i z^i \tag{1.3.75}$$

  with $a \leq z \leq b$
- The previous polynomial must be transformed as

$$P_n(x) = \sum_{i=0}^{n} a_i^* x^i \tag{1.3.76}$$

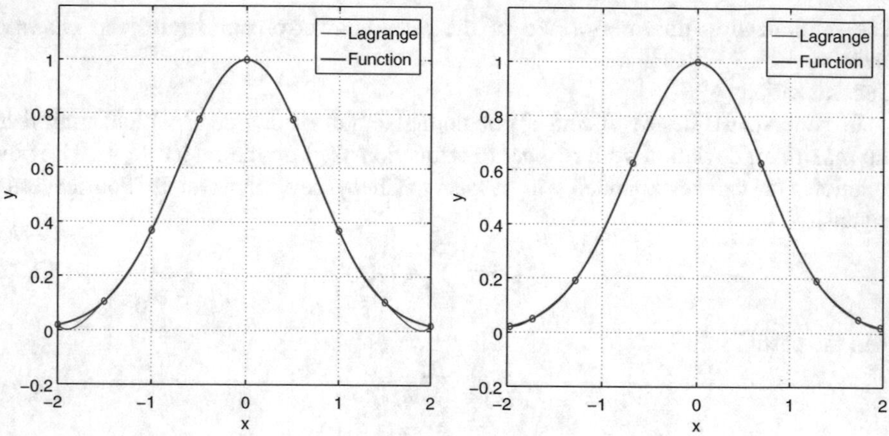

Fig. 1.9 Interpolation of a function by a Lagrange polynomial of degree n with $n+1$ regularly spaced points (left) and with irregularly spaced Chebyshev points (right)

with $-1 \leq x \leq 1$ and $x = \dfrac{2z - a - b}{b - a}$

- The polynomial $P_n(x)$ is expressed under the form of its expansion with respect to Chebyshev polynomials by using the tables

$$P_n(x) = \sum_{i=0}^{n} b_i T_i(x) \tag{1.3.77}$$

- The Chebyshev economization process is done by neglecting the terms whose contribution is lower than a given threshold ϵ. Thus, a truncated polynomial of degree m is obtained

$$P_n(x) = P_m(x) + E = \sum_{i=0}^{m} b_i T_i(x) + E \tag{1.3.78}$$

with the term corresponding to the truncation

$$E = \sum_{i=m+1}^{n} b_i T_i(x) \tag{1.3.79}$$

so that the maximum error is lower than ϵ

$$E_{max} = \sum_{i=m+1}^{n} |b_i| \leq \epsilon \tag{1.3.80}$$

- The economized polynomial

$$P_m(x) = \sum_{i=0}^{m} c_i^* x^i \tag{1.3.81}$$

is written by using the tables.

- Then, the optimized polynomial must be transformed with respect to the original variable z

$$P_m(z) = \sum_{i=0}^{m} c_i z^i \tag{1.3.82}$$

1.3.6.3 Runge Phenomenon

When the number $(n + 1)$ of points used for the interpolation of a function $f(x)$ increases, we can expect the convergence of the Lagrange interpolation polynomial $P_n(x)$ to increase. Difficulties can be highlighted when the function is nearly constant (plane function) or linear on some part of the domain of x and has a totally different behavior in the other parts of the domain.

First, examine the case of regularly spaced points. Consider on the interval $[-2, 2]$ the function $y = \exp(-10\,x^2)$ which tends rapidly to 0 outside a domain close to the origin. The polynomial very well approximates the function around the middle of the interpolation interval $[a, b]$ and thus there is convergence, but on the opposite a divergence occurs close to the endpoints of the interval (Figure 1.10) when the degree n of the interpolation polynomial increases. This is Runge phenomenon (Hairer 1993).

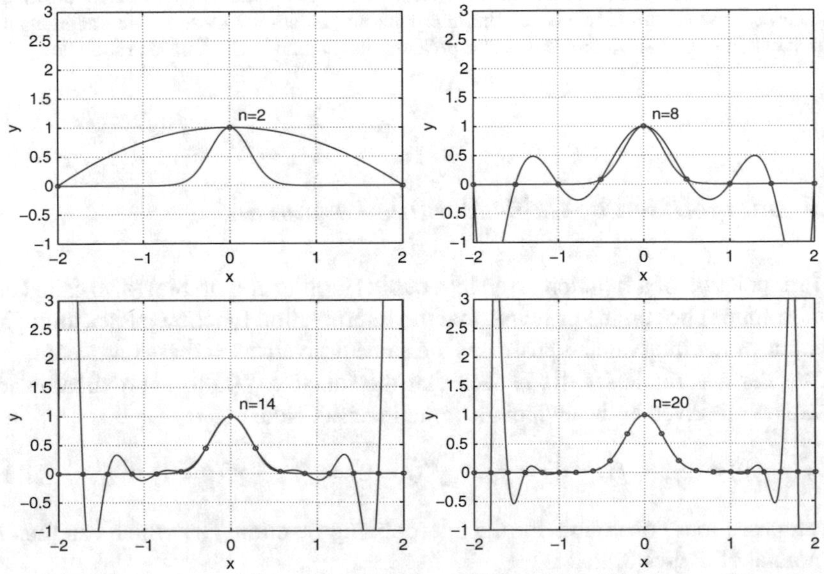

Fig. 1.10 Runge phenomenon and regularly spaced points: highlighting of the divergence between the function and the interpolation polynomial close to the endpoints of the interpolation interval. The degree of the interpolation polynomial is given in each subfigure

When irregularly spaced points are used for interpolation according to the equation of Chebyshev polynomial roots (Figure 1.11), the convergence is neatly improved. With respect to Figure 1.10, the same Lagrange interpolation polynomial was used, only the position of the interpolation points changed.

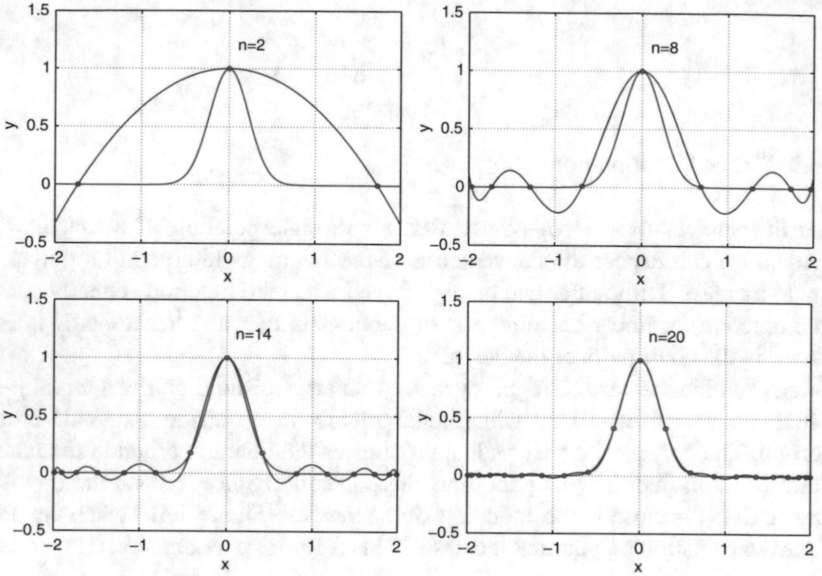

Fig. 1.11 Runge phenomenon and irregularly spaced Chebyshev points: highlighting of the improved convergence between the function and the interpolation polynomial close to the endpoints of the interpolation interval. The degree of the interpolation polynomial is given in each subfigure

1.3.7 Interpolation by Cubic Hermite Polynomial

The interpolation of a function given by a cubic Hermite polynomial (also called cubic Hermite spline) lies on the principle that the interpolation function passes through the endpoints of the interval and possesses the same derivatives at these endpoints.

Consider a variable $t \in [0, 1]$ such that the function $f(t)$ to approximate and its derivatives are given at the endpoints of the interval. Note

$$f(t=0) = f_0\,, \quad f(t=1) = f_1\,, \quad f'(t=0) = f'_0\,, \quad f'(t=1) = f'_1 \tag{1.3.83}$$

This imposes four constraints for the interpolating function $\tilde{f}(t)$ which can thus be a polynomial of degree 3, that is

$$\tilde{f}(t) = a_0 + a_1\,t + a_2\,t^2 + a_3\,t^3 \tag{1.3.84}$$

The respect of the constraints results in the following linear system:

$$
\begin{aligned}
&a_0 = f_0 \\
&a_0 + a_1 + a_2 + a_3 = f_1 \\
&a_1 = f_0' \\
&a_1 + 2\,a_2 + 3\,a_3 = f_1'
\end{aligned}
\tag{1.3.85}
$$

from which the values of the polynomial coefficients are drawn

$$
\begin{aligned}
&a_0 = f_0 \\
&a_1 = f_0' \\
&a_2 = -3\,f_0 + 3\,f_1 - 2\,f_0' - f_1' \\
&a_3 = 2\,f_0 - 2\,f_1 + f_0' + f_1'
\end{aligned}
\tag{1.3.86}
$$

It is possible to reorder the interpolating function under the form

$$
\tilde{f}(t) = f_0(1 - 3\,t^2 + 2\,t^3) + f_1(3\,t^2 - 2\,t^3) + f_0'(t - 2\,t^2 + t^3) + f_0'(-t^2 + t^3) \tag{1.3.87}
$$

which is noted as

$$
\tilde{f}(t) = f_0\,h_{00}(t) + f_1\,h_{01}(t) + f_0'\,h_{10}(t) + f_1'\,h_{11}(t) \tag{1.3.88}
$$

where the polynomials $h_{ij}(t)$ are Hermite basis polynomials defined by

$$
\begin{aligned}
&h_{00}(t) = 1 - 3\,t^2 + 2\,t^3 \\
&h_{01}(t) = 3\,t^2 - 2\,t^3 \\
&h_{10}(t) = t - 2\,t^2 + t^3 \\
&h_{11}(t) = -t^2 + t^3
\end{aligned}
\tag{1.3.89}
$$

which are displayed in Figure 1.12. The interpolation here defined on $[0, 1]$ of course can be extended to subintervals $[x_i, x_{i+1}]$ of any domain $[a, b]$. This type of interpolation by Hermite cubic polynomials is frequently used to simulate ordinary differential equations with boundary conditions (Section 6.6) or partial differential equations (Section 7.10.2).

In the case of an interval $[x_i, x_{i+1}]$, by using the linear relation

$$
t = \frac{x - x_i}{x_{i+1} - x_i} \tag{1.3.90}
$$

and the transformation

$$
g(x) = f(t) \quad \text{and} \quad g'(x) = f'(t)\frac{1}{x_{i+1} - x_i} \tag{1.3.91}
$$

we obtain the interpolation of a function $g(x)$ by an interpolating function $\tilde{g}(x)$

$$
\begin{aligned}
\tilde{g}(x) = {} & g(x_i)\,h_{00}\left(\frac{x - x_i}{x_{i+1} - x_i}\right) + g(x_{i+1})\,h_{01}\left(\frac{x - x_i}{x_{i+1} - x_i}\right) \\
& + g'(x_i)\,(x_{i+1} - x_i)\,h_{10}\left(\frac{x - x_i}{x_{i+1} - x_i}\right) \\
& + g'(x_{i+1})\,(x_{i+1} - x_i)\,h_{11}\left(\frac{x - x_i}{x_{i+1} - x_i}\right)
\end{aligned}
\tag{1.3.92}
$$

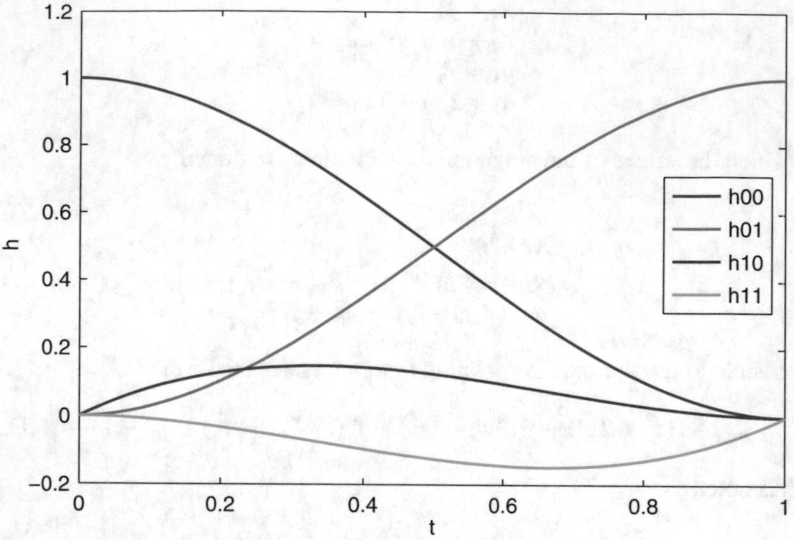

Fig. 1.12 Basis Hermite polynomials

with the basis polynomials h, of Equation (1.3.89), that can be calculated as

$$h(t) = h\left(\frac{x - x_i}{x_{i+1} - x_i}\right) \tag{1.3.93}$$

Noting $h = x_{i+1} - x_i$ and $s = x - x_i$, Equation (1.3.92) becomes

$$\begin{aligned}\tilde{g}(x) = g(x_i)\,\frac{h^3 - 3hs^2 + 2s^3}{h^3} + g(x_{i+1})\,\frac{3hs^2 - 2s^3}{h^3} \\ + g'(x_i)\,\frac{s(s-h)^2}{h^2} + g'(x_{i+1})\,\frac{s^2(s-h)}{h^2}\end{aligned} \tag{1.3.94}$$

When the tangents $g'(x)$ are not known at points (x_i, y_i), they can be estimated by the approximation m_i defined by

$$m_i = \frac{\Delta_{i-1} + \Delta_i}{2} \quad \text{with} \quad \Delta_i = \frac{y_{i+1} - y_i}{x_{i+1} - x_i} \tag{1.3.95}$$

i.e. the tangent at a point is approximated as the half-sum of the slopes of two segments situated on both sides of this point. Another possible approximation of the tangent is the harmonic mean as

$$\frac{1}{m_i} = \frac{1}{2}\left(\frac{1}{\Delta_{i-1}} + \frac{1}{\Delta_i}\right) \tag{1.3.96}$$

By using Equations (1.3.94) and (1.3.95), it is thus possible to approximate any function only known by the endpoints of the subintervals composing the domain by means of cubic Hermite polynomials.

The approximation by cubic Hermite polynomials (or Hermite splines) is local while that by the cubic splines (Section 1.3.8) is global, as a slight modification of the position of a point (x_i, y_i) influences only the calculation on the interval $[x_{i-1}, x_{i+1}]$ by cubic Hermite polynomials whereas it influences all the coefficients by the cubic spline of Section 1.3.8. The cubic splines of Section 1.3.8 are slightly more accurate than cubic Hermite polynomials. They take into account the second derivative, which is not done by cubic Hermite polynomials which are only of class C^1, i.e. continuously differentiable.

Example 1.2 :
Cubic Hermite polynomials: interpolation on subintervals

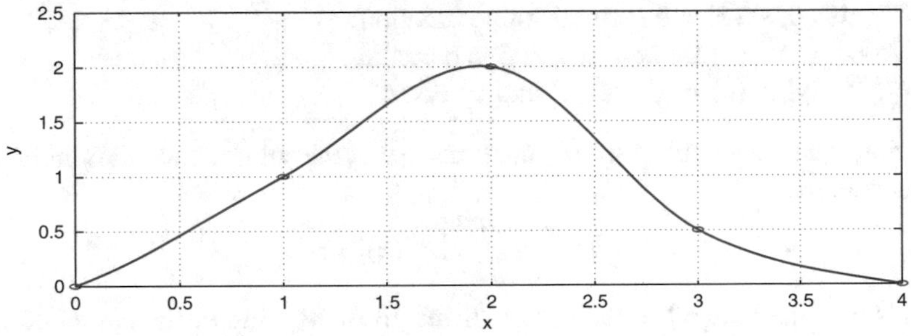

Fig. 1.13 Cubic Hermite polynomials: interpolation on subintervals

For the set of points of Table 1.3, we have plotted the interpolation on subintervals by cubic Hermite polynomials obtained by Equation (1.3.92) (Figure 1.13). The values of the derivatives were specified at the endpoints (at $x = 0$, $y' = 0.7$ and at $x = 4$, $y' = -0.3$). Other values could have been specified to be in a case similar to the natural spline. The values of the tangents at the internal points were estimated by Equation (1.3.95). The result can be compared to that of Figure 1.16 obtained by a cubic spline.

1.3.8 Interpolation by Spline Functions

The spline functions[1] are extremely appreciated for their quality of graphical interpolation (de Boor 1978). There exist several types of spline functions, B-splines, cubic splines, and exponential splines.

The spline functions are produced by associating to a partition $[a = x_1, x_2, \ldots, x_{n-1}, b = x_n]$ a set of piecewise polynomial functions, i.e. to each subinterval $[x_i, x_{i+1}]$ a different polynomial $P_i(x)$ is associated. Nevertheless, all polynomials will have the same degree. Moreover, it is possible to add a condition such that the interpolated functions f coincide at right with the polynomials $P_i(x_i) = f(x_i)$ except at b.

[1] A spline is a thin wood or metal strip used in building construction (Webster Dictionary).

In this section, only the cubic splines which are the most used are described. A spline function $S(x)$ is a real function defined on the partition $[a = x_1, x_2, \ldots, x_{n-1}, b = x_n]$ which possesses the following properties:

- S is twice continuously differentiable on $[a, b]$.
- S coincides on each subinterval $[x_i, x_{i+1}]$ of $[a, b]$ with a polynomial of degree 3.

Thus, a spline function is composed of pieces of cubic polynomials connected together so that, on one side, the first derivatives and, on another side, the second derivatives coincide at the nodes.

Frequently, one of the following conditions is added to ensure the unicity of the spline function S:

(a) $S''(a) = 0, S''(b) = 0$ (natural spline function).
(b) $S^{(k)}(a) = S^{(k)}(b)$ for $k = 0, 1, 2$; S is periodic.
(c) $S'(a) = y'_1, S'(b) = y'_n$ with y'_1 and y'_n given.

An important property of spline functions is that the spline function minimizes the following norm:

$$\|f\|^2 = \int_a^b |f''(x)|^2 dx \tag{1.3.97}$$

which is thus the integral of the square of the absolute value of the curvature on the considered interval. This integral represents the energy of the spline.

The spline function can be compared to the curve drawn by a designer who takes his French curve ruler to draw by eye the best curve passing through a set of points.

On each subinterval $[x_i, x_{i+1}]$, a function $S_i(x)$ is defined so that the function $S(x)$ can be considered as the family of functions $S_i(x)$ connected on the subintervals composing the interval $[a, b]$.

Determination of spline functions:

Let $\Delta = \{x_i, \quad i = 1, \ldots, n\}$ be a partition of the interval $[a, b]$ by the ordered nodes $a = x_1 < x_2 < \quad \ldots \quad < x_n = b$ and $Y = \{y_i, \quad i = 1, \ldots, n\}$ a set of n real numbers.

Let $h_i = x_{i+1} - x_i, \quad i = 1, \ldots, n-1$.

The values of the second derivatives at the nodes x_i are called moments

$$M_i = S''_i(x_i), \quad i = 1, \ldots, n \tag{1.3.98}$$

The spline function being a polynomial of degree 3, the second derivative is a linear function on the subinterval $[x_i, x_{i+1}]$. It results on that subinterval

$$S''_i(x) = M_i \frac{x_{i+1} - x}{h_i} + M_{i+1} \frac{x - x_i}{h_i}, \quad x \in [x_i, x_{i+1}], \quad i = 1, \ldots, n-1 \tag{1.3.99}$$

Then, we proceed to an integration, hence

$$S_i(x) = M_i \frac{(x_{i+1} - x)^3}{6h_i} + M_{i+1} \frac{(x - x_i)^3}{6h_i} + a_i(x - x_i) + b_i, \quad x \in [x_i, x_{i+1}] \tag{1.3.100}$$

- The values at the extremities are imposed, giving the relations

$$\begin{aligned} S_i(x_i) &= M_i \frac{h_i^2}{6} && +b_i = y_i\,, \quad i = 1, \ldots, n-1 \\ S_i(x_{i+1}) &= M_{i+1} \frac{h_i^2}{6} + a_i h_i && +b_i = y_{i+1}\,, \quad i = 1, \ldots, n-1 \end{aligned} \tag{1.3.101}$$

hence the values of the coefficients

$$\begin{aligned} a_i &= \frac{y_{i+1} - y_i}{h_i} - \frac{h_i}{6}(M_{i+1} - M_i) &&, \ i = 1, \ldots, n-1 \\ b_i &= y_i - M_i \frac{h_i^2}{6} &&, \ i = 1, \ldots, n-1 \end{aligned} \tag{1.3.102}$$

- The equality of the first left and right derivatives is imposed, that is

$$S'_{i-1}(x_i) = S'_i(x_i)\,, \quad i = 2, \ldots, n-1 \tag{1.3.103}$$

giving the relations

$$M_i \frac{h_{i-1}}{2} + a_{i-1} = -M_i \frac{h_i}{2} + a_i\,, \quad i = 2, \ldots, n-1 \tag{1.3.104}$$

In total, there are $3n - 2$ unknowns, n moments M_i, $n - 1$ coefficients a_i, and $n - 1$ coefficients b_i. We have $3n - 4$ equations by means of Equations (1.3.102) and (1.3.104). Thus, 2 equations are missing which are provided by the conditions at the extremities a and b. Therefore, one of the three previously mentioned conditions (a), (b), (c) will be exploited.

Indeed, according to Equations (1.3.102), the coefficients a_i and b_i are directly calculable from the moments M_i, thus there remains a problem of calculation of moments by replacing the coefficients a_i and b_i given by Equations (1.3.102) in Equation (1.3.104), which gives

$$\begin{aligned} &\frac{h_{i-1}}{h_{i-1} + h_i} M_{i-1} + 2\,M_i + \frac{h_i}{h_{i-1} + h_i} M_{i+1} = \frac{6}{h_{i-1} + h_i} \left(\frac{y_{i+1} - y_i}{h_i} - \frac{y_i - y_{i-1}}{h_{i-1}} \right) \\ &\quad \text{for} \quad i = 2, \ldots, n-1 \end{aligned} \tag{1.3.105}$$

which can be transformed under the form

$$\mu_i\, M_{i-1} + 2\,M_i + \lambda_i\, M_{i+1} = d_i\,, \quad i = 2, \ldots, n-1 \tag{1.3.106}$$

by posing the auxiliary variables

$$\mu_i = \frac{h_{i-1}}{h_{i-1} + h_i}\,, \quad \lambda_i = \frac{h_i}{h_{i-1} + h_i}\,, \quad d_i = \frac{6}{h_{i-1} + h_i} \left(\frac{y_{i+1} - y_i}{h_i} - \frac{y_i - y_{i-1}}{h_{i-1}} \right) \tag{1.3.107}$$

In addition to $n - 2$ equations (1.3.107), there remains to consider the conditions at the extremities:

- Case (a): $S''(a) = S''(b) = 0 \Longrightarrow M_1 = M_n = 0$.
- Case (b): $S^{(k)}(a) = S^{(k)}(b)$ for $k = 0, 1, 2$. It gives $M_1 = M_n$. Thus, there remain $n - 1$ unknowns to find, the moments M_1 to M_{n-1}.
  Moreover, the condition $S'(a) = S'(b)$ joined to $y_1 = y_n$ gives after transformation

$$2\,M_1+\frac{h_1}{h_1+h_{n-1}}\,M_2+\frac{h_{n-1}}{h_1+h_{n-1}}\,M_{n-1}=\frac{6}{h_1+h_{n-1}}\left(\frac{y_2-y_1}{h_1}-\frac{y_n-y_{n-1}}{h_{n-1}}\right) \tag{1.3.108}$$

Finally, Equation (1.3.105) for $i = n-1$, with $M_1 = M_n$, gives the relation

$$\begin{aligned}\frac{h_{n-1}}{h_{n-1}+h_{n-2}}\,M_1+\frac{h_{n-2}}{h_{n-1}+h_{n-2}}\,M_{n-2}+2\,M_{n-1}=\\ \frac{6}{h_{n-1}+h_{n-2}}\left(\frac{y_n-y_{n-1}}{h_{n-1}}-\frac{y_{n-1}-y_{n-2}}{h_{n-2}}\right)\end{aligned} \tag{1.3.109}$$

- Case (c): $S'(x_1) = y'_1$ known and $S'(x_n) = y'_n$ known. These are the two additional equations that give

$$\begin{aligned}S'_1(x_1)&=-M_1\,\frac{h_1}{2}+a_1=y'_1\\ S'_{n-1}(x_n)&=M_n\,\frac{h_{n-1}}{2}+a_{n-1}=y'_n\end{aligned} \tag{1.3.110}$$

hence

$$\begin{aligned}2\,M_1+M_2&=\frac{6}{h_1}\left(\frac{y_2-y_1}{h_1}-y'_1\right)\\ M_{n-1}+2\,M_n&=\frac{6}{h_{n-1}}\left(y'_n-\frac{y_n-y_{n-1}}{h_{n-1}}\right)\end{aligned} \tag{1.3.111}$$

In all cases, it amounts to solving a linear system.

- In the case (a), the moments are obtained by solving the following linear system:

$$\begin{bmatrix}2 & \lambda_2 & 0 & \dots & & 0\\ \mu_3 & 2 & \lambda_3 & 0 & \dots & \vdots\\ 0 & \mu_4 & 2 & \lambda_4 & \ddots & \\ \vdots & \ddots & \ddots & \ddots & \ddots & 0\\ & & 0 & \mu_{n-2} & 2 & \lambda_{n-2}\\ 0 & \dots & & 0 & \mu_{n-1} & 2\end{bmatrix}\begin{bmatrix}M_2\\ M_3\\ \vdots\\ \\ \\ M_{n-1}\end{bmatrix}=\begin{bmatrix}d_2\\ d_3\\ \vdots\\ \\ \\ d_{n-1}\end{bmatrix} \tag{1.3.112}$$

- In the case (b), the moments are obtained by solving the following linear system:

$$\begin{bmatrix}2 & \lambda_1 & 0 & \dots & 0 & \mu_1\\ \mu_2 & 2 & \lambda_2 & 0 & \dots & 0\\ 0 & \mu_3 & \ddots & \ddots & \ddots & \vdots\\ \vdots & \ddots & \ddots & \ddots & \ddots & 0\\ 0 & \ddots & 0 & \mu_{n-2} & 2 & \lambda_{n-2}\\ \lambda_{n-1} & 0 & \dots & 0 & \mu_{n-1} & 2\end{bmatrix}\begin{bmatrix}M_1\\ M_2\\ \vdots\\ \\ \\ M_{n-1}\end{bmatrix}=\begin{bmatrix}d_1\\ d_2\\ \vdots\\ \\ \\ d_{n-1}\end{bmatrix} \tag{1.3.113}$$

by adding to the variables defined by Equation (1.3.107) the following auxiliary variables:

$$\mu_1=\frac{h_{n-1}}{h_1+h_{n-1}},\quad \lambda_1=\frac{h_1}{h_1+h_{n-1}},\quad d_1=\frac{6}{h_1+h_{n-1}}\left(\frac{y_2-y_1}{h_1}-\frac{y_n-y_{n-1}}{h_{n-1}}\right) \tag{1.3.114}$$

- In the case (c), the moments are obtained by solving the following linear system:

$$\begin{bmatrix} 2 & \lambda_1 & 0 & \dots & & 0 \\ \mu_2 & 2 & \lambda_2 & 0 & \dots & \vdots \\ 0 & \mu_3 & 2 & \lambda_3 & \ddots & \\ \vdots & \ddots & \ddots & \ddots & \ddots & 0 \\ & & 0 & \mu_{n-1} & 2 & \lambda_{n-1} \\ 0 & \dots & & 0 & \mu_n & 2 \end{bmatrix} \begin{bmatrix} M_1 \\ M_2 \\ \vdots \\ \\ \\ M_n \end{bmatrix} = \begin{bmatrix} d_1 \\ d_2 \\ \vdots \\ \\ \\ d_n \end{bmatrix} \tag{1.3.115}$$

by adding to the variables defined by Equation (1.3.107) the following auxiliary variables:

$$\mu_n = 1\,, \quad \lambda_1 = 1\,, \quad d_1 = \frac{6}{h_1}\left(\frac{y_2 - y_1}{h_1} - y_1'\right), \quad d_n = \frac{6}{h_{n-1}}\left(y_n' - \frac{y_n - y_{n-1}}{h_{n-1}}\right) \tag{1.3.116}$$

It is possible to demonstrate that the previous square matrices are not singular.

Example 1.3 :
Interpolation by spline functions

For the set of points of Table 1.3, the three types of splines have been drawn (Figures 1.14, 1.15, and 1.16).

In the case (c), the values of the first derivatives had to be specified at the extremities (at $x = 0$, $y' = 0.7$ and at $x = 4$, $y' = -0.3$); these values were chosen arbitrarily. The two other cases (natural and periodic splines) require no specification. Different behaviors can be noticed close to the extremities a and b related to the conditions chosen at these points.

Table 1.3 Spline functions: experimental points for interpolation

x	0	1	2	3	4
y	0	1	2	0.5	0

1.3.9 Interpolation by Parametric Splines

It may occur that the points (x_i, y_i) of a curve cannot be represented by a function of the form $y = f(x)$. In this case, a parameter t must be used so that the curve is represented by

$$x = f(t)\,, \quad y = g(t) \tag{1.3.117}$$

It is called a parametric curve.

To build this parametric curve, Fortin (2008) proposes to define the parameter t_i such that

$$t_i = t_{i-1} + \|\overrightarrow{M_{i-1}M_i}\|_2 \qquad \forall\, i \geq 1 \qquad \text{with } t_0 = 1 \tag{1.3.118}$$

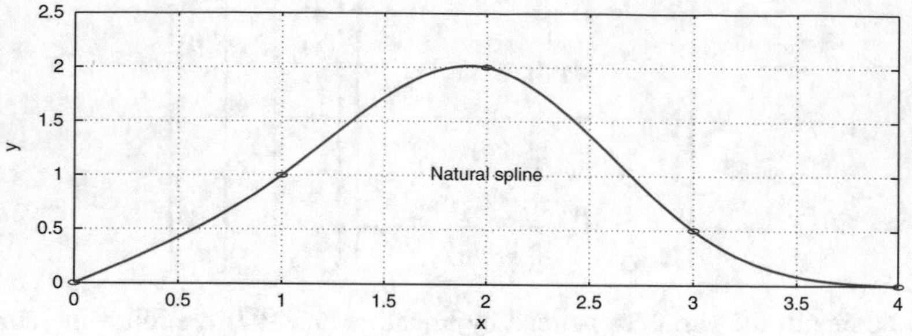

Fig. 1.14 Natural spline

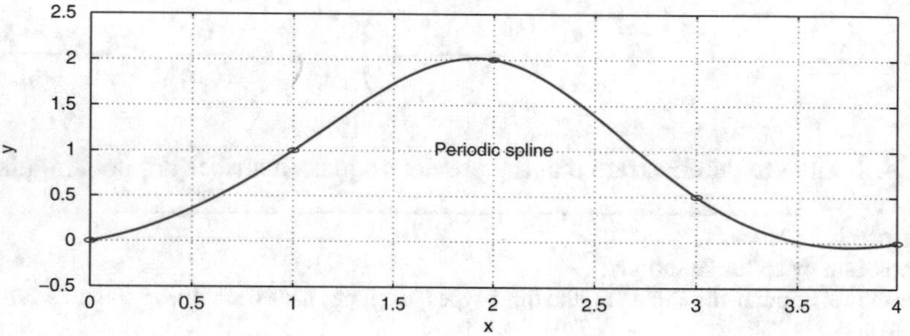

Fig. 1.15 Periodic spline

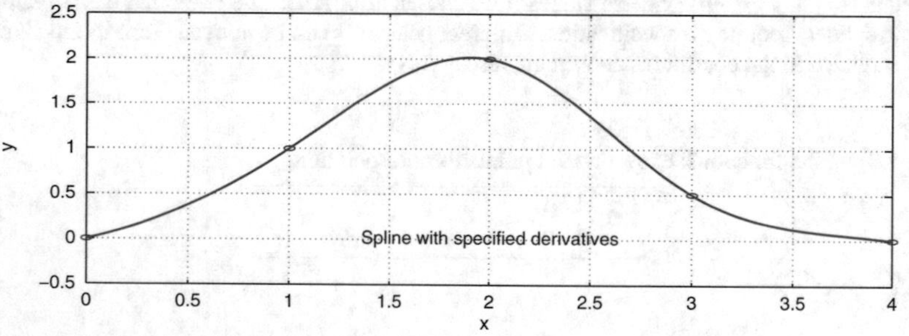

Fig. 1.16 Spline with derivatives specified at the extremities

Thus $\Delta t = t_i - t_{i-1}$ represents the length of the vector formed by two consecutive points. Then, two splines are determined, one passing through the points (t_i, x_i), the other one passing through the points (t_i, y_i), proceeding in the following way:

- Supposing that n points exist, $t_{max} = t_n$ is defined and the set $[t_0, \dots, t_n]$ is defined.
- The interval $[0, t_{max}]$ is divided in steps dt of small size to obtain a fine graphical representation. Let τ be the new variable $dt = \tau_i - \tau_{i-1}$.
- We calculate the splines passing on one hand through the points (t_i, x_i), on the other hand through the points (t_i, y_i), and the splines are evaluated at all points defined by τ_i.

Example 1.4 :
Interpolation by parametric spline functions

Let the cardioid of equation

$$\begin{aligned} x &= a\,\cos^2(\phi) + l\,\cos(\phi) \\ y &= a\,\cos(\phi)\,\sin(\phi) + l\,\sin(\phi) \end{aligned} \tag{1.3.119}$$

with the parameters $a = 1.5$, $l = 1$, $\phi \in [0, 2\pi)$. We consider 20 values of ϕ regularly spaced between 0 and 2π, which gives the couples of points (xc_i, yc_i) to represent the cardioid by means of parametric splines. Let (xc_i, yc_i) be the set issued from the corresponding parameters ϕ_i. We follow the procedure previously described, first the t_i are calculated, then two natural splines are calculated on the sets (t, xc) and (t, yc), and an interpolation is performed on these splines with the parameter τ (Figure 1.17). Then, we simply draw the interpolated set, which represents in a very satisfactory way the cardioid passing through the couples (xc, yc) (Figure 1.18).

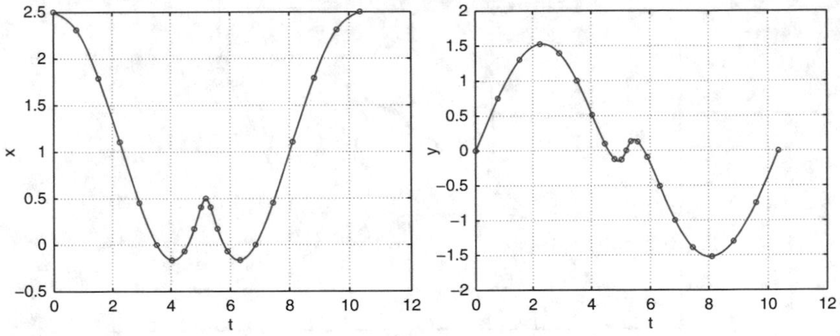

Fig. 1.17 Splines built on (t, xc) (left) and on (t, yc) (right)

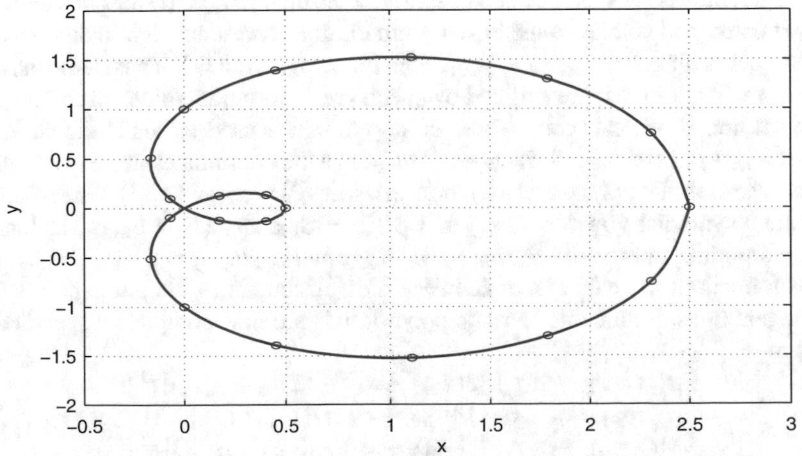

Fig. 1.18 Parametric spline: rebuilding of the cardioid with the points (xc, yc) (o)

1.4 Bézier Curves

When we desire to represent a set of points (x_i, y_i) by means of parametric curves, a possibility often used is Bézier curves. This technique consists in determining a couple of cubic Hermite polynomials depending on a parameter t for each couple of consecutive points, a polynomial for $x(t)$ and a polynomial for $y(t)$. Let (x_k, y_k) and (x_{k+1}, y_{k+1}) be two such points (Figure 1.19). Any point of the curve is written as $(x(t), y(t))$. Thus, $t = 0$ at the beginning point of the curve and $t = 1$ at the endpoint, so that $x_k = x(0)$ and $x_{k+1} = x(1)$, and similarly $y_k = y(0)$ and $y_{k+1} = y(1)$.

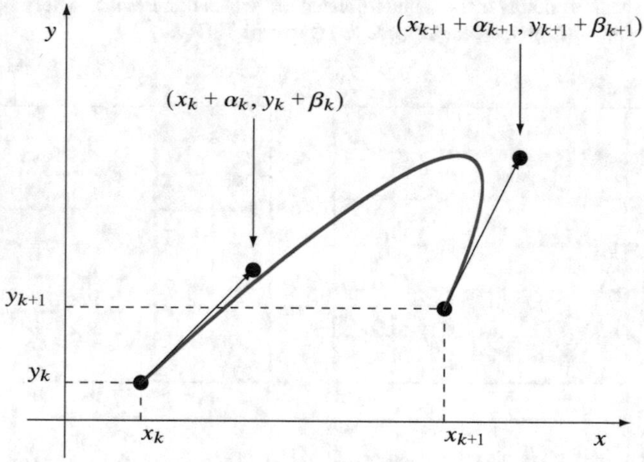

Fig. 1.19 Bézier curve: interpolation with guide-points

The derivatives are specified at both extremities, thus $y'(t)/x'(t)$ for $t = 0$ and $t = 1$. As two cubic polynomials must be determined, this represents eight unknowns with only six constraints $\{x_k, y_k, x_{k+1}, y_{k+1}, dy_k/dx_k, dy_{k+1}/dx_{k+1}\}$. Thus, there exist two degrees of freedom that are fulfilled by specifying two guide-points, each one along a tangent line at one extremity. These guide-points are used to "pull" the curve. Let $(x_k + \alpha_k, y_k + \beta_k)$ and $(x_{k+1} + \alpha_{k+1}, y_{k+1} + \beta_{k+1})$ be the coordinates of these two guide-points. The Hermite polynomial $x(t)$ must verify $x'(0) = \alpha_k$ and $x'(1) = \alpha_{k+1}$, and the Hermite polynomial $y(t)$ must also verify $y'(0) = \beta_k$ and $y'(1) = \beta_{k+1}$. The tangents at the extremities must verify $\beta_k/\alpha_k = dy_k/dx_k$ and $\beta_{k+1}/\alpha_{k+1} = dy_{k+1}/dx_{k+1}$, which leaves a freedom for, either α, or β, hence a displacement of the guide-points along the tangent lines. Both cubic Hermite polynomials are now completely specified and equal to

$$\begin{aligned} x(t) &= x_k + \alpha_k\, t + [3(x_{k+1} - x_k) - (2\,\alpha_k + \alpha_{k+1})]\, t^2 + \\ &\quad [2(x_k - x_{k+1}) + (\alpha_k + \alpha_{k+1})]\, t^3\,, \quad t \in [0, 1] \\ y(t) &= y_k + \beta_k\, t + [3(y_{k+1} - y_k) - (2\,\beta_k + \beta_{k+1})]\, t^2 + \\ &\quad [2(y_k - y_{k+1}) + (\beta_k + \beta_{k+1})]\, t^3 \end{aligned} \tag{1.4.1}$$

The form of the parametric equations for Bézier curves is very slightly different from the previous Hermite polynomials, as each term α or β is multiplied by a factor 3, but this is not a fundamental change.

Example 1.5 :
Bézier curves

Two examples are given to show how Bézier curves can be flexible. The extremities are (1, 1) and (3, 2). Three cases have been considered for each example (Tables 1.4 and 1.5).

Table 1.4 Bézier curves: first study

Case 1 $(dy/dx)_0 = 1$ $(dy/dx)_1 = -2$				Case 2 $(dy/dx)_0 = -1$ $(dy/dx)_1 = -2$				Case 3 $(dy/dx)_0 = 1$ $(dy/dx)_1 = 2$			
α_0	β_0	α_1	β_1	α_0	β_0	α_1	β_1	α_0	β_0	α_1	β_1
1	1	1	−2	1	−1	1	−2	1	−1	1	2
3	3	3	−6	3	−3	3	−6	3	−3	3	6
6	6	6	−12	6	−6	6	−12	6	−6	6	12
10	10	10	−20	10	−10	10	−20	10	−10	10	20

For each case, three sets of parameters have been used. The slope specified at x_1 is different between the case 1 and cases 2, 3. The form of the parametric equations is (1.4.1).

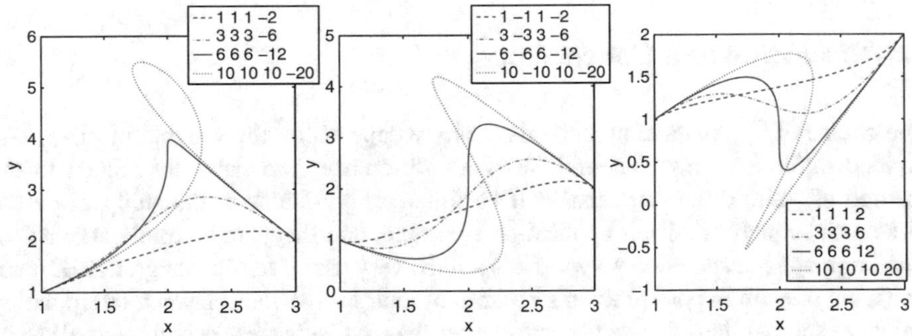

Fig. 1.20 Bézier curves in the three cases of Table 1.4. The values indicated in the legend correspond to $\{\alpha_0, \beta_0, \alpha_1, \beta_1\}$

Figures 1.20 in the three cases of Table 1.4 show both a great diversity and flexibility.

The different choice $\alpha_0 = -\alpha_1$ produces different curves while still respecting the tangent lines at the origin, because of the modification of the guide-points. This is demonstrated in Figures 1.21 which correspond to Table 1.5.

Table 1.5 Bézier curves: second study

Case 1 $(dy/dx)_0 = 1$ $(dy/dx)_1 = -2$				Case 2 $(dy/dx)_0 = -1$ $(dy/dx)_1 = -2$				Case 3 $(dy/dx)_0 = 1$ $(dy/dx)_1 = 2$			
α_0	β_0	α_1	β_1	α_0	β_0	α_1	β_1	α_0	β_0	α_1	β_1
1	1	−1	2	1	−1	−1	2	1	1	−1	−2
3	3	−3	6	3	−3	−3	6	3	3	−3	−6
6	6	−6	12	6	−6	−6	12	6	6	−6	−12
10	10	−10	20	10	−10	−10	20	10	10	−10	−20

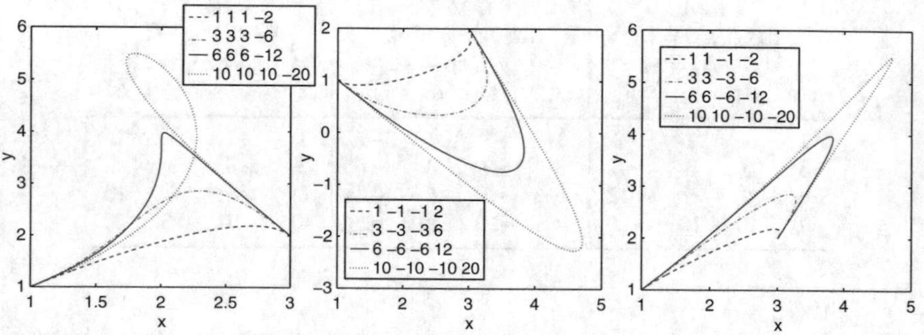

Fig. 1.21 Bézier curves in the three cases of Table 1.5. The values indicated in the legend correspond to $\{\alpha_0, \beta_0, \alpha_1, \beta_1\}$

1.5 Discussion and Conclusion

The choice of approximation methods strongly depends on the sought objective. The divided differences methods and Newton's differences are naturally related to the method of finite differences that will be found in particular in the study of partial differential equations. Their application is simple, and they can be easily automated. The Lagrange interpolation polynomials will be very useful for the integration of functions, but they often pose problems because of their oscillations between interpolation points as soon as their degree becomes larger than 2. Chebyshev polynomials allow us to considerably improve the interpolation quality by using irregularly spaced points. Hermite polynomials will be used to approximate the solution of ordinary differential equations with boundary values. The interpolation of important sets of points by spline functions possessing remarkable smoothness qualities is very important and heavily used in some technical domains. Bézier curves find applications in particular in the graphics domain due to their potential to generate complex forms.

1.6 Exercise Set

Exercise 1.6.1 (Easy)

Figure 1.22 shows a simple countercurrent heat exchanger made of an internal cylindrical tube and an external annulus. The heating fluid is saturated steam. It is assumed that the temperature of the heating fluid remains constant and equal to T_s.

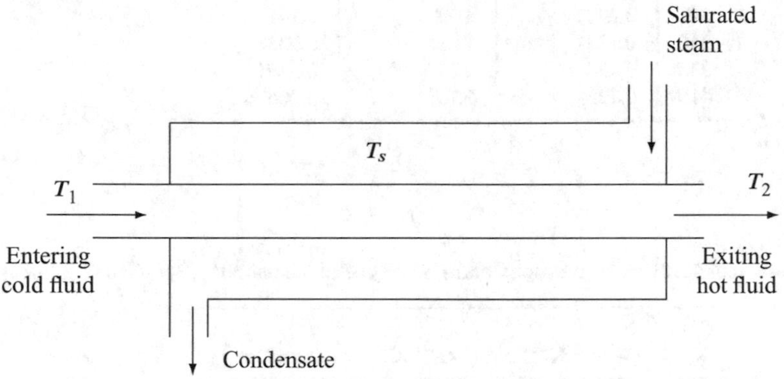

Fig. 1.22 Countercurrent heat exchanger

Expressing the energy conservation leads to the energy balance for a small length increment

$$\dot{m} C_p dT = \pi D h (T_s - T) \tag{1.6.1}$$

then, the length L of the heat exchanger results by integration

$$L = \frac{\dot{m}}{\pi D} \int_{T_1}^{T_2} \frac{C_p dT}{h(T_s - T)} \tag{1.6.2}$$

where D is the internal tube diameter, $\dot{m}$ the mass flow rate of the fluid, and h the local heat transfer coefficient between the inner wall of the tube and the fluid.

A literature correlation (Bergman et al. 2011, p. 595) allows us to calculate the dimensionless Nusselt number with respect to Reynolds and Prandtl dimensionless numbers in turbulent flow

$$Nu = 0.023 Re^{4/5} Pr^{2/5} \qquad \text{with } Re = \frac{4\dot{m}}{\pi D \mu}, \; Pr = \frac{\nu}{\alpha} = \frac{\mu C_p}{\lambda} \tag{1.6.3}$$

hence the local heat transfer coefficient

$$h = \frac{0.0697\,\lambda}{D} \left(\frac{\dot{m}}{\pi D \mu}\right)^{0.8} \left(\frac{\mu C_p}{\lambda}\right)^{0.4} \tag{1.6.4}$$

where C_p, μ, λ are the heat capacity, viscosity, and thermal conductivity of the fluid, respectively. These physical properties can be temperature dependent.

Tables 1.6 and 1.7 give the values of the thermal conductivity and viscosity of oil and ammoniac NH_3 at different temperatures.

Table 1.6 Thermal conductivity, heat capacity, and viscosity of oil (Bergman et al. 2011)

T (K)	λ ($\mathrm{W\,m^{-1}\,K^{-1}}$)	C_p ($\mathrm{J\,kg^{-1}\,K^{-1}}$)	μ ($\mathrm{kg\,m^{-1}\,s^{-1}}$)
290	0.147	1868	0.999
320	0.143	1993	0.141
350	0.138	2118	0.0356
380	0.136	2250	0.0141
410	0.133	2381	0.00698

Table 1.7 Thermal conductivity, heat capacity, and viscosity of ammoniac NH_3 (Bergman et al. 2011)

T (K)	λ ($\mathrm{W\,m^{-1}\,K^{-1}}$)	C_p ($\mathrm{J\,kg^{-1}\,K^{-1}}$)	μ ($\mathrm{kg\,m^{-1}\,s^{-1}}$)
300	0.0247	2158	101.5×10^{-7}
380	0.0340	2254	131×10^{-7}
460	0.0463	2393	159×10^{-7}
540	0.0575	2540	186.5×10^{-7}

1. By means of the finite difference method, find the interpolation polynomial of lowest degree allowing you to approximate in each case the thermal conductivity of NH_3 and the viscosity of oil in the desired temperature range with a precision of 1%. Remark: About viscosity, $\log(\mu)$ is more easily correlated to temperature than μ.
2. Write a program which calculates the length L of the heat exchanger with respect to the data of $\dot{m}$, T_1, T_2, T_s, D (Table 1.8) as well as the physical properties which depend on temperature.

Table 1.8 Characteristics of the heat exchanger

Fluid phase	NH_3 gas	Oil liquid
$\dot{m}$ (kg/s)	3×10^{-3}	6
T_1 (K)	300	290
T_2 (K)	520	360
T_s (K)	560	390
D (cm)	1.27	2.54
C_p (J/kg K)	$1661.6 + 1.606T$	$625.17 + 4.277T$
λ (W/m K)	cf. Table 1.7	$0.1802–1.167 \times 10^{-4}T$
μ (kg/m s)	$\exp(-12.86 + 5.66 \times 10^{-3}T - 3.73 \times 10^{-6}T^2)$	cf. Table 1.6

In general, it is admitted that, as a first approximation, the physical properties are constant. They are evaluated at the mean temperature $\bar{T} = (T_1 + T_2)/2$. Compare the difference of length calculated by both methods.

Exercise 1.6.2 (Easy)
Given the function g defined by

$$g(x) = \text{erf}(x) = \frac{2}{\sqrt{\pi}} \int_0^x \mathrm{e}^{-t^2} dt \tag{1.6.5}$$

we study how we could calculate the values of this function at the points of abscissa $x = 0$, $x = 0.25$, $x = 0.5$, $x = 0.75$, $x = 1$, with a precision better than 10^{-4} by using a nth degree Taylor polynomial in the neighborhood of 0 (thus, a MacLaurin series expansion).

1. To that purpose, you will first use a third degree Taylor polynomial and examine whether the chosen order is sufficient. Give the results at that order with an estimation of the precision achieved.
2. Based on that assessment, explain under an algorithmic form how to proceed to obtain the required precision.

Exercise 1.6.3 (Medium)
A family of quadratic splines is defined for each spline on $[x_i, x_{i+1}]$ by the equation

$$S_i(x) = a_i + b_i(x - x_i) + c_i(x - x_i)^2 \tag{1.6.6}$$

The points of abscissa x_i are not regularly spaced but are ordered by increasing abscissae. The family of splines passes through all the points $(x_i, f(x_i))$. The family is studied on the set from Table 1.9.

Find the relations about the coefficients a_i, b_i, c_i, so that the solution satisfies a maximum of conditions for a high quality interpolation and the solution is consistent. Deduce the numerical solving method. Write a program achieving this interpolation.

Table 1.9 Points x_i, y_i to interpolate by the family of splines

x	−2	−1.2	0	1.2	2
y	−0.964	−0.834	0	0.834	0.964

Exercise 1.6.4 (Medium)
Given a set of pairs (x_i, y_j), a two-dimensional interpolation is desired to estimate the value of the function $f(x, y)$ corresponding to a given value of a pair (x, y). First, a rectangular grid is achieved as in Figure 1.23.

We desire to extend the use of spline functions to this two-dimensional case. We suppose the following general for a spline function:

$$S(t) = M_k \frac{(t_{k+1} - t)^3}{6h_k} + M_{k+1} \frac{(t - t_k)^3}{6h_k} + a_k (t - t_k) + b_k \quad , \ t \in [t_k, t_{k+1}] \tag{1.6.7}$$

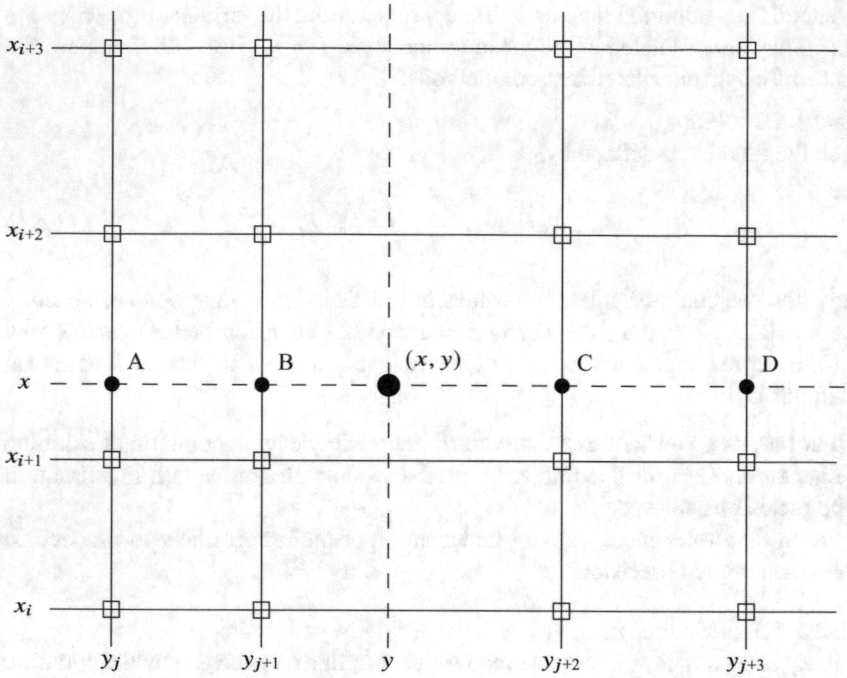

Fig. 1.23 Principle of the two-dimensional interpolation

where t is any variable. We decide to interpolate four times (for each of the values $\{y_i, y_{i+1}, y_{i+2}, y_{i+3}\}$) in the x direction to obtain the points A, B, C, D for the set $\{x_i, x_{i+1}, x_{i+2}, x_{i+3}\}$ and any given value of y, then in the direction y to obtain the desired value $f(x, y)$.

Do a general flow chart and then detail the algorithm of this method by clearly explaining what is known and how you obtain the necessary information at each instant.

Table 1.10 Surface $z = f(x, y)$ to interpolate by the splines in two dimensions

$x(i)$	1	1.25	1.6	2
$y(j)$	3	3.7	4.8	6

$z(i, j)$	10.000	14.690	24.040	37.000
	10.562	15.253	24.602	37.562
	11.560	16.250	25.600	38.560
	13.000	17.690	27.040	40

Write a program achieving this interpolation and verify it on the points of the surface $z = f(x, y)$ (Table 1.10). The element $z(i, j)$ of Table 1.10 must be considered as the element z_{ij} of the matrix z where i is the row index and j the column index.

Exercise 1.6.5 (Medium)

Construction of Bézier curves

The interpolation of curves by classic methods is not possible when several values of y correspond to a given value of x like in parametric curves. Bézier curves are often used in graphical programs on computers when fast continuous plots using the mouse are performed. Bézier curves are a very slight variant of the following construction by Hermite polynomials.

The parametric curve $(x(t), y(t))$ passing through the points (x_0, y_0) and (x_1, y_1) is built (Figure 1.24). Furthermore, at point (x_0, y_0), the tangent line is drawn on which the point $(x_0 + \alpha_0, y_0 + \beta_0)$ is located. This last point is called guide-point for (x_0, y_0). Similarly, the point $(x_1 + \alpha_1, y_1 + \beta_1)$ belongs to the tangent line at (x_1, y_1) and is called guide-point for (x_1, y_1). These guide-points are used to draw the parametric curve in a given direction.

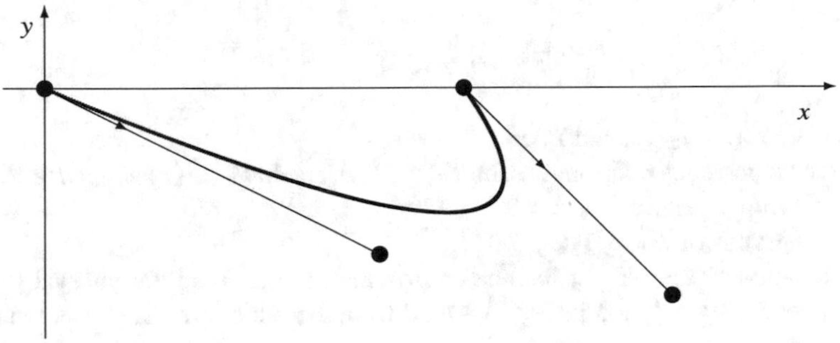

Fig. 1.24 Interpolation by a Bézier curve

Interpolation polynomials are chosen for $x(t)$ and $y(t)$ separately. These are Hermite cubic polynomials which are such that the parameter t varies in the interval $[0, 1]$ and $x(0) = x_0$, $y(0) = y_0$, $x(1) = x_1$, $y(1) = y_1$.

The polynomial $x(t)$ satisfies the derivative conditions at the bounds $x'(0) = \alpha_0$ and $x'(1) = \alpha_1$ and, similarly, the polynomial $y(t)$ satisfies the derivative conditions at the bounds $y'(0) = \beta_0$ and $y'(1) = \beta_1$.

1. Demonstrate that the expression of polynomial $x(t)$ is

$$x(t) = [2(x_0 - x_1) + (\alpha_0 + \alpha_1)]\, t^3 + [3(x_1 - x_0) - (\alpha_1 + 2\,\alpha_0)]\, t^2 + \alpha_0\, t + x_0 \quad (1.6.8)$$

   Find the expression of polynomial $y(t)$.
2. Let C_i be the Bézier cubic curve built parametrically by the relation

$$(x_i(t), y_i(t)) = (a_0^{(i)} + a_1^{(i)}\, t + a_2^{(i)}\, t^2 + a_3^{(i)}\, t^3, b_0^{(i)} + b_1^{(i)}\, t + b_2^{(i)}\, t^2 + b_3^{(i)}\, t^3) \quad (1.6.9)$$

   where (x_i, y_i) is the left interpolation point, (x_{i+1}, y_{i+1}) is the right interpolation point, (x_i^g, y_i^g) is the left guide-point, (x_{i+1}^d, y_{i+1}^d) the right guide-point. Calculate the parameters $a_j^{(i)}$ and $b_j^{(i)}$.

Exercise 1.6.6 (Difficult)

We first consider the usual Chebyshev polynomials $T_n(x)$ for which the coefficient of its highest degree monomial is equal to 2^{n-1}. A polynomial is called monic when the coefficient of its highest degree monomial is equal to 1. From $T_n(x)$, we build the monic Chebyshev polynomials $\tilde{T}_n(x)$ defined by

$$\tilde{T}_n(x) = \frac{T_n(x)}{2^{n-1}} \qquad \text{for } n \geq 1 \qquad \text{and } \tilde{T}_0(x) = 1 \tag{1.6.10}$$

An important property of monic Chebyshev polynomials $\tilde{T}_n(x)$ is that

$$\max_{x \in [-1,1]} |\tilde{T}_n(x)| = \frac{1}{2^{n-1}} \qquad \text{for } n \geq 1 \tag{1.6.11}$$

Furthermore, if all monic polynomials $P_n(x)$ are considered, all of them verify

$$\max_{x \in [-1,1]} |P_n(x)| \geq \frac{1}{2^{n-1}} \qquad \text{for } n \geq 1 \tag{1.6.12}$$

1. Calculate $\tilde{T}_1(x)$, $\tilde{T}_2(x)$, and $\tilde{T}_3(x)$.
2. Give the recurrence relation relating $\tilde{T}_n(x)$, $\tilde{T}_{n-1}(x)$, and $\tilde{T}_{n-2}(x)$ when $n \geq 2$. Verify this recurrence relation for $n = 3$.
3. Plot the graphs of $\tilde{T}_1(x)$, $\tilde{T}_2(x)$, $\tilde{T}_3(x)$.
4. The function $f(x) = \exp(x)$ can be approximated near 0, on the interval $[-1, +1]$, by the following polynomial $P_5(x)$ issued from the nth degree Taylor polynomial:

$$P_5(x) = 1 + x + \frac{x^2}{2} + \frac{x^3}{6} + \frac{x^4}{24} + \frac{x^5}{120} = \sum_{i=0}^{n} a_i x^i \tag{1.6.13}$$

with an error e equal to

$$e = \exp(x) - P_5(x) = x^6 \frac{f^{(6)}(\xi)}{720} \qquad \text{for } x \in [-1, 1] \tag{1.6.14}$$

(a) Build the polynomial $Q(x) = P_5(x) - a_5 \tilde{T}_5(x)$.
(b) What is the degree of $Q(x)$?
(c) Give the expression of $\{P_5(x) - Q(x)\}$? Deduce an upper bound of $|P_5(x) - Q(x)|$.
(d) Find an upper bound of the absolute value of the error e.
(e) What is the error done by choosing the polynomial $Q(x)$ as an approximation of $\exp(x)$?
(f) Compare the previous error to that which would be done if the polynomial $P_4(x)$ issued from the nth degree Taylor polynomial was chosen. Conclude.

Exercise 1.6.7 (Medium)

The Chebyshev series expansion of a function $f(x)$ is

$$f(x) = \sum_{i=0}^{\infty} a_i T_i(x) \qquad , x \in [-1, 1] \tag{1.6.15}$$

where $T_i(x)$ is the Chebyshev polynomial of rank i. When $f(x)$ is a polynomial, this expansion can be easily achieved.

1. Demonstrate that, when $f(x)$ is any function, the orthogonality of Chebyshev polynomials implies

$$a_0 = \frac{1}{\pi}\int_{-1}^{1} \frac{f(x)}{\sqrt{1-x^2}} dx \quad , \quad a_i = \frac{2}{\pi}\int_{-1}^{1} \frac{f(x)T_i(x)}{\sqrt{1-x^2}} dx \quad \text{for } i \geq 1 \quad (1.6.16)$$

2. Carry out Chebyshev expansion of order 2 for the function $f(x) = \exp(-x)$ based on property (1.6.16). The integrals will be calculated by a five-point Gauss–Legendre quadrature.
3. Carry out a second degree Taylor polynomial for $f(x)$ in the neighborhood of 0. Call that approximation $g(x)$. Deduce the Chebyshev expansion of order 2 for the function $g(x)$. Compare this expansion to that of $f(x)$ obtained in question 2.

Exercise 1.6.8 (Medium)

We desire to determine the family of natural cubic splines approximating the function $f(x) = \cos(\pi x)$ by using the points of abscissae $x = 0$; $x = 0.25$; $x = 0.5$; $x = 0.75$; $x = 1$. Each spline approximates the function on $[x_i, x_{i+1}]$.

1. Explain how you calculate the family of splines and their coefficients.
2. Write a program calculating the splines.

Exercise 1.6.9 (Medium)

Integration of an ordinary differential equation by approximation

Let the ordinary differential equation

$$x\,y'' + y' - 4\,y = -4x^2 + 8\,x - 1 \quad , 0 \leq x \leq 1 \qquad (1.6.17)$$

with the conditions $y(0) = 0$, $y(1) = 0$.

We search an approximation of the solution by means of spline functions. For that purpose, we decide to apply the spline functions on subintervals of length $h = 0.2$.

1. Pose the equations so that the problem is *complete*. "Complete" means well posed i.e. the number of equations must be in agreement with the number of unknowns.
2. Transform the problem so that it can be executed by a computer.

Remarks: This requires that the equations must not be posed in any order. Constantly think in terms of unknowns, number of unknowns, and number of available equations. There exist several different ways to pose the problem.

Exercise 1.6.10 (Easy)

Find the third order Lagrange interpolation polynomial for the following function:

$$f(x) = \cos(x) + \sin(x) \qquad (1.6.18)$$

with the nodes $x_0 = 0$, $x_1 = 0,25$, $x_2 = 0,5$, $x_3 = 1$.

Exercise 1.6.11 (Medium)

1. A cubic spline, defined by the specification of the first derivatives at the extremities, approximating a function f, is defined by

$$\begin{aligned} S_0(x) &= 1 + ax + 2x^2 - 2x^3 && \text{for } 0 \le x \le 1 \\ S_1(x) &= 1 + b(x-1) - 4(x-1)^2 + 7(x-1)^3 && \text{for } 1 \le x \le 2 \end{aligned} \tag{1.6.19}$$

   Deduce $f'(0)$ and $f'(2)$ that were necessary.
2. Let $f(x)$ be a function, indeed a third order polynomial on $[a, b]$ that will be noted $P_3(x)$. Demonstrate that $f(x)$ can be considered as a cubic spline defined by its first derivatives at the extremities, but that it is neither a natural spline nor a periodic spline. Which conditions should be fulfilled so that $f(x)$ could be considered as a natural spline or a periodic spline?

Exercise 1.6.12 (Easy)

A quadratic spline is defined by

$$S_i(x) = a_i + b_i(x - x_i) + c_i(x - x_i)^2 \,, \quad x \in [x_i, x_{i+1}] \tag{1.6.20}$$

Find the quadratic splines that interpolate the data $f(0) = 2$, $f(1) = 5$, $f(2) = 1$, with $f'(0) = 1$.

Exercise 1.6.13 (Difficult)

Consider the following ordinary differential equation:

$$y'' + \frac{1}{4} y(x) - \frac{3}{32} \sin\left(\frac{x}{4}\right) = 0 \tag{1.6.21}$$

with both boundary conditions $y(0) = 1/3$ and $y(\pi) = \sqrt{2}/4$.

A manner to find an approximate solution of that ordinary differential equation is to use a family of cubic splines. The nodes are as follows $x_1 = 0$, $x_2 = 0.25\pi$, $x_3 = 0.5\pi$, $x_4 = 0.75\pi$, $x_5 = \pi$.

Let $S_i(x)$ be a curve of that family defined on $[x_i, x_{i+1}]$, $i = 1, \ldots, 4$, by the following equation:

$$S_i(x) = M_i \frac{(x_{i+1} - x)^3}{6h_i} + M_{i+1} \frac{(x - x_i)^3}{6h_i} + a_i(x - x_i) + b_i \,, \quad x \in [x_i, x_{i+1}] \tag{1.6.22}$$

with $h_i = x_{i+1} - x_i$. The parameters M_i are called moments. The set of parameters to find is thus formed by the M_i, the parameters a_i and b_i.

The following conditions are imposed:

(a) The continuity of splines at x_i, $i = 2, \ldots, 4$.
(b) The continuity of derivatives at x_i, $i = 2, \ldots, 4$.
(c) The fact that the splines pass through the extreme points at 0 and π.
(d) The splines are solution of the ordinary differential equation at points x_i.

1. Find the number of unknown parameters.

2. Explain each of these conditions from (a) to (d) by indicating the number of equations that each condition represents.
3. Verify if the system is perfectly determined (same number of equations and parameters).
4. Write a program that solves the problem.

Exercise 1.6.14 (Medium)

Fourier series expansion

Let $\{f_0, f_1, \dots, f_{2n-1}\}$ be the set of functions defined by

$$\begin{aligned} f_0(x) &= \frac{1}{2} \\ f_k(x) &= \cos(kx)\,, \quad \forall\, k = 1, 2, \dots, n \\ f_{n+k}(x) &= \sin(kx)\,, \quad \forall\, k = 1, 2, \dots, n-1 \end{aligned} \tag{1.6.23}$$

1. Demonstrate that the previous functions are orthogonal on the interval $[-\pi, +\pi]$ with the weight function $w(x) = 1$, i.e. they should verify

$$\begin{aligned} &\int_{-\pi}^{+\pi} w(x) f_i(x) f_j(x) dx = 0\,, \quad \forall\, i, j = 1, 2, \dots, n\,, \quad i \neq j \\ &\int_{-\pi}^{+\pi} w(x) f_{n+i}(x) f_{n+j}(x) dx = 0\,, \quad \forall\, i, j = 1, 2, \dots, n-1\,, \quad i \neq j \end{aligned} \tag{1.6.24}$$

The following trigonometric relations can be used:

$$\begin{aligned} \sin(p)\sin(q) &= \frac{1}{2}\left(\cos(p-q) - \cos(p+q)\right) \\ \cos(p)\cos(q) &= \frac{1}{2}\left(\cos(p+q) + \cos(p-q)\right) \\ \sin(p)\cos(q) &= \frac{1}{2}\left(\sin(p+q) + \sin(p-q)\right) \end{aligned} \tag{1.6.25}$$

2. We desire to approximate any function $g(x)$ on the interval $[-\pi, +\pi]$ by a function $S(x)$ which is a Fourier expansion by means of functions $f_i(x)$, $i = 0, 1, \dots, 2n-1$, according to

$$S(x) = \frac{a_0}{2} + \sum_{i=1}^{n-1} [a_i \cos(ix) + b_i \sin(ix)] + a_n \cos(nx) \tag{1.6.26}$$

Demonstrate that the coefficients a_i and b_i are given by the following relations:

$$\begin{aligned} a_i &= \frac{1}{\pi} \int_{-\pi}^{+\pi} g(x) \cos(ix)\, dx\,, \quad \forall\, i = 0, 1, \dots, n \\ b_i &= \frac{1}{\pi} \int_{-\pi}^{+\pi} g(x) \sin(ix)\, dx\,, \quad \forall\, i = 0, 1, \dots, n-1 \end{aligned} \tag{1.6.27}$$

The following trigonometric relations can be used:

$$\begin{aligned} \cos^2(x) &= \frac{1}{2}[1 + \cos(2x)] \\ \sin^2(x) &= \frac{1}{2}[1 - \cos(2x)] \end{aligned} \tag{1.6.28}$$

References

G. Allaire. *Numerical Analysis and optimization*. Oxford University Press, Oxford, 2007.

J. J. Beers. *Numerical methods in chemical engineering - Applications in Matlab*. Cambridge University Press, Cambridge, 2007.

T. L. Bergman, A. S. Lavine, F. P. Incropera, and D. P. DeWitt. *Fundamentals of Heat and Mass Transfer*. John Wiley, New York, 7th edition, 2011.

I. N. Bronstein, K. A. Semendjajew, G. Musiol, and H. Mühlig. *Taschenbuch der Mathematik*. Verlag Harri Deutsch, 2005.

R. L. Burden and J. D. Faires. *Numerical Analysis*. Brooks/Cole, Boston, 9th edition, 2011.

B. Carnahan, H. A. Luther, and J. O. Wilkes. *Applied Numerical Methods*. Wiley, New York, 1969.

S. D. Conte and C. De Boor. *Elementary numerical analysis*. Mc Graw Hill, Singapore, 1981.

C. de Boor. *A practical guide to splines*. Springer-Verlag, Berlin, 1978.

B. Demidovitch and L. Maron. *Elements de calcul numérique*. Mir, Moscou, 1973.

J. F. Epperson. *An Introduction to Numerical Methods and Analysis*. Wiley, Hoboken, 2nd edition, 2013.

A. Fortin. *Analyse numérique pour ingénieurs*. Presses Internationales Polytechnique, Canada, 3ème edition, 2008.

W. Gautschi. *Numerical Analysis*. Birhäuser-Springer, New York, 2nd edition, 2012.

C. F. Gerald and P. O. Wheatley. *Applied Numerical Analysis*. Pearson, Boston-Addison Wesley, 7th edition, 2004.

E. Hairer. *Introduction à l'analyse numérique*. Université de Genève, 1993.

G. I. Marchuk. *Methods of numerical mathematics*. Springer-Verlag, New York, 1980.

W. H. Press, B. P. Flannery, S. A. Teukolsky, and W. T. Vetterling. *Numerical recipes in Fortran. The art of scientific computing*. Cambridge University Press, Cambridge, 1992.

S. S. Sastry. *Introductory Methods of Numerical Analysis*. Prentice-Hall of India, New Delhi, 4th edition, 2006.

T. Sauer. *Numerical Analysis*. Pearson, Boston, 2nd edition, 2012.

J. Stoer and R. Bulirsch. *Introduction to numerical analysis*. Springer Verlag, New York, second edition, 1996.

J. J. Tuma. *Handbook of Numerical Calculations in Engineering*. Mc Graw Hill, New York, 1989.

Chapter 2
Numerical Integration

2.1 Introduction

In some simple cases, the calculation of the definite integral

$$\int_a^b f(x)dx \tag{2.1.1}$$

is directly possible when the primitive (or antiderivative) function $F(x)$ is known

$$\int f(x)dx = F(x) \tag{2.1.2}$$

hence

$$\int_a^b f(x)dx = F(b) - F(a) \tag{2.1.3}$$

Most often, this is impossible and the only possible solution is numerical. Frequently, moreover, the function $f(x)$ is only known at a given number of points x_i, $i = 0, 1, \dots, n$. In this case, it is possible to search an approximation $g(x)$ of the function $f(x)$ and to proceed to a formal integration.

The interpolation polynomials $P_n(x)$ possess the required approximation properties and are easily integrable. Thus, they will be largely used in numerical integration (also called quadrature).

2.2 Newton and Cotes Closed Integration Formulas

The following integration formulas are called "closed" as they use the two basis points a and b to determine the approximation polynomial.

J.-P. Corriou, *Numerical Methods and Optimization*, Springer Optimization and Its Applications 187, https://doi.org/10.1007/978-3-030-89366-8_2

2.2.1 Global Integration on Interval $[a, b]$

Consider basis points uniformly distributed on interval $[a, b]$

$$x_i = a + ih\,, \quad i = 0, 1, \ldots, n \quad \text{with} \quad h = \frac{b-a}{n} \tag{2.2.1}$$

Note that n is the degree of the interpolation polynomial $P_n(x)$ such that

$$P_n(x_i) = f(x_i) = f_i\,, \quad i = 0, 1, \ldots, n \tag{2.2.2}$$

For example, a Lagrange polynomial can be chosen as an interpolation polynomial. In this case

$$P_n(x) = \sum_{i=0}^{n} L_i(x) f_i \tag{2.2.3}$$

with

$$L_i(x) = \prod_{\substack{k=0\\k\neq i}}^{n} \frac{x - x_k}{x_i - x_k} \tag{2.2.4}$$

The variable $t \in [0, n]$ is introduced such that $x = a + ht$. The polynomial $L_i(x)$ becomes

$$L_i(x) = \phi_i(t) = \prod_{\substack{k=0\\k\neq i}}^{n} \frac{t - k}{i - k} \tag{2.2.5}$$

By integrating, we get

$$\begin{aligned}\int_a^b P_n(x)dx &= \sum_{i=0}^{n} f_i \int_a^b L_i(x)dx \\ &= h \sum_{i=0}^{n} f_i \int_0^n \phi_i(t)dt \\ &= h \sum_{i=0}^{n} f_i w_i\end{aligned} \tag{2.2.6}$$

The coefficients w_i are called *weights*; they depend only on n, thus they neither depend on the function f nor on the integration limits a and b. Recall that $h = (b - a)/n$.
Example: $n = 1$

$$w_0 = \int_0^1 \frac{t-1}{0-1} dt = \int_0^1 (1-t)dt = \frac{1}{2} \tag{2.2.7}$$

$$w_1 = \int_0^1 \frac{t-0}{1-0} dt = \int_0^1 t dt = \frac{1}{2} \tag{2.2.8}$$

which gives the following result:

$$\int_a^b P_1(x)dx = \frac{h}{2}(f_0 + f_1) = \frac{h}{2}[f(a) + f(b)] \tag{2.2.9}$$

corresponding to the trapezoidal rule with $h = (b - a)$ (Figure 2.2).
Example: $n = 2$

$$w_0 = \int_0^2 \frac{t-1}{0-1}\frac{t-2}{0-2}dt = \frac{1}{2}\int_0^2 (t^2 - 3t + 2)dt = \frac{1}{3} \tag{2.2.10}$$

$$w_1 = \int_0^2 \frac{t-0}{1-0}\frac{t-2}{1-2}dt = -\int_0^2 (t^2 - 2t)dt = \frac{4}{3} \tag{2.2.11}$$

$$w_2 = \int_0^2 \frac{t-0}{2-0}\frac{t-1}{2-1}dt = \frac{1}{2}\int_0^2 (t^2 - t)dt = \frac{1}{3} \tag{2.2.12}$$

which gives the following result:

$$\int_a^b P_2(x)dx = \frac{h}{3}(f_0 + 4f_1 + f_2) = \frac{h}{3}[f(a) + 4f(\frac{a+b}{2}) + f(b)] \tag{2.2.13}$$

which is Simpson rule with $h = (b - a)/2$ (Figure 2.1).

By continuing, Table 2.1 results for different values of the degree n of the interpolation polynomial. From the degree n, the value of s results, then the weights w_i. The values σ_i are introduced only to display a table of integer values instead of fractional weights.

Table 2.1 Integration rules with respect to the degree n of the interpolation polynomials

n	ns	$\sigma_i = w_i s$	Error	Name of the rule
1	2	1 1	$h^3 1/12 f^{(2)}(\xi)$	Trapezoidal rule
2	6	1 4 1	$h^5 1/90 f^{(4)}(\xi)$	Simpson's rule
3	8	1 3 3 1	$h^5 3/80 f^{(4)}(\xi)$	Newton's rule
4	90	7 32 12 32 7	$h^7 8/845 f^{(6)}(\xi)$	Milne's rule (or Boole)

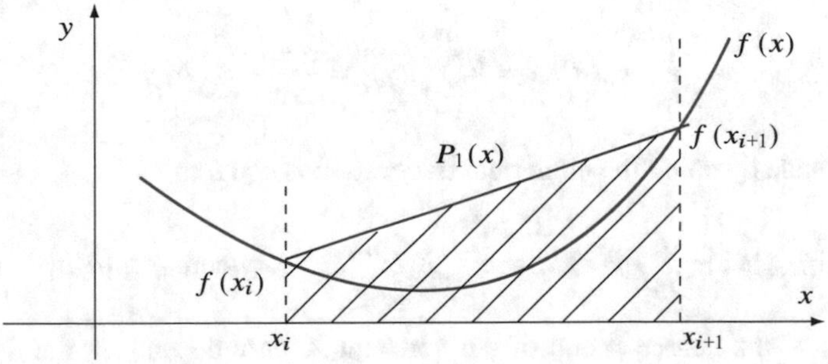

Fig. 2.1 Simpson's rule

Newton–Cotes formulas thus give the approximation of the integral

$$\int_a^b P_n(x)dx = h\sum_{i=0}^{n} w_i f_i \tag{2.2.14}$$

with

$$h = \frac{b-a}{n} \tag{2.2.15}$$

The weights w_i are such that their sum is equal to the degree n of interpolation polynomial

$$\sum_{i=0}^{n} w_i = n \tag{2.2.16}$$

Let s be the lowest common denominator of the weights w_i. The integer numerators σ_i are such that

$$\sigma_i = s w_i \tag{2.2.17}$$

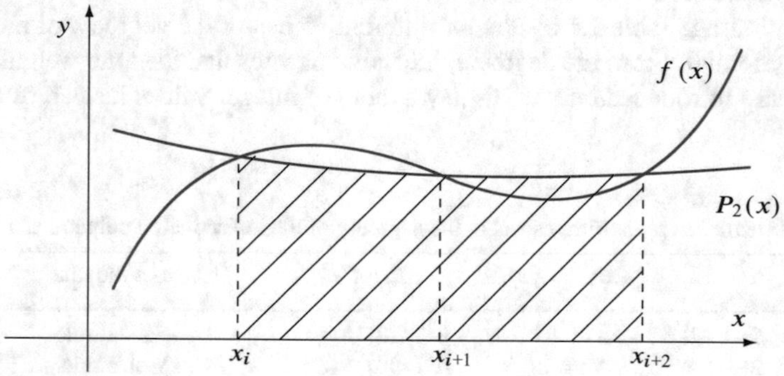

Fig. 2.2 Trapezoidal rule

Example: For Simpson's rule, according to Table 2.1, we have $n = 2$, $ns = 6$, thus $s = 3$. The weights result $w_0 = 1/3$, $w_1 = 4/3$, $w_2 = 1/3$.

Newton–Cotes are now expressed as

$$\int_a^b P_n(x)dx = h\sum_{i=0}^{n} w_i f_i = \frac{b-a}{ns}\sum_{i=0}^{n} \sigma_i f_i \tag{2.2.18}$$

The error made by doing the numerical integration is equal to

$$\int_a^b P_n(x)dx - \int_a^b f(x)dx = h^{p+1} K f^{(p)}(\xi) \qquad \text{where } \xi \in [a, b] \tag{2.2.19}$$

The values of the degree p and of the constant K only depend on the degree n of the interpolation polynomial. The error being of order p, any polynomial function of degree lower than p will be exactly integrated as the derivative of order p will be zero.

2.2.2 Integration on Subintervals

In general, Newton–Cotes formulas are not applied on all the interval $[a, b]$, but on the sequence of subintervals composing $[a, b]$. The type of subinterval depends on the order of the chosen method. The points x_i composing the interval $[a, b]$ are defined by

$$x_i = a + ih\,, \quad i = 0, 1, ..., N \quad \text{with } h = \frac{b-a}{N} \tag{2.2.20}$$

It can be noticed that, in the previous formula, the definition of h is different from Equation (2.2.1). N must be chosen in agreement with the order n of the integration formula.

- For the trapezoidal rule, a subinterval is defined by $[x_i, x_{i+1}]$.
- For Simpson's rule, N is chosen even (the number of calculation points x_i is odd), a subinterval is defined by $[x_{2i}, x_{2i+1}, x_{2i+2}]$, $i = 0, 1, ..., N/2 - 1$.
- For the 3/8 rule, N is a multiple of 3 and a subinterval will be defined by $[x_{3i}, x_{3i+1}, x_{3i+2}, x_{3i+3}]$, $i = 0, 1, ..., N/3 - 1$.
- Application of the *trapezoidal rule*:
  On a subinterval, the trapezoidal rule gives

$$I_i = \frac{h}{2}[f(x_i) + f(x_{i+1})] \tag{2.2.21}$$

Applying it to all the interval $[a, b]$, we get

$$\begin{aligned} I(h) = \sum_{i=0}^{N-1} I_i &= h\left[\frac{f(a)}{2} + f(a+h) + \cdots + f(b-h) + \frac{f(b)}{2}\right] \\ &= \frac{b-a}{2N}\left[f(a) + f(b) + 2\sum_{i=1}^{N-1} f\left(a + i\frac{b-a}{N}\right)\right] \end{aligned} \tag{2.2.22}$$

The function f is assumed to be continuously differentiable. On each subinterval, the error is equal to

$$I_i - \int_{x_i}^{x_{i+1}} f(x)dx = \frac{h^3}{12} f^{(2)}(\xi_i) \tag{2.2.23}$$

Then the sum of the individual errors is

$$I(h) - \int_a^b f(x)dx = \frac{h^3}{12}\sum_{i=0}^{N-1} f^{(2)}(\xi_i) = \frac{h^2}{12}\frac{b-a}{N}\sum_{i=0}^{N-1} f^{(2)}(\xi_i) \tag{2.2.24}$$

The summation term can be bounded

$$\min_i f^{(2)}(\xi_i) \leq \frac{1}{N}\sum_{i=0}^{N-1} f^{(2)}(\xi_i) \leq \max_i f^{(2)}(\xi_i) \tag{2.2.25}$$

As $f^{(2)}$ is continuous, $\exists\, \xi \in [\min_i \xi_i, \max_i \xi_i] \subset [a, b]$ such that

$$f^{(2)}(\xi) = \frac{1}{N}\sum_{i=0}^{N-1} f^{(2)}(\xi_i) \tag{2.2.26}$$

hence

$$I(h) - \int_a^b f(x)dx = \frac{b-a}{12}h^2 f^{(2)}(\xi), \quad \xi \in [a, b] \tag{2.2.27}$$

This result means that the error done when a trapezoidal rule is used decreases like the square of h and thus the method is of order 2.

- Application of *Simpson's rule*:
  With N even, on each subinterval $[x_{2i}, x_{2i+1}, x_{2i+2}]$, $i = 0, 1, \ldots, N/2 - 1$. It gives

$$I_i = \frac{h}{3}[f(x_{2i}) + 4f(x_{2i+1}) + f(x_{2i+2})] \qquad \text{with} \quad h = \frac{b-a}{N} \tag{2.2.28}$$

By summing these $N/2$ values, the approximation on $[a, b]$ results

$$\begin{aligned} I(h) &= \sum_{i=0}^{N/2-1} I_i \\ &= \tfrac{h}{3}[f(a) + 4f(a+h) + 2f(a+2h) + 4f(a+3h) + \cdots + \\ &\quad 2f(b-2h) + 4f(b-h) + f(b)] \\ &= \tfrac{h}{3}\left[f(a) + f(b) + 2\sum_{i=1}^{N/2-1} f(a+2ih) + 4\sum_{i=0}^{N/2-1} f(a+(2i+1)h)\right] \end{aligned} \tag{2.2.29}$$

The error done is

$$S(h) - \int_a^b f(x)dx = \frac{h^5}{90}\sum_{i=0}^{N/2-1} f^{(4)}(\xi_i) = \frac{h^4}{90}\frac{b-a}{2}\frac{2}{N}\sum_{i=0}^{N/2-1} f^{(4)}(\xi_i) \tag{2.2.30}$$

In the same way as for the trapezoidal rule, provided that f is 4 times continuously differentiable, it results that

$$S(h) - \int_a^b f(x)dx = \frac{b-a}{180}h^4 f^{(4)}(\xi) \tag{2.2.31}$$

Thus Simpson's rule is a method of order 4.

2.3 Open Newton and Cotes Integration Formulas

The following integration formulas are called "open" as they do not demand one or the other one of the bounds of the integration interval. The interpolation polynomial is of order $n - 2$. Consider $n - 1$ base points regularly spaced $x_1, \ldots, x_{n-1}$. It is supposed that the lower integration limit a coincides with $x_0 = x_1 - h$ where h is the spacing between adjacent points. The upper limit b is not fixed. The integration formula is

$$\int_a^b f(x)dx \approx \int_a^b P_{n-2}(x)dx \tag{2.3.1}$$

Thus, by defining

$$\bar{\alpha} = \frac{b - x_0}{h} \tag{2.3.2}$$

we get
for $\bar{\alpha} = 2$

$$\int_{x_0}^{x_2} f(x)dx = 2hf(x_1) + \frac{h^3}{3} f^{(2)}(\xi) \tag{2.3.3}$$

$\bar{\alpha} = 3$

$$\int_{x_0}^{x_3} f(x)dx = \frac{3h}{2}[f(x_1) + f(x_2)] + \frac{3h^3}{4} f^{(2)}(\xi) \tag{2.3.4}$$

$\bar{\alpha} = 4$

$$\int_{x_0}^{x_4} f(x)dx = \frac{4h}{3}[2f(x_1) - f(x_2) + 2f(x_3)] + \frac{14h^5}{45} f^{(4)}(\xi) \tag{2.3.5}$$

$\bar{\alpha} = 5$

$$\int_{x_0}^{x_5} f(x)dx = \frac{5h}{24}[11f(x_1) + f(x_2) + f(x_3) + 11f(x_4)] + \frac{95h^5}{144} f^{(4)}(\xi) \tag{2.3.6}$$

The closed Newton–Cotes formulas are more accurate than the open formulas as soon as a number of points larger than 2 or 3 is used. Thus, in general, it is better to use the closed formulas.

2.4 Conclusions on Newton and Cotes Integration Formulas

The formulas with m points for m odd have the same order of accuracy as the formulas with $m + 1$ points. Their degree of precision is equal to m. A formula of degree of precision m exactly integrates all the polynomials of degree lower than or equal to m. The polynomials of larger degree are not exactly integrated. Thus, Simpson's rule exactly integrates polynomials of degree lower than or equal to 3. Except for the trapezoidal rule used because of its simplicity, it is preferable to use formulas with an odd number of base points than with an even number.

The formulas with a number of base points larger than 8 are rarely used. Indeed, the rounding errors become large because of large weight factors with alternate signs.

A way to reduce the error is the use of composite integration formulas. Rather than using a formula of a high order, it is often better to choose a formula having a low order, divide the integration interval $[a, b]$ into subintervals, and use the formula of low order separately on each subinterval.

2.5 Repeated Integration by Dichotomy and Romberg's Integration

Let $I_{N,1}$ be the estimation of the integral

$$\int_a^b f(x)dx \tag{2.5.1}$$

obtained by using the composite trapezoidal rule with a number n of subintervals such that $n = 2^N$. $I_{0,1}$ is the estimation of the integral obtained by using the simple trapezoidal rule (step = h)

$$I_{0,1} = \frac{(b-a)}{1}\left\{\frac{1}{2}\left[f(a)+f(b)\right]\right\} \tag{2.5.2}$$

$I_{1,1}$ is the estimation of the integral obtained by using the simple trapezoidal rule applied two times (step = $h/2$)

$$\begin{aligned} I_{1,1} &= \frac{(b-a)}{2}\left\{\frac{1}{2}\left[f(a)+f(b)\right]+f\left(a+\frac{(b-a)}{2}\right)\right\} \\ &= \frac{1}{2}\left\{I_{0,1}+(b-a)f\left(a+\frac{(b-a)}{2}\right)\right\} \end{aligned} \tag{2.5.3}$$

$I_{2,1}$ is the estimation of the integral obtained by using the simple trapezoidal rule applied four times (step = $h/2^2$)

$$\begin{aligned} I_{2,1} &= \frac{(b-a)}{4}\left\{\frac{1}{2}\left[f(a)+f(b)\right]+\sum_{i=1}^{3} f\left(a+i\frac{(b-a)}{4}\right)\right\} \\ &= \frac{1}{2}\left\{I_{1,1}+\frac{(b-a)}{2}\sum_{\substack{i=1 \\ \Delta i=2}}^{3} f\left(a+i\frac{(b-a)}{4}\right)\right\} \end{aligned} \tag{2.5.4}$$

The recurrence relation relating $I_{n,1}$ (step = $h/2^n$) to $I_{n-1,1}$ (step = $h/2^{n-1}$) is thus expressed as

$$I_{n,1} = \frac{1}{2}\left\{I_{n-1,1}+\frac{(b-a)}{2^{n-1}}\sum_{\substack{i=1 \\ \Delta i=2}}^{2^n-1} f\left(a+i\frac{(b-a)}{2^n}\right)\right\} \tag{2.5.5}$$

The error term corresponding to $I_{n,1}$ is equal to

$$-\frac{(b-a)^3}{12(2)^{2n}}f^{(2)}(\xi)\,, \quad \xi \in [a,b] \tag{2.5.6}$$

Provided that the function $f^{(2)}$ be continuous and bounded, $I_{n,1}$ converges to the exact value of the integral.

Richardson's Extrapolation

Now, introduce the general technique of Richardson's extrapolation. Given a quantity

g_{approx} obtained by means of a discretization step h, g can be an integral, a derivative, ..., approximating the exact value g_{exact}. Suppose that the approximation is of order n, hence

$$g_{exact} = g_{approx}(h) + 0(h^n) \tag{2.5.7}$$

which could be written as

$$g_{exact} = g_{approx}(h) + a_n\, h^n + a_{n+1}\, h^{n+1} + 0(h^{n+2}) \tag{2.5.8}$$

where the coefficients a_i depend on the approximation method used. If instead of using a step h, we use $h/2$, Equation (2.5.8) becomes

$$g_{exact} = g_{approx}(\frac{h}{2}) + a_n\, \frac{h^n}{2^n} + a_{n+1}\, \frac{h^{n+1}}{2^{n+1}} + 0(h^{n+2}) \tag{2.5.9}$$

where $g_{approx}(\frac{h}{2})$ is in general a better approximation of g_{exact} than $g_{approx}(h)$. If Equations (2.5.8) and (2.5.9) are combined in order to eliminate the term about h^n, the approximation is then improved as the error will be at least about h^{n+1}. We thus get

$$(2^n - 1)\, g_{exact} = 2^n\, g_{approx}(\frac{h}{2}) - g_{approx}(h) - \frac{a_{n+1}}{2}\, h^{n+1} + 0(h^{n+2}) \tag{2.5.10}$$

Richardson's extrapolation formula results

$$g_{exact} = \frac{2^n\, g_{approx}(\frac{h}{2}) - g_{approx}(h)}{2^n - 1} + 0(h^{n+1}) \tag{2.5.11}$$

demonstrating that this new formula gives a better result with an approximation of order $(n + 1)$.

Application of Richardson's Extrapolation

Let us apply Richardson's extrapolation technique to a pair of adjacent elements of the sequence $I_{i,1}$ to obtain a better approximation of the integral. Each integral $I_{k,1}$ resulting from the trapezoidal rule is obtained with an error of order 2, thus $n = 2$. The application of Richardson's extrapolation then gives the relation

$$I = \frac{2^2\, I_{k+1,1} - I_{k,1}}{2^2 - 1} + 0(h^3) \tag{2.5.12}$$

as $I_{k+1,1}$ uses a step two times lower than $I_{k,1}$. The approximation results

$$I \approx \frac{2^2\, I_{k+1,1} - I_{k,1}}{2^2 - 1} \tag{2.5.13}$$

Applied to the previous sequence, Richardson's extrapolation (Figure 2.3) gives for two adjacent elements

$$I_{k,2} = \frac{4 I_{k+1,1} - I_{k,1}}{3} \tag{2.5.14}$$

For $k = 0$ corresponding to the trapezoidal rule applied once for $I_{0,1}$ on the interval $[a, b]$ and twice for $I_{1,1}$, Richardson's extrapolation yields

$$\begin{aligned} I_{0,2} &= \frac{4I_{1,1} - I_{0,1}}{3} \\ &= \frac{(b-a)}{6}\left\{\left[f(a) + 4f\left(a + \frac{(b-a)}{2}\right) + f(b)\right]\right\} \end{aligned} \tag{2.5.15}$$

that is Simpson's rule. The error on $I_{k,2}$ is equal to

$$-\frac{(b-a)^5}{2880(2)^{4k}} f^{(4)}(\xi), \quad \xi \in [a, b] \tag{2.5.16}$$

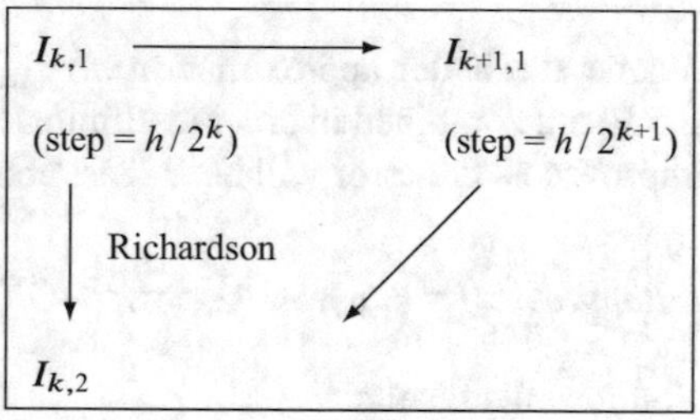

Fig. 2.3 Richardson's extrapolation

By repeating Richardson's extrapolation, Romberg's extrapolation formula results

$$I_{k,j} = \frac{4^{j-1} I_{k+1,j-1} - I_{k,j-1}}{4^{j-1} - 1} \tag{2.5.17}$$

The error corresponding to $I_{k,j}$ is equal to

$$-\frac{c(j)}{2^{2jk}} f^{(2j)}(\xi), \quad \xi \in [a, b] \tag{2.5.18}$$

where $c(j)$ is a constant depending on a, b, j.

It is interesting to note that if $I_{k,j}$ converges to the exact value of the integral when k increases, $I_{0,j}$ also converges when j increases. Thus, in the particular numerical Example 2.1, $I_{0,j}$ is close to the solution to about 10^{-6} for $j = 10$ whereas it is only the case for $I_{13,1}$.

Example 2.1 :
Integral by Richardson's extrapolation
The following integral has been calculated:

$$I = \int_0^3 x \exp(x^2) dx \tag{2.5.19}$$

by three different methods:

(a) Gauss–Legendre quadrature with 5 points gives $I \approx 3963.45$ and with 10 points, it gives $I \approx 4051.04$.
(b) Simpson's rule with 5 base points gives $I \approx 6441.21$, with 11 base points $I \approx 4258.18$, with 101 base points $I \approx 4051.07$.
(c) Richardson's extrapolation gives the following Table 2.2. The values of the calculated integrals follow the notations of Equation (2.5.17) with tabulated values of $I_{k,j}$ obtained by iterative use of Richardson's extrapolation formula. It converges to 4051.042.

Table 2.2 Richardson's extrapolation. Values of $I_{k,j}$ are given

$k \backslash j$	1	2	3	4	5	6	7	8
1	36,463.878	12,183.089	6058.419	4295.426	4061.334	4051.171	4051.042	4051.042
2	18,253.286	6441.211	4322.973	4062.248	4051.181	4051.042	4051.042	
3	9394.230	4455.363	4066.322	4051.224	4051.043	4051.042		
4	5690.080	4090.637	4051.460	4051.043	4051.042			
5	4490.498	4053.908	4051.050	4051.042				
6	4163.056	4051.228	4051.042					
7	4079.185	4051.054						
8	4058.087							

2.6 Numerical Integration with Irregularly Spaced Points

Previously, all developed integration formulas were of the form

$$\int_a^b f(x)dx \approx h \sum_{i=0}^{n} w_i f(x_i) \tag{2.6.1}$$

where the $n+1$ weights w_i are known from the $n+1$ values x_i. When the points x_i are not fixed, there are $2n+2$ unknowns (w_i and x_i), which allows us to determine a polynomial of degree $2n+1$.

2.6.1 Reminder on Orthogonal Polynomials

Two functions $g_m(x)$ and $g_n(x)$ belonging to a family of functions $g_i(x)$ are *orthogonal with respect to a weight function $w(x)$ on the interval* $[a, b]$ when, for all n

$$< g_m|g_n >= \int_a^b w(x)g_m(x)g_n(x)dx = 0 \qquad \text{if } n \neq m \tag{2.6.2}$$

$$< g_n|g_n >= \int_a^b w(x)[g_n(x)]^2 dx = c(n) \neq 0 \tag{2.6.3}$$

where the notation $< f|g >$ is called *scalar product of the functions* f and g relative to the weight function w. The scalar product is a number. Two functions are orthogonal when their scalar product is zero. A function is normalized when the scalar product of the function by itself is equal to 1. If all orthogonal functions two by two of the ensemble are normalized, the ensemble is orthonormal. In general, the value of c depends on n.

A way to generate an ensemble of orthogonal polynomials for a given weight function $w(x)$ is to use the recurrence relation

$$\begin{aligned} P_{-1}(x) &\equiv 0 \\ P_0(x) &\equiv 1 \\ P_{n+1}(x) &= (x - a_n)P_n(x) - b_n P_{n-1}(x)\,, \quad n = 0, 1, 2, \dots \end{aligned} \tag{2.6.4}$$

with the coefficients defined by

$$\begin{aligned} a_n &= \frac{< xP_n|P_n >}{< P_n|P_n >}\,, \quad n = 0, 1, 2, \dots \\ b_n &= \frac{< xP_n|P_{n-1} >}{< P_{n-1}|P_{n-1} >}\,, \quad n = 1, 2, \dots\,, \quad \text{any } b_0 \end{aligned} \tag{2.6.5}$$

To demonstrate Equation (2.6.5), it suffices to consider Equation (2.6.4) and to multiply by $w(x)P_n$ or $w(x)P_{n-1}$ respectively, then to take the integral of the new equation and to use the properties of orthogonal polynomials.

The polynomials defined by (2.6.4) are monic, i.e. the coefficient of the monomial x^n of largest degree of $P_n(x)$ is equal to 1. If each polynomial is divided by $< P_n|P_n >^{1/2}$, the ensemble of polynomials becomes orthonormal.

Other orthogonal polynomials can be met with different normalizations.

Each polynomial $P_n(x)$ has exactly n distinct roots in the interval $[a, b]$.

Among the known families of orthogonal functions, let us cite the family $(\sin kx)$ and the family $(\cos kx)$.

The monomial functions are not orthogonal. On the opposite, there exist several families of orthogonal polynomials.

Legendre polynomials:

Legendre polynomials $P_n(x)$ are orthogonal on the interval $[-1, 1]$ with the unit weight function $w(x) = 1$

$$\int_{-1}^{1} P_m(x)P_n(x)dx = 0 \qquad \text{if } n \neq m \tag{2.6.6}$$

and moreover

$$\int_{-1}^{1} [P_n(x)]^2 dx = \frac{2}{2n+1} \tag{2.6.7}$$

The first Legendre polynomials are

$$\begin{aligned}
P_0(x) &= 1 \\
P_1(x) &= x \\
P_2(x) &= \frac{1}{2}(3x^2 - 1) \\
P_3(x) &= \frac{1}{2}(5x^3 - 3x) \\
P_4(x) &= \frac{1}{8}(35x^4 - 30x^2 + 3) \\
\dots \\
P_n(x) &= \frac{2n-1}{n} x P_{n-1}(x) - \frac{n-1}{n} P_{n-2}(x)
\end{aligned} \tag{2.6.8}$$

Example 2.2 :

Orthogonal Legendre polynomials

To determine the orthogonal Legendre polynomials, consider the recurrence (2.6.4)

$$P_{n+1}(x) = (x - a_n)P_n(x) - b_n P_{n-1}(x)\,, \quad n = 0, 1, 2, \dots \tag{2.6.9}$$

with $P_0(x) = 1$, $P_1(x) = x$ and the equations of the coefficients (2.6.5), that is

$$\begin{aligned}
a_n &= \frac{\int_{-1}^{1} x P_n^2(x) dx}{\int_{-1}^{1} P_n^2(x) dx}\,, \quad n = 1, 2, \dots \\
b_n &= \frac{\int_{-1}^{1} x P_n(x) P_{n-1}(x) dx}{\int_{-1}^{1} P_{n-1}^2(x) dx}\,, \quad n = 1, 2, \dots
\end{aligned} \tag{2.6.10}$$

If $P_{n+1}(x)$ is calculated by Equation (2.6.9), we get a monic polynomial (whose coefficient of monomial x^{n+1} of highest degree is equal to 1) which does not satisfy Equation (2.6.7), thus we must set the Legendre polynomial equal to

$$PLeg_{n+1}(x) = c_{n+1} P_{n+1}(x) \tag{2.6.11}$$

where c_{n+1} is a coefficient such that

$$\int_{-1}^{1} [PLeg_{n+1}(x)]^2 dx = \frac{2}{2(n+1)+1} = \int_{-1}^{1} [c_{n+1} P_{n+1}(x)]^2 dx \tag{2.6.12}$$

hence

$$c_{n+1} = \sqrt{\frac{2}{2(n+1)+1} \frac{1}{\int_{-1}^{1} [P_{n+1}(x)]^2 dx}} \tag{2.6.13}$$

Thus, Table 2.3 results.

Table 2.3 Orthogonal monic polynomials P and Legendre polynomials $PLeg$

n	a_n	b_n	$P_{n+1}(x)$	$PLeg_{n+1}(x)$
1	0	0.3333	$x^2 - 0.3333$	$1.5x^2 - 0.5$
2	0	0.2666	$x^3 - 0.6x$	$2.5x^3 - 1.5x$
3	0	0.2570	$x^4 - 0.8570x^2 + 0.0857$	$4.375x^4 - 3.750x^2 + 0.3750$

Laguerre polynomials:

Laguerre polynomials $\mathcal{L}_n(x)$ are orthogonal on the interval $[0, +\infty[$ with the weight function $w(x) = \exp(-x)$

$$\int_0^{+\infty} \exp(-x)\mathcal{L}_m(x)\mathcal{L}_n(x)dx = 0 \qquad \text{if } n \neq m \tag{2.6.14}$$

$$\int_0^{+\infty} \exp(-x)[\mathcal{L}_n(x)]^2 dx = \frac{\Gamma(n+1)}{n!} \tag{2.6.15}$$

The first Laguerre polynomials are

$$\begin{aligned}
\mathcal{L}_0(x) &= 1 \\
\mathcal{L}_1(x) &= -x + 1 \\
\mathcal{L}_2(x) &= x^2 - 4x + 2 \\
\mathcal{L}_3(x) &= -x^3 + 9x^2 - 18x + 6 \\
&\dots \\
\mathcal{L}_n(x) &= (2n - 1 - x)\mathcal{L}_{n-1}(x) - (n-1)^2 \mathcal{L}_{n-2}(x)
\end{aligned} \tag{2.6.16}$$

Chebyshev polynomials of the first kind:

Chebyshev polynomials of the first kind $T_n(x)$ are orthogonal on the interval $[-1, 1]$ with the weight function $w(x) = 1/\sqrt{1-x^2}$

$$\begin{aligned}
&\int_{-1}^{+1} \frac{1}{\sqrt{1-x^2}} T_m(x)T_n(x)dx = 0 \qquad \text{if } n \neq m \\
&\int_{-1}^{+1} \frac{1}{\sqrt{1-x^2}} [T_n(x)]^2 dx = \begin{cases} \frac{\pi}{2} & \text{if } n \neq 0 \\ \pi & \text{if } n = 0 \end{cases}
\end{aligned} \tag{2.6.17}$$

The first Chebyshev polynomials of the first kind are

$$\begin{aligned}
T_0(x) &= 1 \\
T_1(x) &= x \\
T_2(x) &= 2x^2 - 1 \\
T_3(x) &= 4x^3 - 3x \\
&\dots \\
T_n(x) &= 2xT_{n-1}(x) - T_{n-2}(x)
\end{aligned} \tag{2.6.18}$$

Hermite polynomials:

Hermite polynomials $H_n(x)$ are orthogonal on the interval $]-\infty, +\infty[$ with the weight function $w(x) = \exp(-x^2)$

$$\int_{-\infty}^{+\infty} \exp(-x^2)H_m(x)H_n(x)dx = 0 \qquad \text{if } n \neq m \tag{2.6.19}$$

$$\int_{-\infty}^{+\infty} \exp(-x^2)[H_n(x)]^2 dx = 2^n n! \sqrt{n} \tag{2.6.20}$$

The first Hermite polynomials are

$$
\begin{aligned}
H_0(x) &= 1\\
H_1(x) &= 2x\\
H_2(x) &= 4x^2 - 2\\
H_3(x) &= 8x^3 - 12x\\
&\dots\\
H_n(x) &= 2xH_{n-1}(x) - 2(n-1)H_{n-2}(x)
\end{aligned}
\tag{2.6.21}
$$

Each of these orthogonal polynomials of degree n with real coefficients has n distinct roots in its definition interval. Any polynomial of degree n can be represented as a linear combination of functions of any of the previous families.

2.6.2 Gauss–Legendre Quadrature

We estimate the integral in the same way as previously by integration of an approximation polynomial of degree n

$$
\int_a^b f(x)dx = \int_a^b P_n(x)dx + \int_a^b R_n(x)dx \tag{2.6.22}
$$

$R_n(x)$ being the error term.

Let us use Lagrange interpolation polynomial

$$
f(x) = \sum_{i=0}^{n} L_i(x) f(x_i) + \left[\prod_{i=0}^{n}(x - x_i)\right] \frac{f^{(n+1)}(\xi)}{(n+1)!}, \quad a < \xi < b \tag{2.6.23}
$$

with

$$
L_i(x) = \prod_{\substack{j=0\\ j\neq i}}^{n} \left(\frac{x - x_j}{x_i - x_j}\right) \tag{2.6.24}
$$

The integration interval $[a, b]$ is transformed into $[-1, 1]$ by a change of variable

$$
z = \frac{2x - (a+b)}{b - a} \tag{2.6.25}
$$

and we define $F(z) = f(x)$, hence

$$
F(z) = \sum_{i=0}^{n} L_i(z) F(z_i) + \left[\prod_{i=0}^{n}(z - z_i)\right] \frac{F^{(n+1)}(\zeta)}{(n+1)!}, \quad -1 < \zeta < 1 \tag{2.6.26}
$$

with

$$
L_i(z) = \prod_{\substack{j=0\\ j\neq i}}^{n} \left(\frac{z - z_j}{z_i - z_j}\right) \tag{2.6.27}
$$

Supposing that $f(x)$ is a polynomial of degree $2n+1$, then

$$\frac{F^{(n+1)}(\zeta)}{(n+1)!} = q_n(\zeta) \tag{2.6.28}$$

is a polynomial of degree n. ζ belonging to the interval $[-1, 1]$, but having no known value in this interval, to be able to pursue the calculations, $q_n(\zeta)$ is transformed into a polynomial $q_n(z)$ on which it will be possible to work.

An estimation of the integral is then given by

$$\begin{aligned}\int_{-1}^{1} F(z)dz &\approx \sum_{i=0}^{n} F(z_i) \int_{-1}^{1} L_i(z)dz \\ &= \sum_{i=0}^{n} w_i F(z_i)\end{aligned} \tag{2.6.29}$$

with the weights

$$w_i = \int_{-1}^{1} L_i(z)dz = \int_{-1}^{1} \prod_{\substack{j=0 \\ j\neq i}}^{n} \left(\frac{z - z_j}{z_i - z_j}\right) dz \tag{2.6.30}$$

The integral to be calculated on $[a, b]$ is related to the integral calculated after change of variable by

$$\int_{a}^{b} f(x)dx = \frac{b-a}{2} \int_{-1}^{1} F(z)dz \approx \frac{b-a}{2} \sum_{i=0}^{n} w_i F(z_i) \tag{2.6.31}$$

Taking into account the previous remark about ζ and $q_n(\zeta)$, the error term of the quadrature formula takes the form

$$\int_{-1}^{1} \left[\prod_{i=0}^{n} (z - z_i)\right] q_n(z)dz \tag{2.6.32}$$

The abscissas z_i must be chosen in order to minimize the error term.

Consider the particular case of Gauss–Legendre quadrature. The polynomials $\prod_{i=0}^{n}(z - z_i)$ and $q_n(z)$ are expressed by means of Legendre polynomials P_i

$$\prod_{i=0}^{n} (z - z_i) = \sum_{i=0}^{n+1} b_i P_i(z) \tag{2.6.33}$$

$$q_n(z) = \sum_{i=0}^{n} c_i P_i(z) \tag{2.6.34}$$

The integral to minimize becomes

$$
\begin{aligned}
&\int_{-1}^{1}\left[\prod_{i=0}^{n}(z-z_i)\right]q_n(z)dz = \\
&\int_{-1}^{1}\left[\sum_{i=0}^{n}\sum_{j=0}^{n} b_i c_j P_i(z)P_j(z) + b_{n+1}\sum_{i=0}^{n} c_i P_i(z)P_{n+1}(z)\right]dz = \\
&\int_{-1}^{1}\sum_{i=0}^{n} b_i c_i\,[P_i(z)]^2\,dz = \\
&\sum_{i=0}^{n} b_i c_i \int_{-1}^{1}[P_i(z)]^2\,dz
\end{aligned}
\tag{2.6.35}
$$

as a result of orthogonality properties.

The error term can be rendered equal to zero by imposing that the first $n+1$ coefficients b_i be zero. There remains only one coefficient b_{n+1} different from zero so that, from Equation (2.6.33)

$$
\prod_{i=0}^{n}(z-z_i) = b_{n+1}P_{n+1}(z)
\tag{2.6.36}
$$

As the coefficient of the term of degree n of the left-hand polynomial is equal to 1, it results that b_{n+1} is equal to the inverse of the coefficient of the term of highest degree of P_{n+1}.

From the previous equality, it is obvious that *the* $n+1$ *base points* z_i *used in the integration formula are the* $n+1$ *roots of Legendre polynomial of degree* $n+1$. Now that the base points are determined, the weights w_i can be calculated by Equation (2.6.30). Several different methods have been developed to efficiently calculate the weights. In the particular case of Legendre polynomials (Abramowitz and Stegun 1972), it is possible to use

$$
w_i = \frac{2}{(1-z_i^2)[P'_{n+1}(z_i)]^2}
\tag{2.6.37}
$$

This constitutes Gauss–Legendre quadrature. Gauss–Legendre quadrature gives an exact integration result when the integrated function f is a polynomial of maximum degree $(2n+1)$.

The values of the roots z_i and the corresponding weights w_i for a family of given orthogonal polynomials are tabulated (Table 2.4).

Remark:

The calculation of the integral

$$
I = \int_a^b f(x)dx
\tag{2.6.38}
$$

is brought back to the calculation of the integral approximated by the sum

$$
I = \frac{b-a}{2}\int_{-1}^{1}F(z)dz \approx \frac{b-a}{2}\sum_{i=0}^{n} w_i F(z_i)
\tag{2.6.39}
$$

Rather than making the change of variable $x \rightarrow z$ to determine the function $F(z)$ from $f(x)$, it is simpler to calculate the roots x_i on $[a, b]$ corresponding to the roots z_i on $[-1, 1]$ and to use the equality $f(x_i) = F(z_i)$. Thus, we get

$$I = \frac{b-a}{2} \sum_{i=0}^{n} w_i f(x_i) \tag{2.6.40}$$

Table 2.4 Gauss–Legendre quadrature formulas

Gauss–Legendre quadrature

$$\int_{-1}^{1} F(z)dz \approx \sum_{i=0}^{n} w_i F(z_i)$$

Roots z_i	Type	Weights w_i
$\pm 1/\sqrt{3}$	Formula with 2 points ($n = 1$)	1.000000000000000
0.000000000000000	Formula with 3 points ($n = 2$)	8/9
$\pm\sqrt{15}/5$		5/9
$\pm\sqrt{525 - 70\sqrt{30}}/35$	Formula with 4 points ($n = 3$)	$(18 + \sqrt{30})/36$
$\pm\sqrt{525 + 70\sqrt{30}}/35$		$(18 - \sqrt{30})/36$
0.000000000000000	Formula with 5 points ($n = 4$)	0.568888888888888
±0.538469310105683		0.478628670499366
±0.906179845938664		0.236926885056189
±0.238619186083197	Formula with 6 points ($n = 5$)	0.467913934572691
±0.661209386466265		0.360761573048139
±0.932469514203152		0.171324492379170
±0.148874338981631	Formula with 10 points ($n = 9$)	0.295524224714753
±0.433395394129247		0.269266719309996
±0.679409568299024		0.219086362515982
±0.865063366688985		0.149451349150581
±0.973906528517172		0.066671344308688

Example 2.3 :
Gauss–Legendre quadrature

Calculate numerically the following integral:

$$I = \int_{-2}^{3} x^4 \, dx$$

If it is integrated analytically, the exact value of this integral is I=55.

We perform an integration according to Gauss–Legendre quadrature with 3 points ($n = 2$).

We proceed in the following way:

- Calculation of the abscissas x_i on interval $[-2, 3]$ corresponding to the tabulated zeros z_i which are in $[-1, 1]$.
- Calculation of the values of the function $f(x_i)$.
- Evaluation of the integral according to Equation (2.6.40).

Thus, Table 2.5 results.

Table 2.5 Gauss–Legendre quadrature

z_i	x_i	$f(x_i)$	w_i	$w_i f(x_i)$
0	0.5	0.0625	$\frac{8}{9}$	0.05555
$\frac{\sqrt{15}}{5}$	2.4364915	35.2419	$\frac{5}{9}$	19.5788
$-\frac{\sqrt{15}}{5}$	−1.4364915	4.2580	$\frac{5}{9}$	2.36555

Finally

$$I \approx \frac{5}{2}(0.05555 + 19.5788 + 2.36555) \approx 55$$

Notice that, as the polynomial function to integrate had a degree lower than $(2n + 1)$ with $n = 2$, the integration is exact.

Comparison with the trapezoidal rule and Simpson's rule without subintervals:

Trapezoidal rule:

$$I \approx 5\,[\frac{1}{2}(-2)^4 + \frac{1}{2}(3)^4) \approx 242.5$$

Simpson's rule:

$$I \approx \frac{5}{2}\,[\frac{1}{3}(-2)^4 + \frac{4}{3}(0.5)^4 + \frac{1}{3}(3)^4)] \approx 81.04$$

The interest of Gauss–Legendre quadrature is obvious with respect to the trapezoidal rule and Simpson's rule.

2.6.3 Gauss–Laguerre Quadrature

Gauss–Laguerre quadrature is based on the same principle as Gauss–Legendre quadrature. However, the integration formula takes into account the weight function under the form

$$\int_0^{+\infty} \exp(-z)F(z)dz = \sum_{i=0}^{n} w_i F(z_i) \Longrightarrow \int_0^{+\infty} G(z)dz = \sum_{i=0}^{n} w_i \exp(z_i)G(z_i) \tag{2.6.41}$$

where the z_i are the roots of Laguerre polynomial $\mathcal{L}_n$ and the weights are equal to

$$w_i = \frac{(n!)^2 z_i}{(n+1)^2)[\mathcal{L}_{n+1}(z_i)]^2} \tag{2.6.42}$$

2.6.4 Gauss–Chebyshev Quadrature

In the case of Chebyshev polynomials of the first kind, Gauss–Chebyshev quadrature gives

$$\int_{-1}^{+1} \frac{F(z)}{\sqrt{1-z^2}} dz = \sum_{i=0}^{n} w_i F(z_i) \Longrightarrow \int_{-\infty}^{+\infty} G(z) dz = \sum_{i=0}^{n} w_i \sqrt{1-z_i^2} G(z_i) \tag{2.6.43}$$

where the z_i are the roots of Chebyshev polynomial of the first kind T_n equal to

$$z_i = \cos\left(\frac{(2i-1)\pi}{2n}\right) \tag{2.6.44}$$

and the weights are equal to

$$w_i = \frac{\pi}{n} \tag{2.6.45}$$

2.6.5 Gauss–Hermite Quadrature

Gauss–Hermite quadrature gives

$$\int_{-\infty}^{+\infty} \exp(-z^2) F(z) dz = \sum_{i=0}^{n} w_i F(z_i) \Longrightarrow \int_{-\infty}^{+\infty} G(z) dz = \sum_{i=0}^{n} w_i \exp(z_i^2) G(z_i) \tag{2.6.46}$$

where the z_i are the roots of Hermite polynomial H_n and the weights are equal to (by using the orthonormal ensemble of Hermite polynomials)

$$w_i = \frac{2}{[H_n'(z_i)]^2} \tag{2.6.47}$$

2.7 Discussion and Conclusion

Just as in approximation methods, the use of irregularly spaced points imposed by the quadrature method clearly demonstrates its advantage with respect to the accuracy of the result of numerical integration at the expense of a slightly more important work of understanding and design. Gauss–Legendre quadrature is the most frequently used. At a comparable level of precision, the number of calculations required to evaluate the integral is considerably lower than for the rules based on regularly spaced points. These latter are still used by many users who do not want to invest time, do a simple program, have the impression to master the method, in particular the trapezoidal rule. This shows the interest to use numerical libraries which are available and would provide the precision to them with a relatively reduced investment.

2.8 Exercise Set

Exercise 2.8.1 (Easy)

Calculate the following integral:

$$\int_{-2}^{3} x^4 dx \tag{2.8.1}$$

first by the trapezoidal method, then Simpson's method, and finally Gauss–Legendre quadrature with 3 points ($n = 2$), by using in the three cases the bounds of the integration interval as calculation interval. Comment.

Exercise 2.8.2 (Easy)

The error function erf(x) is defined by

$$\text{erf}(x) = \frac{2}{\sqrt{\pi}} \int_0^x \exp(-t^2)dt = 1 - \frac{2}{\sqrt{\pi}} \int_x^{+\infty} \exp(-t^2)dt \tag{2.8.2}$$

1. Using Gauss–Legendre quadrature with 4 points, give an approximation of both integrals for $x = 1$. For the calculation of the second integral, it will be useful to think about the choice of the bound to use to replace $+\infty$ and the influence of the value of the term $\exp(-t^2)$. It may be useful to make a few trials to better understand this influence. Discuss the results thus obtained. Deduce an approximation of erf(x).
2. Do the same estimation with the first integral by Simpson's method with a step $h = 0.1$. Compare the result thus obtained.

Remark: The error function is used in many problems of physics. Consider a solid plate where the one-dimensional heat transfer is ruled by Fourier's law

$$\frac{\partial T}{\partial t} = \alpha \frac{\partial^2 T}{\partial x^2} \tag{2.8.3}$$

subjected to a constant temperature at the interface $T(x = 0, t) = T_s$. Let $T(x, t = 0) = T_0$ be the initial temperature. The temperature along time (Incropera and DeWitt 1996) is expressed as

$$\frac{T(x, t) - T_s}{T_0 - T_s} = \text{erf}\left(\frac{x}{2\sqrt{\alpha\, t}}\right) \tag{2.8.4}$$

where α is the thermal diffusivity.

Exercise 2.8.3 (Medium)

The fugacity f (atm) of a gas at a pressure P (atm) and at temperature T is given by

$$\ln \frac{f}{P} = \int_0^P \frac{Z-1}{P} \, dP \tag{2.8.5}$$

with the compressibility factor $Z = PV/(RT)$, R gas constant, and V molar volume (Smith et al. 2018; Vidal 1997).

For methane, the experimental data of the compressibility factor Z with respect to pressure are given in Table 2.6.

Table 2.6 Compressibility factor Z of methane with respect to pressure P

P	Z
1	0.9940
10	0.9370
20	0.8683
30	0.7928
40	0.7034
50	0.5936
60	0.4515
80	0.3429
100	0.3767
120	0.4259
140	0.4753
160	0.5252
180	0.5752
200	0.6246

We desire to calculate the fugacity at $P = 200$ atm.

1. A first crude method would consist in using the trapezoidal method by using only the experimental data. Calculate in this way the fugacity with detailed calculation.
2. A second method would consist in using Gauss–Legendre quadrature. For that purpose, we propose an approximation function of the form

$$Z = 1 - 0.00858\, P - 0.000463\, P \ln(P+1) + 0.0000475\, P^2 \tag{2.8.6}$$

By explaining the steps of the calculation, without fully explaining Gauss–Legendre quadrature, calculate the fugacity.

Exercise 2.8.4 (Easy)

Calculate the following integral:

$$I = \int_0^{+\infty} \frac{1}{1+x^4} dx \tag{2.8.7}$$

by both methods,

1. Simpson's method with the step $h = 0.25$.
2. Gauss–Legendre quadrature with five points.

Remark: To calculate this integral, it is recommended to divide the integration domain as $[0, 1]$ and $[1, +\infty[$, and then to do a change of variable $t = 1/x$ on the second domain.

Exercise 2.8.5 (Medium)
Calculate the following integral:

$$I = \int_{-1}^{1} \int_{-1+x^2}^{1-x^2} \sqrt{x^2 + y^2}\, dx\, dy \tag{2.8.8}$$

by Gauss–Legendre quadrature with three points, clearly explaining the technique used and giving intermediate results.

Exercise 2.8.6 (Easy)
Calculate the following integral:

$$I = \int_{-2}^{+2} \exp(-x^2) dx \tag{2.8.9}$$

by 5-point Gauss–Legendre quadrature.

References

M. Abramowitz and I. A. Stegun. *Handbook of Mathematical Functions with Formulas, Graphs and Mathematical Tables*. Dover, New York, 1972.

F. P. Incropera and D. P. DeWitt. *Fundamentals of Heat and Mass Transfer*. John Wiley, New York, 4th edition, 1996.

J. M. Smith, H. C. Van Ness, M. M. Abbott, and M. T. Swihart. *Introduction to Chemical Engineering Thermodynamics*. McGraw-Hill, New York, 8th edition, 2018.

J. Vidal. *Thermodynamique*. Technip, Paris, 1997.

Chapter 3
Equation Solving by Iterative Methods

3.1 Introduction

The problem is to develop adequate methods to find the solutions of the general equation

$$f(x) = 0 \tag{3.1.1}$$

The roots will be noted α_i. In a given number of cases, $f(x)$ will be supposed to be a polynomial of degree n

$$f(x) = x^n + a_1 x^{n-1} + \cdots + a_{n-1} x^1 + a_n \tag{3.1.2}$$

but some methods are applicable to any type of function.

There exist four main classes of methods (Gritton et al. 2001) to find the roots of a nonlinear equation of the form

$$f(x) = 0 \tag{3.1.3}$$

1. *Local methods* that require an initial estimation of the root (e.g. successive substitutions, Newton). Frequently, local methods are designed to search for only one real root of the nonlinear equation, even if multiple roots exist. Nevertheless, they are often very robust and nearly always converge (e.g. Newton, quasi-Newton), but they present the drawback to need to provide an initial estimation sufficiently close to the root.
2. *Global methods* that find a root from an arbitrary initial value (e.g. homotopy). They are adapted to the search of multiple roots.
3. *Interval methods* that find all the roots in a specified domain of x (e.g. dichotomy, regula falsi). They are robust but slow.
4. *Graphical methods* or spreadsheet that uses a graphical view of $f(x)$ in a specified domain of x.

Some of the methods presented below (Graeffe, Bernoulli, Bairstow) are more interesting from a mathematical point of view than for real applications, but they present a historical interest and their exposure may promote future ideas.

J.-P. Corriou, *Numerical Methods and Optimization*, Springer Optimization and Its Applications 187, https://doi.org/10.1007/978-3-030-89366-8_3

3.2 Graeffe's Method

Graeffe's method is a global method, as it gives a simultaneous approximation of all roots.

Consider a monic polynomial f of type (3.1.2). To f, the following adjoint function ϕ is associated

$$\begin{aligned}\phi(x) &= (-1)^n f(x) f(-x)\\ &= (x^2-\alpha_1^2)(x^2-\alpha_2^2)\dots(x^2-\alpha_n^2),\end{aligned} \tag{3.2.1}$$

where α_i are the searched roots, ordered by decreasing modulus. As $\phi(x)$ contains only even powers, a new function is defined

$$f_2(x) = \phi(\sqrt{x}) = (x-\alpha_1^2)(x-\alpha_2^2)\dots(x-\alpha_n^2) \tag{3.2.2}$$

which has the property that its roots are the squares of the roots of f. The operation can be repeated, and we obtain a sequence of polynomials $f_2, f_4, f_8\dots$ such that

$$f_m(x) = (x-\alpha_1^m)(x-\alpha_2^m)\dots(x-\alpha_n^m) \tag{3.2.3}$$

where m is an integer positive of 2 and f_m has roots $\alpha_1^m, \alpha_2^m, \dots, \alpha_n^m$. The aim of this sequence is to form an equation whose roots have very different orders of magnitude, that is, if the roots are real, the ratios $|\alpha_{i-1}^m/\alpha_i^m|$ can be made as small as desired when m becomes large.

$f_m(x)$ can be developed

$$\begin{aligned}f_m(x) &= x^n - (\alpha_1^m + \dots)x^{n-1} + (\alpha_1^m\alpha_2^m + \dots)x^{n-2}\\ &\quad -(\alpha_1^m\alpha_2^m\alpha_3^m + \dots)x^{n-3} + \dots + (-1)^n(\alpha_1^m\alpha_2^m\dots\alpha_n^m)\\ &= x^n - A_1x^{n-1} + \dots + (-1)^i A_i x^{n-i} + \dots + (-1)^n A_n\end{aligned} \tag{3.2.4}$$

the approximations result

$$\alpha_1^m \doteq A_1\,, \quad \alpha_2^m \doteq \frac{A_2}{A_1}\quad ,\dots \quad \alpha_n^m \doteq \frac{A_n}{A_{n-1}} \tag{3.2.5}$$

hence an approximation of the absolute values or of the moduli of the searched roots by taking the mth root.

The sign of the roots is not determined by this method. It must be verified by substitution in the original equation.

If multiple or complex roots exist, $|\alpha_i| = |\alpha_{i+1}|$, the equation

$$A_{i-1}x^2 - A_ix + A_{i+1} = 0 \tag{3.2.6}$$

gives approximations of α_i^m and α_{i+1}^m.

Graeffe's method presents some numerical difficulties and is not commonly used.

Example 3.1 :
Graeffe's method

Consider a polynomial with real roots, equal to 1, −2, 3, and −4. This polynomial is equal to

$$P(x) = x^4 + 2x^3 - 13x^2 - 14x + 24 \tag{3.2.7}$$

Table 3.1 Graeffe's method: root finding for a polynomial having real roots

i	m	$A_1^{1/m}$	$(A_2/A_1)^{1/m}$	$(A_3/A_2)^{1/m}$	$(A_4/A_3)^{1/m}$
1	2	5.4772	3.0166	1.7331	0.8381
2	4	4.3376	2.9409	1.9173	0.9812
3	8	4.0497	2.9787	1.9904	0.9994
4	16	4.0024	2.9984	1.9998	0.9999
5	32	4.0000	3.0000	2.0000	1.0000

Using Graeffe's method, the successive polynomials are found

$$\begin{aligned}
f_2(x) &= x^4 - 30\,x^3 + 273\,x^2 - 820\,x + 576\\
f_4(x) &= x^4 - 354\,x^3 + 26481\,x^2 - 357904\,x + 331776\\
f_8(x) &= x^4 - 72354\,x^3 + 448510881\,x^2 - 110523752704\,x + 110075314176
\end{aligned} \tag{3.2.8}$$

This shows that the polynomial coefficients increase very rapidly. In Table 3.1, the values of $A_1^{1/m}$, $(A_2/A_1)^{1/m}, \dots,$ have been gathered to highlight the limits that are the ordered root moduli. The convergence is fast. However, it must be noted that the coefficients A_i very rapidly take very large values, which poses huge numerical problems.

In the case of a polynomial with complex roots, Graeffe's method is difficult to use. At the best, it allows to have an estimation of α_i^m.

3.3 Bernoulli's Method

Bernoulli's method to find a root α_k of the polynomial

$$P(x) = \sum_{i=0}^{n} a_i x^{n-i} \quad \text{with } a_0 = 1 \tag{3.3.1}$$

first consists of building a sequence $\{u_i\}$ by associating to each monomial x^{n-k} a term u_{i-k} (thus $i \geq n$).

To understand the interest of the building of the sequence u_i, first consider the fact that the roots α_i of the polynomial $P(x)$ are supposed to be ordered according to their modulus $|\alpha_1| > \cdots > |\alpha_n|$.

Express that α_i is a root

$$\begin{aligned}
&a_0\alpha_1^n + a_1\alpha_1^{n-1} + \cdots + a_{n-1}\alpha_1 + a_n = 0\\
&\vdots\\
&a_0\alpha_n^n + a_1\alpha_n^{n-1} + \cdots + a_{n-1}\alpha_n + a_n = 0
\end{aligned} \tag{3.3.2}$$

Each of the previous equations is then multiplied by an arbitrary coefficient c_i, we sum the rows, that is

$$c_1(a_0\alpha_1^n + a_1\alpha_1^{n-1} + \cdots + a_{n-1}\alpha_1 + a_n) + \cdots + c_n(a_0\alpha_n^n + a_1\alpha_n^{n-1} + \cdots + a_{n-1}\alpha_n + a_n) = 0 \tag{3.3.3}$$

and then we order again with respect to the coefficients a_i, i.e.

$$a_0(c_1\alpha_1^n + \cdots + c_n\alpha_n^n) + a_1(c_1\alpha_1^{n-1} + \cdots + c_n\alpha_n^{n-1}) + \cdots + a_n(c_1 + \cdots + c_n) = 0 \tag{3.3.4}$$

We set

$$u_i = c_1\alpha_1^i + \cdots + c_n\alpha_n^i, \quad 0 \le i \le n \tag{3.3.5}$$

hence

$$a_0u_n + a_1u_{n-1} + \cdots + a_nu_0 = 0 \tag{3.3.6}$$

To simplify the writing, we consider $a_0 = 1$, hence

$$u_n = -a_1u_{n-1} - \cdots - a_nu_0 = -\sum_{i=1}^{n} a_iu_{n-i} \tag{3.3.7}$$

By extension, the sequence is defined

$$u_i = -\sum_{j=1}^{n} a_ju_{i-j} \qquad \text{for } i \ge n \tag{3.3.8}$$

Thus, it can be seen that to define this sequence, it suffices to arbitrarily choose the real numbers u_i for $0 \le i < n$. From the form of the solutions u_i previously given

$$u_i = c_1\alpha_1^i + \cdots + c_n\alpha_n^i, \quad 0 \le i \le n-1 \tag{3.3.9}$$

a system of n equations with n unknowns α_j^i results (Vandermonde determinant). To find the solution, an initial vector u_i as simple as possible can be chosen

$$u_i = 0, \quad 0 \le i < n-1 \qquad \text{and } u_{n-1} \ne 0 \tag{3.3.10}$$

Equation (3.3.8) can also be written as

$$u_{n+k} = -a_1u_{n+k-1} - \cdots - a_nu_k \tag{3.3.11}$$

Now consider the ratio of two successive terms

$$\frac{u_{n+k}}{u_{n+k-1}} = -a_1 - a_2\frac{u_{n+k-2}}{u_{n+k-1}} \cdots - a_n\frac{u_k}{u_{n+k-1}} \tag{3.3.12}$$

and suppose that this ratio admits a limit l when $k \to \infty$

$$\lim_{k\to\infty} \frac{u_{n+k}}{u_{n+k-1}} = l \tag{3.3.13}$$

From Equation (3.3.12), we deduce

$$l = -a_1 - a_2 \frac{1}{l} \cdots - a_n \frac{1}{l^{n-1}} \Rightarrow l^n + a_1 l^{n-1} + \cdots + a_n = 0 \tag{3.3.14}$$

hence the limit l is a root of the polynomial.

Theorem:

$$\text{If } |\alpha_1| > |\alpha_2| , \quad \alpha_1 = \lim_{i \to \infty} \frac{u_i}{u_{i-1}}, \tag{3.3.15}$$

thus the limit l corresponds to the root of larger modulus. Bernoulli's method remains valid even when multiple roots exist: $\alpha_1 = \alpha_2 = \cdots = \alpha_j$ provided we have: $|\alpha_j| > |\alpha_{j+1}|$.

If there are no multiple roots, the unique solution for $i \geq 0$ is of the form

$$u_i = c_1 \alpha_1^i + \cdots + c_n \alpha_n^i , \quad 0 \leq i \leq n-1 \tag{3.3.16}$$

the values of the coefficients c_i depending on the initial value taken for u_i.

In the case where a double root exists: $\alpha_1 = \alpha_2$ with $|\alpha_2| > |\alpha_3|$, the solution is of the form

$$u_i = c_1 \alpha_1^i + i c_2 \alpha_1^i + c_3 \alpha_3^i + \cdots + c_n \alpha_n^i \tag{3.3.17}$$

To find the solution, among all possible sequences, it may be useful to associate the two following sequences $\{v_i\}$ and $\{t_i\}$ to the sequence $\{u_i\}$

$$v_i = u_i^2 - u_{i+1} u_{i-1} \tag{3.3.18}$$

and

$$t_i = u_i u_{i-1} - u_{i+1} u_{i-2} \tag{3.3.19}$$

Thus, in the case (in particular, if α_1 and α_2 are complex) where

$$|\alpha_1| = |\alpha_2| > |\alpha_3| \tag{3.3.20}$$

we obtain the following results

$$\lim_{i \to \infty} \frac{v_i}{v_{i-1}} = \alpha_1 \alpha_2 \tag{3.3.21}$$

and

$$\lim_{i \to \infty} \frac{t_{i+1}}{v_i} = \alpha_1 + \alpha_2 \tag{3.3.22}$$

Bernoulli's method is rarely employed since the intensive use of computers.

Example 3.2 :

Bernoulli's method

Consider a polynomial with real roots, equal to 1, −2, 3, and −4. This polynomial is

$$P(x) = x^4 + 2\,x^3 - 13\,x^2 - 14\,x + 24 \tag{3.3.23}$$

First, the sequence $\{u_i\}$ is calculated by initializing with $u_0 = 0$, $u_1 = 0$, $u_2 = 0$, $u_3 = 1$. The other terms of this sequence obey Equation (3.3.16). The sequences $\{v_i\}$ and $\{t_i\}$ are also calculated according to Equations (3.3.18) and (3.3.19), respectively. The value of α_1 results as the root of largest modulus. Equations (3.3.21) and (3.3.22) are also used to calculate the product and the sum of α_1 and α_2.

In Table 3.2, the results of the calculations are gathered. It appears that the ratio u_i/u_{i-1} tends toward the root α_1 of largest modulus, the ratio v_i/v_{i-1} tends toward the product of both roots $(\alpha_1\alpha_2)$ of larger modulus, and the ratio t_{i+1}/v_i tends toward their sum. However, the convergence is slow.

Table 3.2 Bernoulli's method: root finding for a polynomial in the case of real roots

i	u_i	u_i/u_{i-1}	v_i/v_{i-1}	t_{i+1}/v_i
0	0			
1	0			
2	0			
3	1			
4	−2	−2		
5	17	−8.5	−15.1538	−1.3197
6	−46	−2.7058	−11.7817	−0.9358
7	261	−5.6739	−12.8207	−1.1177
8	−834	−3.1954	−11.7885	−0.9610
9	4009	−4.8069	−12.2893	−1.0459
10	−14102	−3.5175	−11.8809	−0.9796
11	62381	−4.4235	−12.1143	−1.0187
12	−231946	−3.7182	−11.9411	−0.9901
13	981201	−4.2302	−12.0477	−1.0079
14	−3765918	−3.8380	−11.9724	−0.9953
15	15543061	−4.1272	−12.0204	−1.0034
16	−60739538	−3.9078	−11.9873	−0.9978
17	247267193	−4.0709	−12.0089	−1.0014
18	−976163494	−3.9478	−11.9943	−0.9990
19	3943413501	−4.0397	−12.0039	
20	−15657462810	−3.9705		

Example 3.3 :
Bernoulli's method

Consider the following real polynomial

$$P(x) = x^4 + 2x^3 + 3x^2 + 4x + 5 \tag{3.3.24}$$

This polynomial has complex conjugate roots, equal to $0.2878 \pm 1.4160i$ and -1.2878 ± 0.8578. The respective moduli of these roots are 1.4450 and 1.5474.

The sequence $\{u_i\}$ is initialized exactly like in previous Example 3.2 with $u_0 = 0, u_1 = 0, u_2 = 0, u_3 = 1$. The other terms of this sequence are calculated according to Equation (3.3.16). The sequences $\{v_i\}$ and $\{t_i\}$ are also calculated from Equations (3.3.18) and (3.3.19), respectively. In Table 3.3, it can be noticed that, opposite to the previous case where the roots were real, the ratio u_i/u_{i-1} is not stabilized, which indicates that the root of largest modulus is complex. Thus, we must deal with the two complex conjugate roots of largest modulus. Equations (3.3.21) and (3.3.22) are used to calculate the product and the sum of α_1 and α_2 that are conjugate. It appears that the ratio v_i/v_{i-1} tends toward the product of both roots $(\alpha_1\alpha_2)$ (equal to 2.3944) of largest modulus and that the ratio t_{i+1}/v_i tends toward their sum (equal to −2.5756). The number of iterations is very large with respect to a very limited convergence.

Table 3.3 Bernoulli's method: root finding for a real polynomial in the case of complex conjugate roots

i	u_i	u_i/u_{i-1}	v_i/v_{i-1}	t_{i+1}/v_i
0	0			
1	0			
2	0			
3	1			
4	−1	−2		
5	1	−0.5	0.3333	0
6	0	id.	id.	id.
7	0	id.	id.	id.
8	6	id.	id.	−2.8333
9	−17	−2.8333	5.3611	−1.2538
10	16	−0.9411	0.8860	−0.4678
20	1746	−1.7962	2.6906	−1.3435
30	126,576	−2.6744	4.0406	−2.1753
40	7,484,506	−4.6352	2.9735	−2.6754
50	340,135,536	6.5020	2.4605	−2.6521
60		0.2271	2.4442	−2.6057
70		−0.6304	2.4000	−2.6199
80		−1.0668	2.3533	−2.6013
90		−1.4228	2.3650	−2.5732
100		−1.8325	2.3912	−2.5670

3.4 Bairstow's Method

Bairstow's method allows us to find the complex conjugate roots of a real polynomial by noticing that they correspond to the factorization of a real polynomial of degree 2. To calculate these roots by Newton's method, it would be necessary to use numerical complex calculation.

In a general manner, a polynomial $P(x)$

$$P(x) = \sum_{i=0}^{n} a_i x^{n-i} \tag{3.4.1}$$

by setting $P_0(x) \equiv P(x)$ can be written under the form

$$P_0(x) = P_1(x)(x^2 - rx - q) + A_0 x + B_0 \tag{3.4.2}$$

where the polynomial $P_1(x)$ is of degree $n - 2$ and the remainder of the division of $P_0(x)$ by $P_1(x)$ is equal to $A_0 x + B_0$. The coefficients A_0 and B_0 depend on the values of r and q; thus the issue is to find the values of these coefficients that make $A_0(r, q)$ and $B_0(r, q)$ equal to zero.

Indeed, Bairstow's method makes use of Newton–Raphson method that will be examined in Section 5.12, in the solution of systems of nonlinear equations. The recurrence relation giving r and q is

$$\begin{bmatrix} r_{i+1} \\ q_{i+1} \end{bmatrix} = \begin{bmatrix} r_i \\ q_i \end{bmatrix} - \begin{bmatrix} \frac{\partial A_0}{\partial r} & \frac{\partial A_0}{\partial q} \\ \frac{\partial B_0}{\partial r} & \frac{\partial B_0}{\partial q} \end{bmatrix}^{-1}_{\substack{r=r_i \\ q=q_i}} \begin{bmatrix} A_0(r_i, q_i) \\ B_0(r_i, q_i) \end{bmatrix} \tag{3.4.3}$$

However, $A_0(r, q)$ and $B_0(r, q)$ are unknown, thus the elements of the Jacobian matrix also. Consequently, it is necessary to determine those four partial derivatives that are present in the following identities:

$$\frac{\partial P_0(x)}{\partial r} = (x^2 - rx - q)\frac{\partial P_1(x)}{\partial r} - xP_1(x) + x\frac{\partial A_0}{\partial r} + \frac{\partial B_0}{\partial r} \equiv 0 \tag{3.4.4}$$

$$\frac{\partial P_0(x)}{\partial q} = (x^2 - rx - q)\frac{\partial P_1(x)}{\partial q} - P_1(x) + x\frac{\partial A_0}{\partial q} + \frac{\partial B_0}{\partial q} \equiv 0 \tag{3.4.5}$$

After this first division by $(x^2 - rx - q)$, the operation can be repeated; hence

$$P_1(x) = P_2(x)(x^2 - rx - q) + A_1 x + B_1 \tag{3.4.6}$$

Supposing that $(x^2 - rx - q = 0)$ has two distinct roots x_0, x_1, we get

$$P_1(x_i) = A_1 x_i + B_1 \,, \quad i = 0, 1 \tag{3.4.7}$$

and both identities (3.4.4) and (3.4.5) become

$$\left.\begin{cases} -x_i(A_1 x_i + B_1) + \frac{\partial A_0}{\partial r} x_i + \frac{\partial B_0}{\partial r} = 0 \\ -(A_1 x_i + B_1) + \frac{\partial A_0}{\partial q} x_i + \frac{\partial B_0}{\partial q} = 0 \end{cases}\right\} , \quad i = 0, 1 \tag{3.4.8}$$

From the second equation of Equation (3.4.8) and from Equation (3.4.7), we draw

$$\frac{\partial A_0}{\partial q} = A_1 \,, \quad \frac{\partial B_0}{\partial q} = B_1 \tag{3.4.9}$$

and consequently the first equation of (3.4.8) becomes

$$-x_i^2 \frac{\partial A_0}{\partial q} + x_i\left(\frac{\partial A_0}{\partial r} - \frac{\partial B_0}{\partial q}\right) + \frac{\partial B_0}{\partial r} = 0 \,, \quad i = 0, 1 \tag{3.4.10}$$

As $x_i^2 = rx_i + q$, it gives

$$x_i\left(\frac{\partial A_0}{\partial r} - \frac{\partial B_0}{\partial q} - \frac{\partial A_0}{\partial q} r\right) + \frac{\partial B_0}{\partial r} - \frac{\partial A_0}{\partial q} q = 0 \,, \quad i = 0, 1 \tag{3.4.11}$$

hence

$$\frac{\partial A_0}{\partial r} - \frac{\partial B_0}{\partial q} - \frac{\partial A_0}{\partial q} r = 0 \tag{3.4.12}$$

$$\frac{\partial B_0}{\partial r} - \frac{\partial A_0}{\partial q} q = 0 \tag{3.4.13}$$

and finally

$$\frac{\partial A_0}{\partial q} = A_1\,, \qquad \frac{\partial B_0}{\partial q} = B_1$$
$$\frac{\partial A_0}{\partial r} = rA_1 + B_1\,, \qquad \frac{\partial B_0}{\partial r} = qA_1 \tag{3.4.14}$$

By taking

$$P_0(x) = \sum_{i=0}^{n} a_i x^{n-i} \quad \text{and} \quad P_1(x) = \sum_{i=0}^{n-2} b_i x^{n-2-i} \tag{3.4.15}$$

it results

$$\sum_{i=0}^{n} a_i x^{n-i} = (x^2 - rx - q)\sum_{i=0}^{n-2} b_i x^{n-2-i} + A_0 x + B_0 \tag{3.4.16}$$

By identifying with respect to the successive powers of x by decreasing order, we get the relations

$$\begin{aligned}
a_0 &= b_0\\
a_1 &= b_1 - r\,b_0\\
a_2 &= b_2 - r\,b_1 - q\,b_0\\
&\ldots\\
a_i &= b_i - r\,b_{i-1} - q\,b_{i-2}\,, \quad i = 2, \ldots, n-2\\
&\ldots\\
a_{n-1} &= -r\,b_{n-2} - q\,b_{n-3} + A_0\\
a_n &= -q\,b_{n-2} + B_0
\end{aligned} \tag{3.4.17}$$

and the values of A_0 and B_0 result by means of Horner' scheme

$$\begin{aligned}
b_0 &= a_0\\
b_1 &= b_0\,r + a_1\\
b_2 &= b_0\,q + b_1\,r + a_2\\
&\ldots\\
b_i &= b_{i-2}\,q + b_{i-1}\,r + a_i\,, \quad i = 2, \ldots, n-2\\
&\ldots\\
A_0 &= b_{n-3}\,q + b_{n-2}\,r + a_{n-1}\\
B_0 &= b_{n-2}\,q + a_n
\end{aligned} \tag{3.4.18}$$

The process is repeated with P_1 and so on until exhaustion.

Example 3.4 :
Bairstow's method

Consider the following real polynomial

$$P(x) = x^4 + 2\,x^3 + 3\,x^2 + 4\,x + 5 \tag{3.4.19}$$

having complex conjugate roots, equal to $0.2878 \pm 1.4160i$ and $-1.2878 \pm 0.8578i$.

By applying Bairstow's method, the values of $A_0(r, q)$ and $B_0(r, q)$ are found by applying Equation (3.4.18)

$$\begin{aligned}
b_0 &= 1 \\
b_1 &= r + 2 \\
b_2 &= q + r\,(r + 2) + 3 \\
A_0 &= (r + 2)\,q + (q + (r + 2)\,r + 3)\,r + 4 = 2\,q\,r + 2\,q + r^3 + 2\,r^2 + 3\,r + 4 \\
B_0 &= (q + (r + 2)r + 3)\,q + 5 = q^2 + q\,r^2 + 2\,q\,r + 3\,q + 5
\end{aligned} \tag{3.4.20}$$

Applying Equation (3.4.3), Newton–Raphson's algorithm follows at iteration i

$$\begin{bmatrix} r \\ q \end{bmatrix}_{i+1} = \begin{bmatrix} r \\ q \end{bmatrix}_i - \begin{bmatrix} 2\,q + 3\,r^2 + 4\,r + 3 & 2\,r + 2 \\ 2\,q\,r + 2\,q & 2\,q + r^2 + 2\,r + 3 \end{bmatrix}_i^{-1} \begin{bmatrix} A_0 \\ B_0 \end{bmatrix}_i \tag{3.4.21}$$

The initialization is simply done with $(r, q) = (0, 0)$. After a limited number of iterations, Table 3.4 results.

Table 3.4 Bairstow's method: roots solving for a real polynomial having complex conjugate roots

Iteration i	r	q	A_0	B_0
1	0	0	4	5
2	−0.2222	−1.6667	0.8285	3.4362
3	1.0132	−1.3463	4.7119	−1.3364
4	0.5601	−1.6794	1.2433	0.374
5	0.5514	−2.0700	0.0071	0.1630
6	0.5760	−2.0880	0.0013	−0.0023
7	0.5756	−2.0882	0.62×10^{-6}	-0.029×10^{-6}

Having found $(A_0, B_0) \approx (0, 0)$, the searched polynomial results

$$x^2 - r\,x - q = x^2 - 0.5756\,x + 2.0882 \tag{3.4.22}$$

whose roots are $0.2878 \pm 1.4160\,i$. In a general way, then it would be necessary to continue the method by applying it to the polynomial P_1 obtained by exact division of $P(x)$ by $(x^2 - r\,x - q)$. In the present case, as $P_1(x)$ is of degree 2, the solution is immediate.

3.5 Existence of a Root of a Function

Before searching for the root α of a continuous function f by an iterative method such that $f(\alpha) = 0$, it is essential to find an interval $[a, b]$ such that

$$f(a)\,f(b) < 0 \tag{3.5.1}$$

This guarantees the existence of a solution α in the interval $[a, b]$. Finding the interval $[a, b]$ is the initial step of bracketing.

To compare different root-finding methods, the following equation will be considered

$$f(x) = \exp(x) - x^2 \tag{3.5.2}$$

represented in Figure 3.1. This function relatively flat in the neighborhood of its zero is well adapted to highlight the characteristics of the studied methods.

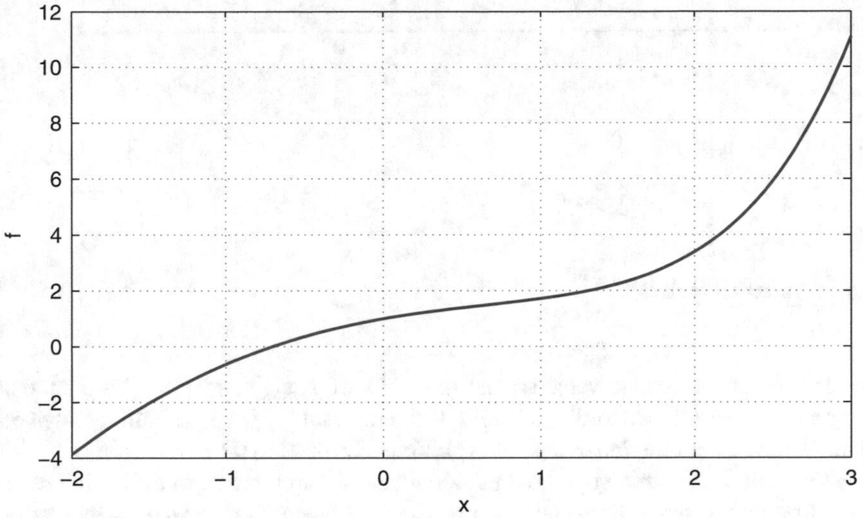

Fig. 3.1 Root finding: function of a single variable used

3.6 Bisection and Regula Falsi Methods

In Newton's method, the tangent line to the curve $f(x)$ at the point of abscissa x_k is drawn and its intersection with the x-axis is equal to x_{k+1}. This method works well when the sign of the concavity of the curve does not change between x_k and α like in Figure 3.6.

If the sign of the concavity of the curve changes, other methods by repeated bracketing the root can be better adapted such as the bisection and regula falsi methods.

3.6.1 Bisection Method

The bisection method is also called dichotomy or interval halving.

If the function $f(x)$ is continuous and if two values a and b are known such that $f(a)f(b) < 0$, then these values bracket the sought root α.

Let $x_{L0} = a$ (L = Lower) and $x_{U0} = b$ (U = Upper) (Figure 3.2). We calculate the value of the function at the midpoint $f((x_{L0} + x_{U0})/2)$. If this value is equal to 0, a root has been found. Otherwise, we compare its sign to the signs of $f(x_{L0})$ and $f(x_{U0})$. We choose the new pair (x_{L1}, x_{U1}) by taking $(x_{L0} + x_{U0})/2$ for one of these two values and

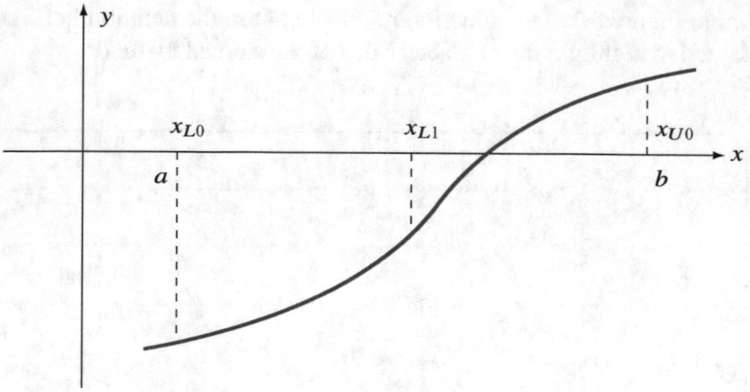

Fig. 3.2 Bisection method

x_{L0} or x_{U0} for the second value so that the signs of $f(x_{L1})$ and $f(x_{U1})$ are opposite. The method is continued until the length of the interval $[x_{Lk}, x_{Uk}]$ is sufficiently small or the value of the function at a point is sufficiently close to 0.

This method is extremely robust but slow, and it can be used to determine several roots. Its convergence is linear and the rate of convergence is equal to 0.5. Its accuracy is very good.

The number of necessary iterations to obtain a given result with an error ϵ starting from an initial interval $[a, b]$ is

$$r \approx \frac{\ln|a - b| - \ln\epsilon}{\ln 2} \tag{3.6.1}$$

Example 3.5 :
Bisection method

Bisection is used to find the root of Equation (3.5.2) also solved by Newton's method (Section 3.8). The initial search interval is $[-2, 2]$ after checking that a root exists in this interval. The results are gathered in Table 3.5. The convergence is slow, in particular with respect to Newton's method (Table 3.8).

3.6.2 Regula Falsi Method

This method derived from the false position is also based on the bracketing, i.e. two numbers, one x_{Lk} lower bound of the solution, the other one x_{Uk} upper bound of the solution, are determined at each iteration so that

$$f(x_{Lk})f(x_{Uk}) < 0 \tag{3.6.2}$$

Table 3.5 Bisection method: root finding of Equation (3.5.2)

Iteration	x_L	x_U	$f(x_L)$	$f(x_U)$
0	−2	2	−0.38647e+01	0.33891e+01
1	−2	0	−0.38647e+01	0.10000e+01
2	−1	0	−0.63212e+00	0.10000e+01
3	−1	−0.5	−0.63212e+00	0.35653e+00
4	−0.75000	−0.50000	−0.90133e−01	0.35653e+00
5	−0.75000	−0.62500	−0.90133e−01	0.14464e+00
6	−0.75000	−0.68750	−0.90133e−01	0.30175e−01
7	−0.71875	−0.68750	−0.29240e−01	0.30175e−01
8	−0.71875	−0.70313	−0.29240e−01	0.65113e−03
9	−0.71094	−0.70313	−0.14249e−01	0.65113e−03
10	−0.70703	−0.70313	−0.67873e−02	0.65113e−03
11	−0.70508	−0.70313	−0.30652e−02	0.65113e−03
12	−0.70410	−0.70313	−0.12063e−02	0.65113e−03
13	−0.70361	−0.70313	−0.27741e−03	0.65113e−03

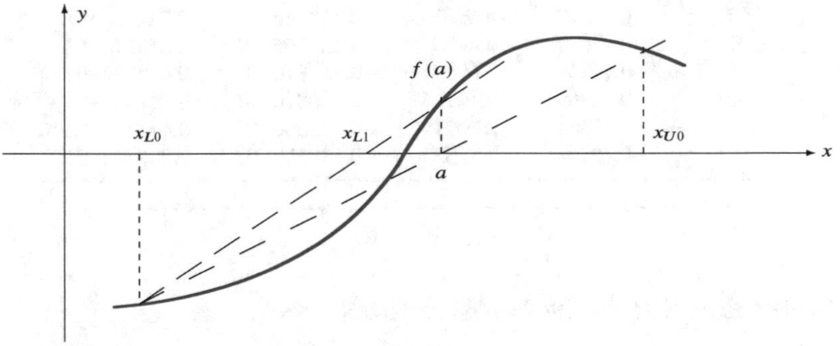

Fig. 3.3 Regula falsi method

Let a be the intersection of the x-axis and the chord joining the points $(x_{Lk}, f(x_{Lk}))$, $(x_{Uk}, f(x_{Uk}))$ (Figure 3.3). a is equal to

$$a = \frac{x_{Lk} f(x_{Uk}) - x_{Uk} f(x_{Lk})}{f(x_{Uk}) - f(x_{Lk})} \tag{3.6.3}$$

We calculate $f(a)$. If its value is zero, a is a root; otherwise,

$$\begin{cases} \text{if } f(a)f(x_{Lk}) > 0 & x_{Uk+1} = x_{Uk}\,, \quad x_{Lk+1} = a \\ \text{if } f(a)f(x_{Uk}) > 0 & x_{Lk+1} = x_{Lk}\,, \quad x_{Uk+1} = a \end{cases} \tag{3.6.4}$$

The iterations are continued until a suitable value of a is found so that $|f(a)|$ is sufficiently small or the interval $[x_{Lk}, x_{Uk}]$ is sufficiently small.

The convergence of regula falsi method is linear.

Example 3.6 :
Regula falsi method

Regula falsi method is used to find the root of Equation (3.5.2) also solved by Newton's method and bisection. The initial search interval is $[-2, 2]$ after checking that a root exists in this interval. The results are gathered in Table 3.6. The convergence is slow, hardly better than bisection (Table 3.5) and bad with respect to Newton's method (Table 3.8).

Table 3.6 Regula falsi method: root finding of Equation (3.5.2)

Iteration	x_L	x_U	$f(x_L)$	$f(x_U)$
0	−2	2	−0.38647e+01	0.33891e+01
1	−2	0.13114	−0.38647e+01	0.11229e+01
2	−0.86457	0.13114	−0.32624e+00	0.11229e+01
3	−0.86457	−0.09304	−0.32624e+00	0.90250e+00
4	−0.86457	−0.69088	−0.32624e+00	0.23823e−01
5	−0.81845	−0.69088	−0.22875e+00	0.23823e−01
6	−0.81845	−0.69956	−0.22875e+00	0.74197e−02
7	−0.71077	−0.69956	−0.13936e−01	0.74197e−02
8	−0.71077	−0.69991	−0.13936e−01	0.67517e−02
9	−0.70369	−0.69991	−0.41581e−03	0.67517e−02
10	−0.70369	−0.70114	−0.41581e−03	0.44148e−02
11	−0.70354	−0.70114	−0.13530e−03	0.44148e−02
12	−0.70354	−0.70333	−0.13530e−03	0.25670e−03

3.7 Method of Successive Substitutions

The method of successive substitutions can be applied to any type of function. The original equation

$$f(x) = 0 \tag{3.7.1}$$

is written under the modified form

$$x = F(x) \tag{3.7.2}$$

Supposing that α is a root of Equation (3.7.2), α is called a fixed point of F and the method is also named as fixed point iteration method (Burden and Faires 2011; Fortin 2008).

We use the recurrence relation

$$x_{i+1} = F(x_i) \tag{3.7.3}$$

hoping that the sequence x_i converges to the desired solution α of the equation. Cauchy's criterion is applicable:
A sequence u_i of R^n converges $\iff \forall\, \epsilon > 0,\ \exists N(\epsilon)$ such that $\forall\, j, k > N(\epsilon),\ \|u_j - u_k\| < \epsilon$.

Take a given point α of F. For any vector x_0 taken in the neighborhood on $\mathcal{N}(\alpha)$, suppose that an inequality of the form

$$\|x_{i+1} - \alpha\| \leq C\|x_i - \alpha\|^p \tag{3.7.4}$$

is verified $\forall\, i$, with $0 < C < 1$ if $p = 1$. In this case, the iteration method defined by F is said to be a convergent method of order at least p to determine α. Any method of order p to determine α is locally convergent (on $\mathcal{N}(\alpha)$). If $\mathcal{N}(\alpha)$ extends to $\mathcal{R}^n$, the method is globally convergent.

Let us reason on $\mathcal{R}$, α is such that $\alpha = F(\alpha)$. Suppose that we can write

$$|F(x) - F(\alpha)| \leq C|x - \alpha| \tag{3.7.5}$$

when $|x - \alpha| \leq |x_1 - \alpha|$. It results

$$|x_2 - \alpha| = |F(x_1) - \alpha| = |F(x_1) - F(\alpha)| \leq C|x_1 - \alpha| \tag{3.7.6}$$

by iterating

$$|x_3 - \alpha| = |F(x_2) - \alpha| \leq C|x_2 - \alpha| \leq C^2|x_1 - \alpha| \tag{3.7.7}$$

and in general

$$|x_i - \alpha| \leq C^{i-1}|x_1 - \alpha| \tag{3.7.8}$$

hence

$$\lim_{i\to\infty} x_i = \alpha \tag{3.7.9}$$

provided that $0 \leq C < 1$.

This method of successive substitutions (Figure 3.4) works only when the first derivative F' is such that

$$|F'(x)| \leq C < 1\,, \quad |x - \alpha| < |x_1 - \alpha| \tag{3.7.10}$$

and thus the method does not work when $|F'(x)| > 1$ in the studied region (Figure 3.5). In a general way, when x_i is close to α, the relation

$$x_{i+1} - \alpha \doteq F'(\alpha)(x_i - \alpha) \tag{3.7.11}$$

is verified and $F'(\alpha)$ is the asymptotic convergence factor.

The condition (3.7.10) is necessary for the convergence of the method but is not sufficient. Indeed, the initial value x_0 must be adequately placed. Thus the attraction basin of the root α is defined as the ensemble of points x_0 such that the method converges.

It must be noticed that the equation $x = F(x)$ can be formed from equation $f(x) = 0$ in many ways. For example, if the sought solution is α such that $\alpha = F(\alpha)$, among different possibilities, we can use the iterative equation

$$x_{j+1} = (1 - k)x_j + kF(x_j) \tag{3.7.12}$$

to try to improve the convergence.

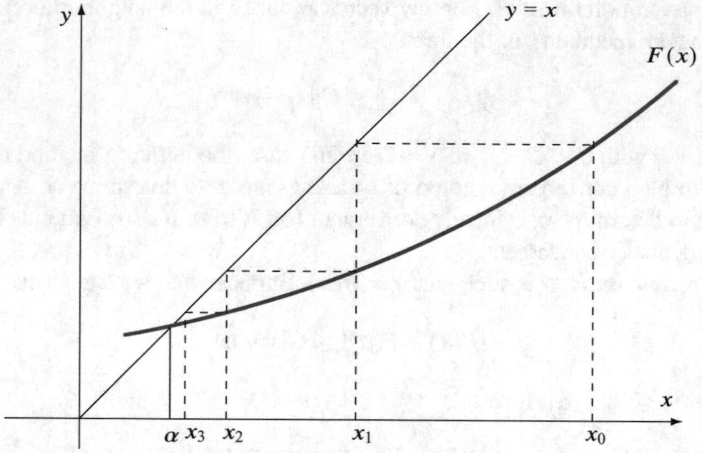

Fig. 3.4 Successive substitutions: Convergence in the case where $|F'(x)| \leq C < 1$ near α

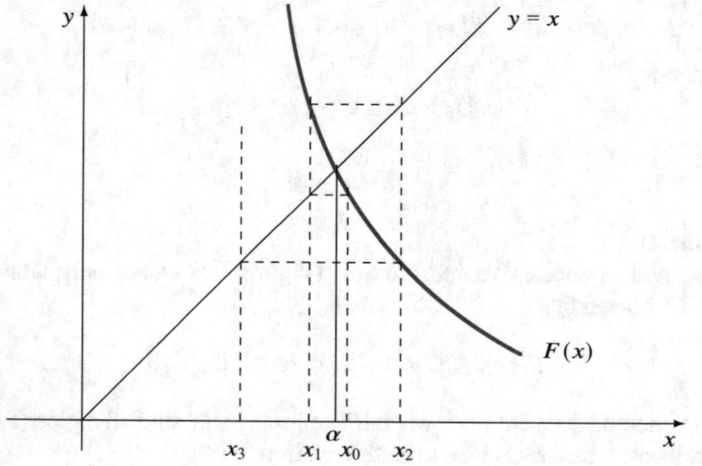

Fig. 3.5 Successive substitutions: Divergence in the case where $|F'(x)| > 1$ near α

Example 3.7 :
Method of successive substitutions

Using the method of successive substitutions, we search for the root of Equation (3.5.2)

$$f(x) = \exp(x) - x^2 \tag{3.7.13}$$

The equation is first transformed under the form

$$x = F(x) = x - (\exp(x) - x^2) \tag{3.7.14}$$

The initial value $x = 0$ is chosen. The results are given in Table 3.7. It can be noticed that the convergence is slow. Obviously, the method of successive substitutions is ill adapted to this equation.

Table 3.7 Method of successive substitutions: root finding of Equation (3.7.14)

Iteration	x	f
1	0	1.0000
2	−1.0000	−0.6321
3	−0.3679	0.5569
4	−0.9247	−0.4585
5	−0.4662	0.4100
6	−0.8762	−0.3514
7	−0.5248	0.3162
8	−0.8411	−0.2761
9	−0.5649	0.2492
10	−0.8142	−0.2199
11	−0.5943	0.1988
12	−0.7931	−0.1765
13	−0.6166	0.1596
14	−0.7762	−0.1423
15	−0.6339	0.1288
16	−0.7626	−0.1151
17	−0.6475	0.1041
18	−0.7516	−0.0933
19	−0.6583	0.0844
20	−0.7427	−0.0757
30	−0.7175	−0.0269
40	−0.7085	−0.0096
50	−0.7053	−0.0034
60	−0.7041	−0.0012
70	−0.7037	−0.0004
80	−0.7035	−0.0002
90	−0.7035	−0.0001
100	−0.7035	−0.0000

3.8 Newton's Method and Derived Methods

3.8.1 Newton's Method

Newton's method can be applied to functions of a real variable as well as to functions of a complex variable.

Consider a second degree Taylor polynomial for the function f at point α, sought solution, belonging to the neighborhood of a given point x

$$f(\alpha) = f(x)+(\alpha-x)f'(x)+\frac{1}{2}(\alpha-x)^2 f''(x)+\frac{1}{3!}(\alpha-x)^3 f^{(3)}(\xi)\,, \quad \xi \text{ between } x \text{ and } \alpha \tag{3.8.1}$$

We know that $f(\alpha) = 0$. If the series expansion is truncated at first order, it gives

$$f(x) = (x - \hat{\alpha})f'(x) \qquad \text{or } \hat{\alpha} = x - \frac{f(x)}{f'(x)} \tag{3.8.2}$$

where $\hat{\alpha}$ is an estimation of the root α. This requires that $f'(\alpha) \neq 0$, i.e. α is not a multiple root of $f(x) = 0$.

Newton's iterative formula results

$$x_{k+1} = x_k - \frac{f(x_k)}{f'(x_k)} \tag{3.8.3}$$

Thus, this formula amounts to drawing the tangent line to the curve $f(x)$ at the point of abscissa x_k (Figure 3.6). Its intersection with the x-axis is equal to x_{k+1}.

Using the framework of the method of successive substitutions (Section 3.7) and setting

$$F(x) = x - \frac{f(x)}{f'(x)} \tag{3.8.4}$$

Newton's method can be considered as a fixed point method.

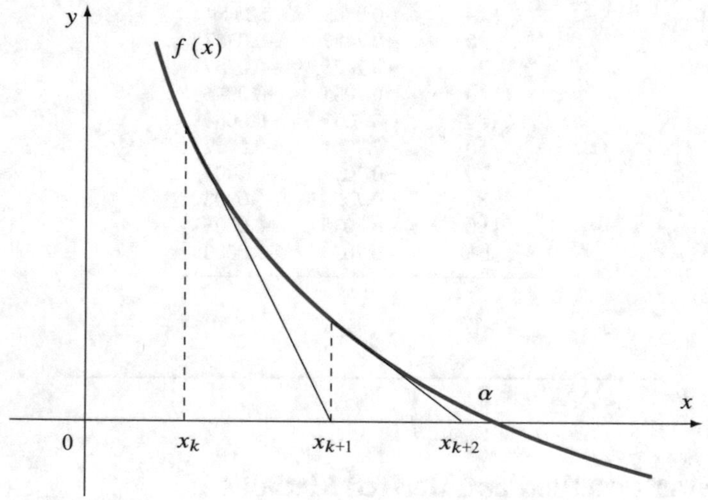

Fig. 3.6 Newton's method: finding the root α

To study the convergence of Newton's method, let us introduce the error e_k

$$e_k = x_k - \alpha \tag{3.8.5}$$

The derivative of F is

$$F'(x) = 1 - \frac{f'(x)}{f'(x)} + \frac{f(x)}{f'(x)^2} = \frac{f(x)}{f'(x)^2} \tag{3.8.6}$$

It verifies $F'(\alpha) = 0$. Now, consider the first degree Taylor polynomial of F near α

$$F(\alpha + e_k) = F(\alpha) + e_k F'(\alpha) + \frac{e_k^2}{2} F''(\xi), \quad \xi \text{ between } \alpha \text{ and } x_k \tag{3.8.7}$$

As F corresponds to a fixed point method, $F(\alpha) = \alpha$, so that the first degree Taylor polynomial of F becomes

$$F(\alpha + e_k) = \alpha + \frac{e_k^2}{2} F''(\xi), \quad \xi \text{ between } \alpha \text{ and } x_k \tag{3.8.8}$$

As $F(\alpha + e_k) = x_{k+1}$, it comes

$$\frac{x_{k+1} - \alpha}{(x_k - \alpha)^2} = \frac{F''(\xi_k)}{2}, \quad \xi \text{ between } \alpha \text{ and } x_k \tag{3.8.9}$$

It can be shown that $|F''(\xi_k)|$ is bounded in the neighborhood of α or $|F''(\xi_k)| < M$ so that

$$\left| \frac{x_{k+1} - \alpha}{(x_k - \alpha)^2} \right| < \frac{M}{2}, \quad \text{when } k \to \infty \tag{3.8.10}$$

As the error made on x_{k+1} is proportional to the square of the error made on x_k, the convergence is *quadratic* and it is neatly faster than the linear convergence of other methods.

The convergence issue can be shown differently using

$$x_{k+1} - \alpha = x_k - \alpha - \frac{f(x_k)}{f'(x_k)} \quad \text{or} \quad e_{k+1} = e_k - \frac{f(\alpha + e_k)}{f'(\alpha + e_k)} \tag{3.8.11}$$

Now, introduce the first degree Taylor polynomial of $f(\alpha)$ around x_k as

$$f(\alpha) = f(x_k - e_k) = f(x_k) - e_k f'(x_k) + \frac{e_k^2}{2} f''(\xi), \ \xi \text{ between } \alpha \text{ and } x_k \tag{3.8.12}$$

As $f(\alpha) = 0$, provided that $f'(x_k) \neq 0$ verified as α is a single root, it gives

$$\frac{f(x_k)}{f'(x_k)} = e_k - \frac{e_k^2}{2} \frac{f''(\xi)}{f'(x_k)} \tag{3.8.13}$$

hence, in Equation (3.8.11)

$$e_{k+1} = \frac{e_k^2}{2} \frac{f''(\xi)}{f'(x_k)} \tag{3.8.14}$$

thus

$$\lim_{k \to \infty} \frac{e_{k+1}}{e_k^2} = \frac{f''(\alpha)}{2 f'(\alpha)} \tag{3.8.15}$$

Considering that $|f''|$ is upper bounded and $|f'|$ is lower bounded, this demonstrates the quadratic convergence.

A sufficient condition of convergence of Newton's method is

$$\left| \frac{f(x_k) f''(x_k)}{(f'(x_k))^2} \right| < 1 \tag{3.8.16}$$

More complex conditions can be found in the literature (Ezquerro et al. 2009; Kantorovich and Akilov 1982). Theoretically, if the function is monotonous, if the derivative is calculable and nonzero, Newton's method converges whatever the initial point x_0 (globally convergent method of order 2). Otherwise, Newton's method works if the initial value x_0 is sufficiently close to the solution (locally convergent method of order 2).

Newton's method can be applied to the function of the complex variable

$$f(z) = f(x + iy) = u(x, y) + iv(x, y) \tag{3.8.17}$$

for which the iteration formula will be

$$z_{k+1} = z_k - \frac{f(z_k)}{f'(z_k)} \tag{3.8.18}$$

and is equivalent to both formulas

$$x_{k+1} = x_k + \left(\frac{v u_y - u u_x}{u_x^2 + u_y^2} \right)_{x_k, y_k} \tag{3.8.19}$$

$$y_{k+1} = y_k + \left(\frac{-v u_x - u u_y}{u_x^2 + u_y^2} \right)_{x_k, y_k} \tag{3.8.20}$$

where u_x represents the partial derivative $\partial u / \partial x$. Cauchy–Riemann equations, $u_x = v_y$ and $u_y = -v_x$, were used in the previous calculation. These formulas can also be found by using Newton–Raphson method.

Example 3.8 :
Newton's method

Newton's method is used to search for the root of Equation (3.5.2)

$$f(x) = \exp(x) - x^2 \tag{3.8.21}$$

The initial value is $x = 0$. Thus, the algorithm is

$$x_{n+1} = x_n - \frac{f(x_n)}{f'(x_n)} = x_n - \frac{\exp(x_n) - x_n^2}{\exp(x_n) - 2x_n} \tag{3.8.22}$$

The results are gathered in Table 3.8. They demonstrate the fast quadratic convergence.

Table 3.8 Newton's method: root finding of Equation (3.5.2)

Iteration	x	f
0	0.0000	0.10000e+01
1	−1.0000	−0.63212e+00
2	−0.73304	−0.56908e−01
3	−0.70381	−0.64739e−03
4	−0.70347	−0.87166e−07

3.8.2 Secant Method

Newton's method requires the knowledge of the derivative f' of the function f, which may be tedious. By using the definition of a derivative

$$f'(x_0) = \lim_{x \to x_0} \frac{f(x) - f(x_0)}{x - x_0} \tag{3.8.23}$$

and setting $x = x_{k-1}$, $x_0 = x_k$, the following approximation of the derivative results

$$f'(x_k) \approx \frac{f(x_{k-1}) - f(x_k)}{x_{k-1} - x_k} \tag{3.8.24}$$

By implementing this approximation of the derivative in Newton's method, the secant method results

$$x_{k+1} = x_k - f(x_k)\frac{x_k - x_{k-1}}{f(x_k) - f(x_{k-1})} \tag{3.8.25}$$

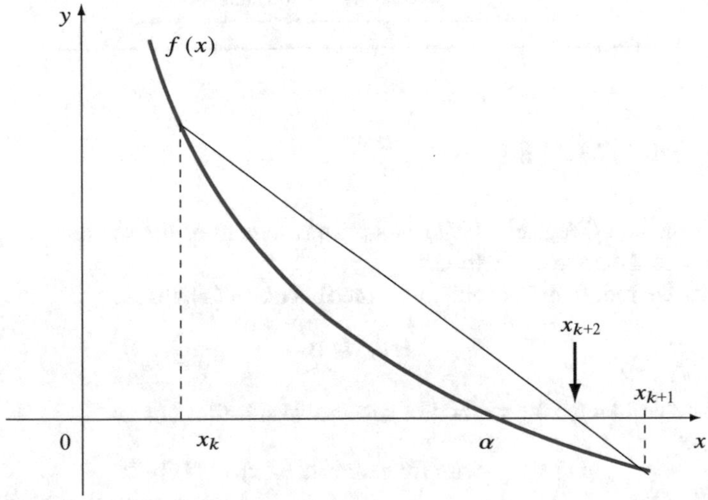

Fig. 3.7 Secant method: finding the root α

The secant method can be interpreted graphically (Figure 3.7). We start with two points x_0 and x_1 bracketing the solution α. The point x_2 is the intersection of the straight line joining the points $(x_0, f(x_0))$ and $(x_1, f(x_1))$ with the x-axis. Then, we continue with the straight line joining the points $(x_1, f(x_1))$ and $(x_2, f(x_2))$. It can be shown that the convergence of the secant method is superlinear, not quadratic.

Example 3.9 :
Secant method

The secant method is used to find the root of Equation (3.5.2) already solved by Newton's method. The initial search interval is $[-2, 2]$ ($x_0 = -2$ and $x_1 = 2$) by noticing that $f(x_0)f(x_1) < 0$; thus there exists a root in this interval. The results are gathered in Table 3.9. The method starts when the iteration index is equal to 2. The convergence is a little slower than that for Newton's method (Table 3.8), but much better than that for dichotomy or regula falsi (Section 3.6.1).

Table 3.9 Secant method: root finding of Equation (3.5.2)

Iteration	x	f
0	−2	−0.38647e+01
1	2	0.33891e+01
2	0.13114	0.11229e+01
3	−0.79493	−0.18031e+00
4	−0.66681	0.68713e−01
5	−0.70216	0.24824e−02
6	−0.70349	−0.36425e−04
7	−0.70347	0.18831e−07

3.9 Wegstein's Method

Wegstein's method (Wegstein 1958) is an improvement of the method of successive substitutions and the secant method.

Consider the root finding problem for the function $f(x)$, that is

$$f(x) = 0 \tag{3.9.1}$$

that could be written for the method of successive substitutions as

$$g(x) = x \quad \text{with for example} \quad g(x) = f(x) + x \tag{3.9.2}$$

Consider the two first iterations of the substitution method, with x_0 initial value. It gives

$$x_1 = g(x_0) \tag{3.9.3}$$

then

$$x_2 = g(x_1) \tag{3.9.4}$$

then the secant method is applied as

$$\frac{y - g(x_2)}{x - x_2} = \frac{g(x_2) - g(x_1)}{x_2 - x_1} \tag{3.9.5}$$

thus the points $(x_1, g(x_1))$, $(x_2, g(x_2))$, and (x, y) are lined up. By imposing the substitution $y = g(x)$ and $x_3 = x = g(x)$ (Figure 3.8), we obtain, by generalizing

$$x_{n+1}\left[1-\frac{g(x_n)-g(x_{n-1})}{x_n-x_{n-1}}\right]=g(x_n)-x_n\frac{g(x_n)-g(x_{n-1})}{x_n-x_{n-1}} \tag{3.9.6}$$

which gives Wegstein's formula

$$x_{n+1}=\frac{x_n g(x_{n-1})-x_{n-1}g(x_n)}{x_n-x_{n-1}-(g(x_n)-g(x_{n-1}))} \tag{3.9.7}$$

We can set

$$s_n=\frac{g(x_n)-g(x_{n-1})}{x_n-x_{n-1}} \tag{3.9.8}$$

so that Equation (3.9.6) becomes

$$x_{n+1}=\frac{g(x_n)}{1-s_n}-x_n\frac{s_n}{1-s_n} \tag{3.9.9}$$

By setting

$$\omega_n=\frac{1}{1-s_n} \tag{3.9.10}$$

Wegstein's formula results under a simple form

$$x_{n+1}=\omega_n\, g(x_n)+(1-\omega_n)\, x_n \tag{3.9.11}$$

where ω_n appears as a relaxation coefficient.

Wegstein's formula can be applied under the strict form, that is by calculating ω_n from the previous formulas, which amounts to

$$\omega_n=\frac{x_n-x_{n-1}}{x_n-x_{n-1}-(g(x_n)-g(x_{n-1}))} \tag{3.9.12}$$

By using the method of dominant eigenvalues, Orbach and Crowe (1971), Crowe and Nishio (1975) propose to survey the ratio

$$|\lambda|_{max}\approx\frac{\|\Delta x_n\|}{\|\Delta x_{n-1}\|} \tag{3.9.13}$$

and the optimum value of the relaxation factor would be

$$\omega=\frac{1}{1-|\lambda|_{max}} \tag{3.9.14}$$

which amounts to a form that is no more strictly Equation (3.9.6). This technique can be used after a small number of iterations (for example 5) according to the strict method.

Wegstein's method is frequently used in chemical process simulators when the process flowsheets present one or several recyclings and it is needed to proceed to iterations to determine the values of the variables influencing the recycling (Finlayson 2014; Orbach and Crowe 1971).

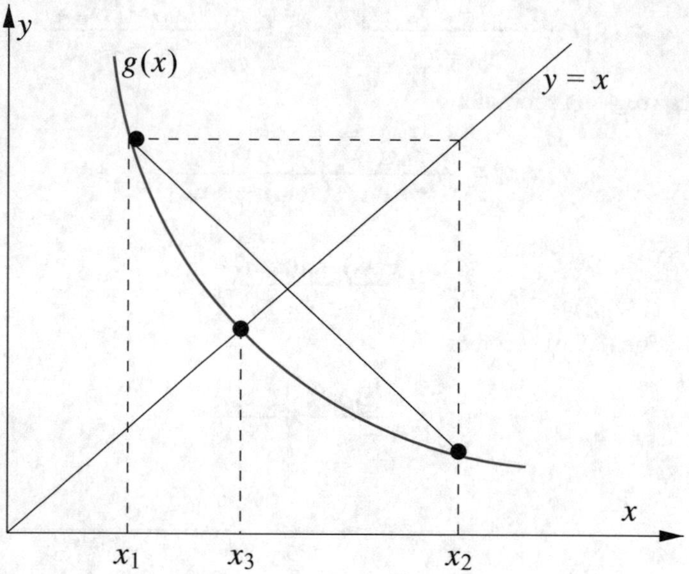

Fig. 3.8 Wegstein's method: Determination of the point x_3 from x_1 and x_2. The sought solution x is the intersection of the first bisector and of $g(x)$

Example 3.10 :
Wegstein's method

We desire to compare the convergence of the method of successive substitutions and Wegstein's method in the root finding of Equation (3.5.2)

$$f(x) = \exp(x) - x^2 \tag{3.9.15}$$

transformed for Wegstein's method under the form adapted to the method of successive substitutions

$$g(x) = (\exp(x) - x^2) + x \tag{3.9.16}$$

The initial value is $x = 0$. The results are gathered in Table 3.10. It can be noticed that Wegstein's method notably improves the convergence with respect to the method of successive substitutions, but also with respect to to Aitken's method, but does not reach the performance of Newton type or secant methods.

3.10 Aitken's Method

The previously cited methods differ by their rate of convergence, either linear, or more rarely superlinear and quadratic. Aitken's method (Aitken 1925) differs in that it is designed to accelerate the convergence of methods converging linearly to the origin.

Consider a sequence u_n that converges to a limit l. The forward difference is defined as

$$\Delta u_n = u_{n+1} - u_n \tag{3.10.1}$$

Table 3.10 Wegstein's method compared to the method of successive substitutions: root finding of Equation (3.5.2)

Iteration	Successive substitutions		Wegstein	
	x	$f(x)$	x	$f(x)$
1	0	1.0000	0	1.0000
2	−1.0000	−0.6321	1	1.718281
3	−0.3679	0.5569	0.511736	1.4063112
4	−0.9247	−0.4585	0.0023744	1.023464
5	−0.4662	0.4100	−1.280806	−1.362652
6	−0.8762	−0.3514	−0.535809	0.298103
7	−0.5248	0.3162	−0.669535	6.366837×10^{-2}
8	−0.8411	−0.2761	−0.705853	-4.541520×10^{-3}
9	−0.5649	0.2492	−0.703435	6.147391×10^{-5}
10	−0.8142	−0.2199	−0.703467	5.799760×10^{-8}
11	−0.5943	0.1988	−0.703467	$-7.415179 \times 10^{-13}$

Aitken's method consists of defining a new sequence v_n such that

$$v_n = u_n - \frac{(\Delta u_n)^2}{\Delta^2 u_n} = u_n - \frac{(u_{n+1} - u_n)^2}{u_{n+2} - 2u_{n+1} + u_n} \tag{3.10.2}$$

which in principle converges more rapidly to l than the sequence u_n. Aitken's method can also be written under the form

$$v_n = \frac{u_{n+2}\, u_n - u_{n+1}^2}{u_{n+2} - 2u_{n+1} + u_n} \tag{3.10.3}$$

This method can be applied to many linearly convergent methods such as solving an equation, or the search of eigenvalues by the power method (Section 4.4).

However, Aitken's method can also be applied to Newton's or secant methods.

Aitken's method applied to a fixed point is called Steffensen's method. To find the solution of $f(x) = 0$, Steffensen's method is written as

$$x_{n+1} = x_n - \frac{f(x_n)}{g(x_n)} \quad \text{with} \quad g(x_n) = \frac{f(x_n + f(x_n))}{f(x_n)} \tag{3.10.4}$$

A condition of convergence is that the function $f(x)$ verifies $1 < f'(\alpha) < 0$, where α is the sought solution. Clearly, $g(x)$ is an approximation of the derivative of $f(x)$ at x_n. With respect to the secant method, Steffensen's method presents a relatively small advantage and, like Newton's method, it requires a good initialization.

Example 3.11 :
Aitken's method

In order to improve the convergence of the method of successive substitutions, Aitken's method is applied to find the root of Equation (3.5.2)

$$f(x) = \exp(x) - x^2 \tag{3.10.5}$$

transformed under the form adapted to the method of successive substitutions

$$x = F(x) = x - (\exp(x) - x^2) \tag{3.10.6}$$

The initial chosen value is $x = 0$. The results are gathered in Table 3.11. We notice that Aitken's method neatly improves the convergence with respect to the method of successive substitutions. However, as the base method used here is little efficient for that problem, Aitken's method applied to that method of successive substitutions does not reach the performance of Newton type or secant methods.

Table 3.11 Aitken's method to improve the convergence of the method of successive substitutions: root finding of Equation (3.5.2)

	Successive substitutions		Aitken	
Iteration	x	$f(x)$	v	$f(v)$
3	−0.3679	0.5569	−0.6127	0.1665
4	−0.9247	−0.4585	−0.6639	0.0740
5	−0.4662	0.4100	−0.6733	0.0567
6	−0.8762	−0.3514	−0.6827	0.0392
7	−0.5248	0.3162	−0.6870	0.0311
8	−0.8411	−0.2761	−0.6913	0.0231
9	−0.5649	0.2492	−0.6936	0.0186
10	−0.8142	−0.2199	−0.6959	0.0143
11	−0.5943	0.1988	−0.6974	0.0116
12	−0.7931	−0.1765	−0.6987	0.0091
13	−0.6166	0.1596	−0.6996	0.0074
14	−0.7762	−0.1423	−0.7004	0.0058
15	−0.6339	0.1288	−0.7010	0.0048
16	−0.7626	−0.1151	−0.7015	0.0038
17	−0.6475	0.1041	−0.7018	0.0031
18	−0.7516	−0.0933	−0.7022	0.0025
19	−0.6583	0.0844	−0.7024	0.0020
20	−0.7427	−0.0757	−0.7026	0.0016
30	−0.7175	−0.0269	−0.7034	0.0002
40	−0.7085	−0.0096	−0.7035	0.0000
50	−0.7053	−0.0034	−0.7035	0.0000

3.11 Homotopy Method

3.11.1 Introduction

The continuation method by homotopy (Bondy 1991; Ficken 1951; Watson 1986) allows us to find the solution of a nonlinear equation by following a path from an initial point that is the solution of a simpler function. The original function and the simpler function are incorporated in a parameterized initial value problem of ordinary differential equations that is solved by a predictor–corrector method.

The case of the search of real and complex roots of non-polynomial functions $f(x)$ or with transcendental terms is the most difficult. There also exist many cases where the homotopy method fails (Nocedal and Wright 1999).

3.11.2 Continuation Method

An often used homotopy consists of generating a function $h(x,t)$ that is a linear combination of the function $f(x)$ whose roots are searched for and a second function $g(x)$ for which a root is known or can be easily obtained

$$h(x,t) = t\, f(x) + (1-t)\, g(x) \tag{3.11.1}$$

where t is the homotopy parameter that allows us to follow the path that joins an arbitrary initial point x^0 to x^*, solution of $f(x) = 0$. By varying t from 0 to 1, a sequence of solutions is obtained, which draws a path toward the solution x^*.

The two functions $g(x)$ mostly used in homotopy are Newton's homotopy

$$g(x) = f(x) - f(x^0) \tag{3.11.2}$$

and fixed point homotopy

$$g(x) = x - x^0 \tag{3.11.3}$$

The differential form of Equation (3.11.1) with respect to curvilinear abscissa s (s is the length of the arc along the curve joining x^0 to x) is obtained by differentiating $h(x,t)$ with respect to s

$$\frac{\partial h}{\partial x}\frac{dx}{ds} + \frac{\partial h}{\partial t}\frac{dt}{ds} = 0 \tag{3.11.4}$$

with moreover the equation verified by ds

$$ds^2 = dx^2 + dt^2 \Rightarrow \left(\frac{dx}{ds}\right)^2 + \left(\frac{dt}{ds}\right)^2 = 1 \tag{3.11.5}$$

For a real valued function, Equations (3.11.1), (3.11.4), and (3.11.5) can be solved by Kubicek's method (Kubicek 1976) that uses an Adams–Bashforth predictor to solve the ordinary differential equations (3.11.4) and (3.11.5) and a Newton corrector to solve the homotopy (3.11.1).

Gritton et al. (2001) extended this method to functions of a complex variable.

Solving this type of problem can be performed by integrating ordinary differential equations like in Example 3.12. It can also be done by an algebraic predictor–corrector (see for Example 3.13) functioning in the following way. Let a point (x, t) belong to the path. It is possible to build a predicted point according to equation

$$(x^p, t^p) = (x, t) + \epsilon \frac{1}{dl} \left(\frac{dx}{ds}, \frac{dt}{ds} \right) \qquad \text{with } dl = \pm \sqrt{\left(\frac{dx}{ds} \right)^2 + \left(\frac{dt}{ds} \right)^2} \tag{3.11.6}$$

where ϵ will influence the step length. According to Nocedal and Wright (1999), the sign of dl can be determined by using the heuristics that the tangent vector $\left(\frac{dx}{ds}, \frac{dt}{ds} \right)$ to the new point must form an angle lower than $\pi/2$ with the tangent vector at the old point. As the predicted point (x^p, t^p) does not belong exactly to the sought path, correcting iterations must be done (algebraic predictor–corrector scheme) until converging to a new point (x^n, t^n) really belonging to the sought path such that $h(x^n, t^n) = 0$. Nocedal and Wright (1999) propose to calculate the corrections $(\delta x, \delta t)$ according to a Newton–Raphson scheme

$$\begin{bmatrix} \delta x \\ \delta t \end{bmatrix} = - \begin{bmatrix} \frac{\partial h}{\partial x} & \frac{\partial h}{\partial t} \\ & e_i \end{bmatrix}^{-1} \begin{bmatrix} h \\ 0 \end{bmatrix} \tag{3.11.7}$$

where e_i is a unit vector of dimension $n + 1$ (in the case where $x \in \mathbb{R}^n$) whose ith coordinate is 1 (the other ones being equal to 0), and where the choice of i is done by selecting the ith component of (x, t) that varies the most during the corrections. To avoid brutal passing from a variation with respect to a coordinate to another one, it may be preferable to choose a unit vector that has no zero coordinates and will act like a weighting.

It must be noted that the continuation methods can fail in some cases. Thus, Nocedal and Wright (1999) show that there exists no path allowing us to go from $t = 0$ to $t = 1$ for the function

$$h(x, t) = t\,(x^2 - 1) + (1 - t)\,(x + 2) \tag{3.11.8}$$

with $f(x) = x^2 - 1$, $g(x) = x - x_0$, $x_0 = -2$. Indeed, the path starting from $(x_0, t) = (-2, 0)$ has no limit (see for Example 3.13).

Example 3.12 :

Continuation method by homotopy

Shacham (1989) gives several examples of problems in chemical engineering that he solves by a continuation method by homotopy. One such example deals with the chemical equilibrium of the system N_2-H_2-NH_3, for which the function $f(x)$ is

$$f(x) = \frac{8\,(4 - x)^2\, x^2}{(6 - 3\,x)^2\,(2 - x)} - 0.1886 = 0 \tag{3.11.9}$$

which is transformed under the form

$$f(x) = 8\,(4 - x)^2\, x^2 - 0.1886\,(6 - 3\,x)^2\,(2 - x) = 0 \tag{3.11.10}$$

We choose the start point $x^0 = 1$ at $t = 0$, and the homotopy

$$h(x, t) = t\, f(x) + (1 - t)\,(x - x^0) \tag{3.11.11}$$

From a practical point of view (numerical solving), Equations (3.11.4) and (3.11.5) are not present under the simple form of an explicit system of two first order ordinary differential equations. For that reason, we performed the following transformations:

$$\frac{dx}{ds} = \epsilon_1 \sqrt{1 - \left(\frac{dt}{ds}\right)^2} \tag{3.11.12}$$

with $\epsilon_1 = \pm 1$. Equation (3.11.4) then becomes

$$\frac{\partial h}{\partial x} \epsilon_1 \sqrt{1 - \left(\frac{dt}{ds}\right)^2} + \frac{\partial h}{\partial t}\frac{dt}{ds} = 0 \tag{3.11.13}$$

hence the expression

$$\frac{dt}{ds} = \epsilon_2 \frac{\dfrac{\partial h}{\partial x}}{\sqrt{\left(\dfrac{\partial h}{\partial x}\right)^2 + \left(\dfrac{\partial h}{\partial t}\right)^2}} \tag{3.11.14}$$

that can be simultaneously integrated with (3.11.12) knowing the partial derivatives $\frac{\partial h}{\partial x}$ and $\frac{\partial h}{\partial t}$. It can be shown that

$$\epsilon_2 = -\epsilon_1 . \text{sign}\left(\frac{\partial h}{\partial t}\right) \tag{3.11.15}$$

An hypothesis must be done about the sign of ϵ_1 that can be verified from the final value of $f(x)$ that should be zero. If it is not the case, just change the sign of ϵ_1.

On another side, it can be noticed that the integration of the system of ordinary differential equations (3.11.12) and (3.11.14) deals with the variable s, whereas what prevails is in fact the variable t that must vary from 0 to 1. Thus s starts at 0, but its final value is unknown. An integration must thus be done while surveying the t that will give the end signal. Figure 3.9 illustrates the dependence of t and of x with respect to s. The advancement of s was stopped at $s = 1.578$ when t reached the value 1. Furthermore, Figure 3.10 shows the variations of x and $f(x)$ versus t. The sought value x^* that makes $f(x)$ equal to zero is found at $t = 1$ and is equal to $x^* = 0.2794$.

Example 3.13 :
Continuation method by homotopy

The example given by Nocedal and Wright (1999) is treated

$$f(x) = x^2 - 1 \tag{3.11.16}$$

The function has two roots -1 and $+1$. We use $g(x) = x - x_0$ with three different values of x_0, that is $x_0 = -2$, or $x_0 = -0.5$, or $x_0 = 0.5$. It results

$$h(x, t) = t\,(x^2 - 1) + (1 - t)\,(x - x_0) \tag{3.11.17}$$

If we use the code lsoda designed to solve stiff systems of ordinary differential equations, the present system does not converge for any of the three tested values of x_0. Instead of using the method based on the integration of ordinary differential equations, we decide to use the algebraic predictor–corrector described by Equation (3.11.7) that allows the convergence in some cases (Figures 3.11 and 3.12) but does not ensure it whatever x_0 (Figure 3.11).

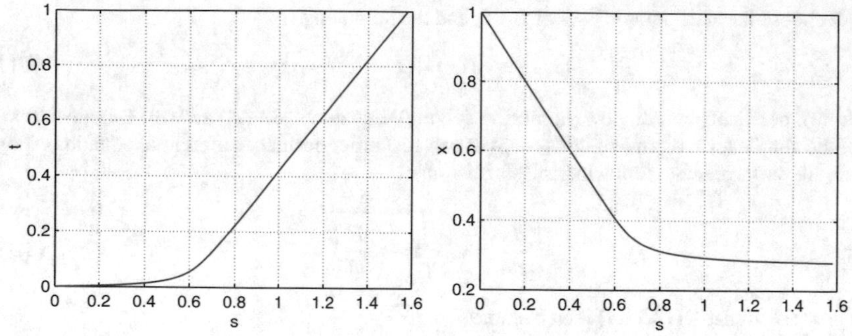

Fig. 3.9 Homotopy method: Evolution of t versus s (left). Evolution of x versus s (right). Integration with lsoda

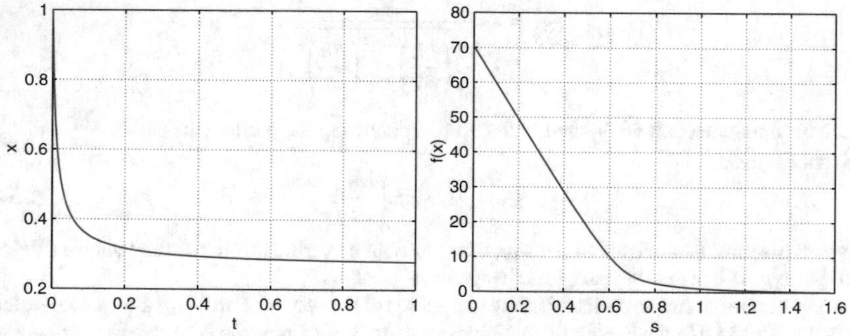

Fig. 3.10 Homotopy method: Evolution of x versus t (left). Evolution of f versus t (right). Integration with lsoda

Figure 3.11 shows that x continuously decreases from $x = -2$ and that the function $f(x)$ continuously increases without converging to 0. If a different value of x_0 is used, for example $x_0 = -0.5$, x converges to +1 (Figure 3.12), or if $x_0 = +0.5$, x converges to +1 (Figure 3.13). We found no value for x_0 such that x converges to -1; thus only one of the two roots was found.

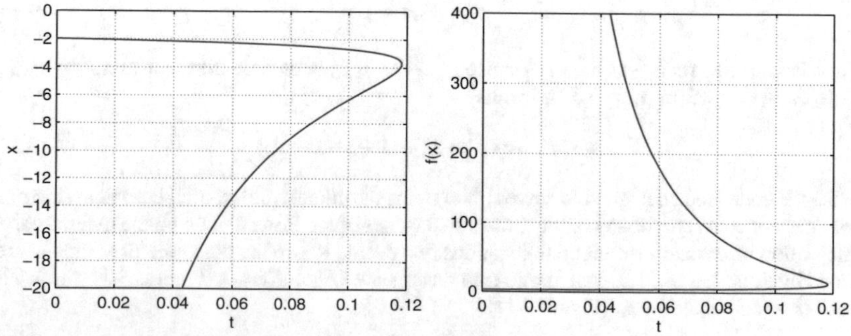

Fig. 3.11 Homotopy method: case of non-convergence with $x_0 = -2$. Evolution of x versus t (left). Evolution of f versus t (right). Solving by the algebraic method (3.11.7)

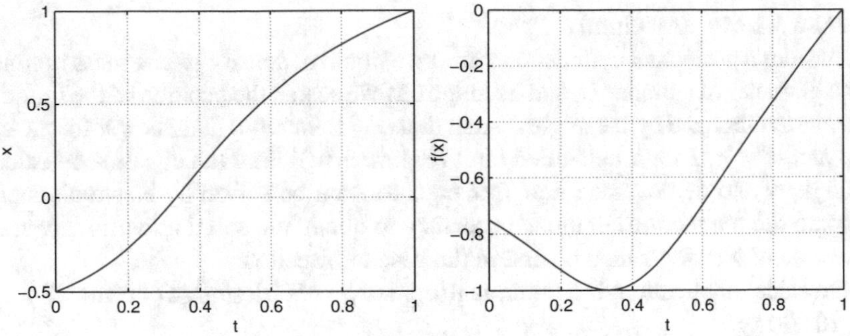

Fig. 3.12 Homotopy method: case of convergence to +1 with $x_0 = -0.5$. Evolution of x versus t (left). Evolution of f versus t (right). Solving by the algebraic method (3.11.7)

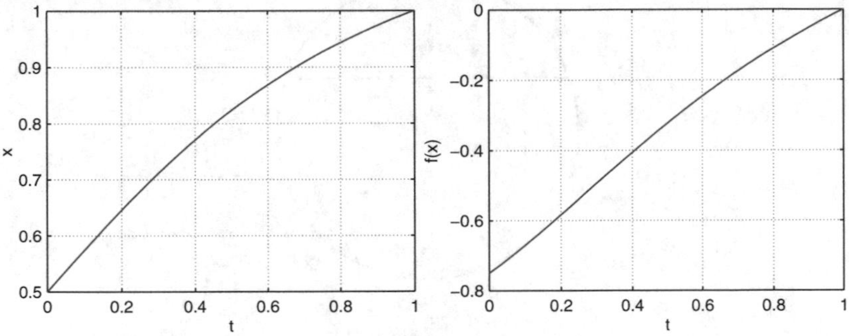

Fig. 3.13 Homotopy method: case of convergence to +1 with $x_0 = +0.5$. Evolution of x versus t (left). Evolution of f versus t (right). Solving by the algebraic method (3.11.7)

3.12 Discussion and Conclusion

Some of the presented methods (Graeffe, Bernoulli, and Bairstow) only offer a theoretical interest. Many methods are available and usable, some of them are robust like bisection or regula falsi, but the rate of convergence of the method must be examined. Wegstein's method gives an excellent convergence. Aitken's method that allows us to accelerate the convergence rate of linear convergence methods should be more often used. Newton's method possesses a quadratic convergence and must be privileged, even if the analytic expression of the derivative must be replaced by its numerical approximation. The homotopy method is delicate and not frequently used.

3.13 Exercise Set

Preamble: For many exercises, students can write simple general programs to solve the problems.

Exercise 3.13.1 (Medium)

Consider a horizontal cylinder containing a liquid of density ρ. The mass of liquid is m, the diameter of cylinder D, and its length L. We search the position of the liquid–gas interface symbolized by the angle α such that $\alpha = 0$ for a full tank, $\alpha = \pi$ for an empty tank, and $\alpha = \pi/2$ for a half-filled tank (Figure 3.14). Find the equation to calculate the angle α and demonstrate how this equation can be solved by Newton's method. Compare the number of iterations necessary to obtain the solution with a given error in the case of Newton's method and in the case of bisection.

Numerical application: the liquid is propane. $\rho = 500\,\text{kg/m}^3$, $D = 2\,\text{m}$, $L = 10\,\text{m}$, $m = 10{,}000\,\text{kg}$.

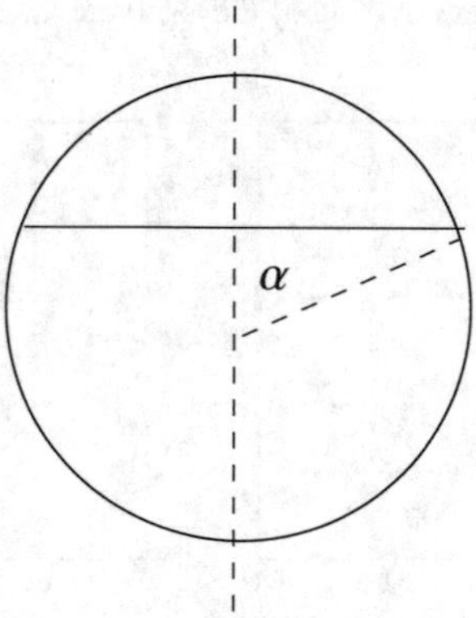

Fig. 3.14 Position of the liquid–gas interface identified by α in the horizontal cylinder

Exercise 3.13.2 (Medium)

The exercise is similar to Example 3.13.1, but in the present case, the tank is a sphere of diameter D instead of being a cylinder. The mass of liquid is m. We search the position of the liquid–gas interface symbolized by the angle α such that $\alpha = 0$ for a full tank, $\alpha = \pi$ for an empty tank, and $\alpha = \pi/2$ for a half-filled tank (Figure 3.15). Find the equation to calculate the angle α and demonstrate how this equation can be solved by Newton's method. Compare the number of iterations necessary to obtain the solution with a given error in the case of Newton's method and in the case of bisection. Information:

(a) If V_g is the volume occupied by the gas phase and h the height of gas in the vertical axis, R the radius of the sphere, h responds to the following equation:

$$h^3 - 3\,R\,h^2 + 3\,\frac{V_g}{\pi} = 0 \tag{3.13.1}$$

(b) A general cubic polynomial

$$x^3 + b\,x^2 + c\,x + d = 0 \tag{3.13.2}$$

is transformed into an equation of the form

$$X^3 + p\,X + q = 0 \tag{3.13.3}$$

by posing $X = x + b/3$ with $p = c - b^3/3$ and $q = 2b^3/27 + d - bc/3$.
Let $D = 4\,p^3 + 27\,q^2$.
If $D > 0$, this equation has two complex roots and a real root equal to

$$X = \left(-\frac{q}{2} + \sqrt{\left(\frac{q}{2}\right)^2 + \left(\frac{p}{3}\right)^3}\right)^{1/3} + \left(-\frac{q}{2} - \sqrt{\left(\frac{q}{2}\right)^2 + \left(\frac{p}{3}\right)^3}\right)^{1/3} \tag{3.13.4}$$

If $D < 0$, this equation has three distinct real roots.
If $D = 0$, this equation has three roots $X_1 = -\sqrt{-p/3}$, $X_2 = -X_1$, $X_3 = 3p/q$.

The general solution of Equation (3.13.3) is obtained by posing: $X = \sqrt{-\frac{4p}{3}}\cos\theta$.
The following equation results

$$4\cos^3\theta - 3\cos\theta = \cos(3\,\theta) = -q\sqrt{-\frac{27}{4p^3}} \tag{3.13.5}$$

hence the general solution

$$X = \sqrt{-\frac{4p}{3}}\cos\left(\theta + \frac{2\,k\,\pi}{3}\right) \quad , k = 0, 1, 2 \tag{3.13.6}$$

(c) The angle α is equal to

$$\alpha = \sin\left(\frac{1}{R}\sqrt{\frac{\frac{6\,V_g}{\pi\,h} - h^2}{3}}\right) \tag{3.13.7}$$

Numerical application: the liquid is propane. $\rho = 500\,\text{kg/m}^3$, $D = 10\,\text{m}$, $m = 160{,}000\,\text{kg}$.

Exercise 3.13.3 (Medium)

The pressure drop for an incompressible fluid (Bird et al. 2002, page 180) flowing in a rough pipe of length L and diameter D is given by

$$\Delta P = \frac{4\rho u_m^2 L f}{D} \tag{3.13.8}$$

where u_m is the mean flow velocity and f is the friction factor. In laminar regime, according to Hagen–Poiseuille equation, the friction factor depends only on the dimensionless Reynolds number Re

$$f = \frac{16}{Re} \quad \text{with} \quad Re = \frac{\rho u_m D}{\mu} \quad \text{when } Re < 2100 \tag{3.13.9}$$

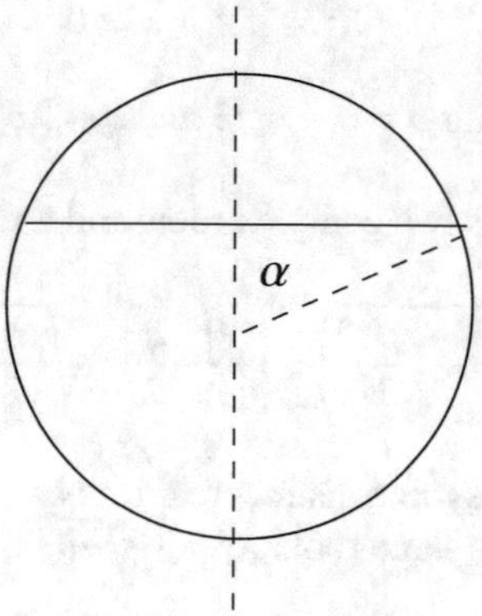

Fig. 3.15 Position of the liquid–gas interface identified by α in the sphere

On the contrary, in turbulent regime, the friction factor depends both on the Reynolds number and on the tube rugosity ϵ. When the Reynolds number is larger than 2100, Churchill's equation can be used

$$\frac{1}{\sqrt{f}} = -1.737 \ln\left(\frac{0.27\epsilon}{D} + \frac{1.256}{Re\sqrt{f}}\right) \quad \text{when } Re > 2100 \tag{3.13.10}$$

A good initial approximation for the value of f is obtained by means of Blasius relation for a flat tube ($\epsilon = 0$)

$$f = 0.0791\, Re^{-0.25} \tag{3.13.11}$$

Write a program allowing you to calculate the pressure drop ΔP from the known values (Table 3.12), flow rate Q, tube diameter D, length L, rugosity ϵ, fluid density ρ, and viscosity μ.

Table 3.12 Data for two different cases

Data	Case 1	Case 2
Q [m^3/h]	39	1
D [cm]	7.8	1.6
L [m]	3050	30
ρ [kg/m^3]	1000	1290
μ [kg/m s]	1.04×10^{-3}	7.44×10^{-2}
ϵ [m]	5.1×10^{-5}	1.27×10^{-5}

Exercise 3.13.4 (Easy)

A correlation allowing us to calculate the saturated vapor pressure of a liquid with respect to temperature is extended Antoine equation (Yaws and Yang 1989):

$$\ln P = a_1 + \frac{a_2}{a_3 + T} + a_4 T + a_5 T^2 + a_6 \ln T \qquad (3.13.12)$$

For propane, when the pressure is expressed in Pa and temperature in K, the parameters a_i are equal to

$$\begin{aligned} a_1 &= -0.128 \times 10^3 \\ a_2 &= 0.835 \times 10^{-3} \\ a_3 &= -0.338 \times 10^3 \\ a_4 &= -0.125 \\ a_5 &= 0.829 \times 10^{-4} \\ a_6 &= 0.302 \times 10^2 \end{aligned} \qquad (3.13.13)$$

Calculate the saturated vapor temperature corresponding to a saturated vapor pressure equal to: $P = 16 \times 10^5$ Pa, by using Newton's method and taking for initial value $T = 300$ K. Present the results (with intermediate calculations) in a table.

Exercise 3.13.5 (Medium)

Van der Waals equation for a real fluid (Smith et al. 2018) is a cubic equation of the form

$$\left(P + \frac{a}{V^2}\right)(V - b) = RT \qquad (3.13.14)$$

where P is the pressure, T the temperature, V the molar volume, and R the universal gas constant. Let us consider ethane for which the critical temperature and pressure are $T_c = 305.4$ K and $P_c = 48.2 \times 10^5$ Pa, respectively. The parameters a and b are equal to

$$a = \frac{27}{64}\frac{R^2 T_c^2}{P_c}, \quad b = \frac{RT_c}{8P_c} \qquad (3.13.15)$$

Furthermore, we know that the critical volume is $V_c = 0.148 \times 10^{-3}$ m^3/mol, which can give an order of magnitude for initialization values. At a pressure of 33×10^5 Pa and a temperature of 280 K, calculate the roots (molar volumes) of this equation by detailing the used method. Comment the results obtained.

Data: $R = 8.314$ Pa m^3 mol^{-1} K^{-1}

Exercise 3.13.6 (Difficult)

This exercise is an extension of Exercise 3.13.5. The objective is to calculate the liquid–vapor equilibrium and the corresponding saturated vapor pressure (Smith et al. 2018). Van der Waals equation for a real fluid is a cubic equation of the form

$$\left(P + \frac{a}{V^2}\right)(V - b) = RT \qquad (3.13.16)$$

Let us consider ethane for which the critical temperature and pressure are $T_c = 305.4$ K and $P_c = 48.2 \times 10^5$ Pa, respectively. The parameters a and b are equal to

$$a = \frac{27}{64}\frac{R^2 T_c^2}{P_c}, \quad b = \frac{RT_c}{8P_c} \qquad (3.13.17)$$

Furthermore, we know that $V_c = 0.148 \times 10^{-3}$ m^3/mol, which can give an order of magnitude for initialization values. At a pressure of 33×10^5 Pa and a temperature of 280 K, calculate the roots (molar volumes V_l, V_i, V_g) of this equation by a method such as Newton–Raphson. The molar volumes of the liquid, intermediate phase (without physical signification), and gas are ordered by increasing values. The condition of liquid–vapor equilibrium for a saturated vapor pressure P_{sat} corresponds to the equality of the areas included between the curve $P = f(V)$ and the horizontal straight line $P = P_{sat}$ according to

$$\int_{V_l}^{V_i} (P - P_{sat})dV = \int_{V_i}^{V_g} (P - P_{sat})dV \tag{3.13.18}$$

Thus, to find that liquid–vapor equilibrium, the pressure must be swept so that Equation (3.13.18) is satisfied. Calculate the saturated vapor pressure and the corresponding liquid and gas molar volumes.

Data: $R = 8.314$ Pa m^3 mol^{-1} K^{-1}

Exercise 3.13.7 (Medium)

Consider a fourth degree polynomial equal to

$$P(x) = x^4 - x^3 - 72\,x^2 + 516\,x + 1396 \tag{3.13.19}$$

1. Find the expansion of this polynomial with respect to Chebyshev polynomials.
2. Find the roots of this polynomial by Bairstow's method. Choose $(r, q) = (-10, -20)$ as initial couple to limit the number of iterations.

Exercise 3.13.8 (Medium)

Let the sequence be defined by

$$x_0 = 0\,, \quad x_{n+1} = \sqrt{2 + x_n} \quad \text{for } n \geq 0 \tag{3.13.20}$$

1. Find in two ways (mathematical and numerical) the limit of that sequence.
2. Determine the numerical type of convergence (linear, superlinear, quadratic, ...) of this sequence.

Exercise 3.13.9 (Difficult)

In the case of the isentropic flow of an ideal gas from a tank through a convergent–divergent nozzle (Anderson 2003; Zucker and Biblarz 2002), operating at the sound speed at the throat, the following formula is applied

$$\frac{A_g^2}{A^2} = \left(\frac{\gamma+1}{2}\right)^{\frac{\gamma+1}{\gamma-1}} \left(\frac{2}{\gamma-1}\right) \left[\left(\frac{P}{P_1}\right)^{\frac{2}{\gamma}} - \left(\frac{P}{P_1}\right)^{\frac{\gamma+1}{\gamma}}\right] \tag{3.13.21}$$

where P is the pressure at any point where the nozzle has a cross section area A, while P_1 is the pressure of the tank, A_g the cross section at the throat (gorge), and γ the ratio of the heat capacity at constant pressure over the heat capacity at constant volume. A_g and P_1 are data of the system.

Knowing that, for any given cross section area A, there exist two possible pressures P that satisfy Equation (3.13.21), we propose to think in the following manner:

- First, search analytically the minimum of an associated function

$$g(x) = x^{\frac{2}{\gamma}} - x^{\frac{\gamma+1}{\gamma}} \tag{3.13.22}$$

- Deduce two domains that can contain the roots.
- Calculate these roots by a convenient numerical method. Give the results along iterations in a table.
  Values of parameters:
  $A_g = 0.01\ \text{m}^2$, $\gamma = 1.4$, $P_1 = 7$ atm, $A = 0.012\ \text{m}^2$.

Exercise 3.13.10 (Medium)
Let the polynomial

$$P(x) = x^3 - 3.6\,x^2 - 23.31\,x + 39.68 \tag{3.13.23}$$

1. Find a root of this polynomial by Newton's method starting from $x = 0$. Deduce the other roots.
2. Find the roots of this polynomial by Bairstow's method.

References

A. C. Aitken. On Bernoulli's numerical solution of algebraic equations. *Proc. Royal Soc. Edinburgh*, 46:289, 1925.

J. D. Anderson. *Modern Compressible Flow with Historical Perspective*. McGraw Hill, Boston, 3rd edition, 2003.

R. B. Bird, W. E. Stewart, and E. N. Lightfoot. *Transport phenomena*. Wiley, New York, second edition, 2002.

R. W. Bondy. Physical continuation approaches to solving reactive distillation problems. In *AIChE Meeting, Los Angeles*, 1991.

R. L. Burden and J. D. Faires. *Numerical Analysis*. Brooks/Cole, Boston, 9th edition, 2011.

C. M. Crowe and M. Nishio. Convergence promotion in the simulation of chemical processes. *AIChE J.*, 21(3):520–533, 1975.

J. A. Ezquerro, M. A. Hernandez, and N. Romero. Newton-type methods of high order and domains of semilocal and global convergence. *Applied Mathematics and Computation*, 214:142–154, 2009.

F. A. Ficken. The continuation method for functional equations. *Communications on Pure and Applied Mathematics*, 4(4):435–456, 1951.

B. A. Finlayson. *Introduction to Chemical Engineering Computing*. Wiley, Hoboken, 2014.

A. Fortin. *Analyse numérique pour ingénieurs*. Presses Internationales Polytechnique, Canada, 3ème edition, 2008.

K. S. Gritton, J. D. Seader, and W. J. Lin. Global homotopy continuation procedures for seeking all roots of a nonlinear equation. *Comp. Chem. Engng.*, 25:1003–1019, 2001.

L. V. Kantorovich and G. P. Akilov. *Functional Analysis*. Pergamon Press, Oxford, 1982.

M. Kubicek. Algorithm 502. Dependence of solution of nonlinear systems on a parameter. *ACM Transactions on Mathematical Software*, 2(1):98, 1976.

J. Nocedal and S. J. Wright. *Numerical optimization*. Springer-Verlag, New York, 1999.

O. Orbach and C. M. Crowe. Convergence promotion in the simulation of chemical processes with recycle - the dominant eigenvalue method. *Can. J. Chem. Eng.*, 49: 509, 1971.

M. Shacham. An improved memory method for the solution of a nonlinear equation. *Chem. Eng. Sci.*, 27:2099–2101, 1989.

J. M. Smith, H. C. Van Ness, M. M. Abbott, and M. T. Swihart. *Introduction to Chemical Engineering Thermodynamics*. McGraw-Hill, New York, 8th edition, 2018.

L. T. Watson. Numerical linear algebra aspects of globally convergent homotopy methods. *SIAM Review*, 28(4):529–544, 1986.

J. H. Wegstein. Accelerating convergence of iterative processes. *Comm. Assoc. Comput. Mach.*, 1(6):9, 1958.

R. D. Zucker and O. Biblarz. *Fundamentals of Gas Dynamics*. Wiley, Hoboken, 2nd edition, 2002.

C. L. Yaws and H. C. Yang. To estimate vapor pressure easily. Antoine coefficients relate vapor pressure to temperature for almost 700 major organic compounds. *Hydrocarbon Processing*, 68(10):65–68, 1989

Chapter 4
Numerical Operations on Matrices

4.1 Introduction

A very large part of numerical algorithms makes use of more or less intricate matrix operations. For this reason, a chapter dealing only on reminders about matrices and most common transformations is necessary (Golub and Van Loan 1991; Rotella and Borne 1995). Furthermore, many specific techniques about matrix use, transformation, and computation are exposed.

4.2 Reminder About Matrices

Matrices with real or complex elements are considered. The matrix $\boldsymbol{A}$ is an array whose typical element is a_{ij} where i is the row index and j the column index. It is an $m \times n$ matrix if it has m rows and n columns

$$\boldsymbol{A} = \begin{bmatrix} a_{11} & \dots & a_{1n} \\ \vdots & a_{ij} & \vdots \\ a_{m1} & \dots & a_{mn} \end{bmatrix} \tag{4.2.1}$$

Matrices $\boldsymbol{A}$ and $\boldsymbol{B}$ of same size $m \times n$, thus having the same number of rows and columns, can be added

$$\boldsymbol{S} = \boldsymbol{A} + \boldsymbol{B} \iff s_{ij} = a_{ij} + b_{ij} \tag{4.2.2}$$

The *addition* is a commutative operation. The *product* of two matrices, $\boldsymbol{A}$ of size $m \times n$ and $\boldsymbol{B}$ of size $n \times p$, is equal to the $m \times p$ matrix $\boldsymbol{P}$

$$\boldsymbol{P} = \boldsymbol{A}\boldsymbol{B} \iff p_{ij} = \sum_{k=1}^{n} a_{ik} b_{kj} \tag{4.2.3}$$

In general, the product of two matrices is not commutative.

J.-P. Corriou, *Numerical Methods and Optimization*, Springer Optimization and Its Applications 187, https://doi.org/10.1007/978-3-030-89366-8_4

A matrix $\boldsymbol{A}$ is *square* if it has the same number of rows and columns, $m = n$. The *order* of the matrix is equal to n.

The $n \times n$ *identity* matrix $\boldsymbol{I}$ is such that all its elements are zero except its diagonal elements ($i = j$) that are all equal to 1. A *square* matrix $\boldsymbol{A}$ possesses the property

$$\boldsymbol{A}\boldsymbol{I} = \boldsymbol{I}\boldsymbol{A} = \boldsymbol{A} \tag{4.2.4}$$

The *transpose* matrix $\boldsymbol{A}^T$ of size $n \times m$ of the matrix $\boldsymbol{A}$ of size $m \times n$ is such that its typical element $a^T_{ij} = a_{ji}$. It possesses the following properties:

$$(\boldsymbol{A} + \boldsymbol{B})^T = \boldsymbol{A}^T + \boldsymbol{B}^T \quad \text{and} \quad (\boldsymbol{A}\boldsymbol{B})^T = \boldsymbol{B}^T \boldsymbol{A}^T \tag{4.2.5}$$

A matrix $\boldsymbol{A}$ is *symmetric* if and only if $\boldsymbol{A} = \boldsymbol{A}^T$.

If we note $\bar{a}_{ij}$ the complex conjugate number of a_{ij}, and $\bar{\boldsymbol{A}}$ the matrix obtained by replacing a_{ij} by its complex conjugate $\bar{a}_{ij}$, the following property is obtained:

$$\overline{\boldsymbol{A}\boldsymbol{B}} = \bar{\boldsymbol{A}}\bar{\boldsymbol{B}} \tag{4.2.6}$$

Let $\boldsymbol{A}^*$ be the transpose conjugate matrix of $\boldsymbol{A}$, that is $\boldsymbol{A}^* = (\bar{\boldsymbol{A}})^T$. The following properties are observed:

$$(\bar{\boldsymbol{A}})^T = \overline{(\boldsymbol{A}^T)} \tag{4.2.7}$$

$$(\boldsymbol{A}\boldsymbol{B})^* = \boldsymbol{B}^* \boldsymbol{A}^* \tag{4.2.8}$$

A matrix $\boldsymbol{A}$ is *Hermitian* if and only if $\boldsymbol{A}^* = \boldsymbol{A}$. A real matrix is Hermitian if and only if it is symmetric.

A matrix $\boldsymbol{A}$ is *diagonal* if it is square ($n \times n$) and if all its elements a_{ij} with $i \neq j$ are zero (off-diagonal elements).

A matrix is *scalar* if it is diagonal and all its diagonal elements are equal, $a_{ii} = a$. The identity matrix is a particular scalar matrix.

If there exists a matrix $\boldsymbol{B}$ such that $\boldsymbol{A}\boldsymbol{B} = \boldsymbol{I}$, then $\boldsymbol{B}$ is called *inverse* of $\boldsymbol{A}$ and noted $\boldsymbol{A}^{-1}$. The following properties are verified:

$$\boldsymbol{A}\boldsymbol{A}^{-1} = \boldsymbol{A}^{-1}\boldsymbol{A} = \boldsymbol{I} \tag{4.2.9}$$

In the case where $\boldsymbol{A}^{-1}$ exists, $\boldsymbol{A}$ is *nonsingular*.
If the inverses of matrices $\boldsymbol{A}$ and $\boldsymbol{B}$ exist, then we have the property

$$(\boldsymbol{A}\boldsymbol{B})^{-1} = \boldsymbol{B}^{-1}\boldsymbol{A}^{-1} \tag{4.2.10}$$

The square matrix $\boldsymbol{B}$ is *congruent* to the square matrix $\boldsymbol{A}$ of same order if there exists a nonsingular matrix $\boldsymbol{Q}$ such that

$$\boldsymbol{B} = \boldsymbol{Q}^T \boldsymbol{A} \boldsymbol{Q} \tag{4.2.11}$$

It results that:

1. If $\boldsymbol{A}$ is symmetric, $\boldsymbol{B}$ is symmetric.
2. If $\boldsymbol{A}$ and $\boldsymbol{B}$ are unitary, $\boldsymbol{Q}^T = \boldsymbol{Q}^{-1}$.

A square matrix is *unitary* if $\boldsymbol{A}^* = \boldsymbol{A}^{-1}$
A square matrix is *orthogonal* if $\boldsymbol{A}^* = \boldsymbol{A}^{-1} = \boldsymbol{A}^T$
The diagonal elements of a square matrix have the same row and column indexes, a_{ii}. If all elements below the diagonal are zero, the matrix is *upper triangular* and if moreover, the diagonal elements are zero, the matrix is strictly upper triangular. In the same way, a *lower triangular* matrix and a strictly lower triangular matrix are defined.

The *determinant* associated to the square matrix $\boldsymbol{A}$ is the number noted $\det(\boldsymbol{A})$. It can be shown that

$$\det(\boldsymbol{A}^T) = \det(\boldsymbol{A}), \quad \det(\boldsymbol{AB}) = \det(\boldsymbol{A})\det(\boldsymbol{B}), \quad \det(\boldsymbol{A}^{-1}) = [\det(\boldsymbol{A})]^{-1} \quad (4.2.12)$$

$$\det(\bar{\boldsymbol{A}}) = \overline{\det(\boldsymbol{A})}, \quad \det(\boldsymbol{A}^*) = \overline{\det(\boldsymbol{A})} \quad (4.2.13)$$

If $\boldsymbol{A}$ is a triangular matrix, its determinant is equal to the product of the diagonal terms.

The *minor* of a_{ij} is the determinant of the submatrix obtained by removing the row i and the column j from the matrix $\boldsymbol{A}$. The *cofactor* of a_{ij} is equal to the product of the minor of a_{ij} by $(-1)^{i+j}$. Let $\boldsymbol{A}_{ij}$ be the cofactor of a_{ij}. The determinant of $\boldsymbol{A}$ is equal to

$$\det(\boldsymbol{A}) = \sum_{j=1}^{n} a_{lj}\boldsymbol{A}_{lj} = \sum_{i=1}^{n} a_{ic}\boldsymbol{A}_{ic} \quad (4.2.14)$$

where l is an index of any row and c an index of any column.

The *adjoint* matrix of $\boldsymbol{A}$ noted $\text{adj}(\boldsymbol{A})$ is the transpose matrix of the matrix of cofactors. The adjoint matrix of $\boldsymbol{A}$ noted $\text{adj}(\boldsymbol{A})$ is such that

$$\boldsymbol{A}\,\text{adj}(\boldsymbol{A}) = \det(\boldsymbol{A})\boldsymbol{I} \quad (4.2.15)$$

The typical element adj_{ij} of the adjoint matrix is equal to the cofactor $\boldsymbol{A}_{ji}$ of a_{ji}

$$adj_{ij} = \boldsymbol{A}_{ji} \quad (4.2.16)$$

If the determinant of $\boldsymbol{A}$ is nonzero, the matrix $\boldsymbol{A}$ is *invertible*. As $\det(\det(\boldsymbol{A})\boldsymbol{I}) = [\det(\boldsymbol{A})]^n$, it results that $\det(\text{adj}(\boldsymbol{A})) = [\det(\boldsymbol{A})]^{n-1}$.

Consider $\boldsymbol{A}$ of size $m \times n$. From that matrix $\boldsymbol{A}$, it is possible to form square submatrices whose determinant can be calculated. The *rank* of $\boldsymbol{A}$ is the order of the nonsingular submatrix of higher order.

A matrix is *sparse* if most of its elements are zero. On the opposite, a matrix is *dense* if most of its elements are nonzero. A *band* matrix is formed of nonzero elements which are diagonal and close to the diagonal, all the other elements are zero.

4.3 Reminder on Vectors

A vector $\boldsymbol{v}$ is a column matrix of size $n \times 1$.

The *scalar product* of two vectors $\boldsymbol{u}$ and $\boldsymbol{v}$ is defined by

$$(\boldsymbol{u}, \boldsymbol{v}) = \boldsymbol{u}^*\boldsymbol{v} \quad (4.3.1)$$

Example: $\boldsymbol{u}^T = [1 + 2i, 3i, -1 - i, 4 - 2i]$, $\boldsymbol{v}^T = [2 + i, 4 - 2i, 1 + 7i, i]$

The scalar product is thus equal to

$$\begin{aligned}(\boldsymbol{u}, \boldsymbol{v}) &= [1 - 2i, -3i, -1 + i, 4 + 2i][2 + i, 4 - 2i, 1 + 7i, i]^T \\ &= (1 - 2i)(2 + i) + (-3i)(4 - 2i) + (-1 + i)(1 + 7i) + (4 + 2i)(i) \\ &= -12 - 17i\end{aligned} \tag{4.3.2}$$

The scalar product verifies the following properties:

- With respect to commutativity: $(\boldsymbol{u}, \boldsymbol{v}) = \overline{(\boldsymbol{v}, \boldsymbol{u})}$
- Multiplication by a scalar: $(\lambda \boldsymbol{u}, \boldsymbol{v}) = \bar{\lambda}(\boldsymbol{u}, \boldsymbol{v})$
- Associativity: $(\boldsymbol{u} + \boldsymbol{v}, \boldsymbol{w}) = (\boldsymbol{u}, \boldsymbol{w}) + (\boldsymbol{v}, \boldsymbol{w})$
- Norm or length of $\boldsymbol{u}$: $\|\boldsymbol{u}\|$ such that $\|\boldsymbol{u}\|^2 = (\boldsymbol{u}, \boldsymbol{u})$
- Null vector: $(\boldsymbol{u}, \boldsymbol{u}) = \boldsymbol{0} \iff \boldsymbol{u} = \boldsymbol{0}$

Two vectors are *orthogonal* if $(\boldsymbol{u}, \boldsymbol{v}) = 0$. A vector is *normed* or *unitary* if $\|\boldsymbol{u}\| = 1$. A set of vectors is *orthogonal* if these vectors are orthogonal between themselves. Moreover, this set is *orthonormal* if they are unitary. An orthonormal set $\{\boldsymbol{u}_1, \ldots, \boldsymbol{u}_n\}$ verifies the following property $\overline{[\boldsymbol{u}_1 \ldots \boldsymbol{u}_n]^T} [\boldsymbol{u}_1 \ldots \boldsymbol{u}_n] = \boldsymbol{I}$.

Consider a vector space of dimension n and n linearly independent vectors $\boldsymbol{V}_i$. Then, it is possible to build a set of n orthogonal vectors $\boldsymbol{W}_i$ following Gram–Schmidt orthogonalization procedure

$$\begin{cases} \boldsymbol{W}_1 = \boldsymbol{V}_1 \\ \boldsymbol{W}_k = \boldsymbol{V}_k - \displaystyle\sum_{i=1}^{k-1} \frac{(\boldsymbol{V}_k, \boldsymbol{W}_i)}{(\boldsymbol{W}_i, \boldsymbol{W}_i)} \boldsymbol{W}_i \,, \quad k = 2, \ldots, n \end{cases} \tag{4.3.3}$$

This algorithm can lead to instabilities, and it is preferable to slightly modify the previous algorithm so that the vectors $\boldsymbol{U}_i$ deduced from the orthogonal vectors $\boldsymbol{W}_i$ are orthonormal. We set

$$\boldsymbol{U}_1 = \frac{\boldsymbol{W}_1}{\|\boldsymbol{W}_1\|} \tag{4.3.4}$$

then

$$\begin{aligned} &\boldsymbol{U}_j = \frac{\boldsymbol{W}_j^{(j)}}{\|\boldsymbol{W}_j^{(j)}\|} \quad \text{with} \quad \boldsymbol{W}_k^{(j+1)} = \boldsymbol{W}_k^{(j)} - (\boldsymbol{W}_k^{(j)}, \boldsymbol{U}_j)\boldsymbol{U}_j \\ &\text{for} \quad j = 1, \ldots, n \quad \text{and} \quad k = j + 1, \ldots, n \end{aligned} \tag{4.3.5}$$

where $(\boldsymbol{W}_k^{(j)}, \boldsymbol{U}_j)$ is the scalar product. At iteration 1, we set $\boldsymbol{W}_j^{(1)} = \boldsymbol{W}_j$ for $j = 1, \ldots, n$.

The matrices $\boldsymbol{A}_1, \boldsymbol{A}_2, \ldots, \boldsymbol{A}_p$ $m \times n$ are *linearly dependent* if there exist scalars $\lambda_1, \lambda_2, \ldots, \lambda_p$ such that they form a linear combination equal to zero, $\lambda_1 \boldsymbol{A}_1 + \lambda_2 \boldsymbol{A}_2 + \cdots + \lambda_p \boldsymbol{A}_p = \boldsymbol{0}$. In the opposite case, the set of matrices is *linearly independent*.

A *vector space* is of dimension p if there exist only p linearly independent vectors belonging to that space and if any set of $p+1$ vectors would be linearly dependent. Any vector of the space can be expressed as a linear combination of a set of independent vectors generating the space that thus form a *basis* of the space.

The vector space $\mathcal{V}_n$ of all $n \times 1$ matrices is generated by the orthonormal vectors $\boldsymbol{\epsilon}_i = [0, \ldots, 1, \ldots, 0]^T$ where the only nonzero value occupies the ith position and is equal to 1. Any vector $\boldsymbol{v}$ of $\mathcal{V}_n$ can be written as a linear combination of the vectors of the basis

$$\boldsymbol{v} = \sum_{i=1}^{n} c_i \boldsymbol{\epsilon}_i \tag{4.3.6}$$

If we consider $n+1$ vectors $\boldsymbol{v}_i$ of $\mathcal{V}_n$, each of these vectors can be written as a linear combination of the vectors of the basis

$$\boldsymbol{v}_j = \sum_{i=1}^{n} a_{ji} \boldsymbol{\epsilon}_i \tag{4.3.7}$$

We can prove that it is possible to find constants α_i, all of them nonzero, such that

$$\sum_{j=1}^{n+1} \alpha_j \boldsymbol{v}_j = \sum_{i=1}^{n} \boldsymbol{\epsilon}_i \sum_{j=1}^{n+1} \alpha_j a_{ji} = 0 \tag{4.3.8}$$

The dimension of the vector space $\mathcal{V}_n$ is n, and the vectors $\boldsymbol{\epsilon}_i$ constitute a basis of that space.

An $m \times n$ matrix $\boldsymbol{A}$ can be represented as

$$\boldsymbol{A} = [\boldsymbol{v}_1 \ldots \boldsymbol{v}_n] \tag{4.3.9}$$

each vector of dimension m being a column of the matrix. The maximum number of columns linearly independent is equal to the maximum number of rows linearly independent and is the rank of the matrix.

4.4 Linear Transformations and Subspaces

A set of vectors $\mathcal{W}$ of a vector space $\mathcal{V}$ forms a vector subspace of $\mathcal{V}$ if and only if, whatever the scalar α and any two vectors $\boldsymbol{v}_1$ and $\boldsymbol{v}_2$ of $\mathcal{W}$, $\alpha \boldsymbol{v}_1$ and $\boldsymbol{v}_1 + \boldsymbol{v}_2$ belong to $\mathcal{W}$. Let $\boldsymbol{A}$ be a given $n \times n$ matrix. Let $\mathcal{V}_n$ be a vector space of dimension n. The set of vectors of the form $\boldsymbol{A}\boldsymbol{v}$ where $\boldsymbol{v}$ is any vector of $\mathcal{V}_n$ constitutes a subspace of $\mathcal{V}_n$, thus a set of linearly independent vectors $\boldsymbol{A}\boldsymbol{v}$ can constitute a basis for the space of the vectors thus formed. It can be noticed that the ith column of the matrix $\boldsymbol{A}$ is equal to $\boldsymbol{A}\boldsymbol{\epsilon}_i$, thus

$$\boldsymbol{v} = \sum_{j=1}^{n} c_j \boldsymbol{\epsilon}_j \Rightarrow \boldsymbol{A}\boldsymbol{v} = \sum_{j=1}^{n} c_j \boldsymbol{A}\boldsymbol{\epsilon}_j \tag{4.4.1}$$

The dimension of the subspace $\mathcal{W}$ of vectors $\boldsymbol{A}\boldsymbol{v}$ is the rank of $\boldsymbol{A}$.

A transformation $\mathcal{T}$ is linear if and only if, whatever the vectors $\boldsymbol{u}$ and $\boldsymbol{v}$ of $\mathcal{V}$ and the scalar α

$$\mathcal{T}(\boldsymbol{u} + \boldsymbol{v}) = \mathcal{T}\boldsymbol{u} + \mathcal{T}\boldsymbol{v} \tag{4.4.2}$$

$$\mathcal{T}(\alpha \boldsymbol{u}) = \alpha \mathcal{T}\boldsymbol{u} \tag{4.4.3}$$

The square matrices of order n can thus be considered as linear operators for the vector space $\mathcal{V}_n$.

Let $\boldsymbol{A}$ be an $n \times n$ matrix and $\boldsymbol{v}$ a vector of $\mathcal{V}_n$ such that

$$\boldsymbol{A}\boldsymbol{v} = 0 \tag{4.4.4}$$

If $\boldsymbol{A}$ is nonsingular, its determinant is nonzero and the solution of the system is the zero vector $\boldsymbol{v} = \boldsymbol{0}$. If $\boldsymbol{A}$ is singular, its determinant is zero and there exist several solutions $\boldsymbol{v}$ different from 0. The set of vectors whose image through a given matrix $\boldsymbol{A}$ is the zero vector is a subspace of $\mathcal{V}_n$ called *kernel*, also called *nullspace*. Its rank is the dimension of the kernel of $\boldsymbol{A}$, and the dimension of the vector space $\mathcal{V}_n$ is the sum of the dimension of the kernel of $\boldsymbol{A}$ and of the rank of $\boldsymbol{A}$. The dimension of the kernel of $\boldsymbol{A}$ is the nullity of $\boldsymbol{A}$.

According to the transfer rule, the scalar product $(\boldsymbol{u}, \boldsymbol{A}\boldsymbol{v})$, also noted as $\boldsymbol{u}^*(\boldsymbol{A}\boldsymbol{v})$ (matrix product) is given by

$$(\boldsymbol{u}, \boldsymbol{A}\boldsymbol{v}) = (\boldsymbol{A}^*\boldsymbol{u}, \boldsymbol{v}) \tag{4.4.5}$$

indeed as $(\boldsymbol{A}^*\boldsymbol{u})^* = \boldsymbol{u}^*\boldsymbol{A}$, we have $\boldsymbol{u}^*(\boldsymbol{A}\boldsymbol{v}) = (\boldsymbol{A}^*\boldsymbol{u})^*\boldsymbol{v}$.

A nonzero vector $\boldsymbol{u}$ is an *eigenvector* of $\boldsymbol{A}$ if and only if there exists a scalar λ (which may be equal to zero), such that

$$\boldsymbol{A}\boldsymbol{u} = \lambda\boldsymbol{u} \tag{4.4.6}$$

the scalar λ is the associated *eigenvalue*. The eigenvectors $\boldsymbol{u}$ and the eigenvalues verify the relation

$$(\boldsymbol{A} - \lambda\boldsymbol{I})\boldsymbol{u} = \boldsymbol{0} \tag{4.4.7}$$

Let

$$F(\lambda) = \det(\boldsymbol{A} - \lambda\boldsymbol{I}) = 0 \tag{4.4.8}$$

the function $F(\lambda)$ is called *characteristic equation or characteristic polynomial* of the matrix $\boldsymbol{A}$. If the coefficients of the matrix $\boldsymbol{A}$ are a_{ij}, the matrix $(\boldsymbol{A} - \lambda\boldsymbol{I})$ is equal to

$$\boldsymbol{A} - \lambda\boldsymbol{I} = \begin{bmatrix} a_{11} - \lambda & a_{12} & \dots & a_{1n} \\ a_{21} & a_{22} - \lambda & \dots & a_{2n} \\ & & \ddots & \\ \dots & \dots & & \dots \\ a_{n1} & a_{n2} & \dots & a_{nn} - \lambda \end{bmatrix} \tag{4.4.9}$$

A matrix and its transpose have the same characteristic equation.

We could simply think that, to calculate the eigenvalues of a matrix, it thus suffices to calculate the coefficients of the characteristic polynomial, then to determine the roots of this polynomial. Actually, when the order n is large, a small perturbation in the coefficients of the polynomial induces an important error in the roots of the polynomial. Thus, this technique is not recommended (Hairer 1993).

If a matrix $\boldsymbol{A}$ is real, and if λ_1 and λ_2 two eigenvalues such that $\bar{\lambda}_1 \neq \lambda_2$, an eigenvector $\boldsymbol{u}$ associated to λ_1 such that $\boldsymbol{A}\boldsymbol{u} = \lambda_1\boldsymbol{u}$ and an eigenvector $\boldsymbol{v}$ associated to λ_2 such that $\boldsymbol{A}^T\boldsymbol{v} = \lambda_2\boldsymbol{v}$ are orthogonal.

Demonstration:

On one side, we have $(\boldsymbol{v}, \boldsymbol{A}\boldsymbol{u}) = (\boldsymbol{v}, \lambda_1 \boldsymbol{u}) = \lambda_1(\boldsymbol{v}, \boldsymbol{u})$
and on the other side $(\boldsymbol{v}, \boldsymbol{A}\boldsymbol{u}) = (\boldsymbol{A}^T \boldsymbol{v}, \boldsymbol{u}) = (\lambda_2 \boldsymbol{v}, \boldsymbol{u}) = \bar{\lambda}_2(\boldsymbol{v}, \boldsymbol{u})$
hence $(\boldsymbol{v}, \boldsymbol{u}) = 0$

The *trace* of a matrix is defined by

$$\mathrm{Tr}(\boldsymbol{A}) = \sum_{i=1}^{n} \lambda_i = \sum_{i=1}^{n} a_{ii} \tag{4.4.10}$$

and is thus equal to the sum of the diagonal elements.

The product of the eigenvalues is equal to the determinant of the matrix

$$\prod_{i=1}^{n} \lambda_i = \det(\boldsymbol{A}) \tag{4.4.11}$$

The *spectrum* of a matrix is the set of the eigenvalues of the matrix

$$\mathrm{Sp}(\boldsymbol{A}) = \{\lambda\ ,\ \lambda \text{ eigenvalue of } \boldsymbol{A}\} \tag{4.4.12}$$

The *spectral radius* $\rho(\boldsymbol{A})$ of a matrix $\boldsymbol{A}$ is defined as the maximum of the modulus of the eigenvalues of $\boldsymbol{A}$

$$\rho(\boldsymbol{A}) = \max\{|\lambda|\ ,\ \lambda \in \mathrm{Sp}(\boldsymbol{A})\} \tag{4.4.13}$$

4.4.1 Gershgorin Theorem

A theorem allowing to check the possible singularity of a square matrix of order n is as follows:

$$\text{If} \quad |a_{ii}| > \sum_{\substack{j=1 \\ j \neq i}}^{n} |a_{ij}| \quad \text{or if} \quad |a_{ii}| > \sum_{\substack{j=1 \\ j \neq i}}^{n} |a_{ji}| \tag{4.4.14}$$

then $\boldsymbol{A}$ is nonsingular. Such a matrix is named *diagonally dominant* and is frequently met in the models of physical systems.

This theorem finds an application to locate in the complex plane the eigenvalues of a matrix with potentially complex elements. The eigenvalues are situated in the union $\cup \mathcal{D}_i$ of the Gershgorin disks (often called circles) defined by

$$\mathcal{D}_i = \{\lambda \in \mathbb{C} \text{ such that} \quad |\lambda - a_{ii}| \leq \sum_{\substack{j=1 \\ j \neq i}}^{n} |a_{ij}|\ , \quad i = 1, \ldots, n\} \tag{4.4.15}$$

and in the union $\cup \mathcal{D}_j$ of the disks defined by

$$\mathcal{D}_j = \{\lambda \in \mathbb{C} \text{ such that } |\lambda - a_{jj}| \leq \sum_{\substack{i=1 \\ i \neq j}}^{n} |a_{ij}|\ , \quad j = 1, \ldots, n\} \tag{4.4.16}$$

The condition on the sum of the off-diagonal elements in each row is often the only one expressed. Indeed, as the transpose matrix A^T has the same eigenvalues as matrix A, the condition (4.4.16) on the columns obviously derives from the condition (4.4.15) on the rows.

In summary, Gershgorin theorem says that the largest eigenvalue in terms of modulus of the matrix A cannot be larger than the largest sum of the moduli of the coefficients of any row or column of A.

Example 4.1 :
Gershgorin theorem: Location of eigenvalues

Consider the following matrix:

$$\begin{bmatrix} 1 & -5 & 2 \\ 3 & 6 & -1 \\ 2 & 4 & -3 \end{bmatrix} \tag{4.4.17}$$

The eigenvalues of that matrix are $\lambda_1 = -4.1217$ and $\lambda_{2\,or\,3} = 4.0608 \pm 2.9974i$. The application of Gershgorin theorem gives the following results to the disks defined by the sum of the off-diagonal elements in each row:

$$\begin{array}{ll} \text{(a)} & |\lambda - 1| \le 7 \\ \text{(b)} & |\lambda - 6| \le 4 \\ \text{(c)} & |\lambda + 3| \le 6 \end{array} \tag{4.4.18}$$

Gershgorin theorem gives the following results to the disks defined by the sum of the off-diagonal elements in each column:

$$\begin{array}{ll} \text{(a)} & |\lambda - 1| \le 5 \\ \text{(b)} & |\lambda - 6| \le 9 \\ \text{(c)} & |\lambda + 3| \le 3 \end{array} \tag{4.4.19}$$

Gershgorin circles resulting on one hand of the union of the disks defined by condition (4.4.18) and on the other hand of the union of the disks defined by condition (4.4.19) have been represented on Figure 4.1. It can be noticed that the eigenvalues belong to the intersection of both unions.

4.4.2 Cayley–Hamilton Theorem and Consequences

Given a square matrix A of dimension $n \times n$, the characteristic equation associated to that matrix is

$$P_n(\lambda) = b_0\,\lambda^n + b_1\,\lambda^{n-1} + \cdots + b_{n-1}\,\lambda + b_n = 0 \tag{4.4.20}$$

with $b_0 = (-1)^n$. $P_n(\lambda)$ is the determinant of the matrix $(A - \lambda I)$.

According to Cayley–Hamilton theorem, the matrix A itself verifies the polynomial equation with respect to matrices

$$P_n(A) = b_0\,A^n + b_1\,A^{n-1} + \cdots + b_{n-1}\,A + b_n\,I = 0 \tag{4.4.21}$$

which is expressed as "any square matrix itself satisfies its own characteristic equation."

Suppose now $b_0 = 1$, otherwise all coefficients can be multiplied by (-1).

Cayley–Hamilton theorem has several consequences:

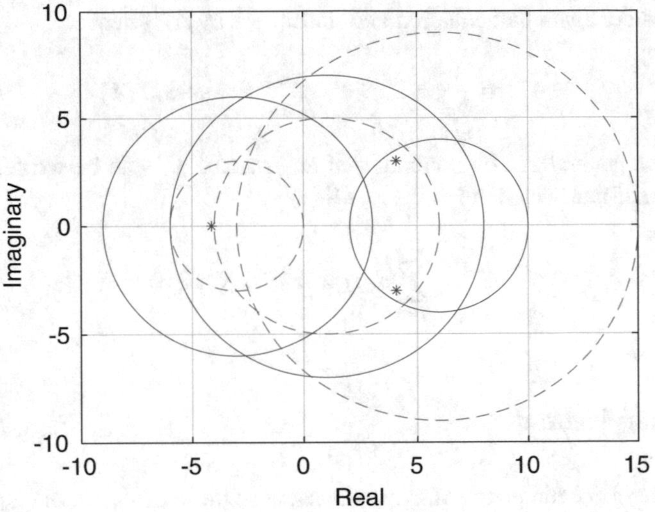

Fig. 4.1 Gershgorin circles and eigenvalues for the matrix (4.4.17). The union resulting from condition (4.4.18) corresponds to disks of radius 4, 6, 7. The union resulting from condition (4.4.19) corresponds to disks of radius 3, 5, 9. Both unions are represented by a different symbol

- The matrix $\boldsymbol{A}^n$ is a linear combination of lower power matrices

$$\boldsymbol{A}^n = -[b_1\,\boldsymbol{A}^{n-1} + \cdots + b_{n-1}\,\boldsymbol{A} + b_n\,\boldsymbol{I}] \tag{4.4.22}$$

- By multiplying Equation (4.4.22) by $\boldsymbol{A}$, we get

$$\begin{aligned} \boldsymbol{A}^{n+1} &= -[b_1\,\boldsymbol{A}^n + \cdots + b_{n-1}\,\boldsymbol{A}^2 + b_n\,\boldsymbol{A}] \\ &= -b_1(-[b_1\,\boldsymbol{A}^{n-1} + \cdots + b_{n-1}\,\boldsymbol{A} + b_n\,\boldsymbol{I}]) \\ &\quad -[b_2\,\boldsymbol{A}^{n-1} + \cdots + b_{n-1}\,\boldsymbol{A}^2 + b_n\,\boldsymbol{A}] \\ &= c_0\,\boldsymbol{A}^{n-1} + \cdots + c_{n-3}\,\boldsymbol{A}^2 + c_{n-2}\,\boldsymbol{A} + c_{n-1}\,\boldsymbol{I} \end{aligned} \tag{4.4.23}$$

which shows iteratively that any matrix $\boldsymbol{A}^{n+i}$ can be written as a linear combination of matrices $\boldsymbol{A}^j$ $(j = 0, \ldots, n-1)$

$$\boldsymbol{A}^{n+i} = \sum_{j=0}^{n-1} c_{n-1-j}\,\boldsymbol{A}^j \qquad \forall\ i \geq 0 \tag{4.4.24}$$

- Taking Equation (4.4.22) in the other direction, we have

$$\boldsymbol{I} = \frac{1}{b_n}[\boldsymbol{A}^n + b_1\,\boldsymbol{A}^{n-1} + \cdots + b_{n-1}\,\boldsymbol{A}] \tag{4.4.25}$$

We can consider again that equation and multiply it by $\boldsymbol{A}^{-1}$, hence

$$\boldsymbol{A}^{-1} = \frac{1}{b_n}[\boldsymbol{A}^{n-1} + b_1\,\boldsymbol{A}^{n-2} + \cdots + b_{n-1}\,\boldsymbol{I}] \tag{4.4.26}$$

which can be generalized by noticing that any matrix $\boldsymbol{A}^{-i}$ can be written as a linear combination of matrices $\boldsymbol{A}^j$ ($j = 0, \ldots, n-1$)

$$\boldsymbol{A}^{-i} = \sum_{j=0}^{n-1} d_{n-1-j}\,\boldsymbol{A}^j \qquad \forall\ i \geq 0 \tag{4.4.27}$$

4.4.3 Power Method

The eigenvectors of the matrix $\boldsymbol{A}^{-1}$ are the same as those of $\boldsymbol{A}$, but the eigenvalues of the matrix $\boldsymbol{A}^{-1}$ are the inverse of the eigenvalues of $\boldsymbol{A}$.

To calculate the eigenvalues of a matrix, the recommended algorithm is based on the power method which consists in posing the following iterative algorithm:

$$\boldsymbol{y}_{k+1} = \boldsymbol{A}\,\boldsymbol{y}_k \tag{4.4.28}$$

where $\boldsymbol{y}_0$ is an arbitrary vector. The following properties (Hairer 1993) can be shown

(a) $\boldsymbol{y}_k$ tends to an eigenvector of $\boldsymbol{A}$.
(b) $\boldsymbol{y}_k^*\boldsymbol{A}\boldsymbol{y}_k/\boldsymbol{y}_k^*\boldsymbol{y}_k$ tends to the eigenvalue of $\boldsymbol{A}$ of largest modulus.

Indeed, consider the eigenvalues ordered as $|\lambda_1| > |\lambda_2| \geq |\lambda_3| \cdots \geq |\lambda_n|$, and the associated normed eigenvectors $\boldsymbol{V}_1, \ldots, \boldsymbol{V}_n$ (with $\|\boldsymbol{V}_i\|_2 = 1$). Supposing that $\boldsymbol{A}$ is diagonalizable, choose as initial vector $\boldsymbol{y}_0 = \sum_{i=1}^n \alpha_i\,\boldsymbol{V}_i$. The algorithm (4.4.28) gives at iteration k

$$\begin{aligned}\boldsymbol{y}_k &= \boldsymbol{A}^k\,\boldsymbol{y}_0 \\ &= \textstyle\sum_{i=1}^n \alpha_i\,\lambda_i^k\,\boldsymbol{V}_i = \lambda_1^k\,(\alpha_1\,\boldsymbol{V}_1 + \alpha_2\left(\frac{\lambda_2}{\lambda_1}\right)^k\,\boldsymbol{V}_2 + \cdots + \alpha_n\left(\frac{\lambda_n}{\lambda_1}\right)^k\,\boldsymbol{V}_n)\end{aligned} \tag{4.4.29}$$

This result can be symbolized under the form

$$\boldsymbol{y}_k = \lambda_1^k(\alpha_1\,\boldsymbol{V}_1 + 0\left(\left|\frac{\lambda_2}{\lambda_1}\right|^k\right) \tag{4.4.30}$$

As, along the iterations, the modulus of $\boldsymbol{y}_k$ rapidly increases, it is recommended at the following iteration to replace the vector $\boldsymbol{y}_k$ by $\boldsymbol{y}_k/\|\boldsymbol{y}_k\|$.

On another side, we have the following equalities:

$$\begin{aligned}
\boldsymbol{y}_k^* \boldsymbol{A}\, \boldsymbol{y}_k &= \boldsymbol{y}_k^* \boldsymbol{y}_{k+1} = \sum_{i=1}^n |\alpha_i|^2\, |\lambda_i|^{2k} \lambda_i + \sum_{i \neq j} \bar{\alpha}_i\, \alpha_j \bar{\lambda}_i^k\, \lambda_j^{k+1}\, V_i^*\, V_j \\
&= |\alpha_1|^2\, |\lambda_1|^{2k}\, \lambda_1\, (1 + 0\left(|\frac{\lambda_2}{\lambda_1}|^k\right)) \\
\boldsymbol{y}_k^* \boldsymbol{y}_k &= \sum_{i=1}^n |\alpha_i|^2\, |\lambda_i|^{2k} + \sum_{i \neq j} \bar{\alpha}_i\, \bar{\alpha}_j \bar{\lambda}_i^k\, \lambda_j^k\, \boldsymbol{V}_i^*\, \boldsymbol{V}_j \\
&= |\alpha_1|^2\, |\lambda_1|^{2k}\, (1 + 0\left(|\frac{\lambda_2}{\lambda_1}|^k\right))
\end{aligned} \tag{4.4.31}$$

hence the result announced in (b).

When the eigenvalues λ_1 and λ_2 have very close moduli, the convergence is slow and the algorithm (4.4.28) must be modified according to Wielandt method (Hairer 1993) by replacing the matrix $\boldsymbol{A}$ in Equation (4.4.28) by $(\boldsymbol{A} - \beta \boldsymbol{I})$ where β is an approximation of the sought eigenvalue λ_1.

The power method can be generalized and allows to motivate the iterative QR factorization which is in general used to calculate the eigenvalues of a matrix (Golub and Van Loan 1991).

Example 4.2 :

Power method: calculation of eigenvalue of largest modulus and corresponding eigenvector

Consider the matrix

$$\boldsymbol{A} = \begin{bmatrix} 1 & 3 & 5 \\ 2 & 6 & 4 \\ 3 & 1 & 7 \end{bmatrix} \tag{4.4.32}$$

whose eigenvalues ordered by decreasing modulus are $\lambda_1 = 10.816$, $\lambda_2 = 4.2325$, $\lambda_3 = -1.0485$ and the corresponding eigenvectors

$$\boldsymbol{V}_1 = \begin{bmatrix} -0.4895 \\ -0.6684 \\ -0.5600 \end{bmatrix}, \quad \boldsymbol{V}_2 = \begin{bmatrix} -0.9375 \\ -0.0728 \\ 0.3404 \end{bmatrix}, \quad \boldsymbol{V}_3 = \begin{bmatrix} -0.1219 \\ -0.8839 \\ 0.4515 \end{bmatrix}$$

The initial vector chosen is a very simple vector which favors no direction. If the vector $\boldsymbol{y}_k$ is normed at each iteration, the iterative vectors are in Table 4.1.

Table 4.1 Power method: calculation of the eigenvalue of largest modulus and corresponding eigenvector

Iteration k	1	2	3	4	5	6
Vector y_k normed when $k > 1$	$\begin{bmatrix} 1 \\ 1 \\ 1 \end{bmatrix}$	$\begin{bmatrix} 0.4838 \\ 0.6451 \\ 0.5914 \end{bmatrix}$	$\begin{bmatrix} 0.4914 \\ 0.6585 \\ 0.5700 \end{bmatrix}$	$\begin{bmatrix} 0.4899 \\ 0.6647 \\ 0.5641 \end{bmatrix}$	$\begin{bmatrix} 0.4897 \\ 0.6669 \\ 0.5616 \end{bmatrix}$	$\begin{bmatrix} 0.4896 \\ 0.6678 \\ 0.5606 \end{bmatrix}$
$\boldsymbol{y}_k^* \boldsymbol{A} \boldsymbol{y}_k / \boldsymbol{y}_k^* \boldsymbol{y}_k$		10.9364	10.8531	10.8320	10.8222	10.8185

It shows that $\boldsymbol{y}_k^* \boldsymbol{A} \boldsymbol{y}_k / \boldsymbol{y}_k^* \boldsymbol{y}_k$ converges to the eigenvalue of $\boldsymbol{A}$ of largest modulus, and also that the vector y tends to the eigenvector V_1.

4.5 Similar Matrices and Matrix Polynomials

A matrix $\boldsymbol{B}$ is *similar* to a matrix $\boldsymbol{A}$ if there exists a matrix $\boldsymbol{P}$ such that $\boldsymbol{B} = \boldsymbol{P}^{-1}\boldsymbol{A}\boldsymbol{P}$. $\boldsymbol{P}$ is called a *change of basis* matrix. The matrices $\boldsymbol{A}$ and $\boldsymbol{B}$ have the same eigenvalues. We will pass from the eigenvectors $\boldsymbol{u}_i$ of $\boldsymbol{A}$ to the eigenvectors $\boldsymbol{v}_i$ of $\boldsymbol{B}$ by a square nonsingular matrix $\boldsymbol{P}$, $\boldsymbol{v}_i = \boldsymbol{P}^{-1}\boldsymbol{u}_i$ so that $\boldsymbol{U} = \boldsymbol{P}\boldsymbol{V}$.

Suppose that all the eigenvalues of the matrix $\boldsymbol{A}$ are distinct. Let $\boldsymbol{u}_i$ be the eigenvectors corresponding to the eigenvalues λ_i and $\boldsymbol{U}$ the matrix formed by the eigenvectors (its ith column is the vector $\boldsymbol{u}_i$). In this case

$$\boldsymbol{A}\,\boldsymbol{U} = \boldsymbol{U}\,\mathrm{diag}(\lambda_1, \ldots, \lambda_n) \tag{4.5.1}$$

and

$$\boldsymbol{U}^{-1}\,\boldsymbol{A}\,\boldsymbol{U} = \mathrm{diag}(\lambda_1, \ldots, \lambda_n) \tag{4.5.2}$$

The diagonal matrix formed by the eigenvalues of $\boldsymbol{A}$ is thus similar to $\boldsymbol{A}$.

Whatever the matrix $\boldsymbol{A}$, it is also possible to determine a nonsingular unitary matrix $\boldsymbol{U}$ such that

$$\boldsymbol{U}^{-1}\,\boldsymbol{A}\,\boldsymbol{U} = \boldsymbol{T} \tag{4.5.3}$$

where $\boldsymbol{T}$ is an upper triangular matrix whose diagonal elements are the eigenvalues of $\boldsymbol{A}$. $\boldsymbol{T}$ is thus similar to $\boldsymbol{A}$.

If the matrix $f(\boldsymbol{A})$ is defined as the polynomial of matrix $\boldsymbol{A}$ equal to

$$f(\boldsymbol{A}) = \sum_{i=1}^{p} c_i\,\boldsymbol{A}^{p-i} \tag{4.5.4}$$

where $\boldsymbol{A}^0$ is the identity matrix, and if $\boldsymbol{B}$ is a nonsingular matrix

$$f(\boldsymbol{B}^{-1}\,\boldsymbol{A}\,\boldsymbol{B}) = \boldsymbol{B}^{-1}\,f(\boldsymbol{A})\,\boldsymbol{B} \tag{4.5.5}$$

the eigenvalues of the matrix $f(\boldsymbol{A})$ are equal to $f(\lambda_i)$ where λ_i are the eigenvalues of $\boldsymbol{A}$. The definition of polynomials of matrices allows to define sequences and to obtain important properties.

4.6 Symmetric Matrices and Hermitian Matrices

A *Hermitian* form is an expression of type

$$\sum_{i,j=1}^{n} a_{ij}\bar{u}_i u_j \quad \text{with } \bar{a}_{ij} = a_{ji} \tag{4.6.1}$$

defined with respect the complex variables u_i, the a_{ij} being complex constants. Such a form can be written as $(\boldsymbol{u}, \boldsymbol{A}\boldsymbol{u})$ where the matrix $\boldsymbol{A}$ is a Hermitian matrix ($\boldsymbol{A} = \boldsymbol{A}^*$)

A *quadratic* form is an expression of type

$$\sum_{i,j=1}^{n} a_{ij}u_i u_j \quad \text{with } a_{ij} = a_{ji} \tag{4.6.2}$$

defined with respect the complex variables u_i, the a_{ij} being complex constants. In general, the quadratic form is real and the matrix $\boldsymbol{A}$ is a symmetric matrix with real values. If only quadratic real forms are considered (that will be the case here), the quadratic form is equal to the scalar product

$$(\boldsymbol{u}, \boldsymbol{A}\boldsymbol{u}) \quad \text{with } \boldsymbol{A} = \bar{\boldsymbol{A}} = \boldsymbol{A}^T \text{ and } \boldsymbol{u} \text{ real} \tag{4.6.3}$$

If $\boldsymbol{u}$ and $\boldsymbol{A}$ satisfy one of the two previous conditions, then,

(a) Whatever the vector $\boldsymbol{u}$, the scalar product $(\boldsymbol{u}, \boldsymbol{A}\boldsymbol{u})$ is real.
(b) All the eigenvalues of $\boldsymbol{A}$ are real.
(c) The eigenvectors corresponding to distinct eigenvalues are orthogonal.

A matrix $\boldsymbol{A}$ is *unitary* $\iff \boldsymbol{A}^* = \boldsymbol{A}^{-1}$.
A real matrix $\boldsymbol{A}$ real is *orthogonal* $\iff \boldsymbol{A}^T = \boldsymbol{A}^{-1}$.
A matrix $\boldsymbol{A}$ is *isometric* $\iff$ whatever the vectors $\boldsymbol{u}$ and $\boldsymbol{v}$ of dimension n, the scalar products are equal $(\boldsymbol{u}, \boldsymbol{v}) = (\boldsymbol{A}\boldsymbol{u}, \boldsymbol{A}\boldsymbol{v})$

A real unitary matrix is orthogonal. A matrix $\boldsymbol{A}$ is isometric if and only if it is unitary.

If $\boldsymbol{A}$ is any Hermitian matrix, there exists a unitary matrix $\boldsymbol{P}$ such that

$$\boldsymbol{A} = \boldsymbol{P}\boldsymbol{\Lambda}\boldsymbol{P}^* \tag{4.6.4}$$

where $\boldsymbol{\Lambda}$ is the diagonal matrix formed by the eigenvalues of $\boldsymbol{A}$, $\boldsymbol{\Lambda} = [\lambda_1 \dots \lambda_n]$.

A Hermitian matrix $\boldsymbol{A}$ is *positive definite* $\iff$ whatever $\boldsymbol{u} \neq 0$, the scalar product $(\boldsymbol{u}, \boldsymbol{A}\boldsymbol{u}) > 0$ and *positive semidefinite* if the scalar product $(\boldsymbol{u}, \boldsymbol{A}\boldsymbol{u}) \geq 0$. In the same manner, a negative definite matrix is defined. Any Hermitian matrix that cannot be defined in this way is indefinite.

The leading submatrices $\boldsymbol{A}^{(m)}$ of a square matrix $\boldsymbol{A}$ are the square matrices of order m obtained by taking the m first rows and columns of $\boldsymbol{A}$. Their determinants are noted Δ_m. A theorem says that between two distinct eigenvalues of the submatrix $\boldsymbol{A}^{(k+1)}$, there is at least one eigenvalue of $\boldsymbol{A}^{(k)}$ which may coincide with one of both eigenvalues. Moreover, if λ is a multiple eigenvalue of order p for $\boldsymbol{A}^{(k+1)}$, there exists a multiple eigenvalue of order at least $p-1$ for $\boldsymbol{A}^{(k)}$.

For a Hermitian matrix $\boldsymbol{A}$ to be positive definite, one of the following conditions is sufficient:

(a) All its eigenvalues are positive.
(b) The coefficients of the characteristic equation of $\boldsymbol{A}$ have alternate signs.
(c) Each submatrix of $\boldsymbol{A}$ is positive definite.
(d) The determinant of the leading submatrices $(1 \leq i \leq n)$ is strictly positive.

If the eigenvalues of a Hermitian matrix $\boldsymbol{A}$ are ordered, we can show that

$$\lambda_n \leq \frac{(\boldsymbol{u}, \boldsymbol{A}\boldsymbol{u})}{(\boldsymbol{u}, \boldsymbol{u})} \leq \frac{(\boldsymbol{A}\boldsymbol{u}, \boldsymbol{A}\boldsymbol{u})}{(\boldsymbol{u}, \boldsymbol{A}\boldsymbol{u})} \leq \lambda_1 \tag{4.6.5}$$

This property is useful to find the eigenvalues of a Hermitian matrix as soon as one of them has been found.

To find the eigenvalues and the eigenvectors of a matrix, the original matrix will be transformed into a simpler matrix by a series of similar transformations.

If the matrix $\boldsymbol{A}$ is positive definite,

(a) $\boldsymbol{A}$ is nonsingular.
(b) $a_{ii} > 0 \quad \forall\ 1 \le i \le n.$
(c) $\max_{1 \le j \ne k \le n} |a_{jk}| \le \max_{1 \le i \le n} |a_{ii}|.$
(d) $a_{ij}^2 < a_{ii} a_{jj} \quad \forall\ i \ne j.$

A matrix $\boldsymbol{A}$ is *normal* $\Longleftrightarrow$ there exists a unitary matrix $\boldsymbol{u}$ such that

$$\boldsymbol{u}^{-1} \boldsymbol{A} \boldsymbol{u} = \begin{bmatrix} \lambda_1 & & 0 \\ & \ddots & \\ 0 & & \lambda_n \end{bmatrix} \tag{4.6.6}$$

Given any $m \times n$ matrix $\boldsymbol{A}$, the matrix $\boldsymbol{A}^* \boldsymbol{A}$ is positive semidefinite. The eigenvalues of $\boldsymbol{A}^* \boldsymbol{A}$ are positive or zero and they can be written under the form $\lambda_i(\boldsymbol{A}^* \boldsymbol{A}) = \lambda_i = \sigma_i^2$ with $\sigma_i \ge 0$. The numbers σ_i are *singular values* of $\boldsymbol{A}$ that are ordered as $\sigma_1 \ge \sigma_2 \ge \cdots \ge \sigma_n$.

Let $\mathcal{V}$ be a vector space on the field of scalars. The *norm* of a vector is an application of $\mathcal{V}$ in $\mathcal{R}$ which must verify the following properties:

- $\|\boldsymbol{v}\| > 0$ if $\boldsymbol{v} \ne 0$ (positivity)
- $\|\alpha \boldsymbol{v}\| = |\alpha| \, \|\boldsymbol{v}\|$ (multiplication by a scalar)
- $\|\boldsymbol{u} + \boldsymbol{v}\| \le \|\boldsymbol{u}\| + \|\boldsymbol{v}\|$ (triangular inequality)

In a general way, the norm $\|\boldsymbol{v}\|_p$ of a vector is defined by

$$\|\boldsymbol{v}\|_p = \left(\sum_i |v_i|^p \right)^{1/p} \tag{4.6.7}$$

The three following norms are commonly used:

The norm $\|\boldsymbol{v}\|_1$

$$\|\boldsymbol{v}\|_1 = \sum_{i=1}^{n} |v_i| \tag{4.6.8}$$

The Euclidean norm

$$\|\boldsymbol{v}\|_2 = \sqrt{\boldsymbol{v}^* \boldsymbol{v}} = \sqrt{\sum_{i=1}^{n} |v_i|^2} \tag{4.6.9}$$

The maximum norm

$$\|\boldsymbol{v}\|_\infty = \max_i |v_i| \tag{4.6.10}$$

A matrix norm noted $\|\boldsymbol{A}\|$ must verify the four following properties:

- $\|A\| > 0$ if $A \neq 0$ (positivity)
- $\|\alpha A\| = |\alpha| \, \|A\|$ (multiplication by a scalar)
- $\|A + B\| \leq \|A\| + \|B\|$ (triangular inequality)
- $\|A\, B\| \leq \|A\| \, \|B\|$ (product of norms)

The fourth property is new with respect to the properties required by the vector norm.

Given a vector space $\mathcal{V}$ and a square matrix A, the norm of matrix *subordinate* to or *induced* by the given vector norm is defined as

$$\text{lub}(A) = \max_{\|v\|_p \neq 0} \frac{\|Av\|_p}{\|v\|_p} \tag{4.6.11}$$

or still as

$$\text{lub}(A) = \max_{\|v\|_p = 1} \|Av\|_p \tag{4.6.12}$$

This matrix norm is compatible with the vector norm

$$\|Av\| \leq \text{lub}(A)\|v\| \tag{4.6.13}$$

Moreover, lub(A) is the smallest of matrix norms compatible with the vector norm

$$\|Av\| \leq \|A\| \, \|v\| \quad \forall\, v \quad \Rightarrow \frac{\|Av\|}{\|v\|} \leq \|A\| \tag{4.6.14}$$

As

$$\text{lub}(A) = \max \frac{\|Av\|}{\|v\|} \tag{4.6.15}$$

the previous inequality implies

$$\text{lub}(A) \leq \|A\| \tag{4.6.16}$$

In the same manner as vector norms, different norms are used to characterize the matrices,

(a) Maximum absolute column sum norm or 1-norm

$$\|A\|_1 = \max \frac{\|Av\|_1}{\|v\|_1} = \max_j \sum_{i=1}^{n} |a_{ij}| \tag{4.6.17}$$

(b) Schur norm

$$\|A\|_s = \sqrt{\sum_{i=1}^{m} \sum_{j=1}^{n} |a_{ij}|^2} = \sqrt{\text{trace de}\,(A\, A^T)} \tag{4.6.18}$$

This norm is also called trace norm or Frobenius norm.

The 2-norm of A is the square root of the largest eigenvalue of $A^T\, A$

$$\|A\|_2 = \sqrt{\max(\lambda_{A^T\, A})} = \sqrt{\rho(A^T\, A)} \tag{4.6.19}$$

On another side, the spectral radius is

$$\rho(\boldsymbol{A}) = \max \frac{\|\boldsymbol{A}\boldsymbol{v}\|_2}{\|\boldsymbol{v}\|_2} \tag{4.6.20}$$

and

$$\rho(\boldsymbol{A}) \le \|\boldsymbol{A}\| \tag{4.6.21}$$

whatever the matrix norm used.

The following inequality is valid:

$$\|\boldsymbol{A}\|_2 \le \|\boldsymbol{A}\|_s \tag{4.6.22}$$

(c) Maximum absolute row sum norm or ∞-norm

$$\|\boldsymbol{A}\|_\infty = \max \frac{\|\boldsymbol{A}\boldsymbol{v}\|_\infty}{\|\boldsymbol{v}\|_\infty} = \max_i \sum_{j=1}^{n} |a_{ij}| = \text{lub}_\infty(\boldsymbol{A}) \tag{4.6.23}$$

The norms (a) and (b) are submultiplicative, that is $\|\boldsymbol{A}\boldsymbol{B}\| \le \|\boldsymbol{A}\|\|\boldsymbol{B}\|$

The *condition number* of the matrix $\boldsymbol{A}$ is noted cond($\boldsymbol{A}$) and is defined as follows. Suppose the system of linear equations is to be solved

$$\boldsymbol{A}\boldsymbol{x} = \boldsymbol{b} \tag{4.6.24}$$

where $\boldsymbol{x}$ is a vector. If a small variation $\Delta\boldsymbol{x}$ with respect to the solution is done, a displacement $\Delta\boldsymbol{b}$ results such that

$$\boldsymbol{A}(\boldsymbol{x} + \Delta\boldsymbol{x}) = \boldsymbol{b} + \Delta\boldsymbol{b} \tag{4.6.25}$$

hence

$$\frac{\|\Delta\boldsymbol{x}\|}{\|\boldsymbol{x}\|} \le \|\boldsymbol{A}\|\,\|\boldsymbol{A}^{-1}\|\,\frac{\|\Delta\boldsymbol{b}\|}{\|\boldsymbol{b}\|} = \text{cond}(\boldsymbol{A})\,\frac{\|\Delta\boldsymbol{b}\|}{\|\boldsymbol{b}\|} \tag{4.6.26}$$

with

$$\text{cond}(\boldsymbol{A}) = \|\boldsymbol{A}\|\,\|\boldsymbol{A}^{-1}\| \tag{4.6.27}$$

The condition number measures the sensitivity of the relative error on the solution $\boldsymbol{x}$ for variations of the right-hand member $\boldsymbol{b}$.

The largest singular value is the subordinate norm

$$\sigma_1 = \max_{\boldsymbol{v}\neq 0} \frac{\|\boldsymbol{A}\boldsymbol{v}\|_2}{\|\boldsymbol{v}\|_2} = \text{lub}_2(\boldsymbol{A}) \tag{4.6.28}$$

and on the opposite

$$\sigma_n = \min_{\boldsymbol{v}\neq 0} \frac{\|\boldsymbol{A}\boldsymbol{v}\|_2}{\|\boldsymbol{v}\|_2} \tag{4.6.29}$$

In the case where $m = n$ and $\boldsymbol{A}$ is nonsingular, we have

$$\frac{1}{\sigma_n} = \text{lub}_2(\boldsymbol{A}^{-1}) \tag{4.6.30}$$

$$\text{cond}_2(\boldsymbol{A}) = \text{lub}_2(\boldsymbol{A})\,\text{lub}_2(\boldsymbol{A}^{-1}) = \frac{\sigma_1}{\sigma_n} \tag{4.6.31}$$

The ratio of the largest singular value to the smallest singular value thus characterizes the condition number of the nonsingular square matrix $\boldsymbol{A}$. It may occur that the condition number is defined as the inverse of the previous value, i.e. the ratio of the smallest singular value to the largest singular value. In both cases, what matters is the comparison of the condition number to the reference value equal to 1.

Note that if the rank of a matrix is a strict notion at the mathematical level, numerically the rank of a matrix $\boldsymbol{A}$ can be calculated as the number of singular values of $\boldsymbol{A}$ larger than a given scalar threshold $\epsilon > 0$.

Example 4.3 :
Ill conditioning: Hilbert matrices

Hilbert matrices are symmetric square matrices of order n whose typical element is equal to $1/(i+j-1)$. They are well known to be ill-conditioned.

Let $n = 4$. The corresponding Hilbert matrix is equal to

$$\boldsymbol{A} = \begin{bmatrix} 1.0000 & 0.5000 & 0.3333 & 0.2500 \\ 0.5000 & 0.3333 & 0.2500 & 0.2000 \\ 0.3333 & 0.2500 & 0.2000 & 0.1667 \\ 0.2500 & 0.2000 & 0.1667 & 0.1429 \end{bmatrix}$$

The eigenvalues of this matrix are equal to 0.0001, 0.0067, 0.1691, 1.5002, and the condition number is thus equal to $\text{cond}(\boldsymbol{A}) \approx 15514$.

For $n = 12$, the condition number becomes equal to $\text{cond}(\boldsymbol{A}) \approx 1.66 \times 10^{16}$, and the matrix is no more invertible by most computing codes.

4.7 Reduction of Matrices Under a Simple Form

Consider the recurrent sequence $\boldsymbol{A}_i$ of matrices obtained by transformation of the matrix $\boldsymbol{A} = \boldsymbol{A}_0$ as

$$\boldsymbol{A}_i = \boldsymbol{T}_i^{-1}\boldsymbol{A}_{i-1}\boldsymbol{T}_i \tag{4.7.1}$$

After m transformations, a simpler matrix results

$$\boldsymbol{B} = \boldsymbol{A}_m = \boldsymbol{T}^{-1}\boldsymbol{A}\boldsymbol{T} \quad \text{with } \boldsymbol{T} = \boldsymbol{T}_1 \ldots \boldsymbol{T}_m \tag{4.7.2}$$

whose values and eigenvectors remain to be determined.

A condition is that the eigenvalue problem is better conditioned for $\boldsymbol{B}$ than for $\boldsymbol{A}$. It would not be the case if $\text{cond}(\boldsymbol{A}) \gg 1$.

We know that $\text{cond}(\boldsymbol{T}) = \text{cond}(\boldsymbol{T}_1 \ldots \boldsymbol{T}_n) \leq \text{cond}(\boldsymbol{T}_1) \ldots \text{cond}(\boldsymbol{T}_m)$. The condition number will be well ensured if the matrices $\boldsymbol{T}_i$ are taken such that it is not too large. In particular, for the norm $\|x\|_\infty$ and the Euclidean norm $\|x\|_2$, $\boldsymbol{T}_i$ can be chosen such that

$$\boldsymbol{T}_i = \begin{bmatrix} 1 & & & & 0 \\ & \ddots & & & \\ & & 1 & & \\ & & l_{i+1,i} & \ddots & \\ & & \vdots & & \ddots \\ 0 & & l_{n,i} & 0 & 1 \end{bmatrix} \tag{4.7.3}$$

where the element l_{ij} is obtained by a pivot method (cf. Chapter 5, about solution of systems of linear equations). The unitary matrices (e.g. Householder matrices) have a condition number equal to 1 and thus are adequate as elimination matrices.

Thus, the resulting matrix $\boldsymbol{B}$ contains only zeros below the diagonal.

4.8 Rutishauser's LR Method

Rutishauser's LR method (Rutishauser 1958, 1970; Rutishauser and Schwarz 1963) allows to calculate the eigenvalues and the eigenvectors of a matrix. In this part, a lower triangular matrix will be noted $\boldsymbol{L}$ ("Left") with its diagonal elements all equal to 1 and $\boldsymbol{R}$ ("Right") an upper triangular matrix.

Let $\boldsymbol{A} = \boldsymbol{A}_1$ be a square real matrix such that $\boldsymbol{A}_1 = \boldsymbol{L}_1\boldsymbol{R}_1$ (this is possible if $\boldsymbol{A}$ is positive definite. Refer later to the decomposition of a matrix $\boldsymbol{B}$ into lower $\boldsymbol{C}$ and upper $\boldsymbol{D}$). Then, we define $\boldsymbol{A}_2 = \boldsymbol{R}_1\boldsymbol{L}_1$ that is decomposed as $\boldsymbol{A}_2 = \boldsymbol{L}_2\boldsymbol{R}_2$, then, we define $\boldsymbol{A}_3 = \boldsymbol{R}_2\boldsymbol{L}_2$ and so on. In a general manner

$$\boldsymbol{A}_k = \boldsymbol{L}_k\boldsymbol{R}_k\,, \quad \boldsymbol{A}_{k+1} = \boldsymbol{R}_k\boldsymbol{L}_k \tag{4.8.1}$$

Notice that

$$\boldsymbol{A}_{k+1} = \boldsymbol{L}_k^{-1}\boldsymbol{A}_k\boldsymbol{L}_k = \boldsymbol{R}_k\boldsymbol{A}_k\boldsymbol{R}_k^{-1} \tag{4.8.2}$$

If we note

$$\boldsymbol{L}'_k = \boldsymbol{L}_1 \dots \boldsymbol{L}_k\,, \quad \boldsymbol{R}'_k = \boldsymbol{R}_k \dots \boldsymbol{R}_1 \tag{4.8.3}$$

we get

$$\boldsymbol{A}_{k+1} = \boldsymbol{L}'^{-1}_k\boldsymbol{A}\boldsymbol{L}'_k = \boldsymbol{R}'_k\boldsymbol{A}\boldsymbol{R}'^{-1}_k \tag{4.8.4}$$

thus, the matrix $\boldsymbol{A}_k$ is similar to $\boldsymbol{A}$. Consequently, it has the same eigenvalues and corresponding eigenvectors.

Suppose that

$$\lim_{k\to\infty} \boldsymbol{L}'_k = \boldsymbol{L}' \tag{4.8.5}$$

it means that

$$\lim_{k\to\infty} \boldsymbol{L}_k = \boldsymbol{I}\,, \quad \lim_{k\to\infty} \boldsymbol{A}_k = \lim_{k\to\infty} \boldsymbol{R}_k = \boldsymbol{R} \tag{4.8.6}$$

It results that

$$\boldsymbol{L}'^{-1}\boldsymbol{A}\boldsymbol{L}' = \boldsymbol{R} \Longrightarrow \boldsymbol{L}'\boldsymbol{R}\boldsymbol{L}'^{-1} = \boldsymbol{A} \tag{4.8.7}$$

$\boldsymbol{R}$ has the same eigenvalues as $\boldsymbol{A}$, and moreover they are its diagonal elements. Thus, when there will be convergence (it will be ensured for the definite positive matrices), the knowledge of $\boldsymbol{L}'$ and $\boldsymbol{R}$ will provide the eigenvalues and the eigenvectors of $\boldsymbol{A}$. Indeed, suppose that $\boldsymbol{v}$ is an eigenvector of $\boldsymbol{R}$. It implies that $\boldsymbol{R}\,\boldsymbol{v} = \lambda\,\boldsymbol{v}$; on another side, the equality $\boldsymbol{A}\,\boldsymbol{L}' = \boldsymbol{L}'\,\boldsymbol{R}$ leads to $\boldsymbol{L}'\,\boldsymbol{R}\,\boldsymbol{v} = \boldsymbol{L}'\,\lambda\,\boldsymbol{v} = \boldsymbol{A}\,\boldsymbol{L}'\,\boldsymbol{v}$, hence $\lambda\,\boldsymbol{L}'\,\boldsymbol{v} = \boldsymbol{A}\,\boldsymbol{L}'\,\boldsymbol{v}$, thus $\boldsymbol{L}'\,\boldsymbol{v}$ is an eigenvector of $\boldsymbol{A}$.

Description of LR decomposition method:

Let $\boldsymbol{A} = [a_{ij}]$ be a matrix that is decomposed as $\boldsymbol{A} = \boldsymbol{LR}$ where $\boldsymbol{L} = [l_{ij}]$ is lower triangular with 1 on the diagonal and $\boldsymbol{R} = [r_{ij}]$ upper triangular. By hypothesis

$$l_{ij} = 0 \quad \text{for } i < j \leq n,\ l_{ii} = 1 \tag{4.8.8}$$

$$r_{ij} = 0 \quad \text{for } i > j \tag{4.8.9}$$

As $\boldsymbol{A} = \boldsymbol{LR}$

$$a_{ij} = \sum_{k=1}^{n} l_{ik} r_{kj} \tag{4.8.10}$$

by using the definitions of l_{ij} and r_{ij}, we get

$$a_{ij} = \sum_{k=1}^{i-1} l_{ik} r_{kj} + r_{ij} \quad \text{for } i \leq j \tag{4.8.11}$$

$$a_{ij} = \sum_{k=1}^{j-1} l_{ik} r_{kj} + l_{ij} r_{jj} \quad \text{for } i > j \tag{4.8.12}$$

thus

$$r_{1j} = a_{1j} \quad \forall\, j \tag{4.8.13}$$

$$r_{ij} = a_{ij} - \sum_{k=1}^{i-1} l_{ik} r_{kj} \quad \text{for } 1 < i \leq j \tag{4.8.14}$$

$$l_{i1} = \frac{a_{i1}}{r_{11}} \quad \text{for } 1 < i \tag{4.8.15}$$

$$l_{ij} = \frac{1}{r_{jj}} \left[a_{ij} - \sum_{k=1}^{j-1} l_{ik} r_{kj} \right] \quad \text{for } i > j > 1 \tag{4.8.16}$$

Given $\boldsymbol{A}$, the previous calculation of $\boldsymbol{L}$ and $\boldsymbol{R}$ is done in an alternate manner, we calculate the first row of $\boldsymbol{R}$, then the first column of $\boldsymbol{L}$, then the second row of $\boldsymbol{R}$, then the second column of $\boldsymbol{L}$, and so on (cf. Crout's method, Section 5.3).

Suppose that $\boldsymbol{A}$ is real and that the eigenvalues are ordered with respect to their absolute value as $|\lambda_1| > \cdots > |\lambda_n|$ and are real. Noticing that

$$\boldsymbol{A}_k = \boldsymbol{L}_k \boldsymbol{R}_k \text{ and } \boldsymbol{A}_{k+1} = \boldsymbol{R}_k \boldsymbol{L}_k \Rightarrow \boldsymbol{L}_k \boldsymbol{R}_k = \boldsymbol{R}_{k-1} \boldsymbol{L}_{k-1} \tag{4.8.17}$$

we get

$$\begin{aligned} \boldsymbol{L}'_k \boldsymbol{R}'_k &= (\boldsymbol{L}_1 \dots \boldsymbol{L}_{k-2}) \boldsymbol{L}_{k-1} \boldsymbol{L}_k \boldsymbol{R}_k \boldsymbol{R}_{k-1} (\boldsymbol{R}_{k-2} \dots \boldsymbol{R}_1) \\ &= (\boldsymbol{L}_1 \dots \boldsymbol{L}_{k-2}) \boldsymbol{L}_{k-1} \boldsymbol{R}_{k-1} \boldsymbol{L}_{k-1} \boldsymbol{R}_{k-1} (\boldsymbol{R}_{k-2} \dots \boldsymbol{R}_1) \\ &= (\boldsymbol{L}_1 \dots \boldsymbol{L}_{k-2}) \boldsymbol{R}_{k-2} \boldsymbol{L}_{k-2} \boldsymbol{R}_{k-2} \boldsymbol{L}_{k-2} (\boldsymbol{R}_{k-2} \dots \boldsymbol{R}_1) \\ &= \dots \\ &= (\boldsymbol{L}_1 \boldsymbol{R}_1)^k \\ &= \boldsymbol{A}^k \end{aligned} \tag{4.8.18}$$

Under that form, Rutishauser's method appears as a power method as in Equation (4.4.28).

LR Rutishauser's method presents several drawbacks,

(a) It is expensive relatively to the number of operations.
(b) It fails if a matrix $\boldsymbol{A}_i$ does not admit a triangular decomposition.
(c) The convergence is very slow if two eigenvalues are very close in absolute value or modulus.

To avoid (a), it is advised to apply this method only to reduced matrices, for example under a tridiagonal form.

To avoid (c), there exist acceleration techniques such as applying the LR method to the matrix $\boldsymbol{A} - p\boldsymbol{I}$ where p is an approximation of λ_n.

Among later developments of Rutishauser's method, we can cite QR Francis's method.

Example 4.4 :
Rutishauser's method: Application

Given the matrix

$$\boldsymbol{A} = \begin{bmatrix} 7 & 1 & 3 \\ 4 & 8 & 5 \\ 2 & 3 & 6 \end{bmatrix} \tag{4.8.19}$$

- This matrix is decomposed under the form

$$\boldsymbol{A}_1 = \boldsymbol{A} = \boldsymbol{L}_1 \boldsymbol{R}_1 \tag{4.8.20}$$

where $\boldsymbol{L}_1$ is lower triangular and all its diagonal elements are equal to 1 while $\boldsymbol{R}_1$ is upper triangular.
We get

$$\boldsymbol{L}_1 = \begin{bmatrix} 1 & 0 & 0 \\ 0.5714 & 1 & 0 \\ 0.2857 & 0.3654 & 1 \end{bmatrix}, \quad \boldsymbol{R}_1 = \begin{bmatrix} 7 & 1 & 3 \\ 0 & 7.4286 & 3.2857 \\ 0 & 0 & 3.9423 \end{bmatrix} \tag{4.8.21}$$

- Then, we calculate the matrix

$$\boldsymbol{A}_2 = \boldsymbol{R}_1 \boldsymbol{L}_1 \tag{4.8.22}$$

equal to

$$\boldsymbol{A}_2 = \begin{bmatrix} 8.4286 & 2.0962 & 3 \\ 5.1837 & 8.6291 & 3.2857 \\ 1.1264 & 1.4405 & 3.9423 \end{bmatrix} \tag{4.8.23}$$

- The matrix $\boldsymbol{A}_2$ is decomposed under the form

$$\boldsymbol{A}_2 = \boldsymbol{L}_2 \, \boldsymbol{R}_2 \tag{4.8.24}$$

and we get

$$\boldsymbol{L}_2 = \begin{bmatrix} 1 & 0 & 0 \\ 0.6150 & 1 & 0 \\ 0.1336 & 0.1581 & 1 \end{bmatrix}, \quad \boldsymbol{R}_2 = \begin{bmatrix} 8.4286 & 2.0962 & 3 \\ 0 & 7.3400 & 1.4407 \\ 0 & 0 & 3.3136 \end{bmatrix} \tag{4.8.25}$$

- Then we calculate $\boldsymbol{A}_3 = \boldsymbol{R}_2 \, \boldsymbol{L}_2$, we decompose $\boldsymbol{A}_3 = \boldsymbol{L}_3 \, \boldsymbol{R}_3$ and so on. Moreover, we simultaneously calculate $\boldsymbol{L}'_k = \boldsymbol{L}_1 \, \dots \, \boldsymbol{L}_k$.
- At the 10th iteration, we find

$$\boldsymbol{L}_{10} = \begin{bmatrix} 1 & 0 & 0 \\ 0.0012 & 1 & 0 \\ 0.0000 & 0.0015 & 1 \end{bmatrix}, \quad \boldsymbol{R}_{10} = \begin{bmatrix} 12.7898 & 2.9700 & 3 \\ 0 & 5.0080 & -1.3387 \\ 0 & 0 & 3.2005 \end{bmatrix} \tag{4.8.26}$$

  The limit of $\boldsymbol{R}_k$ equal to the limit of $\boldsymbol{A}_k$ is a matrix which has the same eigenvalues as $\boldsymbol{A}$ and they are its diagonal elements. Thus $\{12.7898, 5.0080, 3.2005\}$ are estimations of the eigenvalues of $\boldsymbol{A}$. In fact, the eigenvalues of $\boldsymbol{A}$ are $\{12.7958, 5.0000, 3.2042\}$. Thus, this shows that the eigenvalues are well approximated.
- Then, we search the eigenvectors $\boldsymbol{v}$ of $\boldsymbol{R}_k$ (noted in the following $\boldsymbol{R}$ which is their limit). For that, the n following systems must be solved:

$$\boldsymbol{R} \, \boldsymbol{v}_i = \lambda_i \, \boldsymbol{v}_i \,, \quad i = 1, \dots, n \tag{4.8.27}$$

  where λ_i is an estimated eigenvalue of $\boldsymbol{R}$. Each vector $\boldsymbol{v}_i$ presents the following components:

$$\boldsymbol{v}_i = \begin{bmatrix} x_1 \\ \vdots \\ x_{i-1} \\ 1 \\ 0 \\ \vdots \\ 0 \end{bmatrix} \tag{4.8.28}$$

  where all the components of rank lower than i are nonzero, the component of rank i is equal to 1, and the following components are zero. If a system (4.8.27) is explained for a given eigenvalue λ_i (in fact, approximated by r_{ii}, diagonal element of $\boldsymbol{R}_k$ and thus eigenvalue of $\boldsymbol{R}_k$), we get

$$\begin{array}{ll} (r_{11} - r_{ii})x_1 + r_{12}x_2 + \dots + r_{1,i-1}x_{i-1} & = -r_{1i} \\ (r_{22} - r_{ii})x_2 + r_{23}x_3 + \dots + r_{2,i-1}x_{i-1} & = -r_{2i} \\ \dots & \\ (r_{i-1,i-1} - r_{ii})x_{i-1} & = -r_{i-1,i} \end{array} \tag{4.8.29}$$

  Solving this upper triangular system of the form (5.2.1), and considering the set of eigenvalues of $\boldsymbol{R}_k$, the eigenvectors $\boldsymbol{v}_i$ of $\boldsymbol{R}$ (in fact, $\boldsymbol{R}_k$) result, that is

$$\boldsymbol{v}_1 = \begin{bmatrix} 1 \\ 0 \\ 0 \end{bmatrix}, \quad \boldsymbol{v}_2 = \begin{bmatrix} -0.3817 \\ 1 \\ 0 \end{bmatrix}, \quad \boldsymbol{v}_3 = \begin{bmatrix} -0.5422 \\ 0.7406 \\ 1 \end{bmatrix} \tag{4.8.30}$$

- If $\boldsymbol{v}$ is an eigenvector of $\boldsymbol{R}$, then $\boldsymbol{L}'\boldsymbol{v}$ is an eigenvector of $\boldsymbol{A}$. Thus, as $\boldsymbol{L}'$ has been stored, the estimations of the eigenvectors of $\boldsymbol{A}$ are easily deduced.

  Knowing $\boldsymbol{L}'$ after 10 iterations

$$L'_{10} = \begin{bmatrix} 1 & 0 & 0 \\ 2.1129 & 1 & 0 \\ 1.2269 & 0.6582 & 1 \end{bmatrix} \tag{4.8.31}$$

the following estimations $\boldsymbol{w}_i$ are found (the numerical results correspond to vectors that have been normed after the calculation $\boldsymbol{L}'\boldsymbol{v}$) of the eigenvectors of $\boldsymbol{A}$, according to the formula $\boldsymbol{w}_i = \boldsymbol{L}'\boldsymbol{v}_i$, hence

$$\boldsymbol{w}_1 = \begin{bmatrix} 0.3788 \\ 0.8003 \\ 0.4647 \end{bmatrix}, \quad \boldsymbol{w}_2 = \begin{bmatrix} -0.8151 \\ 0.4135 \\ 0.4057 \end{bmatrix}, \quad \boldsymbol{w}_3 = \begin{bmatrix} -0.5092 \\ -0.3803 \\ 0.7721 \end{bmatrix} \tag{4.8.32}$$

that can be compared to the eigenvectors of $\boldsymbol{A}$ given by Matlab in columns

$$\begin{bmatrix} -0.3787 & -0.8165 & -0.5102 \\ -0.8004 & 0.4082 & -0.3793 \\ -0.4648 & 0.4082 & 0.7719 \end{bmatrix} \tag{4.8.33}$$

The convergence is relatively good for this simple example.

4.9 Householder's Method

Householder's method (Householder 1958) allows to obtain a symmetric tridiagonal matrix similar to a given symmetric matrix $\boldsymbol{A}$. From the definition of similar matrices (Section 4.5), and in the case where $\boldsymbol{A}$ is symmetric, there exists a unitary matrix $\boldsymbol{U}$ such that $\boldsymbol{U}^{-1}\boldsymbol{A}\boldsymbol{U} = \boldsymbol{D}$ where $\boldsymbol{D}$ is diagonal and its diagonal elements are the eigenvalues of $\boldsymbol{A}$.

Householder's matrices are unitary matrices defined by

$$\boldsymbol{H} = \boldsymbol{I} - 2\boldsymbol{w}\boldsymbol{w}^* \quad \text{with } \boldsymbol{w}^*\boldsymbol{w} = 1 \tag{4.9.1}$$

where $\boldsymbol{w}$ is a vector of dimension n. The matrix $\boldsymbol{H}$ thus defined is Hermitian, that is $\boldsymbol{H}^* = \boldsymbol{H}$, unitary as $\boldsymbol{H}^*\boldsymbol{H} = \boldsymbol{I}$ and *involutory* as $\boldsymbol{H}^2 = \boldsymbol{I}$. These Householder's matrices are studied to carry out the orthogonalization of a matrix $\boldsymbol{A}$ and thus the choice of the vector $\boldsymbol{w}$ will have to be done adequately.

Householder's method starts in the following manner:

- Set $\boldsymbol{A}^{(1)} = \boldsymbol{A}$ with $\boldsymbol{A}$ symmetric.
- Determine Householder's matrix $\boldsymbol{H}^{(1)}$ such that

$$\boldsymbol{A}^{(2)} = \boldsymbol{H}^{(1)}\boldsymbol{A}^{(1)}\boldsymbol{H}^{(1)} \tag{4.9.2}$$

with

$$\begin{aligned} a_{11}^{(2)} &= a_{11}^{(1)} \\ a_{i1}^{(2)} &= 0\,,\ \forall\, i = 3, \ldots, n \\ a_{1j}^{(2)} &= 0\,,\ \forall\, j = 3, \ldots, n \quad \text{(as each matrix } \boldsymbol{A}^{(k)} \text{is symmetric)} \end{aligned} \tag{4.9.3}$$

As we desire to obtain a tridiagonal matrix at the end of the algorithm, it is required that

$$\boldsymbol{H}^{(1)} \begin{bmatrix} a_{11} \\ a_{21} \\ a_{31} \\ \dots \\ a_{n1} \end{bmatrix} = \begin{bmatrix} a_{11} \\ \alpha \\ 0 \\ \dots \\ 0 \end{bmatrix} \tag{4.9.4}$$

where α is to be determined. The solving gives

$$\begin{aligned} \alpha &= -(\text{sign}(a_{21})) \left(\sum_{i=2}^{n} a_{i1}^2 \right)^{1/2} \\ r &= \left(\frac{1}{2}\alpha(\alpha - a_{21}) \right)^{1/2} \\ w_1 &= 0 \\ w_2 &= \frac{a_{21} - \alpha}{2r} \\ w_i &= \frac{a_{i1}}{2r} \quad \forall i = 3, \dots, n \end{aligned} \tag{4.9.5}$$

We thus obtain a matrix $\boldsymbol{A}^{(2)}$ which is tridiagonal with respect to the first row and the first column

$$\boldsymbol{A}^{(2)} = \boldsymbol{H}^{(1)} \boldsymbol{A}^{(1)} \boldsymbol{H}^{(1)} = \begin{bmatrix} a_{11}^{(2)} & a_{12}^{(2)} & 0 & \dots & 0 \\ a_{21}^{(2)} & a_{22}^{(2)} & a_{23}^{(2)} & \dots & a_{2n}^{(2)} \\ 0 & a_{32}^{(2)} & a_{33}^{(2)} & \dots & a_{3n}^{(2)} \\ \vdots & \vdots & \vdots & \dots & \vdots \\ 0 & a_{n3}^{(2)} & \dots & \dots & a_{nn}^{(2)} \end{bmatrix} \tag{4.9.6}$$

By doing an additional iteration, a matrix $\boldsymbol{A}^{(3)}$ results that is tridiagonal with respect to the two first rows and the two first columns

$$\boldsymbol{A}^{(3)} = \boldsymbol{H}^{(2)} \boldsymbol{A}^{(2)} \boldsymbol{H}^{(2)} = \begin{bmatrix} a_{11}^{(2)} & a_{12}^{(2)} & 0 & 0 & \dots & 0 \\ a_{21}^{(2)} & a_{22}^{(2)} & a_{23}^{(2)} & 0 & \dots & 0 \\ 0 & a_{32}^{(2)} & a_{33}^{(2)} & a_{34}^{(2)} & \dots & a_{3n}^{(2)} \\ 0 & 0 & a_{43}^{(2)} & \vdots & \dots & a_{4n}^{(2)} \\ \vdots & \vdots & \vdots & \vdots & \vdots & \vdots \\ 0 & 0 & a_{n3}^{(2)} & \vdots & \dots & a_{nn}^{(2)} \end{bmatrix} \tag{4.9.7}$$

The process is repeated until obtaining a symmetric tridiagonal matrix $\boldsymbol{A}^{(n-1)}$ equal to

$$\boldsymbol{A}^{(n-1)} = \boldsymbol{H}^{(n-2)} \dots \boldsymbol{H}^{(1)} \boldsymbol{A}_1 \boldsymbol{H}^{(1)} \dots \boldsymbol{H}^{(n-2)} \tag{4.9.8}$$

The full algorithm contains the stages 1 to 9,

1. For $k = 1, \dots, n-2$, do the following stages,
2. Let

$$q = \sum_{i=k+1}^{n} (a_{ik}^{(k)})^2 \tag{4.9.9}$$

3.

$$\begin{cases} \text{If } a_{k+1,k}^{(k)} = 0, \text{ set } \alpha = -\sqrt{q}, \\ \text{otherwise}, \quad \text{set } \alpha = -\dfrac{\sqrt{q}\, a_{k+1,k}^{(k)}}{|a_{k+1,k}^{(k)}|} \end{cases} \tag{4.9.10}$$

4. Calculate $r = \sqrt{\frac{\alpha}{2}(\alpha - a_{k+1,k}^{(k)})}$.
5. Set

$$\boldsymbol{w} = \frac{1}{2r}\boldsymbol{v} \tag{4.9.11}$$

that is

$$\begin{aligned} v_i &= 0 && \forall\, i = 1, \ldots, k \\ v_{k+1} &= a_{k+1,k}^{(k)} - \alpha && \\ v_i &= a_{ik}^{(k)} && \forall\, i = k+2, \ldots, n \end{aligned} \tag{4.9.12}$$

6. Set

$$\boldsymbol{u} = \frac{1}{2r^2}\boldsymbol{A}^{(k)}\,\boldsymbol{v} = \frac{1}{r}\boldsymbol{A}^{(k)}\,\boldsymbol{w} \tag{4.9.13}$$

that is

$$u_i = \frac{1}{2r^2} \sum_{j=k+1}^{n} a_{ij}^{(k)} v_j \qquad \forall\, i = k, \ldots, n \tag{4.9.14}$$

7. Set

$$p = \boldsymbol{v}^T \boldsymbol{u} = \frac{1}{2r^2}\boldsymbol{v}^T\,\boldsymbol{A}^{(k)}\,\boldsymbol{v} \tag{4.9.15}$$

that is

$$p = \sum_{i=k+1}^{n} v_i\, u_i \tag{4.9.16}$$

8. Set

$$\boldsymbol{z} = \boldsymbol{u} - \frac{1}{4r^2}\boldsymbol{v}^T\,\boldsymbol{u}\,\boldsymbol{v} = \boldsymbol{u} - \boldsymbol{w}\,\boldsymbol{w}^T\,\boldsymbol{u} = \frac{1}{r}\boldsymbol{A}^{(k)}\,\boldsymbol{w} - \boldsymbol{w}\,\boldsymbol{w}^T\,\frac{1}{r}\boldsymbol{A}^{(k)}\,\boldsymbol{w} \tag{4.9.17}$$

that is

$$z_i = u_i - \frac{p}{4r^2}v_i \qquad \forall\, i = k, \ldots, n \tag{4.9.18}$$

9. We get

$$\boldsymbol{H}^{(k)} = \boldsymbol{I} - 2\boldsymbol{w}\,\boldsymbol{w}^T \tag{4.9.19}$$

and

$$\boldsymbol{A}^{(k+1)} = \boldsymbol{A}^{(k)} - \boldsymbol{v}\,\boldsymbol{z}^T - \boldsymbol{z}\,\boldsymbol{v}^T = \boldsymbol{H}^{(k)}\,\boldsymbol{A}^{(k)}\,\boldsymbol{H}^{(k)} \tag{4.9.20}$$

that is, for $i = k+1, \ldots, n-1$, do

$$\begin{aligned} a_{ji}^{(k+1)} &= a_{ji}^{(k)} - v_i\, z_j - v_j\, z_i \qquad \forall j = i+1, \dots, n \\ a_{ji}^{(k+1)} &= a_{ij}^{(k+1)} \qquad \forall j = i+1, \dots, n \\ a_{ii}^{(k+1)} &= a_{ii}^{(k)} - 2\, v_i\, z_i \end{aligned} \tag{4.9.21}$$

and

$$a_{nn}^{(k+1)} = a_{nn}^{(k)} - 2\, v_n\, z_n \tag{4.9.22}$$

$$a_{ik}^{(k+1)} = a_{ki}^{(k+1)} = 0 \quad \text{for } i = k+2, \dots, n \tag{4.9.23}$$

$$\begin{cases} a_{k+1,k}^{(k+1)} = a_{k+1,k}^{(k)} - v_{k+1} z_k \\ a_{k,k+1}^{(k+1)} = a_{k+1,k}^{(k+1)} \end{cases} \tag{4.9.24}$$

Return to 1. When all iterations are done, the matrix $\boldsymbol{A}^{(n-1)}$ is symmetric, tridiagonal, and similar to $\boldsymbol{A}$.

It is possible to apply a modified Householder's algorithm when $\boldsymbol{A}$ is any matrix, nonsymmetric. From a vector and matrix point of view, nothing is changed. However, the detail of calculations differs and the stages 6 to 9 are replaced by

6.

$$\begin{aligned} u_i &= \frac{1}{2r^2} \sum_{j=k+1}^{n} a_{ij}^{(k)} v_j \qquad \forall i = k, \dots, n \\ y_i &= \frac{1}{2r^2} \sum_{j=k+1}^{n} a_{ji}^{(k)} v_j \end{aligned} \tag{4.9.25}$$

7.

$$z_i = u_i - \frac{p}{2r^2} v_i \qquad \forall i = 1, \dots, n \tag{4.9.26}$$

8. For $i = k+1, \dots, n$, do

$$\begin{aligned} a_{ji}^{(k+1)} &= a_{ji}^{(k)} - v_i\, z_j \qquad \forall j = 1, \dots, k \\ a_{ij}^{(k+1)} &= a_{ij}^{(k+1)} - v_i\, y_j \qquad \forall j = 1, \dots, k \\ a_{ji}^{(k+1)} &= a_{ji}^{(k)} - v_i\, z_j - v_i\, y_j \qquad \forall j = k+1, \dots, n \end{aligned} \tag{4.9.27}$$

The resulting matrix $\boldsymbol{A}^{(n-1)}$ is not tridiagonal except if $\boldsymbol{A}$ is symmetric. However, it has zeros below the subdiagonal and is called upper Hessenberg's matrix.

A matrix $\boldsymbol{H}$ is said to be upper Hessenberg's matrix when $h_{ij} = 0,\ \forall i > j+1$. For a square matrix, the elements below the first subdiagonal are zero.

Example 4.5 :
Householder's method: Application to a symmetric matrix
Let the symmetric matrix $\boldsymbol{A}$ be equal to

$$\begin{bmatrix} 3 & -2 & 5 & 1 \\ -2 & 7 & -8 & 6 \\ 5 & -8 & 4 & -3 \\ 1 & 6 & -3 & 2 \end{bmatrix} \tag{4.9.28}$$

For the first iteration, $k = 1$, we set $\boldsymbol{A}^{(1)} = \boldsymbol{A}$. We obtain successively $q = 30$; $\alpha = 5.4772$; $r = 4.5252$

$$\boldsymbol{w}^T = \begin{bmatrix} 0 & -0.8262 & 0.5525 & 0.1105 \end{bmatrix}$$

$$\boldsymbol{u}^T = \begin{bmatrix} 1.0000 & -2.1082 & 1.8757 & -1.4129 \end{bmatrix}$$

$p = 23.7291$

$$\boldsymbol{z}^T = \begin{bmatrix} 1.0000 & 0.0580 & 0.4272 & -1.7026 \end{bmatrix}$$

$$\boldsymbol{H}^{(1)} = \begin{bmatrix} 1 & 0 & 0 & 0 \\ 0 & -0.3651 & 0.9129 & 0.1826 \\ 0 & 0.9129 & 0.3896 & -0.1221 \\ 0 & 0.1826 & -0.1221 & 0.9756 \end{bmatrix}$$

and finally $\boldsymbol{A}^{(2)}$ that ends the first iteration

$$\boldsymbol{A}^{(2)} = \boldsymbol{H}^{(1)}\boldsymbol{A}^{(1)}\boldsymbol{H}^{(1)} = \begin{bmatrix} 3.0000 & 5.4772 & 0 & 0 \\ 5.4772 & 7.8667 & -5.0956 & -6.7885 \\ 0 & -5.0956 & -0.2718 & 5.0857 \\ 0 & -6.7885 & 5.0857 & 5.4051 \end{bmatrix}$$

Then, we go to iteration 2, $k = 2$. The operations are repeated and we finally obtain

$$\boldsymbol{H}^{(2)} = \begin{bmatrix} 1 & 0 & 0 & 0 \\ 0 & 1 & 0 & 0 \\ 0 & 0 & -0.6003 & -0.7998 \\ 0 & 0 & -0.7998 & 0.6003 \end{bmatrix}$$

and

$$\begin{aligned} \boldsymbol{A}^{(3)} &= \boldsymbol{H}^{(2)}\boldsymbol{A}^{(2)}\boldsymbol{H}^{(2)} = \boldsymbol{H}^{(2)}\boldsymbol{H}^{(1)}\boldsymbol{A}^{(1)}\boldsymbol{H}^{(1)}\boldsymbol{H}^{(2)} \\ &= \begin{bmatrix} 3.0000 & 5.4772 & 0 & 0 \\ 5.4772 & 7.8667 & 8.4882 & 0 \\ 0 & 8.4882 & 8.2426 & -1.3055 \\ 0 & 0 & -1.3055 & -3.1093 \end{bmatrix} \end{aligned}$$

which is the symmetric, tridiagonal matrix, similar to $\boldsymbol{A}$ that we desired to obtain.

Example 4.6 :

Householder's method: Application to a nonsymmetric matrix

Let the nonsymmetric matrix $\boldsymbol{A}$ be equal to

$$\begin{bmatrix} 3 & -2 & 5 & 1 \\ 1 & 7 & -8 & 6 \\ 9 & 2 & 4 & -3 \\ -5 & 4 & -2 & 3 \end{bmatrix} \tag{4.9.29}$$

For the first iteration, $k = 1$, we set $\boldsymbol{A}^{(1)} = \boldsymbol{A}$. We successively obtain $q = 107$; $\alpha = -10.3441$; $r = 7.6598$

$$\boldsymbol{w}^T = \begin{bmatrix} 0 & 0.7405 & 0.5875 & -0.3264 \end{bmatrix}$$

$$\boldsymbol{u}^T = \begin{bmatrix} 0.1475 & -0.1925 & 0.6280 & 0.1055 \end{bmatrix}$$

$p = 2.9404$

$$\boldsymbol{z}^T = \begin{bmatrix} 0.1475 & -0.3347 & 0.5152 & 0.1681 \end{bmatrix}$$

$$H^{(1)} = \begin{bmatrix} 1 & 0 & 0 & 0 \\ 0 & -0.0967 & -0.8701 & 0.4834 \\ 0 & -0.8701 & 0.3097 & 0.3835 \\ 0 & 0.4834 & 0.3835 & 0.7870 \end{bmatrix}$$

and finally $A^{(2)}$ that ends the first iteration

$$A^{(2)} = H^{(1)}A^{(1)}H^{(1)} = \begin{bmatrix} 3.0000 & -3.6736 & 3.6722 & 1.7377 \\ -10.3441 & 4.9252 & 0.6174 & 1.0964 \\ 0 & -8.5025 & 3.8104 & -2.9868 \\ 0 & 4.6806 & -5.9837 & 5.2644 \end{bmatrix}$$

We then go to iteration 2, $k = 2$. The operations are repeated and finally we obtain

$$H^{(2)} = \begin{bmatrix} 1 & 0 & 0 & 0 \\ 0 & 1 & 0 & 0 \\ 0 & 0 & -0.8760 & 0.4823 \\ 0 & 0 & 0.4823 & 0.8760 \end{bmatrix}$$

and

$$\begin{aligned} A^{(3)} &= H^{(2)}A^{(2)}H^{(2)} = H^{(2)}H^{(1)}A^{(1)}H^{(1)}H^{(2)} \\ &= \begin{bmatrix} 3.0000 & -3.6736 & -2.3790 & 3.2932 \\ -10.3441 & 4.9252 & -0.0121 & 1.2582 \\ 0 & 9.7057 & 7.9383 & 1.5148 \\ 0 & 0 & 4.5117 & 1.1364 \end{bmatrix} \end{aligned}$$

which is an upper Hessenberg's matrix, the elements below the subdiagonal are zero.

4.10 Francis's QR Method

The QR method by Francis (1961–62a,b) is a development of the previous Rutishauser's method. The QR transformation preserves the symmetry of the matrix A, preserves the tridiagonal form of A, and preserves Hessenberg's form of A. The QR method is a method of matrix reduction whose objective is to determine the set of eigenvalues of a symmetric tridiagonal matrix A. In the case where A is any matrix, the method is not well adapted and its convergence is slow. Let A of dimension $n \times n$ be the initial matrix. If A is symmetric, but not tridiagonal, it is necessary to start applying a Householder's transformation to obtain a symmetric tridiagonal matrix similar to the initial matrix A. The algorithm is numerically stable as it proceeds by similar orthogonal transformations.

We set $A^{(1)} = A$ and we form matrices $Q^{(i)}, R^{(i)}, A^{(i)}$ responding to the following conditions:

$$A^{(i)} = Q^{(i)} R^{(i)} \quad \text{with } Q^{(i)*} Q^{(i)} = I \tag{4.10.1}$$

$$R^{(i)} = \begin{bmatrix} * & \dots & * \\ & \ddots & \vdots \\ 0 & & * \end{bmatrix} \tag{4.10.2}$$

The matrix $\boldsymbol{A}^{(i)}$ is decomposed as a product of a unitary (orthogonal if real) matrix $\boldsymbol{Q}^{(i)}$ and an upper triangular matrix $\boldsymbol{R}^{(i)}$

$$\boldsymbol{A}^{(i+1)} = \boldsymbol{R}^{(i)}\,\boldsymbol{Q}^{(i)} = \boldsymbol{Q}^{(i)*}\,\boldsymbol{Q}^{(i)}\,\boldsymbol{R}^{(i)}\,\boldsymbol{Q}^{(i)} = \boldsymbol{Q}^{(i)*}\,\boldsymbol{A}^{(i)}\,\boldsymbol{Q}^{(i)} \tag{4.10.3}$$

Thus, if $\boldsymbol{A}^{(i)}$ is symmetric, the matrix $\boldsymbol{A}^{(i+1)}$ is symmetric and similar to $\boldsymbol{A}^{(i)}$, it has the same eigenvalues as $\boldsymbol{A}^{(i)}$. We thus get

$$\boldsymbol{A}^{(i+1)} = (\boldsymbol{Q}^{(1)}\ldots\boldsymbol{Q}^{(i)})^*\boldsymbol{A}^{(1)}(\boldsymbol{Q}^{(1)}\ldots\boldsymbol{Q}^{(i)}) \tag{4.10.4}$$

and $\boldsymbol{A}^{(i+1)}$ has the same eigenvalues as $\boldsymbol{A}$. If $\boldsymbol{A}$ is not symmetric, $\boldsymbol{A}^{(i)}$ converges to an upper triangular matrix with the eigenvalues along the diagonal when the eigenvalues are distinct. Moreover, the eigenvalues are ordered by increasing modulus. In the case of a multiple eigenvalue of order p, there appears a diagonal block of order p possessing that eigenvalue. If $\boldsymbol{A}$ is symmetric, $\boldsymbol{A}^{(i+1)}$ tends to a diagonal matrix with the eigenvalues along the diagonal.

We set the matrices $\boldsymbol{P}^{(i)}, \boldsymbol{R}^{(i)}$ such that

$$\boldsymbol{P}^{(i)} = \boldsymbol{Q}^{(1)}\,\boldsymbol{Q}^{(2)}\ldots\boldsymbol{Q}^{(i)}, \quad \boldsymbol{U}^{(i)} = \boldsymbol{R}^{(i)}\,\boldsymbol{R}^{(i-1)}\ldots\boldsymbol{R}^{(1)} \tag{4.10.5}$$

where $\boldsymbol{P}^{(i)}$ is unitary and $\boldsymbol{U}^{(i)}$ is upper triangular. It results

$$\boldsymbol{A}^{(i+1)} = \boldsymbol{P}^{(i)}\,\boldsymbol{A}^{(1)}\,\boldsymbol{U}^{(i)} \tag{4.10.6}$$

In the case of distinct eigenvalues, the convergence rate to zero of an off-diagonal element of the matrix $\boldsymbol{A}^{(i+1)}$, i.e. $\boldsymbol{A}^{(i+1)}_{j,j-1}$, depends on the ratio $|\lambda_{j+1}/\lambda_j|$, which can result in a slow convergence when the ratio is close to 1. Then, there exists an acceleration technique of the convergence by setting

$$\boldsymbol{A}^{(i)} = s\,\boldsymbol{I} + \boldsymbol{Q}^{(i)}\,\boldsymbol{R}^{(i)} \quad \text{and} \quad \boldsymbol{A}^{(i+1)} = s\,\boldsymbol{I} + \boldsymbol{R}^{(i)}\,\boldsymbol{Q}^{(i)} \tag{4.10.7}$$

where s is a scalar chosen close to an eigenvalue of $\boldsymbol{A}$. The convergence rate then depends on the ratio $|(\lambda_{j+1}-s)/(\lambda_j-s)|$. s should be chosen close to λ_{j+1}. There exist techniques of this type where s is modified at each iteration i.

It remains to describe the construction of the factorization matrices $\boldsymbol{Q}$ and $\boldsymbol{R}$. The QR decomposition (or factorization) always exists and can be numerically calculated in a stable way. The QR decomposition is not unique. Three methods of QR factorization are presented in the following.

4.10.0.1 QR Factorization by Gram–Schmidt

This factorization is valid for any matrix $\boldsymbol{A}$. The matrix $\boldsymbol{A}$ is defined by means of its column vectors as

$$\boldsymbol{A} = \left[\boldsymbol{a}_1\ \boldsymbol{a}_2\ \ldots\ \boldsymbol{a}_n\right] \tag{4.10.8}$$

and the vector projection of $\boldsymbol{v}_1$ on $\boldsymbol{v}_2$ defined from two arbitrary vectors $\boldsymbol{v}_1$ and $\boldsymbol{v}_2$

$$\text{proj}_{\boldsymbol{v}_2}(\boldsymbol{v}_1) = \frac{<\boldsymbol{v}_2, \boldsymbol{v}_1>}{<\boldsymbol{v}_2, \boldsymbol{v}_2>} \boldsymbol{v}_2 \tag{4.10.9}$$

where $<\boldsymbol{v}_1, \boldsymbol{v}_2>$ is the scalar product of $\boldsymbol{v}_1$ by $\boldsymbol{v}_2$.

Gram–Schmidt procedure successively defines the orthogonal vectors $\boldsymbol{u}_i$ and the corresponding unit vectors $\boldsymbol{e}_i$

$$\begin{aligned}
&\boldsymbol{u}_1 = \boldsymbol{a}_1\,, && \boldsymbol{e}_1 = \frac{\boldsymbol{u}_1}{||\boldsymbol{u}_1||}\\
&\boldsymbol{u}_2 = \boldsymbol{a}_2 - \text{proj}_{\boldsymbol{e}_1}(\boldsymbol{a}_2)\,, && \boldsymbol{e}_2 = \frac{\boldsymbol{u}_2}{||\boldsymbol{u}_2||}\\
&\vdots\\
&\boldsymbol{u}_i = \boldsymbol{a}_i - \sum_{j=1}^{i-1} \text{proj}_{\boldsymbol{e}_j}(\boldsymbol{a}_i)\,, \; \boldsymbol{e}_i = \frac{\boldsymbol{u}_i}{||\boldsymbol{u}_i||} && \forall i = 1, \dots, n
\end{aligned} \tag{4.10.10}$$

The vectors $\boldsymbol{a}_i$ can be represented on the basis $\boldsymbol{e}$ as

$$a_i = \sum_{j=1}^{i} <\boldsymbol{e}_j, \boldsymbol{a}_i> \boldsymbol{e}_j \tag{4.10.11}$$

with $<\boldsymbol{e}_i, \boldsymbol{a}_i> = ||\boldsymbol{u}_i||$. Equation (4.10.11) is written under matrix form

$$\boldsymbol{A} = \boldsymbol{Q}\,\boldsymbol{R} \tag{4.10.12}$$

with

$$\begin{aligned}
\boldsymbol{Q} &= \left[\boldsymbol{e}_1\ \boldsymbol{e}_2 \dots \boldsymbol{e}_n\right]\\
\boldsymbol{R} = \boldsymbol{Q}^T \boldsymbol{A} &= \begin{bmatrix} <\boldsymbol{e}_1, \boldsymbol{a}_1> & <\boldsymbol{e}_1, \boldsymbol{a}_2> & \dots & <\boldsymbol{e}_1, \boldsymbol{a}_n>\\ 0 & <\boldsymbol{e}_2, \boldsymbol{a}_2> & \dots & <\boldsymbol{e}_2, \boldsymbol{a}_n>\\ \vdots & & \ddots & \vdots\\ 0 & \dots & 0 & <\boldsymbol{e}_n, \boldsymbol{a}_n> \end{bmatrix}
\end{aligned} \tag{4.10.13}$$

which ends that QR decomposition.

Example 4.7 :

QR decomposition by Gram–Schmidt procedure: Application to any matrix

The matrix

$$\boldsymbol{A} = \begin{bmatrix} 3 & -2 & 5 & 1\\ 1 & 7 & -8 & 6\\ 9 & 2 & 4 & -3\\ -5 & 4 & -2 & 3 \end{bmatrix} \tag{4.10.14}$$

gives the following result:

$$\boldsymbol{Q} = \begin{bmatrix} 0.2785 & -0.2311 & 0.3646 & 0.8580\\ 0.0928 & 0.8203 & -0.4249 & 0.3714\\ 0.8356 & 0.2432 & 0.3444 & -0.3521\\ -0.4642 & 0.4631 & 0.7536 & -0.0448 \end{bmatrix}$$

$$\boldsymbol{R} = \begin{bmatrix} 10.7703 & -0.0928 & 4.9209 & -3.0640\\ 0 & 8.5435 & -7.6717 & 5.3509\\ 0 & 0 & 5.0921 & -0.9567\\ 0 & 0 & 0 & 4.0080 \end{bmatrix}$$

4.10.0.2 QR Householder's Factorization

Consider any real matrix $\boldsymbol{A}$ (not necessarily square, of dimension $m \times n$) that is represented by means of its column vectors as

$$\boldsymbol{A} = \begin{bmatrix} \boldsymbol{a}_1 \ \boldsymbol{a}_2 \ \dots \ \boldsymbol{a}_n \end{bmatrix} \tag{4.10.15}$$

We set $\boldsymbol{A}_1 = \boldsymbol{A}$. First, a Householder's transformation (reflection) must be made that transforms the first column $\boldsymbol{a}_1$ of $\boldsymbol{A}$ into a vector $||\boldsymbol{a}_1|| \boldsymbol{e}_1$ where $\boldsymbol{e}_1$ is the unit vector $[1 \ 0 \ \dots \ 0]$. Let $\alpha = ||\boldsymbol{a}_1||$. We set

$$\boldsymbol{u} = \boldsymbol{a}_1 - \alpha \, \boldsymbol{e}_1 \quad , \qquad \boldsymbol{v} = \frac{\boldsymbol{u}}{||\boldsymbol{u}||} \tag{4.10.16}$$

Householder's matrix is

$$\boldsymbol{H}_1 = \boldsymbol{I} - 2 \, \boldsymbol{v} \, \boldsymbol{v}^T \tag{4.10.17}$$

The product $\boldsymbol{H}_1 \boldsymbol{A}_1$ is a matrix containing zeros in the first column, except the first row, that is

$$\boldsymbol{H}_1 \, \boldsymbol{A}_1 = \begin{bmatrix} \alpha_1 & * \ \dots & * \\ 0 & & \\ \vdots & \boldsymbol{A}_2 & \\ 0 & & \end{bmatrix} \tag{4.10.18}$$

The matrix $\boldsymbol{A}_2$ is of dimension lower than $\boldsymbol{A}_1$ by one unit with respect to the number of rows and of columns. The operation can be exactly repeated on that matrix $\boldsymbol{A}_2$. Note $\boldsymbol{H}'_2$ Householder's matrix that intervenes. The reason of the notation $\boldsymbol{H}'_2$ comes from its dimension lower than $\boldsymbol{H}_1$ by one unit. For that reason, we note $\boldsymbol{H}_2$ deduced from $\boldsymbol{H}'_2$ by

$$\boldsymbol{H}_2 = \begin{bmatrix} 1 & \boldsymbol{0} \\ \boldsymbol{0} & \boldsymbol{H}'_2 \end{bmatrix} \tag{4.10.19}$$

and in a general way

$$\boldsymbol{H}_k = \begin{bmatrix} \boldsymbol{I}_{k-1} & \boldsymbol{0} \\ \boldsymbol{0} & \boldsymbol{H}'_k \end{bmatrix} \tag{4.10.20}$$

where $\boldsymbol{I}_{k-1}$ is the identity matrix of dimension $k-1$. After a number of iterations equal to $p = \min(m-1, n)$, the matrices of the QR factorization result

$$\boldsymbol{Q} = \boldsymbol{H}_1^T \ \dots \ \boldsymbol{H}_p^T \, , \quad \boldsymbol{R} = \boldsymbol{H}_p \ \dots \ \boldsymbol{H}_1 \, \boldsymbol{A} \tag{4.10.21}$$

where $\boldsymbol{R}$ is upper triangular. This method has a better stability than Gram–Schmidt factorization.

Example 4.8 :
QR decomposition by Householder's procedure: Application to any matrix
The matrix

$$A = \begin{bmatrix} 3 & -2 & 5 & 1 \\ 1 & 7 & -8 & 6 \\ 9 & 2 & 4 & -3 \\ -5 & 4 & -2 & 3 \end{bmatrix} \tag{4.10.22}$$

gives in the first stage $A_1 = A$, $\alpha = 10.7703$, $u = [-7.7703\ 1\ 9\ -5]$, $v = [-0.6006\ 0.0773\ 0.6957\ -0.3865]$, the first Householder matrix H_1

$$H_1 = \begin{bmatrix} 0.2785 & 0.0928 & 0.8356 & -0.4642 \\ 0.0928 & 0.9881 & -0.1075 & 0.0597 \\ 0.8356 & -0.1075 & 0.0321 & 0.5377 \\ -0.4642 & 0.0597 & 0.5377 & 0.7013 \end{bmatrix}$$

It results

$$H_1 A_1 = \begin{bmatrix} 10.7703 & -0.0928 & 4.9209 & -3.0640 \\ 0 & 6.7546 & -7.9898 & 6.5230 \\ 0 & -0.2090 & 4.0916 & 1.7071 \\ 0 & 5.2272 & -2.0509 & 0.3849 \end{bmatrix}$$

hence

$$A_2 = \begin{bmatrix} 6.7546 & -7.9898 & 6.5230 \\ -0.2090 & 4.0916 & 1.7071 \\ 5.2272 & -2.0509 & 0.3849 \end{bmatrix}$$

The operation is repeated on that matrix A_2 giving

$$H'_2 = \begin{bmatrix} 0.7906 & -0.0245 & 0.6118 \\ -0.0245 & 0.9971 & 0.0715 \\ 0.6118 & 0.0715 & -0.7878 \end{bmatrix}, \quad H_2 = \begin{bmatrix} 1 & 0 & 0 & 0 \\ 0 & 0.7906 & -0.0245 & 0.6118 \\ 0 & -0.0245 & 0.9971 & 0.0715 \\ 0 & 0.6118 & 0.0715 & -0.7878 \end{bmatrix}$$

It results

$$H'_2 A_2 = \begin{bmatrix} 8.5435 & -7.6717 & 5.3509 \\ 0 & 4.1287 & 1.5702 \\ 0 & -2.9804 & 3.8098 \end{bmatrix} \quad \text{hence} \quad A_3 = \begin{bmatrix} 4.1287 & 1.5702 \\ -2.9804 & 3.8098 \end{bmatrix}$$

$$H'_3 = \begin{bmatrix} 0.8108 & -0.5853 \\ -0.5853 & -0.8108 \end{bmatrix}, \quad H_3 = \begin{bmatrix} 1 & 0 & 0 & 0 \\ 0 & 1 & 0 & 0 \\ 0 & 0 & 0.8108 & -0.5853 \\ 0 & 0 & -0.5853 & -0.8108 \end{bmatrix}$$

Finally, the result of the factorization is

$$Q = H_1^T H_2^T H_3^T = \begin{bmatrix} 0.2785 & -0.2311 & 0.3646 & -0.8580 \\ 0.0928 & 0.8203 & -0.4249 & -0.3714 \\ 0.8356 & 0.2432 & 0.3444 & 0.3521 \\ -0.4642 & 0.4631 & 0.7536 & 0.0448 \end{bmatrix}$$

$$R = H_3 H_2 H_1 A = \begin{bmatrix} 10.7703 & -0.0928 & 4.9209 & -3.0640 \\ 0 & 8.5435 & -7.6717 & 5.3509 \\ 0 & 0 & 5.0921 & -0.9567 \\ 0 & 0 & 0 & -4.0080 \end{bmatrix}$$

4.10.0.3 QR Factorization by Rotation Matrices

The Jacobi rotation matrix was retained by Givens (1958) and Golub and Van Loan (1991) in the framework of the QR factorization. It is the same matrix. A Givens rotation matrix G is such that

$$
G(i, j, \theta) =
\begin{array}{c}
\qquad\qquad\qquad\quad \downarrow i \quad\; \downarrow j \\
\begin{bmatrix}
1 & 0 & \cdots\cdots\cdots\cdots\cdots\cdots\cdots\cdots\cdots & 0 \\
0 & \ddots & 0 & & & & & & & \vdots \\
\vdots & & \ddots & 1 & 0 & & & & & \\
\vdots & & \cdots & 0 & \cos(\theta) & 0 & \cdots & \sin(\theta) & 0 & \\
\vdots & & & & 0 & 1 & 0 & & & \\
\vdots & & & & & 0 & \ddots & 0 & & \\
\vdots & & \cdots & 0 & -\sin(\theta) & 0 & \cdots & \cos(\theta) & 0 & \\
\vdots & & & & & & & 0 & 1 & 0 \\
\vdots & & & & & & & & 0 & \ddots & 0 \\
0 & \cdots & & & & & & & & 0 & 1
\end{bmatrix}
\begin{array}{l}
\\ \\ \\ \leftarrow i \\ \\ \\ \leftarrow j \\ \\ \\ \\
\end{array}
\end{array}
\tag{4.10.23}
$$

The notation $G(i, j, \theta)$ means that the matrix is specified by the indexes i and j and the value of θ. A rotation matrix differs from the identity matrix only by four elements at the intersections of the rows i, j and the columns i, j so that $G_{ii} = G_{jj} = \cos(\theta)$ and $G_{ij} = -G_{ji} = \sin(\theta)$. The multiplication of $G(i, j, \theta)$ by a matrix A under the form $G^T A$ affects the rows i and j, while under the form AG, it affects the columns i and j.

The idea of Jacobi's method (note $J = G$ in his honor) is to search a matrix J such that the matrix $B = J^T(i, j, \theta) A J(i, j, \theta)$ is diagonal with respect to elements b_{ii}, b_{jj} (as $b_{ij} = b_{ji} = 0$). The other elements of B are identical to the elements of A. Jacobi's cyclic algorithm per row to solve the eigenvalue problem is simpler, but less performing, than QP factorization (Golub and Van Loan 1991).

To understand the rotation operation, this latter is explained in the case of the multiplication $G(i, j, \theta) X$ where X is a vector

$$
Y = G(i, j, \theta)^T X \quad, \quad X \in \mathbb{R}^n
\tag{4.10.24}
$$

Frequently, we note $c = \cos(\theta)$, $s = \sin(\theta)$. In this case, we find

$$
Y_k = \begin{cases}
c\, X_i - s\, X_j & \text{if } k = i \\
s\, X_i + c\, X_j & \text{if } k = j \\
X_k & \text{if } k \neq i, j
\end{cases}
\tag{4.10.25}
$$

We wish the element j of the vector Y to be zero. It results

$$
\frac{c}{s} = -\frac{X_i}{X_j}
\tag{4.10.26}
$$

and we use $c^2 + s^2 = 1$. Two pairs of solutions are possible; we choose

$$c = \cos(\theta) = \frac{X_i}{\sqrt{X_i^2 + X_j^2}}, \quad s = \sin(\theta) = -\frac{X_j}{\sqrt{X_i^2 + X_j^2}} \tag{4.10.27}$$

The QR factorization algorithm is described in the case of any matrix $\boldsymbol{A}$, not necessarily square (of dimension $m \times n$, however assuming $m \geq n$). We set $\boldsymbol{A}_1 = \boldsymbol{A}$. The objective is, by a sequence of rotations, to obtain an upper triangular matrix $\boldsymbol{R}$. The first transformation by means of the first rotation matrix $\boldsymbol{G}_1$ has the objective that $\boldsymbol{A}_2 = \boldsymbol{G}_1\,\boldsymbol{A}_1$ has the zero element at position $(m, 1)$, then $\boldsymbol{A}_2$ will have the zero element at position $(m-1, 1)$, and thus the first column will be swept under the diagonal, then the second column under diagonal, and so on. $(m-1) \times (n-1)$ rotation operations are thus required. In the case of a tridiagonal symmetric matrix $(n \times n)$, $(n-1)$ rotation operations are required.

Thus, the rotation matrices $\boldsymbol{G}_1, \ldots, \boldsymbol{G}_{n-1}$ are determined such that the factorization matrices are

$$\boldsymbol{Q} = \boldsymbol{G}_1 \ldots \boldsymbol{G}_t\,, \quad \boldsymbol{R} = \boldsymbol{G}_t^T \ldots \boldsymbol{G}_1^T\,\boldsymbol{A}\,, \quad \boldsymbol{A} = \boldsymbol{Q}\,\boldsymbol{R} \tag{4.10.28}$$

where $\boldsymbol{R}$ is upper triangular. t is the number of rotation operations.

The Matlab algorithm can be summarized as:

```
for j=1:na-1
  for i=ma:-1:j+1
    xi = A(j,j);
    xj = A(i,j);
% elements of simple Givens rotation matrix (Golub, p. 202)
    c = xi / sqrt(xi^2 + xj^2);
    s = - xj / sqrt(xi^2 + xj^2);
% rotation matrix G
    G = eye(ma,ma);
    G(i,i) = c;
    G(j,j) = c;
    G(i,j) = -s;
    G(j,i) = s;
    A = G' * A;
    Q = Q * G;
    R = G' * R;
  end
end
```

where the columns under the diagonal are swept as indicated, and a choice is made for elements X_i and X_j. The matrix $\boldsymbol{A}$ is renewed at each step.

Golub and Van Loan (1991) slightly modifies algorithm (4.10.27) to avoid calculation overflow as

$$
\begin{array}{lll}
\text{if } X_j = 0, & c = \cos(\theta) = 1, & s = \sin(\theta) = 0 \\
\text{if } |X_i| > |X_j|, & c = \cos(\theta) = \dfrac{1}{\sqrt{1 + (X_j/X_i)^2}}, & s = \sin(\theta) = -\dfrac{X_j}{X_i} c \\
\text{if } |X_j| > |X_i|, & c = \cos(\theta) = -\dfrac{X_i}{X_j} s, & s = \sin(\theta) = \dfrac{1}{\sqrt{1 + (X_i/X_j)^2}}
\end{array}
\tag{4.10.29}
$$

Example 4.9 :

QR factorization by Givens rotation matrices: Application to any matrix

The QR factorization by Givens rotation matrices is applied to the following arbitrary matrix:

$$
\boldsymbol{A} = \begin{bmatrix} 3 & -2 & 5 & 1 \\ 1 & 7 & -8 & 6 \\ 9 & 2 & 4 & -3 \\ -5 & 4 & -2 & 3 \end{bmatrix}
\tag{4.10.30}
$$

We set $\boldsymbol{A}_1 = \boldsymbol{A}$, initially $\boldsymbol{Q} = \boldsymbol{I}_4$, $\boldsymbol{R} = \boldsymbol{A}_1$. Consider the first stage of the algorithm, $j = 1$, $i = 4$, hence $x_i = 3$, $x_j = -5$. The rotation matrix of the first stage is

$$
\boldsymbol{G}_1 = \begin{bmatrix} 0.5145 & 0 & 0 & 0.8575 \\ 0 & 1 & 0 & 0 \\ 0 & 0 & 1 & 0 \\ -0.8575 & 0 & 0 & 0.5145 \end{bmatrix}
$$

hence for this first stage

$$
\boldsymbol{Q} = G_1 \,, \quad \boldsymbol{R} = \begin{bmatrix} 5.8310 & -4.4590 & 4.2875 & -2.0580 \\ 1 & 7 & -8 & 6 \\ 9 & 2 & 4 & -3 \\ 0 & 0.3430 & 3.2585 & 2.4010 \end{bmatrix}
$$

and

$$
\boldsymbol{A}_2 = \boldsymbol{G}_1^T \boldsymbol{A}_1 = \boldsymbol{R}
$$

It is noted that the element at position (4, 1) of $\boldsymbol{A}_2$ is zero. The operations are repeated as described in the algorithm. Finally, the matrices of the QR factorization QR are obtained

$$
\boldsymbol{Q} = \boldsymbol{G}_1 \ldots \boldsymbol{G}_t = \begin{bmatrix} 0.2785 & -0.2311 & 0.3646 & 0.8580 \\ 0.0928 & 0.8203 & -0.4249 & 0.3714 \\ 0.8356 & 0.2432 & 0.3444 & -0.3521 \\ -0.4642 & 0.4631 & 0.7536 & -0.0448 \end{bmatrix}
$$

$$
\boldsymbol{R} = \boldsymbol{G}_t^T \ldots \boldsymbol{G}_1^T \boldsymbol{A} = \begin{bmatrix} 10.7703 & -0.0928 & 4.9209 & -3.0640 \\ 0 & 8.5435 & -7.6717 & 5.3509 \\ 0 & 0 & 5.0921 & -0.9567 \\ 0 & 0 & 0 & 4.0080 \end{bmatrix}
$$

where $\boldsymbol{R}$ is upper triangular. We can verify $\boldsymbol{Q}\,\boldsymbol{R} = \boldsymbol{A}$.

The three methods previously described, QR factorization by Gram–Schmidt, QR factorization by Householder, QR factorization by Givens rotation matrices, are compared in the following to determine, according to QR Francis's method, the eigenvalues of a symmetric tridiagonal matrix.

Example 4.10 :
QR Francis's method: Determination of the eigenvalues of a symmetric tridiagonal matrix

QR Francis's method is applied to determine the eigenvalues of a symmetric tridiagonal matrix $\boldsymbol{A}$ equal to

$$\boldsymbol{A} = \begin{bmatrix} 1 & 1 & 0 & 0 \\ 1 & 2 & -1 & 0 \\ 0 & -1 & 3 & 1 \\ 0 & 0 & 1 & 4 \end{bmatrix} \tag{4.10.31}$$

whose eigenvalues are

$$\boldsymbol{\lambda} = \left[0.2547\ 1.8227\ 3.1773\ 4.7453\right]^T \tag{4.10.32}$$

1. *QR factorization by Gram–Schmidt*
We set $\boldsymbol{A}^{(1)} = \boldsymbol{A}$. We first perform the QR factorization of $\boldsymbol{A}$ by QR decomposition of Gram–Schmidt. We thus obtain

$$\boldsymbol{Q}^{(1)} = \begin{bmatrix} 0.7071 & -0.4082 & -0.4364 & 0.3780 \\ 0.7071 & 0.4082 & 0.4364 & -0.3780 \\ 0 & -0.8165 & 0.4364 & -0.3780 \\ 0 & 0 & 0.6547 & 0.7559 \end{bmatrix}$$

$$\boldsymbol{R}^{(1)} = \begin{bmatrix} 1.4142 & 2.1213 & -0.7071 & 0 \\ 0 & 1.2247 & -2.8577 & -0.8165 \\ 0 & 0 & 1.5275 & 3.0551 \\ 0 & 0 & 0 & 2.6458 \end{bmatrix}$$

Then, we calculate $\boldsymbol{A}^{(2)} = \boldsymbol{R}^{(1)}\boldsymbol{Q}^{(1)}$, thus

$$\boldsymbol{A}^{(2)} = \begin{bmatrix} 2.5000 & 0.8660 & 0 & 0 \\ 0.8660 & 2.8333 & -1.2472 & 0 \\ 0 & -1.2472 & 2.6667 & 1.7321 \\ 0 & 0 & 1.7321 & 2.0000 \end{bmatrix}$$

and this operation is continued until most nondiagonal elements (except the case of a multiple eigenvalue) tend to 0. Thus, we find after k iterations (k = 5, 10, 20)

$$\boldsymbol{A}^{(6)} = \begin{bmatrix} 4.2928 & 0.7213 & 0 & 0 \\ 0.7213 & 3.5561 & -0.3350 & 0 \\ 0 & -0.3350 & 1.8964 & 0.0004 \\ 0 & 0 & 0.0004 & 0.2547 \end{bmatrix}$$

$$\boldsymbol{A}^{(11)} = \begin{bmatrix} 4.7342 & 0.1314 & 0 & 0 \\ 0.1314 & 3.1881 & -0.0186 & 0 \\ 0 & -0.0186 & 1.8230 & 0.0000 \\ 0 & 0 & 0.0000 & 0.2547 \end{bmatrix}$$

$$\boldsymbol{A}^{(21)} = \begin{bmatrix} 4.7453 & 0.0024 & 0 & 0 \\ 0.0024 & 3.1773 & -0.0001 & 0 \\ 0 & -0.0001 & 1.8227 & 0.0000 \\ 0 & 0 & 0.0000 & 0.2547 \end{bmatrix}$$

Through the nondiagonal elements of $\boldsymbol{A}^{(k)}$, we note the convergence when the number of iterations k increases and the diagonal elements of $\boldsymbol{A}^{(21)}$ approximate the eigenvalues of $\boldsymbol{A}$ to within 10^{-4} in absolute value and are ordered by decreasing absolute value.

2. *QR factorization by Householder.*
The procedure is the same as that used in 1. except that QR factorization by Householder is used instead of QR factorization by Gram–Schmidt. We thus get

$$\boldsymbol{Q}^{(1)} = \begin{bmatrix} 0.7071 & -0.4082 & -0.4364 & -0.3780 \\ 0.7071 & 0.4082 & 0.4364 & 0.3780 \\ 0 & -0.8165 & 0.4364 & 0.3780 \\ 0 & 0 & 0.6547 & -0.7559 \end{bmatrix}$$

$$\boldsymbol{R}^{(1)} = \begin{bmatrix} 1.4142 & 2.1213 & -0.7071 & 0 \\ 0 & 1.2247 & -2.8577 & -0.8165 \\ 0 & 0 & 1.5275 & 3.0551 \\ 0 & 0 & 0 & -2.6458 \end{bmatrix}$$

Then, we calculate $\boldsymbol{A}^{(2)} = \boldsymbol{R}^{(1)}\boldsymbol{Q}^{(1)}$, that is

$$\boldsymbol{A}^{(2)} = \begin{bmatrix} 2.5000 & 0.8660 & 0 & 0 \\ 0.8660 & 2.8333 & -1.2472 & 0 \\ 0 & -1.2472 & 2.6667 & -1.7321 \\ 0 & 0 & -1.7321 & 2.0000 \end{bmatrix}$$

Then, the operation is repeated until most nondiagonal elements (except the case of a multiple eigenvalue) tend to zero. After k iterations ($k = 5,\ 10,\ 20$), we find

$$\boldsymbol{A}^{(6)} = \begin{bmatrix} 4.2928 & 0.7213 & 0 & 0 \\ 0.7213 & 3.5561 & -0.3350 & 0 \\ 0 & -0.3350 & 1.8964 & -0.0004 \\ 0 & 0 & -0.0004 & 0.2547 \end{bmatrix}$$

$$\boldsymbol{A}^{(11)} = \begin{bmatrix} 4.7342 & 0.1314 & 0 & 0 \\ 0.1314 & 3.1881 & -0.0186 & 0 \\ 0 & -0.0186 & 1.8230 & 0.0000 \\ 0 & 0 & 0.0000 & 0.2547 \end{bmatrix}$$

$$\boldsymbol{A}^{(21)} = \begin{bmatrix} 4.7453 & 0.0024 & 0 & 0 \\ 0.0024 & 3.1773 & -0.0001 & 0 \\ 0 & -0.0001 & 1.8227 & -0.0000 \\ 0 & 0 & -0.0000 & 0.2547 \end{bmatrix}$$

The convergence is the same as for QR factorization by Gram–Schmidt, but the present method is reputed to be more stable.

3. *QR factorization QR by Givens rotation matrices.*
We set $\boldsymbol{A}^{(1)} = \boldsymbol{A}$. QR factorization of $\boldsymbol{A}$ is first performed by QR decomposition by Givens rotation matrices. We thus get

$$\boldsymbol{Q}^{(1)} = \begin{bmatrix} 0.7071 & -0.4082 & -0.4364 & 0.3780 \\ 0.7071 & 0.4082 & 0.4364 & -0.3780 \\ 0 & -0.8165 & 0.4364 & -0.3780 \\ 0 & 0 & 0.6547 & 0.7559 \end{bmatrix}$$

$$\boldsymbol{R}^{(1)} = \begin{bmatrix} 1.4142 & 2.1213 & -0.7071 & 0 \\ 0 & 1.2247 & -2.8577 & -0.8165 \\ 0 & 0.0000 & 1.5275 & 3.0551 \\ 0 & -0.0000 & 0.0000 & 2.6458 \end{bmatrix}$$

Then, we calculate $\boldsymbol{A}^{(2)} = \boldsymbol{R}^{(1)}\boldsymbol{Q}^{(1)}$, that is

$$\boldsymbol{A}^{(2)} = \begin{bmatrix} 2.5000 & 0.8660 & 0 & 0 \\ 0.8660 & 2.8333 & -1.2472 & 0 \\ 0 & -1.2472 & 2.6667 & 1.7321 \\ 0 & 0 & 1.7321 & 2.0000 \end{bmatrix}$$

Then, the operation is repeated. After k iterations ($k = 5,\ 10,\ 20$), we find

$$A^{(6)} = \begin{bmatrix} 4.2928 & 0.7213 & 0 & 0 \\ 0.7213 & 3.5561 & -0.3350 & 0 \\ 0 & -0.3350 & 1.8964 & 0.0004 \\ 0 & 0 & 0.0004 & 0.2547 \end{bmatrix}$$

$$A^{(11)} = \begin{bmatrix} 4.7342 & 0.1314 & 0 & 0 \\ 0.1314 & 3.1881 & -0.0186 & 0 \\ 0 & -0.0186 & 1.8230 & 0.0000 \\ 0 & 0 & 0.0000 & 0.2547 \end{bmatrix}$$

$$A^{(21)} = \begin{bmatrix} 4.7453 & 0.0024 & 0 & 0 \\ 0.0024 & 3.1773 & -0.0001 & 0 \\ 0 & -0.0001 & 1.8227 & 0.0000 \\ 0 & 0 & 0.0000 & 0.2547 \end{bmatrix}$$

In this simple example, the convergence of the diagonal elements of $A^{(i)}$ to the real eigenvalues of $A^{(i)}$ is identical for the three described QR factorization methods.

4.10.0.4 Application of QR Factorization

The QR factorization allows to find a least squares solution of an overdetermined linear system by doing a QR factorization.

Consider the overdetermined system

$$\begin{aligned} a_{11}x_1 + a_{12}x_2 + \cdots + a_{1n}x_n &= b_1 \\ a_{21}x_1 + a_{22}x_2 + \cdots + a_{2n}x_n &= b_2 \\ \vdots \\ a_{m1}x_1 + a_{m2}x_2 + \cdots + a_{mn}x_n &= b_m \end{aligned} \tag{4.10.33}$$

with $m \geq n$ (number of equations larger than the number of unknowns), that is

$$A\,x = b \tag{4.10.34}$$

We do the QR factorization of A, that is

$$Q\,R = A \tag{4.10.35}$$

hence

$$Q\,R\,x = b \Rightarrow Q^T\,Q\,R\,x = Q^T\,b \Rightarrow R\,x = Q^T\,b \tag{4.10.36}$$

As R is triangular, the system can thus be solved without any matrix inversion (see for Example 5.8).

4.11 Discussion and Conclusion

Numerical methods make an extensive use of matrices. Their correct handling is thus essential, and the matrix definitions and properties cited at the beginning of this chapter must be mastered. The transformation of matrices to obtain similar matrices possessing better properties often occurs in numerical calculation, even if the end-user is not conscious of it. The eigenvalues represent a summary of the information transported by a matrix, but their calculation remains a delicate numerical operation. For that reason, an important part of the chapter has been devoted to Rutishauser's method and more recent related methods.

4.12 Exercise Set

Preamble: Writing general codes can be very helpful for the students to learn and understand the techniques. They also can use existing codes.

Exercise 4.12.1 (Medium)

Decomposition of a matrix

Write a Fortran subroutine or a Matlab function allowing you to decompose a square matrix $\boldsymbol{A}$ under the form $\boldsymbol{A} = \boldsymbol{B}\,\boldsymbol{C}$ where $\boldsymbol{B}$ is a lower triangular matrix whose diagonal elements are all equal to 1 and $\boldsymbol{C}$ is an upper triangular matrix.

Exercise 4.12.2 (Easy)

Consider the matrix $\boldsymbol{A}$ equal to

$$\begin{bmatrix} 3 & 1.1 & -0.1 \\ 0.9 & 15 & -2.1 \\ 0.1 & -1.9 & 19.5 \end{bmatrix} \tag{4.12.1}$$

1. Determine domains as small as possible where the eigenvalues of this matrix are located.
2. Calculate an upper bound of $\text{cond}_2(\boldsymbol{A})$ by using the previous estimations. What is your conclusion?

Exercise 4.12.3 (Easy)

Consider the matrix

$$M = \begin{bmatrix} 1 & -5 & 2 & 7 \\ 3 & 4 & -2 & 8 \\ 2 & -4 & 1 & -6 \\ 6 & 3 & 2 & -5 \end{bmatrix} \tag{4.12.2}$$

1. Find the eigenvalue of largest modulus by the power method by taking as initial vector $V^{(0)} = [0.75 \quad 0.25 \quad -0.5 \quad -0.75]^T$. This initialization allows you to find the result in less than 20 iterations.
2. Find a bracketing of the eigenvalues. Represent graphically this bracketing. Confirm the result of question 1.

Exercise 4.12.4 (Easy)
Consider the following matrix $\boldsymbol{A}$:

$$\boldsymbol{A} = \begin{bmatrix} 1 & 0 & 2 \\ 0 & 1 & -1 \\ -1 & 1 & 1 \end{bmatrix} \tag{4.12.3}$$

1. Calculate the norms $\|\boldsymbol{A}\|_1$, $\|\boldsymbol{A}\|_2$, $\|\boldsymbol{A}\|_\infty$.
2. Calculate the eigenvalues and eigenvectors of the matrix according to the usual mathematical manner by explaining the calculations.
3. By Gershgorin's method, give a bracketing of the eigenvalues and verify the consistence with the result of question 2.
4. Give an interval for the condition number of $\boldsymbol{A}$.

Exercise 4.12.5 (Medium)
We consider the following algorithm that factorizes the matrix $\boldsymbol{A}$ under the form $\boldsymbol{L}\boldsymbol{D}\boldsymbol{L}^T$ where $\boldsymbol{L}$ is a lower triangular matrix with its diagonal elements equal to 1 and $\boldsymbol{D}$ is a diagonal matrix whose nonzero elements are positive,

1. For $i = 1, \ldots, n$, do the steps 2 to 4.
2. For $j = 1, \ldots, i-1$, set $v_j = l_{ij}\, d_j$.
3. Set $d_i = a_{ii} - \sum_{k=1}^{i-1} l_{ik}\, v_k$.
4. For $j = i+1, \ldots, n$, set $l_{ji} = (a_{ji} - \sum_{k=1}^{i-1} l_{jk}\, v_k)/d_i$.

Apply this algorithm to the following matrix $\boldsymbol{A}$:

$$\boldsymbol{A} = \begin{bmatrix} 4 & -1 & 1 \\ -1 & 4.25 & 2.75 \\ 1 & 2.75 & 3.5 \end{bmatrix} \tag{4.12.4}$$

by detailing the intermediate results of calculation. Give the final factorization.

Exercise 4.12.6 (Medium)
Consider the following system:

$$\begin{aligned} 3.333\, x_1 + 8.920\, x_2 + 10.333\, x_3 &= -2.540 \\ 2.222\, x_1 + 16.710\, x_2 + 9.612\, x_3 &= 0.965 \\ -1.5611\, x_1 + 5.1792\, x_2 - 1.6855\, x_3 &= 2.714 \end{aligned} \tag{4.12.5}$$

that can be symbolized under the form $\boldsymbol{A}\,\boldsymbol{x} = \boldsymbol{b}$.

1. Solve this system by Gauss method by explaining the calculations and giving the results of each step.
2. Solve the same system by Gauss–Jordan method by explaining the calculations and giving the results of each step.
3. The following questions use all or part of the previous results. Answer these questions by justifying all the calculations.

   (a) Calculate the determinant of matrix $\boldsymbol{A}$.

(b) Do the decomposition $\boldsymbol{LU}$ of matrix $\boldsymbol{A}$. Verify that the decomposition is correct.
(c) Calculate the inverse of matrix $\boldsymbol{A}$. Verify that the inverse is correct.

Exercise 4.12.7 (Medium)

Three questions of mathematics

1. Let $\boldsymbol{w}$ be a vector of R^n such that $\boldsymbol{w}^T\boldsymbol{w} = 1$. We define the matrix $\boldsymbol{P} = \boldsymbol{I} - 2\boldsymbol{w}\boldsymbol{w}^T$. The matrix $\boldsymbol{P}$ is called a Householder transformation.
   (a) How to qualify this property of $\boldsymbol{w}$?
   (b) Prove that the matrix $\boldsymbol{P}$ is symmetrical.
   (c) Prove that $\boldsymbol{P}^{-1} = \boldsymbol{P}$.
   (d) Prove that $\boldsymbol{P}$ is orthogonal.
2. Any matrix $\boldsymbol{A}$ is said stable for the differential problem $\dot{\boldsymbol{x}} = \boldsymbol{A}\boldsymbol{x}$ when its eigenvalues are negative or with a negative real part. Explain this property by thinking about the problem $\dot{x} = ax$ when x and a are scalars.
3. Let the following matrix of dimension $n \times n$:

$$\boldsymbol{A} = \begin{bmatrix} 1+2\alpha & -\alpha & 0 & \dots & 0 \\ -\alpha & 1+2\alpha & -\alpha & \ddots & \vdots \\ 0 & \ddots & \ddots & \ddots & \vdots \\ \vdots & & -\alpha & 1+2\alpha & -\alpha \\ 0 & \dots & 0 & -\alpha & 1+2\alpha \end{bmatrix} \tag{4.12.6}$$

   (a) How do you qualify such a matrix?
   (b) The eigenvalues of $\boldsymbol{A}$ are equal to $\lambda_i = 1 + 4\alpha \left(\sin\left(\frac{\pi i}{2(n+1)}\right)\right)^2$, $i = 1, \dots, n$. When is the matrix $\boldsymbol{A}$ stable?
   (c) Verify the formula of the eigenvalues when $n = 2$.

Exercise 4.12.8 (Easy)

1. With a precision of 10^{-1}, find the eigenvalue of largest modulus of matrix $\boldsymbol{A}$ without solving the characteristic equation

$$\boldsymbol{A} = \begin{bmatrix} 4 & 1 & -1 & 0.5 \\ 1 & 3.2 & 0.7 & -0.3 \\ -1 & 0.7 & -6.4 & 2.1 \\ 0.5 & -0.3 & 2.1 & 3.7 \end{bmatrix} \tag{4.12.7}$$

   The solution must be found in less than ten iterations.
2. Verify mathematically that this calculation corresponds to an eigenvalue (with a given uncertainty).

Exercise 4.12.9 (Easy)
Consider the following matrix $\boldsymbol{A}$:

$$A = \begin{bmatrix} 1 & 0 & 2 \\ 0 & 1 & -1 \\ -1 & 1 & 1 \end{bmatrix} \tag{4.12.8}$$

1. Calculate the norms $\|A\|_1$, $\|A\|_2$, $\|A\|_\infty$.
2. Calculate the eigenvalues and the eigenvectors of the matrix using the usual mathematical manner and explain your calculations.
3. Using Gershgorin's method, give a domain for the eigenvalues and verify the consistency with the result of question 2.
4. Give an interval for the condition number A.

Exercise 4.12.10 (Easy)
Consider the following matrix A:

$$A = \begin{bmatrix} \alpha & 1 & 0 \\ \beta & 2 & 1 \\ 0 & 1 & 2 \end{bmatrix} \tag{4.12.9}$$

Calculate all the values of α and β for which

1. The matrix A is singular.
2. The matrix A is strictly diagonally dominant.
3. The matrix A is symmetrical.
4. The matrix A is positive definite.

Exercise 4.12.11 (Medium)
Consider the following matrix:

$$\begin{bmatrix} 4 & -2 & 1 \\ 1 & 8 & 3 \\ -1 & 3 & 9 \end{bmatrix} \tag{4.12.10}$$

1. Calculate a domain for the eigenvalues and make a graphical representation.
2. Determine the characteristic polynomial associated to that matrix.
3. Use the following technique to seek an eigenvalue:
   - Find the interval for the real eigenvalues and make a bisection from that interval until the length of the interval is lower than 1.
   - Continue the search by Newton's method initialized at the midpoint of the last interval.

   Then, deduce the two other eigenvalues by a simple calculation to be detailed. Verify graphically that all the eigenvalues indeed belong to the domain of question 1.

Exercise 4.12.12 (Easy)
Write a program achieving the LU decomposition by Crout's method and apply it on the following matrix A:

$$A = \begin{bmatrix} 2 & 1 & 3 & -5 \\ 1 & -2 & 7 & -4 \\ -3 & 5 & 1 & 6 \\ 4 & -9 & 5 & 3 \end{bmatrix} \tag{4.12.11}$$

Calculate the determinant of A.

Exercise 4.12.13 (Medium)
Consider the matrix

$$A = \begin{bmatrix} 7 & 1 & 3 \\ 4 & 8 & 5 \\ 2 & 3 & 6 \end{bmatrix} \tag{4.12.12}$$

1. Make the decomposition of the matrix under the form

$$A = L_1 R_1 \tag{4.12.13}$$

where L_1 is lower triangular and has its diagonal elements equal to 1 while R_1 is upper triangular (refer to Rutishauser's method, Section 4.8).
2. Apply Rutishauser's method.
3. Deduce an estimation of the eigenvalues of A.
4. Deduce an estimation of the determinant of A.

References

J. G. F. Francis. The QR transformation. *Computer Journal*, 4, Part I:265–271, 1961–62a.

J. G. F. Francis. The QR transformation. *Computer Journal*, 4, Part II:332–345, 1961–62b.

W. Givens. Computation of plane unitary rotations transforming a general matrix to triangular form. *SIAM J. App. Math.*, 6:26–50, 1958.

G. H. Golub and C. F. Van Loan. *Matrix computations.* John Hopkins University Press, Baltimore, 1991.

E. Hairer. *Introduction à l'analyse numérique*. Université de Genève, 1993.

A. S. Householder. Unitary triangularization of a nonsymmetric matrix. *Journal of the ACM*, 5(4):339–342, 1958.

F. Rotella and P. Borne. *Théorie et pratique du calcul matriciel*. Technip, Paris, 1995.

H. Rutishauser. Solutions of eigenvalues problems with the LR-transformation. *National Bureau of Standards Applied Mathematics Series*, 49:47–81, 1958.

H. Rutishauser. Simultaneous iteration method for symmetric matrices. *Numerische Mathematik*, 16:205–223, 1970.

H. Rutishauser and H. R. Schwarz. The LR transformation method for symmetric matrices. *Numerische Mathematik*, 5:273–289, 1963.

Chapter 5
Numerical Solution of Systems of Algebraic Equations

5.1 Introduction

The system of n algebraic equations with n unknowns to be solved can be written under the general form

$$\begin{aligned} f_1(x_1, \ldots, x_n) &= 0 \\ &\vdots \\ f_n(x_1, \ldots, x_n) &= 0 \end{aligned} \tag{5.1.1}$$

In the first part of the present chapter, we will consider the particular case of systems of linear algebraic equations with respect to the variable x which can be set under the form

$$\begin{aligned} a_{11}x_1 + a_{12}x_2 + \cdots + a_{1n}x_n &= b_1 \\ a_{21}x_1 + a_{22}x_2 + \cdots + a_{2n}x_n &= b_2 \\ &\vdots \\ a_{n1}x_1 + a_{n2}x_2 + \cdots + a_{nn}x_n &= b_n \end{aligned} \tag{5.1.2}$$

or under the matrix form

$$\boldsymbol{A x} = \boldsymbol{b} \tag{5.1.3}$$

In the second part of the present chapter, the general case of a system of nonlinear equations (5.1.1) will be studied.

5.2 Solution of Linear Triangular Systems

Two particular cases of linear systems are simple to solve, when the matrix $\boldsymbol{A}$ is triangular.

- Case where $\boldsymbol{A}$ is upper triangular.
  We note $\boldsymbol{U} = \boldsymbol{A}$ (U="Upper") and the system is written as

$$\boldsymbol{U}\,\boldsymbol{x} = \boldsymbol{b} \tag{5.2.1}$$

J.-P. Corriou, *Numerical Methods and Optimization*, Springer Optimization and Its Applications 187, https://doi.org/10.1007/978-3-030-89366-8_5

Its solution is given by

$$x_i = \frac{1}{u_{ii}}(b_i - \sum_{j=i+1}^{n} u_{ij}x_j)\,, \quad i = n, \ldots, 1 \tag{5.2.2}$$

- Case where $\boldsymbol{A}$ is lower triangular.
  We note $\boldsymbol{L} = \boldsymbol{A}$ (L="Lower") and the system is written as

$$\boldsymbol{L}\,\boldsymbol{x} = \boldsymbol{b} \tag{5.2.3}$$

Its solution is given by

$$x_i = \frac{1}{l_{ii}}(b_i - \sum_{j=1}^{i-1} l_{ij}x_j)\,, \quad i = 1, \ldots, n \tag{5.2.4}$$

5.3 Solution of Linear Systems: Gauss Elimination Method

The system of linear equations described by

$$\boldsymbol{Ax} = \boldsymbol{b} \tag{5.3.1}$$

where $\boldsymbol{A}$ is an $n \times n$ dense matrix and $\boldsymbol{b}$ a vector of $\mathcal{R}^n$, will be transformed by steps into a linear system, having the same solution vector, of the form

$$\boldsymbol{Ux} = \boldsymbol{c} \tag{5.3.2}$$

where $\boldsymbol{U}$ is an $n \times n$ upper triangular matrix.

Suppose that $a_{11} \neq 0$. If it were not the case, it would be sufficient to reorder the equations so that $a_{11} \neq 0$. In the first step, we subtract from all equations except the first one a multiple of that first equation so that the coefficient of x_1 disappears in each of these equations going from 2 to n. Indeed, these operations are performed on the augmented matrix $(\boldsymbol{A}, \boldsymbol{b})$ defined by

$$(\boldsymbol{A}, \boldsymbol{b}) = \begin{bmatrix} a_{11} & a_{12} & \ldots & a_{1n} & b_1 \\ a_{21} & a_{22} & \ldots & a_{2n} & b_2 \\ \vdots & \vdots & & \vdots & \vdots \\ a_{n1} & a_{n2} & \ldots & a_{nn} & b_n \end{bmatrix} \tag{5.3.3}$$

- **Gauss algorithm with partial pivoting**:

  (a) *Determine an element* $a_{r1} \neq 0$ *and do step b/. If such an index r does not exist,* $\boldsymbol{A}$ *is singular, pose* $(\boldsymbol{A}', \boldsymbol{b}') = (\boldsymbol{A}, \boldsymbol{b})$ *, stop.*
  (b) *Interchange the rows r and* 1 *of the matrix* $(\boldsymbol{A}, \boldsymbol{b})$*. The result is the matrix* $(\bar{\boldsymbol{A}}, \bar{\boldsymbol{b}})$.
  (c) *For the indexes* $i = 2, 3, \ldots, n$*, subtract from the row i of the matrix* $(\bar{\boldsymbol{A}}, \bar{\boldsymbol{b}})$ *the multiple by* l_{ij} *of row 1, with*

$$l_{ij} = \bar{a}_{ij}/\bar{a}_{11}\,, \quad j = 1, \ldots, n+1 \tag{5.3.4}$$

The result of the first sequence of steps (for $j = 1$) is the matrix $(\boldsymbol{A}', \boldsymbol{b}')$ such that

$$\begin{aligned} a'_{1j} &= \bar{a}_{1j} \\ a'_{ij} &= \bar{a}_{ij} - \frac{\bar{a}_{i1}}{\bar{a}_{11}}\,\bar{a}_{1j} \\ b'_1 &= \bar{b}_1 \\ b'_i &= \bar{b}_i - \frac{\bar{a}_{i1}}{\bar{a}_{11}}\,\bar{b}_1 \end{aligned} \tag{5.3.5}$$

The element $a_{r1} = \bar{a}_{11}$ is called pivot element or *pivot* and the step a/ pivot selection step.

It thus leads to a matrix $(\boldsymbol{A}', \boldsymbol{b}')$ equal to

$$(\boldsymbol{A}', \boldsymbol{b}') = \begin{bmatrix} a'_{11} & a'_{12} & \cdots & a'_{1n} & b'_1 \\ 0 & a'_{22} & \cdots & a'_{2n} & b'_2 \\ \vdots & \vdots & & \vdots & \vdots \\ 0 & a'_{n2} & \cdots & a'_{nn} & b'_n \end{bmatrix} \tag{5.3.6}$$

This transformation $(\boldsymbol{A}, \boldsymbol{b}) \rightarrow (\bar{\boldsymbol{A}}, \bar{\boldsymbol{b}}) \rightarrow (\boldsymbol{A}', \boldsymbol{b}')$ can be described by the following matrix operations:

$$(\bar{\boldsymbol{A}}, \bar{\boldsymbol{b}}) = \boldsymbol{P}_1\,(\boldsymbol{A}, \boldsymbol{b})\,, \quad (\boldsymbol{A}', \boldsymbol{b}') = \boldsymbol{G}_1\,(\bar{\boldsymbol{A}}, \bar{\boldsymbol{b}}) = \boldsymbol{G}_1\,\boldsymbol{P}_1\,(\boldsymbol{A}, \boldsymbol{b}) \tag{5.3.7}$$

where $\boldsymbol{P}_1$ is an interchange matrix and $\boldsymbol{G}_1$ a lower triangular matrix defined by

$$\boldsymbol{P}_1 = \begin{bmatrix} 0 & & & 1 & & & 0 \\ & 1 & & & & & \\ & & \ddots & & & & \\ & & & 1 & & & \\ 1 & & & 0 & & & \\ & & & & 1 & & \\ & & & & & \ddots & \\ 0 & & & & & & 1 \end{bmatrix} \leftarrow r \qquad \boldsymbol{G}_1 = \begin{bmatrix} 1 & & & 0 \\ -l_{21} & 1 & & \\ \vdots & 0 & \ddots & \\ -l_{n1} & & 0 & 1 \end{bmatrix} \tag{5.3.8}$$

A matrix such that $\boldsymbol{G}_1$, being different from the identity matrix by one column at the most, is called the *Frobenius matrix*. Both matrices $\boldsymbol{P}_1$ and $\boldsymbol{G}_1$ are nonsingular and have the following properties:

$$\boldsymbol{P}_1^{-1} = \boldsymbol{P}_1\,, \quad \boldsymbol{G}_1^{-1} = \begin{bmatrix} 1 & & & 0 \\ l_{21} & 1 & & \\ \vdots & 0 & \ddots & \\ l_{n1} & & 0 & 1 \end{bmatrix} \tag{5.3.9}$$

It can be easily shown that both systems $\boldsymbol{A}\boldsymbol{x} = \boldsymbol{b}$ and $\boldsymbol{A}'\boldsymbol{x} = \boldsymbol{b}'$ have the same solution

$$\boldsymbol{A}\boldsymbol{x} = \boldsymbol{b} \Rightarrow \boldsymbol{G}_1\,\boldsymbol{P}_1\,\boldsymbol{A}\,\boldsymbol{x} = \boldsymbol{A}'\,\boldsymbol{x} = \boldsymbol{b}' = \boldsymbol{G}_1\,\boldsymbol{P}_1\boldsymbol{b} \tag{5.3.10}$$

$$\boldsymbol{A}'\boldsymbol{x} = \boldsymbol{b}' \Rightarrow \boldsymbol{P}_1^{-1}\,\boldsymbol{G}_1^{-1}\boldsymbol{A}'\,\boldsymbol{x} = \boldsymbol{A}\,\boldsymbol{x} = \boldsymbol{b} = \boldsymbol{P}_1^{-1}\,\boldsymbol{G}_1^{-1}\boldsymbol{b}' \tag{5.3.11}$$

For numerical stability reasons, the pivot is taken as the largest in absolute value of the elements of the first column

$$|a_{r1}| = \max_i |a_{i1}| \tag{5.3.12}$$

This pivot selection method is called *partial pivoting* in contrast to *complete pivoting* where the search of the pivot is done on all elements of the matrix. In this case, the steps (a) and (b) become (a') and (b') as follows:

- **Gauss algorithm with complete pivoting**:

(a') Determine r and s such that

$$|a_{rs}| = \max_{i,j} |a_{ij}| \tag{5.3.13}$$

and continue with (b') if $a_{rs} \neq 0$. Otherwise, $\boldsymbol{A}$ is singular, set $(\boldsymbol{A}', \boldsymbol{b}') = (\boldsymbol{A}, \boldsymbol{b})$, stop.

(b') Interchange the rows 1 and r of $(\boldsymbol{A}, \boldsymbol{b})$ as well as the columns 1 and s. The resulting matrix is $(\bar{\boldsymbol{A}}, \bar{\boldsymbol{b}})$.

(c') Identical to (c).

After the first elimination step, the resulting matrix is of the form

$$(\boldsymbol{A}', \boldsymbol{b}') = \left[\begin{array}{c|c|c} a'_{11} & a'^T & b'_1 \\ \hline 0 & \tilde{\boldsymbol{A}} & \tilde{\boldsymbol{b}} \end{array}\right] \tag{5.3.14}$$

the matrix $\tilde{\boldsymbol{A}}$ having $n-1$ rows. The following elimination step consists in applying to the reduced matrix $(\tilde{\boldsymbol{A}}, \tilde{\boldsymbol{b}})$ the processes previously described for the matrix $(\boldsymbol{A}, \boldsymbol{b})$. By continuing in this way, we obtain a sequence of matrices

$$(\boldsymbol{A}, \boldsymbol{b}) \equiv (\boldsymbol{A}^{(0)}, \boldsymbol{b}^{(0)}) \to (\boldsymbol{A}^{(1)}, \boldsymbol{b}^{(1)}) \to \cdots \to (\boldsymbol{A}^{(n-1)}, \boldsymbol{b}^{(n-1)}) \equiv (\boldsymbol{U}, \boldsymbol{c}) \tag{5.3.15}$$

An intermediate matrix $(\boldsymbol{A}^{(i)}, \boldsymbol{b}^{(i)})$ of this sequence is of the form

$$(\boldsymbol{A}^{(i)}, \boldsymbol{b}^{(i)}) = \left[\begin{array}{cccc|ccc} * & * & \ldots & * & * & * \ldots & * \\ 0 & \ddots & & \vdots & \vdots & & \vdots \\ \vdots & \ldots & \ddots & & & & \\ 0 & \ldots & 0 & * & * & * \ldots & * \\ \hline 0 & \ldots & & 0 & * & * \ldots & * \\ \vdots & & & \vdots & \vdots & & \vdots \\ 0 & \ldots & & 0 & * & * \ldots & * \end{array}\right] \tag{5.3.16}$$

or still

$$(A^{(j)}, b^{(j)}) = \left[\begin{array}{c|c|c} A_{11}^{(j)} & A_{12}^{(j)} & b_1^{(j)} \\ \hline 0 & A_{22}^{(j)} & b_2^{(j)} \end{array}\right] \tag{5.3.17}$$

Thus, the matrix $A_{11}^{(j)}$ is upper triangular. We still have the transformation

$$(A^{(j)}, b^{(j)}) = G_j\, P_j\, (A^{(j-1)}, b^{(j-1)}) \tag{5.3.18}$$

$$(U, c) = G_{n-1}\, P_{n-1}\, G_{n-2}\, P_{n-2}\, \ldots\, G_1\, P_1 (A, b) \tag{5.3.19}$$

where the interchange matrices P_i have been previously defined, whereas the nonsingular Frobenius matrix is of the form

$$G_j = \begin{bmatrix} 1 & & & & & 0 \\ 0 & \ddots & & & & \\ \vdots & & 1 & & & \\ & & -l_{j+1,j} & 1 & & \\ & & \vdots & 0 & \ddots & \\ 0 & & -l_{n,j} & \vdots & & 1 \end{bmatrix} \tag{5.3.20}$$

In the jth elimination step, $(A^{(j-1)}, b^{(j-1)}) \rightarrow (A^{(j)}, b^{(j)})$, the elements below the diagonal in the jth column are no more useful. When this algorithm is implemented on a computer, these places can be used to store the elements l_{ij} of the matrix G_j. Thus, we work on a matrix of the form

$$T^{(j)} = \left[\begin{array}{ccccc|cccc} u_{11} & u_{12} & & \ldots & u_{1j} & u_{1,j+1} & \ldots & u_{1,n} & c_1 \\ \lambda_{21} & u_{22} & & \ldots & u_{2j} & \vdots & & \vdots & \vdots \\ \lambda_{31} & \lambda_{32} & & \vdots & & & & & \\ \vdots & \vdots & & \vdots & & \vdots & & \vdots & \vdots \\ \vdots & \vdots & & & u_{jj} & u_{j,j+1} & \ldots & u_{j,n} & c_j \\ \hline \vdots & \vdots & & & \lambda_{j+1,j} & a_{j+1,j+1}^{(j)} & \ldots & a_{j+1,n}^{(j)} & b_{j+1}^{(j)} \\ \vdots & \vdots & & & \vdots & \vdots & & & \vdots \\ \lambda_{n1} & \lambda_{n2} & \ldots & & \lambda_{nj} & a_{n,j+1}^{(j)} & \ldots & a_{n,n}^{(j)} & b_n^{(j)} \end{array}\right] \tag{5.3.21}$$

The subdiagonal elements $\lambda_{k+1,k}$, $\lambda_{k+2,k}$, and $\lambda_{n,k}$ of the kth column are an interchange of the elements $l_{k+1,k}, \ldots, l_{n,k}$ of G_k.

The jth step $T^{(j-1)} \rightarrow T^{(j)}$ $(j = 1, \ldots, n-1)$ is described as follows:

(a) Choice of the partial pivot: determine r such that

$$|t_{rj}^{(j-1)}| = \max_{i \geq j} |t_{ij}^{(j-1)}| \tag{5.3.22}$$

If $t_{rj}^{(j-1)} = 0$, *set* $\boldsymbol{T}^{(j)} \equiv \boldsymbol{T}^{(j-1)}$; $\boldsymbol{A}$ *is singular, stop. Otherwise, continue in step (b).*

(b) *Interchange the rows r and j of* $\boldsymbol{T}^{(j-1)}$ *and store the result under* $\bar{\boldsymbol{T}} = (\bar{t}_{ik})$.

(c) *Replace*

$$t_{ij}^{(j)} \equiv l_{ij} \equiv \frac{\bar{t}_{ij}}{\bar{t}_{jj}}, \qquad i = j+1, \ldots, n \tag{5.3.23}$$

$$t_{ik}^{(j)} \equiv \bar{t}_{ik} - l_{ij}\bar{t}_{jk}, \quad i = j+1, \ldots, n \text{ and } k = j+1, \ldots, n+1 \tag{5.3.24}$$

$$t_{ik}^{(j)} \equiv \bar{t}_{ik} \text{ otherwise} \tag{5.3.25}$$

In step (c), the elements $l_{j+1,j}, \ldots, l_{nj}$ of $\boldsymbol{G}^{(j)}$ are stored in the natural order under the form $t_{j+1,j}^{(j)}, \ldots, t_{nj}^{(j)}$. However, the order can be modified in a later step, $\boldsymbol{T}^{(k)} \to \boldsymbol{T}^{(k+1)}$, $k \geq j$, as in b/ the rows of the matrix $\boldsymbol{T}^{(k)}$ are reordered. It follows that the following lower triangular $\boldsymbol{L}$ and upper triangular $\boldsymbol{U}$ matrices

$$\boldsymbol{L} = \begin{bmatrix} 1 & & & 0 \\ t_{21} & \ddots & & \\ \vdots & \ddots & \ddots & \\ t_{n1} & \ldots & t_{n,n-1} & 1 \end{bmatrix}, \quad \boldsymbol{U} = \begin{bmatrix} t_{11} & \ldots & & t_{1n} \\ & \ddots & & \vdots \\ & & \ddots & \vdots \\ 0 & & & t_{nn} \end{bmatrix} \tag{5.3.26}$$

which are contained in the final matrix $\boldsymbol{T}^{(n-1)} = (t_{ik})$ yield a decomposition of the matrix $\boldsymbol{PA}$

$$\boldsymbol{L}\,\boldsymbol{U} = \boldsymbol{P}\,\boldsymbol{A} \tag{5.3.27}$$

where $\boldsymbol{P}$ is the product of all interchanges

$$\boldsymbol{P} = \boldsymbol{P}_{n-1}\,\boldsymbol{P}_{n-2}\,\ldots\,\boldsymbol{P}_1 \tag{5.3.28}$$

Suppose that no interchange is necessary during the elimination process. In this case, $\boldsymbol{P}_1 = \cdots = \boldsymbol{P}_n = I$ and

$$\boldsymbol{L} = \begin{bmatrix} 1 & & & 0 \\ l_{21} & \ddots & & \\ \vdots & & \ddots & \\ l_{n1} & \ldots & l_{n,n-1} & 1 \end{bmatrix} \tag{5.3.29}$$

As

$$\boldsymbol{U} = \boldsymbol{G}_{n-1}\,\ldots\,\boldsymbol{G}_1\,\boldsymbol{A} \tag{5.3.30}$$

it results

$$\boldsymbol{G}_1^{-1}\,\ldots\,\boldsymbol{G}_{n-1}^{-1}\,\boldsymbol{U} = \boldsymbol{A} \tag{5.3.31}$$

From the value of $\boldsymbol{G}_i^{-1}$, we deduce that the product of these matrices is a lower triangular matrix with the diagonal exclusively constituted by 1, hence

$$\boldsymbol{G}_1^{-1}\,\ldots\,\boldsymbol{G}_{n-1}^{-1} = \boldsymbol{L} \tag{5.3.32}$$

Thus, Gauss method allows to solve a system

$$\boldsymbol{Ax} = \boldsymbol{b} \tag{5.3.33}$$

by doing the following triangular decomposition:

$$\boldsymbol{LU} = \boldsymbol{PA}(= \boldsymbol{A} \text{ if } \boldsymbol{P} = \boldsymbol{I}) \tag{5.3.34}$$

provided that the matrix $\boldsymbol{A}$ is nonsingular.

In the case where no interchanges are needed, the matrix $\boldsymbol{A}$ is then decomposed as $\boldsymbol{A} = \boldsymbol{LU}$ (this is the *LU decomposition*), that is,

$$a_{ij} = \sum_{k=1}^{min(i,j)} l_{ik} u_{kj} \quad \text{with } l_{ii} = 1 \tag{5.3.35}$$

- **Crout's method** does the LU decomposition by using an alternate partitioning of rows and columns:

$$a_{1i} = \sum_{k=1}^{1} l_{1k} u_{ki} \Rightarrow u_{1i} = a_{1i} \quad , i = 1, \dots, n \qquad \text{(Calculation of 1st row of } U) \tag{5.3.36}$$

$$a_{i1} = \sum_{k=1}^{1} l_{ik} u_{k1} \Rightarrow l_{i1} = \frac{a_{i1}}{u_{11}}, \quad i = 2, \dots, n \qquad \text{(Calculation of 1st column of } L) \tag{5.3.37}$$

If $u_{11} = 0$, the factorization is impossible.

$$a_{2i} = \sum_{k=1}^{2} l_{2k} u_{ki} \Rightarrow u_{2i} = a_{2i} - l_{21} u_{1i}, \quad i = 2, \dots, n \qquad \text{(2nd row of } U) \tag{5.3.38}$$

then

$$a_{i2} = \sum_{k=1}^{2} l_{ik} u_{k2} \Rightarrow l_{i2} = \frac{1}{u_{22}}(a_{i2} - l_{i1} u_{12}), \quad i = 3, \dots, n \qquad \text{(2nd column of } L) \tag{5.3.39}$$

and so on.

In a general manner, we have for $i = 1, 2, \dots, n$

$$\begin{cases} l_{ii} = 1 \\ u_{ij} = a_{ij} - \displaystyle\sum_{k=1}^{i-1} l_{ik} u_{kj}, \quad j = i, i+1, \dots, n \qquad (i\text{th row of } U) \\ l_{ji} = \dfrac{1}{u_{ii}}(a_{ji} - \displaystyle\sum_{k=1}^{i-1} l_{jk} u_{ki}), \quad j = i+1, i+2, \dots, n \qquad (i\text{th column of } L) \end{cases} \tag{5.3.40}$$

which are in fact formulas identical to Equations (4.8.13) to (4.8.16). If $u_{ii} = 0$, the factorization is impossible. It can be noticed that, in algorithm (5.3.40), the calculation of the rows of U and the columns of L is done alternately.

Example 5.1 :
Gauss method: partial pivoting

Using Gauss method with partial pivoting (Section 5.3), the following system will be solved:

$$\begin{aligned} 2\,x_1 + x_2 - 3\,x_3 &= 2.1 \\ x_1 - 4\,x_2 + 7\,x_3 &= -2.4 \\ 5\,x_1 + 3\,x_2 + 8\,x_3 &= 0.7 \end{aligned} \tag{5.3.41}$$

First iteration:
We set

$$(\boldsymbol{A}^{(0)}, \boldsymbol{b}^{(0)}) = \begin{bmatrix} 2 & 1 & -3 & 2.1 \\ 1 & -4 & 7 & -2.4 \\ 5 & 3 & 8 & 0.7 \end{bmatrix} \tag{5.3.42}$$

The pivot is element 5 of the first column. The rows 1 and 3 are interchanged (step (b) of algorithm), hence

$$(\bar{\boldsymbol{A}}, \bar{\boldsymbol{b}}) = \begin{bmatrix} 5 & 3 & 8 & 0.7 \\ 1 & -4 & 7 & -2.4 \\ 2 & 1 & -3 & 2.1 \end{bmatrix} \tag{5.3.43}$$

Step (c) of algorithm is performed with

$$l_{ij} = \bar{a}_{ij}/\bar{a}_{11} = \bar{a}_{ij}/5\,, \quad i = 1, \ldots, n\,, \quad j = 1, \ldots, n+1 \tag{5.3.44}$$

Then, the following matrix results:

$$(\boldsymbol{A}', \boldsymbol{b}') = (\boldsymbol{A}^{(1)}, \boldsymbol{b}^{(1)}) = \begin{bmatrix} 5 & 3 & 8 & 0.7 \\ 0 & -4.6 & 5.4 & -2.54 \\ 0 & -0.2 & -6.2 & 1.82 \end{bmatrix} \tag{5.3.45}$$

with the interchange matrix and the Frobenius matrix

$$\boldsymbol{P}_1 = \begin{bmatrix} 0 & 0 & 1 \\ 0 & 1 & 0 \\ 1 & 0 & 0 \end{bmatrix}, \quad \boldsymbol{G}_1 = \begin{bmatrix} 1 & 0 & 0 \\ -0.2 & 1 & 0 \\ -0.4 & 0 & 1 \end{bmatrix} \tag{5.3.46}$$

Second iteration:

From matrix $(\boldsymbol{A}^{(1)}, \boldsymbol{b}^{(1)})$, the reduced matrix is considered

$$(\tilde{\boldsymbol{A}}, \tilde{\boldsymbol{b}}) = \begin{bmatrix} -4.6 & 5.4 & -2.54 \\ -0.2 & -6.2 & 1.82 \end{bmatrix} \tag{5.3.47}$$

The partial pivot is -4.6. It is not necessary to do an interchange, thus $\boldsymbol{P}_2 = \boldsymbol{I}$ and $(\bar{\boldsymbol{A}}, \bar{\boldsymbol{b}}) = (\boldsymbol{A}^{(1)}, \boldsymbol{b}^{(1)})$. The step c/ of the algorithm is done with

$$l_{ij} = \bar{a}_{ij}/\bar{a}_{22} = \bar{a}_{ij}/(-4.6)\,, \quad i = 2, \ldots, n\,, \quad j = 2, \ldots, n+1 \tag{5.3.48}$$

Then, the matrix results

$$(\boldsymbol{A}', \boldsymbol{b}') = (\boldsymbol{A}^{(2)}, \boldsymbol{b}^{(2)}) = \begin{bmatrix} 5 & 3 & 8 & 0.7 \\ 0 & -4.6 & 5.4 & -2.54 \\ 0 & 0 & -6.435 & 1.930 \end{bmatrix} \tag{5.3.49}$$

with the interchange matrix and the Frobenius matrix

$$P_2 = \begin{bmatrix} 1 & 0 & 0 \\ 0 & 1 & 0 \\ 0 & 0 & 1 \end{bmatrix}, \quad G_2 = \begin{bmatrix} 1 & 0 & 0 \\ 0 & 1 & 0 \\ 0 & -0.0435 & 1 \end{bmatrix} \tag{5.3.50}$$

All steps are achieved and we get the system

$$\begin{aligned} 5\,x_1 + 3\,x_2 + 8\,x_3 &= 0.7 \\ -4.6\,x_2 + 5.4\,x_3 &= -2.54 \\ -6.435\,x_3 &= 1.930 \end{aligned} \tag{5.3.51}$$

of the form $\boldsymbol{U}\,\boldsymbol{x} = \boldsymbol{b}$ corresponding to system (5.2.1) that is easily solved to get the final solution

$$\boldsymbol{x} = \begin{bmatrix} 0.5 \\ 0.2 \\ -0.3 \end{bmatrix} \tag{5.3.52}$$

It can be noted that we have found the matrix $\boldsymbol{U}$ equal to

$$\boldsymbol{U} = \begin{bmatrix} 5 & 3 & 8 \\ 0 & -4.6 & 5.4 \\ 0 & 0 & -6.435 \end{bmatrix} \tag{5.3.53}$$

and the matrix $\boldsymbol{L}$ equal to

$$\boldsymbol{L} = \boldsymbol{G}_2^{-1}\,\boldsymbol{G}_1^{-1} = \begin{bmatrix} 1 & 0 & 0 \\ 0.2 & 1 & 0 \\ 0.4 & 0.0435 & 1 \end{bmatrix} \tag{5.3.54}$$

such that

$$\boldsymbol{L}\boldsymbol{U} = \boldsymbol{P}\,\boldsymbol{A} = \boldsymbol{P}_1\,\boldsymbol{A} \tag{5.3.55}$$

Example 5.2 :

Crout's method: LU decomposition without interchange

We desire to do an LU decomposition without interchange for a matrix $\boldsymbol{A}$. Using Crout's method (Equation (5.3.40)), we find

$$\begin{aligned} \boldsymbol{A} &= \begin{bmatrix} 2 & 1 & 3 & -5 \\ 1 & -2 & 7 & -4 \\ -3 & 5 & 1 & 6 \\ 4 & -9 & 5 & 3 \end{bmatrix} \\ &= \begin{bmatrix} 1 & 0 & 0 & 0 \\ 0.5 & 1 & 0 & 0 \\ -1.5 & -2.6 & 1 & 0 \\ 2 & 4.4 & -1.2727 & 1 \end{bmatrix} \begin{bmatrix} 2 & 1 & 3 & -5 \\ 0 & -2.5 & 5.5 & -1.5 \\ 0 & 0 & 19.8 & -5.4 \\ 0 & 0 & 0 & 12.7273 \end{bmatrix} \\ &= \boldsymbol{L}\boldsymbol{U} \end{aligned} \tag{5.3.56}$$

which is the desired LU decomposition.

5.4 Calculation of a Matrix Determinant

Let us use the relation obtained during the operations for solving the system

$$\boldsymbol{P}\,\boldsymbol{A} = \boldsymbol{L}\,\boldsymbol{U} \tag{5.4.1}$$

As the matrix $\boldsymbol{P}$ is the product of interchange matrices, it results

$$\det \boldsymbol{P} = (-1)^p \tag{5.4.2}$$

where p is the number of interchanges performed. As $\boldsymbol{L}$ and $\boldsymbol{U}$ are triangular matrices and $\det \boldsymbol{L} = 1$, there comes

$$\det \boldsymbol{A} = (-1)^p \, u_{11} \, u_{22} \, \dots \, u_{nn} \tag{5.4.3}$$

Example 5.3 :
Determinant of a matrix

From Example 5.1, about the matrix $\boldsymbol{A}$

$$\boldsymbol{A} = \begin{bmatrix} 2 & 1 & -3 \\ 1 & -4 & 7 \\ 5 & 3 & 8 \end{bmatrix} \tag{5.4.4}$$

for which we found

$$\boldsymbol{U} = \begin{bmatrix} 5 & 3 & 8 \\ 0 & -4.6 & 5.4 \\ 0 & 0 & -6.435 \end{bmatrix} \tag{5.4.5}$$

The determinant of $\boldsymbol{A}$ results

$$\det \boldsymbol{A} = (-1)^p \, u_{11} \, u_{22} \, u_{33} = (-1) \, 5 \, (-4.6) \, (-6.435) \approx -148.005 \tag{5.4.6}$$

when the exact value is 148.

5.5 Gauss–Jordan Algorithm

Gauss elimination method allows to calculate the inverse of a matrix, but less simply than Gauss–Jordan algorithm that will be described below. Thus, if the decomposition is done under the previous Gauss form $\boldsymbol{PA} = \boldsymbol{LU}$, the ith column of $\boldsymbol{A}^{-1}$ (of typical element $\bar{a}_{ij}$) is obtained as the solution of the system

$$\boldsymbol{LU} \, \bar{\boldsymbol{a}}_i = \boldsymbol{P} \boldsymbol{e}_i \tag{5.5.1}$$

where e_i is the ith vector of coordinates. Gauss–Jordan method would require the same calculation effort but is preferable with respect to organization.

Gauss–Jordan method consists in progressively passing from the system $\boldsymbol{Ax} = \boldsymbol{y}$ to a system $\boldsymbol{A'y} = \boldsymbol{x}$. The system to be solved is posed as

$$\begin{aligned} a_{11}x_1 + \dots + a_{1n}x_n &= y_1 \\ &\vdots \\ a_{n1}x_1 + \dots + a_{nn}x_n &= y_n \end{aligned} \tag{5.5.2}$$

In the first step of Gauss–Jordan method, the variable x_1 is interchanged with one of the variables y_r. To do it, we search the partial pivot

$$|a_{r1}| = \max_i |a_{i1}| \tag{5.5.3}$$

and the equations r and 1 are interchanged. The following system results:

$$\begin{aligned} \bar{a}_{11}x_1 + \cdots + \bar{a}_{1n}x_n &= \bar{y}_1 \\ &\vdots \\ \bar{a}_{n1}x_1 + \cdots + \bar{a}_{nn}x_n &= \bar{y}_n \end{aligned} \tag{5.5.4}$$

for which we have $\bar{a}_{1i} = a_{ri}$, $\bar{y}_1 = y_r$. As $\boldsymbol{A}$ is nonsingular, $\bar{a}_{11}$ is nonzero. The first equation is solved to get x_1 and a substitution is done by replacing x_1 by its value in the other equations, hence the new system

$$\begin{aligned} a'_{11}\bar{y}_1 + a'_{12}x_2 + \cdots + a'_{1n}x_n &= x_1 \\ a'_{21}\bar{y}_1 + a'_{22}x_2 + \cdots + a'_{2n}x_n &= \bar{y}_2 \\ &\vdots \\ a'_{n1}\bar{y}_1 + a'_{n2}x_2 \cdots + a'_{nn}x_n &= \bar{y}_n \end{aligned} \tag{5.5.5}$$

where the coefficients are equal to

$$a'_{11} = \frac{1}{\bar{a}_{11}}, \quad a'_{1j} = -\frac{\bar{a}_{1j}}{\bar{a}_{11}}, \tag{5.5.6}$$

$$a'_{i1} = \frac{\bar{a}_{i1}}{\bar{a}_{11}}, \quad a'_{ij} = \bar{a}_{ij} - \frac{\bar{a}_{i1}\bar{a}_{1j}}{\bar{a}_{11}}, \quad \text{for } i, j = 2, 3, \ldots, n \tag{5.5.7}$$

At the following step, x_2 is interchanged with one of the variables $\bar{y}_2, \ldots, \bar{y}_n$ and so on.

This sequence of transformations can be represented by means of successive matrices

$$\boldsymbol{A} = \boldsymbol{A}^{(0)} \rightarrow \boldsymbol{A}^{(1)} \rightarrow \cdots \rightarrow \boldsymbol{A}^{(n)} \tag{5.5.8}$$

a typical matrix $\boldsymbol{A}^{(j)} = [a_{ik}^{(j)}]$ being related to the system of equations of the form

$$\begin{aligned} a_{11}^{(j)}\tilde{y}_1 + \cdots + a_{1j}^{(j)}\tilde{y}_j + a_{1,j+1}^{(j)}x_{j+1} + \cdots + a_{1n}^{(j)}x_n &= x_1 \\ &\vdots \\ a_{j1}^{(j)}\tilde{y}_1 + \cdots + a_{jj}^{(j)}\tilde{y}_j + a_{j,j+1}^{(j)}x_{j+1} + \cdots + a_{jn}^{(j)}x_n &= x_j \\ a_{j+1,1}^{(j)}\tilde{y}_1 + \cdots + a_{j+1,j}^{(j)}\tilde{y}_j + a_{j+1,j+1}^{(j)}x_{j+1} + \cdots + a_{j+1,n}^{(j)}x_n &= \tilde{y}_{j+1} \\ &\vdots \\ a_{n1}^{(j)}\tilde{y}_1 + \cdots + a_{nj}^{(j)}\tilde{y}_j + a_{n,j+1}^{(j)}x_{j+1} + \cdots + a_{nn}^{(j)}x_n &= \tilde{y}_n \end{aligned} \tag{5.5.9}$$

In the following summary of the algorithm, we note

$$\boldsymbol{A}^{(j-1)} = [a_{ik}], \quad \boldsymbol{A}^{(j)} = [a'_{ik}] \tag{5.5.10}$$

At iteration j, the following sequence will be executed:

(a) Choice of the partial pivot: determine r such that

$$|a_{rj}| = \max_{i \geq j} |a_{ij}| \tag{5.5.11}$$

If $a_{rj} = 0$, the matrix is singular, stop. Otherwise, continue in (b).

(b) Interchange the rows r and j of $\boldsymbol{A}^{(j-1)}$ and store the result in $\bar{\boldsymbol{A}} = [\bar{a}_{ik}]$

(c) Calculate $\boldsymbol{A}^{(j)} = [a'_{ik}]$ according to the following formulas:

$$a'_{jj} = \frac{1}{\bar{a}_{jj}} \tag{5.5.12}$$

$$a'_{jk} = -\frac{\bar{a}_{jk}}{\bar{a}_{jj}}, \quad a'_{ij} = \frac{\bar{a}_{ij}}{\bar{a}_{jj}}, \quad \text{for } i, k \neq j \tag{5.5.13}$$

$$a'_{ik} = \bar{a}_{ik} - \frac{\bar{a}_{ij}\bar{a}_{jk}}{\bar{a}_{jj}}, \quad \text{for } i, k \neq j \tag{5.5.14}$$

Finally, we obtain the equation

$$\boldsymbol{A}^{(n)} \tilde{\boldsymbol{y}} = \boldsymbol{x} \tag{5.5.15}$$

where the vector $\tilde{\boldsymbol{y}}$ is an interchange of the original vector $\boldsymbol{y}$. It results that

$$\boldsymbol{A}^{(n)} \boldsymbol{P}\, \boldsymbol{y} = \boldsymbol{x} \tag{5.5.16}$$

and, as $\boldsymbol{A}\boldsymbol{x} = \boldsymbol{y}$, the inverse of matrix $\boldsymbol{A}$ is

$$\boldsymbol{A}^{-1} = \boldsymbol{A}^{(n)} \boldsymbol{P} \tag{5.5.17}$$

Example 5.4 :

Gauss–Jordan algorithm

Instead of Gauss elimination method, Gauss–Jordan algorithm is used on the same Example 5.1, that is,

$$\begin{aligned} 2\,x_1 + x_2 - 3\,x_3 &= 2.1 \\ x_1 - 4\,x_2 + 7\,x_3 &= -2.4 \\ 5\,x_1 + 3\,x_2 + 8\,x_3 &= 0.7 \end{aligned} \tag{5.5.18}$$

or $\boldsymbol{A}\,\boldsymbol{x} = \boldsymbol{y}$ by setting the vector

$$\boldsymbol{y} = \begin{bmatrix} 2.1 \\ -2.4 \\ 0.7 \end{bmatrix} \tag{5.5.19}$$

which will be used until the end of the steps of the algorithm so as to make explicitly appear the inverse matrix of $\boldsymbol{A}$ at the end of the algorithm. The y_i should especially not be replaced by their numerical values along the calculation, but only when all iterations are ended, or otherwise the inverse matrix will not appear.

First iteration:

We set $\boldsymbol{A}^{(0)} = \boldsymbol{A}$. We look for the partial pivot in the first column of $\boldsymbol{A}^{(0)}$, that is, $a_{31} = 5$, and we interchange the rows 1 and 3, hence the new system

$$\begin{aligned} 5\,x_1 + 3\,x_2 + 8\,x_3 &= y_3 \\ x_1 - 4\,x_2 + 7\,x_3 &= y_2 \\ 2\,x_1 + x_2 - 3\,x_3 &= y_1 \end{aligned} \tag{5.5.20}$$

which is symbolized as

$$\bar{\boldsymbol{A}}\boldsymbol{x} = \bar{\boldsymbol{y}} \tag{5.5.21}$$

with moreover the interchange matrix

$$\boldsymbol{P}_1 = \begin{bmatrix} 0 & 0 & 1 \\ 0 & 1 & 0 \\ 1 & 0 & 0 \end{bmatrix} \tag{5.5.22}$$

The operations (5.5.12-5.5.14) are done (equivalent to the replacement of x_1 by its expression with respect to x_2, x_3, and $\bar{y}_1$ in the last two rows of the system (5.5.21)), hence the matrix $\boldsymbol{A}^{(1)}$

$$\boldsymbol{A}^{(1)} = \begin{bmatrix} 0.2 & -0.6 & -1.6 \\ 0.2 & -4.6 & 5.4 \\ 0.4 & -0.2 & -6.2 \end{bmatrix} \tag{5.5.23}$$

<u>Second iteration</u>:
We look for the pivot in the elements $a_{i2}^{(1)}$ with $i \geq 2$. The pivot is -4.6. It is not necessary to do an interchange, thus $\boldsymbol{P} = \boldsymbol{I}$. Then, we obtain the matrix $\boldsymbol{A}^{(2)}$

$$\boldsymbol{A}^{(2)} = \begin{bmatrix} 0.1739 & 0.1304 & -2.3043 \\ 0.0434 & -0.2174 & 1.1739 \\ 0.3913 & 0.4348 & -6.435 \end{bmatrix} \tag{5.5.24}$$

<u>Third iteration</u>:
The pivot is $a_{33}^{(3)} = -6.435$. We obtain the matrix $\boldsymbol{A}^{(3)}$

$$\boldsymbol{A}^{(3)} = \begin{bmatrix} 0.0338 & 0.1149 & 0.3581 \\ 0.1149 & -0.2095 & -0.1824 \\ 0.0608 & 0.0068 & -0.1554 \end{bmatrix} \tag{5.5.25}$$

such that

$$\boldsymbol{A}^{-1} = \boldsymbol{A}^{(n)}\boldsymbol{P} \tag{5.5.26}$$

As $\boldsymbol{P} = \boldsymbol{P}_1$, the inverse matrix results

$$\boldsymbol{A}^{-1} = \begin{bmatrix} 0.3581 & 0.1149 & 0.0338 \\ -0.1824 & -0.2095 & 0.1149 \\ -0.1554 & 0.0068 & 0.0608 \end{bmatrix} \tag{5.5.27}$$

and the solution of the initial system as

$$\boldsymbol{x} = \boldsymbol{A}^{-1}\boldsymbol{y} \tag{5.5.28}$$

by replacing $\boldsymbol{y}$ by its value, hence

$$\boldsymbol{x} = \begin{bmatrix} 0.5 \\ 0.2 \\ -0.3 \end{bmatrix} \tag{5.5.29}$$

5.6 $\boldsymbol{LDL^T}$ Factorization

We consider a definite positive matrix $\boldsymbol{A}$. This matrix can be decomposed under the form $\boldsymbol{LDL^T}$, where $\boldsymbol{L}$ is a lower triangular matrix with 1 on the diagonal and $\boldsymbol{D}$ is a diagonal matrix with positive elements.

To do the $\boldsymbol{LDL^T}$ factorization, it suffices to execute the following algorithm: $\forall\ i = 1, \ldots, n$, set $l_{ii} = 1$, then repeat steps 1 to 3,

1. $c_j = l_{ij}\, d_{jj} \quad \forall\ j = 1, \ldots, i-1,$
2. $d_{ii} = a_{ii} - \sum_{j=1}^{i-1} l_{ij}\, c_j,$
3. $l_{ji} = \left(a_{ji} - \sum_{k=1}^{i-1} l_{jk}\, c_k \right) / d_{ii} \quad \forall\ j = i+1, \ldots, n.$

As $\boldsymbol{D}$ is a diagonal matrix with positive elements, the relation

$$\boldsymbol{A} = \boldsymbol{LDL^T} \tag{5.6.1}$$

can be written under the form

$$\boldsymbol{A} = \boldsymbol{L}\,\boldsymbol{D}^{\frac{1}{2}}\,\boldsymbol{D}^{\frac{1}{2}}\,\boldsymbol{L}^T = \boldsymbol{L}'\boldsymbol{L}'^T = \boldsymbol{R}^T\boldsymbol{R} \tag{5.6.2}$$

which is Cholesky's factorization by setting $\boldsymbol{L}' = \boldsymbol{L}\,\boldsymbol{D}^{\frac{1}{2}}$ and $\boldsymbol{R} = \boldsymbol{L}'^T$.

Example 5.5 :
$\boldsymbol{LDL^T}$ factorization

Consider the definite positive matrix $\boldsymbol{A}$ of Example 5.6.

$$\boldsymbol{A} = \begin{bmatrix} 3 & 0.6 & -1.5 \\ 0.6 & 1.12 & 0.3 \\ -1.5 & 0.3 & 3.11 \end{bmatrix} \tag{5.6.3}$$

Applying the previous algorithm, we find again the matrices of Example 5.6 having been used to build the matrix $\boldsymbol{A}$

$$\boldsymbol{L} = \begin{bmatrix} 1 & 0 & 0 \\ 0.2 & 1 & 0 \\ -0.5 & 0.6 & 1 \end{bmatrix} \quad \text{and } \boldsymbol{D} = \begin{bmatrix} 3 & 0 & 0 \\ 0 & 1 & 0 \\ 0 & 0 & 2 \end{bmatrix} \tag{5.6.4}$$

5.7 Cholesky's Decomposition

Cholesky's decomposition is related to symmetrical definite positive matrices. That could appear as a restricted case, but indeed, in practical applications, such as linear least squares, this case is relatively frequent.

Cholesky's decomposition is a particular type of $\boldsymbol{LDL^T}$ factorization in which the matrix $\boldsymbol{D}$ is the identity matrix. In the case of definite positive matrices, the choice of the pivot does not matter anymore.

Recall that a matrix is definite positive if

- it is Hermitian $\boldsymbol{A} = \boldsymbol{A}^*$
- $\boldsymbol{u}^*\boldsymbol{A}\boldsymbol{u} > 0 \quad , \forall \boldsymbol{u} \in C^n \, , \boldsymbol{u} \neq 0$

On another side, if $\boldsymbol{A}$ is definite positive, $\boldsymbol{A}^{-1}$ exists and is also definite positive. Moreover, all the main submatrices of a definite positive matrix are definite positive, and all the principal minors of a definite positive matrix are positive.

Theorem: For any $n \times n$ definite positive matrix $\boldsymbol{A}$, there exists an $n \times n$ unique lower triangular matrix $\boldsymbol{L}$ whose all diagonal elements l_{ii} are strictly positive and which satisfies $\boldsymbol{A} = \boldsymbol{L}\,\boldsymbol{L}^*$. If $\boldsymbol{A}$ is real, $\boldsymbol{L}$ is also real

$$\boldsymbol{A} = \boldsymbol{L}\,\boldsymbol{L}^* \,, \quad l_{ik} = 0 \;\forall\, k > i \,,\, l_{ii} > 0 \;\forall\, i \tag{5.7.1}$$

This factorization $\boldsymbol{A} = \boldsymbol{L}\,\boldsymbol{L}^*$ is sometimes called "taking the square root of the matrix."

Demonstration by induction:

If $n = 1$, $\boldsymbol{A} = a > 0$. We can write $a = l_{11} l_{11}^*$, hence $l_{11} = \sqrt{a}$.

Suppose that the theorem is true until order $n - 1$ and we demonstrate it at order n. The $n \times n$ matrix $\boldsymbol{A}$ can be partitioned as

$$\boldsymbol{A} = \begin{bmatrix} \boldsymbol{A}_{n-1} & \boldsymbol{b} \\ \boldsymbol{b}^* & a_{nn} \end{bmatrix} \tag{5.7.2}$$

$\boldsymbol{b}$ is a vector of C^{n-1} and $\boldsymbol{A}_{n-1}$ is an $(n-1) \times (n-1)$ definite positive matrix. Due to the recurrence property, we know that there exists an $(n-1) \times (n-1)$ unique lower triangular matrix $\boldsymbol{L}_{n-1}$ and with positive diagonal elements such that

$$\boldsymbol{A}_{n-1} = \boldsymbol{L}_{n-1}\,\boldsymbol{L}_{n-1}^* \,, \quad l_{ik} = 0 \quad \forall\, k > i \,,\, l_{ii} > 0 \quad \forall\, i \tag{5.7.3}$$

Suppose that there exists a matrix $\boldsymbol{L}$ such that

$$\boldsymbol{L} = \begin{bmatrix} \boldsymbol{L}_{n-1} & 0 \\ \boldsymbol{c}^* & \alpha \end{bmatrix} \tag{5.7.4}$$

and we try to find the vector $\boldsymbol{c} \in C^{n-1}$ and the scalar $\alpha > 0$ such that

$$\begin{bmatrix} \boldsymbol{L}_{n-1} & 0 \\ \boldsymbol{c}^* & \alpha \end{bmatrix} \begin{bmatrix} \boldsymbol{L}_{n-1}^* & \boldsymbol{c} \\ 0 & \alpha \end{bmatrix} = \begin{bmatrix} \boldsymbol{A}_{n-1} & \boldsymbol{b} \\ \boldsymbol{b}^* & a_{nn} \end{bmatrix} = \boldsymbol{A} \tag{5.7.5}$$

thus

$$\begin{aligned} &\boldsymbol{L}_{n-1}\boldsymbol{c} = \boldsymbol{b} \\ &\boldsymbol{c}^*\,\boldsymbol{c} + \alpha^2 = a_{nn} \quad \Rightarrow \alpha^2 = a_{nn} - \boldsymbol{c}^*\,\boldsymbol{c} \quad \text{with } \alpha > 0 \end{aligned} \tag{5.7.6}$$

We draw

$$\boldsymbol{c} = \boldsymbol{L}_{n-1}^{-1}\boldsymbol{b} \tag{5.7.7}$$

as $\boldsymbol{L}_{n-1}$ is invertible. On another side, the relation

$$\det(\boldsymbol{A}) = |\det(\boldsymbol{L}_{n-1})|^2 \alpha^2 \tag{5.7.8}$$

is consistent as the determinant of $\boldsymbol{A}$ is positive. Consequently, there exists only one value of α corresponding to the equality $\boldsymbol{L}\,\boldsymbol{L}^* = \boldsymbol{A}$ and which is equal to

$$\alpha = \sqrt{a_{nn} - \boldsymbol{c}^*\,\boldsymbol{c}} \tag{5.7.9}$$

The decomposition is done in the same manner as for Gauss method. Supposing that all elements l_{ij} are known until column $k-1$ included ($j \leq k-1$), the equations giving the following elements l_{kk} and l_{ik} ($i \geq k+1$) are

$$\begin{aligned} a_{kk} &= |l_{k1}|^2 + |l_{k2}|^2 + \cdots + |l_{kk}|^2\,, \quad (l_{kk} > 0) \\ a_{ik} &= l_{i1}\bar{l}_{k1} + l_{i2}\bar{l}_{k2} + \cdots + l_{ik}\bar{l}_{kk} \end{aligned} \tag{5.7.10}$$

The elements of the matrix $\boldsymbol{L}$ are equal to

$$\begin{aligned} l_{ii} &= \left(a_{ii} - \sum_{k=1}^{i-1} l_{ik}^2\right)^{1/2} \\ l_{ji} &= \frac{1}{l_{ii}}\left(a_{ij} - \sum_{k=1}^{i-1} l_{ik}\, l_{jk}\right), \quad j = i+1, \ldots, n \end{aligned} \tag{5.7.11}$$

Example 5.6 :
Cholesky's decomposition
We build the matrix $\boldsymbol{A}$ equal to

$$\boldsymbol{A} = \boldsymbol{L}_1\,\boldsymbol{D}_1\,\boldsymbol{L}_1^T \qquad \text{with } \boldsymbol{L}_1 = \begin{bmatrix} 1 & 0 & 0 \\ 0.2 & 1 & 0 \\ -0.5 & 0.6 & 1 \end{bmatrix} \qquad \text{and } \boldsymbol{D}_1 = \begin{bmatrix} 3 & 0 & 0 \\ 0 & 1 & 0 \\ 0 & 0 & 2 \end{bmatrix} \tag{5.7.12}$$

We know that the resulting matrix $\boldsymbol{A}$ is definite positive by construction

$$\boldsymbol{A} = \begin{bmatrix} 3 & 0.6 & -1.5 \\ 0.6 & 1.12 & 0.3 \\ -1.5 & 0.3 & 3.11 \end{bmatrix} \tag{5.7.13}$$

Let us apply Equation (5.7.11) of Cholesky's decomposition to the matrix $\boldsymbol{A}$. We thus obtain the matrix $\boldsymbol{L}$ equal to

$$\boldsymbol{L} = \begin{bmatrix} 1.7321 & 0 & 0 \\ 0.3464 & 1 & 0 \\ -0.8660 & 0.6000 & 1.3748 \end{bmatrix} \tag{5.7.14}$$

such that $\boldsymbol{L}\boldsymbol{L}^T = \boldsymbol{A}$.

5.8 Singular Value Decomposition (SVD)

When a set of equations or a matrix is either singular or numerically close to singularity, the techniques of Gaussian elimination or LU decomposition do not give satisfactory results, and the singular value decomposition (SVD) method allows either to solve the problem or to identify the cause of the difficulties encountered. Moreover, this is the recommended method to solve most least squares problems (Golub and Van Loan 1991; Mandel 1982).

The singular value decomposition can be applied to any matrix $\boldsymbol{A}$ of dimension $m \times n$, not necessarily square, which can be real or complex. In general, the number of rows m is larger than or equal to the number of columns n, as the system comprises more equations than unknowns. Otherwise, there does not exist a unique solution to the system.

Theorem: Any real or complex matrix $\boldsymbol{A}$ of dimension $m \times n$ can be decomposed as

$$\boldsymbol{A} = \boldsymbol{U}\,\boldsymbol{\Sigma}\,\boldsymbol{V}^* \tag{5.8.1}$$

where $\boldsymbol{U}$ is an $m \times m$ unitary matrix (orthogonal if $\boldsymbol{A}$ is real), $\boldsymbol{V}$ is an $n \times n$ unitary matrix ($\boldsymbol{V}^T$ is orthogonal if $\boldsymbol{A}$ is real), and $\boldsymbol{\Sigma}$ is an $m \times n$ matrix, including a diagonal submatrix of dimension $m \times m$ if $m \leq n$ or of dimension $n \times n$ if $n \leq m$, of typical element $\sigma_{ij} = 0$ if $i \neq j$ and $\sigma_{ii} = \sigma_i \geq 0$. The quantities σ_i are the singular values of $\boldsymbol{A}$ and the columns U_i of $\boldsymbol{U}$, V_i of $\boldsymbol{V}$ respectively, are the left singular vectors, and right singular vectors, such that

$$\boldsymbol{A} V_i = \sigma_i U_i \qquad \text{and} \qquad \boldsymbol{A}^* U_i = \sigma_i V_i \qquad \forall i \tag{5.8.2}$$

The singular value decomposition is not unique.

The left singular vectors U_i are the orthonormal eigenvectors of $\boldsymbol{A}\boldsymbol{A}^*$. The right singular vectors V_i are the orthonormal eigenvectors of $\boldsymbol{A}^*\boldsymbol{A}$. Recall that in the case of a real matrix, $\boldsymbol{A}^* = \boldsymbol{A}^T$.

If $\boldsymbol{A}$ is an $n \times n$ square matrix, the matrices $\boldsymbol{A}^*\,\boldsymbol{A}$ and $\boldsymbol{A}\,\boldsymbol{A}^*$ have the same nonzero eigenvalues. Moreover, the matrices $\boldsymbol{U}$, $\boldsymbol{\Sigma}$ and $\boldsymbol{V}$ are square and the inverse of the matrix $\boldsymbol{A}$ can be easily calculated

$$\boldsymbol{A}^{-1} = \boldsymbol{V}\,\mathbf{diag}(1/\sigma_i)\,\boldsymbol{U}^T \tag{5.8.3}$$

The only difficulty that may come is that a singular value be zero or numerically very close to zero. Thus, the diagnostic is very simple. Numerically, the matrix is quasi-singular when the condition number (here taken as the magnitude of the ratio of the smallest to the largest singular value of the matrix) is close to the computer error.

Example 5.7 :
Singular Value Decomposition

(a) The system contains more columns than rows

$$\boldsymbol{A} = \begin{bmatrix} 5 & 4 & 6 \\ 1 & 7 & 3 \end{bmatrix}$$
$$= \begin{bmatrix} 0.7661 & -0.6427 \\ 0.6427 & 0.7661 \end{bmatrix} \begin{bmatrix} 10.9448 & 0 & 0 \\ 0 & 4.0264 & 0 \end{bmatrix} \begin{bmatrix} 0.4087 & -0.6079 & 0.6808 \\ 0.6911 & 0.6933 & 0.2042 \\ 0.5962 & -0.3870 & -0.7035 \end{bmatrix}$$

The singular values of $\boldsymbol{A}$ are 4.0264 and 10.9448. The square roots of the eigenvalues of $(\boldsymbol{A}^*\boldsymbol{A})$ are equal to 0, 4.0264, and 10.9448, and there exist $n - m$ eigenvalues equal to 0 for a system with a number m of equations lower than the number n of unknowns.

(b) The system contains more rows than columns

$$\boldsymbol{A} = \begin{bmatrix} 5 & 4 \\ 6 & 1 \\ 7 & 3 \end{bmatrix}$$
$$= \begin{bmatrix} 0.5414 & 0.7218 & -0.4311 \\ 0.5141 & -0.6900 & -0.5095 \\ 0.6652 & -0.0542 & 0.7447 \end{bmatrix} \begin{bmatrix} 11.4469 & 0 \\ 0 & 2.2290 \\ 0 & 0 \end{bmatrix} \begin{bmatrix} 0.9128 & -0.4085 \\ 0.4085 & 0.9128 \end{bmatrix}$$

The singular values of $\boldsymbol{A}$ are 2.2290 and 11.4469. The square roots of the eigenvalues of $(\boldsymbol{A}^*\boldsymbol{A})$ are equal to 2.2290 and 11.4469.

(c) The matrix $\boldsymbol{A}$ is square

$$\boldsymbol{A} = \begin{bmatrix} 5 & 4 \\ 6 & 1 \end{bmatrix} = \begin{bmatrix} 0.7276 & 0.6860 \\ 0.6860 & -0.7276 \end{bmatrix} \begin{bmatrix} 8.5474 & 0 \\ 0 & 2.2229 \end{bmatrix} \begin{bmatrix} 0.9072 & -0.4207 \\ 0.4207 & 0.9072 \end{bmatrix}$$

The singular values of $\boldsymbol{A}$ are 2.2229 and 8.5474. They are equal to the square roots of the eigenvalues of $(\boldsymbol{A}^*\boldsymbol{A})$.

5.9 Least Squares Method for Linear Overdetermined Systems

Consider the following system of linear equations, with real coefficients:

$$\begin{aligned} a_{11}x_1 + a_{12}x_2 + \cdots + a_{1n}x_n &= b_1 \\ a_{21}x_1 + a_{22}x_2 + \cdots + a_{2n}x_n &= b_2 \\ &\vdots \\ a_{m1}x_1 + a_{m2}x_2 + \cdots + a_{mn}x_n &= b_m \end{aligned} \tag{5.9.1}$$

in the case $m \geq n$, that is, when the number of equations is larger than the number of unknowns. The system is called *overdetermined*. This happens very often in the treatment of experimental data. In general, this system has no solution and this problem can be solved in the least squares sense by setting the following optimization problem:

$$\min_{\boldsymbol{x}} \|\boldsymbol{A}\,\boldsymbol{x} - \boldsymbol{b}\|_2 \tag{5.9.2}$$

of minimization of the Euclidean norm expressed as the quadratic form

$$f(\boldsymbol{x}) = (\boldsymbol{A}\,\boldsymbol{x} - \boldsymbol{b})^T\,(\boldsymbol{A}\,\boldsymbol{x} - \boldsymbol{b}) = \boldsymbol{x}^T\,\boldsymbol{A}^T\,\boldsymbol{A}\,\boldsymbol{x} - 2\,\boldsymbol{x}^T\,\boldsymbol{A}^T\,\boldsymbol{b} + \boldsymbol{b}^T\,\boldsymbol{b} \tag{5.9.3}$$

Its minimization implies that its gradient is zero

$$f_x(\boldsymbol{x}) = 0 = 2\,(\boldsymbol{A}^T\,\boldsymbol{A}\,\boldsymbol{x} - \boldsymbol{A}^T\,\boldsymbol{b}) \tag{5.9.4}$$

The solution is thus obtained by solving the square linear system

$$\boldsymbol{A}^T\,\boldsymbol{A}\,\boldsymbol{x} = \boldsymbol{A}^T\,\boldsymbol{b} \tag{5.9.5}$$

Note that $A^T A$ was present in the singular value decomposition of Section 5.8. Formally, if the matrix $A^T A$ is invertible, the solution of system (5.9.5) is thus

$$x = [A^T A]^{-1} A^T b \tag{5.9.6}$$

However, it happens frequently that the matrix $A^T A$ is ill-conditioned, and then it is not recommended to invert that matrix.

The solution of system (5.9.5) can be done while avoiding to invert the matrix $A^T A$, which increases the solving robustness. The process must be the following:

- Note that the matrix $A^T A$ is definite positive. We do Cholesky's decomposition of that matrix which gives
$$L L^T = A^T A \tag{5.9.7}$$
where the matrix L is lower triangular. Equation (5.9.5) then becomes
$$L L^T x = A^T b \tag{5.9.8}$$
- Posing in Equation (5.9.8)
$$y = L^T x \tag{5.9.9}$$
we get the linear system
$$L y = A^T b \tag{5.9.10}$$
which can be solved without inverting as L is simply lower triangular, cf. Equation (5.2.3).
- y being now known, we solve the system (5.9.9)
$$L^T x = y \tag{5.9.11}$$
according to Equation (5.2.1), hence the sought solution.

Example 5.8 :
Linear least squares: parametric estimation for a linear model

Consider the model of heat capacity

$$C_p = \theta_1 + \theta_2 (T/100) + \theta_3 (T/100)^2 + \theta_4 (T/100)^3 \tag{5.9.12}$$

whose parameters are to be determined. It is noted that the model is linear with respect to these parameters. Six experiments are performed (more experiments than parameters) and given in Table 5.1.

Table 5.1 Experiments of heat capacity

Experiment ♯	1	2	3	4	5	6
T (K)	300	350	400	450	500	550
C_p (J mol^{-1} K^{-1})	33.67	34.05	34.47	34.91	35.39	35.89

The temperature and the heat capacity of experiment ♯ i are noted T_i and $C_{p,i}$, respectively.

It could be written that the experiments verify the model, giving the set of equations

$$\theta_1 + \theta_2\,(T_i/100) + \theta_3\,(T_i/100)^2 + \theta_4\,(T_i/100)^3 = C_{pi}\,, \quad i = 1, \ldots, 6 \tag{5.9.13}$$

This system is overdetermined (more equations than unknowns), which can only be solved by searching an optimal solution by the least squares method.

The matrix $\boldsymbol{A}$ and the vector $\boldsymbol{b}$ of Equation (5.9.5) are then equal to

$$\boldsymbol{A} = \begin{bmatrix} 1 & 3.0 & 9.00 & 27.000 \\ 1 & 3.5 & 12.25 & 42.875 \\ 1 & 4.0 & 16.00 & 64.000 \\ 1 & 4.5 & 20.25 & 91.125 \\ 1 & 5.0 & 25.00 & 125.000 \\ 1 & 5.5 & 30.25 & 166.375 \end{bmatrix}, \quad \boldsymbol{b} = \begin{bmatrix} 33.67 \\ 34.05 \\ 34.47 \\ 34.91 \\ 35.39 \\ 35.89 \end{bmatrix} \tag{5.9.14}$$

The matrix $\boldsymbol{A}^T \boldsymbol{A}$ being definite positive, Cholesky's decomposition of that matrix is done, which gives the matrix $\boldsymbol{L}$

$$\begin{bmatrix} 2.4495 & 0 & 0 & 0 \\ 10.4103 & 2.0917 & 0 & 0 \\ 46.0300 & 17.7790 & 1.5275 & 0 \\ 210.8092 & 115.9820 & 19.4759 & 1.0062 \end{bmatrix} \tag{5.9.15}$$

such that

$$\boldsymbol{L}\,\boldsymbol{L}^T = \boldsymbol{A}^T\,\boldsymbol{A} \tag{5.9.16}$$

which allows to transform relation (5.9.5) as

$$\boldsymbol{L}\,\boldsymbol{L}^T\,\boldsymbol{\theta} = \boldsymbol{A}^T\,\boldsymbol{b} \tag{5.9.17}$$

Setting $\boldsymbol{y} = \boldsymbol{L}^T\,\boldsymbol{\theta}$, we get

$$\boldsymbol{L}\,\boldsymbol{y} = \boldsymbol{A}^T\,\boldsymbol{b} \tag{5.9.18}$$

which can be solved easily by recognizing a system of the form (5.2.3), hence

$$\boldsymbol{y} = [85.0708 \quad 1.8598 \quad 0.0917 \quad -0.003]^T \tag{5.9.19}$$

and then we solve the system

$$\boldsymbol{y} = \boldsymbol{L}^T\,\boldsymbol{\theta} \tag{5.9.20}$$

which is of the form (5.2.1), hence the solution of the vector of model parameters

$$\boldsymbol{\theta} = [32.2027 \quad 0.2223 \quad 0.0978 \quad -0.0030]^T \tag{5.9.21}$$

Thus, this solution was obtained without inverting any matrix. Figure 5.1 confirms the determination of the model parameters.

5.10 Iterative Solution of Large Linear Systems (Jacobi and Gauss–Seidel)

In some types of problems, for example, solving partial differential equations discretized by methods of finite differences, finite volumes, finite elements, it is common that a large number of large size systems need to be solved many times. Frequently, the occurring matrices will be sparse.

In these conditions, different methods are recommended or even necessary. Indeed, an iterative method is used which consists in generating a sequence of vectors $\boldsymbol{x}_i$ which converges to the solution. When the matrix is sparse, the search of the solution is easier. Such methods should not be used on dense matrices.

The system to solve is

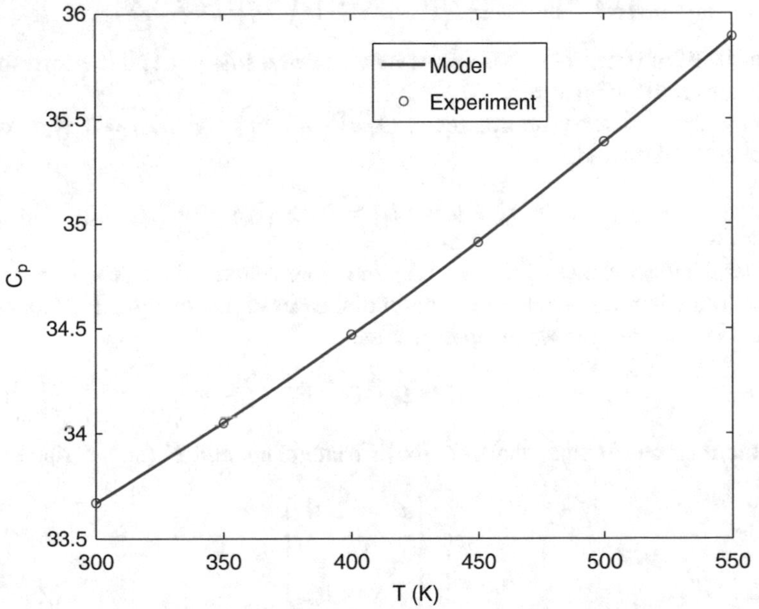

Fig. 5.1 Least squares with Cholesky's decomposition: experimental points and model of heat capacity

$$\boldsymbol{A}\boldsymbol{x} = \boldsymbol{b} \tag{5.10.1}$$

where $\boldsymbol{A}$ is an $n \times n$ nonsingular matrix.

Consider an iterative method of the form

$$\boldsymbol{x}^{(i+1)} = \boldsymbol{\phi}(\boldsymbol{x}^{(i)}) \tag{5.10.2}$$

By introducing an arbitrary nonsingular matrix $\boldsymbol{B}$ of dimension $n \times n$, algorithms can be built, which satisfy the previous recurrence according to

$$\boldsymbol{B}\,\boldsymbol{x} + (\boldsymbol{A} - \boldsymbol{B})\,\boldsymbol{x} = \boldsymbol{b} \tag{5.10.3}$$

by writing

$$\boldsymbol{B}\,\boldsymbol{x}^{(i+1)} + (\boldsymbol{A} - \boldsymbol{B})\,\boldsymbol{x}^{(i)} = \boldsymbol{b} \tag{5.10.4}$$

and expressing $\boldsymbol{x}^{(i+1)}$ that is searched

$$\boldsymbol{x}^{(i+1)} = \boldsymbol{x}^{(i)} - \boldsymbol{B}^{-1}\,(\boldsymbol{A}\,\boldsymbol{x}^{(i)} - \boldsymbol{b}) = (\boldsymbol{I} - \boldsymbol{B}^{-1}\,\boldsymbol{A})\,\boldsymbol{x}^{(i)} + \boldsymbol{B}^{-1}\,\boldsymbol{b} \tag{5.10.5}$$

The iteration can be written as

$$\begin{bmatrix} 1 \\ \boldsymbol{x}^{(i+1)} \end{bmatrix} = \boldsymbol{W} \begin{bmatrix} 1 \\ \boldsymbol{x}^{(i)} \end{bmatrix}, \quad \boldsymbol{W} = \left[\begin{array}{c|c} 1 & 0 \\ \hline \boldsymbol{B}^{-1}\,\boldsymbol{b} & \boldsymbol{I} - \boldsymbol{B}^{-1}\,\boldsymbol{A} \end{array}\right] \tag{5.10.6}$$

The matrix $\boldsymbol{W}$ of order $n+1$ has one of its eigenvalues $\lambda_0 = 1$ and the corresponding eigenvector $[1, \boldsymbol{x}]^T$ with $\boldsymbol{x} = \boldsymbol{A}^{-1}\,\boldsymbol{b}$.

The sequence $[1, \boldsymbol{x}^{(i)}]^T$ converges to $[1, \boldsymbol{x}]^T$ if and only if $\lambda_0 = 1$ is a dominant eigenvalue of $\boldsymbol{W}$, that is,

$$\lambda_0 = 1 > |\lambda_1| \geq \cdots \geq |\lambda_n| \tag{5.10.7}$$

The lower the eigenvalues of $(\boldsymbol{I} - \boldsymbol{B}^{-1}\,\boldsymbol{A})$, the more adapted the method.

In the following, we indicate a few particular cases of this method. For that purpose, it is necessary to decompose the matrix $\boldsymbol{A}$ as

$$\boldsymbol{A} = \boldsymbol{D} - \boldsymbol{E} - \boldsymbol{F} \tag{5.10.8}$$

where the matrices $\boldsymbol{D}$ (diagonal), $\boldsymbol{E}$ (lower triangular), and $\boldsymbol{F}$ (upper triangular) are defined as

$$\boldsymbol{D} = \begin{bmatrix} a_{11} & & 0 \\ & \ddots & \\ 0 & & a_{nn} \end{bmatrix} \tag{5.10.9}$$

$$\boldsymbol{E} = -\begin{bmatrix} 0 & & & 0 \\ a_{21} & \ddots & & \vdots \\ \vdots & & \ddots & \vdots \\ a_{n1} & \dots & a_{n,n-1} & 0 \end{bmatrix} \tag{5.10.10}$$

$$\boldsymbol{F} = -\begin{bmatrix} 0 & a_{12} & \dots & a_{1n} \\ \dots & \ddots & & \vdots \\ \vdots & & \ddots & a_{n-1,n} \\ 0 & \dots & \dots & 0 \end{bmatrix} \tag{5.10.11}$$

as well as the associated matrices

$$\boldsymbol{L} = \boldsymbol{D}^{-1}\boldsymbol{E}, \quad \boldsymbol{U} = \boldsymbol{D}^{-1}\boldsymbol{F}, \quad \boldsymbol{J} = \boldsymbol{L} + \boldsymbol{U}, \quad \boldsymbol{H} = (\boldsymbol{I} - \boldsymbol{L})^{-1}\boldsymbol{U} \tag{5.10.12}$$

by supposing

$$a_{ii} \neq 0 \quad \forall i \tag{5.10.13}$$

- In *Jacobi's method*, the matrices are

$$\boldsymbol{B} = \boldsymbol{D}, \quad \boldsymbol{I} - \boldsymbol{B}^{-1}\boldsymbol{A} = \boldsymbol{J} \tag{5.10.14}$$

hence the algorithm

$$\boldsymbol{B}\,\boldsymbol{x}^{(i+1)} - (\boldsymbol{E} + \boldsymbol{F})\,\boldsymbol{x}^{(i)} = \boldsymbol{b} \Longleftrightarrow \boldsymbol{x}^{(i+1)} = (\boldsymbol{L} + \boldsymbol{U})\,\boldsymbol{x}^{(i)} + \boldsymbol{D}^{-1}\boldsymbol{b} \tag{5.10.15}$$

Under explicit form, the iteration sequence (indexed i) is

$$a_{jj}x_j^{(i+1)} + \sum_{k \neq j} a_{jk}x_k^{(i)} = b_j\,, \quad j = 1, \ldots, n \tag{5.10.16}$$

the iteration vector being equal to $\boldsymbol{x}^{(i)} = \left[x_1^{(i)}, \ldots, x_n^{(i)}\right]^T$.

- In *Gauss–Seidel method*, the matrices are

$$\boldsymbol{B} = \boldsymbol{D} - \boldsymbol{E}, \quad \boldsymbol{I} - \boldsymbol{B}^{-1}\boldsymbol{A} = (\boldsymbol{I} - \boldsymbol{L})^{-1}\boldsymbol{U} = \boldsymbol{H} \tag{5.10.17}$$

hence the algorithm

$$(\boldsymbol{D} - \boldsymbol{E})\,\boldsymbol{x}^{(i+1)} - \boldsymbol{F}\,\boldsymbol{x}^{(i)} = \boldsymbol{b} \tag{5.10.18}$$

Under explicit form, the iteration sequence (indexed i) is

$$\sum_{k<j} a_{jk}x_k^{(i+1)} + a_{jj}x_j^{(i+1)} + \sum_{k>j} a_{jk}x_k^{(i)} = b_j\,, \quad j = 1, \ldots, n \tag{5.10.19}$$

Remark:
The system to solve can be transformed by dividing each row by the corresponding diagonal coefficient a_{ii}, so that the new system to solve is

$$\begin{bmatrix} 1 & r_{12} & r_{13} & \ldots & r_{1n} \\ r_{21} & 1 & r_{23} & \ldots & r_{2n} \\ r_{31} & r_{32} & 1 & & r_{3n} \\ \vdots & \vdots & & \ddots & \vdots \\ r_{n1} & \ldots & & r_{n-1,n} & 1 \end{bmatrix} \begin{bmatrix} x_1 \\ x_2 \\ \vdots \\ \\ x_n \end{bmatrix} = \begin{bmatrix} m_1 \\ m_2 \\ \vdots \\ \\ m_n \end{bmatrix} \tag{5.10.20}$$

where the new elements are such that

$$r_{ij} = \frac{a_{ij}}{a_{ii}} \text{ with } i \neq j\,, \; m_i = \frac{b_i}{a_{ii}} \tag{5.10.21}$$

This system is called strictly diagonally *dominant* system if the sum of the absolute values of the off-diagonal elements in each row is lower than 1

$$\sum_{j \neq i} |r_{ij}| < 1 \quad \forall i \tag{5.10.22}$$

A strictly diagonally dominant matrix is nonsingular. When the condition (5.10.22) is satisfied, the system can be solved by an iterative method such as Jacobi or Gauss–Seidel. This condition is very often fulfilled in physical cases where a diagonal effect overcomes the nondiagonal effects, as discovered by Onsager's reciprocal relations in thermodynamics of systems out of equilibrium represented by the matrix of phenomenological coefficients. For example, a pressure or a stress causes on a material a deformation (main or diagonal effect), but it can cause a piezoelectric effect (secondary or off-diagonal effect). In most cases, Gauss–Seidel method converges faster than Jacobi's method.

In some cases, it can be more interesting to modify Gauss–Seidel method (5.10.19) by introducing an *under-* or an *over-relaxation* according to the following equation:

$$x_j^{(i+1)} = (1-\omega)x_j^{(i)} + \frac{\omega}{a_{jj}} \left[b_j - \sum_{k<j} a_{jk} x_k^{(i+1)} - \sum_{k>j} a_{jk} x_k^{(i)} \right], \quad j = 1, \dots, n \tag{5.10.23}$$

where ω is the under-relaxation ($0 < \omega < 1$) or over-relaxation ($\omega > 1$) parameter. The under-relaxation is used in the case of systems not converging by Gauss–Seidel method, while the over-relaxation is used in the case of systems that converge by Gauss–Seidel method. Sometimes, it is called successive over-relaxation.

Example 5.9 :

Jacobi and Gauss–Seidel methods: iterative solution of a linear system

The following linear system is solved to compare the convergence performances of Jacobi and Gauss–Seidel methods:

$$\begin{cases} 7\,x_1 + 2\,x_2 + x_3 = 21 \\ 3\,x_1 - 8\,x_2 - x_3 = 15 \\ x_1 - 2\,x_2 + 6\,x_3 = 17 \end{cases} \tag{5.10.24}$$

The same initial condition $\boldsymbol{x} = (0, 0, 0)$ is used for both methods. First, it is verified that the system is diagonally dominant.

Jacobi's method corresponds to the following algorithm:

$$\begin{aligned} 7\,x_1^{(i+1)} + 2\,x_2^{(i)} + x_3^{(i)} &= 21 \\ 3\,x_1^{(i)} - 8\,x_2^{(i+1)} - x_3^{(i)} &= 15 \\ x_1^{(i)} - 2\,x_2^{(i)} + 6\,x_3^{(i+1)} &= 17 \end{aligned} \tag{5.10.25}$$

while Gauss–Seidel method corresponds to the following algorithm:

$$\begin{aligned} 7\,x_1^{(i+1)} + 2\,x_2^{(i)} + x_3^{(i)} &= 21 \\ 3\,x_1^{(i+1)} - 8\,x_2^{(i+1)} - x_3^{(i)} &= 15 \\ x_1^{(i+1)} - 2\,x_2^{(i+1)} + 6\,x_3^{(i+1)} &= 17 \end{aligned} \tag{5.10.26}$$

It can be noticed in Table 5.2 that Gauss–Seidel method converges faster than Jacobi's, as expected from the design of both algorithms.

Table 5.2 Jacobi's and Gauss–Seidel methods: solution of system (5.10.24)

Iteration	Jacobi's method x_1	x_2	x_3	Gauss–Seidel method x_1	x_2	x_3
0	0.00000	0.00000	0.00000	0.00000	0.00000	0.00000
1	3.0000	−1.8750	2.8333	3.0000	−0.75000	2.0833
2	3.1310	−1.1042	1.7083	2.9167	−1.0417	2.0000
3	3.0714	−0.91443	1.9435	3.0119	−0.99554	1.9995
4	2.9836	−0.96615	2.0166	2.9988	−1.0004	2.0001
5	2.9880	−1.0082	2.0140	3.0001	−0.99997	2.0000
6	3.0003	−1.0063	1.9993	3.0000	−1.0000	2.0000
7	3.0019	−0.99978	1.9979			
8	3.0002	−0.99902	1.9998			
9	2.9998	−0.99988	2.0003			
10	2.9999	−1.0001	2.0001			
11	3.0000	−1.0000	2.0000			

5.11 Solution of Tridiagonal Linear Systems of Equations

In the simulation of physical processes, it frequently occurs that the system to solve is of the form

$$\boldsymbol{A}\,\boldsymbol{X} = \boldsymbol{D} \tag{5.11.1}$$

where $\boldsymbol{A}$ is a tridiagonal matrix such that

$$\begin{bmatrix} b_1 & c_1 & 0 & \dots & 0 \\ a_2 & b_2 & c_2 & 0 & \vdots \\ 0 & \ddots & \ddots & \ddots & \\ \vdots & & a_{n-1} & b_{n-1} & c_{n-1} \\ 0 & \dots & 0 & a_n & b_n \end{bmatrix} \begin{bmatrix} x_1 \\ x_2 \\ \vdots \\ x_n \end{bmatrix} = \begin{bmatrix} d_1 \\ d_2 \\ \vdots \\ d_n \end{bmatrix} \tag{5.11.2}$$

The solution method (also known as Thomas algorithm, also similar to the LU decomposition) consists in transforming that system under the following upper triangular form that will be easy to solve:

$$\begin{bmatrix} 1 & \alpha_1 & 0 & \dots & 0 \\ 0 & 1 & \alpha_2 & 0 & \vdots \\ 0 & \ddots & \ddots & \ddots & \\ \vdots & & \ddots & \ddots & \alpha_{n-1} \\ 0 & \dots & & 0 & 1 \end{bmatrix} \begin{bmatrix} x_1 \\ x_2 \\ \vdots \\ x_n \end{bmatrix} = \begin{bmatrix} \beta_1 \\ \beta_2 \\ \vdots \\ \beta_n \end{bmatrix} \tag{5.11.3}$$

Suppose that the coefficients α and β are known. The solution is obtained under recurrent form, starting by the nth row

$$\begin{cases} x_n &= \beta_n \\ x_{n-1} &= \beta_{n-1} - \alpha_{n-1}\, x_n \\ \vdots & \\ x_i &= \beta_i - \alpha_i\, x_{i+1} \\ \vdots & \\ x_1 &= \beta_1 - \alpha_1\, x_2 \end{cases} \tag{5.11.4}$$

The initial system gives for the ith row equation

$$a_i\, x_{i-1} + b_i\, x_i + c_i\, x_{i+1} = d_i \quad \forall\ 1 \leq i \leq n \tag{5.11.5}$$

with $a_1 = 0$ and $c_n = 0$. The value of x_{i-1} given by the transformed system can be introduced in this equation, which gives the new equation

$$\begin{aligned} & a_i\,(\beta_{i-1} - \alpha_{i-1}\, x_i) + b_i\, x_i + c_i\, x_{i+1} = d_i \Longleftrightarrow \\ & (b_i - a_i\, \alpha_{i-1})\, x_i + c_i\, x_{i+1} = d_i - a_i\, \beta_{i-1} \quad \forall\ 2 \leq i \leq n \end{aligned} \tag{5.11.6}$$

which is identified with the corresponding equation of the transformed system $x_i + \alpha_i\, x_{i+1} = \beta_i$. The recurrence relations are

$$\begin{cases} \alpha_i = \dfrac{c_i}{b_i - a_i\, \alpha_{i-1}} & \forall\ 2 \leq i \leq n-1 \\ \beta_i = \dfrac{d_i - a_i\, \beta_{i-1}}{b_i - a_i\, \alpha_{i-1}} & \forall\ 2 \leq i \leq n \end{cases} \tag{5.11.7}$$

Moreover, we identify the equation $b_1\, x_1 + c_1\, x_2 = d_1$ and the equation of the transformed system $x_1 = \beta_1 - \alpha_1\, x_2$, which allows to initialize the sequences α_i and β_i

$$\begin{cases} \alpha_1 = \dfrac{c_1}{b_1} \\ \beta_1 = \dfrac{d_1}{b_1} \end{cases} \tag{5.11.8}$$

5.12 Solution of Nonlinear Systems: Newton–Raphson Method

In this case, the system of equations to solve is no more linear and can be set under the general form

$$\boldsymbol{f}(\boldsymbol{x}) = \begin{bmatrix} f_1(x_1, \ldots, x_n) \\ \vdots \\ f_n(x_1, \ldots, x_n) \end{bmatrix} = \mathbf{0} \tag{5.12.1}$$

Newton–Raphson method is an extension of Newton's method designed for root-finding in the case of a single variable function (Section 3.8). Similarly, we proceed to a linearization of $\boldsymbol{f}$. Assuming that $\boldsymbol{x} = \boldsymbol{\alpha}$ is a solution vector of the system ($\boldsymbol{f}(\boldsymbol{\alpha}) = 0$) and that $\boldsymbol{x}^{(0)}$ is an approximation of $\boldsymbol{\alpha}$, if $\boldsymbol{f}$ differentiable, it can be written at first order

$$\boldsymbol{f}(\boldsymbol{\alpha}) \approx \boldsymbol{f}(\boldsymbol{x}^{(0)}) + \boldsymbol{D}\boldsymbol{f}(\boldsymbol{x}^{(0)})(\boldsymbol{\alpha} - \boldsymbol{x}^{(0)}) \qquad \text{with} \quad \boldsymbol{f}(\boldsymbol{\alpha}) = 0 \tag{5.12.2}$$

where $\boldsymbol{D}\boldsymbol{f}(\boldsymbol{x}^{(0)})$ is the Jacobian matrix equal to

$$\boldsymbol{D}\boldsymbol{f}(\boldsymbol{x}^{(0)}) = \begin{bmatrix} \frac{\partial f_1}{\partial x_1} & \cdots & \frac{\partial f_1}{\partial x_n} \\ \vdots & & \vdots \\ \frac{\partial f_n}{\partial x_1} & \cdots & \frac{\partial f_n}{\partial x_n} \end{bmatrix}_{\boldsymbol{x}=\boldsymbol{x}^{(0)}} \tag{5.12.3}$$

and the deviation vector equal to

$$\boldsymbol{\alpha} - \boldsymbol{x}^{(0)} = \begin{bmatrix} \alpha_1 - (x_1)^{(0)} \\ \vdots \\ \alpha_n - (x_n)^{(0)} \end{bmatrix} \tag{5.12.4}$$

If the Jacobian is nonsingular, it is possible to solve the equation

$$\boldsymbol{f}(\boldsymbol{x}^{(0)}) + \boldsymbol{D}\boldsymbol{f}(\boldsymbol{x}^{(0)})(\boldsymbol{x}^{(1)} - \boldsymbol{x}^{(0)}) = 0 \tag{5.12.5}$$

with the result of the first iteration

$$\boldsymbol{x}^{(1)} = \boldsymbol{x}^{(0)} - (\boldsymbol{D}\boldsymbol{f}(\boldsymbol{x}^{(0)}))^{-1}\boldsymbol{f}(\boldsymbol{x}^{(0)}) \tag{5.12.6}$$

hence the recurrence formula

$$\boldsymbol{x}^{(i+1)} = \boldsymbol{x}^{(i)} - (\boldsymbol{D}\boldsymbol{f}(\boldsymbol{x}^{(i)}))^{-1}\boldsymbol{f}(\boldsymbol{x}^{(i)}) \tag{5.12.7}$$

In the same way as for Newton's method, the convergence of Newton–Raphson method is quadratic. The choice of the initial vector $\boldsymbol{x}^{(0)}$ for Newton–Raphson method is very important and essential simply for obtaining the solution and for convergence. This remark is still more crucial in this case than for simple Newton's method.

Very often, it is not possible to obtain the Jacobian under an analytic form. In this case, an approximation can be achieved for each partial derivative for example under the form of a central difference quotient (or a forward or backward difference quotient if the central difference quotient is not possible)

$$\frac{\partial f_i}{\partial x_j} \approx \frac{f_i(x_j(1+\epsilon)) - f_i(x_j(1-\epsilon))}{2\,x_j\,\epsilon} \tag{5.12.8}$$

where ϵ is small compared to 1. Then, the convergence will only be superlinear.

According to Equation (5.12.7), the Jacobian matrix $\boldsymbol{D}\boldsymbol{f}$ should be inverted. Indeed, Newton–Raphson method can be presented in the much more general frame of gradient methods (Section 9.5) corresponding to the following algorithm:

$$\boldsymbol{x}^{(i+1)} = \boldsymbol{x}^{(i)} + \alpha_i\, \boldsymbol{p}^{(i)} \tag{5.12.9}$$

where $\boldsymbol{p}^{(i)}$ indicates the displacement direction and α_i the displacement intensity. The displacement direction must be such that the scalar product with the gradient $\boldsymbol{g}$ of the function to minimize is negative

$$\boldsymbol{p}^{(i)}\, \boldsymbol{g}(\boldsymbol{x}^{(i)}) < 0 \tag{5.12.10}$$

In the present case, the problem (5.12.1) is not posed as a minimization problem. However, the vector $\boldsymbol{f}(\boldsymbol{x})$ plays the same role as the gradient in the minimization problem.

Thus, in the case of Newton–Raphson method, it is possible to avoid the explicit inversion of the Jacobian matrix $\boldsymbol{Df}$ by solving at each iteration k the linear system

$$Df(\boldsymbol{x}^{(i)})\, \boldsymbol{\eta}^{(i)} = -f(\boldsymbol{x}^{(i)}) \tag{5.12.11}$$

which gives both the displacement direction $\boldsymbol{p}^{(i)}$ and intensity $\alpha^{(i)}$ under the form $\boldsymbol{\eta}^{(i)} = \alpha^{(i)} \boldsymbol{p}^{(i)}$ and then use the gradient formula

$$\boldsymbol{x}^{(i+1)} = \boldsymbol{x}^{(i)} + \boldsymbol{\eta}^{(i)} \tag{5.12.12}$$

The matrix inversion is thus replaced by the solution of a linear problem which only moves the difficulties. In solving (5.12.11), the issue of the condition number of the Jacobian matrix $Df(\boldsymbol{x}^{(i)})$ remains important. Different methods (Section 5.13) will be proposed to avoid these pitfalls.

Example 5.10 :
Newton–Raphson method for a nonlinear system

An example of application of Newton–Raphson method is treated in Section 3.4. Consider the second example corresponding to the following system of equations:

$$\begin{aligned} f_1(\boldsymbol{x}) &= 3\, x_1 - \cos(x_2\, x_3) - 0.5 \\ f_2(\boldsymbol{x}) &= 4\, x_1^2 - 625\, x_2^2 + 2\, x_2 - 1 \\ f_3(\boldsymbol{x}) &= \exp(-x_1\, x_2) + 20\, x_3 + (10\pi - 3)/3 \end{aligned} \tag{5.12.13}$$

The Jacobian of this system is equal to

$$\boldsymbol{Df}(\boldsymbol{x}) = \begin{bmatrix} 3 & x_3 \sin(x_2\, x_3) & x_2 \sin(x_2\, x_3) \\ 8\, x_1 & -1250\, x_2 + 2 & 0 \\ -x_2 \exp(-x_1\, x_2) & -x_1 \exp(-x_1\, x_2) & 20 \end{bmatrix} \tag{5.12.14}$$

Starting from the initial point $\boldsymbol{x} = (0, 0, 0)$, during solving by the Newton–Raphson method, iterations yield the results of Table 5.3. The iterations are stopped when simultaneously the norm of vector $\delta\boldsymbol{x}$ and that of vector $\boldsymbol{f}$ become lower than given tolerances, respectively.

5.13 Solution of Nonlinear Systems by Optimization

Solving a system of nonlinear equations of the form

$$f_i(\boldsymbol{x}) = 0\,, \quad i = 1, \ldots, n \tag{5.13.1}$$

can be considered as an optimization problem where the function $F(\boldsymbol{x})$ to minimize is the sum of squares of the nonlinear functions

$$F(\boldsymbol{x}) = \frac{1}{2} \sum_{i=1}^{m} f_i(\boldsymbol{x})^2 \tag{5.13.2}$$

Often, it will be necessary to use weights because of different orders of magnitude of the nonlinear functions as

$$F(\boldsymbol{x}) = \frac{1}{2} \sum_{i=1}^{m} w_i\, f_i(\boldsymbol{x})^2 \tag{5.13.3}$$

In the absence of constraints, the minimum condition is that the gradient of F is equal to zero

Table 5.3 Newton–Raphson method: solution of system (5.12.13)

Iteration	Vector $\boldsymbol{x}$	Vector $\boldsymbol{f}$
0	$\begin{bmatrix} 0 \\ 0 \\ 0 \end{bmatrix}$	$\begin{bmatrix} -1.5000 \\ -1.0000 \\ 10.4720 \end{bmatrix}$
1	$\begin{bmatrix} 0.5000 \\ 0.5000 \\ -0.5235 \end{bmatrix}$	$\begin{bmatrix} 0.03407 \\ -155.25 \\ -0.2211 \end{bmatrix}$
2	$\begin{bmatrix} 0.50016 \\ 0.2508 \\ 0.5173 \end{bmatrix}$	$\begin{bmatrix} 0.0089 \\ -38.81 \\ 0.0063 \end{bmatrix}$
3	$\begin{bmatrix} 0.499945 \\ 0.1262 \\ -0.5204 \end{bmatrix}$	$\begin{bmatrix} 0.0019 \\ -9.70 \\ 0.0017 \end{bmatrix}$
4	$\begin{bmatrix} 0.499986 \\ 0.0639 \\ -0.5220 \end{bmatrix}$	$\begin{bmatrix} 0.51\,10^{-3} \\ -2.42 \\ 0.46\,10^{-3} \end{bmatrix}$
5	$\begin{bmatrix} 0.499995 \\ 0.03277 \\ -0.522780 \end{bmatrix}$	$\begin{bmatrix} 0.32\,10^{-4} \\ -0.151 \\ 0.29\,10^{-4} \end{bmatrix}$
8	$\begin{bmatrix} 0.499999 \\ 0.00571 \\ -0.523456 \end{bmatrix}$	$\begin{bmatrix} 0.19\,10^{-5} \\ -0.89\,10^{-2} \\ 0.17\,10^{-5} \end{bmatrix}$
12	$\begin{bmatrix} 0.500000 \\ 0.003199 \\ -0.523518 \end{bmatrix}$	$\begin{bmatrix} 0.27\,10^{-11} \\ -0.12\,10^{-7} \\ 0.24\,10^{-11} \end{bmatrix}$

$$\begin{cases} \dfrac{\partial F}{\partial x_1} = 0 \\ \vdots \\ \dfrac{\partial F}{\partial x_n} = 0 \end{cases} \tag{5.13.4}$$

which yields a system of n nonlinear equations with n unknowns. Theoretically, that system could be solved by Newton–Raphson method. However, the inversion of the Hessian matrix of F, equal to the Jacobian of the gradient, often poses numerical problems, in particular when the system scale is large.

For that reason, it is preferable to apply other methods to such a problem, such as Gauss–Newton method, quasi-Newton method, and Levenberg–Marquardt method, which are frequently used in optimization. The problem being posed as an optimization problem, to have a complete development of these methods, it will be useful to refer to Chapter 9, in particular in Section 9.5 and especially in Sections 9.5.7, 9.5.8, 9.5.9, and 9.5.10.

5.14 Discussion and Conclusion

When solving a system of linear equations $\boldsymbol{A}\,\boldsymbol{x} = \boldsymbol{b}$, first the type of matrix $\boldsymbol{A}$ must be analyzed. Is this matrix dense, triangular, sparse, band, and tridiagonal? Specialized methods for some types of matrices such as sparse matrices exist in the numerical libraries, whether they are open (LAPACK) or commercial (GAMS, NAG, and IMSL). In such a case, it is essential to avoid using a Gauss method designed for dense matrices. For a dense matrix, opposite to Gauss method, Gauss–Jordan method gives the inverse matrix and thus allows to calculate the condition number of the matrix, which indicates the confidence that we can have with respect to the solution. For least squares parameter estimation of a linear model, use Cholesky's decomposition. When a system of nonlinear equations is to be solved, Newton–Raphson method seems a priori obvious. However, it presents many drawbacks and it is useful to think over another way to solve the system. In some cases, in steady-state process modelling, when the steady state is searched, it is often simpler to solve the associated dynamic model constituted by a system of ordinary differential equations and then integrate that differential system to deduce the asymptotic solution (when $t \to \infty$), which corresponds to the steady state. In large size systems of nonlinear equations, as this is the case for process simulators, the problem is transformed into an optimization problem (Section 5.13) and an approximate solution is given by some method of quasi-Newton type or close.

5.15 Exercise Set

Preamble: In many cases, the students can write their own programs to solve linear or nonlinear systems of algebraic equations. In addition to the proposed methods, they can write programs corresponding to more advanced methods. They can also use existing codes.

Exercise 5.15.1 (Medium)

1. Solve the following linear system by Jacobi's method:

$$\begin{cases} x_1 - 2\,x_2 + 2\,x_3 = -5 \\ -3\,x_1 + 2\,x_2 - 5\,x_3 = 6 \\ 7\,x_1 + 4\,x_2 + 6\,x_3 = 9 \end{cases} \tag{5.15.1}$$

   starting from point (0, 0, 0) and comment.
2. Solve the following linear system first by Jacobi's method and then by Gauss–Seidel method:

$$\begin{cases} 10\,x_1 - x_2 + 2\,x_3 = 35 \\ -x_1 + 7\,x_2 - 3\,x_3 = -16 \\ 2\,x_1 + 4\,x_2 + 9\,x_3 = 20 \end{cases} \tag{5.15.2}$$

   starting from point (0, 0, 0) and comment.

Exercise 5.15.2 (Difficult)

Solving a system of nonlinear equations

Introduction

A gas mixture is initially composed of 2 moles of CH_4 and 3 moles of H_2O. At thermodynamic equilibrium, the present species are CH_4, H_2O, CO, CO_2, and H_2. We desire to determine the composition at equilibrium (Vidal 1997; Zeggeren and Storey 1970).

Equilibrium condition for a chemical reaction

Consider the reaction

$$\nu_1\,\mathrm{A}_1 + \nu_2\,\mathrm{A}_2 + \cdots \rightleftharpoons \nu_1'\,\mathrm{A}_1' + \nu_2'\,\mathrm{A}_2' + \ldots \tag{5.15.3}$$

The condition of thermodynamic equilibrium is

$$dG_{T,P} = 0 \tag{5.15.4}$$

where G is the Gibbs free energy of the reaction medium, equal to

$$G = \sum_i n_i\,\mu_i \tag{5.15.5}$$

where the chemical potential μ_i of species i is equal to

$$\mu_i = \mu_i^0 + RT \ln y_i \tag{5.15.6}$$

As the chemical potential μ_i is an intensive variable, it does not depend on n_i. A variation of G (extensity) related to a variation of the number of moles n_i is expressed under the form

$$dG = \sum_i \mu_i dn_i + \sum_i n_i d\mu_i = \sum_i \mu_i dn_i \tag{5.15.7}$$

using Gibbs–Duhem relation

$$S dT - V dP + \sum_i n_i d\mu_i = 0 \tag{5.15.8}$$

at constant T and P. By introducing the degree of advancement ξ of the reaction such that

$$-\frac{dn_1}{\nu_1} = -\frac{dn_2}{\nu_2} = \dots = \frac{dn'_1}{\nu'_1} = \frac{dn'_2}{\nu'_2} = \dots = d\xi \tag{5.15.9}$$

the variation of Gibbs free energy can be expressed under the form

$$dG = \sum_i \mu_i dn_i = [(\nu'_1\mu'_1 + \nu'_2\mu'_2 + \dots) - (\nu_1\mu_1 + \nu_2\mu_2 + \dots)]d\xi = \Delta G\, d\xi \tag{5.15.10}$$

by posing

$$\Delta G = \sum_i \nu_i \mu_i \tag{5.15.11}$$

where ν_i is positive for a product and negative for a reactant. The equilibrium condition (5.15.4) thus becomes

$$\Delta G = 0 \tag{5.15.12}$$

Moreover, the evolution can only occur in the direction of a decrease of Gibbs free energy such that

$$\left(\frac{\partial \Delta G}{\partial \xi}\right)_{T,P} > 0 \tag{5.15.13}$$

and the equilibrium conditions are finally Equations (5.15.12) and (5.15.13).

In practice, we determine the minimum of G.

Mathematical position of the problem

The problem is set in the framework of optimization, here the minimization of a function subject to the constraints

$$\min(G) = \min\left(\sum_i n_i\, \mu_i\right) \tag{5.15.14}$$

subject to the following equality constraints g_i:

$$\begin{array}{ll} \text{C:} & n_{CH_4} + n_{CO} + n_{CO_2} - 2 = 0 \\ \text{H:} & 4n_{CH_4} + 2n_{H_2O} + 2n_{H_2} - 14 = 0 \\ \text{O:} & n_{H_2O} + n_{CO} + 2n_{CO_2} - 3 = 0 \end{array} \tag{5.15.15}$$

To solve this problem, we form the Lagrangian function (or Lagrangian)

$$
\begin{aligned}
\mathcal{L} = G \quad & + \sum_i \lambda_i g_i \\
= \sum_i n_i \, \mu_i \; & + \lambda_1 (n_{CH_4} + n_{CO} + n_{CO_2} - 2) + \lambda_2 (4 n_{CH_4} + 2 n_{H_2O} + 2 n_{H_2} - 14) \\
& + \lambda_3 (n_{H_2O} + n_{CO} + 2 n_{CO_2} - 3)
\end{aligned}
\tag{5.15.16}
$$

where the λ_i are Lagrange multipliers.

The extremum condition imposes that the gradient of $\mathcal{L}$ is zero, that is,

$$
\begin{aligned}
n_{CH_4}: \quad & \mu_{CH_4} + \lambda_1 + 4\lambda_2 = 0 \\
n_{H_2O}: \quad & \mu_{H_2O} + 2\lambda_2 + \lambda_3 = 0 \\
n_{CO}: \quad & \mu_{CO} + \lambda_1 + \lambda_3 = 0 \\
n_{CO_2}: \quad & \mu_{CO_2} + \lambda_1 + 2\lambda_3 = 0 \\
n_{H_2}: \quad & \mu_{H_2} + 2\lambda_2 = 0 \\
\lambda_1: \quad & n_{CH_4} + n_{CO} + n_{CO_2} - 2 = 0 \\
\lambda_2: \quad & 4 n_{CH_4} + 2 n_{H_2O} + 2 n_{H_2} - 14 = 0 \\
\lambda_3: \quad & n_{H_2O} + n_{CO} + 2 n_{CO_2} - 3 = 0
\end{aligned}
\tag{5.15.17}
$$

thus forming a nonlinear system of 8 equations with 8 unknowns. Indeed, the problem is not solved under that form because of the following remark.

As the chemical potentials μ_i depend on mole fractions y_i (intensity), the problem is solved with respect to these mole fractions y_i as

$$\mu^0_{CH_4} + RT \ln(y_{CH_4}) + \lambda_1 + 4\lambda_2 = 0 \tag{5.15.18}$$

$$\mu^0_{H_2O} + RT \ln(y_{H_2O}) + 2\lambda_2 + \lambda_3 = 0 \tag{5.15.19}$$

$$\mu^0_{CO} + RT \ln(y_{CO}) + \lambda_1 + \lambda_3 = 0 \tag{5.15.20}$$

$$\mu^0_{CO_2} + RT \ln(y_{CO_2}) + \lambda_1 + 2\lambda_3 = 0 \tag{5.15.21}$$

$$\mu^0_{H_2} + RT \ln(y_{H_2}) + 2\lambda_2 = 0 \tag{5.15.22}$$

$$y_{CH_4} + y_{CO} + y_{CO_2} - \frac{2}{\sum n_i} = 0 \tag{5.15.23}$$

$$4 y_{CH_4} + 2 y_{H_2O} + 2 y_{H_2} - \frac{14}{\sum n_i} = 0 \tag{5.15.24}$$

$$y_{H_2O} + y_{CO} + 2 y_{CO_2} - \frac{3}{\sum n_i} = 0 \tag{5.15.25}$$

$$y_{CH_4} + y_{H_2O} + y_{CO} + y_{CO_2} + y_{H_2} - 1 = 0 \tag{5.15.26}$$

which forms a nonlinear system of 9 equations with 9 unknowns. These latter are

$$y_{CH_4}, \; y_{H_2O}, \; y_{CO}, \; y_{CO_2}, \; y_{H_2}, \; \lambda_1, \; \lambda_2, \; \lambda_3, \; \sum n_i \tag{5.15.27}$$

as the total number of moles is a variable.

Numerical data

The equilibrium temperature is 1000K and the atmospheric pressure is 1.01325 10^5Pa. Use Newton–Raphson method to calculate the equilibrium composition of the mixture (Poling et al. 2001).
$\mu^0_{CH_4} = -50.45$ kJ.mol^{-1}, $\mu^0_{H_2O} = -228.42$ kJ.mol^{-1}, $\mu^0_{CO} = -137.16$ kJ.mol^{-1}, $\mu^0_{CO_2} = -394.38$ kJ.mol^{-1}, $\mu^0_{H_2} = 0$ kJ.mol^{-1}, $R = 8.314$ J.mol^{-1}.K^{-1}.

Exercise 5.15.3 (Easy)
Solve the following system by Gauss–Seidel method:

$$\begin{cases} 7\,x_1 + 2\,x_2 - x_3 = 21 \\ x_1 - 5\,x_2 + 2\,x_3 = 4 \\ -2\,x_1 - x_2 + 4\,x_3 = -13 \end{cases} \tag{5.15.28}$$

starting from the initial point (1,1,1). Do 5 iterations and give all the intermediate results in a table.

Exercise 5.15.4 (Easy)
Solve the following system by Newton–Raphson method:

$$\begin{cases} x_1 \exp(-0.001\,x_2) = 4.0937\ 10^{-4} \\ x_1^{0.5}\,x_2 = 4.4721 \end{cases} \tag{5.15.29}$$

starting from the initial point (1,1). Do 5 iterations and give all the intermediate results in a table.

Exercise 5.15.5 (Medium)
Solve the following linear system $\boldsymbol{A}\,\boldsymbol{X} = \boldsymbol{B}$ by Gauss–Jordan method:

$$\begin{cases} x_1 + 2\,x_2 + 8\,x_3 \quad = -3 \\ 2\,x_1 - 4\,x_2 - 6\,x_3 \quad = 0 \\ -2\,x_1 + 4\,x_2 + 2\,x_3 = 4 \end{cases} \tag{5.15.30}$$

by detailing the intermediate matrices. Calculate the matrix $\boldsymbol{A}^{-1}$ inverse of $\boldsymbol{A}$.

Exercise 5.15.6 (Difficult)
Consider the following linear system:

$$\boldsymbol{A}\,\boldsymbol{X} = \boldsymbol{B} \tag{5.15.31}$$

1. Demonstrate that the iteration sequence of Jacobi

$$a_{jj} x_j^{(i+1)} + \sum_{k \neq j} a_{jk} x_k^{(i)} = b_j\,, \quad j = 1, \ldots, n \tag{5.15.32}$$

can be written under the form

$$\boldsymbol{X}^{(k+1)} = \boldsymbol{D}^{-1}\,\boldsymbol{B} - \boldsymbol{D}^{-1}\,(\boldsymbol{L} + \boldsymbol{U})\,\boldsymbol{X}^{(k)} \tag{5.15.33}$$

where $\boldsymbol{X}^{(k)}$ and $\boldsymbol{X}^{(k+1)}$ are vectors of the estimated solution at the kth and the $(k+1)$th iteration, respectively. $\boldsymbol{U}$ is a strictly upper triangular matrix, $\boldsymbol{L}$ a strictly lower triangular matrix, and $\boldsymbol{D}$ a diagonal matrix.

2. Let

$$\boldsymbol{C} = -\boldsymbol{D}^{-1}\,(\boldsymbol{L} + \boldsymbol{U}) \tag{5.15.34}$$

so that the iteration can be rewritten as

$$\boldsymbol{X}^{(k+1)} = \boldsymbol{D}^{-1}\,\boldsymbol{B} + \boldsymbol{C}\,\boldsymbol{X}^{(k)} \tag{5.15.35}$$

Noting α the exact solution of Equation (5.15.31) and ϵ_k the error vector

$$\epsilon_k = \alpha - \boldsymbol{X}^{(k)} \tag{5.15.36}$$

demonstrate that if $\boldsymbol{X}^{(0)}$ is the initial vector, then the error vector for $\boldsymbol{X}^{(k+1)}$ is given by

$$\epsilon_{k+1} = \boldsymbol{C}^{k+1}\,\epsilon_0 \tag{5.15.37}$$

Deduce the condition of convergence of the sequence $\boldsymbol{X}^{(k)}$.

Exercise 5.15.7 (Difficult)

Suppose that the sequence $\{x_i\}$ has the limit l. On another side, suppose that the sequence $\{x_i\}$ converges geometrically to l with a common ratio k such that $|k| < 1$, that is,

$$x_{i+1} - l = k(x_i - l) \tag{5.15.38}$$

1. Deduce the expressions of the limit l and the common ratio k with respect to x_i, x_{i+1}, x_{i+2}.
2. We pose

$$x'_i = x_i - \frac{(x_{i+1} - x_i)^2}{x_{i+2} - 2\,x_{i+1} + x_i} \tag{5.15.39}$$

Justify the interest of this expression. This is Aitken's method.

3. Suppose that the function $f(x)$ has only one root α such that $f(\alpha) = 0$. We pose $\Phi(x) = x - f(x)$. Let the recurrence according to the successive substitution method $x_{i+1} = \Phi(x_i)$. Demonstrate that it leads to an expression of the form

$$x'_i = x_i - \frac{f(x_i)^2}{f(x_i) - f[x_i - f(x_i)]} \tag{5.15.40}$$

Justify the interest of this expression. This is a quasi-Newton method.

4. Consider the following iteration:

$$\begin{cases} y = x_n - \dfrac{f(x_n)}{f'(x_n)} \\ x_{n+1} = y - \dfrac{f(y)}{f'(x_n)} \end{cases} \tag{5.15.41}$$

Find the numerical type of convergence of that iteration to the solution α.

5. Application:
   Let the function

$$f(x) = x^3 - x^2 - x - 1 \tag{5.15.42}$$

   We search the root α starting from $x = 0$ by means of several iterative methods. Often the calculation needs a rather large number of iterations but will be stopped after a maximum of 40 iterations (beyond 20 iterations, present the results every 5 iterations).

   Present the results in a table with 5 columns as
column 1: iteration number,
column 2: Newton's method,
column 3: quasi-Newton method (corresponding to question 3),
column 4: method of question 4 starting from $x = 0$, and
column 5: method of question 4 starting from $x = 0.02$.
   From the point of view of numerical analysis, comment the different methods used.

Exercise 5.15.8 (Medium)
Solve the following system by Gauss–Jordan method and detail the calculations step by step:

$$\begin{aligned}
3.333\,x_1 + 15920\,x_2 + 10.333\,x_3 &= 7953\\
2.222\,x_1 + 16.71\,x_2 + 9.612\,x_3 &= 0.965\\
-1.5611\,x_1 + 5.1792\,x_2 - 1.6855\,x_3 &= 2.714
\end{aligned} \tag{5.15.43}$$

Exercise 5.15.9 (Medium)
Solve the following system:

$$\begin{aligned}
&3\,x_1 - \cos(x_2\,x_3) - \frac{1}{2} = 0\\
&4\,x_1^2 - 625\,x_2^2 + 2\,x_2 - 1 = 0\\
&\exp(-x_1\,x_2) + 20\,x_3 + \frac{10\,\pi - 3}{3} = 0
\end{aligned} \tag{5.15.44}$$

starting from the point (0, 0, 0). Detail the method used and give the iterative results in a table.

Exercise 5.15.10 (Medium)
Solve the following system:

$$\begin{aligned}
&6\,x_1 - 2\,\cos(x_2\,x_3) = 1\\
&9\,x_2 + \sqrt{x_1^2 + \sin(x_3) + 1.06} = -0.9\\
&60\,x_3 + 3\,\exp(-x_1\,x_2) = -10\,\pi + 3
\end{aligned} \tag{5.15.45}$$

starting from the point (0, 0, 0). In principle, the convergence to the solution is obtained in less than 4 iterations.

Exercise 5.15.11 (Medium)
The pressure P required to sink a heavy and large object in a homogeneous soft ground lying on a hard basis is modelled by considering a circular plate of radius R sunk by a depth z according to a first empirical law by Bernstein and later Goriatchkin

$$P = a_1 z^{a_2} \tag{5.15.46}$$

where the coefficients a_i depend on the depth z and the ground consistency, but not the radius R. a_1 is called a soil stiffness constant and a_2 the exponent of sinkage.

The second following empirical law by Bekker is also used

$$P = \left(\frac{a_1}{R} + a_2\right) z^{a_3} \tag{5.15.47}$$

where a_1 and a_2 are soil stiffness constants and a_3 an exponent of deformation.

Other more complex models are available (Rashidi et al. 2012; Salman and Kiss 2019).

Experiments give Table 5.4. To find the coefficients a_i, we set the criterion

$$S = \sum_j (P_{exp,j} - P_{mod,j})^2 \tag{5.15.48}$$

where $P_{exp,j}$ is the experimental value of Table 5.4, while $P_{mod,j}$ is the value of model (5.15.47) depending on the radius R_j and the depth z.

The minimum of S with respect to coefficients a_i is obtained by writing that its gradient is zero

$$f_i = \frac{\partial S}{\partial a_i} = 0 \quad \forall i \tag{5.15.49}$$

1. Give the equations in the present case that will be called "zero gradient conditions."
2. Explain in detail the necessary equations to solve the equations of zero gradient conditions to numerically find the coefficients a_i.
3. Give the algorithm of the calculation.
4. Numerically solve the problem, starting from the initial point (1,1,1).

Table 5.4 Sinking of the plate

Radius R (cm)	Depth z (m)	Pressure P (kPa)
0.04	0.01	190
0.04	0.025	287
0.04	0.05	392
0.04	0.1	535
0.08	0.01	221
0.08	0.025	334
0.08	0.05	457
0.08	0.1	624

Exercise 5.15.12 (Medium)

Solve the following system:

$$
\begin{aligned}
&2\,x_1\,x_3^2 + 1.5\,\cos(x_1 x_2 x_3) = -13.8156\\
&x_1^2\,x_2 + \sqrt{x_1^2 + x_3^2} = 3.9111\\
&\exp\left(\frac{x_1}{x_2}\right) + x_3^2 = 7.0494
\end{aligned} \tag{5.15.50}
$$

starting from point $(-1; 1; 2)$.

Exercise 5.15.13 (Medium)

Solve the following system:

$$
\begin{aligned}
&\frac{1}{1 + x_1^2} + x_1\,x_2\,x_3 = -12.8121\\
&x_2 + \sqrt{x_1^2 + x_3^2} = 3.0654\\
&\frac{x_2}{x_1 + x_3} = -0.2258
\end{aligned} \tag{5.15.51}
$$

starting from point $(1, 1, 1)$.

Exercise 5.15.14 (Medium)

Solve the following system:

$$
\begin{aligned}
&x_1^1 + 2\,x_2^2 - x_2 - 2\,x_3 = 0\\
&x_1^2 - 8\,x_2^2 + 10\,x_3 = 0\\
&\frac{x_1^2}{7\,x_2\,x_3} = 1
\end{aligned} \tag{5.15.52}
$$

detailing your method and give the iterative results in a table. The initial point is free.

Exercise 5.15.15 (Easy)

Solve the following system:

$$
\begin{aligned}
&5\,x_1^2 - x_2^2 = 0\\
&x_2 - 0.25\,(\sin x_1 + \cos x_2) = 0
\end{aligned} \tag{5.15.53}
$$

Exercise 5.15.16 (Medium)

Solve the following system:

$$
\begin{aligned}
&6\,x_1 - 2\,\cos(x_2\,x_3) = 1\\
&9\,x_2 + \sqrt{x_1^2 + \sin(x_3) + 1.06} = -0.9\\
&60\,x_3 + 3\,\exp(-x_1\,x_2) = -10\,\pi + 3
\end{aligned} \tag{5.15.54}
$$

starting from the point $(0, 0, 0)$. Normally, the convergence to the solution is obtained in less than 4 iterations.

Exercise 5.15.17 (Medium)

Solve the following system:

$$
\begin{aligned}
&\ln(x_1^2 + x_2^2) + 0.5\,\sin(x_1\,x_3) = 0.45\\
&x_1\sqrt{(x_1^2 + x_3^2)} = 0.5\\
&\frac{1}{x_3}\,\exp(x_1\,x_2) = 0.3
\end{aligned} \tag{5.15.55}
$$

starting from the point $(0.1, -0.5, 1)$. The convergence can be obtained in about 5 iterations with a good precision.

Exercise 5.15.18 (Medium)
Solve the following system:

$$\frac{x_1}{x_1^2 + x_2^2 + x_3^2} = 0.1333 \tag{5.15.56}$$

$$\exp(-x_1\, x_2\, x_3) = 0.9656 \tag{5.15.57}$$

$$2\frac{x_1}{x_2} + \frac{x_2}{x_3} = 1.1143 \tag{5.15.58}$$

starting from the point $\boldsymbol{x} = [0.2 \quad 0.7 \quad 0.9]$.

References

G. H. Golub and C. F. Van Loan. *Matrix computations*. John Hopkins University Press, Baltimore, 1991.

J. Mandel. Use of the singular value decomposition in regression analysis. *The American Statistician*, 36(1):15–24, 1982.

B. E. Poling, J. M. Prausnitz, and J. P. O'Connel. *The properties of gases and liquids*. McGraw-Hill, New York, 5th edition, 2001.

M. Rashidi, M. Fakhri, M. A. Sheikhi, S. Azadeh, and S. Razavi. Evaluation of Bekker model in predicting soil pressure-sinkage behaviour under field conditions. *Middle-East Journal of Scientific Research*, 12(10):1364–1369, 2012.

N. D. Salman and P. Kiss. A study of pressure-sinkage relationship used in a tyre terrain interaction. *International Journal of Engineering and Management Science*, 4(1):186–199, 2019.

J. Vidal. *Thermodynamique*. Technip, Paris, 1997.

Z. Van Zeggeren and S. H. Storey. *The Computation of Chemical Equilibria*. University Press, Cambridge, 1970.

Chapter 6
Numerical Integration of Ordinary Differential Equations

6.1 Introduction

Many physical models make use of ordinary differential equations. This is the case for lumped parameter systems considered in transient regime and distributed parameter systems in steady state, when modelled with respect to one dimension. The number of numerical methods available to users is very important, and these methods are far from being equivalent.

It must be noted that solving a system of differential and algebraic equations (Brenan et al. 1989) is more tricky than that of a simple system of ordinary differential equations.

Essential remark:

When a system of ordinary differential equations is numerically integrated, the solution depends on the integration step chosen. Normally, the step must be chosen as large as possible without having any influence on the solution while ensuring the fulfillment of the calculation in a minimum time.

First of all, consider an ordinary differential equation of order n under *implicit* form

$$F(x, y, y^{(1)}(x), y^{(2)}(x), \dots, y^{(n)}(x)) = 0 \tag{6.1.1}$$

that can often be brought back to an *explicit* equation of the form

$$y^{(n)}(x) = G(x, y, y^{(1)}(x), \dots, y^{(n-1)}(x)) \tag{6.1.2}$$

The term of highest order of this equation is of order n. Only derivatives with respect to the x variable intervene by opposite to partial differential equations. The solution to that equation is not unique, and it is necessary to provide additional conditions. If the n conditions are given at a same value of x (e.g. x_0), the problem is called *initial condition* or *initial value* problem. If several values of x intervene, the problem is called *boundary condition* or *boundary value* problem (Section 6.6).

Any ordinary differential equation of order n can be brought back to a system of first order ordinary differential equations by introducing auxiliary functions

J.-P. Corriou, *Numerical Methods and Optimization*, Springer Optimization and Its Applications 187, https://doi.org/10.1007/978-3-030-89366-8_6

$$\begin{aligned} u_1(x) &= y(x) \\ u_2(x) &= y^{(1)}(x) \\ &\dots \\ u_n(x) &= y^{(n-1)}(x) \end{aligned} \tag{6.1.3}$$

which allow to write the system under an equivalent form

$$\boldsymbol{u}' = \begin{bmatrix} u_1' \\ \vdots \\ u_{n-1}' \\ u_n' \end{bmatrix} = \begin{bmatrix} u_2 \\ \vdots \\ u_n \\ G(x, u_1, u_2, \dots, u_n) \end{bmatrix} \tag{6.1.4}$$

In the following, only first order ordinary differential equations will be considered

$$y'(x) = f(x, y) \tag{6.1.5}$$

Furthermore, the explanations are given for one ordinary differential equation, but *all the techniques described can be applied in the same manner and with the same ease to a system of ordinary differential equations. In this case, we make the integration of the ordinary differential equations progress simultaneously. The system is qualified as a vector of ordinary differential equations*. Especially, it must never be tried to separately integrate the ordinary differential equations of the system, which would unavoidably lead to an erroneous solution.

6.1.1 Linear and Nonlinear Ordinary Differential Equations

The following second order ordinary differential equation

$$y\, y'' + y'^2 = 0 \tag{6.1.6}$$

is *nonlinear*.

On the opposite, in the following second order ordinary differential equation (6.1.7), y and its derivatives are only present under a *linear* form

$$y'' + f(x)\, y' + g(x)\, y = h(x) \tag{6.1.7}$$

Consider the second order *homogeneous* ordinary differential equation (6.1.8) associated to the previous second order linear ordinary differential equation

$$y'' + f(x)\, y' + g(x)\, y = 0 \tag{6.1.8}$$

This equation has two independent solutions y_1 and y_2; thus any linear combination of these solutions will be the general form of the solution of that equation (6.1.8)

$$y = \alpha\, y_1 + \beta\, y_2 \tag{6.1.9}$$

α and β being particular constants. To find the general solution of the original equation (6.1.7), it suffices to find a particular solution y_p, so that the general solution of (6.1.7) will be of the form

$$y = y_p + \alpha\, y_1 + \beta\, y_2 \tag{6.1.10}$$

This analytical procedure can be used only for linear ordinary differential equations.

6.1.2 *Uniqueness of the Solution*

The initial value problem

$$y^{(n)} = f(x, y, y', \ldots, y^{(n-1)}) \tag{6.1.11}$$

where the function y and its $n-1$ first derivatives are specified at a point a admits a unique solution if f is sufficiently smooth with respect to y, y', ... , $y^{(n-1)}$.

Consider the system of n ordinary differential equations

$$\boldsymbol{y}' = \boldsymbol{f}(x, \boldsymbol{y}) \tag{6.1.12}$$

Theorem:
Let $\boldsymbol{f}$ be defined and continuous on the band defined by $B = \{x \in [a, b],\ \boldsymbol{y} \in \mathcal{R}^n\}$. Let a constant C such that Lipschitz condition is verified

$$\|\boldsymbol{f}(x, \boldsymbol{y}_1) - \boldsymbol{f}(x, \boldsymbol{y}_2)\| \le C\|\boldsymbol{y}_1 - \boldsymbol{y}_2\|\,, \quad \forall\, x \in [a, b], \forall\, \boldsymbol{y}_1, \boldsymbol{y}_2 \in \mathcal{R}^n \tag{6.1.13}$$

$\forall\ x_0 \in [a, b]$, $\forall\ \boldsymbol{y}_0 \in \mathcal{R}^n$, there exists only one function $\boldsymbol{y}(x)$ such that:

(a) $\boldsymbol{y}(x)$ is continuous and continuously differentiable for $x \in [a, b]$.
(b) $\boldsymbol{y}'(x) = \boldsymbol{f}(x, \boldsymbol{y}(x)), \quad x \in [a, b]$.
(c) $\boldsymbol{y}(x_0) = \boldsymbol{y}_0$.

Lipschitz condition is verified if the partial derivatives $\partial f_i / \partial y_j$ exist on the band B, are continuous, and bounded on it.

6.2 Initial Value Problems

We search the solution of a first order ordinary differential equation of the form

$$\frac{dy}{dx} = f(x, y) \tag{6.2.1}$$

The solution $y(x)$ must satisfy that ordinary differential equation and an initial condition such as $y(x_0) = y_0$. The solution is searched on an interval $[a = x_0, b = x_n]$ that is divided into *steps* of fixed length

$$h = \frac{b-a}{n} \tag{6.2.2}$$

The values of x on the integration interval are regularly spaced $\{x_0, x_1, \ldots, x_i, \ldots, x_n\}$ that is

$$x_i = x_0 + ih\,, \quad i = 0, 1, \ldots, n \tag{6.2.3}$$

and the approximation of the solution will be given under the form of a vector of dimensions $n+1$, $\{y_i\,,\ i = 0, 1, \ldots, n\}$, while the true solution is noted $y(x_i)$. Similarly, the true derivative is $f(x_i, y(x_i))$ approximated by $f(x_i, y_i)$. The *discretization error* or *truncation error* is equal to

$$\epsilon_i = y_i - y(x_i) \tag{6.2.4}$$

and mainly depends on the type of numerical method in use (the computer rounding error is not considered).

6.2.1 One-Step Methods

The one-step methods allow the calculation of y_{i+1} from the ordinary differential equation and the value of x_i alone, thus a single value of y_i according to the following relation

$$y_{i+1} = y_i + h_i \Phi(x_i, y_i, h_i) \tag{6.2.5}$$

where $\Phi(x_i, y_i, h_i)$ is the increment function, and h_i the step retained for the subinterval $[x_i, x_{i+1}]$. This integration formula is said to be consistent with the initial value problem if

$$\Phi(x_i, y_i, 0) = f(x, y) \tag{6.2.6}$$

6.2.1.1 Euler's Method

The nth degree Taylor polynomial allows to approximate the solution near a point x_0

$$y(x_0 + h) = y(x_0) + hf(x_0, y(x_0)) + \frac{h^2}{2!} f'(x_0, y(x_0)) + \frac{h^3}{3!} f''(x_0, y(x_0)) + \ldots \tag{6.2.7}$$

provided the derivatives of f of order larger than or equal to 1 are known, for example for the first derivative

$$\frac{df}{dx} = \frac{\partial f}{\partial x} + \frac{\partial f}{\partial y}\frac{dy}{dx} \tag{6.2.8}$$

An iterative algorithm allowing to pass from a value x_i to a value x_{i+1} will be written according to the nth degree Taylor polynomial

$$
\begin{aligned}
y(x_{i+1}) &= y(x_i + h) \\
&= y(x_i) + hf(x_i, y(x_i)) + \frac{h^2}{2!} f'(x_i, y(x_i)) + \frac{h^3}{3!} f''(x_i, y(x_i)) + \dots \qquad (6.2.9) \\
&+ \frac{h^n}{n!} f^{(n-1)}(x_i, y(x_i)) + \frac{h^{n+1}}{(n+1)!} f^{(n)}(\xi, y(\xi)), \quad \xi \in [x_i, x_{i+1}]
\end{aligned}
$$

This kind of algorithm is said to be of order h^n. The error is of order h^{n+1}, and the local truncation error e_i is bounded by

$$
|e_i| \leq \frac{h^{n+1}}{(n+1)!} C \qquad (6.2.10)
$$

with

$$
C \geq \max |f^{(n)}(x, y(x))|, \quad x \in [x_i, x_{i+1}] \qquad (6.2.11)
$$

Due to the difficulty of differentiation, this method is very little used except at first order

$$
y(x_{i+1}) = y(x_i) + hf(x_i, y(x_i)) + 0(h^2) \qquad (6.2.12)
$$

The algorithm of Euler's method is thus formulated

$$
y_{i+1} = y_i + hf(x_i, y_i), \quad i \geq 1 \qquad (6.2.13)
$$

The Euler's method consists of approximating the solution at x_{i+1} by the point taken at x_{i+1} on the tangent line to the curve $y(x)$ at x_i (Figure 6.1).

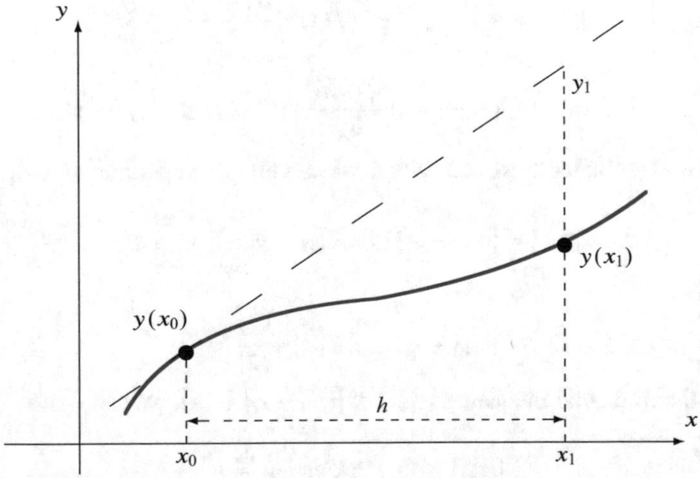

Fig. 6.1 Euler's method

The term of local error in Euler's method is equal to

$$
\epsilon_i = y_i - y(x_i) \qquad (6.2.14)
$$

which implies that the error related to passing through the step from x_i to x_{i+1} is equal to

$$\Delta\epsilon_i = \Delta y_i - \Delta y(x_i) \tag{6.2.15}$$

that is

$$\epsilon_{i+1} - \epsilon_i = y_{i+1} - y_i - (y(x_{i+1}) - y(x_i)) \tag{6.2.16}$$

with the condition

$$\epsilon_0 = 0 \tag{6.2.17}$$

Both contributions to the error can be evaluated, first that of explicit Euler's algorithm

$$y_{i+1} - y_i = hf(x_i, y_i) \tag{6.2.18}$$

then the error related to the second degree Taylor polynomial

$$y(x_{i+1}) - y(x_i) = hf(x_i, y(x_i)) + \frac{h^2}{2!} f'(\xi, y(\xi)), \quad \xi \in [x_i, x_{i+1}] \tag{6.2.19}$$

hence

$$\Delta\epsilon_i = h[f(x_i, y_i) - f(x_i, y(x_i))] - \frac{h^2}{2!} f'(\xi, y(\xi)) \tag{6.2.20}$$

From the properties of the function $f(x, y)$ and its first partial derivatives continuous and bounded on $x \in [a, b]$, assuming that a solution $y(x)$ exists, then constants K and M exist such that

$$|y''(x)| = |\frac{\partial f(x, y)}{\partial x} + f(x, y)\frac{\partial f(x, y)}{\partial y}| \leq M \tag{6.2.21}$$

$$|f(x, y^*) - f(x, y)| = |\frac{\partial f(x, \alpha)}{\partial y}|\,|y^* - y| \leq K\,|y^* - y| \tag{6.2.22}$$

with $y^* < \alpha < y$. Under these conditions, we can find an upper bound $|\Delta\epsilon_i|$

$$|\Delta\epsilon_i| = |\epsilon_{i+1} - \epsilon_i| \leq hK|y_i - y(x_i)| + \frac{M}{2!} h^2 \tag{6.2.23}$$

that is

$$|\epsilon_{i+1} - \epsilon_i| \leq hK|\epsilon_i| + \frac{M}{2!} h^2 \tag{6.2.24}$$

and using the triangular inequality $|\epsilon_{i+1}| \leq |\epsilon_{i+1} - \epsilon_i| + |\epsilon_i|$, we can write

$$|\epsilon_{i+1}| \leq (1 + hK)|\epsilon_i| + \frac{M}{2!} h^2 \tag{6.2.25}$$

Applying this relation i times, by induction, it comes

$$|\epsilon_i| \leq \frac{Mh}{2K}[(1 + hK)^i - 1] \tag{6.2.26}$$

Using the relation (coming from the nth degree Taylor polynomial for the exponential function)

$$(1 + hK) < \exp(hK) \tag{6.2.27}$$

and $x_n - x_0 = nh = L$, we get for the total integration interval

$$|\epsilon_n| \leq \frac{Mh}{2K} \exp(LK) \tag{6.2.28}$$

The error tends to 0 when h tends to 0

$$\lim_{h \to 0} |\epsilon_n| \leq \lim_{h \to 0} \frac{Mh}{2K} \exp(LK) = 0 \tag{6.2.29}$$

Euler's method is *convergent* as the error tends to 0 when the step tends to 0

$$\lim_{h \to 0} |\epsilon_i| = 0 \tag{6.2.30}$$

Euler's method converges with a global error defined by

$$|\epsilon_i| = |y_i - y(x_i)| = 0(h) \tag{6.2.31}$$

On another side, even if the local truncation error (estimated from the nth degree Taylor polynomial) is of order h^2 for Euler's method

$$|e_i| = \frac{h^2}{2!} |f'(\xi, y(\xi))| \,, \quad \xi \in [x_i, x_{i+1}] \tag{6.2.32}$$

the previous calculation demonstrates that the total truncation error is of order h.

Euler's method is thus a first order method. This method is not recommended because of its lack of precision. In particular, it must absolutely be avoided for *stiff* systems (where the derivative changes rapidly). A Runge–Kutta method of larger order hardly more difficult to operate is much better (even if it should not be also used for stiff systems).

Euler's method up to now is explicit, as x_i and y_i are only needed to calculate y_{i+1}. An implicit Euler's method exists

$$y_{i+1} = y_i + hf(x_{i+1}, y_{i+1}) \tag{6.2.33}$$

It gives better results than the explicit method.

6.2.1.2 A Few Ideas of Numerical Calculation

When we want to assess a procedure using any numerical method, it is important to validate it on a simple example for which the analytical solution is known.

For ordinary differential systems, consider a few simple cases. A first example can be

$$y^{(2)}(t) = -y(t) \quad \text{with} \quad y(0) = 1 \,, \quad y'(0) = 0 \tag{6.2.34}$$

whose solution is $y(t) = \cos(t)$. To numerically solve that second order ordinary differential equation, first it must be transformed into a system of first order ordinary differential equations.

A second example could be a system with a polynomial solution

$$y'(x) = x^3 - 2x + 1\,, \quad y(0) = 0 \tag{6.2.35}$$

Indeed, the previous examples are very easy to numerically solve on condition of choosing an adequate time step.

A third example could be a system with an exponential solution, as

$$y'(t) = 2y(t)\,, \quad y(0) = 1 \tag{6.2.36}$$

a difficulty lies in the rapid variations of the solution $y(t) = \exp(2t)$ on a time interval like $[0, 20]$ as $\exp(2 \times 20) = 2.3\,10^{17}$.

Example 6.1 :
Numerical comparison of integration methods

Consider now a fourth example of an ordinary differential system

$$\begin{aligned} y_1'(t) &= -3y_2(t) + 2y_3(t) &, \quad y_1(0) &= 2 \\ y_2'(t) &= -3y_2(t) &, \quad y_2(0) &= 1 \\ y_3'(t) &= 2y_3(t) &, \quad y_3(0) &= 1 \end{aligned} \tag{6.2.37}$$

for which the analytical solution is

$$\begin{aligned} y_1(t) &= \exp(2t) + \exp(-3t) \\ y_2(t) &= \exp(-3t) \\ y_3(t) &= \exp(2t) \end{aligned} \tag{6.2.38}$$

These solutions vary very differently in the interval $t \in [0, 20]$. Nevertheless, with relatively simple methods of Runge–Kutta–Merson type and a time step not too large like $h = 0.1$, we obtain results (Table 6.1) of excellent quality compared to Euler's method using the same step. Given the form of function $y_1(t)$, its values are very close to those of $y_3(t)$ when t becomes large. For that reason, $y_3(t)$ was not reported in the table.

Table 6.1 Euler's and Runge–Kutta–Merson (Runge-K-M) methods: comparison of integration results of system (6.2.37) to the theoretical results (integration step $h = 0.1$)

	y_1			y_2		
t	Theoretical	Euler	Runge-K-M	Theoretical	Euler	Runge
0	2.000	2.000	2.000	1.000	1.000	1.000
1	$0.7439\,10^{1}$	$0.6220\,10^{1}$	$0.7439\,10^{1}$	$0.4979\,10^{-1}$	$0.2825\,10^{-1}$	$0.4979\,10^{-1}$
2	$0.5460\,10^{2}$	$0.3834\,10^{2}$	$0.5460\,10^{2}$	$0.2479\,10^{-2}$	$0.7979\,10^{-3}$	$0.2479\,10^{-2}$
3	$0.4034\,10^{3}$	$0.2374\,10^{3}$	$0.4034\,10^{3}$	$0.1234\,10^{-3}$	$0.2254\,10^{-4}$	$0.1234\,10^{-3}$
4	$0.2981\,10^{4}$	$0.1470\,10^{4}$	$0.2981\,10^{4}$	$0.6144\,10^{-5}$	$0.6367\,10^{-6}$	$0.6145\,10^{-5}$
5	$0.2203\,10^{5}$	$0.9100\,10^{4}$	$0.2203\,10^{5}$	$0.3059\,10^{-6}$	$0.1798\,10^{-7}$	$0.3060\,10^{-6}$
6	$0.1628\,10^{6}$	$0.5635\,10^{5}$	$0.1628\,10^{6}$	$0.1523\,10^{-7}$	$0.5080\,10^{-9}$	$0.1523\,10^{-7}$
8	$0.8886\,10^{7}$	$0.2160\,10^{7}$	$0.8886\,10^{7}$	$0.3775\,10^{-10}$	$0.4054\,10^{-12}$	$0.3776\,10^{-10}$
10	$0.4852\,10^{9}$	$0.8282\,10^{8}$	$0.4851\,10^{9}$	$0.9358\,10^{-13}$	$0.3234\,10^{-15}$	$0.9361\,10^{-13}$
15	$0.1069\,10^{14}$	$0.7537\,10^{12}$	$0.1069\,10^{14}$	$0.2863\,10^{-19}$	$0.5817\,10^{-23}$	$0.2864\,10^{-19}$
20	$0.2354\,10^{18}$	$0.6859\,10^{16}$	$0.2354\,10^{18}$	$0.8757\,10^{-26}$	$0.1046\,10^{-30}$	$0.8762\,10^{-26}$

Thus, using different examples, it will be possible to compare the performances of algorithms designed to solve the same type of problem. The systems of ordinary differential equations can present integration difficulties when time constants of various orders of magnitude are present.

6.2.1.3 Runge–Kutta Methods

Different Runge–Kutta algorithms (Fatunla 1988) exist, all of which only use the evaluations of the first derivative. These algorithms of orders 2, 3, or 4 (or more) require the estimation of $f(x, y)$ at 2, 3, or 4 points (or more) on the interval $x \in [x_i, x_{i+1}]$. They give approximations with precisions equivalent to the Taylor polynomial up to the degrees 2, 3, or 4 (or more).

Indeed, a Runge–Kutta method of order p can be considered as an extension of a Taylor polynomial where the evaluation of the derivatives is replaced by n evaluations of the function inside each integration interval $[x_i, x_{i+1}]$ (Figure 6.2).

The Runge–Kutta procedure appears as a substitution method of the form

$$y_{i+1} = y_i + h\Phi_{RK}(x_i, y_i, h) \tag{6.2.39}$$

where the function Φ_{RK} is a weighted average of the slopes at particular points z_r of the subinterval

$$\{z_r \quad | \quad z_r = x_i + c_r h\,, \quad 0 \le c_r \le 1, \quad r = 1, \dots, s\} \tag{6.2.40}$$

with

$$\Phi_{RK} = \sum_{j=1}^{s} b_j Y_j\,,\ s \ge 1 \quad (s - \text{number of steps of the method}) \tag{6.2.41}$$

The coefficients b_j are related by the constraint

$$\sum_{j=1}^{s} b_j = 1 \tag{6.2.42}$$

The slope Y at point z_r depends on the slopes Y_j ($j = 1, \dots, s$) and is defined by

$$Y_r = f(z_r, y_i + h \sum_{j=1}^{s} a_{rj} Y_j) \tag{6.2.43}$$

where z_r has been previously defined and

$$c_r = \sum_{j=1}^{s} a_{rj}\,, \quad r = 1, \dots, s \tag{6.2.44}$$

Several types of Runge–Kutta methods (Gear 1971b) exist:

- Explicit: $a_{rj} = 0$ for $j \geq r$ (**A** is lower triangular with diagonal elements equal to zero); $s(s+1)/2$ coefficients
- Semi-implicit: $a_{rj} = 0$ for $j > r$ (**A** is lower triangular with diagonal elements nonzero); $s(s+3)/2$ coefficients
- Implicit: $a_{rj} \neq 0$ for at least one $j > r$ (**A** is non-triangular); $s(s+1)$ coefficients

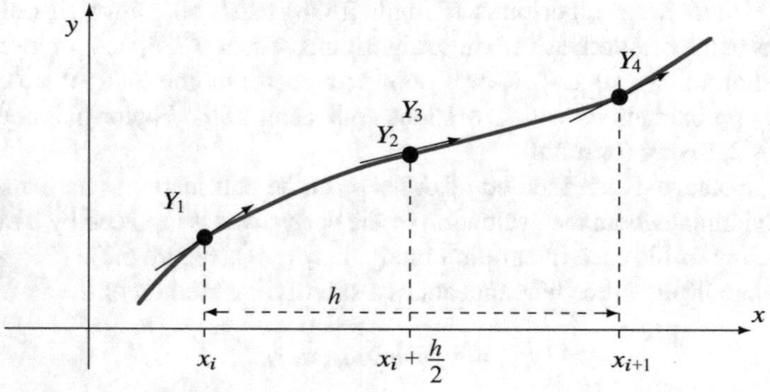

Fig. 6.2 Explicit classic Runge–Kutta method: principle

The unknown coefficients $\{b_j, c_j, a_{rj}\}$ are determined from a set of nonlinear equations according to the following procedure:

Step 1:
Obtain the nth degree Taylor polynomial for Y_r near the point (x_i, y_i) for $r = 1, \ldots, s$.
Step 2:
Insert these expansions and the condition

$$c_r = \sum_{j=1}^{s} a_{rj}\,, \quad r = 1, \ldots, s \tag{6.2.45}$$

in equation

$$\Phi_{RK} = \sum_{j=1}^{s} b_j Y_j\,,\ s \geq 1 \tag{6.2.46}$$

Step 3:
Compare the coefficients of the powers of h both for the function Φ_{RK} equal to

$$\Phi_{RK} = \sum_{j=1}^{s} b_j Y_j\,,\ s \geq 1 \tag{6.2.47}$$

and defined as

$$y_{i+1} = y_i + h\Phi_{RK}(x_i, y_i, h) \tag{6.2.48}$$

and the function Φ_T of the Taylor polynomial of degree p equal to

$$\Phi_T(x_i, y_i, h) = \sum_{r=0}^{p-1} \frac{h^r}{(r+1)!} f^{(r)}(x_i, y_i) \tag{6.2.49}$$

As, in general, the number of unknowns $\{b_j, c_j, a_{rj}\}$ is larger than the number of equations, some coefficients can be chosen with respect to an objective:

- Minimize a limit of the local truncation error.
- Maximize the possible order of the system.
- Optimize the interval of absolute stability.
- Reduce the requirements of memory storage.

Application:
Obtaining the explicit two-step Runge–Kutta method: (Butcher and Tracogna 1997; Hairer and Wanner 1997; Skvortsov 2010)

$$y_{i+1} = y_i + h\Phi_{RK2}(x_i, y_i, h) \tag{6.2.50}$$

The increment function is equal to

$$\Phi_{RK2} = b_1 Y_1 + b_2 Y_2 \tag{6.2.51}$$

with the constraint

$$b_1 + b_2 = 1 \tag{6.2.52}$$

The slopes Y_1 and Y_2 are given by

$$Y_1 = f(x_i, y_i) \quad \text{and} \quad Y_2 = f(x_i + c_2 h, y_i + h a_{21} Y_1) \tag{6.2.53}$$

with the constraint

$$c_2 = a_{21} \tag{6.2.54}$$

The second degree Taylor polynomial near point (x_i, y_i) gives

$$\begin{aligned} Y_2 = {} & f + h\,(c_2\, f_x + a_{21}\, f\, f_y) + \\ & \frac{1}{2} h^2\,(c_2^2\, f_{xx} + 2 c_2\, a_{21}\, f\, f_{xy} + a_{21}^2\, f^2\, f_{yy}) + 0(h^3) \end{aligned} \tag{6.2.55}$$

all the terms being evaluated at (x_i, y_i). By substitution, we get

$$\begin{aligned} \Phi_{RK2} = {} & (b_1 + b_2)\, f + h\,(b_2\, c_2\, f_x + b_2\, a_{21}\, f\, f_y) + \\ & \frac{1}{2}\, h^2\,(b_2\, c_2^2\, f_{xx} + 2\, b_2\, c_2\, a_{21}\, f\, f_{xy} + b_2\, a_{21}^2\, f^2\, f_{yy}) + 0(h^3) \end{aligned} \tag{6.2.56}$$

The equivalent increment function issued from the second degree Taylor polynomial is equal to

$$\begin{aligned} \Phi_{T2} = {} & f + \frac{1}{2}\, h\,(f_x + f\, f_y) + \\ & \frac{1}{6}\, h^2\,(f_{xx} + 2\, f\, f_{xy} + f^2\, f_{yy} + f_x\, f_y + f\, f_y^2) + 0(h^3) \end{aligned} \tag{6.2.57}$$

Both the following equations result

$$b_1 + b_2 = 1 \quad \text{and} \quad b_2\, c_2 = \frac{1}{2} \tag{6.2.58}$$

As there are three unknowns, the system has one degree of freedom. We must take

$$c_2 = a_{21} \tag{6.2.59}$$

otherwise, there is no solution.

The error function of this explicit 2nd order Runge–Kutta scheme is evaluated from the previous developments

$$\begin{aligned}\Psi(x_i, y_i; h) &= \Phi_{RK2} - \Phi_{T2} \\ &= \frac{1}{6}\, h^2\, [(3b_2\, c_2^2\, f_{xx} + 6b_2\, c_2\, a_{21}\, f\, f_{xy} + 3b_2 a_{21}^2\, f^2 f_{yy} \\ &-f_{xx} - 2f\, f_{xy} - f^2\, f_{yy} - f_x\, f_y - f\, f_y^2]_{(x_i, y_i)}\end{aligned} \tag{6.2.60}$$

Taking into account the previous results

$$b_1 + b_2 = 1\,, \quad b_2 c_2 = \frac{1}{2}\,, \quad c_2 = a_{21} \tag{6.2.61}$$

and taking $\alpha = c_2$, the error function becomes

$$\begin{aligned}\Psi(x_i, y_i; h) = \frac{1}{6}\, h^2\, [\, &\alpha(\frac{3}{2} f_{xx} + 3f\, f_{xy} + \frac{3}{2} f^2\, f_{yy}) - \\ &(f_{xx} + 2f\, f_{xy} + f^2\, f_{yy}) - (f_x\, f_y + f\, f_y^2)]_{(x_i, y_i)}\end{aligned} \tag{6.2.62}$$

It is necessary to bound the partial derivatives. Supposing that the differential system satisfies

$$|f(x, y)| \le M\,, \quad |\frac{\partial f}{\partial y}| \le L\,, \quad x \in [a, b] \text{ and } y \text{ finite} \tag{6.2.63}$$

Lotkin supposes the inequality (which is only a theoretical tool)

$$|\frac{\partial^{i+j} f(x, y)}{\partial x^i \partial y^j}| < L^{i+j} M^{1-j} \tag{6.2.64}$$

This type of bound will be used for the explicit second order Runge–Kutta scheme with respect to the partial derivatives

$$|f_{xx}| < ML^2\,, \quad |f_{xy}| < L^2\,, \quad |f_{yy}| < M^{-1}\, L^2 \tag{6.2.65}$$

hence the bound on the error

$$|\Psi(x_i, y_i; h)| < (|\alpha - \frac{2}{3}| + \frac{1}{3})\, h^2\, M\, L^2 \tag{6.2.66}$$

which is minimum for $\alpha = \frac{2}{3}$; hence

$$\alpha = c_2 = a_{21} = \frac{2}{3}\,, \quad b_1 = \frac{1}{4}\,, \quad b_2 = \frac{3}{4} \tag{6.2.67}$$

In summary, the *optimal explicit 2nd order Runge–Kutta scheme* (Heun) is described as

$$\begin{aligned} y_{i+1} &= y_i + \frac{h}{4}(Y_1 + 3Y_2) \\ Y_1 &= f(x_i, y_i) \\ Y_2 &= f(x_i + \tfrac{2}{3}h, y_i + \tfrac{2}{3}hY_1) \end{aligned} \tag{6.2.68}$$

According to the choice of parameter α, other second order Runge–Kutta methods exist. Thus, we can cite the semi-implicit scheme

$$y_{i+1} = y_i + h\, f(x_{i+1/2}, \frac{1}{2}(y_i + y_{i+1})) \tag{6.2.69}$$

Runge–Kutta schemes can be *represented under matrix form* known as Butcher's tableau.

$$\begin{bmatrix} \boldsymbol{A} & \boldsymbol{c} \\ \boldsymbol{b}^T & 0 \end{bmatrix} = \begin{bmatrix} a'_{11} & \dots & a'_{1s} & a'_{1,s+1} \\ \vdots & & & \vdots \\ a'_{s1} & \dots & a'_{ss} & a'_{s,s+1} \\ a'_{s+1,1} & \dots & a'_{s+1,s} & 0 \end{bmatrix} \tag{6.2.70}$$

The coefficients of the resulting matrix $\boldsymbol{A'}$ equal to the matrix $\boldsymbol{A}$ augmented by one row and one column are defined by

$$\begin{aligned} a'_{i,s+1} &= c_r\,, \quad i = 1, \dots, s \\ a'_{s+1,j} &= b_j\,, \quad j = 1, \dots, s \\ a'_{s+1,s+1} &= 0 \\ a'_{ij} &= a_{ij}\,, \quad i, j = 1, \dots, s \end{aligned} \tag{6.2.71}$$

(cf. Eqs. (6.2.39)–(6.2.44)).

Thus, we get:

1-step schemes:

$$\text{Euler's method (1 step):} \quad \begin{bmatrix} 0 & 0 \\ 1 & 0 \end{bmatrix} \tag{6.2.72}$$

2-step schemes:

$$\text{Explicit trapezoidal rule (2 steps):} \quad \begin{bmatrix} 0 & 0 & 0 \\ 1 & 0 & 1 \\ \frac{1}{2} & \frac{1}{2} & 0 \end{bmatrix} \tag{6.2.73}$$

$$\text{Optimal Heun's scheme (2 steps):} \quad \begin{bmatrix} 0 & 0 & 0 \\ \frac{2}{3} & 0 & \frac{2}{3} \\ \frac{1}{4} & \frac{3}{4} & 0 \end{bmatrix} \tag{6.2.74}$$

3-step schemes:

Optimal Ralston's scheme (3 steps): $$\begin{bmatrix} 0 & 0 & 0 & 0 \\ \frac{1}{2} & 0 & 0 & \frac{1}{2} \\ 0 & \frac{3}{4} & 0 & \frac{3}{4} \\ \frac{2}{9} & \frac{1}{3} & \frac{4}{9} & 0 \end{bmatrix} \quad (6.2.75)$$

4-step schemes:

Classic scheme (4 steps): $$\begin{bmatrix} 0 & 0 & 0 & 0 & 0 \\ \frac{1}{2} & 0 & 0 & 0 & \frac{1}{2} \\ 0 & \frac{1}{2} & 0 & 0 & \frac{1}{2} \\ 0 & 0 & 1 & 0 & 1 \\ \frac{1}{6} & \frac{2}{6} & \frac{2}{6} & \frac{1}{6} & 0 \end{bmatrix} \quad (6.2.76)$$

that is

$$\begin{aligned} Y_1 &= f(x, y) \\ Y_2 &= f(x + \tfrac{1}{2}h, y + \tfrac{1}{2}hY_1) \\ Y_3 &= f(x + \tfrac{1}{2}h, y + \tfrac{1}{2}hY_2) \\ Y_4 &= f(x + h, y + hY_3) \\ y_{i+1} &= y_i + h(\tfrac{1}{6}Y_1 + \tfrac{2}{6}Y_2 + \tfrac{2}{6}Y_3 + \tfrac{1}{6}Y_4) \end{aligned} \quad (6.2.77)$$

Kutta's scheme (4 steps): $$\begin{bmatrix} 0 & 0 & 0 & 0 & 0 \\ \frac{1}{3} & 0 & 0 & 0 & \frac{1}{3} \\ -\frac{1}{3} & 1 & 0 & 0 & \frac{2}{3} \\ 1 & -1 & 1 & 0 & 1 \\ \frac{1}{8} & \frac{3}{8} & \frac{3}{8} & \frac{1}{8} & 0 \end{bmatrix} \quad (6.2.78)$$

Gill's scheme (4 steps): $$\begin{bmatrix} 0 & 0 & 0 & 0 & 0 \\ \frac{1}{2} & 0 & 0 & 0 & \frac{1}{2} \\ \frac{\sqrt{2}-1}{2} & \frac{2-\sqrt{2}}{2} & 0 & 0 & \frac{1}{2} \\ 0 & -\frac{\sqrt{2}}{2} & \frac{2+\sqrt{2}}{2} & 0 & 1 \\ \frac{1}{6} & \frac{2-\sqrt{2}}{6} & \frac{2+\sqrt{2}}{6} & \frac{1}{6} & 0 \end{bmatrix} \quad (6.2.79)$$

5-step schemes:

Kutta–Merson scheme (5 steps): $$\begin{bmatrix} 0 & 0 & 0 & 0 & 0 & 0 \\ \frac{1}{3} & 0 & 0 & 0 & 0 & \frac{1}{3} \\ \frac{1}{6} & \frac{1}{6} & 0 & 0 & 0 & \frac{1}{3} \\ \frac{1}{8} & 0 & \frac{3}{8} & 0 & 0 & \frac{1}{2} \\ \frac{1}{2} & 0 & -\frac{3}{2} & 2 & 0 & 1 \\ \frac{1}{6} & 0 & 0 & \frac{2}{3} & \frac{1}{6} & 0 \end{bmatrix} \quad (6.2.80)$$

Thus, Runge–Kutta–Merson method is expressed as

$$\begin{aligned} Y_1 &= f(x, y) \\ Y_2 &= f(x + \tfrac{1}{3}h, y + \tfrac{1}{3}hY_1) \\ Y_3 &= f(x + \tfrac{1}{3}h, y + \tfrac{1}{6}hY_1 + \tfrac{1}{6}hY_2) \\ Y_4 &= f(x + \tfrac{1}{2}h, y + \tfrac{1}{8}hY_1 + \tfrac{3}{8}hY_3) \\ Y_5 &= f(x + h, y + \tfrac{1}{2}hY_1 - \tfrac{3}{2}hY_3 + 2hY_4) \\ y_{i+1} &= y_i + h(\tfrac{1}{6}Y_1 + \tfrac{2}{3}Y_4 + \tfrac{1}{6}Y_5) \end{aligned} \tag{6.2.81}$$

Example 6.2 :

Classic Runge–Kutta method

Consider the following ordinary differential equation

$$y'' = -4\,y \quad \text{with} \quad y(0) = 1 \quad \text{and} \quad y'(0) = 0 \tag{6.2.82}$$

Setting

$$\begin{aligned} u_1 &= y \\ u_2 &= y' \end{aligned} \tag{6.2.83}$$

this equation is transformed into a system of two first order ordinary differential equations that are solved by the classic Runge–Kutta method

$$\begin{aligned} u_1' &= u_2 \quad \text{with} \quad u_1(0) = 1 \\ u_2' &= -4\,u_1 \quad \text{with} \quad u_2(0) = 0 \end{aligned} \tag{6.2.84}$$

As the solution is known $y(t) = \cos(2t)$, the result of the numerical integration can be compared to the theoretical solution. The results together with the details along integration are given in Table 6.2.

Table 6.2 Classic Runge–Kutta method: detail of derivatives and variables during the integration of system (6.2.84) (integration step $h = 0.1$)

t	Y_1	Y_2	Y_3	Y_4	u	u Theoretical
0	$\begin{bmatrix} 0 \\ -4 \end{bmatrix}$				$\begin{bmatrix} 1 \\ 0 \end{bmatrix}$	$\begin{bmatrix} 1 \\ 0 \end{bmatrix}$
0.05		$\begin{bmatrix} -0.2 \\ -4 \end{bmatrix}$				
0.05			$\begin{bmatrix} -0.2 \\ -3.96 \end{bmatrix}$			
0.1				$\begin{bmatrix} -0.396 \\ -3.92 \end{bmatrix}$	$\begin{bmatrix} 0.980066 \\ -0.397333 \end{bmatrix}$	$\begin{bmatrix} 0.980066 \\ -0.397334 \end{bmatrix}$
0.1	$\begin{bmatrix} -0.3973 \\ -3.9202 \end{bmatrix}$					
0.15		$\begin{bmatrix} -0.5933 \\ -3.8408 \end{bmatrix}$				
0.15			$\begin{bmatrix} -0.5893 \\ -3.8015 \end{bmatrix}$			
0.2				$\begin{bmatrix} -0.7774 \\ -3.6845 \end{bmatrix}$	$\begin{bmatrix} 0.92106 \\ -0.77883 \end{bmatrix}$	$\begin{bmatrix} 0.92106 \\ -0.77883 \end{bmatrix}$

6.2.1.4 Semi-Implicit and Implicit Runge–Kutta Schemes

With respect to explicit Runge–Kutta schemes, the *implicit Runge–Kutta schemes* possess strong stability properties. However, they require solving systems of nonlinear equations to calculate the slopes Y_i.

The implicit 2-step Runge–Kutta can be obtained with their coefficients determined from the following system of 4 nonlinear equations with 6 unknowns

$$\begin{aligned}
&b_1 + b_2 &&= 1 \\
&b_1 c_1 + b_2 c_2 &&= \frac{1}{2} \\
&b_1 c_1^2 + b_2 c_2^2 &&= \frac{1}{3} \\
&b_1(a_{11}\, c_1 + c_2(c_1 - a_{11})) + b_2(c_1(c_2 - a_{22}) + a_{22} c_2) &&= \frac{1}{6}
\end{aligned} \tag{6.2.85}$$

Thus, this system has two degrees of freedom that will be chosen with respect to the desired objective. From the three first equations, the relation comes

$$c_2 = \frac{-\frac{1}{2} c_1 + \frac{1}{3}}{-c_1 + \frac{1}{2}} \tag{6.2.86}$$

The value $c_1 = 0$ gives the 2-step implicit Runge–Kutta scheme of order $p = 3$

$$\begin{bmatrix} -1 & 1 & 0 \\ \frac{2}{3} & 0 & \frac{2}{3} \\ \frac{1}{4} & \frac{3}{4} & 0 \end{bmatrix} \tag{6.2.87}$$

that corresponds (refer to Eqs. (6.2.39)–(6.2.44)) to the following equations:

$$\begin{aligned}
Y_1 &= f(x_i, y_i - hY_1 + hY_2) \\
Y_2 &= f(x_i + \tfrac{2}{3}h, y_i + \tfrac{2}{3}hY_1) \\
y_{i+1} &= y_i + h(\tfrac{1}{4}Y_1 + \tfrac{3}{4}Y_2)
\end{aligned} \tag{6.2.88}$$

In general, an s-step implicit scheme has $s(s+1)$ distinct coefficients.

Among implicit Runge–Kutta schemes, let us cite the following schemes:

$$\text{Backward Euler } (s = 1,\ p = 1): \quad \begin{bmatrix} 1 & 1 \\ 1 & 0 \end{bmatrix} \tag{6.2.89}$$

that is

$$\begin{aligned}
Y_1 &= f(x_i + h, y_i + hY_1) \\
y_{i+1} &= y_i + hY_1
\end{aligned} \tag{6.2.90}$$

$$\text{Butcher } (s = 3,\, p = 4): \quad \begin{bmatrix} \frac{5}{36} & \frac{2}{9} - \frac{\sqrt{15}}{15} & \frac{5}{36} - \frac{\sqrt{15}}{30} & \frac{1}{2} - \frac{\sqrt{15}}{10} \\ \frac{5}{36} + \frac{\sqrt{15}}{24} & \frac{2}{9} & \frac{5}{36} - \frac{\sqrt{15}}{14} & \frac{1}{2} \\ \frac{5}{36} + \frac{\sqrt{15}}{30} & \frac{2}{9} + \frac{\sqrt{15}}{15} & \frac{5}{36} & \frac{1}{2} + \frac{\sqrt{15}}{10} \\ \frac{5}{18} & \frac{4}{9} & \frac{5}{18} & 0 \end{bmatrix} \tag{6.2.91}$$

The disadvantage of implicit schemes lies in the important calculation effort necessary to solving the system of nonlinear equations to find the slopes Y_r.

The semi-implicit schemes are decoupled into a system of s uncoupled simultaneous nonlinear equations.

If the matrix $\boldsymbol{A}$ of coefficients (corresponding to the above matrices minus the last row and the last column) is lower triangular, the calculation is much simplified, and we are brought back to *semi-implicit schemes*

$$Y_r = f(x_i + c_r h, y_i + h \sum_{j=1}^{s} a_{rj} Y_j), \quad r = 1, \ldots, s \tag{6.2.92}$$

with respect to the $(r-1)$th equation, where all Y_j are assumed to be known, and the rth equation uses only Y_r that can thus be determined.

Semi-implicit 2-step Runge–Kutta schemes are reduced to the following system of 4 equations with 5 unknowns:

$$\begin{aligned} & b_1 + b_2 && = 1 \\ & b_1 c_1 + b_2 c_2 && = \frac{1}{2} \\ & b_1 c_1^2 + b_2 c_2^2 && = \frac{1}{3} \\ & b_1 c_1^2 + b_2 (c_1 (c_2 - a_{22}) + b_2 a_{22} c_2) && = \frac{1}{6} \end{aligned} \tag{6.2.93}$$

by using the relations

$$c_1 = a_{11}\,, \quad a_{21} + a_{22} = c_2 \tag{6.2.94}$$

Among semi-implicit schemes, let us cite

$$\text{Semi-implicit trapezoidal rule } (s = 2,\, p = 2): \quad \begin{bmatrix} 1 & 0 & 1 \\ 0 & 1 & 1 \\ \frac{1}{2} & \frac{1}{2} & 0 \end{bmatrix} \tag{6.2.95}$$

that is

$$\begin{aligned} Y_1 &= f(x_i + h, y_i + h Y_1) \\ Y_2 &= f(x_i + h, y_i + h Y_2) \\ y_{i+1} &= y_i + h(\tfrac{1}{2} Y_1 + \tfrac{1}{2} Y_2) \end{aligned} \tag{6.2.96}$$

$$\text{Rosenbrock } (s = 2, p = 3): \begin{bmatrix} 1 + \frac{\sqrt{6}}{6} & 0 & 1 + \frac{\sqrt{6}}{6} \\ 0 & 1 - \frac{\sqrt{6}}{6} & 1 - \frac{\sqrt{6}}{6} \\ -0.41315432 & 1.41315432 & 0 \end{bmatrix} \tag{6.2.97}$$

$$\text{Norsett } (s = 2): \begin{bmatrix} \frac{3+\sqrt{3}}{6} & 0 & \frac{3+\sqrt{3}}{6} \\ -\frac{\sqrt{3}}{3} & \frac{3+\sqrt{3}}{6} & \frac{3-\sqrt{3}}{6} \\ \frac{1}{2} & \frac{1}{2} & 0 \end{bmatrix} \tag{6.2.98}$$

6.2.1.5 Variable-Step Runge–Kutta–Fehlberg Method

The control of the integration step is fundamental when a system of ordinary differential equations is integrated. Through variable-step Runge–Kutta–Fehlberg method (Hairer et al. 1993), we try to explain how this control is automatically done. The method is qualified as adaptive as the number and position of the nodes are adapted in the approximation to ensure that the truncation error remains inside a given limit.

An ideal integration method should make use of a minimum number of points so that the global error does not exceed a threshold ϵ. Although the global error cannot be determined, a link exists between the local truncation error and the global error.

Suppose that two integration methods of different orders are available. The first Taylor polynomial of degree n is of the form

$$y(x_{i+1}) = y(x_i) + h\Phi_T(x_i, y(x_i), h) + 0(h^{n+1}) \tag{6.2.99}$$

which gives as approximation

$$y_{i+1} = y_i + h\Phi_T(x_i, y_i, h) \tag{6.2.100}$$

with the local truncation error $\tau_{i+1} = 0(h^n)$.

The second method is supposed to be of higher order. For example, it could be a Runge–Kutta method. Suppose that it is a Taylor polynomial of degree $n + 1$ of the form

$$y(x_{i+1}) = y(x_i) + h\tilde{\Phi}_T(x_i, y(x_i), h) + 0(h^{n+2}) \tag{6.2.101}$$

which gives as approximation

$$\tilde{y}_{i+1} = \tilde{y}_i + h\tilde{\Phi}_T(x_i, \tilde{y}_i, h) \tag{6.2.102}$$

with the local truncation error $\tilde{\tau}_{i+1} = 0(h^{n+1})$.

Suppose that the initial point of local integration is the same, $y_i \approx \tilde{y}_i \approx y(x_i)$, and choose a fixed step h that gives the approximations y_{i+1} and $\tilde{y}_{i+1}$ of $y(x_{i+1})$.

Under these conditions, we get

$$\begin{aligned}\tau_{i+1}(h) &= \frac{y(x_{i+1}) - y(x_i)}{h} - \Phi_T(x_i, y(x_i), h) \\ &= \frac{y(x_{i+1}) - y_i}{h} - \Phi_T(x_i, y(x_i), h) \\ &= \frac{y(x_{i+1}) - [y_i + h\,\Phi_T(x_i, y(x_i), h)]}{h} \\ &= \frac{1}{h}[y(x_{i+1}) - y_{i+1}]\end{aligned} \tag{6.2.103}$$

Similarly

$$\tilde{\tau}_{i+1}(h) = \frac{1}{h}[y(x_{i+1}) - \tilde{y}_{i+1}] \tag{6.2.104}$$

It results

$$\begin{aligned}\tau_{i+1}(h) &= \frac{1}{h}[y(x_{i+1}) - \tilde{y}_{i+1} + \tilde{y}_{i+1} - y_{i+1}] \\ &= \tilde{\tau}_{i+1}(h) + \frac{1}{h}[\tilde{y}_{i+1} - y_{i+1}]\end{aligned} \tag{6.2.105}$$

As we know that the orders of the local errors $\tau_{i+1}(h)$ and $\tilde{\tau}_{i+1}(h)$ are n and $n+1$, respectively, the main contribution to the most important error, that is $\tau_{i+1}(h)$, then comes from $[\tilde{y}_{i+1} - y_{i+1}]/h$, which can be written as

$$\tau_{i+1}(h) \approx \frac{\tilde{y}_{i+1} - y_{i+1}}{h} \tag{6.2.106}$$

On another side, the local error can be formulated as

$$\tau_{i+1}(h) \approx K h^n \tag{6.2.107}$$

where K is a constant independent of h. If a different step αh was applied, the error would be

$$\tau_{i+1}(\alpha h) \approx K\,(\alpha h)^n = K\,\alpha^n\,h^n = \alpha^n\,\tau_{i+1}(h) \approx \frac{\alpha^n}{h}[\tilde{y}_{i+1} - y_{i+1}] \tag{6.2.108}$$

It is desired that the error with this different step αh is bounded by ϵ. This gives q such that

$$\tau_{i+1}(\alpha h) = \frac{\alpha^n}{h}|\tilde{y}_{i+1} - y_{i+1}| \le \epsilon \quad \Longrightarrow \quad \alpha \le \left(\frac{\epsilon\,h}{|\tilde{y}_{i+1} - y_{i+1}|}\right)^{\frac{1}{n}} \tag{6.2.109}$$

The Runge–Kutta–Fehlberg method is based on this type of technique. It is a Runge–Kutta method of order 4 with a local truncation of order 5. It is defined by the following matrix:

$$\text{Runge–Kutta–Fehlberg (6 steps)}: \begin{bmatrix} 0 & 0 & 0 & 0 & 0 & 0 & 0 \\ \frac{1}{4} & 0 & 0 & 0 & 0 & 0 & \frac{1}{4} \\ \frac{3}{32} & \frac{9}{32} & 0 & 0 & 0 & 0 & \frac{3}{8} \\ \frac{1932}{2197} & -\frac{7200}{2197} & \frac{7296}{2197} & 0 & 0 & 0 & \frac{12}{13} \\ \frac{439}{216} & -8 & \frac{3680}{513} & -\frac{845}{4104} & 0 & 0 & 1 \\ -\frac{8}{27} & 2 & -\frac{3544}{2565} & \frac{1859}{4104} & -\frac{11}{40} & 0 & \frac{1}{2} \\ \frac{25}{216} & 0 & \frac{1408}{2565} & \frac{2197}{4104} & -\frac{1}{5} & 0 & 0 \end{bmatrix} \tag{6.2.110}$$

which is expressed by the calculation of the derivatives as

$$\begin{aligned} Y_1 &= f(x, y) \\ Y_2 &= f(x + \tfrac{1}{4}h, y + \tfrac{1}{4}hY_1) \\ Y_3 &= f(x + \tfrac{3}{8}h, y + \tfrac{3}{32}hY_1 + \tfrac{9}{32}hY_2) \\ Y_4 &= f(x + \tfrac{12}{13}h, y + \tfrac{1932}{2197}hY_1 - \tfrac{7200}{2197}hY_2 + \tfrac{7296}{2197}hY_3) \\ Y_5 &= f(x + h, y + \tfrac{439}{216}hY_1 - 8hY_2 + \tfrac{3680}{513}hY_3 - \tfrac{845}{4104}hY_4) \\ Y_6 &= f(x + \tfrac{1}{2}h, y - \tfrac{8}{27}hY_1 + 2hY_2 - \tfrac{3544}{2565}hY_3 + \tfrac{1859}{4104}hY_4 - \tfrac{11}{40}hY_5) \end{aligned} \tag{6.2.111}$$

and the following calculation of functions

$$y_{i+1} = y_i + h(\frac{25}{216}Y_1 + \frac{1408}{2565}Y_3 + \frac{2197}{4104}Y_4 - \frac{1}{5}Y_5) \tag{6.2.112}$$

Thus, at each step, the Runge–Kutta–Fehlberg method requires 6 calculations of derivatives.

The control of the error is done by using a Runge–Kutta method having a local truncation error of order 5 given by

$$\tilde{y}_{i+1} - y_i = h(\frac{16}{135}Y_1 + \frac{6656}{12825}Y_3 + \frac{28561}{56430}Y_4 - \frac{9}{50}Y_5 + \frac{2}{55}Y_6) \tag{6.2.113}$$

to estimate the local error in a Runge–Kutta method of order 4 (see Equation (6.2.112)) given by

$$y_{i+1} - y_i = h(\frac{25}{216}Y_1 + \frac{1408}{2565}Y_3 + \frac{2197}{4104}Y_4 - \frac{1}{5}Y_5) \tag{6.2.114}$$

which gives

$$\tilde{y}_{i+1} - y_{i+1} = h(\frac{1}{360}Y_1 - \frac{128}{4275}Y_3 - \frac{2197}{75240}Y_4 + \frac{1}{50}Y_5 + \frac{2}{55}Y_6) \tag{6.2.115}$$

The value of the parameter α is chosen in a conservative way for the usual method with $n = 4$

$$\alpha = \left(\frac{\epsilon\, h}{2\, |\tilde{y}_{i+1} - y_{i+1}|} \right)^{\frac{1}{4}} \tag{6.2.116}$$

where the difference $\tilde{y}_{i+1} - y_{i+1}$ is obtained by difference of Eqs. (6.2.113) and (6.2.114). On another side, a test is done to avoid too small values of the step in some regions presenting important variations of the derivatives and also to avoid using a too large step

$$\begin{array}{ll} \text{if } \alpha \leq 0.1 & \text{then fix } h = 0.1\, h \\ \text{if } \alpha \geq 4 & \text{then fix } h = 4\, h \\ \text{otherwise} & \text{fix } h = \alpha\, h \end{array} \tag{6.2.117}$$

At initial time, the chosen step is the maximum step. A minimum bound of the step is also fixed to avoid too small steps and too long calculations.

6.2.2 Multi-Step Methods

One-step methods integrate the ordinary differential equation

$$y' = f(x, y) \tag{6.2.118}$$

with the initial condition $y_0 = y(x_0)$. In the case of Euler's method, we proceed by successive integration in order to obtain evaluations of the form

$$\begin{aligned} y_1 &= y_0 + \int_{x_0}^{x_1} f_0 dx \\ &\vdots \\ y_{i+1} &= y_i + \int_{x_i}^{x_{i+1}} f_i dx \end{aligned} \tag{6.2.119}$$

so that, for an integration occurring on $k + 1$ intervals, we can write

$$y_{i+1} = y_{i-k} + \int_{x_{i-k}}^{x_{i+1}} \psi_i(x) dx \tag{6.2.120}$$

$\psi_i(x)$ being a step function determined by its ordinates $f_j = f(x_j, y_j)$ on the half-open intervals $[x_j, x_{j+1}[\,,\quad j = i - k, \ldots, i$. If the function $\psi_i(x)$ were replaced by an interpolation polynomial, it can be imagined that the result would be better. This is the base of *multi-step methods* that can be described by

$$y_{i+1} = y_{i-k} + \int_{x_{i-k}}^{x_{i+1}} \phi(x) dx \tag{6.2.121}$$

with $\phi(x) = \sum_{j=0}^{r} a_j x^j$. These methods are not suitable for problems where the solutions contain singularities. Thus, when solving physical problems, frequently, conditions are posed, which lead to a change of model, thus to a discontinuity of the derivatives at a point and the local non-applicability of a multi-step integration formula. Then, the method must be initialized again at that point.

6.2.3 Open Integration Formulas

The x_j being regularly spaced, the previous values $f_i, f_{i-1}, \dots, f_{i-r} (f_i = f(x_i, y_i))$ being known, the interpolation polynomial (Figure 6.3) can be chosen as Newton's interpolation polynomial based on backward differences (defined by $\nabla f_i = f(x_i, y_i) - f(x_{i-1}, y_{i-1})$)

$$\phi(x_i + \alpha h) = f_i + \alpha \nabla f_i + \dots + \alpha(\alpha+1)\dots(\alpha+r-1)\frac{\nabla^r f_i}{r!} \tag{6.2.122}$$

with $\alpha = (x - x_i)/h$. This formula is explicit.

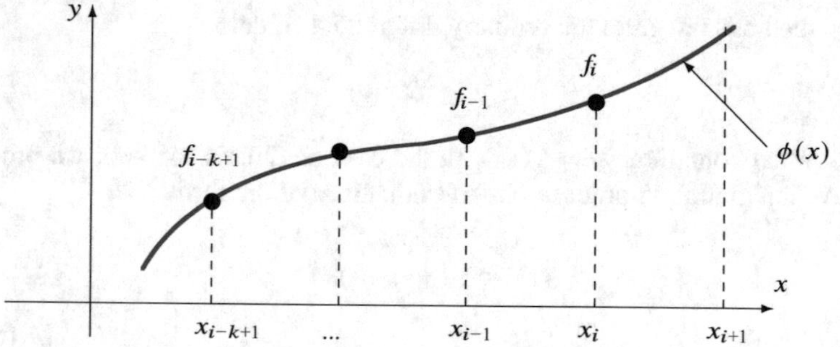

Fig. 6.3 Open multi-step method: principle

The integral term of Equation (6.2.121) is equal to

$$\int_{x_{i-k}}^{x_{i+1}} \phi_i(x)dx = h\left[\alpha f_i + \frac{\alpha^2}{2}\nabla f_i + \alpha^2\left(\frac{\alpha}{3} + \frac{1}{2}\right)\frac{\nabla^2 f_i}{2!} + \dots\right] \tag{6.2.123}$$

Thus, we find

$$k = 0: \quad y_{i+1} = y_i + h\left(f_i + \frac{1}{2}\nabla f_i + \frac{5}{12}\nabla^2 f_i + \frac{3}{8}\nabla^3 f_i + \frac{251}{720}\nabla^4 f_i + \dots\right) \tag{6.2.124}$$

$$k = 1: \quad y_{i+1} = y_{i-1} + h\left(2f_i + 0\nabla f_i + \frac{1}{3}\nabla^2 f_i + \frac{1}{3}\nabla^3 f_i + \frac{29}{90}\nabla^4 f_i + \dots\right) \tag{6.2.125}$$

$$k = 2: \quad y_{i+1} = y_{i-2} + h\left(3f_i - \frac{3}{2}\nabla f_i + \frac{3}{4}\nabla^2 f_i + \frac{3}{8}\nabla^3 f_i + \frac{27}{80}\nabla^4 f_i + \dots\right) \tag{6.2.126}$$

$$k = 3: \quad y_{i+1} = y_{i-3} + h\left(4f_i - 4\nabla f_i + \frac{8}{3}\nabla^2 f_i + 0\nabla^3 f_i + \frac{14}{45}\nabla^4 f_i + \dots\right) \tag{6.2.127}$$

When k is even, the kth coefficient of the kth backward difference is zero. For that reason, the most frequently used formulas are for even k by keeping only $k+1$ terms. In this case, the degree r of the interpolation polynomial is thus apparently equal to k, but as the $(k+1)$th is zero, its degree is indeed $r-1$.

The most important formulas are the following (R, error term):

$$\begin{cases} k = 0,\ r = 3: \\ y_{i+1} = y_i + h(f_i + \frac{1}{2}\nabla f_i + \frac{5}{12}\nabla^2 f_i + \frac{3}{8}\nabla^3 f_i), \qquad R = \frac{251}{720} h^5 f^{(4)}(\xi) \end{cases} \tag{6.2.128}$$

$$\begin{cases} k = 1,\ r = 1: \\ y_{i+1} = y_{i-1} + h(2f_i + 0\nabla f_i), \qquad R = \frac{1}{3} h^3 f^{(2)}(\xi) \end{cases} \tag{6.2.129}$$

$$\begin{cases} k = 3,\ r = 3: \\ y_{i+1} = y_{i-3} + h(4f_i - 4\nabla f_i + \frac{8}{3}\nabla^2 f_i + 0\nabla^3 f_i), \qquad R = \frac{14}{45} h^5 f^{(4)}(\xi) \end{cases} \tag{6.2.130}$$

$$\begin{cases} k = 5,\ r = 5: \\ y_{i+1} = y_{i-5} + h(6f_i - 12\nabla f_i + 15\nabla^2 f_i - 9\nabla^3 f_i \\ \quad + \frac{33}{10}\nabla^4 f_i + 0\nabla^5 f_i), \quad R = \frac{41}{140} h^7 f^{(6)}(\xi) \end{cases} \tag{6.2.131}$$

Now, if the open formulas are expressed with respect to the derivatives and no more the backward differences, we get

$$\begin{cases} k = 0,\ r = 3,\ \text{(Adams–Bashforth formula)}: \\ y_{i+1} = y_i + \frac{h}{24}(55f_i - 59f_{i-1} + 37f_{i-2} - 9f_{i-3}), \quad R = 0(h^5) \end{cases} \tag{6.2.132}$$

$$\begin{cases} k = 1,\ r = 1: \\ y_{i+1} = y_{i-1} + 2hf_i\,, \quad R = 0(h^3) \end{cases} \tag{6.2.133}$$

$$\begin{cases} k = 3,\ r = 3: \\ y_{i+1} = y_{i-3} + \frac{4h}{3}(2f_i - f_{i-1} + 2f_{i-2})\,, \quad R = 0(h^5) \end{cases} \tag{6.2.134}$$

$$\begin{cases} k = 5,\ r = 5: \\ y_{i+1} = y_{i-5} + \frac{3h}{10}(11f_i - 14f_{i-1} + 26f_{i-2} - 14f_{i-3} + 11f_{i-4})\,, \quad R = 0(h^7) \end{cases} \tag{6.2.135}$$

These formulas make use of an interpolation polynomial that passes through the known points $(x_j, f_j), (j = i, i-1, \ldots, i-r)$. As the integration covers all the interval $[x_{i-k}, x_{i+1}]$, indeed the polynomial extrapolates on the interval $[x_i, x_{i+1}]$.

The open integration formulas are directly usable as long as the initial values are known. For example, if the formula (6.2.134) is used, for $i = 1, 2, 3$, it will be necessary to use a different scheme such as the explicit fourth order Runge–Kutta scheme that gives an error of the same order. The values y_1, y_2, y_3 thus calculated serve as an initialization to Equation (6.2.134) that can be used since $i = 4$. The formula (6.2.134) is explicit.

Example 6.3 :
Construction of Adams–Bashforth 3-step method

This method is written under the general form as

$$y_{i+1} = y_i + a\,h\,f_i + b\,h\,f_{i-1} + c\,h\,f_{i-2} \tag{6.2.136}$$

and the problem is to find the coefficients a, b, c.

As the method is built by referring to the first degree Taylor polynomial, it can be written under the general form

$$y(x_{i+1}) = y(x_i) + a\,h\,f(x_i, y(x_i)) + b\,h\,f(x_{i-1}, y(x_{i-1})) + c\,h\,f(x_{i-2}, y(x_{i-2})) \tag{6.2.137}$$

adapted to a later identification.

First, the third degree Taylor polynomial for $y(x_{i+1})$ near x_i is considered, that is

$$\begin{aligned} y(x_{i+1}) &= y(x_i) + h\,y'(x_i) + \frac{h^2}{2} y^{(2)}(x_i) + \frac{h^3}{3!} y^{(3)}(x_i) + 0(h^4) \\ &= y(x_i) + h\,f + \frac{h^2}{2}\left[f_x + f f_y\right] + \frac{h^3}{6}\left[f_{xx} + f_x f_y + 2 f f_{xy} + f f_y^2 + f^2 f_{yy}\right] + 0(h^4) \end{aligned} \tag{6.2.138}$$

where, in the interest of abbreviation, all terms $f, f_x, f_y, \ldots$ are evaluated at $(x_i, y(x_i))$.

Then, the second degree Taylor polynomial for $f(x_{i-1}, y(x_{i-1}))$ near x_i is written, that is,

$$\begin{aligned} f(x_{i-1}, y(x_{i-1})) &= f(x_i, y(x_i)) - h\left[f_x + f f_y\right] + \\ &\quad \frac{h^2}{2}\left[f_{xx} + f_x f_y + 2 f f_{xy} + f f_y^2 + f^2 f_{yy}\right] + 0(h^3) \\ \text{that is} \quad y'(x_{i-1}) &= y'(x_i) - h y^{(2)}(x_i) + \frac{h^2}{2} y^{(3)}(x_i) + 0(h^3) \end{aligned} \tag{6.2.139}$$

and the second degree Taylor polynomial for $f(x_{i-2}, y(x_{i-2}))$ near x_i, that is,

$$\begin{aligned} f(x_{i-2}, y(x_{i-2})) &= f(x_i, y(x_i)) - 2h\left[f_x + f f_y\right] + \\ &\quad \frac{(2h)^2}{2}\left[f_{xx} + f_x f_y + 2 f f_{xy} + f f_y^2 + f^2 f_{yy}\right] + 0(h^3) \\ \text{that is} \quad y'(x_{i-2}) &= y'(x_i) - 2h y^{(2)}(x_i) + \frac{(2h)^2}{2} y^{(3)}(x_i) + 0(h^3) \end{aligned} \tag{6.2.140}$$

The new form of Equation (6.2.137) then comes

$$\begin{aligned} y(x_{i+1}) &= y(x_i) + a\,h\,f + \\ &\quad b\,h\left\{f - h\left[f_x + f f_y\right] + \frac{h^2}{2}\left[f_{xx} + f_x f_y + 2 f f_{xy} + f f_y^2 + f^2 f_{yy}\right]\right\} + \\ &\quad c\,h\left\{f - 2h\left[f_x + f f_y\right] + 2h^2\left[f_{xx} + f_x f_y + 2 f f_{xy} + f f_y^2 + f^2 f_{yy}\right]\right\} + 0(h^3) \\ \text{that is} \quad & y(x_{i+1}) = y(x_i) + (a + b + c)\,h y'(x_i) + (-b - 2c)\,h^2 y^{(2)}(x_i) + \\ &\quad (\frac{b}{2} + 2c)\,h^3 y^{(3)}(x_i) + 0(h^3) \end{aligned} \tag{6.2.141}$$

that is identified with respect to the powers of h with Equation (6.2.138); hence, the system

$$\begin{aligned} a + b + c &= 1 \\ -b - 2c &= \frac{1}{2} \\ \frac{b}{2} + 2c &= \frac{1}{6} \end{aligned} \tag{6.2.142}$$

which gives the coefficients $a = 23/12$, $b = -4/3$, $c = 5/12$; hence, the relation

$$y(x_{i+1}) = y(x_i) + h\left[\frac{23}{12} f(x_i, y(x_i)) - \frac{4}{3} f(x_{i-1}, y(x_{i-1})) + \frac{5}{12} f(x_{i-2}, y(x_{i-2}))\right] \tag{6.2.143}$$

The algorithm of 3-step Adams–Bashforth method results

$$y_{i+1} = y_i + h\left[\frac{23}{12}f_i - \frac{4}{3}f_{i-1} + \frac{5}{12}f_{i-2}\right] \quad (6.2.144)$$

6.2.3.1 Closed Integration Formulas

The closed integration formulas are formed in the same manner

$$y_{i+1} = y_{i-k} + \int_{x_{i-k}}^{x_{i+1}} \phi(x)dx \quad (6.2.145)$$

but the interpolation polynomial ϕ, in addition to the previous known points $(x_j, f_j), (j = i, i-1, \ldots, i-r)$, passes through the unknown point (x_{i+1}, f_{i+1}) (Figure 6.4). Such a formula is thus implicit.

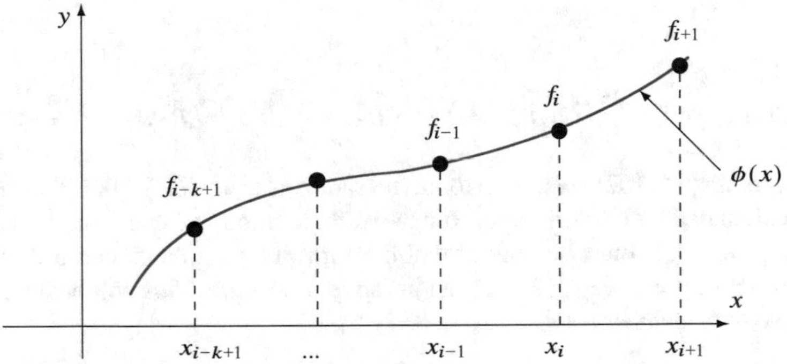

Fig. 6.4 Closed multi-step method: principle

The transformation $x = x_i + \alpha h$ is still used, which gives

$$y_{i+1} = y_{i-k} + h\int_{-k}^{1} \phi(x_i + \alpha h)d\alpha \quad (6.2.146)$$

By using the backward difference at x_{i+1}, we get

$$y_{i+1} = y_{i-k} + h\int_{-k}^{1}\left[f_{i+1} + (\alpha - 1)\nabla f_{i+1} + \frac{(\alpha-1)\alpha}{2!}\nabla^2 f_{i+1} + \ldots\right] \quad (6.2.147)$$

The three most important closed integration formulas are

$$\begin{cases} k = 0, r = 3 : \\ y_{i+1} = y_i + h(f_{i+1} - \frac{1}{2}\nabla f_{i+1} - \frac{1}{12}\nabla^2 f_{i+1} - \frac{1}{24}\nabla^3 f_{i+1}), \quad R = -\frac{19}{720}h^5 f^{(4)}(\xi) \end{cases} \quad (6.2.148)$$

$$\begin{cases} k = 1,\ r = 3: \\ y_{i+1} = y_{i-1} + h(2f_{i+1} - 2\nabla f_{i+1} + \frac{1}{3}\nabla^2 f_{i+1} + 0\nabla^3 f_{i+1}), \quad R = -\frac{1}{90}h^5 f^{(4)}(\xi) \end{cases} \tag{6.2.149}$$

$$\begin{cases} k = 3,\ r = 5: \\ y_{i+1} = y_{i-3} + h(4f_{i+1} - 8\nabla f_{i+1} + \frac{20}{3}\nabla^2 f_{i+1} \\ \quad -\frac{8}{3}\nabla^3 f_{i+1} + \frac{14}{45}\nabla^4 f_{i+1}), \quad R = -\frac{8}{945}h^7 f^{(6)}(\xi) \end{cases} \tag{6.2.150}$$

Expressed with respect to the derivatives, the previous formulas become

$$\begin{cases} k = 0,\ r = 3, \text{ (Adams–Moulton formula)}: \\ y_{i+1} = y_i + \frac{h}{24}(9f_{i+1} + 19f_i - 5f_{i-1} + f_{i-2}), \quad R = 0(h^5) \end{cases} \tag{6.2.151}$$

$$\begin{cases} k = 1,\ r = 3, \text{ (Milne's formula)}: \\ y_{i+1} = y_{i-1} + \frac{h}{3}(f_{i+1} + 4f_i + f_{i-1}), \quad R = 0(h^5) \end{cases} \tag{6.2.152}$$

$$\begin{cases} k = 3,\ r = 5: \\ y_{i+1} = y_{i-3} + \frac{2h}{45}(7f_{i+1} + 32f_i + 12f_{i-1} + 32f_{i-2} + 7f_{i-3}), \quad R = 0(h^7) \end{cases} \tag{6.2.153}$$

The formula (6.2.152) gives an error of the same order as (6.2.134). If it is initialized with y_1 calculated by a fourth order Runge–Kutta method, we could think that we are calculating y_2, but it must be noticed that the formula is implicit, as it uses f_2 that is unknown. Thus, it is necessary to use an iterative procedure. This will be the objective of predictor–corrector methods.

6.2.3.2 Predictor–Corrector Methods

Before using the predictor–corrector algorithm, it is necessary to start on a given number of steps with a one-step method, such as a fourth order Runge–Kutta method, to calculate the past variables y_{n+j} or the derivatives y'_{n+j} used by the predictor–corrector method that calculates y_{n+k} ($k > j$, Equation (6.2.154)). Thus, we start on a given number of steps with a Runge–Kutta method, and then, as soon as possible, the predictor–corrector method is used. It must be noticed that, for a physical problem, it is also necessary to make a similar initialization each time a change of model occurs that influences the derivatives. This often happens due to typical conditions such as "If ... Then" present in a computing code.

The implementation of an implicit linear multi-step method includes 3 stages:

- **Prediction** of the initial value of the variable y_{n+k} noted $y_{n+k}^{[0]}$

$$y_{n+k}^{[0]} = \sum_{j=0}^{k-1}(-\alpha_{1j}y_{n+j} + \beta_{1j}hy'_{n+j}) \tag{6.2.154}$$

- **Evaluation** of the derivative at $(x_{n+k}, y^{[0]}_{n+k})$, that is,

$$f^{[0]}_{n+k} = f(x_{n+k}, y^{[0]}_{n+k}) \tag{6.2.155}$$

- **Correction**

$$y^{[1]}_{n+k} = \sum_{j=0}^{k-1}(-\alpha_{2j} y_{n+j} + \beta_{2j} h y'_{n+j}) + \beta_{2k} h f^{[0]}_{n+k} \tag{6.2.156}$$

Frequently, the stages of evaluation (6.2.155) and correction (6.2.156) are repeated until satisfaction of a predefined convergence criterion.

Let us cite a few predictor–corrector algorithms,

Milne's method of order 4 (nonstable):

Predictor:

$$y_{i+1} = y_{i-3} + \frac{4h}{3}(2f_i - f_{i-1} + 2f_{i-2}), \quad R = 0(h^5) \tag{6.2.157}$$

Corrector:

$$y_{i+1} = y_{i-1} + \frac{h}{3}(f_{i+1} + 4f_i + f_{i-1}), \quad R = 0(h^5) \tag{6.2.158}$$

Milne's method of order 6:

Predictor:

$$y_{i+1} = y_{i-5} + \frac{3h}{10}(11f_i - 14f_{i-1} + 26f_{i-2} - 14f_{i-3} + 11f_{i-4}), \quad R = 0(h^7) \tag{6.2.159}$$

Corrector:

$$y_{i+1} = y_{i-3} + \frac{2h}{45}(7f_{i+1} + 32f_i + 12f_{i-1} + 32f_{i-2} + 7f_{i-3}), \quad R = 0(h^7) \tag{6.2.160}$$

Adams–Bashforth–Moulton method :

Predictor:

$$y_{i+1} = y_i + \frac{h}{24}(55f_i - 59f_{i-1} + 37f_{i-2} - 9f_{i-3}), \quad R = 0(h^5) \tag{6.2.161}$$

Corrector:

$$y_{i+1} = y_i + \frac{h}{24}(9f_{i+1} + 19f_i - 5f_{i-1} + f_{i-2}), \quad R = 0(h^5) \tag{6.2.162}$$

The error constants of the correctors are in general lower than those of the predictors.

To use a predictor–corrector method, we can proceed in the following way, for example in the case of Adams–Bashforth–Moulton method:

- Calculate $y^{[0]}_{i+1}$ by using the predictor

$$y^{[0]}_{i+1} = y_i + \frac{h}{24}(55f_i - 59f_{i-1} + 37f_{i-2} - 9f_{i-3}) \tag{6.2.163}$$

- Calculate $f(x_{i+1}, y_{i+1}^{[0]})$, and apply the corrector iteratively ($j = 0, 1, \dots$) by using the following equation:

$$y_{i+1}^{[j+1]} = y_i + \frac{h}{24}(9f(x_{i+1}, y_{i+1}^{[j]}) + 19f_i - 5f_{i-1} + f_{i-2}) \tag{6.2.164}$$

until a convergence criterion is verified, for example

$$\left|\frac{y_{i+1}^{[j+1]} - y_{i+1}^{[j]}}{y_{i+1}^{[j+1]}}\right| \leq \epsilon \tag{6.2.165}$$

If the number of iterations is too large, it is then necessary to decrease the integration step h.

While the previous methods need to make iterations to ensure the convergence, Hamming's method does not require that iteration. Moreover, it is stable opposite to Milne's method.

Hamming's method:

Predictor:

$$p_{i+1} = y_{i-3} + \frac{4h}{3}(2f_i - f_{i-1} + 2f_{i-2}) \tag{6.2.166}$$

Modifier:

$$m_{i+1} = p_{i+1} - \frac{112}{121}(p_i - c_i) \quad \text{with} \quad m'_{i+1} = f(x_{i+1}, m_{i+1}) \tag{6.2.167}$$

Corrector:

$$c_{i+1} = \frac{1}{8}\left[9y_i - y_{i-2} + 3h(m'_{i+1} + 2f_i - f_{i-1})\right] \tag{6.2.168}$$

Final value:

$$y_{i+1} = c_{i+1} + \frac{9}{121}(p_{i+1} - c_{i+1}) \tag{6.2.169}$$

6.2.3.3 Backward Differentiation Methods (BDF)

In predictor–corrector methods, the calculation of y_{i+1} makes use of the values of derivatives $f(x, y)$ at the past points. The backward differentiation methods are based on the direct use of the ordinates y_j of the past points. Suppose that a polynomial $P(x)$ passes through a set of $(k + 1)$ points going from $i - k + 1$ to $i + 1$, that is

$$P(x_j) = y_j \qquad \text{for} \quad j = i - k + 1, \dots, i + 1 \tag{6.2.170}$$

according to Figure 6.5. The backward differentiation method consists of determining y_{i+1} such that

$$P'(x_{i+1}) = f(x_{i+1}, P(x_{i+1})) \tag{6.2.171}$$

The same Newton interpolation polynomial (Equation 6.2.122) as previously is used to calculate $P'(x_{i+1})$. If the points are regularly spaced, the following equation results

$$\sum_{j=1}^{k} \nabla^j y_{i+1} = h\, f_{i+1} \tag{6.2.172}$$

which thus yields

$$\begin{aligned}
i = 1: &\quad y_{i+1} - y_i = h\, f_{i+1} \\
i = 2: &\quad \frac{3}{2}\, y_{i+1} - 2\, y_i + \frac{1}{2}\, y_{i-1} = h\, f_{i+1} \\
i = 3: &\quad \frac{11}{6}\, y_{i+1} - 3\, y_i + \frac{3}{2}\, y_{i-1} - \frac{1}{3}\, y_{i-2} = h\, f_{i+1} \\
i = 4: &\quad \frac{25}{12}\, y_{i+1} - 4\, y_i + 3\, y_{i-1} - \frac{4}{3}\, y_{i-2} - \frac{1}{4}\, y_{i-3} = h\, f_{i+1}
\end{aligned} \tag{6.2.173}$$

The backward differentiation method is only stable when $k \leq 6$ (Figure 6.5). These backward differentiation formulas are used in particular in the case of systems of stiff ordinary differential equations.

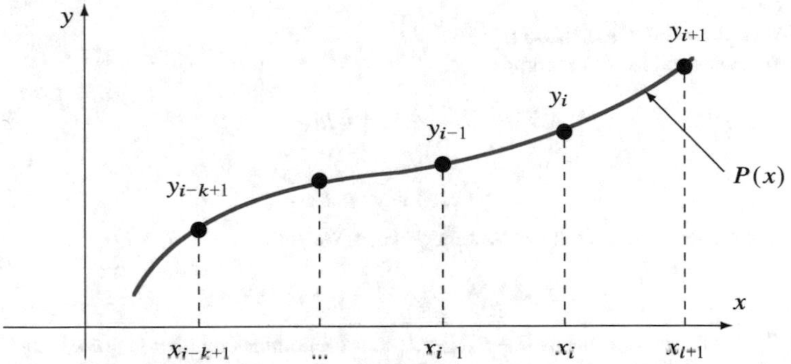

Fig. 6.5 Backward differentiation method: principle

6.3 Stability of Numerical Integration Methods

Approximately, it can be said that a method is unstable when the errors issued from the calculations increase exponentially along the calculation. Thus, the predictor–corrector methods can be either unstable whatever the step, or stable for a domain of step values.

To judge the stability of predictor–corrector methods, the following equation is used

$$y' = \lambda\, y \tag{6.3.1}$$

for which the solution is an exponential function of the form

$$y(t) = y_0 \exp(\lambda_R + i\, \lambda_I)t \tag{6.3.2}$$

supposing that λ is complex. We desire that the modulus of the solution decreases exponentially, that is $|y_{j+1}| < |y_j|$. The *time constant* of the first order Equation (6.3.1) is the reciprocal of the absolute value of λ_R. The larger the time constant, the more rapidly the solution varies, increasing if $\lambda_R > 0$ (assuming $y_0 > 0$), and decreasing if $\lambda_R < 0$.

We form the homogeneous difference equation for which we then search the roots $\alpha_i (i = 1, \ldots, k)$. The general solution of the homogeneous difference equation is then

$$y_n = c_1\,\alpha_1^n + \cdots + c_k\,\alpha_k^n \tag{6.3.3}$$

One of the solutions, for example α_1^n, tends to the exact solution of Equation (6.3.1) when $h \to 0$, while all the other solutions are not natural. A multi-step method is called *strongly stable* if all the nonnatural roots satisfy the condition

$$\text{When } h \to 0 \quad |\alpha_i| < 1\,, \quad i = 2, \ldots, k \tag{6.3.4}$$

Example 6.4 :
Stability of explicit Euler's method

Consider an explicit Euler's method

$$y_{i+1} = y_i + h\,f_i \tag{6.3.5}$$

Set

$$f(x, y) = y' = \lambda\,y \tag{6.3.6}$$

where λ is a complex number in the most general case. We get

$$y_{i+1} = y_i + h\,\lambda\,y_i = (1 + h\,\lambda)y_i \tag{6.3.7}$$

For the method to be stable, it needs $|1 + h\,\lambda| < 1$; thus the complex number $(1 + h\,\lambda)$ must be strictly in a disk of center (0, 0) and unit radius, or equivalently, the complex number $(h\,\lambda)$ must be in a disk of center $(-1, 0)$ and unit radius.

If λ is real, it must be negative and such that $-2 < h\,\lambda < 0$, thus $h < -2/\lambda$.

If λ is complex, $\lambda = \lambda_R + i\,\lambda_I$. The constraint thus deals only with the real part λ_R, which must verify $-1 < 1 + h\,\lambda_R < -1$, thus $h < -2/\lambda_R$.

An illustration of the stability problem has been done in the case of the ordinary differential equation

$$y'(x) = \lambda\,y \qquad \text{with } \lambda = -2 \tag{6.3.8}$$

The stability criterion imposes $h < 1$. Two steps have been tested $h = 0.8$ and $h = 1.2$ (Figure 6.6). The right figure clearly shows the instability, the left figure shows the stability, but also that the chosen step is not satisfactory, being too large. A step equal to 0.1 gives an apparently satisfying result with respect to the stability considered alone (Figure 6.7), but, in reality, the numerical error remains very important and is related to the explicit Euler's method that gives a modest precision.

Example 6.5 :
Stability of Milne's method of order 4

Consider Milne's method of order 4. The equation of the corrector is

$$y_{i+1} = y_{i-1} + \frac{h}{3}(f_{i+1} + 4 f_i + f_{i-1}) \tag{6.3.9}$$

Set

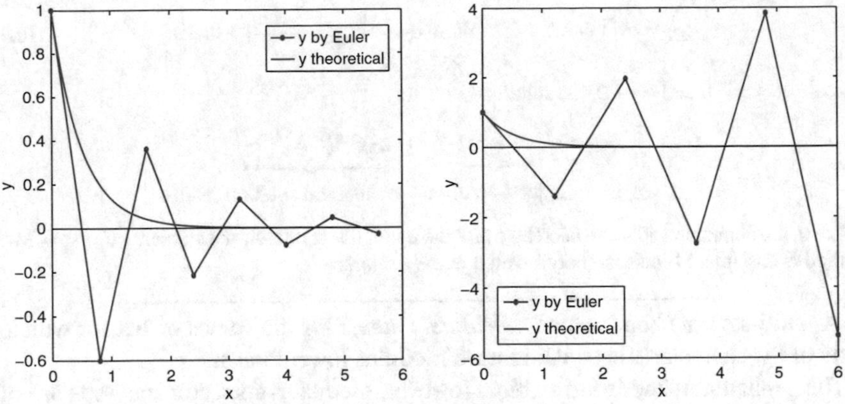

Fig. 6.6 Stability illustration with Euler's method. Integration step = 0.8 (left) and 1.2 (right)

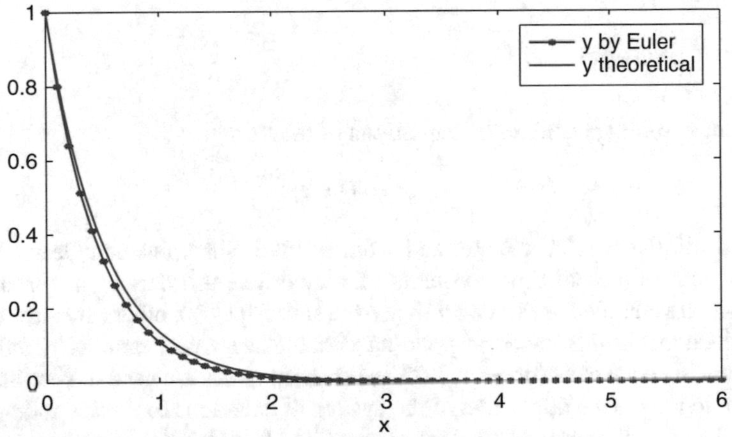

Fig. 6.7 Integration by Euler's method with a step equal to 0.1

$$f(x, y) = y' = \lambda y \tag{6.3.10}$$

The equation becomes

$$y_{i+1} - y_{i-1} - \frac{h\lambda}{3}(y_{i+1} + 4y_i + y_{i-1}) = 0 \tag{6.3.11}$$

which gives the following characteristic equation:

$$\alpha^2 - 1 - \frac{h\lambda}{3}(\alpha^2 + 4\alpha + 1) = 0 \tag{6.3.12}$$

When $h \rightarrow 0$, the roots are $\alpha = 1$ and $\alpha = -1$; thus Milne's method is not strongly stable. For h small, the roots are

$$\begin{aligned} \alpha_1 &= 1 + \lambda h + 0(h^2) \\ \alpha_2 &= -1 + \frac{\lambda h}{3} + 0(h^2) \end{aligned} \tag{6.3.13}$$

hence, the general solution

$$y_n = c_1[1 + \lambda h + 0(h^2)]^n + c_2(-1)^n \left[[1 - \frac{\lambda h}{3}] + 0(h^2)\right]^n \tag{6.3.14}$$

If we set $n = x_n/h$ and $h \to 0$, the solution tends to

$$y_n = \underbrace{c_1\, e^{\lambda x_n}}_{\text{natural solution}} + \underbrace{c_2\,(-1)^n\, e^{-\lambda x_n/3}}_{\text{nonnatural solution}} \tag{6.3.15}$$

If $\lambda > 0$, the nonnatural solution decreases and the method is stable. At the opposite, if $\lambda < 0$, Milne's method is unstable. Milne's method is called *weakly stable*.

A multi-step method is called *absolutely stable* for the values of $h\,\lambda$ for which the roots of the characteristic equation have modulus lower than 1.

Runge–Kutta methods are stable. However, predictor–corrector methods are often more accurate and allow a good estimation of the error.

6.4 Stiff Systems

A system of ordinary differential equations of the form

$$\dot{\boldsymbol{x}} = \boldsymbol{f}(\boldsymbol{x}, t) \tag{6.4.1}$$

is called stiff (Gear 1971b; Hairer and Wanner 1996; Shampine and Gear 1979) when these equations present time constants of very different orders of magnitude. Some equations are qualified as fast (we will use the subscript "f"), others are slow (subscript "s"), which means that the corresponding variables vary very rapidly (noted $\boldsymbol{x}_f$) or at the opposite very slowly (noted $\boldsymbol{x}_s$). These characteristics are used in simulation when we want to reduce the dimension of the system of equations to extract an approximate model. The fast differential equations are replaced by algebraic equations by setting to zero the differential term of the left-hand member, and the only remaining differential equations are the slow equations, yielding the approximate model

$$\begin{cases} 0 = \boldsymbol{f}_f(\boldsymbol{x}, t) \\ \dot{\boldsymbol{x}}_s = \boldsymbol{f}_s(\boldsymbol{x}, t) \end{cases} \tag{6.4.2}$$

Many systems of ordinary differential equations representing chemical kinetics are stiff. If the dimension of the system is not reduced, the integration is tricky and requires to use very efficient integration methods.

Example 6.6:
Integration of stiff systems

The code "lsoda" proposes the following example to illustrate the problem of different time scales:

$$\begin{array}{lll} \dot{y}_1 = -0.04\, y_1 + 10^4\, y_2\, y_3 & , & y_1(0) = 1 \\ \dot{y}_2 = 0.04\, y_1 - 10^4\, y2\, y3 - 3\, 10^7\, y_2^2 & , & y_2(0) = 0 \\ \dot{y}_3 = 3\, 10^7\, y_2^2 & , & y_3(0) = 0 \end{array} \tag{6.4.3}$$

This system must be integrated on the interval [0, 4 10^{10}]. To obtain excellent results, it is much better to use efficient subroutines such as "lsoda" that gives the results of Table 6.3. This code partly lies on Gear's method (Hindmarsh 1983), which is a predictor–corrector method of high order and variable step. Note that the precision of the computer must be specified in the calculation, and that in the case of a bad specification, the results may become abnormal in particular when t increases.

Table 6.3 Stiff system: results of the integration of system (6.4.3) by an improved Gear's method with the variable-step "lsoda" subroutine

t	y_1	y_2	y_3
4.0000×10^{-1}	y = 9.851712×10^{-1}	3.386380×10^{-5}	1.479493×10^{-2}
4.0000	y = 9.055333×10^{-1}	2.240655×10^{-5}	9.444430×10^{-2}
4.0000 10^{1}	y = 7.158403×10^{-1}	9.186334×10^{-6}	2.841505×10^{-1}
4.0000×10^{2}	y = 4.505250×10^{-1}	3.222964×10^{-6}	5.494717×10^{-1}
4.0000×10^{3}	y = 1.831975×10^{-1}	8.941774×10^{-7}	8.168016×10^{-1}
4.0000×10^{4}	y = 3.898730×10^{-2}	1.621940×10^{-7}	9.610125×10^{-1}
4.0000×10^{5}	y = 4.936363×10^{-3}	1.984221×10^{-8}	9.950636×10^{-1}
4.0000×10^{6}	y = 5.161831×10^{-4}	2.065786×10^{-9}	9.994838×10^{-1}
4.0000×10^{7}	y = 5.179817×10^{-5}	2.072032×10^{-10}	9.999482×10^{-1}
4.0000×10^{8}	y = 5.283401×10^{-6}	2.113371×10^{-11}	9.999947×10^{-1}
4.0000×10^{9}	y = 4.659031×10^{-7}	1.863613×10^{-12}	9.999995×10^{-1}
4.0000×10^{10}	y = 1.404280×10^{-8}	5.617126×10^{-14}	1.000000

6.5 Differential–Algebraic Systems

In the previous sections, we were concerned with purely differential systems. These later can present integration difficulties, in particular, those related to time constants of very different orders of magnitude.

Another type of difficulty is the presence of algebraic equations that must be solved simultaneously with the ordinary differential equations (Brenan et al. 1989). Many physical problems lead to differential–algebraic systems such as modelling of real processes, calculus of variations, optimal control, theory of singular perturbations, and discretization of partial differential equations by the method of lines. Typically, a transient process with steady thermodynamic equations will be modelled as a differential–algebraic system.

Some apparently insignificant systems can pose tricky issues. Consider the following differential–algebraic system:

$$\begin{aligned} y_1'(t) &= y_1(t) + y_2(t) + y_3(t) \\ y_2'(t) &= y_1(t) - y_2(t) - y_3(t) \\ y_3(t) &= y_1(t) + 2\,y_2(t) \end{aligned} \tag{6.5.1}$$

where y_1 and y_2 are the differential variables, whereas y_3 is a variable algebraic. The solution of system (6.5.1) is

$$\begin{aligned} y_1(t) &= (y_1(0) + 0.6\, y_2(0)) \exp(2t) - 0.6\, y_2(0) \exp(-3t) \\ y_2(t) &= y_2(0) \exp(-3t) \\ y_3(t) &= y_1(t) + 2\, y_2(t) \end{aligned} \tag{6.5.2}$$

Due to the simultaneous presence of positive and negative exponential (which results in rapidly increasing and decreasing exponentials), a good quality solution on a time interval of some amplitude (for example [0,20]) is difficult to obtain.

What should absolutely not be done is first solving the differential system, then the algebraic problem. To efficiently solve differential–algebraic systems (refer to DAEs), let us cite the subroutines called "ddassl" or more recently "ddasspk" largely used in the domain of chemical engineering. "ddassl" lies on a Gear's predictor–corrector method of order 5 (Gear 1971a; Gear and Petzold 1984; Petzold 1982a,b). To explain how the algebraic problem is simultaneously solved, a simple reasoning will be presented below.

A differential–algebraic system is presented under the following form (Brenan et al. 1989):

$$\begin{aligned} F(y, y', z, t) &= 0 \\ G(y, z, t) &= 0 \end{aligned} \tag{6.5.3}$$

where the system $F(y, y', z, t) = 0$ is formed by first order implicit differential equations and the system $G(y, z, t) = 0$ represents algebraic constraints. In general, the Jacobian $\partial F / \partial y'$ is nonsingular.

It can be presented under a slightly different form that is explicit with respect to the differential equations

$$\begin{aligned} y' &= f(t, y, z)\,, \quad y(t^0) = y^0 \\ 0 &= g(t, y, z) \end{aligned} \tag{6.5.4}$$

where y are called differential variables and z algebraic variables. Often, the system (6.5.4) is called semi-explicit.

The differential–algebraic system can also be written under the form

$$E\, y' = f(y)\,, \quad y(t^0) = y^0 \tag{6.5.5}$$

where the mass matrix E is singular when the system contains purely algebraic equations.

To solve the problem (6.5.4), if the variables at time $n-1$ are assumed to be known, the derivatives can be discretized to get the following algebraic system:

$$\begin{aligned} \frac{y^n - y^{n-1}}{\Delta t} &= f(t^n, y^n, z^n)\,, \quad y(t^0) = y^0 \\ 0 &= g(t^n, y^n, z^n) \end{aligned} \tag{6.5.6}$$

which is implicit and can be solved by a nonlinear programming method such as Newton–Raphson or quasi-Newton. If, in the right-hand members of the equations, the variables at time $n - 1$ instead of n had been used, the resulting explicit system would have been obvious to solve, but would probably have diverged. This presentation of the solving is on purpose extremely simplified, and its only objective is to illustrate the consequence of discretization. Indeed, as previously commented, multi-step methods are used.

The case of differential–algebraic systems (Petzold 1982a) is also more complex as they refer to the notion of index (Gani and Cameron 1992; Gritsis et al. 1995; Unger et al. 1995) related to the notion of tractability of the differential–algebraic system. Consider the semi-explicit differential–algebraic

$$\begin{aligned} y' &= f(t, y, z) \\ 0 &= g(t, y, z) \end{aligned} \tag{6.5.7}$$

The algebraic constraint equation is differentiated with respect to time; hence,

$$\begin{aligned} y' &= f(t, y, z) \\ g_y(t, y, z)\, y' + g_z(t, y, z)\, z' &= -g_t(t, y, z) \end{aligned} \tag{6.5.8}$$

If g_z is nonsingular, the system (6.5.8) is an implicit system of differential equations and (6.5.7) has an index equal to 1. In a simplified manner, the index is defined as the number of times that the differential–algebraic system or a part of it must be differentiated with respect to t to make z' appear as a continuous function of t and z. Then, the system becomes a purely differential system. A system of ordinary differential equations has an index 0. A system of index 1 is easily solved. When the index is larger than 1, the usual calculation codes (Gear and Petzold 1984; Petzold 1982b) frequently meet issues that become nearly insurmontable from a numerical point of view if the index is larger than 2. In this case, the best solution (Bachmann et al. 1990) consists of modifying the model in order to obtain an equivalent model of index lower than or equal to 1, which sometimes implies to introduce additional variables.

6.6 Ordinary Differential Equations with Boundary Conditions

It may occur that ordinary differential equations have conditions at more than one boundary, thus called boundary conditions. Partial differential equations (Chapter 7) present such boundary conditions, but some ordinary differential equations also present these characteristics. A vibrating string obeys the partial differential equation

$$\frac{\partial^2 u}{\partial t^2} - c^2 \frac{\partial^2 u}{\partial x^2} = 0 \tag{6.6.1}$$

where $u(t, x)$ represents the displacement at any point of the string that is fixed at both extremities; hence, the conditions

$$u(t, x = 0) = 0\,, \quad u(t, x = L) = 0 \tag{6.6.2}$$

After spatial discretization, for example by the method of lines (Section 7.7), Equation (6.6.1) becomes a system of ordinary differential equations with boundary conditions.

This problem of boundary conditions systematically occurs in dynamic optimization (Chapter 12) (Corriou 2012; Bryson 1999; Corriou 2018) where a system of ordinary differential equations of the form must be solved

$$\begin{aligned} \dot{\boldsymbol{x}} &= \boldsymbol{f}(\boldsymbol{x}, \boldsymbol{u}, t) \quad \text{with } \boldsymbol{x}(t = 0) = \boldsymbol{x}_0 \\ \dot{\boldsymbol{\psi}} &= \boldsymbol{g}(\psi, \boldsymbol{x}, \boldsymbol{u}, t) \text{ with } \boldsymbol{\psi}(t = t_f) = \boldsymbol{\psi}_f \end{aligned} \tag{6.6.3}$$

where the state $\boldsymbol{x}$ is defined by its initial value $\boldsymbol{x}_0$, while the adjoint variables are defined by their final value $\boldsymbol{\psi}_f$. In the case of dynamic optimization, different techniques are proposed in the literature, some of them leading, after parameterization of the states and the adjoint variables, to a problem of nonlinear algebraic optimization.

In simpler cases, such as a second order ordinary differential equation with two boundaries, it is possible to imagine using a given function that represents the solution and to search the parameters of that function (cf. Example 6.8.6). It is also possible to use a shooting method by making a variable hypothesis at one boundary to satisfy the condition at the other boundary (cf. Example 6.8.4). Other solutions such as the discretization of ordinary differential equations are possible (cf. Example 6.8.11).

In a general manner, several methods adapted to this type of problem are described in Section 7.10 devoted to spectral methods of Chapter 7 and more especially Section 7.10.4. Such methods are detailed in Chapter 12 about dynamic optimization.

6.7 Discussion and Conclusion

The understanding of the integration methods for ordinary differential equations is extremely important to solve a given system, if only to have criteria at our disposal allowing to choose the most adapted method. As often, the system needs to be examined. Is it a stiff system with time constants of different orders of magnitude? Is it a purely differential system or an algebraic–differential one? Are there model changes (condition "If... then") during the integration? In most cases, it is recommended to use an efficient numerical library. Comparing the calculation speed and the results given by different methods for a given problem is always very informative. In easy cases, fourth or fifth order Runge–Kutta methods can be satisfactory, but a variable-step method is always interesting to accelerate the calculation speed. For stiff differential systems, predictor–corrector methods like lsoda, lsode (ODEPACK, in Fortran) (Hindmarsh 1983), or ode15s (for Matlab) are necessary. It must not be forgotten to inform an integrator such as lsoda about events creating a change of model along time. If the system is differential–algebraic, DDASSL, DASPK, DAEPAK (in Fortran), and other recent methods are very efficient. However, we must remain conscious and vigilant with respect to potential index issues. In general, boundary value problems must be examined one by one and solved by different specialized available techniques.

6.8 Exercise Set

Preamble: Most exercises consider the integration on a limited interval to check if the students have understood the details of the integration method. Of course, the same equations can be solved numerically on more realistic longer intervals with dedicated codes. It is an excellent learning process if the students write codes capable of integrating general ordinary differential equations.

Exercise 6.8.1 (Medium)
Integrate the following ordinary differential equation:

$$y'' - 2\,x\,y' - 2\,y - (4\,x^2 + 2)\exp(x^2 + 2) = 0 \tag{6.8.1}$$

with the initial conditions

$$y(0) = \exp(2) \qquad \text{and} \qquad y'(0) = 3\,\exp(2) \tag{6.8.2}$$

by third order optimal Ralston's method. The integration step $h = 0.1$ and the integration will be done on two steps while giving all the intermediate results allowing you to verify the integration.

Exercise 6.8.2 (Medium)
Integrate the following system of ordinary differential equations:

$$\begin{cases} \dot{x}_1 = -3\sqrt{x_1}\,x_2^2 \\ \dot{x}_2 = x_1\,x_2 \end{cases} \tag{6.8.3}$$

with the initial conditions

$$x_1(0) = \exp(2) \qquad \text{and} \qquad x_2(0) = 3\,\exp(2) \tag{6.8.4}$$

on two time steps ($h = 0.1$) by third order Heun's method. Write a subroutine or a function for integrating ordinary differential equations as general as possible, corresponding to Heun's method. Explain how this subroutine or function communicates with the main program, receives or contains information about derivatives.

Exercise 6.8.3 (Difficult)
Under dimensionless form, the ordinary differential equation describing the deflection y, with respect to abscissa x, of a beam fixed at both extremities and subjected at both extremities to a load P directed toward the center of the beam is

$$\frac{d^2y}{dx^2} + \gamma\left[1 + \left(\frac{dy}{dx}\right)^2\right]^{\frac{3}{2}} y = 0 \tag{6.8.5}$$

valid for small curvatures (Boucard et al. 2020; Goodno and Gere 2018). Assuming that the curvature is equal to $\gamma = 10^{-2}\,\pi^2$, starting from $x = 0$ where $y = 0$ and

$dy/dx = 10^{-2}\,\pi$, first apply Heun's method on three steps with a step equal to 0.1 and then continue the integration until $x = 1$ with Adams–Bashforth formula (6.2.132).

Exercise 6.8.4 (Difficult)

Consider the flow of a liquid close to a plane plate (Bird et al. 2002, page 137) in order to describe the boundary layer. This problem has been solved by (Blasius 1908; Boyd 1998; Parand and Taghavi 2009). Far from the plate, the liquid velocity is v_∞. Let ν the kinematic viscosity. The components of the velocity in the directions x and y noted v_x and v_y, respectively, vary in the boundary layer close to the plate (Figure 6.8).

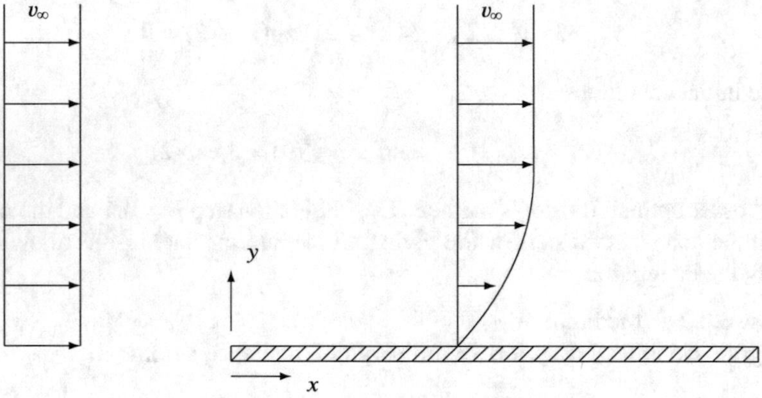

Fig. 6.8 Velocity profiles before the plane plate and above the plane plate

The momentum and continuity equations are

$$\begin{aligned} v_x \frac{\partial v_x}{\partial x} + v_y \frac{\partial v_x}{\partial y} &= \nu \frac{\partial^2 v_x}{\partial y^2} \\ \frac{\partial v_x}{\partial x} + \frac{\partial v_y}{\partial y} &= 0 \end{aligned} \tag{6.8.6}$$

subject to the following boundary conditions:

$$\begin{array}{lll} v_x = v_y = 0 & \text{for } y = 0 & \forall\, x \geq 0 \\ v_x = v_\infty & \text{for } y \to \infty & \forall\, x \geq 0 \\ v_x = v_\infty & \text{for } x = 0 & \forall\, y \geq 0 \end{array} \tag{6.8.7}$$

The stream function $\psi(x, y)$ is defined by

$$v_x = -\frac{\partial \psi}{\partial y} \quad , \quad v_y = \frac{\partial \psi}{\partial x} \tag{6.8.8}$$

Introduce the dimensionless coordinate

$$\Pi(\eta) = \frac{v_x}{v_\infty} \qquad \text{with} \quad \eta = y \sqrt{\frac{v_\infty}{2\nu\, x}} \tag{6.8.9}$$

On another side, the stream function that gives the previous velocity distribution is equal to

$$\psi = -\sqrt{2\nu\, x\, v_\infty}\, f(\eta) \qquad \text{with} \quad f(\eta) = \int_0^\eta \Pi(\zeta)d\zeta \tag{6.8.10}$$

where $f(\eta)$ is a dimensionless stream function.

1. Demonstrate the following partial differential equation:

$$\frac{\partial\psi}{\partial y}\frac{\partial^2\psi}{\partial x\partial y} - \frac{\partial\psi}{\partial x}\frac{\partial^2\psi}{\partial y^2} = -\nu\frac{\partial^3\psi}{\partial y^3} \tag{6.8.11}$$

2. Demonstrate that the dimensionless velocities are given by the following equations:

$$\begin{aligned} V_x &= \frac{v_x}{v_\infty} = f' \\ V_y &= \frac{v_y}{\sqrt{\dfrac{\nu\, v_\infty}{2x}}} = \eta\, f' - f \end{aligned} \tag{6.8.12}$$

3. Demonstrate that Equation (6.8.11) is then reduced to a single ordinary differential equation with respect to the dimensionless stream function $f(\eta)$

$$f\, f^{(2)} + f^{(3)} = 0 \tag{6.8.13}$$

4. Justify the boundary conditions

$$\begin{aligned} &f = 0\,,\ f' = 0 \quad \text{for } \eta = 0 \\ &f' = 1 \quad \text{for } \eta \longrightarrow \infty \end{aligned} \tag{6.8.14}$$

5. Numerical solution of the problem:
First, use third order Runge–Kutta method defined by the following matrix:

$$\begin{bmatrix} 0 & 0 & 0 & 0 \\ \frac{2}{3} & 0 & 0 & \frac{2}{3} \\ 0 & \frac{2}{3} & 0 & \frac{2}{3} \\ \frac{2}{8} & \frac{3}{8} & \frac{3}{8} & 0 \end{bmatrix} \tag{6.8.15}$$

and then, as soon as possible (explain at which moment), use Adams–Bashforth–Moulton predictor.
Remark: as the equation of problem (6.8.13) is subject to boundary conditions at $\eta = 0$ and $\eta = \infty$, this is a two-boundary problem. Equation (6.8.13) can be solved by posing some value of $f^{(2)}$ at $\eta = 0$. In the complete solution of the problem, we would search the optimal value $f^{(2)}$ at $\eta = 0$ to satisfy the boundary condition at $\eta = \infty$.
In the present case, we take $f^{(2)} = 0.1$ at $\eta = 0$. An integration step $h = 0.5$ is chosen, and instead of η_∞, an actual value $\eta_{\max} = 10$ will be taken. Integrate until $\eta = 10$.

In a table, give the results of function f and its derivatives f' and $f^{(2)}$ for all the integration points between 0 and 10.

6. Conclude about the initial choice of the derivative $f^{(2)}$ and explain how to exactly solve this two-boundary problem.

Exercise 6.8.5 (Difficult)

Integrate the following ordinary differential equation:

$$y'' + \frac{\pi^2}{4} y = \frac{\pi^2}{16} \cos(\frac{\pi}{4} x) , \quad 0 \le x \le 1 \tag{6.8.16}$$

with the initial condition $y(0) = 0$, $y'(0) = 1$. The integration step is $h = 0.1$ and the integration interval $[0, 20]$. Heun's optimal method is used for the two first steps, and the following Adams–Bashforth method for two following steps:

$$y_{i+1} = y_i + \frac{h}{12} (23 f_i - 16 f_{i-1} + 5 f_{i-2}) \tag{6.8.17}$$

Exercise 6.8.6 (Difficult)

Given the following ordinary differential equation:

$$y'' + \frac{\pi^2}{4} y = \frac{\pi^2}{16} \cos(\frac{\pi}{4} x) , \quad 0 \le x \le 1 \tag{6.8.18}$$

such that $y(0) = 0$ and $y(1) = 0$, this is a two-boundary problem for which the numerical solution is relatively difficult to determine. We choose as the approximation function a spline function defined by

$$S(x) = M_i \frac{(x_{i+1} - x)^3}{6h_{i+1}} + M_{i+1} \frac{(x - x_i)^3}{6h_{i+1}} + a_i(x - x_i) + b_i , \quad x \in [x_i, x_{i+1}] \tag{6.8.19}$$

with $h_{i+1} = x_{i+1} - x_i$ and the following points x_i, $x_0 = 0$, $x_1 = 0.4$, $x_2 = 0.7$, $x_3 = 1$.

1. Find the equations verified by the coefficients M_i, a_i, and b_i.
2. Explain clearly, precisely, and completely the mathematic problem, how the values of these coefficients would be found numerically. Do not try to determine these numerical values.
3. Explain the fundamental difference between Exercise 6.8.5 and Exercise 6.8.6. How is it possible to modify Exercise 6.8.5 so that its solution coincides with that of Exercise 6.8.6?

Exercise 6.8.7 (Medium)

In natural life, two common competitive phenomena occur. On one side, a species A possesses an abundant reserve of natural food; on another side, a species B totally depends for its food on eating species A. Let N_A and N_B be the populations of A and B at a given time. Many models describe this type of evolution (Gauze 1934; Hassell et al. 1991; Holt 1977; Strogatz 1994; Volterra 1926).

Among different models, the following simplified model describing the evolution of two competing species can be obtained

$$\begin{aligned} \text{Prey equation :} \quad & \frac{dN_A}{dt} = \alpha\, N_A - \beta\, N_A\, N_B \\ \text{Predator equation :} \quad & \frac{dN_B}{dt} = -\gamma\, N_B + \delta\, N_A\, N_B \end{aligned} \tag{6.8.20}$$

This is called a Lotka–Volterra predator–prey model.

The constants $\alpha, \beta, \gamma, \delta$ are positive. What is their signification (think about chemical reactions)?

Consider the following values of these constants $\alpha = 0.2$, $\beta = 0.005$, $\gamma = 0.15$, $\delta = 0.001$. The initial values of N_A and N_B are $N_A(0) = 100$, $N_B(0) = 100$.

The integration step is chosen as $h = 1$ and the integration on interval [0, 100] with the Runge–Kutta method described by the following matrix:

$$\begin{bmatrix} 0 & 0 & 0 & 0 & 0 \\ \frac{1}{3} & 0 & 0 & 0 & \frac{1}{3} \\ -\frac{1}{3} & 1 & 0 & 0 & \frac{2}{3} \\ 1 & -1 & 1 & 0 & 1 \\ \frac{1}{8} & \frac{3}{8} & \frac{3}{8} & \frac{1}{8} & 0 \end{bmatrix} \tag{6.8.21}$$

Write a program and plot $N_A = f(t)$, $N_B = f(t)$, $N_B = f(N_A)$. Comment.

Exercise 6.8.8 (Difficult)

An electric system including resistances, a capacitor, and an inductor is ruled by the two following equations:

$$\begin{aligned} & i_1(t) + 3\,[i_1(t) - i_2(t)] + i_1'(t) = 6 \\ & \int_0^t i_2(x)\,dx + 2\,i_2(t) + 3\,[i_2(t) - i_1(t)] = 0 \end{aligned} \tag{6.8.22}$$

with the following initial conditions $i_1(0) = 0$, $i_2(0) = 0$.

Integrate this system by Heun optimal method on three time steps $h = 0.1$, and then continue with Adams–Bashforth method up to $t = 10$.

Exercise 6.8.9 (Difficult)

Consider the ordinary differential equation

$$y'(t) = f(t, y)\,, \quad a \le t \le b\,, \quad y(a) = y_a \tag{6.8.23}$$

1. Demonstrate that

$$y'(t_i) = \frac{-3\,y(t_i) + 4\,y(t_{i+1}) - y(t_{i+2})}{2h} + \frac{h^2}{3} y^{(3)}(\xi)\,, \quad \xi \in [t_i, t_{i+2}] \tag{6.8.24}$$

for $t_{i+1} = t_i + h$, $i = 0, \ldots, N-2$.

Hint: use a third degree Taylor polynomial for $y(t_{i+1})$ in the neighborhood of t_i and a Taylor polynomial for $y(t_{i+2})$ of same degree in the neighborhood of t_i. Then, combine both expansions to eliminate $y^{(2)}(t_i)$.

2. From property (6.8.24), deduce the difference method

$$y_{i+2} = 4\,y_{i+1} - 3\,y_i - 2\,h\,f(t_i, y_i)\,, \quad \text{for } i = 0, \ldots, N-2 \tag{6.8.25}$$

3. Use method (6.8.25) to solve the ordinary differential equation

$$y'' = 4\,y - 4\,t\,, \quad y(0) = 1\,, \quad y'(0) = 3 \tag{6.8.26}$$

on the interval $[0, 1]$ with the step $h = 0.1$. To start the method and find $y(t_1)$, use explicit Euler's method.

4. Study the stability of the proposed method.

Exercise 6.8.10 (Medium)

Integrate the following ordinary differential equation:

$$y'' - (4x^2 + 2)y + (4x^2 + 1)\exp(-x) = 0 \quad \text{with} \quad y(0) = 3\,\exp(1) + 1\,, \quad y'(0) = -1 \tag{6.8.27}$$

on the interval $[0, 1.5]$ with a step $h = 0.05$, using fourth order Runge–Kutta–Gill scheme.

Exercise 6.8.11 (Difficult)

Consider the following linear ordinary differential equation with two boundaries

$$y''(x) + p(x)\,y' + q(x)\,y + r(x) = 0 \tag{6.8.28}$$

on the domain $x \in [a, b]$ with the fixed boundaries $y(a) = \alpha$ and $y(b) = \beta$. The domain $[a, b]$ is divided into regular intervals of length h so that $x_{i+1} = x_i + h$. Take N intervals, and thus $(N + 1)$ points x_i $(i = 0, \ldots, N)$.

1. To solve that problem, finite differences are used in the same way as for solving partial differential equations (except that, here, the time variable is absent, and only the space variable remains). The second derivative y'' is thus approximated by the finite difference taken for the partial derivative $\dfrac{\partial^2 u}{\partial x^2}$ and the first derivative y' approximated by the finite difference taken for the partial derivative $\dfrac{\partial u}{\partial x}$. Note that these two derivatives must be consistent (prove it) from the point of view of the error with respect to step h. Transform the ordinary differential equation (6.8.28) to get a discretized equation of the form

$$a_i\,y_{i-1} + b_i\,y_i + c_i\,y_{i+1} + d_i = 0 \tag{6.8.29}$$

2. Applying the discretized equation at the different concerned points of domain $[a, b]$, demonstrate that it leads to a particular system of linear equations that will be set under the form $\boldsymbol{A}\,\boldsymbol{Y} = \boldsymbol{B}$ and that you will completely define. Discuss the type of resulting system and explain how to solve it.

3. Solve the following example:

$$y''(x) + \frac{2}{x}\,y' - \frac{2}{x^2}\,y - \frac{\sin(\ln(x))}{x^2} = 0 \tag{6.8.30}$$

with $x \in [1, 2]$, $y(1) = 1$, $y(2) = 2$, $N = 5$.

Exercise 6.8.12 (Difficult)
The oscillations of a pendulum of length L are described by the following ordinary differential equation:

$$\frac{d^2\theta}{dt^2} = -\frac{g}{L}\sin(\theta) \tag{6.8.31}$$

It is assumed that at initial time, the pendulum is released with a zero velocity from the angular position $\theta = \frac{\pi}{6}$. The characteristics are $L = 0.5$m and $g = 9.8$ m.s^{-2}. The time step is $dt = 0.1s$, and the solution is searched for $t \in [0, 4]$s. First, the integration method was the explicit classic Runge–Kutta method that gave the results of Table 6.4. Note that only one initial condition is explicitly given. Define the condition that allowed you to obtain the results of Table 6.4 at different instants. This table is only given to help you and, normally, it should result from your calculations.

Table 6.4 Position of the pendulum at different instants

Time (s)	Angle θ
0	0.52360
0.1	0.47530
0.2	0.33872
0.3	0.13822
0.4	−0.088702
0.5	−0.29860
0.6	−0.45192
0.7	−0.52102

We wish to apply as soon as possible Adams–Bashforth–Moulton predictor–corrector method (the corrector will be applied only once). Use as little as possible the results of Table 6.4 and give the detailed results in a table.

Exercise 6.8.13 (Difficult)
The oscillations of a spring are described by the relation

$$\frac{d^2y}{dt^2} = -k\,y \tag{6.8.32}$$

where $y(t)$ is the displacement of the spring with respect to its equilibrium position, and $k = 2$ is the spring stiffness. At initial time $y(t = 0) = 0.4$. The spring is released with a zero velocity $y'(t = 0) = 0$.

Integrate the equation on the interval [0, 10]s with an integration step equal to 0.1s and use the following multi-step Adams–Bashforth algorithm

$$y_{i+1} = y_i + \frac{h}{12}(23 f_i - 16 f_{i-1} + 5 f_{i-2}) \tag{6.8.33}$$

To start, use the midpoint algorithm a sufficient number of times. The midpoint method is of Runge–Kutta type and can be described by the following matrix:

$$\begin{bmatrix} 0 & 0 & 0 \\ \frac{1}{2} & 0 & \frac{1}{2} \\ 0 & 1 & 0 \end{bmatrix} \tag{6.8.34}$$

1. Detail the algorithm of the midpoint method.
2. Clearly explain the procedure. Write a general program.

Exercise 6.8.14 (Difficult)

Obtaining a multi-step algorithm

The points x_i and x_{i+1} are related by the relation $x_{i+1} - x_i = h$ where h is a constant step.

Let $y(x)$ be the function and $y'(x) = f(x, y(x))$ the ordinary differential equation.

1. Write a third degree Taylor polynomial for $y(x_{i+1})$ in the neighborhood of x_i.
2. Write a second degree Taylor polynomial for $f(x_{i-1}, y(x_{i-1}))$ in the neighborhood of x_i.
3. Write a second degree Taylor polynomial for $f(x_{i-2}, y(x_{i-2}))$ in the neighborhood of x_i.
4. Consider the relation

$$y(x_{i+1}) = y(x_i) + a\,h\,f(x_i, y(x_i)) + b\,h\,f(x_{i-1}, y(x_{i-1})) + c\,h\,f(x_{i-2}, y(x_{i-2})) \tag{6.8.35}$$

   From the previous Taylor polynomials, find the values of the parameters a, b, c, by identifying the coefficients of h, h^2, h^3.
5. Deduce an algorithm of multi-step integration of ordinary differential equations.

Exercise 6.8.15 (Difficult)

Integrate the following ordinary differential equation:

$$y^{(3)} + 2\,y^{(2)} - y' - 2\,y = \exp(t) \tag{6.8.36}$$

with the initial condition $y(0) = 1$, $y'(0) = 2$, $y^{(2)}(0) = 0$. The integration step is $h = 0.2$ and the integration interval [0, 1]. Use the trapezoidal rule on the two first steps and then the following two-step Adams–Bashforth method for the following steps:

$$y_{i+1} = y_i + \frac{h}{2}(3\,f_i - f_{i-1}) \tag{6.8.37}$$

Exercise 6.8.16 (Easy)

Deduce Simpson's method of integration of ordinary differential equations by applying Simpson's rule to the following integral:

$$y(t_{i+1}) - y(t_{i-1}) = \int_{t_{i-1}}^{t_{i+1}} f(t, y(t))\, dt \tag{6.8.38}$$

Exercise 6.8.17 (Easy)
Integrate the following ordinary differential equation:

$$y'' - 2y' = 6 - 12x \tag{6.8.39}$$

by Heun's optimal scheme with the initial condition $y(0) = 1, y'(0) = 2$. The integration step is $h = 0.5$ and the integration interval $[0, 2]$.

Exercise 6.8.18 (Easy)
According to Kirchhoff's law, the current $I(t)$ in a closed circuit containing a resistance of R Ohm, a capacitance of C Farad, an inductance of L Henry, and a voltage source of $E(t)$ Volt verifies the following equation valid for each loop:

$$L\frac{dI(t)}{dt} + RI(t) + \frac{1}{C}\int_0^t I(x)dx = E(t) \tag{6.8.40}$$

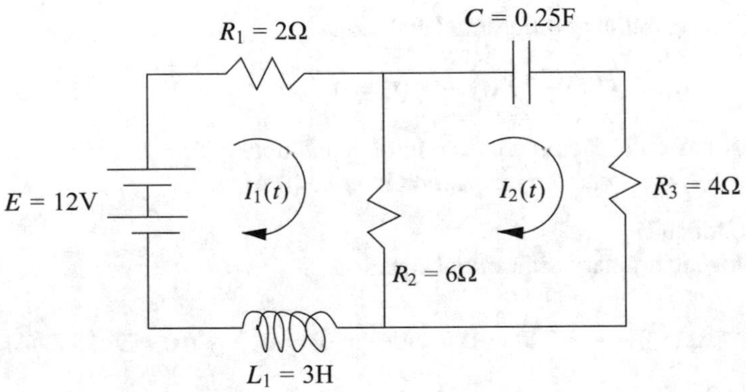

Fig. 6.9 Electrical circuit

Assume that the circuit of Figure 6.9 is closed at time $t = 0$. In this case, the currents I_1 and I_2 are zero at time $t = 0$.

1. Establish the equations for this network.
2. Solve these equations by classic Runge–Kutta method. The integration step is $h = 0.5$ and the integration interval $[0, 10]$.

Exercise 6.8.19 (Medium)

Stability for a fourth order Runge–Kutta method

Consider the usual equation in the integration of ordinary differential equations

$$\frac{dy}{dt} = f(t, y) \tag{6.8.41}$$

Now, consider the following fourth order Runge–Kutta method

$$\begin{aligned} Y_1 &= f(x_i, y_i) \\ Y_2 &= f(x_i + \tfrac{1}{2}h, y_i + \tfrac{1}{2}hY_1) \\ Y_3 &= f(x_i + \tfrac{1}{2}h, y_i + \tfrac{1}{2}hY_2) \\ Y_4 &= f(x_i + h, y_i + hY_3) \\ y_{i+1} &= y_i + \frac{1}{6}\, h(Y_1 + 2\,Y_2 + 2\,Y_3 + Y_4) \end{aligned} \tag{6.8.42}$$

Demonstrate that, when this method is applied to the following ordinary differential equation

$$\frac{dy}{dt} = \lambda\, y \tag{6.8.43}$$

the method can be rewritten under the form

$$y_{i+1} = [1 + h\lambda + \frac{1}{2}(h\lambda)^2 + \frac{1}{6}(h\lambda)^3 + \frac{1}{24}(h\lambda)^4]\, y_i \tag{6.8.44}$$

Exercise 6.8.20 (Easy)

Integrate the following ordinary differential equation:

$$y''(t) - 2y'(t) + 9y(t) = 0 \tag{6.8.45}$$

by optimal Runge–Kutta–Gill scheme with the initial conditions $y(0) = 0$, $y'(0) = 3$. The integration step is $h = 0.1$ and the integration interval $[0, 4]$.

Exercise 6.8.21 (Difficult)

Consider the following ordinary differential equation

$$\frac{1}{6}\, y'' - x\, y' + 4\, x^3\, \exp(x^2) - \frac{1}{3} + 2\, x^2 = 0 \quad \text{with} \quad y(0) = 0\,, \quad y'(0) = 3 \tag{6.8.46}$$

to be integrated on the interval $[0, 1]$ with a step $h = 0.05$.

We want to use a multi-step method, presently 4-point Adams–Bashforth method

$$y_{i+1} = y_i + \frac{h}{24}(55\, f_i - 59\, f_{i-1} + 37\, f_{i-2} - 9\, f_{i-3}) \tag{6.8.47}$$

Start by applying Euler's method on a sufficient number of steps, and then use Adams–Bashforth method as soon as possible.

Exercise 6.8.22 (Easy)

Integrate the following ordinary differential equation:

$$t\,(y''(t) + y(t)) - y'(t) + 3\, t^2\, \sin(t) = 0 \tag{6.8.48}$$

with the initial conditions

$$y(0) = 0\,, \quad y'(0) = 0\,, \quad \lim_{t\to 0} \frac{y'(t)}{t} = 2 \tag{6.8.49}$$

on the interval $[0, 10]$ with the time step $h = 0.1$, using Kutta's scheme.

Exercise 6.8.23 (Medium)
Consider the general ordinary differential equation

$$y'(t) = f(t, y)\,, \quad a \le t \le b\,, \quad y(a) = \alpha \tag{6.8.50}$$

1. Using nth degree Taylor polynomials, demonstrate that we can obtain

$$y'(t_k) = \frac{-3\,y(t_k) + 4\,y(t_{k+1}) - y(t_{k+2})}{2h} + 0(h^2)\,,\ \xi_k \in [t_k, t_{k+2}]\,,\ h = t_{k+1} - t_k\,,\ \forall\, k \tag{6.8.51}$$

2. Deduce the integration algorithm

$$y_{k+2} = 4\,y_{k+1} - 3\,y_k - 2\,h\,f(t_k, y_k)\,, \quad k = 0, 1, \ldots, N-2 \tag{6.8.52}$$

   How do you call this method. Describe its interest.
3. Use the previous algorithm to solve the following ordinary differential equation:

$$y'(t) = 1 - y(t)\,, \quad t \in [0, 0.6]\,, \quad y(0) = 0 \tag{6.8.53}$$

   with the step $h = 0.1$. To initialize the developed method, start with Euler's method and explain when the proposed algorithm can be used.

Exercise 6.8.24 (Difficult)
An electric circuit with a voltage U, having a resistance R, an inductance L, and a capacitance C, is ruled by the following ordinary differential equation that gives the current intensity I

$$\frac{dI}{dt} = C\frac{d^2U}{dt^2} + \frac{1}{R}\frac{dU}{dt} + \frac{1}{L}U \tag{6.8.54}$$

The parameters are $C = 0.2$ Farad, $R = 5\,\Omega$, $L = 2$ Henry. The voltage U is equal to

$$U(t) = \sin(2t - \pi) \tag{6.8.55}$$

The intensity I is zero at time $t = 0$.

1. The step time is $h = 0.1$ and the integration interval $[0, 5]$. Find the solution by using the following three-step Kutta's scheme:

$$\begin{bmatrix} 0 & 0 & 0 & 0 \\ \frac{1}{2} & 0 & 0 & \frac{1}{2} \\ -1 & 2 & 0 & 1 \\ \frac{1}{6} & \frac{2}{3} & \frac{1}{6} & 0 \end{bmatrix} \tag{6.8.56}$$

2. We decide to use Adams–Bashforth multi-step method. To initialize the method, we first use Euler's explicit method and a different step equal to $h = 0.1$. As soon as possible, the multi-step Adams–Bashforth will be used with the step $h = 0.1$ until the final time $t = 5$.
3. Compare the results given by both methods at some selected times.

References

R. Bachmann, L. Brüll, T. Mrziglod, and U. Plaske. On methods for reducing the index of differential algebraic equations. *Comp. Chem. Engng.*, 14(11):1271–1273, 1990.

R. B. Bird, W. E. Stewart, and E. N. Lightfoot. *Transport phenomena.* Wiley, New York, second edition, 2002.

H. Blasius. Grenzschichten in Flüssigkeiten mit kleiner Reibung. *Zeit. Math. Phys.*, 56:1–37, 1908.

P. A. Boucard, J. Lemaitre, and F. Hild. *Résistance mécanique des matériaux et des structures*. Dunod, Paris, second edition, 2020.

J. P. Boyd. The Blasius function in the complex plane. *Experimental Math.*, 8(4): 381–394, 1998.

K. E. Brenan, S. L. Campbell, and L. R. Petzold. *Numerical Solution of Initial-Value Problems in Differential-Algebraic Equations.* North-Holland, 1989.

A. E. Bryson. *Dynamic Optimization*. Addison-Wesley, Menlo Park, California, 1999.

J. C. Butcher and S. Tracogna. Order conditions for two-step Runge-Kutta methods. *Appl. Numer. Math.*, 24(2–3):351–364, 1997.

J. P. Corriou. *Commande des procédés*. Lavoisier, Tec & Doc, Paris, 3rd edition, 2012.

J. P. Corriou. *Process control - Theory and applications*. Springer, London, 2nd edition, 2018.

S. O. Fatunla. *Numerical Methods for Initial Value Problems in Ordinary Differential Equations*. Academic Press, Boston, 1988.

R. Gani and I. T. Cameron. Modelling for dynamic simulation of chemical processes: the index problem. *Chem. Eng. Sci.*, 47(5):1311–1315, 1992.

G. F. Gauze. *The struggle for existence*. The Williams & Wilkins Company, Baltimore, 1934.

C. W. Gear. Simultaneous numerical solution of differential-algebraic equations. *IEEE Transactions on Circuit Theory*, TC-18(4):89–95, 1971a.

C. W. Gear. *Numerical Initial Value Problems in Ordinary Differential Equations*. Prentice-Hall, Englewood Cliffs, N. J., 1971b.

C. W. Gear and L. R. Petzold. ODE methods for the solution of differential/algebraic systems. *SIAM J. Numer. Anal.*, 21(4):716–728, 1984.

B. J. Goodno and J. M. Gere. *Mechanics of Materials*. Cengage Learning, Boston, 9th edition, 2018.

D. M. Gritsis, C. C. Pantelides, and R. W. H. Sargent. Optimal control of systems described by index two differential/algebraic equations. *SIAM J. Sci. Comput.*, 16 (6):1349–1366, 1995.

E. Hairer and G. Wanner. *Solving Ordinary Differential Equations II: Stiff and Differential-Algebraic Problems*. Springer, Berlin, 2nd edition, 1996.

E. Hairer and G. Wanner. Order conditions for general two-step Runge-Kutta methods. *SIAM J. Numer. Anal.*, 34(6):2087–2089, 1997.

E. Hairer, S. Nørsett, and G. Wanner. *Solving Ordinary Differential Equations I: Nonstiff Problems*. Springer, Berlin, 2nd edition, 1993.

M. P. Hassell, H. N. Comins, and R. M. May. Spatial structure and chaos in insect population dynamics. *Nature*, 35:255–258, 1991.

A. C. Hindmarsh. ODEPACK, a systematized collection of ODE solvers. In R. S. Stepleman et al, editor, *Scientific Computing*, pages 55–64. North Holland, Amsterdam, 1983.

R. D. Holt. Predation, apparent competition, and the structure of prey communities. *Theoretical Population Biology*, 12:197–229, 1977.

K. Parand and A. Taghavi. Rational scaled generalized Laguerre function collocation method for solving the Blasius equation. *Journal of Computational and Applied Mathematics*, 233:980–989, 2009.

L. Petzold. Differential/algebraic equations are not ODE's. *SIAM J. Sci. Stat. Comput.*, 3(3):367–384, 1982a.

L. R. Petzold. A description of DASSL: a differential/algebraic system solver. Sand82-8637, Sandia National Laboratories, Livermore, 1982b.

L. F. Shampine and C. W. Gear. A user's view of solving stiff ordinary differential equations. *SIAM Review*, 21(1):1–17, 1979.

L M. Skvortsov. Explicit two-step Runge-Kutta methods. *Mathematical Models and Computer Simulations*, 2:222–231, 2010.

S. Strogatz. *Nonlinear Dynamics and Chaos*. Perseus Books, Reading Mass., 1994.

J. Unger, A. Kröner, and W. Marquardt. Structural analysis of differential-algebraic equation systems - theory and applications. *Comp. Chem. Engng.*, 19(8):867–882, 1995.

V. Volterra. Fluctuations in the abundance of a species considered mathematically. *Nature*, 118:557–559, 1926.

Chapter 7
Numerical Integration of Partial Differential Equations

7.1 Introduction

In many different sectors of scientific activity, to represent systems where time and spatial coordinates intervene simultaneously, scientists develop models based on partial differential equations which may be of very large dimension, for example in meteorology, are often difficult to solve and require advanced numerical techniques. This chapter can only present a first approach for partial differential equations which represent a vast domain covered by general books (Allaire 2007; Ames 1992; Evans et al. 2000; Lapidus and Pinder 1982; LeVeque and Randall 1990; Thomas 1998) in addition to many specialized books and articles which will be cited in the course of the relevant methods. For complex problems such as those of fluid mechanics requiring to solve Navier–Stokes equations, it is recommended to refer to specialized texts. For that reason, the example of Fourier's equation in one-dimensional, two-dimensional, and three-dimensional heat transfer is explained in depth and illustrated. However, examples in domains, such as mass transfer, fluid flow including diffusion or reaction, are also detailed in a few cases. The main presented methods are finite differences, finite volumes, spectral methods, finite elements, boundary elements.

7.2 Some Examples of Physical Systems

The simulation of physical systems described by partial differential equations is often tricky and requires some experience. It must be noted that the model describing the physical process is constituted by a set of partial differential equations, initial conditions, and boundary conditions which are essential for solving the problem. This set of equations constitutes a well-posed problem. In general, when the problem is well-posed, it possesses a unique solution which depends continuously on the variations with respect to the initial and boundary conditions.

J.-P. Corriou, *Numerical Methods and Optimization*, Springer Optimization and Its Applications 187, https://doi.org/10.1007/978-3-030-89366-8_7

7.2.1 Heat Transfer by Conduction

Among such systems, consider the conductive heat transfer in a plane wall supposed to be infinite in height and width, and where the third coordinate (in the thickness direction) is x (Figure 7.1). In the transient case, the conductive heat transfer is described by Fourier's partial differential equation

$$\frac{\partial}{\partial x}\left(\lambda \frac{\partial T}{\partial x}\right) = \rho C_p \frac{\partial T}{\partial t} \tag{7.2.1}$$

with:
$T(x,t)$ temperature
λ thermal conductivity
ρ density
C_p heat capacity
t time
x coordinate in the thickness direction

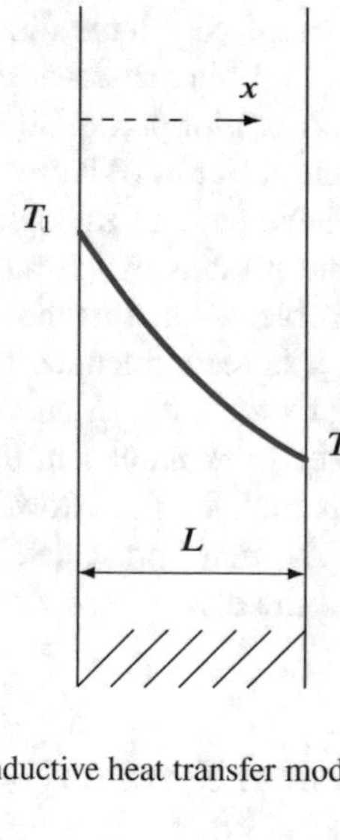

Fig. 7.1 Plane wall subjected to conductive heat transfer modelled in the thickness direction x

The boundary conditions for this one-dimensional problem (Figure 7.1 in the case of a plane wall placed between two isothermal media) can be of different types according to the physical system

- the temperature T is specified at the wall (Dirichlet conditions),
- the thermal flux is specified at the wall, hence the partial derivative of temperature at that point (Neumann condition).

Frequently, the thermal conductivity is assumed to be constant, and Equation (7.2.1) becomes

$$\frac{\partial T}{\partial t} = \alpha \frac{\partial^2 T}{\partial x^2} \tag{7.2.2}$$

using the thermal diffusivity α

$$\alpha = \frac{\lambda}{\rho C_p} \tag{7.2.3}$$

This equation can be transformed by use of the following dimensionless variables (τ is called Fourier's number when t is a characteristic time and L the length through which the conduction takes place)

$$X = \frac{x}{L}, \quad \tau = \frac{\alpha t}{L^2}, \quad \theta = \frac{T - T_{\min}}{T_{\max} - T_{\min}} \tag{7.2.4}$$

and it becomes

$$\frac{\partial \theta}{\partial \tau} = \frac{\partial^2 \theta}{\partial X^2} \tag{7.2.5}$$

7.2.2 Mass Transfer by Diffusion

The mass transfer by diffusion is often described by Fick's law similar to Fourier's law

$$\frac{\partial C}{\partial t} = \mathcal{D} \frac{\partial^2 C}{\partial x^2} \tag{7.2.6}$$

with
$C(x,t)$ concentration
$\mathcal{D}$ diffusion coefficient

Such a case is constituted by a quiescent medium limited laterally by two sources of different concentrations C_1 and C_2 (Figure 7.2). Among such examples, the frontier between two different media can be a membrane (osmosis apparatus) or a gas-liquid interface (evaporation, distillation) or liquid-solid (ion exchange resin) or liquid-liquid (liquid-liquid extraction).

7.2.3 Wave Equation

A string stretched along the axis Ox is moving transversally in a plane containing the axis Ox (Figure 7.3). If the lateral displacements $u(x,t)$ of the string are sufficiently low, the equation ruling the movement is

$$\frac{\partial^2 u}{\partial t^2} - c^2 \frac{\partial^2 u}{\partial x^2} = 0 \tag{7.2.7}$$

with

$u(x, t)$	lateral displacement
$c = \sqrt{T/\rho}$	wave propagation speed
T	string tension
ρ	mass per length unit

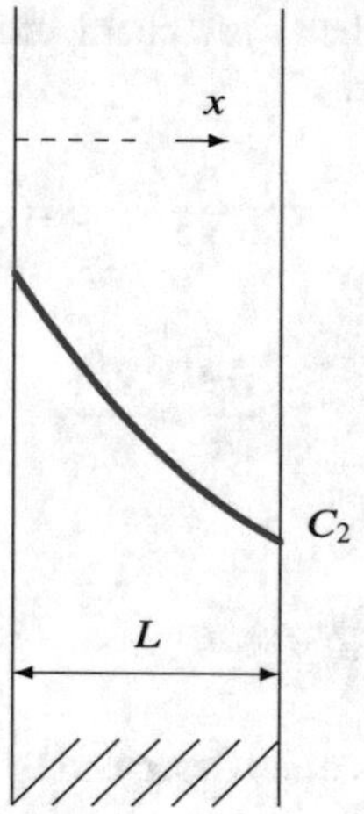

Fig. 7.2 Porous medium subjected to a concentration gradient

Fig. 7.3 Vibrating string

7.2.4 Laplace's Equation

Consider a homogeneous plane wall, of finite dimensions, where heat transfer is purely conductive, in steady state regime. The temperature variation perpendicular to the wall is neglected. Thus, the system is two-dimensional (Figure 7.4), modelled by Fourier's law, without the time dependence. Under these conditions, Laplace's equation describes the temperature field $T(x, y)$ in the wall

$$\frac{\partial^2 T}{\partial x^2} + \frac{\partial^2 T}{\partial y^2} = 0 \tag{7.2.8}$$

7.3 Properties of Partial Differential Equations

7.3.1 Generalities

Consider a partial differential equation where the variable u depends on coordinates x, y, z, i.e.

$$u = u(x, y, z) \tag{7.3.1}$$

The following abbreviations will be frequently used for partial derivatives

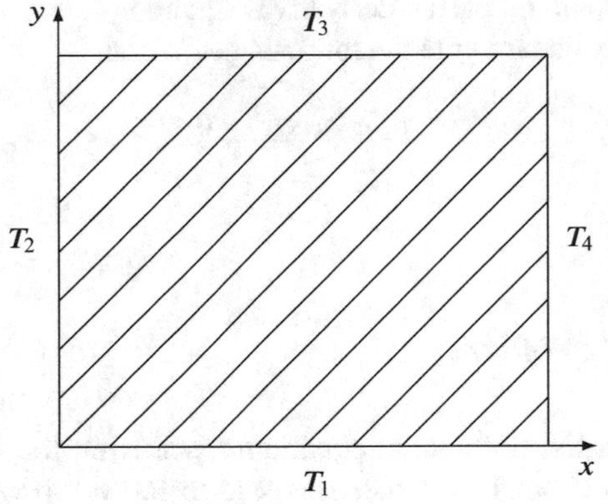

Fig. 7.4 Plane wall subjected to heat transfer

$$u_x = \frac{\partial u}{\partial x}, \quad u_{xx} = \frac{\partial^2 u}{\partial x^2}, \quad u_{xy} = \frac{\partial^2 u}{\partial x \partial y}, \quad \ldots \tag{7.3.2}$$

so that a general partial differential equation can be represented as

$$f(u, u_x, u_y, u_z, u_{xx}, \ldots, u_{xy}, \ldots) = 0 \tag{7.3.3}$$

The order of a partial differential equation is that of the derivative of highest order. Thus, the equation

$$u_x + a u_y = 0 \tag{7.3.4}$$

is of order 1. The equation

$$u_x + a u_{xy} = 0 \tag{7.3.5}$$

is of order 2. When several differential equations are coupled, the order is that obtained by gathering these equations into a single one.

The partial differential equation of the form

$$a(x, y)u_x + b(x, y)u_y = c(x, y) \tag{7.3.6}$$

is linear, as its coefficients depend only on the independent variables x, y.

The partial differential equation of the form

$$a(x, y, u)u_x + b(x, y, u)u_y = c(x, y) \tag{7.3.7}$$

is called quasi-linear, as its coefficients depend on independent variables x, y and on the dependent variable u.

The partial differential equation of the form

$$a(x, y, u, u_x, u_y)u_x + b(x, y, u, u_x, u_y)u_y = c(x, y) \tag{7.3.8}$$

is called nonlinear, as its coefficients depend on independent variables x, y, on the dependent variable u and on partial derivatives u_x and u_y.

Equation (7.3.4) is linear while the following equation

$$u_x + a(u_y)^2 = 0 \tag{7.3.9}$$

is nonlinear.

7.3.2 Well-Posed Problem

The solution of a partial differential equation depends on the boundary conditions and on the initial conditions. A problem is said to be well-posed when its solution is unique and depends continuously on the auxiliary conditions (boundary conditions and initial conditions). When a problem is well-posed, its solution varies smoothly when its auxiliary conditions are subject to small variations.

For example, the wall subjected to heat transfer (Section 7.5.2.5) and modelled by the partial differential equation

$$\frac{\partial T}{\partial t} = \alpha \frac{\partial^2 T}{\partial x^2} \tag{7.3.10}$$

subject to the initial condition

$$T(0, x) = T_0 \,, \quad \forall\ 0 \le x \le L \tag{7.3.11}$$

and to the boundary conditions

$$\begin{aligned} T(t, 0) &= T_A \\ T(t, L) &= T_B \end{aligned} \tag{7.3.12}$$

constitutes a well-posed problem.

7.3.3 Classification

Consider the two-dimensional equation

$$a\frac{\partial^2 u}{\partial x^2} + 2b\frac{\partial^2 u}{\partial x \partial y} + c\frac{\partial^2 u}{\partial y^2} + g_1(x, y, u, \frac{\partial u}{\partial x}, \frac{\partial u}{\partial y}) = 0 \tag{7.3.13}$$

As the function g contains only first order partial derivatives, while Equation (7.3.5) is of order 2, Equation (7.3.13) is called quasi-linear of order 2. The classification of partial differential equations is done by following reasonings similar to those used for the classification of conicals (Lepourhiet 1988) of equation

$$ax^2 + 2bxy + cy^2 + dx + ey + f = 0 \tag{7.3.14}$$

whose asymptotic directions are obtained by considering only the equation keeping the higher order terms

$$ax^2 + 2bxy + cy^2 = 0 \tag{7.3.15}$$

An asymptotic direction corresponds to a slope

$$m = \frac{y}{x} \tag{7.3.16}$$

which gives the equation

$$cm^2 + 2bm + a = 0 \tag{7.3.17}$$

whose discriminant is equal to

$$\Delta = b^2 - ac \tag{7.3.18}$$

This allows us to classify the conicals
$\Delta > 0$ two real asymptotes: the conical is an hyperbola,
$\Delta < 0$ two complex asymptotes: the conical is an ellipse,
$\Delta = 0$ one real double asymptote: the conical is a parabola,

In the same way, a partial differential equation (7.3.13) is
hyperbolic if $b^2 - ac > 0$,
elliptic if $b^2 - ac < 0$,
parabolic if $b^2 - ac = 0$.

The transient one-dimensional conductive heat transfer equation (7.2.2) and the diffusive mass transfer equation (7.2.6) are parabolic equations and of same mathematical nature. The steady-state two-dimensional heat transfer (7.2.8) is an elliptic equation. The wave equation (7.2.7) is hyperbolic.

7.3.4 Characterization of the Solutions

Consider the quasi-linear first order partial differential equation

$$a(x, y, u)\, u_x + b(x, y, u)\, u_y = c(x, y, u) \tag{7.3.19}$$

depending on two independent variables x and y. The solution is a surface of equation $u(x, y) = u$, thus $u(x, y) - u = 0$. At a point of coordinates $M(x, y, u)$, the normal to the surface is given by the gradient vector

$$grad = \begin{bmatrix} u_x \\ u_y \\ -1 \end{bmatrix} \tag{7.3.20}$$

Noting $\theta = [a, b, c]^T$ the vector of coefficients, it can be noticed that the scalar product

$$\theta^T grad = a(x, y, u)\, u_x + b(x, y, u)\, u_y - c(x, y, u) = 0 \tag{7.3.21}$$

corresponding to Equation (7.3.19) is zero and thus the vector of coefficients is parallel to the plane tangent to the surface at point $M(x, y, u)$.

7.4 Method of Characteristics

The property (7.3.21) has an important consequence bearing the name of method of characteristics with the associated characteristic curves (Aris and Amundson 1973; Lapidus and Pinder 1982; Varma and Morbidelli 1997). Typically, the method of characteristics is applicable to first order partial differential equations but also to hyperbolic partial differential equations. It consists in reducing a partial differential equation into a family of ordinary differential equations.

7.4.1 Linear First Order Partial Differential Equation

We still consider the quasi-linear first order partial differential equation

$$a(x, y, u)\, u_x + b(x, y, u)\, u_y = c(x, y, u) \tag{7.4.1}$$

with the initial condition

$$u = u_0(x, y) \quad \text{along the curve } C_0 : \; y = y_0(x) \tag{7.4.2}$$

The curve C_0 expressing the initial condition is contained in the surface $u(x, y)$ solution of Equation (7.4.1), so that the curve C_0 can be parameterized by a parameter r such that

$$x = r\,, \quad y = y_0(r)\,, \quad u = u_0(r) \tag{7.4.3}$$

On one side, the surface solution $u(x, y)$ contains this curve C_0, on another side, it contains characteristic curves crossing the curve C_0 that can be parameterized by a parameter s. Indeed, it can be considered that we start from a curve C obtained by a series of calculations and the previous and following reasonings are identical. In this way, the surface solution $u(x, y)$, parameterized with respect to the parameter s, can be written as $u(x(s), y(s))$ and the following relation results

$$\frac{du}{ds} = \frac{\partial u}{\partial x}\frac{dx}{ds} + \frac{\partial u}{\partial y}\frac{dy}{ds} \tag{7.4.4}$$

Using Equations (7.4.1) and (7.4.4), we deduce

$$\begin{bmatrix} a(x,y,u) & b(x,y,u) \\ \dfrac{dx}{ds} & \dfrac{dy}{ds} \end{bmatrix} \begin{bmatrix} \dfrac{\partial u}{\partial x} \\ \dfrac{\partial u}{\partial y} \end{bmatrix} = \begin{bmatrix} c(x,y,u) \\ \dfrac{du}{ds} \end{bmatrix} \tag{7.4.5}$$

When the determinant of this system is nonzero, the solution $(\partial u/\partial x, \partial u/\partial y)$ allows us to determine u in the neighborhood of C_0 by a first degree Taylor polynomial

$$u(x+dx, y+dy) = u(x,y) + \left(\frac{\partial u}{\partial x}\right)_{x,y} dx + \left(\frac{\partial u}{\partial y}\right)_{x,y} dy \tag{7.4.6}$$

theoretically, by an iterative procedure, thus by operating successive small displacements, the total solution $u(x, y)$ can be determined. This constitutes Cauchy's problem which is well-posed when the determinant is nonzero in any point.

The total differential du is equal to

$$du = \frac{\partial u}{\partial x} dx + \frac{\partial u}{\partial y} dy \tag{7.4.7}$$

Joined to Equation (7.4.1), this relation gives

$$\begin{bmatrix} a(x,y,u) & b(x,y,u) \\ dx & dy \end{bmatrix} \begin{bmatrix} \dfrac{\partial u}{\partial x} \\ \dfrac{\partial u}{\partial y} \end{bmatrix} = \begin{bmatrix} c(x,y,u) \\ du \end{bmatrix} \tag{7.4.8}$$

Whatever the considered point of C_0, there exists a direction called *characteristic direction* defined by (dx, dy) such that

$$\det\left(\begin{bmatrix} a(x,y,u) & b(x,y,u) \\ dx & dy \end{bmatrix}\right) = 0 \Rightarrow a\,dy = b\,dx \tag{7.4.9}$$

The slope of the characteristic direction is thus $(dy/dx = b/a)$ by assuming $a \neq 0$.

Using relation (7.4.9) in (7.4.7) and taking into account (7.4.1), we get (we also suppose $b \neq 0$)

$$\begin{aligned} a\,du &= c\,dx \text{ or } \quad a\frac{du}{ds} = c\frac{dx}{ds} \\ b\,du &= c\,dy \text{ or } \quad b\frac{du}{ds} = c\frac{dy}{ds} \end{aligned} \tag{7.4.10}$$

which means that the variation of u along the characteristics depends only on x or y. Thus, Equations (7.4.10) are ordinary differential equations that have just to be integrated, either one, either the other one of Equations (7.4.10). The solutions of the system formed by Equations (7.4.9) and (7.4.10) are called characteristic curves for Equation (7.4.1).

The method of characteristics thus consists in transforming the problem of a partial differential equation into a problem of ordinary differential equations along the characteristics.

The relations (7.4.9) and (7.4.10) serve as the basis to the solution consisting in two stages:

- determination of the characteristics by Equation (7.4.9),
- calculation of the variation du along the characteristics by Equation (7.4.10).

Frequently, the presentation is done by comparison of Equation (7.4.1) with Equation (7.4.4), which gives the characteristic equations

$$\begin{aligned} \frac{dx}{ds} &= a(x, y, u) \\ \frac{dy}{ds} &= b(x, y, u) \\ \frac{du}{ds} &= c(x, y, u) \end{aligned} \tag{7.4.11}$$

The curve $C = \{x(s), y(s), u(s)\}$ that satisfies these equations is called *integral curve* for the vector field (a, b, c). The integral curves are called *characteristic curves* for Equation (7.4.1) in the space (x, y, u) and depend on the parameter s. The system of ordinary differential equations (7.4.11), by assuming $a \neq 0$, $b \neq 0$, $c \neq 0$, is also often presented under the form

$$\frac{dx}{a} = \frac{dy}{b} = \frac{du}{c} \tag{7.4.12}$$

Thus, we can use Equations (7.4.9) and (7.4.10) or Equation (7.4.11).

A characteristic curve can cross the curve C_0 only in a single point (Figure 7.5).

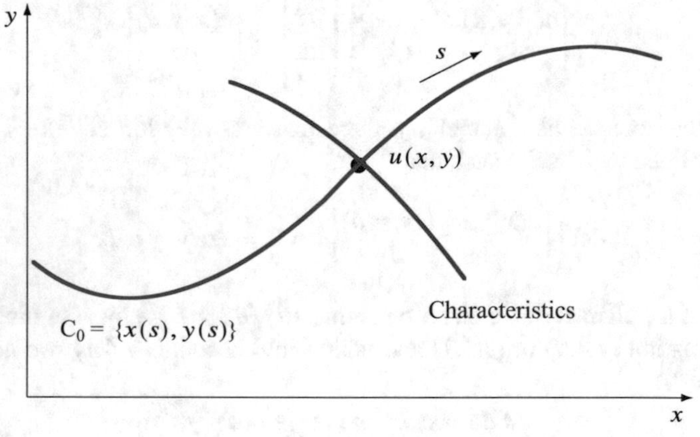

Fig. 7.5 Method of characteristics: Curve C_0 of the surface $u(x, y)$ and a characteristic curve

Finally, the surface characterized by the curve C_0 and the characteristic curves can be represented under the parametric form

$$x = x(r, s) , \quad y = y(r, s) , \quad u = u(r, s) \tag{7.4.13}$$

In principle, the variables r, s, u can reciprocally be expressed with respect to the independent variables x and y, so that the Jacobian

$$J = \det\left(\begin{bmatrix} \frac{\partial x}{\partial r} & \frac{\partial x}{\partial s} \\ \frac{\partial y}{\partial r} & \frac{\partial y}{\partial s} \end{bmatrix}\right) \propto b\,x_r - a y_r \tag{7.4.14}$$

must be nonzero

$$J \neq 0 \tag{7.4.15}$$

If J were zero, C_0 would coincide with a characteristic curve and the problem would admit an infinity of solutions for a single initial condition.

In the case where $c = 0$, we get $du = 0$, thus u is constant along a characteristic and is qualified as *Riemann invariant*.

Aris and Amundson (1973) give a detailed interpretation of the chromatography of two solutes by the method of characteristics. The model is a pair of first order quasi-linear homogeneous partial differential equations coupled by the equilibrium equation, such as Langmuir isotherm.

Anderson (2003) describes the use of the method of characteristics in the domain of flows of compressible fluids, in the supersonic steady-state case for a two- or three-dimensional irrotational and also rotational flow.

Example 7.1 :
Method of characteristics for an isothermal plug-flow reactor

An example frequently studied in chemical engineering dealing with the application of the method of characteristics is the model of the isothermal plug-flow reactor in transient regime with a first order chemical reaction (Figure 7.6).

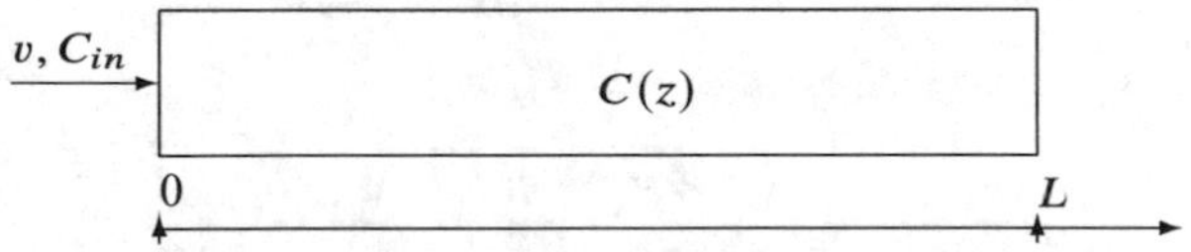

Fig. 7.6 Method of characteristics: Isothermal plug-flow reactor

Noting C the component concentration, v the flow velocity in z direction, and k the kinetic constant, the dynamic model is written as

$$\frac{\partial C}{\partial t} + v\frac{\partial C}{\partial z} + k\,C = 0 \quad \Rightarrow \quad \frac{\partial u}{\partial x} + v\frac{\partial u}{\partial y} + k\,u = 0 \quad \Rightarrow \quad u_x + v\,u_y = -k\,u \tag{7.4.16}$$

with the change of variable $x = t$, $y = z$, $u = C$.

Equation (7.4.16) is a linear first order partial differential equation with constant coefficients that belongs to the class of equations of type (7.4.1).

The initial condition (given concentration profile in the reactor) of the original problem is

$$C = C_0(z) \qquad \text{for } t = 0 \qquad \forall\ 0 \leq z \leq L \tag{7.4.17}$$

and the boundary condition at the inlet is

$$C = C_{in} \qquad \text{for } z = 0 \qquad \forall\ t > 0 \tag{7.4.18}$$

With respect to the new variables, these conditions are transformed. The initial condition becomes

$$u = u_0 \qquad \text{for } x = 0 \qquad \forall\ 0 \leq y \leq L \tag{7.4.19}$$

and the boundary condition at the inlet $C = C_{in}$ at $z = 0$ is

$$u = u_{in} \qquad \text{for } y = 0 \qquad x > 0 \tag{7.4.20}$$

Indeed, instead of the simple change of variables previously proposed to obtain the final equation (7.4.16), it is preferred to introduce the following dimensionless variables

$$\tau = k\,t\,, \quad Z = k\,z/v\,, \quad u = C/C_{in} \tag{7.4.21}$$

which gives the partial differential equation

$$u_\tau + u_Z = -u \tag{7.4.22}$$

associated with the initial condition

$$u(0, Z) = \frac{C_0}{C_{in}} \qquad \text{for } \tau = 0\,, \quad \forall\ 0 \le Z \le \frac{kL}{v} \tag{7.4.23}$$

and the boundary condition at the inlet

$$u(\tau, 0) = 1 \qquad \text{for } \tau > 0\,, \quad Z = 0 \tag{7.4.24}$$

The characteristic direction is given by Equation (7.4.9) which is written here

$$d\tau = dZ \tag{7.4.25}$$

while Equations (7.4.10) give along the characteristics

$$\begin{aligned} du = -u\,d\tau \quad &\text{or} \quad \frac{du}{ds} = -\frac{d\tau}{ds} \\ du = -u\,dZ \quad &\text{or} \quad \frac{du}{ds} = -\frac{dZ}{ds} \end{aligned} \tag{7.4.26}$$

Equation (7.4.25) gives the characteristics

$$Z = \tau + c_1 \tag{7.4.27}$$

parallel to the first bisector.

The integration of Equations (7.4.26) gives

$$\begin{aligned} u &= c_2 \exp(-\tau) \\ u &= c_3 \exp(-Z) \end{aligned} \tag{7.4.28}$$

where the constants c_i must be determined from the initial condition and the inlet condition.

If Equations (7.4.11) had been considered, we would have obtained an equivalent system of relations

$$\begin{cases} \dfrac{dx}{ds} = 1 \\ \dfrac{dy}{ds} = 1 \\ \dfrac{du}{ds} = -u \end{cases} \quad \Rightarrow \quad \begin{cases} \tau = s + c'_1 \\ Z = s + c'_2 \\ u = c'_3 \exp(-s) \end{cases} \tag{7.4.29}$$

where c'_i are constants to determine from the initial condition and the inlet condition.

The initial condition (7.4.23) can be written under parametric form (with respect to parameter r)

$$\tau = 0\,, \quad 0 \le Z = r \le \frac{kL}{v}\,, \quad u = u_0(r) = \frac{C_0(z)}{C_{in}} \tag{7.4.30}$$

where the concentration profile in the reactor $u_0(r)$ depends on r (thus Z). Considering the initial condition amounts to be located along Z axis at $\tau = 0$.

The association of the initial condition (7.4.30) and of Equations (7.4.27)–(7.4.28) gives $r = c_1$, $u_0(r) = c_2$, $u_0(r) = c_3 \exp(-r)$ from which we draw

$$\begin{cases} \tau = s \\ Z = \tau + r \qquad \text{with } 0 \le r \le \frac{kL}{v} \\ u = u_0(r)\exp(-\tau) = u_0(Z-\tau)\exp(-\tau) \qquad \text{for: } Z > \tau \end{cases} \tag{7.4.31}$$

Remark: $u_0(r)$ or $u_0(Z - \tau)$ means that u_0 depends on $r = (Z - \tau)$.

Considering the inlet boundary condition amounts to be located along τ axis at $Z = 0$. The association of the inlet condition (7.4.24) and of Equations (7.4.27)–(7.4.28) gives $\tau = -c_1 < 0$, $c_2 = \exp(\tau)$, $c_3 = 1$, from which we draw

$$\begin{cases} \tau = Z + r \\ Z = s \\ u = \exp(-Z) \qquad \text{for } Z < \tau \end{cases} \tag{7.4.32}$$

This constitutes the second family of characteristic curves (family associated with the inlet boundary condition).

The set of families constitutes the total solution of the problem.

The two families of characteristic curves can be represented as a projection in the plane (τ, Z) (corresponding to (t, z) for the reactor model) or in three dimensions (τ, Z, u) (corresponding to (t, z, C)).

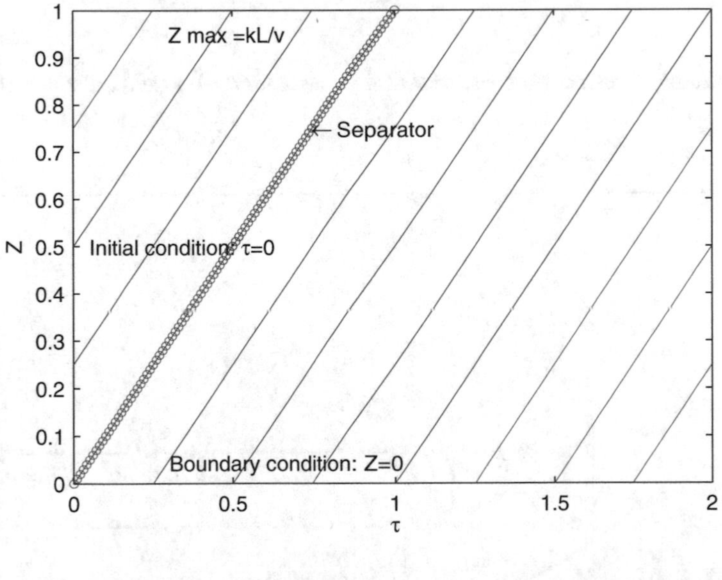

Fig. 7.7 Method of characteristics: Characteristic curves $y = x \pm r$ with indication of the initial condition and the boundary condition

Figure 7.7 shows the two families of characteristics corresponding to parallel lines separated by the separator line $Z = \tau$. Above the separator, the first family lies $(Z > \tau)$, below the separator, the second family $(Z < \tau)$. The value of Z is limited by kL/v, that of τ is unlimited.

Suppose that the concentration in the reactor is zero at time $t = 0$, then $u_0 = 0, \forall Z$. The dimensionless concentration at inlet is equal to $u_{in} = 1$.

Figure 7.8 displays the evolution of dimensionless concentration u versus the dimensionless length Z for two different values of dimensionless time τ. It is noticed for $Z \le \tau$ that the concentration is related to the inlet boundary condition by Equation (7.4.32) while for $Z \ge \tau$ the concentration is related to the initial condition by Equation (7.4.31). As these conditions differ, this causes a discontinuity at $Z = \tau$.

Figure 7.9 in three dimensions displays the evolution of the dimensionless concentration u versus the dimensionless abscissa Z for different values of the dimensionless time τ. In the same way as for the curves of Figure 7.8, it is noticed for $Z \leq \tau$ that the concentration is related to the inlet boundary condition by Equation (7.4.32) whereas for $Z \geq \tau$ the concentration is related to the initial condition by Equation (7.4.31). The same type of discontinuity occurs at $Z = \tau$.

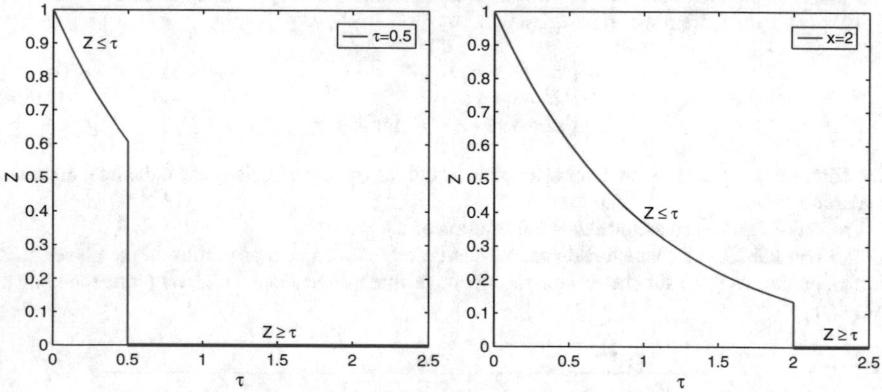

Fig. 7.8 Method of characteristics: Variation of u versus Z for $\tau = 0.5$ (left) and $\tau = 2$ (right)

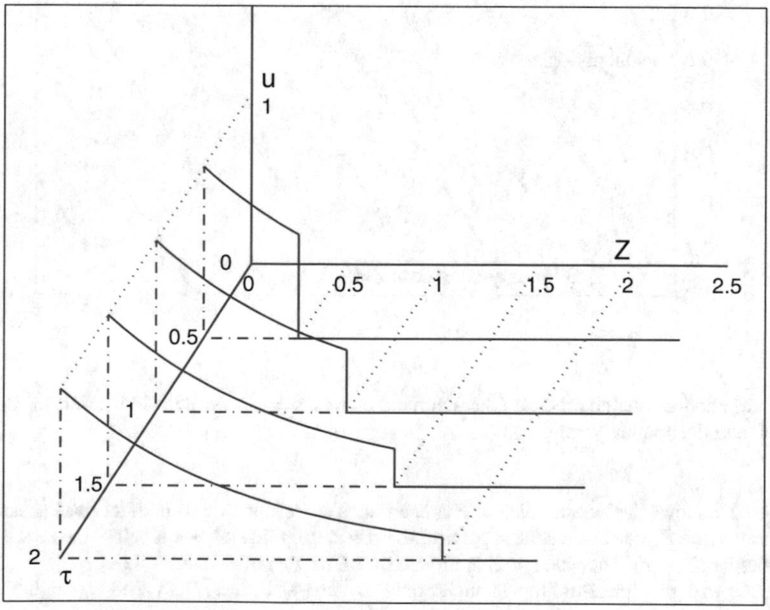

Fig. 7.9 Method of characteristics: Variation of the dimensionless concentration u versus the dimensionless time τ and the dimensionless abscissa Z

Example 7.2 :
Method of characteristics for a quasi-linear first order PDE

An example of quasi-linear first order partial differential equation is given by the phenomena of adsorption of the solute initially contained in a fluid flow and getting adsorbed on a solid disposed in a fixed bed, such as a chromatographic column (Aris and Amundson 1973). The general equation is

$$S\,\epsilon\,\frac{\partial C}{\partial t} + S\,(1-\epsilon)\,\frac{\partial q}{\partial t} + F\,\frac{\partial C}{\partial x} = 0 \tag{7.4.33}$$

where ϵ is the bed porosity, S the column cross-section, C the solute concentration in the fluid, F the volume flow rate of liquid. The fluid velocity is $v(t) = F/(S\epsilon)$, positive or negative according to the flow direction. Equation (7.4.33) can then be written as

$$\frac{\partial C}{\partial t} + \left(\frac{1-\epsilon}{\epsilon}\right)\frac{\partial q}{\partial t} + v(t)\,\frac{\partial C}{\partial x} = 0 \tag{7.4.34}$$

The relation between the solute concentrations in the fluid and the solid in the case of a linear equilibrium valid at low concentrations is

$$q = k\,C \tag{7.4.35}$$

where q is the quantity of solute adsorbed on the solid. In the case where the equilibrium constant k is indeed constant, Equation (7.4.34) becomes

$$\left(1 + \frac{1-\epsilon}{\epsilon}k\right)\frac{\partial C}{\partial t} + v(t)\,\frac{\partial C}{\partial x} = 0 \tag{7.4.36}$$

Setting $V = v/(1 + k(1-\epsilon)/\epsilon)$, this equation is simplified as

$$\frac{\partial C}{\partial t} + V\,\frac{\partial C}{\partial x} = 0 \tag{7.4.37}$$

whose solution is $C = \phi(x - V\,t)$ where ϕ is a Riemann invariant.

The parametric pumping (Pigford et al. 1969) consists in varying the temperature in a synchronous way to the flow and as the equilibrium constant k depends on temperature or pressure, Equation (7.4.34) is transformed under a more complex form

$$\left(1 + \frac{1-\epsilon}{\epsilon}k(t)\right)\frac{\partial C}{\partial t} + \frac{1-\epsilon}{\epsilon}\,C(t)\,\frac{\partial k}{\partial t} + v(t)\,\frac{\partial C}{\partial x} = 0 \tag{7.4.38}$$

Pigford et al. (1969) perform alternate cycles of temperature and velocity in both directions for which the method of characteristics allows an easier interpretation.

7.4.2 Nonlinear First Order Partial Differential Equation

A nonlinear first order partial differential equation is

$$f(x, y, u, u_x, u_y) = 0 \tag{7.4.39}$$

Then, the characteristic equations are (Lapidus and Pinder 1982)

$$
\begin{aligned}
\frac{dx}{ds} &= f_{u_x} \\
\frac{dy}{ds} &= f_{u_y} \\
\frac{du}{ds} &= u_x\, f_{u_x} + u_y\, f_{u_y} \\
\frac{du_x}{ds} &= -f_x - u_x\, f_u \\
\frac{du_y}{ds} &= -f_y - u_y\, f_u
\end{aligned}
\tag{7.4.40}
$$

In this case, the solutions are no more characteristic curves, but characteristic bands (Aris and Amundson 1973).

7.4.3 Quasi-Linear Second Order Partial Differential Equation

A hyperbolic partial differential equation possesses two real characteristic curves, a parabolic partial differential equation possesses one real characteristic curve, an elliptic partial differential equation does not possess a real characteristic curve.

Consider the quasi-linear second order partial differential equation

$$
a\, u_{xx} + b\, u_{xy} + c\, u_{yy} = g \tag{7.4.41}
$$

It is assumed that the curve $C = \{x(s), y(s)\}$ belongs to the surface solution of the partial differential equation (7.4.41) and we aim at propagating the solution u in the neighborhood of this curve, then step by step. C can correspond to the initial profile that must be specified.

To Equation (7.4.41), the two following equations can be associated which simply are properties of the differential forms

$$
\begin{aligned}
u_{xx}\, dx + u_{xy}\, dy &= du_x \\
u_{xy}\, dx + u_{yy}\, dy &= du_y
\end{aligned}
\tag{7.4.42}
$$

These three equations linear with respect to the second order partial derivatives can be written as

$$
\begin{bmatrix} a & b & c \\ dx & dy & 0 \\ 0 & dx & dy \end{bmatrix}
\begin{bmatrix} u_{xx} \\ u_{xy} \\ u_{yy} \end{bmatrix}
=
\begin{bmatrix} g \\ du_x \\ du_y \end{bmatrix}
\tag{7.4.43}
$$

In order to write the linear dependence of these equations, we express that the determinant is zero

$$
\det\left(\begin{bmatrix} a & b & c \\ dx & dy & 0 \\ 0 & dx & dy \end{bmatrix} \right) = 0 \tag{7.4.44}
$$

This yields the relation

$$
a\, dy^2 - b\, dx\, dy + c\, dx^2 = 0 \quad \Rightarrow \quad \frac{dy}{dx} = \frac{b \pm \sqrt{b^2 - 4ac}}{2a} \tag{7.4.45}
$$

which gives two characteristic roots (real in the hyperbolic case, double in the parabolic case, imaginary in the elliptic case). In the case of real roots, thus in the case of a hyperbolic equation, to the limit parabolic, they give the characteristic curves.

To propagate the solution in the neighborhood of C, we desire to use the relation

$$du = u_x\, dx + u_y\, dy \tag{7.4.46}$$

but the first order partial derivatives u_x and u_y must be determined; however, a system such that (7.4.43) gives only the second order partial derivatives, if the determinant is nonzero. Thus it is necessary to calculate the first order partial derivatives from the second order partial derivatives. The determination of $u(x, y)$ in the neighborhood of the characteristics by Equation (7.4.46) is done by using the relations

$$\begin{aligned} du_x &= u_{xx}\, dx + u_{xy}\, dy \\ du_y &= u_{xy}\, dx + u_{yy}\, dy \end{aligned} \tag{7.4.47}$$

where the second order partial derivatives are obtained by solving the system (7.4.43) when the determinant is nonzero along C.

By substitution of columns, other determinants can be considered as zero such as

$$\det\left(\begin{bmatrix} a & g & c \\ dx & du_x & 0 \\ 0 & du_y & dy \end{bmatrix}\right) = 0 \tag{7.4.48}$$

which gives a new relation

$$a\, du_x\, dy - g\, dx\, dy + c\, du_y\, dx = 0 \tag{7.4.49}$$

relating the variations of the first order partial derivatives along a characteristic if dy/dx is imposed by Equation (7.4.45). In the case of a hyperbolic partial differential equation, noting the characteristic slopes λ_1 and λ_2 (Equation (7.4.45)), we thus get

$$\begin{aligned} &dy = \lambda_i\, dx \\ &a\, du_x\, dy - g\, dx\, dy + c\, du_y\, dx = 0 \\ &du = u_x\, dx + u_y\, \lambda_i\, dx \end{aligned} \tag{7.4.50}$$

Consider a point $M_3(x, y)$ in the neighborhood of C. By this point, we make a first characteristics of slope λ_1 pass that crosses C in a close point $M_1(x_1, y_1)$ and a second characteristics of slope λ_2 that crosses C in a second close point $M_2(x_2, y_2)$ (Figure 7.10). The objective is to calculate the solution u at point $M_3(x, y)$. Starting from Equations (7.4.50), along the characteristics that passes through M_1M_3, we can write

$$a\, du_x\, \lambda_1 - g\, \lambda_1\, dx_1 + c\, du_y = 0 \tag{7.4.51}$$

similarly along the characteristics that passes through M_2M_3

$$a\, du_x\, \lambda_2 - g\, \lambda_2\, dx_2 + c\, du_y = 0 \tag{7.4.52}$$

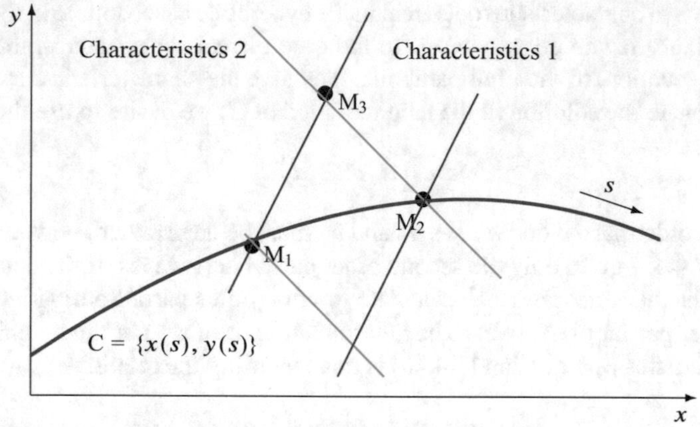

Fig. 7.10 Method of characteristics: Curve C of the surface $u(x, y)$ and two characteristics

As the partial derivatives u_x and u_y are already known in M_1 and M_2, the equations can be integrated, respectively, between M_3 and M_1, and between M_3 and M_2, which allows us to deduce u_x and u_y in M_3 from that system of two equations. Finally, u is calculated by Equation (7.4.50).

The method of characteristics is thus frequently applied to the hyperbolic wave equation (7.2.7) which is homogeneous. On the opposite, for Fourier (7.2.1) and Fick (7.2.6) parabolic equations, the root of Equation (7.4.45) is double, the characteristics are parallel to the horizontal axis and are degenerated, so that is not possible to apply the method of characteristics. For the elliptic Laplace equation (7.2.8), the roots of Equation (7.4.45) are imaginary, and there exist no characteristics.

An advantage of the method of characteristics with respect to the finite difference method is that, by its intrinsic principle, it avoids instabilities close to the limits of the domain for which boundary conditions have been imposed. However, with the progress of computers and advanced numerical techniques such as finite volumes, finite elements and even boundary elements, presently, the method of characteristics is little used.

7.5 Finite Difference Method

7.5.1 Introduction

The finite difference method is old and seems one of the most natural. It must not hide the fact that many other powerful methods exist such as the finite volumes, weighted residuals, spectral methods, finite elements, boundary elements. Thus the finite elements are in general the numerical method used in the professional calculation codes of solid mechanics, heat transfer, and fluid mechanics. The finite volumes are used in several commercial calculation codes available for chemical engineering and also heat transfer and fluid mechanics, possibly with chemical reaction.

7.5.2 *Discretization*

7.5.2.1 Requirement for a Discretization Scheme

An equation

$$f(u) = 0 \tag{7.5.1}$$

will be discretized according to a given numerical scheme and its discretization noted

$$f_\Delta(u) = 0 \tag{7.5.2}$$

the discretization error or truncation error, depending on the discretization scheme, is equal to

$$R_\Delta(u) = f_\Delta(u) - f(u) \tag{7.5.3}$$

Suppose that U_Δ is the solution of Equation (7.5.2). For this solution to be considered as a correct approximation of the exact solution u, three conditions must be fulfilled

- the discretized equation (7.5.2) must be consistent with the original equation (7.5.1), i.e. the truncation error must tend towards zero when the discretization step Δ tends towards 0,
- the discretization scheme must provide a stable solution U_Δ,
- the solution U_Δ must converge towards u.

7.5.2.2 Taylor Series Expansion

The approximation of partial derivatives by divided differences is performed by considering a Taylor series expansion for the relevant variable in the neighborhood of a point. Suppose that the function depends on two variables, i.e. $u(x, y)$. The nth degree Taylor polynomial is given by

$$\begin{aligned} u(x+\Delta x, y+\Delta y) = {} & u(x, y) + (\Delta x\frac{\partial}{\partial x} + \Delta y\frac{\partial}{\partial y})u(x, y) + \\ & \frac{1}{2!}(\Delta x\frac{\partial}{\partial x} + \Delta y\frac{\partial}{\partial y})^2 u(x, y) + \cdots + \\ & \frac{1}{n!}(\Delta x\frac{\partial}{\partial x} + \Delta y\frac{\partial}{\partial y})^n u(x, y) + \\ & \frac{1}{(n+1)!}(\Delta x\frac{\partial}{\partial x} + \Delta y\frac{\partial}{\partial y})^{n+1} u(x+\zeta\Delta x, y+\xi\Delta y) \\ & \text{with } 0 < \zeta < 1 \text{ and } 0 < \xi < 1 \end{aligned} \tag{7.5.4}$$

where

$$(\Delta x\frac{\partial}{\partial x} + \Delta y\frac{\partial}{\partial y}) \tag{7.5.5}$$

is an operator of partial differentiation acting on the function u at point (x, y).

7.5.2.3 Example of Discretization

Consider again the one-dimensional problem of conduction in a rod modelled by

$$\frac{\partial T}{\partial t} = \alpha \frac{\partial^2 T}{\partial x^2} \tag{7.5.6}$$

Two discretizations are necessary for this one-dimensional problem, a time discretization represented by exponent n and a spatial discretization represented by index i. The following problem equivalent to the problem (7.5.6) transformed under the dimensionless form is treated

$$\frac{\partial u}{\partial t} = \frac{\partial^2 u}{\partial x^2} \tag{7.5.7}$$

The variable $u(x, t)$ is symbolized by u_i^n. The axis Ox is indexed by i so that the abscissa is $x = i\Delta x$ where Δx is the length of the increment on Ox. The time t is equal to $t = n\Delta t$.

The variable $u(x, t + \Delta t)$ can be calculated by writing the second degree Taylor polynomial

$$u(x, t + \Delta t) = u_i^{n+1} = u_i^n + \Delta t \left(\frac{\partial u}{\partial t}\right)_i^n + \frac{\Delta t^2}{2} \left(\frac{\partial^2 u}{\partial t^2}\right)_i^n + 0(\Delta t^3) \tag{7.5.8}$$

The approximation of the partial derivative $(\partial u/\partial t)_i^n$ is obtained from Equation (7.5.8), that is

$$\left(\frac{\partial u}{\partial t}\right)_i^n = \frac{u_i^{n+1} - u_i^n}{\Delta t} - \frac{\Delta t}{2} \left(\frac{\partial^2 u}{\partial t^2}\right)_i^n + 0(\Delta t^2) \tag{7.5.9}$$

Patankar (1980) mentions that the difficulty in numerical analysis always comes from the first order partial derivatives. The second order partial derivatives behave well and pose no difficulties. Thus, in the problems of flow simulation, the pressure partial derivatives (in the momentum equations) and velocity partial derivatives (in the continuity equation) are sources for potential problems.

The truncation error produced on the partial derivative approximated as

$$\left(\frac{\partial u}{\partial t}\right)_i^n \approx \frac{u_i^{n+1} - u_i^n}{\Delta t} \tag{7.5.10}$$

is thus equal to

$$\begin{aligned} R_{\Delta t}(u) &= \frac{u_i^{n+1} - u_i^n}{\Delta t} - \left(\frac{\partial u}{\partial t}\right)_i^n \\ &= \frac{\Delta t}{2} \left(\frac{\partial^2 u}{\partial t^2}\right)_i^n + 0(\Delta t^2) \end{aligned} \tag{7.5.11}$$

In a similar way, the variable $u(x + \epsilon\Delta x, t)$ can be calculated by writing the 5th degree Taylor polynomial

$$u(x + \epsilon\Delta x, t) = u^n_{i+\epsilon} = u^n_i + \epsilon\Delta x \left(\frac{\partial u}{\partial x}\right)^n_i + \frac{\epsilon^2\Delta x^2}{2}\left(\frac{\partial^2 u}{\partial x^2}\right)^n_i + \frac{\epsilon^3\Delta x^3}{3!}\left(\frac{\partial^3 u}{\partial x^3}\right)^n_i + \frac{\epsilon^4\Delta x^4}{4!}\left(\frac{\partial^4 u}{\partial x^4}\right)^n_i + \frac{\epsilon^5\Delta x^5}{5!}\left(\frac{\partial^5 u}{\partial x^5}\right)^n_i + 0(\Delta x^6) \tag{7.5.12}$$

with ϵ which can be an integer or not (for example 1, −1, 1/2, −1/2 ...), but not too large with respect to unity.

To obtain the partial derivative $\left(\frac{\partial^2 u}{\partial x^2}\right)^n_i$, it suffices to consider Equation (7.5.12), for two opposite values of ϵ, +1 and −1. It results

$$\left(\frac{\partial^2 u}{\partial x^2}\right)^n_i = \frac{u^n_{i-1} - 2u^n_i + u^n_{i+1}}{\Delta x^2} - \frac{\Delta x^2}{12}\left(\frac{\partial^4 u}{\partial x^4}\right)^n_i + 0(\Delta x^4) \tag{7.5.13}$$

and the partial derivative $\left(\frac{\partial^2 u}{\partial x^2}\right)^n_i$ can be approximated as

$$\left(\frac{\partial^2 u}{\partial x^2}\right)^n_i \approx \frac{u^n_{i-1} - 2u^n_i + u^n_{i+1}}{\Delta x^2} \tag{7.5.14}$$

Consistency of the numerical scheme

The truncation error corresponding to the discretization scheme

$$f_\Lambda(u) = \frac{u^{n+1}_i - u^n_i}{\Delta t} - \frac{u^n_{i-1} - 2u^n_i + u^n_{i+1}}{\Delta x^2} \tag{7.5.15}$$

is obtained by means of Equations (7.5.8) and (7.5.13) and is equal to

$$\begin{aligned} R_{\Delta t,\Delta x}(u) &= \left[\frac{u^{n+1}_i - u^n_i}{\Delta t} - \frac{u^n_{i-1} - 2u^n_i + u^n_{i+1}}{\Delta x^2}\right] - \left[\left(\frac{\partial u}{\partial t}\right)^n_i - \left(\frac{\partial^2 u}{\partial x^2}\right)^n_i\right] \\ &= \frac{\Delta t}{2}\left(\frac{\partial^2 u}{\partial t^2}\right)^n_i - \frac{\Delta x^2}{12}\left(\frac{\partial^4 u}{\partial x^4}\right)^n_i + 0(\Delta t^2) + 0(\Delta x^4) \end{aligned} \tag{7.5.16}$$

Equation (7.5.16) shows that, when the increments Δt and Δx tend towards 0, the truncation error related to the numerical scheme (7.5.15) tends towards 0. The chosen scheme (7.5.15) is thus *consistent* with the partial differential equation (7.3.10).

Noticing that

$$\frac{\partial^2 u}{\partial t^2} = \frac{\partial}{\partial t}\left(\frac{\partial u}{\partial t}\right) = \frac{\partial}{\partial t}\left(\frac{\partial^2 u}{\partial x^2}\right) = \frac{\partial^2}{\partial x^2}\left(\frac{\partial u}{\partial t}\right) = \frac{\partial^4 u}{\partial x^4} \tag{7.5.17}$$

it is possible to write Equation (7.5.16) under the form

$$R_{\Delta t,\Delta x}(u) = \left(\frac{\Delta t}{2} - \frac{\Delta x^2}{12}\right)\left(\frac{\partial^4 u}{\partial x^4}\right)_i^n + 0(\Delta t^2) + 0(\Delta x^4) \qquad (7.5.18)$$

The truncation error $R_{\Delta t,\Delta x}(u)$ is minimized by an adequate choice of the time and spatial discretization steps such that

$$\Delta t - \frac{\Delta x^2}{6} = 0 \qquad (7.5.19)$$

Stability of the numerical scheme

The *stability* implies that the rounding errors unavoidable during an iterative calculation are not amplified by the numerical scheme during solving. The following analysis is known as Fourier or Von Neumann[1] stability analysis (Charnay et al. 1950).

The study of the stability (Lepourhiet 1988) lies on a very general analysis of the numerical scheme where the function of two variables $u(x, t)$ is treated as a function of separated variables and this function is represented by its Fourier series expansion (as it is done in signal processing where any signal of finite duration L is represented as a pseudo-periodic signal) as

$$u(x,t) = \sum_{l=-\infty}^{+\infty} a_l(t) \exp(j\omega_l x)\,, \quad \omega_l = \frac{2\pi l}{2L} \qquad (7.5.20)$$

It can be noticed that the coefficients a_l depend on time (j is the complex number such that $j^2 = -1$).

Replacing $u(x, t)$ by its expression (7.5.20), any numerical scheme, one-dimensional with respect to time and one-dimensional with respect to space, can be resumed to a numerical solution of the form

$$u_i^n = \sum_{l=-N}^{+N} A_l^n \exp(j\omega_l i\Delta x)\,, \quad \omega_l \Delta x = \frac{2\pi l}{2L}\frac{L}{N+1} = l\,\frac{\pi}{N+1} \qquad (7.5.21)$$

where $(N+1)$ is the number of discretization points used to represent the function u on the interval $[0, L]$, $(2N+1)$ being the number of harmonics. A_l^n being the amplitude of the harmonic l at time $n\Delta t$, the study of the stability is concentrated on the evolution of this coefficient along the calculation. The amplification coefficient ξ_l^m is defined by

$$\xi_l^m = \frac{A_l^{n+m}}{A_l^n} \qquad (7.5.22)$$

The discretization scheme can be considered under the form of the more general expression

[1] John Von Neumann (Janos Làjos Neumann) was a brilliant Hungarian-American mathematician and physician (1903–1957). He must not be confused with Neumann that comes from the German mathematician Carl Neumann (1832–1925).

$$f_\Delta(u) = \sum_{k,m} a_k^m u_{i+k}^{n+m} = 0 \tag{7.5.23}$$

In this equation, if u is replaced by its expression (7.5.21), we obtain

$$\begin{aligned}
&\textstyle\sum_{k,m} a_k^m \left[\sum_{l=-N}^{+N} A_l^{n+m} \exp(j\frac{l\pi}{N+1}(i+k)\Delta x)\right] = 0 \Longrightarrow \\
&\textstyle\sum_{k,m} a_k^m \left[\sum_{l=-N}^{+N} \xi_l^m A_l^n \exp(j\frac{l\pi}{N+1} i\Delta x) \exp(j\frac{l\pi}{N+1} k\Delta x)\right] = 0 \Longrightarrow \\
&\textstyle\sum_{l=-N}^{+N} A_l^n \exp(j\frac{l\pi}{N+1} i\Delta x) \left[\sum_{k,m} a_k^m \xi_l^m \exp(j\frac{l\pi}{N+1} k\Delta x)\right] = 0
\end{aligned} \tag{7.5.24}$$

Equation (7.5.24) implies the following relation for each harmonic indexed by l

$$\sum_{k,m} a_k^m \xi_l^m \exp(j\frac{l\pi}{N+1} k\Delta x) = 0 \tag{7.5.25}$$

The solutions of this equation are the amplification coefficients ξ_l^m. So that the numerical solution be stable for that harmonic, it is necessary that the modules of the amplification coefficients be lower than 1

$$|\xi_l^m| \le 1 \tag{7.5.26}$$

Application to the numerical scheme (7.5.15):
The numerical scheme (7.5.15) gives the relation

$$f_\Delta(u) = u_i^{n+1} - u_i^n - r(u_{i-1}^n - 2u_i^n + u_{i+1}^n) = 0 \qquad \text{with } r = \frac{\Delta t}{\Delta x^2} \tag{7.5.27}$$

In the scheme (7.5.27), three points are used: $i-1$, i and $i+1$, corresponding to $2N+1$ harmonics, thus $N = 1$. Thus, the expression

$$u_i^n = \sum_{l=-N}^{+N} A_l^n \exp(j\frac{l\pi}{2} i\Delta x) \tag{7.5.28}$$

is sufficient to represent the scheme (7.5.27)

$$\begin{aligned}
f_\Delta(u) = &\textstyle\sum_{l=-I}^{+I} A_l^{n+1} \exp(j\frac{l\pi}{2} i\Delta x) - \sum_{l=-I}^{+I} A_l^n \exp(j\frac{l\pi}{2} i\Delta x) \\
&\textstyle -r\left[\sum_{l=-I}^{+I} A_l^n \exp(j\frac{l\pi}{2}(i-1)\Delta x) - 2\sum_{l=-I}^{+I} A_l^n \exp(j\frac{l\pi}{2} i\Delta x)\right. \\
&\textstyle \left.+\sum_{l=-I}^{+I} A_l^n \exp(j\frac{l\pi}{2}(i+1)\Delta x)\right] = 0
\end{aligned} \tag{7.5.29}$$

which becomes, using the relation $A_l^{n+1} = \xi_l^1 A_l^n$

$$f_\Delta(u) = \sum_{l=-N}^{+N} A_l^n \exp(j\frac{l\pi}{2} i\Delta x)[\xi_l^1 - 1 - r\underbrace{(\exp(-j\frac{l\pi}{2}\Delta x) - 2 + \exp(j\frac{l\pi}{2}\Delta x))}_{\underbrace{2\cos(\frac{l\pi}{2}\Delta x) - 2}_{-4\sin^2(\frac{l\pi}{4}\Delta x)}}] = 0 \tag{7.5.30}$$

hence the equation

$$\xi_l^1 - 1 + 4r\sin^2(\frac{l\pi}{4}\Delta x) = 0 \tag{7.5.31}$$

whose roots ξ_l^1 must verify

$$|\xi_l^1| \leq 1 \tag{7.5.32}$$

From Equation (7.5.31), it results that

$$\xi_l^1 = 1 - 4r\sin^2(\frac{l\pi}{4}\Delta x) \tag{7.5.33}$$

This root is always lower than 1, on the opposite it is required that

$$-1 \leq 1 - 4r\sin^2(\frac{l\pi}{4}\Delta x) \tag{7.5.34}$$

which gives the stability condition, known as CFL (Courant-Friedrichs-Lewy) condition under a more general form (Courant et al. 1928)

$$r \leq \frac{1}{2} \Longrightarrow \frac{\Delta t}{\Delta x^2} \leq \frac{1}{2} \tag{7.5.35}$$

Thus, the numerical scheme (7.5.15) is stable when the condition (7.5.35) dealing with the respective steps of time and space is respected.

In the case of the *dimensional* following parabolic equation

$$\frac{\partial u}{\partial t} = \nu \frac{\partial^2 u}{\partial x^2} \tag{7.5.36}$$

yielding the explicit discretization

$$\frac{u_i^{n+1} - u_i^n}{\Delta t} = \nu \frac{u_{i-1}^n - 2u_i^n + u_{i+1}^n}{\Delta x^2} \tag{7.5.37}$$

the CFL condition is

$$\frac{\Delta t}{\Delta x^2} \leq \frac{1}{2\nu} \tag{7.5.38}$$

Remark:
If, instead of the dimensionless equation (7.5.7), the dimensional equation (7.5.6) is considered, the discretization of this equation under an explicit form gives

$$\frac{T_i^{n+1} - T_i^n}{\Delta t} = \frac{\alpha}{\Delta x^2}(T_{i+1}^n - 2T_i^n + T_{i-1}^n) \tag{7.5.39}$$

hence

$$\begin{aligned} T_i^{n+1} &= \frac{\alpha \Delta t}{\Delta x^2}(T_{i+1}^n - 2T_i^n + T_{i-1}^n) + T_i^n \\ &= \frac{\alpha \Delta t}{\Delta x^2}(T_{i+1}^n + T_{i-1}^n) + \left(1 - 2\frac{\alpha \Delta t}{\Delta x^2}\right) T_i^n \\ &= \mathrm{Fo}(T_{i+1}^n + T_{i-1}^n) + (1 - 2\mathrm{Fo})T_i^n \end{aligned} \tag{7.5.40}$$

by noting the Fourier number $\mathrm{Fo} = \alpha \Delta t / \Delta x^2$. The stability condition is then $\mathrm{Fo} \leq 1/2$ (Hensen and Nakhi 1994). When the dimensional equation (7.5.6) is considered at the boundaries, in particular in the case of Neumann conditions, the previous discretization must be examined at the boundary and a more restrictive stability condition using both the Fourier and Biot numbers takes place. Note that this stability problem disappears as soon as an implicit discretization scheme is used.

Convergence of the numerical scheme

A finite difference scheme is said to be *convergent* when the difference between the exact analytical solution $u(x, t)$ and the approximate numerical solution u_i^n tends towards 0 when the discretization steps Δt and Δx tend towards 0.

According to Lax theorem, for a well-posed initial value linear problem (i.e. whose solution is unique and continuously depends on the boundary conditions), the consistency and the stability are necessary and sufficient to ensure the convergence.

7.5.2.4 Different Numerical Schemes Applicable to the One-Dimensional Conductive Heat Transfer Equation

In the following, it is proposed to give numerical schemes possible for the same problem of one-dimensional conductive heat transfer modelled by

$$\frac{\partial T}{\partial t} = \alpha \frac{\partial^2 T}{\partial x^2} \tag{7.5.41}$$

which is treated under the dimensionless form (7.5.7).

(a) The scheme (Figure 7.11)

$$\frac{u_i^{n+1} - u_i^n}{\Delta t} = \frac{u_{i-1}^n - 2u_i^n + u_{i+1}^n}{\Delta x^2} \tag{7.5.42}$$

is explicit and stable provided that

$$\frac{\Delta t}{\Delta x^2} \leq \frac{1}{2} \tag{7.5.43}$$

(b) The scheme (Figure 7.12)

$$\frac{u_i^{n+1} - u_i^n}{\Delta t} = \frac{u_{i-1}^{n+1} - 2u_i^{n+1} + u_{i+1}^{n+1}}{\Delta x^2} \tag{7.5.44}$$

is implicit and always stable.

(c) The scheme (Figure 7.13)

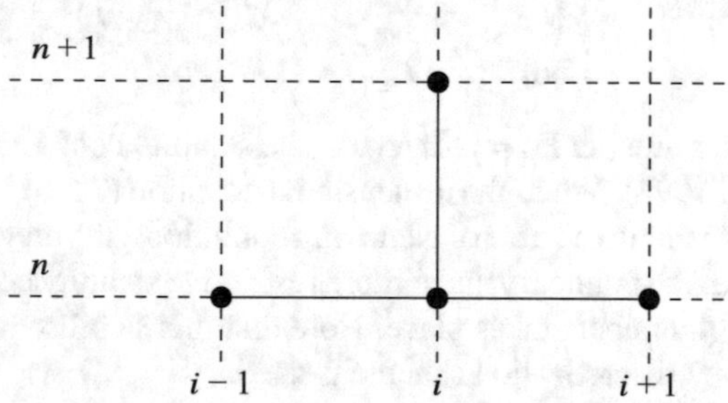

Fig. 7.11 Stencil for the explicit discretization scheme (a)

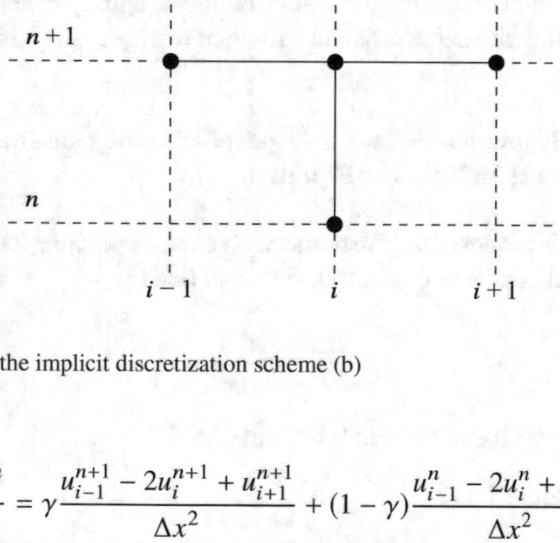

Fig. 7.12 Stencil for the implicit discretization scheme (b)

$$\frac{u_i^{n+1} - u_i^n}{\Delta t} = \gamma \frac{u_{i-1}^{n+1} - 2u_i^{n+1} + u_{i+1}^{n+1}}{\Delta x^2} + (1-\gamma) \frac{u_{i-1}^{n} - 2u_i^{n} + u_{i+1}^{n}}{\Delta x^2} \tag{7.5.45}$$

is implicit and

$$\begin{cases} \text{stable as long as } \dfrac{\Delta t}{\Delta x^2} \leq \dfrac{1}{2(1-2\gamma)} & \text{if } 0 \leq \gamma < 1/2 \\ \text{always stable} & \text{if } 1/2 \leq \gamma \end{cases} \tag{7.5.46}$$

Note that the choice $\gamma = 1/2$ corresponds to the frequently used Crank-Nicolson scheme (Epperson 2013; Hensen and Nakhi 1994). The Crank-Nicolson scheme is thus always stable, meaning that possible oscillations will be damped. The stability does not necessarily mean the convergence towards the right solution. Thus, the stability does not imply that any time step can be used. It will be wise not to use a

too large time step (in practice, the time step should be chosen as large as possible so that a decrease of the time step does not modify the solution).
Now, determine the truncation error for the scheme (7.5.45).
In a similar manner to Equation (7.5.8), we obtain

$$u_i^{n+1} = u_i^n + \Delta t \left(\frac{\partial u}{\partial t}\right)_i^n + \frac{\Delta t^2}{2}\left(\frac{\partial^2 u}{\partial t^2}\right)_i^n + \frac{\Delta t^3}{3!}\left(\frac{\partial^3 u}{\partial t^3}\right)_i^n + 0(\Delta t^4) \tag{7.5.47}$$

hence

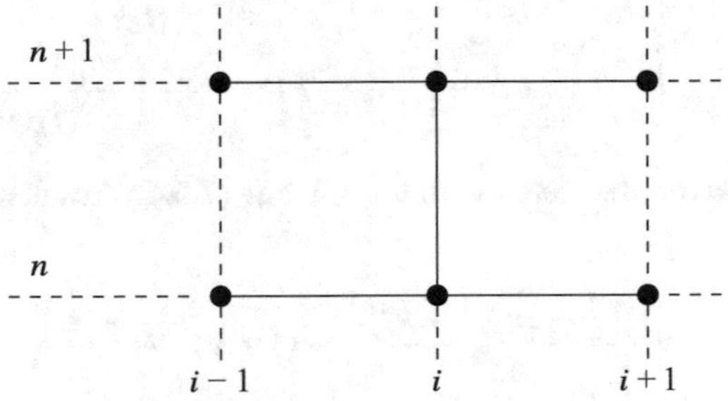

Fig. 7.13 Stencil for the implicit discretization scheme (c)

$$\frac{u_i^{n+1} - u_i^n}{\Delta t} = \left(\frac{\partial u}{\partial t}\right)_i^n + \frac{\Delta t}{2}\left(\frac{\partial^2 u}{\partial t^2}\right)_i^n + \frac{\Delta t^2}{3!}\left(\frac{\partial^3 u}{\partial t^3}\right)_i^n + 0(\Delta t^3) \tag{7.5.48}$$

By using a 6th degree Taylor polynomial similar to (7.5.12), the discretization results

$$\frac{u_{i-1}^n - 2u_i^n + u_{i+1}^n}{\Delta x^2} = \left(\frac{\partial^2 u}{\partial x^2}\right)_i^n + \frac{2\Delta x^2}{4!}\left(\frac{\partial^4 u}{\partial x^4}\right)_i^n + \frac{2\Delta x^4}{6!}\left(\frac{\partial^6 u}{\partial x^6}\right)_i^n + 0(\Delta x^5) \tag{7.5.49}$$

On another side, if we set

$$(\delta^2 u)_i^n = u_{i-1}^n - 2u_i^n + u_{i+1}^n \tag{7.5.50}$$

the second degree Taylor polynomial for $(\delta^2 u)_i^{n+1}$ gives

$$(\delta^2 u)_i^{n+1} = (\delta^2 u)_i^n + \Delta t \frac{\partial}{\partial t}(\delta^2 u)_i^n + \frac{\Delta t^2}{2}\frac{\partial^2}{\partial t^2}(\delta^2 u)_i^n + 0(\Delta t^3) \tag{7.5.51}$$

from which we deduce, keeping only the useful terms,

$$\begin{aligned}\frac{(\delta^2 u)_i^{n+1}}{\Delta x^2} &= \left(\frac{\partial^2 u}{\partial x^2}\right)_i^n + \frac{2\Delta x^2}{4!}\left(\frac{\partial^4 u}{\partial x^4}\right)_i^n + \frac{2\Delta x^4}{6!}\left(\frac{\partial^6 u}{\partial x^6}\right)_i^n \\ &+\Delta t\frac{\partial}{\partial t}\left(\frac{\partial^2 u}{\partial x^2}\right)_i^n + \Delta t\frac{2\Delta x^2}{4!}\frac{\partial}{\partial t}\left(\frac{\partial^4 u}{\partial x^4}\right)_i^n \\ &+\frac{\Delta t^2}{2}\frac{\partial^2}{\partial t^2}\left(\frac{\partial^2 u}{\partial x^2}\right)_i^n + 0(\Delta)\end{aligned} \tag{7.5.52}$$

This expression can be simplified by use of relation (7.5.17) hence

$$\begin{aligned}\frac{(\delta^2 u)_i^{n+1}}{\Delta x^2} &= \left(\frac{\partial^2 u}{\partial x^2}\right)_i^n + \frac{2\Delta x^2}{4!}\left(\frac{\partial^4 u}{\partial x^4}\right)_i^n + \frac{2\Delta x^4}{6!}\left(\frac{\partial^6 u}{\partial x^6}\right)_i^n \\ &+\Delta t\left(\frac{\partial^4 u}{\partial x^4}\right)_i^n + \Delta t\frac{2\Delta x^2}{4!}\left(\frac{\partial^6 u}{\partial x^6}\right)_i^n + \frac{\Delta t^2}{2}\left(\frac{\partial^6 u}{\partial x^6}\right)_i^n + 0(\Delta)\end{aligned} \tag{7.5.53}$$

The truncation error associated with the scheme (7.5.45) results

$$\begin{aligned}&R_{\Delta t,\Delta x}(u) = \\ &\left[\frac{u_i^{n+1}-u_i^n}{\Delta t} - \gamma\frac{u_{i-1}^{n+1}-2u_i^{n+1}+u_{i+1}^{n+1}}{\Delta x^2} - (1-\gamma)\frac{u_{i-1}^n-2u_i^n+u_{i+1}^n}{\Delta x^2}\right] \\ &-\left[\left(\frac{\partial u}{\partial t}\right)_i^n - \left(\frac{\partial^2 u}{\partial x^2}\right)_i^n\right] + 0(\Delta) \\ &= \frac{\Delta t}{2}\left(\frac{\partial^2 u}{\partial t^2}\right)_i^n + \frac{\Delta t^2}{3!}\left(\frac{\partial^3 u}{\partial t^3}\right)_i^n - \frac{2\Delta x^2}{4!}\left(\frac{\partial^4 u}{\partial x^4}\right)_i^n - \frac{2\Delta x^4}{6!}\left(\frac{\partial^6 u}{\partial x^6}\right)_i^n \\ &-\gamma\frac{\Delta t^2}{2}\left(\frac{\partial^6 u}{\partial x^6}\right)_i^n - \gamma\Delta t\left(\frac{\partial^4 u}{\partial x^4}\right)_i^n - \gamma\Delta t\frac{2\Delta x^2}{4!}\left(\frac{\partial^6 u}{\partial x^6}\right)_i^n + 0(\Delta) \\ &= \frac{\Delta t}{2}\left(\frac{\partial^4 u}{\partial x^4}\right)_i^n + \frac{\Delta t^2}{3!}\left(\frac{\partial^6 u}{\partial x^6}\right)_i^n - \frac{2\Delta x^2}{4!}\left(\frac{\partial^4 u}{\partial x^4}\right)_i^n - \frac{2\Delta x^4}{6!}\left(\frac{\partial^6 u}{\partial x^6}\right)_i^n \\ &-\gamma\frac{\Delta t^2}{2}\left(\frac{\partial^6 u}{\partial x^6}\right)_i^n - \gamma\Delta t\left(\frac{\partial^4 u}{\partial x^4}\right)_i^n - \gamma\Delta t\frac{2\Delta x^2}{4!}\left(\frac{\partial^6 u}{\partial x^6}\right)_i^n + 0(\Delta) \\ &= \left(\frac{\partial^4 u}{\partial x^4}\right)_i^n\left[\Delta t(\tfrac{1}{2}-\gamma) - \frac{2\Delta x^2}{4!}\right] \\ &+\left(\frac{\partial^6 u}{\partial x^6}\right)_i^n\left[\frac{\Delta t^2}{3!} - \gamma\frac{\Delta t^2}{2} - \gamma\Delta t\frac{2\Delta x^2}{4!} - \frac{2\Delta x^4}{6!}\right] + 0(\Delta)\end{aligned} \tag{7.5.54}$$

The truncation error is minimized if we choose

$$\Delta t(\frac{1}{2}-\gamma) - \frac{2\Delta x^2}{4!} = 0 \Longrightarrow \gamma = \frac{1}{2} - \frac{2\Delta x^2}{4!\Delta t} \tag{7.5.55}$$

It is noticed that this value of γ implies from (7.5.45) that the corresponding scheme is stable.

On another side, the truncation error for Crank-Nicolson ($\gamma = 1/2$) scheme can be symbolized by $0(\Delta x^2) + 0(\Delta t^2)$ although it is of $0(\Delta x^2) + 0(\Delta t)$ type for the explicit scheme (7.5.42) corresponding to $\gamma = 0$, similarly for the implicit scheme (7.5.44) corresponding to $\gamma = 1$. The Crank-Nicolson scheme should be preferred to those two schemes and will be often used.

(d) The scheme (Figure 7.14)

$$\frac{u_i^{n+1} - u_i^{n-1}}{2\Delta t} = \frac{u_{i-1}^n - u_i^{n+1} - u_i^{n-1} + u_{i+1}^n}{\Delta x^2} \tag{7.5.56}$$

known as Dufort and Frankel scheme is an explicit scheme based on three time steps. The truncation error of this scheme equal to

$$R_{\Delta t,\Delta x}(u) = \left(\frac{(\Delta t)^2}{(\Delta x)^2} - \frac{(\Delta x)^2}{12}\right)\left(\frac{\partial^4 u}{\partial x^4}\right)_i^n + 0(\Delta x^2) + 0\left(\frac{(\Delta t)^2}{(\Delta x)^2}\right) \tag{7.5.57}$$

tends towards 0 only when $\Delta t \ll \Delta x$. This scheme is stable but it is not consistent.

(e) The Gear scheme based on three time steps (Figure 7.15) or BDF scheme (Backward Differentiation Formula)

$$\frac{3u_i^{n+1} - 4u_i^n + u_i^{n-1}}{2\Delta t} = \frac{u_{i-1}^{n+1} - 2u_i^{n+1} + u_{i+1}^{n+1}}{\Delta x^2} \tag{7.5.58}$$

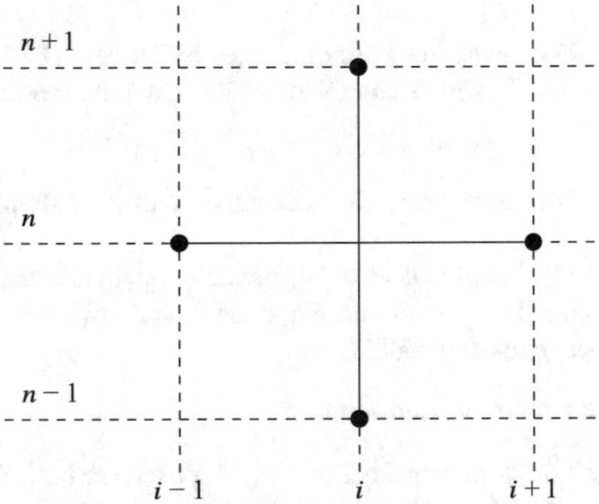

Fig. 7.14 Stencil for the explicit Dufort and Frankel discretization scheme (d) with three time steps

is implicit but makes use of a backward second order difference formula and is qualified as multi-step. The truncation error related to that scheme is equal to

$$R_{\Delta t,\Delta x}(u) = \left(-\frac{(\Delta x)^2}{24}\frac{\partial^4 u}{\partial x^4} + (\Delta t)^2\frac{\partial^6 u}{\partial x^6}\right)_i^{n+1} + 0((\Delta t)^3 + (\Delta x)^3) \tag{7.5.59}$$

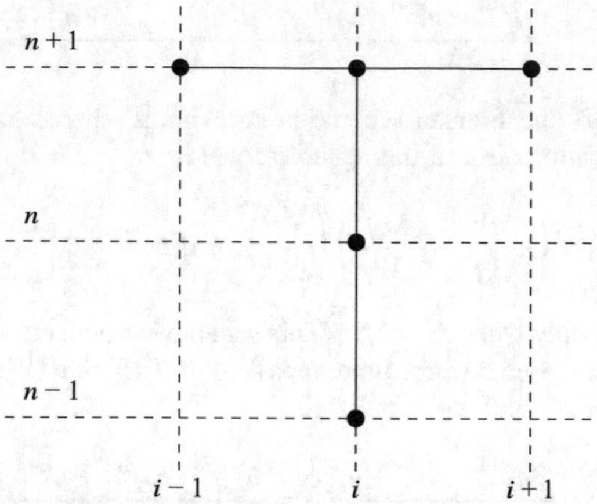

Fig. 7.15 Stencil for the implicit Gear scheme

Higher order methods exist, they are discussed by Iserles (2014), LeVeque (2007), Morton and Mayers (2012), and Wu and White (2004) and their efficiency commented.

7.5.2.5 Influence of the Boundary Conditions on Numerical Solving

In this section, we study the influences of supplementary equations which are the initial conditions and the boundary conditions. Suppose that, at initial time, the temperature of the wall is homogeneous and equal to T_0.

- Case of Dirichlet boundary conditions

It is assumed that the temperature $T(t,x)$ is imposed on both faces (Figure 7.16) since time $t = \epsilon$. The initial and boundary conditions are

$$\begin{array}{ll} T(0,x) = T_0 & \forall\, 0 \le x \le L \\ T(t,0) = T_A & \forall\, 0 < t \\ T(t,L) = T_B & \forall\, 0 < t \end{array} \tag{7.5.60}$$

With N discretization intervals, the space step is

$$\Delta x = \frac{L}{N} \tag{7.5.61}$$

In dimensionless variables (T becomes θ, x becomes xa), the initial and boundary temperatures are, respectively,

$$\begin{array}{ll} \theta_i^0 = \theta_0 & \forall\, 0 \le i \le N \\ \theta_0^n = \theta_A & \forall\, 0 < t \\ \theta_N^n = \theta_B & \forall\, 0 < t \end{array} \tag{7.5.62}$$

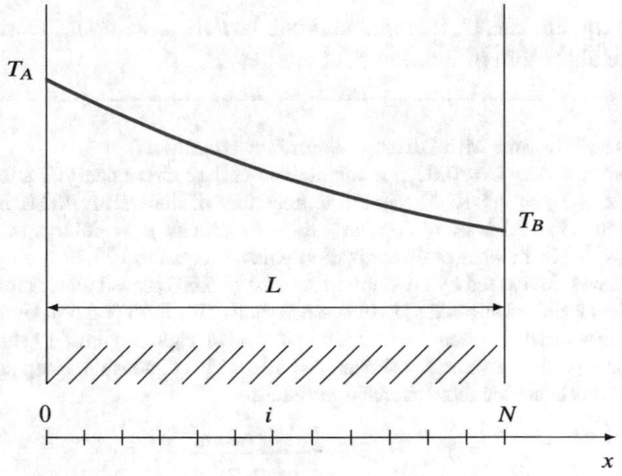

Fig. 7.16 One-dimensional conductive heat transfer: Discretization for a plane wall subjected to Dirichlet boundary conditions

It is assumed that an implicit discretization scheme of the form (7.5.44) is used

$$\frac{\theta_i^{n+1} - \theta_i^n}{\Delta t} = \frac{\theta_{i-1}^{n+1} - 2\theta_i^{n+1} + \theta_{i+1}^{n+1}}{\Delta x^2} \tag{7.5.63}$$

We set

$$r = \frac{\Delta t}{\Delta x^2}$$

hence the system of equations

$$\begin{aligned}
i &= 1, & \theta_1^{n+1} - \theta_1^n &= r(\theta_A - 2\theta_1^{n+1} + \theta_2^{n+1}) \\
i &= 2, & \theta_2^{n+1} - \theta_2^n &= r(\theta_1^{n+1} - 2\theta_2^{n+1} + \theta_3^{n+1}) \\
&\vdots \\
i &= N-2, & \theta_{N-2}^{n+1} - \theta_{N-2}^n &= r(\theta_{N-3}^{n+1} - 2\theta_{N-2}^{n+1} + \theta_{N-1}^{n+1}) \\
i &= N-1, & \theta_{N-1}^{n+1} - \theta_{N-1}^n &= r(\theta_{N-2}^{n+1} - 2\theta_{N-1}^{n+1} + \theta_B)
\end{aligned} \tag{7.5.64}$$

which can be written under matrix form

$$\begin{bmatrix} 1+2r & -r & 0 & \dots & & 0 \\ -r & 1+2r & -r & \ddots & & \vdots \\ 0 & -r & \ddots & \ddots & \ddots & \\ \vdots & & \ddots & \ddots & \ddots & 0 \\ \vdots & & & \ddots & \ddots & -r \\ 0 & \dots & 0 & & -r & 1+2r \end{bmatrix} \begin{bmatrix} \theta_1^{n+1} \\ \vdots \\ \\ \\ \vdots \\ \theta_{N-1}^{n+1} \end{bmatrix} = \begin{bmatrix} \theta_1^n \\ \vdots \\ \\ \\ \vdots \\ \theta_{N-1}^n \end{bmatrix} + r \begin{bmatrix} \theta_A \\ 0 \\ \vdots \\ 0 \\ \theta_B \end{bmatrix} \tag{7.5.65}$$

This system of equations is tridiagonal and can be solved without inverting the matrix by means of the algorithm of Section 5.11.

Example 7.3 :
HT1-FDM-1D: Heat transfer with Dirichlet boundary conditions

A steel wall of thickness $L = 0.02\,\mathrm{m}$ is maintained at the left boundary at a temperature $T_A = 500\,\mathrm{K}$ and at right at $T_B = 400\,\mathrm{K}$. The initial temperature of the wall is 400 K in any point. The thermal conductivity of steel is $\lambda = 26\,\mathrm{W\,m^{-1}\,K^{-1}}$, its density $\rho = 7800\,\mathrm{kg\,m^{-3}}$, heat capacity $C_p = 500\,\mathrm{J\,kg^{-1}\,K^{-1}}$. The number of discretization points is equal to 101, including the boundaries. Fourier's equation was discretized by the implicit scheme (7.5.63) as well under dimensional form as under dimensionless form and allowed us to obtain a system of the form (7.5.65). Under dimensionless form, the temperature varies between $\theta = 1$ at left and $\theta = 0$ at right. Figure 7.17 shows the evolution of the dimensionless profile between $t = 0\,\mathrm{s}$ and $t = 18\,\mathrm{s}$ with a dimensional step equal to 0.6 s. The abscissa and the ordinate are the dimensionless variables.

$$X = \frac{x}{L}, \qquad \theta = \frac{T - T_{min}}{T_{max} - T_{min}} \tag{7.5.66}$$

with $T_{min} = T_B, T_{max} = T_A$. The dimensionless time is

$$\tau = \frac{\alpha t}{L^2} \qquad \text{with } \alpha = \frac{\lambda}{\rho C_p} \tag{7.5.67}$$

Different time steps ($\Delta t = 10^{-2}\,\mathrm{s}, 10^{-3}\,\mathrm{s}, 10^{-4}\,\mathrm{s}$) were used without noticeable influence. Figure 7.18 shows the evolution of profiles between $t = 0\,\mathrm{s}$ and $t = 20\,\mathrm{s}$ every second. The profile is practically stationary at $t = 20\,\mathrm{s}$.

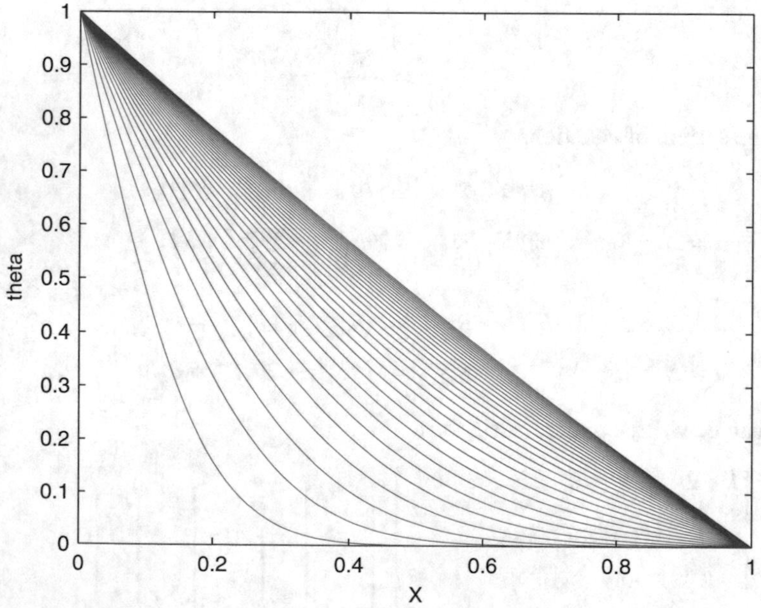

Fig. 7.17 Example HT1-FDM-1D: Dimensionless temperature profile in a wall subject to Dirichlet conditions at both boundaries ($t_{max} = 18\,\mathrm{s}$, visualization step = 0.6 s)

In this simple case of Dirichlet boundary conditions, the theoretical solution of the parabolic dimensional equation can be found

$$\frac{\partial u}{\partial t} = c^2 \frac{\partial^2 u}{\partial x^2} \tag{7.5.68}$$

subject to boundary conditions

$$u(0, t) = u_0 , \quad u(L, t) = u_L \tag{7.5.69}$$

and to the initial condition

$$u(x, 0) = u_{in}(x) \tag{7.5.70}$$

In the case of heat transfer

$$c = \sqrt{\alpha} = \sqrt{\frac{\lambda}{\rho C_p}} \tag{7.5.71}$$

We start by searching a particular solution $u_1(x, t)$ of this problem with the boundary conditions and simplifying the initial condition as $u_1(x, 0) = 0$, hence

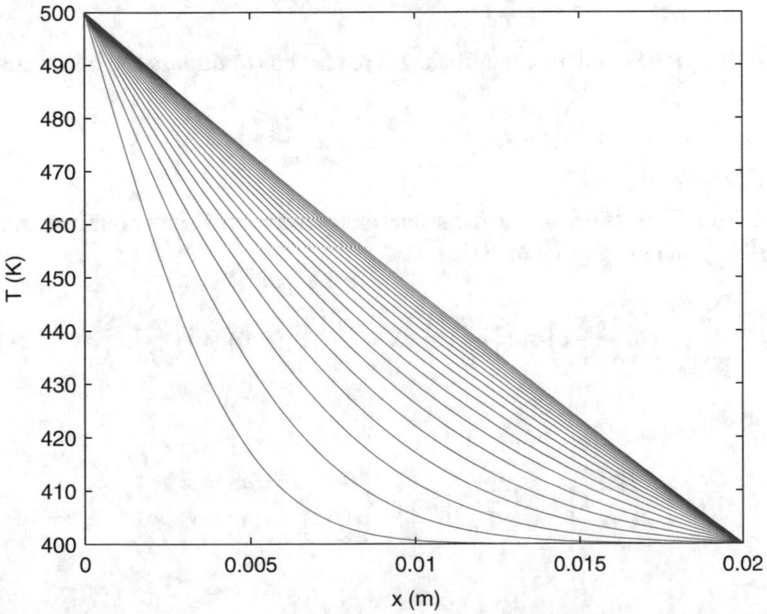

Fig. 7.18 Example HT1-FDM-1D: Dimensional temperature profile in a wall subject to Dirichlet conditions at both boundaries (t_{max} = 20 s, visualization step = 1 s)

$$u_1(x, t) = u_0 + (u_L - u_0)\frac{x}{L} \tag{7.5.72}$$

Setting $u = u_1 + u_2$, we deduce that $u_2(x,t)$ is a solution of the problem

$$\frac{\partial u_2}{\partial t} = c^2 \frac{\partial^2 u_2}{\partial x^2} \tag{7.5.73}$$

subject to boundary conditions

$$u_2(0,t) = 0\,, \quad u_2(L,t) = 0 \tag{7.5.74}$$

and to the initial condition

$$u_2(x,0) = u_{in}(x) - u_0 - (u_L - u_0)\frac{x}{L} \tag{7.5.75}$$

Looking for the solution of the partial differential equation by separation of variables, i.e.

$$u_2(x,t) = f(t)\,g(x) \tag{7.5.76}$$

we find that the solution is the form

$$u_2(x,t) = \sum_{k=1}^{\infty} a_k \exp\left(-\frac{k^2\pi^2c^2}{L^2}t\right) \sin\left(\frac{k\pi}{L}x\right) \tag{7.5.77}$$

which verifies the boundary conditions. There remains to impose the initial condition

$$u_2(x,0) = \sum_{k=1}^{\infty} a_k \sin\left(\frac{k\pi}{L}x\right) \tag{7.5.78}$$

To find the coefficients a_k, we use the properties of orthogonal functions by multiplying by $\sin(\frac{m\pi}{L}x)$ and integrating on $[0, L]$, i.e.

$$\int_0^L \sum_{k=1}^{\infty} a_k \sin\left(\frac{k\pi}{L}x\right) \sin\left(\frac{m\pi}{L}x\right) dx = \int_0^L u_2(x,0) \sin\left(\frac{m\pi}{L}x\right) dx \tag{7.5.79}$$

Knowing that

$$\int_0^L \sin\left(\frac{k\pi}{L}x\right) \sin\left(\frac{m\pi}{L}x\right) dx = \begin{cases} 0 & \text{if } m \neq k \geq 1 \\ \frac{L}{2} & \text{if } m = k \geq 1 \end{cases} \tag{7.5.80}$$

it results

$$a_k = \frac{2}{L}\int_0^L u_2(x,0) \sin\left(\frac{k\pi}{L}x\right) dx\,, \quad k \geq 1 \tag{7.5.81}$$

where $u_2(x,0)$ is given by Equation (7.5.75).

Thus, the general solution is

$$u(x,t) = \sum_{k=1}^{\infty} a_k \exp\left(-\frac{k^2\pi^2c^2}{L^2}t\right) \sin\left(\frac{k\pi}{L}x\right) + u_0 + (u_L - u_0)x \tag{7.5.82}$$

In the case of Example 7.3, the comparison between the theoretical solution and the numerical solution issued from finite differences is performed at $t = 6$ s. In the calculation, the series of the general solution is truncated at $n = 1000$. The comparison is given in Figure 7.19. The agreement is excellent.

- Case of Neumann boundary conditions[2]

When the heat flux is specified at the wall, the boundary condition is called Neumann. The heat flux can be of different natures:

– known numerically. In this case, it is taken into account in the same way as a linear relation which relates two temperatures.
– convective. The heat flux expression is

$$-\lambda \frac{\partial T}{\partial x} = h(T_a - T_b) \tag{7.5.83}$$

which is linear.
– radiative. The heat flux expression is grossly

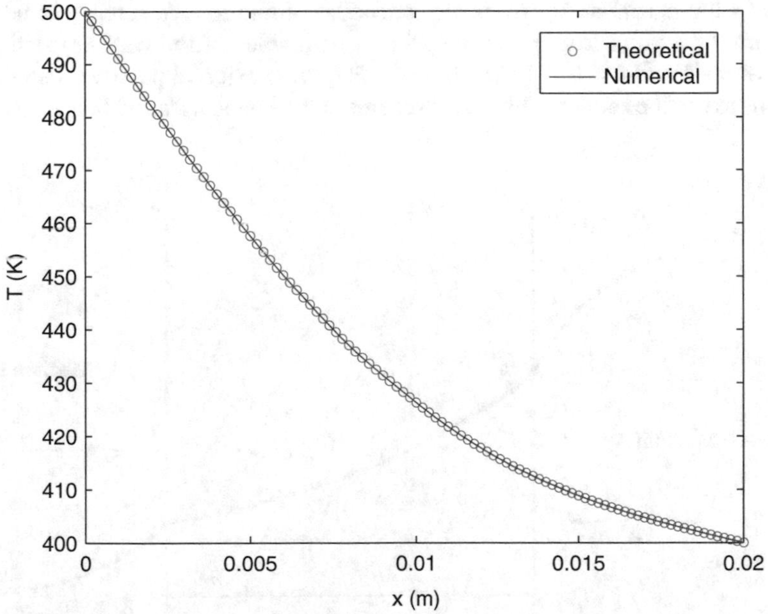

Fig. 7.19 Example HT1-FDM-1D: Comparison between the theoretical and numerical dimensional temperature profiles in a wall subject to Dirichlet conditions at both boundaries ($t = 6$ s)

[2] Neumann comes from the German mathematician Carl Neumann (1832–1925) and must not be confused with John Von Neumann (Janos Làjos Neumann), a brilliant Hungarian-American mathematician and physician (1903–1957).

$$-\lambda \frac{\partial T}{\partial x} = \sigma \epsilon (T_a^4 - T_b^4) \tag{7.5.84}$$

which is nonlinear. Its use in a finite difference scheme is not obvious, and this expression is often approximated by a linear expression

$$-\lambda \frac{\partial T}{\partial x} = \sigma \epsilon (T_a - T_b)(T_a + T_b)(T_a^2 + T_b^2) \approx c(T_a - T_b) \tag{7.5.85}$$

where c would be taken as a constant.

Thus, in all cases, we work as if the expression of the heat flux was linear and we consider the following case of a wall heated by the external medium A on the face $x = 0$ and evacuating the heat towards the external medium B on the face $x = L$ (Figure 7.20)

$$\begin{aligned} \phi_1 &= -\lambda \left(\frac{\partial T}{\partial x}\right)_{x=0} = h_0(T_A - T_0) \\ \phi_2 &= -\lambda \left(\frac{\partial T}{\partial x}\right)_{x=L} = h_L(T_L - T_B) \end{aligned} \tag{7.5.86}$$

where T_B indicates the temperature of the external medium on the side $x = L$ far from the wall, similarly for T_A.

The spatial partial derivative is discretized according to a centered scheme which makes use of a point external to the grid on both sides of the wall, respectively, at indexes $i = -1$ and $i = N + 1$. The fact of using these external points, called dummy, allows us to use the centered difference scheme at the boundaries and thus to guarantee

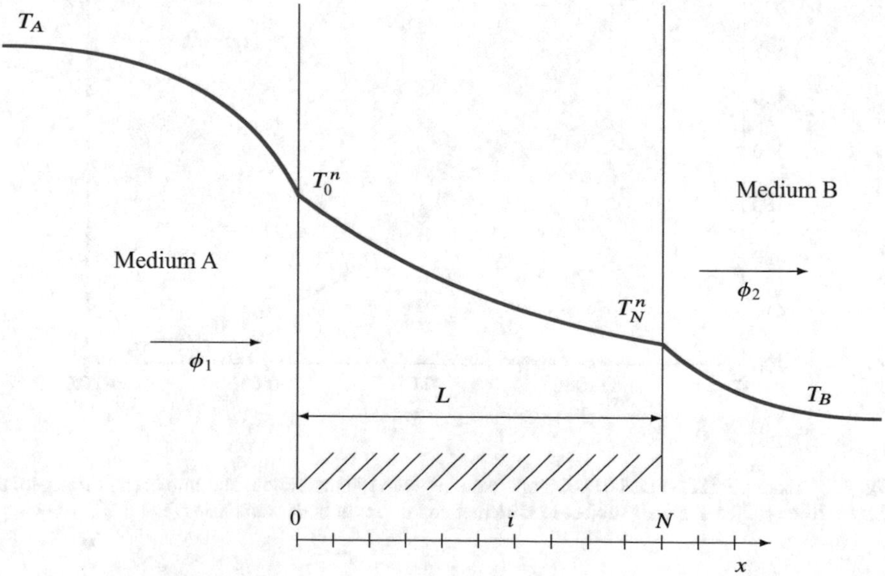

Fig. 7.20 One-dimensional conductive heat transfer: Discretization for a plane wall subjected to Neumann boundary conditions

the consistency of the numerical method with respect to the discretization used in Equation (7.5.63) for Fourier's law. In dimensionless variables, we get

$$\begin{aligned} -\left(\frac{\partial\theta}{\partial xa}\right)_0 &= Bi_A(\theta_A - \theta_0) \approx \frac{\theta_{-1} - \theta_1}{2\Delta x} \\ -\left(\frac{\partial\theta}{\partial xa}\right)_N &= Bi_B(\theta_N - \theta_B) \approx \frac{\theta_{N-1} - \theta_{N+1}}{2\Delta x} \end{aligned} \tag{7.5.87}$$

where $Bi = hL'/\lambda$ is the Biot number, L' being a characteristic dimension. If the wall is insulated, the heat flux is zero and we set $Bi = 0$.

The equations ruling the system are

$$\begin{aligned} &i = 0, &\theta_0^{n+1} - \theta_0^n &= r(\theta_{-1}^{n+1} - 2\theta_0^{n+1} + \theta_1^{n+1}) \\ &i = 1, &\theta_1^{n+1} - \theta_1^n &= r(\theta_0^{n+1} - 2\theta_1^{n+1} + \theta_2^{n+1}) \\ &\vdots \\ &i = N-1, &\theta_{N-1}^{n+1} - \theta_{N-1}^n &= r(\theta_{N-2}^{n+1} - 2\theta_{N-1}^{n+1} + \theta_N^{n+1}) \\ &i = N, &\theta_N^{n+1} - \theta_N^n &= r(\theta_{N-1}^{n+1} - 2\theta_N^{n+1} + \theta_{N+1}^{n+1}) \end{aligned} \tag{7.5.88}$$

Taking into account the boundary conditions (7.5.87), the two previous extreme equations are modified as

$$\begin{aligned} &i = 0, \quad \theta_0^{n+1} - \theta_0^n = r(-2\theta_0^{n+1} + \theta_1^{n+1}) + r\overbrace{[\theta_1^{n+1} + 2Bi_A\Delta x(\theta_A - \theta_0^{n+1})]}^{\theta_{-1}^{n+1}} \\ &i = N, \quad \theta_N^{n+1} - \theta_N^n = r(\theta_{N-1}^{n+1} - 2\theta_N^{n+1}) + r\overbrace{[\theta_{N-1}^{n+1} - 2Bi_B\Delta x(\theta_N^{n+1} - \theta_B)]}^{\theta_{N+1}^{n+1}} \end{aligned} \tag{7.5.89}$$

The system can be written under matrix form

$$\mathcal{A}\begin{bmatrix} \theta_0^{n+1} \\ \vdots \\ \\ \vdots \\ \theta_N^{n+1} \end{bmatrix} = \begin{bmatrix} \theta_0^n \\ \vdots \\ \\ \vdots \\ \theta_N^n \end{bmatrix} + 2r\Delta x \begin{bmatrix} Bi_A\theta_A \\ 0 \\ \vdots \\ 0 \\ Bi_B\theta_B \end{bmatrix} \tag{7.5.90}$$

with

$$\mathcal{A} = \begin{bmatrix} 1+2r(1+Bi_A\Delta x) & -2r & 0 & \dots & & 0 \\ -r & 1+2r & -r & \ddots & & \vdots \\ 0 & -r & \ddots & \ddots & \ddots & \\ \vdots & & \ddots & \ddots & \ddots & 0 \\ \vdots & & & -r & 1+2r & -r \\ 0 & \dots & & 0 & -2r & 1+2r(1+Bi_B\Delta x) \end{bmatrix} \tag{7.5.91}$$

Again, this is a tridiagonal system whose solving is directly possible by the algorithm of Section 5.11.

Example 7.4 :

HT2-FDM-1D: Heat transfer with Neumann boundary conditions

A steel wall of thickness $L = 0.02$ m is subject at the left boundary to a convective flux with temperature $T_A = 500$ K at infinity and at right to a convective flux with temperature $T_B = 400$ K at infinity. The initial wall temperature is 450 K in any point. The steel thermal conductivity is $\lambda = 26\,\mathrm{W\,m^{-1}\,K^{-1}}$, its density $\rho = 7800\,\mathrm{kg\,m^{-3}}$, heat capacity $C_p = 500\,\mathrm{J\,kg^{-1}\,K^{-1}}$. The heat transfer coefficient is $h = 1000\,\mathrm{W\,m^{-2}\,K^{-1}}$. The number of discretization points is equal to 101, including the boundaries. Fourier's equation was discretized by an implicit scheme of type (7.5.63) under dimensional form allowing us to obtain a tridiagonal system similar to (7.5.91). Different time steps ($\Delta t = 10^{-2}$ s, 10^{-3} s, 10^{-4} s) have been used without any noticeable influence. Figure 7.21 shows the evolution of the profiles between $t = 0$ s and $t = 20$ s every second. The final profile is a straight line between a maximum temperature at left where a positive flux is imposed and a minimum temperature at right where heat is yielded to environment.

In the case of the previous Neumann boundary conditions, the theoretical solution cannot be easily found as the boundary conditions are time-dependent. To simplify the problem, we assume that the Neumann boundary conditions are fixed and we attempt to find the theoretical solution of the parabolic dimensional equation

$$\frac{\partial u}{\partial t} = c^2 \frac{\partial^2 u}{\partial x^2} \tag{7.5.92}$$

subject to the boundary conditions

$$-\left(\frac{\partial u}{\partial x}\right)_{x=0} = F_0\,, \quad -\left(\frac{\partial u}{\partial x}\right)_{x=L} = F_L \tag{7.5.93}$$

and to the initial condition

$$u(x,0) = u_0(x) \tag{7.5.94}$$

In the case of heat transfer

$$c = \sqrt{\alpha} = \sqrt{\frac{\lambda}{\rho C_p}} \tag{7.5.95}$$

and

$$-\lambda\left(\frac{\partial u}{\partial x}\right)_{x=0} = \phi_0\,, \quad -\lambda\left(\frac{\partial u}{\partial x}\right)_{x=L} = \phi_L \tag{7.5.96}$$

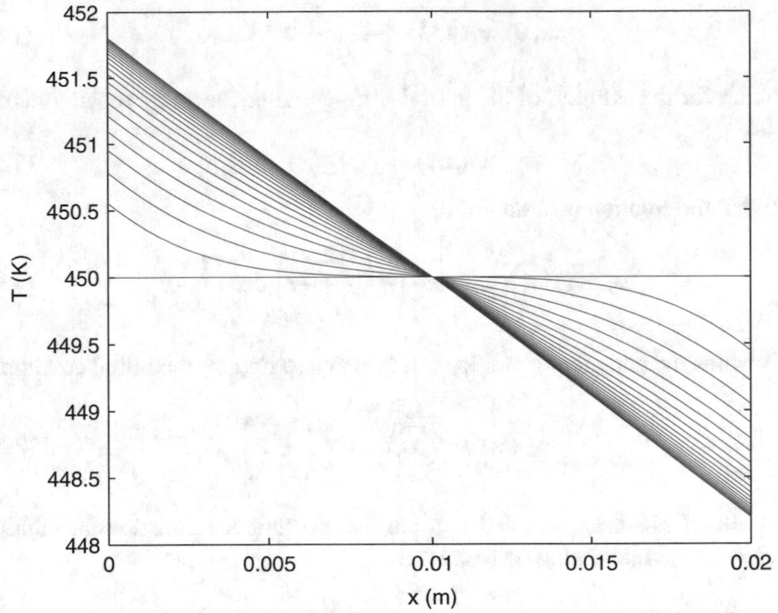

Fig. 7.21 Example HT2-FDM-1D: Temperature profile in a wall subjected to Neumann conditions at both boundaries ($t_{max} = 20$ s, visualization step = 1 s)

so that

$$-\left(\frac{\partial u}{\partial x}\right)_{x=0} = F_0\,, \quad -\left(\frac{\partial u}{\partial x}\right)_{x=L} = F_L \tag{7.5.97}$$

with

$$F_0 = \frac{\phi_0}{\lambda}\,, \quad F_L = \frac{\phi_L}{\lambda} \tag{7.5.98}$$

First, we search a particular solution $u_1(x, t)$ of the problem with the boundary conditions without immediately considering the initial condition, hence

$$u_1(x, t) = \frac{F_0 - F_L}{2L}\,x^2 - F_0\,x + c^2\,\frac{F_0 - F_L}{L}\,t \tag{7.5.99}$$

Setting $u = u_1 + u_2$, we deduce that $u_2(x, t)$ is a solution of the new problem

$$\frac{\partial u_2}{\partial t} = c^2\,\frac{\partial^2 u_2}{\partial x^2} \tag{7.5.100}$$

subject to the boundary conditions

$$\left(\frac{\partial u}{\partial x}\right)_{x=0} = 0\,, \quad \left(\frac{\partial u}{\partial x}\right)_{x=L} = 0 \tag{7.5.101}$$

and to the initial condition

$$u_2(x,0) = u_0(x) - \frac{F_0 - F_L}{2L} x^2 + F_0\, x \tag{7.5.102}$$

Looking for the solution of the partial differential equation by separation of variables, i.e.

$$u_2(x,t) = f(t)\, g(x) \tag{7.5.103}$$

we find that the solution is of the form

$$u_2(x,t) = \sum_{k=0}^{\infty} a_k \exp\left(-\frac{k^2\pi^2c^2}{L^2} t\right) \cos\left(\frac{k\pi}{L} x\right) \tag{7.5.104}$$

which verifies the boundary conditions. It remains to impose the initial condition

$$u_2(x,0) = \sum_{k=0}^{\infty} a_k \cos\left(\frac{k\pi}{L} x\right) \tag{7.5.105}$$

To find the coefficients a_k, we use the properties of orthogonal functions by multiplying by $\cos(\frac{m\pi}{L} x)$ and integrating on $[0, L]$, i.e.

$$\int_0^L \sum_{k=1}^{\infty} a_k \cos\left(\frac{k\pi}{L} x\right) \cos\left(\frac{m\pi}{L} x\right) dx = \int_0^L u_2(x,0) \cos\left(\frac{m\pi}{L} x\right) dx \tag{7.5.106}$$

Knowing that

$$\int_0^L \cos\left(\frac{k\pi}{L} x\right) \cos\left(\frac{m\pi}{L} x\right) dx = \begin{cases} L & \text{if } m = k = 0 \\ 0 & \text{if } m \geq 1 \text{ and } m \neq k \\ \frac{L}{2} & \text{if } m = k \geq 1 \end{cases} \tag{7.5.107}$$

it results

$$\begin{aligned} a_0 &= \frac{1}{L} \int_0^L u_2(x,0)\, dx \\ a_k &= \frac{2}{L} \int_0^L u_2(x,0) \cos\left(\frac{k\pi}{L} x\right) dx \quad , \; k \geq 1 \end{aligned} \tag{7.5.108}$$

Thus the general solution is

$$u(x,t) = \sum_{k=0}^{\infty} a_k \exp\left(-\frac{k^2\pi^2c^2}{L^2} t\right) \cos\left(\frac{k\pi}{L} x\right) + \frac{F_0 - F_L}{2L} x^2 - F_0\, x + c^2 \frac{F_0 - F_L}{L} t \tag{7.5.109}$$

We notice that this solution does not admit a steady state except in the case where $F_0 = F_L$, which corresponds to the equality of the fluxes at left and at right. Considering a heat transfer problem, in the absence of flux equality, the material would be able to indefinitely store heat if $F_0 > F_L$ or indefinitely lose heat if $F_0 < F_L$, which is not physical. Thus, the hypothesis of different fixed fluxes is a mathematical hypothesis, not physical.

In the case of Example 7.4, it is possible to make a comparison between the theoretical solution and the numerical solution only when the steady state is reached, that is when the heat fluxes at left and at right are equal. The time $t = 20$ s was taken as close to the steady state. For the calculations, the series was truncated at $n = 2000$. We found $T_0 = 463.74$ K at left boundary and $T_L = 436.26$ K at right boundary, which corresponds to the equality of the left and right heat fluxes

$$\phi_0 = h(T_A - T_0) = \phi_L = h(T_L - T_B) = 3.6261 \times 10^4 \tag{7.5.110}$$

hence

$$F_0 = F_L = 1.3946 \times 10^3 \tag{7.5.111}$$

With these values, it is possible to perform the comparison and confirm the steady state (Figure 7.22). The agreement between the theoretical and numerical dimensional temperature profiles is very good.

7.5.2.6 Discretization for the One-Dimensional Conductive Heat Transfer Equation in Cylindrical Geometry

We consider a cylindrical rod (Figure 7.23) infinite in the longitudinal direction subject to crosswise one-dimensional conductive heat transfer. The cylindrical geometry leads to the following heat transfer equation

$$\frac{\partial T}{\partial t} = \alpha \left(\frac{\partial^2 T}{\partial r^2} + \frac{1}{r} \frac{\partial T}{\partial r} \right) \tag{7.5.112}$$

Remark: In the case of an isotropic sphere, the heat transfer equation becomes

$$\frac{\partial T}{\partial t} = \alpha \left(\frac{\partial^2 T}{\partial r^2} + \frac{2}{r} \frac{\partial T}{\partial r} \right) \tag{7.5.113}$$

Equation (7.5.112) is first transformed under dimensionless form. The spatial discretization is adopted such that $r = i\Delta r$, i.e. $i = 0$ on the axis and $i = N$ for $r = R$. The following numerical schemes can be obtained, either combining the explicit and implicit forms

$$\begin{aligned} \frac{u_i^{n+1} - u_i^n}{\Delta t} = &\gamma \left(\frac{u_{i-1}^{n+1} - 2u_i^{n+1} + u_{i+1}^{n+1}}{\Delta r^2} + \frac{1}{i\Delta r} \frac{u_{i+1}^{n+1} - u_{i-1}^{n+1}}{2\Delta r} \right) \\ &+(1-\gamma) \left(\frac{u_{i-1}^{n} - 2u_i^{n} + u_{i+1}^{n}}{\Delta r^2} + \frac{1}{i\Delta r} \frac{u_{i+1}^{n} - u_{i-1}^{n}}{2\Delta r} \right), \quad i = 0, \dots, N \end{aligned} \tag{7.5.114}$$

or a totally implicit scheme

$$\frac{u_i^{n+1} - u_i^n}{\Delta t} = \left(\frac{u_{i-1}^{n+1} - 2u_i^{n+1} + u_{i+1}^{n+1}}{\Delta r^2} + \frac{1}{i\Delta r} \frac{u_{i+1}^{n+1} - u_{i-1}^{n+1}}{2\Delta r} \right), \quad i = 0, \dots, N \tag{7.5.115}$$

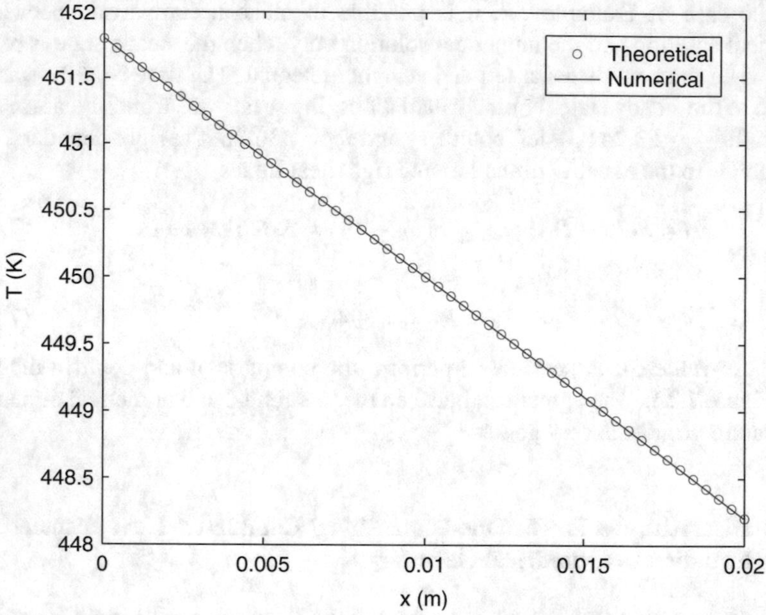

Fig. 7.22 Example HT2-FDM-1D: Comparison between the theoretical and numerical dimensional temperature profiles in a wall subjected to Neumann conditions at both boundaries ($t = 20$ s)

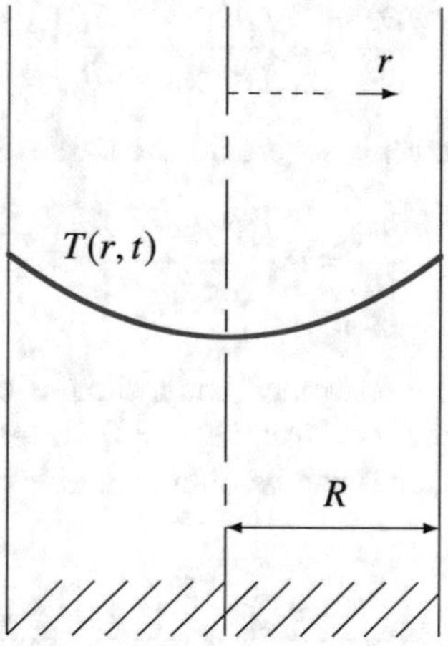

Fig. 7.23 Infinite cylindrical rod subjected to crosswise conductive heat transfer

The first order spatial derivative has been discretized by a centered scheme so that the order of magnitude of the error is the same as that of the second order spatial derivative. When the system is implicit, it is resumed as previously to a tridiagonal linear problem.

However, a small difficulty resides in the denominator term $r = i\Delta r$ which is zero at the cylinder axis (or center in the projected cross section). To solve that point, we consider a first degree Taylor polynomial in the neighborhood of the center as

$$\frac{\partial T}{\partial r} = \left(\frac{\partial T}{\partial r}\right)_{r=0} + r\frac{\partial^2 T}{\partial r^2} + 0(r^2) \tag{7.5.116}$$

and we use the symmetry condition of the temperature profile

$$\left(\frac{\partial T}{\partial r}\right)_{r=0} = 0 \tag{7.5.117}$$

which gives under discretized and dimensionless form

$$\frac{u_{i+1}^{n+1} - u_{i-1}^{n+1}}{2\Delta r} = 0 \qquad \text{for } i = 0 \Rightarrow u_{-1}^{n+1} = u_1^{n+1} \tag{7.5.118}$$

Equation (7.5.112) is then transformed at the center under the form

$$\frac{\partial T}{\partial t} \approx 2\alpha \left(\frac{\partial^2 T}{\partial r^2}\right)_{r=0} \tag{7.5.119}$$

which is expressed under discretized and dimensionless form according to the implicit scheme

$$\frac{u_i^{n+1} - u_i^n}{\Delta t} = 2\frac{2u_{i+1}^{n+1} - 2u_i^{n+1}}{\Delta r^2} \qquad \text{for } i = 0 \tag{7.5.120}$$

which is used in Equation (7.5.118). Thus, Equations (7.5.114) and (7.5.115) are only valid for $1 \le i \le N$.

7.5.2.7 Different Numerical Schemes for the Two-Dimensional Conductive Heat Transfer Equation

We consider, in transient regime, a plane plate subjected to conductive heat transfer requiring to take into account two space directions x and y as

$$\frac{\partial T}{\partial t} = \alpha \left(\frac{\partial^2 T}{\partial x^2} + \frac{\partial^2 T}{\partial y^2}\right) \tag{7.5.121}$$

When the transient phenomenon (term $\partial T/\partial t$) is taken into account, the equation is still parabolic. In the steady state case, it would be an elliptic equation. The problem is transformed under dimensionless form similar to (7.5.7).

We can apply to this equation a scheme including an explicit and an implicit part of the form

$$\frac{u_{i,j}^{n+1} - u_{i,j}^{n}}{\Delta t} = \gamma \left(\frac{u_{i-1,j}^{n+1} - 2u_{i,j}^{n+1} + u_{i+1,j}^{n+1}}{\Delta x^2} + \frac{u_{i,j-1}^{n+1} - 2u_{i,j}^{n+1} + u_{i,j+1}^{n+1}}{\Delta y^2} \right)$$
$$+(1-\gamma) \left(\frac{u_{i-1,j}^{n} - 2u_{i,j}^{n} + u_{i+1,j}^{n}}{\Delta x^2} + \frac{u_{i,j-1}^{n} - 2u_{i,j}^{n} + u_{i,j+1}^{n}}{\Delta y^2} \right) \tag{7.5.122}$$

which requires for the stability to respect the following condition of time and space steps

$$\begin{cases} \text{stable if } r_x + r_y \leq \dfrac{1}{2(1-2\gamma)} & \text{if } 0 \leq \gamma < 1/2 \\ \text{always stable} & \text{if } 1/2 \leq \gamma \leq 1 \end{cases} \tag{7.5.123}$$

with

$$r_x = \frac{\Delta t}{\Delta x^2}, \quad r_y = \frac{\Delta t}{\Delta y^2}$$

The explicit scheme obtained for $\gamma = 0$ poses no problem for solving.

In the case $\gamma = 1$, the scheme is implicit. With respect to the similar case of one-dimensional heat transfer, the problem is more complex as the vector of unknowns is replaced by a matrix of unknowns $u_{i,j}^n$ and the previous tridiagonal matrix has its elements replaced by matrices which are either diagonal or tridiagonal and this new matrix is block-tridiagonal. The solution based on these properties is possible, but complex. A fractional step method is preferred, such as the alternate direction method which is decomposed in two time half-steps

– the first time half-step implicit with respect to x and explicit w.r.t. y

$$\frac{u_{i,j}^{n+1/2} - u_{i,j}^{n}}{\Delta t} = \frac{u_{i-1,j}^{n+1/2} - 2u_{i,j}^{n+1/2} + u_{i+1,j}^{n+1/2}}{\Delta x^2} + \frac{u_{i,j-1}^{n} - 2u_{i,j}^{n} + u_{i,j+1}^{n}}{\Delta y^2} \tag{7.5.124}$$

– the second time half-step explicit w.r.t. x and implicit w.r.t. y

$$\frac{u_{i,j}^{n+1} - u_{i,j}^{n+1/2}}{\Delta t} = \frac{u_{i-1,j}^{n+1/2} - 2u_{i,j}^{n+1/2} + u_{i+1,j}^{n+1/2}}{\Delta x^2} + \frac{u_{i,j-1}^{n+1} - 2u_{i,j}^{n+1} + u_{i,j+1}^{n+1}}{\Delta y^2} \tag{7.5.125}$$

The solution of a time step thus consists in sequentially solving two tridiagonal systems and amounts to a procedure similar to that used in the one-dimensional case.

Example 7.5 :
HT3-FDM-2D: Two-dimensional (2D) steady-state Fourier's equation

We consider the two-dimensional (2D) transient Fourier's equation

$$\frac{\partial T}{\partial t} = \frac{\partial^2 T}{\partial x^2} + \frac{\partial^2 T}{\partial y^2} \tag{7.5.126}$$

which, in the steady case, results in Laplace equation (7.2.8). We consider the steady state example of a plane plate subjected to conductive heat transfer and to Dirichlet conditions on the four sides (Figure 7.24), thus the term $\partial T / \partial t$ is equal to 0. The Dirichlet conditions are noted $T(x=0, y) = T_{x0}$, $T(x=L_x, y) = T_{xL_x}$, $T(x, y=0) = T_{y0}$, $T(x, y=L_y) = T_{yL_y}$.

It is supposed that $x \in [0, L_x]$ and $y \in [0, L_y]$. These discretized coordinates are such that $x_i = (i-1)\Delta x$; $(i = 1, \dots, n_x)$ and $y_j = (j-1)\Delta y$; $(j = 1, \dots, n_y)$ with $\Delta x = L_x/(n_x - 1)$ and $\Delta y = L_y/(n_y - 1)$. We set $r = \Delta x/\Delta y$.

This problem has been solved by two different finite difference methods, the first one being exposed below. The second method based on the method of lines is exposed in Example 7.9.

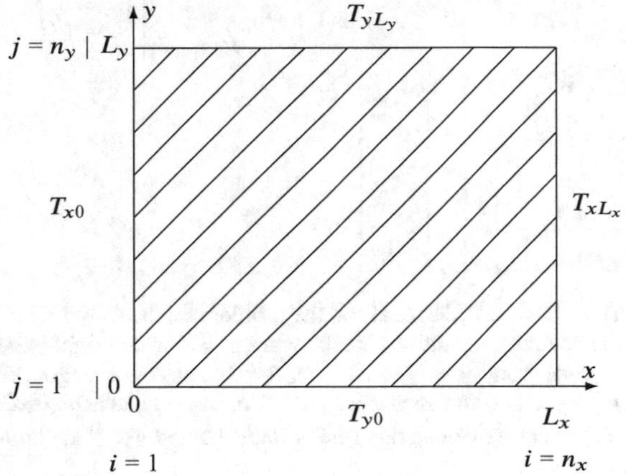

Fig. 7.24 Example HT3-FDM-2D: 2D conductive heat transfer: Plane plate subjected to Dirichlet conditions on the four sides

In this method, the steady state equation (7.5.126) under discretized form becomes

$$\frac{T_{i-1,j} - 2T_{i,j} + T_{i+1,j}}{\Delta x^2} + \frac{T_{i,j-1} - 2T_{i,j} + T_{i,j+1}}{\Delta y^2} = 0 \tag{7.5.127}$$

The solution of this system of linear equations will be found by using a method similar to the alternate directions in two stages. First considering a given index j and a method sweeping the x axis with i increasing from 2 to $(n_x - 1)$, the variable $T_{i-1,j}$ is known, the variables $T_{i,j}$ are unknown as well as $T_{i+1,j}$. Thus we consider an iterative alternate processes of convergence $(k = 0, \dots)$ so that $T_{i-1,j}$ and $T_{i+1,j}$ are known by their previous values and the variables of index i are to be determined.

Equation (7.5.127) thus gives, $\forall\, j = 2, \dots, n_y - 1$, at iteration k of the iterative calculation (the variables noted $y^{(k)}$ are known and those noted $y^{(k+1)}$ are to be determined)

$$\begin{aligned}
&i = 2, \quad T_{x0} - 2T_{2,j}^{(k+1)} + T_{3,j}^{(k+1)} + r^2\left[T_{2,j-1}^{(k)} - 2T_{2,j}^{(k+1)} + T_{2,j+1}^{(k)}\right] = 0\\
&i = 3, \dots, n_x - 2, \quad T_{i-1,j}^{(k+1)} - 2T_{i,j}^{(k+1)} + T_{i+1,j}^{(k+1)} + r^2\left[T_{i,j-1}^{(k)} - 2T_{i,j}^{(k+1)} + T_{i,j+1}^{(k)}\right] = 0\\
&i = n_x - 1, \quad T_{n_x-2,j}^{(k+1)} - 2T_{n_x-1,j}^{(k+1)} + T_{xL_x} + r^2\left[T_{n_x-1,j-1}^{(k)} - 2T_{n_x-1,j}^{(k+1)} + T_{n_x-1,j+1}^{(k)}\right] = 0
\end{aligned} \tag{7.5.128}$$

which can be posed under matrix form

$$\begin{bmatrix} 2(1+r^2) & -1 & 0 & \dots & \dots & 0 \\ -1 & 2(1+r^2) & -1 & 0 & \dots & \vdots \\ 0 & -1 & 2(1+r^2) & -1 & 0 & \vdots \\ \vdots & \ddots & \ddots & \ddots & \ddots & 0 \\ \vdots & \dots & 0 & -1 & 2(1+r^2) & -1 \\ 0 & \dots & \dots & 0 & -1 & 2(1+r^2) \end{bmatrix} \begin{bmatrix} T^{(k+1)}_{2,j} \\ T^{(k+1)}_{3,j} \\ \vdots \\ T^{(k+1)}_{n_x-2,j} \\ T^{(k+1)}_{n_x-1,j} \end{bmatrix} = r^2 \begin{bmatrix} T^{(k)}_{2,j-1} + T^{(k)}_{2,j+1} \\ T^{(k)}_{3,j-1} + T^{(k)}_{3,j+1} \\ \vdots \\ T^{(k)}_{n_x-2,j} + T^{(k)}_{n_x-2,j+1} \\ T^{(k)}_{n_x-1,j} + T^{(k)}_{n_x-1,j+1} \end{bmatrix} + \begin{bmatrix} T_{x0} \\ 0 \\ \vdots \\ 0 \\ T_{xL_x} \end{bmatrix} \tag{7.5.129}$$

It can be noticed that the matrix is tridiagonal and the problem can be solved $\forall\, j = 2, \dots, n_y - 1$ with taking into account the boundary conditions at left (when $j = 2$) and at right (when $j = n_y - 1$).

Then, in the stage done from $(k+1)$ to $(k+2)$, the direction is changed. We consider a given index i and we sweep the y axis with j increasing from 2 to $(n_y - 1)$, which gives $\forall\, i = 2, \dots, n_x - 1$ at iteration $(k+1)$ of the iterative calculation (the variables noted $y^{(k+1)}$ are known and those noted $y^{(k+2)}$ are to be determined)

$$\begin{aligned} &j = 2, \quad \frac{1}{r^2}\left[T^{(k+1)}_{i-1,2} - 2T^{(k+2)}_{i,2} + T^{(k+1)}_{i+1,2}\right] + T_{y0} - 2T^{(k+2)}_{i,2} + T^{(k+2)}_{i,3} = 0 \\ &j = 3, \dots, n_y - 2, \quad \frac{1}{r^2}\left[T^{(k+1)}_{i-1,j} - 2T^{(k+2)}_{i,j} + T^{(k+1)}_{i+1,j}\right] + T^{(k+2)}_{i,j-1} - 2T^{(k+2)}_{i,j} + T^{(k+2)}_{i,j+1} = 0 \\ &j = n_y - 1, \quad \frac{1}{r^2}\left[T^{(k+1)}_{i-1,n_y-1} - 2T^{(k+2)}_{i,n_y-1} + T^{(k+1)}_{i+1,n_y-1}\right] + T^{(k+2)}_{i,n_y-2} - 2T^{(k+2)}_{i,n_y-1} + T_{yL_y} = 0 \end{aligned} \tag{7.5.130}$$

which gives under matrix form

$$\begin{bmatrix} 2(1+1/r^2) & -1 & 0 & \dots & \dots & 0 \\ -1 & 2(1+1/r^2) & -1 & 0 & \dots & \vdots \\ 0 & -1 & 2(1+r^2) & -1 & 0 & \vdots \\ \vdots & \ddots & \ddots & \ddots & \ddots & 0 \\ \vdots & \dots & 0 & -1 & 2(1+1/r^2) & -1 \\ 0 & \dots & \dots & 0 & -1 & 2(1+1/r^2) \end{bmatrix} \begin{bmatrix} T^{(k+2)}_{i,2} \\ T^{(k+2)}_{i,3} \\ \vdots \\ T^{(k+2)}_{i,n_y-2} \\ T^{(k+2)}_{i,n_y-1} \end{bmatrix} = \frac{1}{r^2} \begin{bmatrix} T^{(k+1)}_{i-1,2} + T^{(k+1)}_{i+1,2} \\ T^{(k+1)}_{i-1,3} + T^{(k+1)}_{i+1,3} \\ \vdots \\ T^{(k+1)}_{i-1,n_y-2} + T^{(k+1)}_{i+1,n_y-2} \\ T^{(k+1)}_{i-1,n_y-1} + T^{(k+1)}_{i+1,n_y-1} \end{bmatrix} + \begin{bmatrix} T_{y0} \\ 0 \\ \vdots \\ 0 \\ T_{yL_y} \end{bmatrix} \tag{7.5.131}$$

Again, the matrix is tridiagonal and the problem can be solved $\forall\, i = 2, \dots, n_x - 1$ while taking into account the boundary conditions at bottom (when $i = 2$) and at top (when $i = n_x - 1$).

When we have performed the sweeping w.r.t. Ox, then w.r.t. Oy, we have performed the set of both iterations equivalent to the alternate direction method, but for a steady state problem, and the processes can be repeated until convergence. Of course, the solution can be searched by starting w.r.t. Oy, then Ox.

Numerical application:
$T(x = 0, y) = 400\,\text{K}$, $T(x = L_x, y) = 500\,\text{K}$, $T(x, y = 0) = 450\,\text{K}$, $T(x, y = L_y) = 550\,\text{K}$. $L_x = 0.02\,\text{m}$, $L_y = 0.05\,\text{m}$, $\lambda = 26\,\text{W}\,\text{m}^{-1}\,\text{K}^{-1}$, $\rho = 7800\,\text{kg}\,\text{m}^{-3}$, $C_p = 500\,\text{J}\,\text{kg}^{-1}\,\text{K}^{-1}$. The Dirichlet boundary conditions are shown in Figure 7.24.

We thus get the temperature contours of Figure 7.25 and the temperature surface in Figure 7.26.

A simulation of the same plate based on finite elements has been performed (Section 7.13) by first doing a finite element mesh with the Gmsh code (Figure 7.91), then the simulation to obtain the steady state with the Elmer code (Malinen and Råback 2013) of simulation by finite elements (Figure 7.91). This figure can be compared to the contours of Figure 7.25 obtained by the finite difference method. The same problem has also been solved by the boundary element method (Example 7.36 and Figure 7.106). In this simple case, the results are close.

7.6 Automatic Calculation of Partial Derivatives

The automatic calculation of partial derivatives is often practiced, in particular in the method of lines (Section 7.7). It consists in using a subprogram (procedure or function) calculating systematically the partial derivatives u_x of the variable u at space points x_i distributed on a segment of length L such that

$$x_i = i\Delta x \quad \text{with } \Delta x = \frac{L}{N}, \quad i = 0 \ldots, N \tag{7.6.1}$$

The calculation is simple when it deals with a point interior to the domain. It becomes more intricate when the points are at the boundaries or close to the boundaries.

7.6.1 Calculation of $\left(\frac{\partial u}{\partial x}\right)_0$ and $\left(\frac{\partial u}{\partial x}\right)_N$

We set

$$\begin{aligned}
u(x_0) &= u_0 \\
u(x_0 + \Delta x) &= u_0 + \left(\frac{\partial u}{\partial x}\right)_0 \Delta x + \frac{1}{2!}\left(\frac{\partial^2 u}{\partial x^2}\right)_0 (\Delta x)^2 + \frac{1}{3!}\left(\frac{\partial^3 u}{\partial x^3}\right)_0 (\Delta x)^3 + \ldots \\
u(x_0 + 2\Delta x) &= u_0 + \left(\frac{\partial u}{\partial x}\right)_0 2\Delta x + \frac{1}{2!}\left(\frac{\partial^2 u}{\partial x^2}\right)_0 (2\Delta x)^2 + \frac{1}{3!}\left(\frac{\partial^3 u}{\partial x^3}\right)_0 (2\Delta x)^3 + \ldots
\end{aligned} \tag{7.6.2}$$

The coefficients α and β are chosen such that

$$\alpha u(x_0 + \Delta x) + \beta u(x_0 + 2\Delta x) = c u_0 + \left(\frac{\partial u}{\partial x}\right)_0 \Delta x + 0((\Delta x)^3) \tag{7.6.3}$$

Thus, this sum has a coefficient equal to 1 for $\left(\frac{\partial u}{\partial x}\right)_0$ and a coefficient equal to 0 for $\left(\frac{\partial^2 u}{\partial x^2}\right)_0$, hence

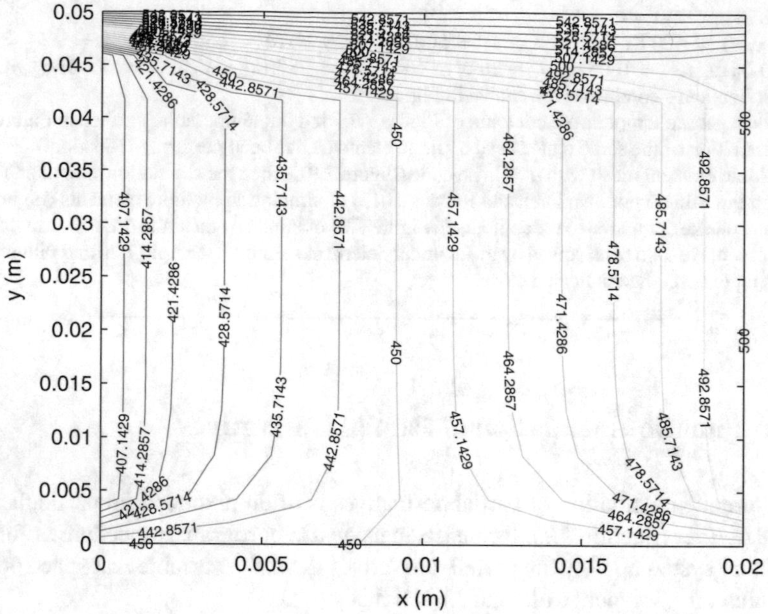

Fig. 7.25 Example HT3-FDM-2D: 2D conductive heat transfer: Temperature contours in the plane plate subjected to Dirichlet conditions on the four sides, by the alternate direction method in steady state

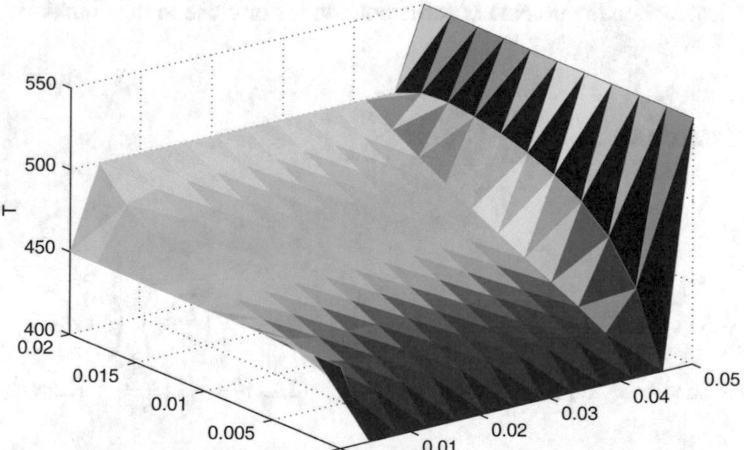

Fig. 7.26 Example HT3-FDM-2D: 2D conductive heat transfer: Representation of the temperature surface in the plane plate subjected to Dirichlet conditions on the four sides, by the alternate direction method in steady state

$$\begin{cases} \alpha + 2\beta = 1 \\ \frac{1}{2!}(\alpha + 4\beta) = 0 \end{cases} \Rightarrow \quad \alpha = 2,\ \beta = -\frac{1}{2} \tag{7.6.4}$$

Then, it comes

$$2u(x_0 + \Delta x) - \frac{1}{2}u(x_0 + 2\Delta x) = 2u_0 - \frac{1}{2}u_0 + \left(\frac{\partial u}{\partial x}\right)_0 \Delta x + 0((\Delta x)^3) \tag{7.6.5}$$

hence the formula

$$\left(\frac{\partial u}{\partial x}\right)_0 \approx \frac{-3u_0 + 4u_1 - u_2}{2\Delta x} \tag{7.6.6}$$

Symmetrically, we obtain

$$\left(\frac{\partial u}{\partial x}\right)_N \approx \frac{3u_N - 4u_{N-1} + u_{N-2}}{2\Delta x} \tag{7.6.7}$$

noting the sign change with respect to Equation (7.6.6).

At any other point x_i, of course we have the usual formula

$$\left(\frac{\partial u}{\partial x}\right)_i \approx \frac{u_{i+1} - u_{i-1}}{2\Delta x} \tag{7.6.8}$$

The previous formulas can be summarized under the following matrix form

$$\begin{bmatrix} u_{x,0} \\ u_{x,1} \\ \vdots \\ u_{x,N-1} \\ u_{x,N} \end{bmatrix} = \frac{1}{2\,\Delta x} \begin{bmatrix} -3 & 4 & -1 & 0 & \dots & 0 \\ -1 & 0 & 1 & 0 & \dots & 0 \\ 0 & -1 & 0 & 1 & 0 & \dots \\ \vdots & & \ddots & \ddots & \ddots & 0 \\ 0 & \dots & & -1 & 0 & 1 \\ 0 & \dots & & 1 & -4 & 3 \end{bmatrix} \begin{bmatrix} u_0 \\ u_1 \\ \vdots \\ u_{N-1} \\ u_N \end{bmatrix} \tag{7.6.9}$$

or still

$$\boldsymbol{u}_x = D_1\, \boldsymbol{u} \tag{7.6.10}$$

where D_1 is the differentiation operator defined by matrix of Equation (7.6.9).

7.6.2 Calculation of $\left(\frac{\partial^2 u}{\partial x^2}\right)_0$ and $\left(\frac{\partial^2 u}{\partial x^2}\right)_N$

We use the same method as previously.

$$\begin{aligned}
u(x_0) &= u_0 \\
u(x_0+\Delta x) &= u_0 + \left(\frac{\partial u}{\partial x}\right)_0 \Delta x + \frac{1}{2!}\left(\frac{\partial^2 u}{\partial x^2}\right)_0 (\Delta x)^2 + \frac{1}{3!}\left(\frac{\partial^3 u}{\partial x^3}\right)_0 (\Delta x)^3 + \dots \\
u(x_0+2\Delta x) &= u_0 + \left(\frac{\partial u}{\partial x}\right)_0 2\Delta x + \frac{1}{2!}\left(\frac{\partial^2 u}{\partial x^2}\right)_0 (2\Delta x)^2 + \frac{1}{3!}\left(\frac{\partial^3 u}{\partial x^3}\right)_0 (2\Delta x)^3 + \dots \\
u(x_0+3\Delta x) &= u_0 + \left(\frac{\partial u}{\partial x}\right)_0 3\Delta x + \frac{1}{2!}\left(\frac{\partial^2 u}{\partial x^2}\right)_0 (3\Delta x)^2 + \frac{1}{3!}\left(\frac{\partial^3 u}{\partial x^3}\right)_0 (3\Delta x)^3 + \dots
\end{aligned} \tag{7.6.11}$$

The coefficients α, β, and γ are chosen such that

$$\alpha u(x_0+\Delta x) + \beta u(x_0+2\Delta x) + \gamma u(x_0+3\Delta x) = c u_0 + \left(\frac{\partial^2 u}{\partial x^2}\right)_0 + 0((\Delta x)^4) \tag{7.6.12}$$

thus this sum has a coefficient equal to 0 for $\left(\frac{\partial u}{\partial x}\right)_0$, a coefficient equal to 1 for $\left(\frac{\partial^2 u}{\partial x^2}\right)_0$ and a coefficient equal to 0 for $\left(\frac{\partial^3 u}{\partial x^3}\right)_0$, hence

$$\begin{cases} \alpha + 2\beta + 3\gamma = 0 \\ \frac{1}{2!}(\alpha + 4\beta + 9\gamma) = 1 \\ \frac{1}{3!}(\alpha + 8\beta + 27\gamma) = 0 \end{cases} \quad \Rightarrow \quad \alpha = -5,\, \beta = 4,\, \gamma = -1 \tag{7.6.13}$$

It results

$$-5u(x_0+\Delta x) + 4u(x_0+2\Delta x) - u(x_0+3\Delta x) = -2u_0 + \left(\frac{\partial^2 u}{\partial x^2}\right)_0 (\Delta x)^2 + 0((\Delta x)^4) \tag{7.6.14}$$

hence the formula

$$\left(\frac{\partial^2 u}{\partial x^2}\right)_0 \approx \frac{2u_0 - 5u_1 + 4u_2 - u_3}{(\Delta x)^2} \tag{7.6.15}$$

and symmetrically

$$\left(\frac{\partial^2 u}{\partial x^2}\right)_N \approx \frac{2u_N - 5u_{N-1} + 4u_{N-2} - u_{N-3}}{(\Delta x)^2} \tag{7.6.16}$$

without the sign change which occurred in the first order derivatives (7.6.6) and (7.6.7).

At any other point x_i, of course the usual formula is valid

$$\left(\frac{\partial^2 u}{\partial x^2}\right)_i \approx \frac{u_{i+1} - 2u_i + u_{i-1}}{(\Delta x)^2} \tag{7.6.17}$$

The previous formulas can be summarized under the following matrix form

$$\begin{bmatrix} u_{xx,0} \\ u_{xx,1} \\ \vdots \\ u_{xx,N-1} \\ u_{xx,N} \end{bmatrix} = \frac{1}{(\Delta x)^2} \begin{bmatrix} 2 & -5 & 4 & -1 & 0 & \dots & 0 \\ 1 & -2 & 1 & 0 & \dots & & 0 \\ 0 & 1 & -2 & 1 & 0 & \dots & \\ \vdots & & \ddots & \ddots & \ddots & & 0 \\ \vdots & & & \ddots & \ddots & \ddots & 0 \\ 0 & \dots & & 0 & 1 & -2 & 1 \\ 0 & \dots & & -1 & 4 & -5 & 2 \end{bmatrix} \begin{bmatrix} u_0 \\ u_1 \\ \vdots \\ u_{N-1} \\ u_N \end{bmatrix} \tag{7.6.18}$$

or still

$$\boldsymbol{u}_{xx} = D_2\, \boldsymbol{u} \tag{7.6.19}$$

where D_2 is the differentiation operator defined by matrix of Equation (7.6.18).

It can be noticed that, instead of using the operator D_2, in some cases, according to the chosen operators, the following operation will be preferred

$$\boldsymbol{u}_{xx} = D_1\,(D_1\, \boldsymbol{u}) \tag{7.6.20}$$

which means that to get the vector of second order derivatives, we use twice the operator of first order differentiation.

7.6.3 *Some Other Differentiation Schemes*

These schemes can be obtained by techniques similar to the previous schemes.

7.6.3.1 Four-Point Scheme

Starting from the general formula

$$\alpha\, u_{i-3} + \beta\, u_{i-2} + \gamma\, u_{i-1} = c\, u_i + \frac{\partial u}{\partial x}_i\, \Delta x \tag{7.6.21}$$

and using nth degree Taylor polynomials, we obtain the four-point scheme formula (Saucez et al. 2001) for the operator D_1 equal to

$$D_1 = \frac{1}{6\,\Delta x} \begin{bmatrix} -11 & 18 & -9 & 2 & 0 & \dots & 0 \\ -2 & -3 & 6 & -1 & 0 & \dots & 0 \\ 1 & -6 & 3 & 2 & 0 & \dots & 0 \\ -2 & 9 & -18 & 11 & 0 & \dots & 0 \\ 0 & -2 & 9 & -18 & 11 & 0 & \dots \\ \vdots & & \ddots & \ddots & \ddots & \ddots & 0 \\ 0 & \dots & & -2 & 9 & -18 & 11 \end{bmatrix} \tag{7.6.22}$$

7.6.3.2 Four-Point Upwind Scheme

Starting from the general formula

$$\alpha\, u_{i-2} + \beta\, u_{i-1} + \gamma\, u_{i+1} = c\, u_i + \frac{\partial u}{\partial x}_i \Delta x \tag{7.6.23}$$

and using nth degree Taylor polynomials, we define the four-point upwind scheme, due to the presence of the term u_{i+1} to obtain $u_{x,i}$, by the operator D_1 equal to

$$D_1 = \frac{1}{6\,\Delta x}\begin{bmatrix} -11 & 18 & -9 & 2 & 0 & \dots & 0 \\ -2 & -3 & 6 & 1 & 0 & \dots & 0 \\ 1 & -6 & 3 & 2 & 0 & \dots & 0 \\ 0 & 1 & -6 & 3 & 2 & 0 & \\ \vdots & & \ddots & \ddots & \ddots & \ddots & \\ 0 & \dots & & 1 & -6 & 3 & 2 \\ 0 & \dots & & -2 & 9 & -18 & 11 \end{bmatrix} \tag{7.6.24}$$

An upwind scheme can be, in particular, useful to take into account convection in partial differential equations of convection-diffusion (Dunnebier et al. 1998).

As the representation of the convection term is the most delicate in a partial differential equation of convection-diffusion, it is usual that only the first order spatial derivative is discretized with such a scheme, whereas the second order spatial derivative will derive from a simple centered difference scheme of type (7.6.18).

7.6.3.3 Five-Point Upwind Scheme

Starting from the general formula

$$\alpha\, u_{i-3} + \beta\, u_{i-2} + \gamma\, u_{i-1} + \delta\, u_{i+1} = c\, u_i + \frac{\partial u}{\partial x}_i \Delta x \tag{7.6.25}$$

and using nth degree Taylor polynomials, we define the five-point upwind scheme (Saucez et al. 2001; Schiesser et al. 1994; Schiesser 1991) by the operator D_1 equal to

$$D_1 = \frac{1}{12\,\Delta x}\begin{bmatrix} -25 & 48 & -36 & 16 & -3 & 0 & \dots & 0 \\ -3 & -10 & 18 & -6 & 1 & 0 & \dots & 0 \\ 1 & -8 & 0 & 8 & -1 & 0 & \dots & 0 \\ -1 & 6 & -18 & 10 & 3 & 0 & \dots & 0 \\ 0 & -1 & 6 & -18 & 10 & 3 & 0 & \\ \vdots & \ddots & & \ddots & \ddots & \ddots & \ddots & \\ 0 & \dots & & -1 & 6 & -18 & 10 & 3 \\ 0 & \dots & & 3 & -16 & 36 & -48 & 25 \end{bmatrix} \tag{7.6.26}$$

This five-point upwind scheme is more accurate than the previous four-point upwind scheme.

7.6.4 Numerical Differentiation by Complex Numbers

The approximation of derivatives by using complex numbers theoretically allows us to get a better accuracy (Martins et al. 2003). A very simple estimation of the derivative of a function f of the real variable x in the neighborhood of x is obtained by the forward divided difference

$$f'(x) = \frac{f(x+h) - f(x)}{h} + 0(h) \tag{7.6.27}$$

with a small step h.

Now consider a complex function g of the complex variable z, that is

$$g(z) = u(x, y) + i\,v(x, y) \qquad \text{with } z = x + i\,y \tag{7.6.28}$$

Assuming g analytic, that is differentiable in the complex plane, Cauchy–Riemann equations are applicable

$$\begin{aligned} \frac{\partial u}{\partial x} &= \frac{\partial v}{\partial y} \\ \frac{\partial u}{\partial y} &= -\frac{\partial v}{\partial x} \end{aligned} \tag{7.6.29}$$

Using the first of these equations, we get

$$\frac{\partial u}{\partial x} = \lim_{h \to 0} \frac{v(x + i(y+h)) - v(x+iy)}{h} \tag{7.6.30}$$

where h is a real number.

Considering that, a priori, we are looking for a derivative of a real function f of a real variable, we can set $y = 0$, $u(x) = f(x)$, $v(x) = 0$, so that we obtain

$$\frac{\partial f}{\partial x} = \lim_{h \to 0} \frac{\mathrm{Im}[f(x+ih)]}{h} \tag{7.6.31}$$

where Im is the imaginary part of f. We can still write the approximation

$$\frac{\partial f}{\partial x} \approx \frac{\mathrm{Im}[f(x+ih)]}{h} \tag{7.6.32}$$

As this approximation of the derivative does not make use of the difference between two values, opposite to approximation (7.6.27), the numerical errors will be decreased.

By expressing a second degree Taylor polynomial in the neighborhood of x, we get

$$f(x+ih) = f(x) + i\,h\,f'(x) - h^2 \frac{f^{(2)}(x)}{2!} - i\,h^3 \frac{f^{(3)}(x)}{3!} + \dots \tag{7.6.33}$$

hence

$$f'(x) = \frac{\mathrm{Im}[f(x+ih)]}{h} + h^2 \frac{f^{(3)}(x)}{3!} + \dots \tag{7.6.34}$$

which shows that the approximation of the derivative by complex numbers is of order 2, and thus better than (7.6.27). An important point is that it is possible to use a step h

as small as desired without numerical consequence. It can also be noticed that

$$f(x) = \mathrm{Re}[f(x+ih)] + h^2 \frac{f^{(2)}(x)}{2!} + \dots \tag{7.6.35}$$

gives an approximation of $f(x)$ as its value on the real axis.

Example 7.6 :
Derivative calculation by means of complex numbers

Martins et al. (2003) gives the following example

$$f(x) = \frac{e^x}{\sqrt{\sin^3 x + \cos^3 x}} \tag{7.6.36}$$

which indicates an idea of some programming difficulties related to this calculation because of the use of complex numbers. We desire to calculate the derivative at $x = 1.5$. For most functions, calculating their value in Fortran poses no major problem, except some functions including the following ones, abs, sin, cos, tan, cosh, sinh, tanh, max, min, sign, log. Martins et al. (2003) gives the code Fortran "complexify.f90" for difficult cases. Of course, the exponential function poses no problem.

In the present case, we have calculated the function sin of the complex variable $z = x + iy$ as

$$\sin(z) = \sin(x)\cosh(y) + i\cos(x)\sinh(y) \tag{7.6.37}$$

and the function cos as

$$\cos(z) = \cos(x)\cosh(y) - i\sin(x)\sinh(y) \tag{7.6.38}$$

where sinh and cosh are the hyperbolic sine and hyperbolic cosine functions, respectively.

By means of these precautions, it is possible to numerically calculate the derivative $f'(x)$ by the complex numbers at $x = 1.5$
as approximately $f'(x) \approx 4.0534278258935892$
while its exact value is $f'(x) = 4.0534278938986201$,
thus a relative error equals to 10^{-8} for a step $h = 10^{-15}$.

7.7 Method of Lines

The method of lines (Shastry and Allen 1998; Vande Wouwer et al. 2004) consists in solving a partial differential equation or a system of partial differential equations by transforming it under the form of ordinary differential equations. In the case where partial derivatives with respect to time and space are present in the model, only the space partial derivatives are discretized. Thus, we consider the conductive heat transfer through a plane wall obeying Fourier's law

$$\frac{\partial T}{\partial t} = \alpha \frac{\partial^2 T}{\partial x^2} \tag{7.7.1}$$

in both cases of Dirichlet and Neumann boundary conditions which were treated above by the finite difference method.

7.7.1 Case of Dirichlet Boundary Conditions

Figure 7.16 represents the studied system. N discretization intervals are assumed:

$$\begin{array}{llll} \text{Positions of the points} & x_i = i\Delta x \quad \text{with } \Delta x = \frac{L}{N}\,, & i = 0,\ldots,N & \\ \text{Initial conditions} & T(x,0) = T_0 & \forall\, 0 \le x \le L & \\ \text{Boundary condition at left} & T(0,t) = T_a & \forall\, 0 < t & \\ \text{Boundary condition at right} & T(L,t) = T_b & \forall\, 0 < t & \end{array} \tag{7.7.2}$$

After discretization of the second order spatial derivative, Fourier's law becomes at any point inside the wall

$$\left.\frac{dT}{dt}\right|_{x=x_i} = \alpha\, \frac{T_{i+1} - 2T_i + T_{i-1}}{\Delta x^2}\,, \qquad \forall\, i = 1,\ldots,N-1 \tag{7.7.3}$$

by noting $T_i = T(x_i,t)$.

The resulting system of $(N-1)$ ordinary differential equations can be solved by any efficient method solving systems of ordinary differential equations according to the following algorithm:

1. Initialization of the temperatures T_i at any point x_i $(i = 0,\ldots,N)$.
2. The Dirichlet boundary conditions $T_0 = T_a$ and $T_N = T_b$ are imposed.
3. The system of ordinary differential equations (7.7.3) is solved on one time step.
4. Return to stage 2 until final time.

By the return to stage 2, we force the system to respect the Dirichlet boundary conditions.

Example 7.7 :
HT1-FDM-1D-b: 1D heat transfer using the method of lines in the case of Dirichlet boundary conditions

The method of lines has been used to solve Example 7.3 and gave extremely close numerical results not represented here.

7.7.2 Case of Boundary Neumann Conditions

We consider the system defined by Figure 7.20. The boundary conditions at left and right of the wall are thus related to the flux entering or leaving the plate. The conditions will be the following:

$$\begin{array}{llll} \text{Positions of the points} & x_i = i\Delta x \quad \text{with } \Delta x = \frac{L}{N}\,, & i = 0,\ldots,N & \\ \text{Initial conditions} & T(x,0) = T_0 \quad \forall\, 0 \le x \le L & & \\ \text{Boundary condition at left} & \left(\frac{\partial T}{\partial x}\right)_{x=0} = -\frac{h_0}{\lambda}(T_A - T_0) & \forall\, 0 < t & \\ \text{Boundary condition at right} & \left(\frac{\partial T}{\partial x}\right)_{x=L} = -\frac{h_0}{\lambda}(T_N - T_B) & \forall\, 0 < t & \end{array} \tag{7.7.4}$$

With respect to the case of Dirichlet boundary conditions, the system of ordinary differential equations differs in the case of Neumann boundary conditions as it includes the derivatives at both boundaries as explained in Equation (7.7.5).

The resulting system of $(N+1)$ ordinary differential equations can then be solved by the following algorithm:

1. Initialization of the temperatures T_i at any point x_i $(i = 0, \dots, N)$.
2. By an automatic procedure, (refer to Equations (7.6.6), (7.6.7), (7.6.8)) the partial derivatives u_x of the variable u are calculated (here, the temperature T) at any point x_i.
3. The boundary Neumann conditions are imposed which specify u_x at $x = 0$ and $x = L$.
4. Knowing u_x in any point, the same automatic procedure can be used to calculate the partial derivatives u_{xx} of the variable u_x at any point x_i.
5. Fourier's law becomes at any point of the wall

$$\left.\frac{dT}{dt}\right|_{x=x_i} = \alpha\, u_{xx,i}\,, \qquad \forall i = 0, \dots, N \tag{7.7.5}$$

   where u_{xx} represents $\dfrac{\partial^2 T}{\partial x^2}$.
6. The system of $(N+1)$ ordinary differential equations (7.7.5) is solved.
7. Return to stage 2 until final time.

Example 7.8 :
HT2-FDM-1D-b: 1D heat transfer using the method of lines in the case of Neumann boundary conditions

The method of lines was used to solve Example 7.4 and gave extremely close numerical results. It must be noted that the method of lines in the Neumann case is largely more complex than in the Dirichlet case. For numerical reasons, instabilities can occur close to the boundaries, in particular just after initialization, when using Equation (7.6.6) and (7.6.7). To avoid these difficulties, it may be interesting to start by using an implicit finite difference method in order to obtain a realistic profile which can be used later as the initial profile to the method of lines.

Example 7.9 :
HT3-FDM-2D-b: Laplace equation by the method of lines

The same problem as Example 7.5 is solved. However, instead of solving the problem under the form of Example 7.5, we propose to use the method of lines. Thus, we transform the steady state problem under transient form as

$$\frac{\partial T}{\partial t} = \alpha \left(\frac{\partial^2 T}{\partial x^2} + \frac{\partial^2 T}{\partial y^2} \right) \tag{7.7.6}$$

and by doing only the discretization of spatial derivatives, which results in the following system of ordinary differential equations

$$\frac{dT_{i,j}}{dt} = \alpha \left(\frac{T_{i-1,j} - 2T_{i,j} + T_{i+1,j}}{\Delta x^2} + \frac{T_{i,j-1} - 2T_{i,j} + T_{i,j+1}}{\Delta y^2} \right) \tag{7.7.7}$$
$$i = 2, \dots, n_x - 1\,, \quad j = 2, \dots, n_y - 1$$

Thus, there remains to integrate that system with the appropriate initial condition and boundary conditions and search the steady state solution (Figure 7.27). To practically solve the system of ordinary differential equations, it may be useful to present it under the form of a vector of ordinary differential equations rather than a matrix. This can be easily done by the transformation of the matrix $T_{i,j}$ into the vector $Tvec_{i+(j-1)n_x}$.

The solution for Figure 7.27 was obtained with $n_x = 11$, $n_y = 11$. The results of Figure 7.27 are close to those of Figure 7.25.

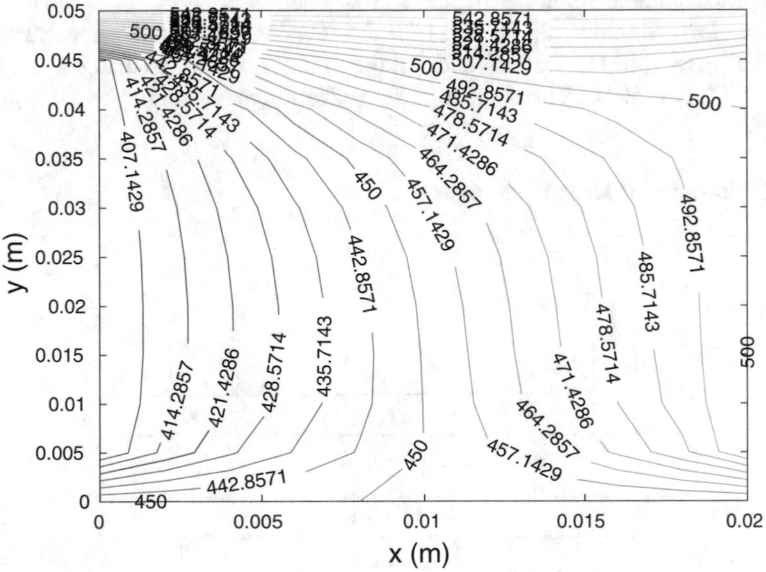

Fig. 7.27 Example HT3-FDM-2D-b: 2D conductive heat transfer: Temperature contours in the plane plate subjected to Dirichlet conditions on the four sides, by the method of lines

---

7.7.3 *Application of the Method of Lines to the Simulation of a Heat Exchanger*

The modelled exchangers are constituted by an external tube in which circulates at velocity v_e a fluid of density ρ_e, heat capacity C_{p_e}, exchanging heat with a fluid of density ρ, heat capacity C_p, circulating in an inner tube at velocity v. It is assumed that both fluids undergo no condensation. The exchanger length is L. The surface areas S_e and S of the internal and external cross sections are assumed to be constant. The surface area subjected to the heat transfer is $\mathcal{S}$ and the heat transfer coefficient is h on the internal side and h_e on the external side.

The energy balances of the heat exchangers make use of first order partial differential equations with respect to time and space related to the convection. For modelling by finite difference method, the length L is discretized into elements of length $\Delta z = L/N$, where N is the number of increments.

The temperature $T(z,t)$ of the internal fluid and the temperature $T_e(z,t)$ of the external fluid depend on time and space along the tube.

The simulation aims at finding the steady state temperature profiles in the tubes. The physical parameters are $a = 2.92\,\text{s}^{-1}$, $v = 1\,\text{m s}^{-1}$, $a_e = 5\,\text{s}^{-1}$, $v_e = 2\,\text{m s}^{-1}$, and $L = 1\,\text{m}$. Moreover $T_0 = T(0,t) = 25\,°\text{C}$. For the co-current exchanger, we impose $T_{e,0} = T_e(0,t) = 50\,°\text{C}$ and for the counter-current exchanger, we impose $T_{e,N} = T_e(L,t) = 50\,°\text{C}$. The number of discretization elements is $N = 50$.

7.7.3.1 Co-current Heat Exchanger

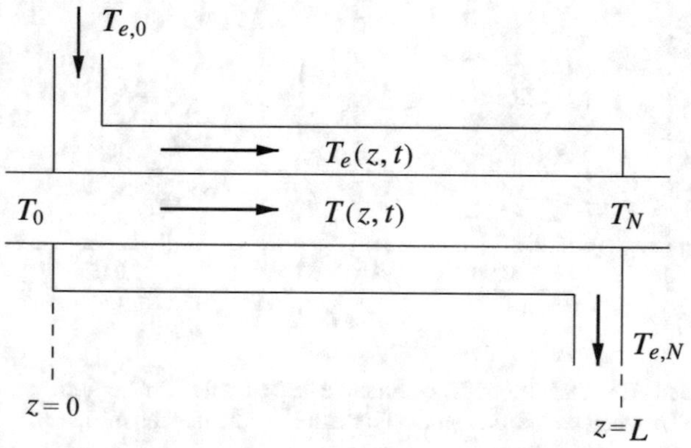

Fig. 7.28 Co-current heat exchanger

In this paragraph, the control point of view is briefly commented. In the modelled co-current heat exchanger (Figure 7.28), the inner fluid enters at temperature $T_0 = T(0,t)$ (considered as a disturbance), the outlet temperature $T_N = T(L,t)$ is the controlled output. The external fluid enters at temperature $T_{e,0} = T_e(0,t)$. This temperature is considered as the manipulated variable.

The energy balance can be written for the internal tube as

$$\frac{\partial T(z,t)}{\partial t} = -v\frac{\partial T(z,t)}{\partial z} + a\,[T_e(z,t) - T(z,t)] \;, \quad a = \frac{hS}{\rho S C_p} \tag{7.7.8}$$

and for the external tube

$$\frac{\partial T_e(z,t)}{\partial t} = -v_e\frac{\partial T_e(z,t)}{\partial z} + a_e\,[T(z,t) - T_e(z,t)] \;, \quad a_e = \frac{h_e S}{\rho_e S_e C_{p_e}} \tag{7.7.9}$$

The initial conditions are profiles $T(z,0) = T^*(z)$ and $T_e(z,0) = T_e^*(z)$. The temperature of the entering external fluid $T_e(0,t)$ must be known and will impose the profiles whereas the temperature of the entering internal fluid $T(0,t)$ is also a known disturbance.

According to the method of lines, the spatial discretization of the energy balances gives the following ordinary differential equations

$$\begin{aligned} \frac{dT_i}{dt} &= -v\frac{T_i - T_{i-1}}{\Delta z} + a(T_{e,i} - T_i)\,, \quad i = 1,\ldots,N \\ \frac{dT_{e,i}}{dt} &= -v_e\frac{T_{e,i} - T_{e,i-1}}{\Delta z} + a_e(T_i - T_{e,i})\,, \quad i = 1,\ldots,N \end{aligned} \tag{7.7.10}$$

The temperature profiles along the co-current exchanger are represented in Figure 7.29.

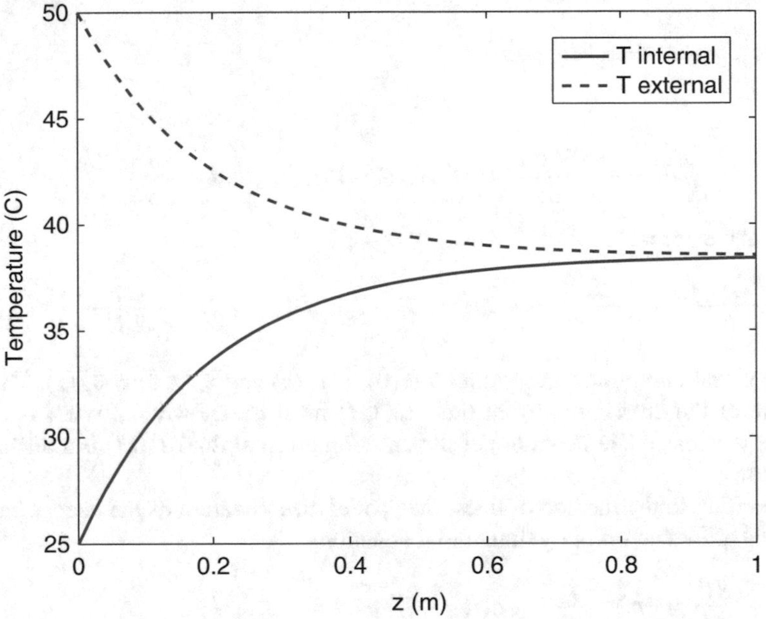

Fig. 7.29 Co-current exchanger: Temperature profiles with $T(0,t) = 25\,°\text{C}$ and $T_e(0,t) = 50\,°\text{C}$

7.7.3.2 Counter-Current Heat Exchanger

From the control point of view, in the modelled counter-current heat exchanger (Figure 7.30), the internal fluid enters at temperature $T_0 = T(0,t)$ (considered as a disturbance), the outlet temperature $T_N = T(L,t)$ is the output to be controlled. Opposite to the co-current exchanger, the external fluid enters at temperature $T_{e,L} = T_e(L,t)$ which is considered as the manipulated variable.

The energy balance can be written for the internal tube as

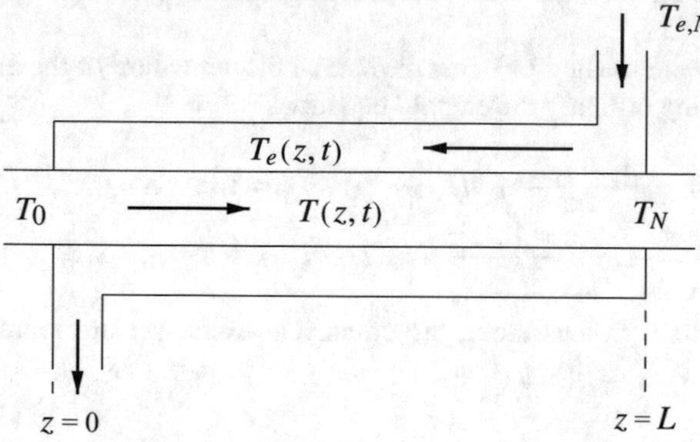

Fig. 7.30 Counter-current heat exchanger

$$\frac{\partial T(z,t)}{\partial t} = -v\frac{\partial T(z,t)}{\partial z} + a\,[T_e(z,t) - T(z,t)]\ , \quad a = \frac{hS}{\rho S C_p} \tag{7.7.11}$$

and for the external tube

$$\frac{\partial T_e(z,t)}{\partial t} = v_e\frac{\partial T_e(z,t)}{\partial z} + a_e\,[T(z,t) - T_e(z,t)]\ , \quad a_e = \frac{h_e S}{\rho_e S_e C_{p_e}} \tag{7.7.12}$$

The initial conditions are profiles $T(z,0) = T^*(z)$ and $T_e(z,0) = T_e^*(z)$. The temperature of the entering external fluid $T_e(L,t)$ must be known and will impose the profiles whereas the temperature of the entering internal fluid $T(0,t)$ is a disturbance also known.

According to the method of lines, the spatial discretization of the energy balances gives the following ordinary differential equations

$$\begin{aligned}
\frac{dT_i}{dt} &= -v\frac{T_i - T_{i-1}}{\Delta z} + a(T_{e,i} - T_i)\,, \quad i = 1,\dots,N \\
\frac{dT_{e,i}}{dt} &= v_e\frac{T_{e,i+1} - T_{e,i}}{\Delta z} + a_e(T_i - T_{e,i})\,, \quad i = 0,\dots,N-1
\end{aligned} \tag{7.7.13}$$

We notice that the spatial discretization is different for the co-current and counter-current heat exchangers (Figure 7.31). The discretization must be done by taking into account the direction of circulation of the fluid. This forces us to use three points for the counter-current while two points are sufficient for the co-current.

The temperature profiles along the exchanger are represented in Figure 7.32.

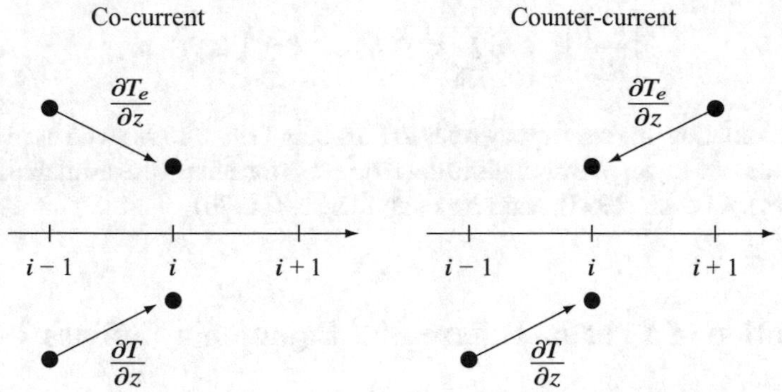

Fig. 7.31 Different spatial discretizations for the co-current and counter-current heat exchangers

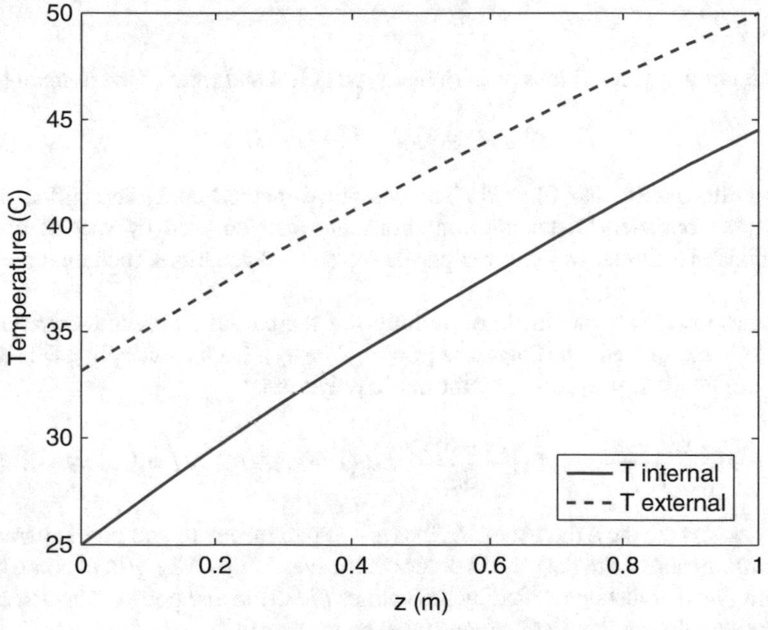

Fig. 7.32 Counter-current heat exchanger: Temperature profiles with $T(0, t) = 25\,°C$ and $T_e(L, t) = 50\,°C$

7.8 Finite Differences on an Irregular Grid

In all previous examples, the finite differences were calculated and applied on a regular grid. In some cases of irregular grids, such as for moving grids (Section 7.11), it is still possible to use finite differences, but the formulas are completely different and make use of a weighting of the form

$$\left[\frac{d^k f}{dx^k}\right]_{x=z} \approx \sum_{i=0}^{n} c_i^k f(x_i)\,, \quad k = 1, 2, \ldots, m \tag{7.8.1}$$

For this calculation, the reader may refer to Fornberg (1988, 1998) who used Lagrange polynomials and Taylor polynomials to develop these formulas, also available in Kelling et al. (2014), Li et al. (1998), and Liszka and Orkisz (1980).

7.9 Solution of a Partial Differential Equation by Splines

The partial differential equation (PDE) chosen is Fourier's law

$$\frac{\partial T(x,t)}{\partial t} = \alpha \frac{\partial^2 T(x,t)}{\partial x^2} \tag{7.9.1}$$

with two boundary conditions of Dirichlet type at left and right of the domain $[a, b]$

$$T(a,t) = T_a\,, \quad T(b,t) = T_b \tag{7.9.2}$$

and an initial profile $T(x, 0) = T_0$. The described method could be applied without difficulty to boundary Neumann conditions and can be used by way of common adaptations in a similar way to other partial differential equations, including nonlinear ones.

It is proposed to approximate the solution of the partial differential equation by a family of cubic splines. The domain is $[a = x_1, b = x_n]$. Each cubic spline S_i is defined by $[x_i, x_{i+1}]$ with $h_i = x_{i+1} - x_i$ by the cubic polynomial

$$S_i(x) = M_i \frac{(x_{i+1} - x)^3}{6h_i} + M_{i+1} \frac{(x - x_i)^3}{6h_i} + a_i(x - x_i) + b_i \quad \forall i = 1, \ldots, n-1 \tag{7.9.3}$$

The unknowns are the n moments M_i, the $(n-1)$ parameters a_i and $(n-1)$ parameters b_i. The moments M_i are the second order derivatives $S''(x_i)$. The grid of x can be any. Equation (7.9.1) must be verified by the splines (7.9.3) in any point of the domain.

Moreover, the continuity of splines must be expressed

$$S_i(x_{i+1}) = S_{i+1}(x_{i+1}) \quad \forall i = 1, \ldots, n-2 \tag{7.9.4}$$

The continuity of the spline derivatives at the nodes must also be expressed

$$S_i'(x_{i+1}) = S_{i+1}'(x_{i+1}) \quad \forall i = 1, \ldots, n-2 \tag{7.9.5}$$

In PDE (7.9.1), $T(x, t)$ can be replaced by the spline $S_i(x)$ where the moments and the parameters are considered as time-dependent. It gives

$$
\begin{aligned}
&M_i' \frac{(x_{i+1} - x)^3}{6h_i} + M_{i+1}' \frac{(x - x_i)^3}{6h_i} + a_i'(x - x_i) + b_i' = \\
&M_i \frac{(x_{i+1} - x)}{h_i} + M_{i+1} \frac{(x - x_i)}{h_i} \quad \forall x \quad , \forall i = 1, \ldots, n-1
\end{aligned}
\tag{7.9.6}
$$

The PDE equation (7.9.6), used at all the nodes x_i, $(i = 1, \ldots, n)$, can be expressed as a linear system of n equations

$$
\boldsymbol{A}_1 \begin{bmatrix} \frac{da_1}{dt} \\ \vdots \\ \frac{da_{n-1}}{dt} \end{bmatrix} + \boldsymbol{B}_1 \begin{bmatrix} \frac{db_1}{dt} \\ \vdots \\ \frac{db_{n-1}}{dt} \end{bmatrix} + \boldsymbol{C}_1 \begin{bmatrix} \frac{dM_1}{dt} \\ \vdots \\ \frac{dM_n}{dt} \end{bmatrix} = 0 \tag{7.9.7}
$$

or

$$
\begin{aligned}
&\begin{bmatrix} 0 & 0 & \ldots & 0 \\ 0 & \ddots & 0 & \\ \vdots & & \ldots & 0 \\ 0 & & 0 & 0 \\ 0 & \ldots & 0 & h_{n-1} \end{bmatrix} \begin{bmatrix} \frac{da_1}{dt} \\ \vdots \\ \\ \frac{da_{n-1}}{dt} \end{bmatrix} + \begin{bmatrix} 1 & 0 & \ldots & 0 \\ 0 & \ddots & 0 & \\ \vdots & & \ddots & 0 \\ 0 & \ldots & 0 & 1 \\ 0 & \ldots & 0 & 1 \end{bmatrix} \begin{bmatrix} \frac{db_1}{dt} \\ \vdots \\ \\ \frac{db_{n-1}}{dt} \end{bmatrix} + \\
&\begin{bmatrix} \frac{h_1^2}{6} & 0 & \ldots & & 0 \\ 0 & \frac{h_2^2}{6} & 0 & \ldots & 0 \\ \vdots & & \ddots & & 0 \\ & & & \frac{h_{n-1}^2}{6} & 0 \\ 0 & \ldots & & 0 & \frac{h_{n-1}^2}{6} \end{bmatrix} \begin{bmatrix} \frac{dM_1}{dt} \\ \vdots \\ \\ \frac{dM_n}{dt} \end{bmatrix} = \begin{bmatrix} M_1 \\ \vdots \\ \\ M_n \end{bmatrix}
\end{aligned}
\tag{7.9.8}
$$

In the case of Dirichlet boundary conditions, the first and the last equations of this system must be eliminated and replaced by two algebraic equations expressing the boundary conditions

$$
\begin{aligned}
&M_1 \frac{h_1^2}{6} + b_1 = u_a \,, \quad \text{for } x = x_a \\
&M_n \frac{h_{n-1}^2}{6} + a_{n-1} h_{n-1} + b_{n-1} = u_b \quad \text{for } x = x_b
\end{aligned}
\tag{7.9.9}
$$

The sizes of matrices $\boldsymbol{A}_1, \boldsymbol{B}_1, \boldsymbol{C}_1$ are $n \times (n-1)$, $n \times (n-1)$, $n \times n$, respectively.

It is also clear that, by differentiating both boundary conditions (7.9.9) with respect to time, we obtain the first and the last equations of the system (7.9.8). The suppression of the first and the last equations of the system (7.9.8) because of Dirichlet boundary conditions reduces the total number of equations by two. It is possible to substitute the equations of the system (7.9.8) at x_1 and x_n by writing similar equations, but at points close to the domain boundaries, for example $(x_1 + x_2)/2$ and $(x_{n-1} + x_n)/2$, which give the following system of n equations

$$\begin{bmatrix} \frac{h_1}{2} & 0 & \dots & 0 \\ 0 & 0 & 0 & \\ \vdots & & \ddots & 0 \\ 0 & & 0 & 0 \\ 0 & \dots & 0 & \frac{h_{n-1}}{2} \end{bmatrix} \begin{bmatrix} \frac{da_1}{dt} \\ \vdots \\ \frac{da_{n-1}}{dt} \end{bmatrix} + \begin{bmatrix} 1 & 0 & \dots & 0 \\ 0 & \ddots & 0 & \\ \vdots & & \ddots & 0 \\ 0 & \dots & 0 & 1 \\ 0 & \dots & 0 & 1 \end{bmatrix} \begin{bmatrix} \frac{db_1}{dt} \\ \vdots \\ \frac{db_{n-1}}{dt} \end{bmatrix} + \begin{bmatrix} \frac{h_1^2}{48} & \frac{h_1^2}{48} & 0 & \dots & & 0 \\ 0 & \frac{h_2^2}{6} & 0 & \dots & & 0 \\ \vdots & & \ddots & & & 0 \\ & & & & \frac{h_{n-1}^2}{6} & 0 \\ 0 & \dots & & & 0\frac{h_{n-1}^2}{48} & \frac{h_{n-1}^2}{48} \end{bmatrix} \begin{bmatrix} \frac{dM_1}{dt} \\ \vdots \\ \frac{dM_n}{dt} \end{bmatrix} = \begin{bmatrix} \frac{M_1 + M_2}{2} \\ M_2 \\ \vdots \\ M_{n-1} \\ \frac{M_{n-1} + M_n}{2} \end{bmatrix} \tag{7.9.10}$$

The continuity of the spline functions can be expressed as the following system of $(n-2)$ linear equations

$$\boldsymbol{A}_2 \begin{bmatrix} a_1 \\ \vdots \\ a_{n-1} \end{bmatrix} + \boldsymbol{B}_2 \begin{bmatrix} b_1 \\ \vdots \\ b_{n-1} \end{bmatrix} + \boldsymbol{C}_2 \begin{bmatrix} M_1 \\ \vdots \\ M_{n-1} \end{bmatrix} = 0 \tag{7.9.11}$$

or

$$\begin{bmatrix} h_1 & 0 & \dots & & 0 \\ 0 & \ddots & 0 & & 0 \\ \vdots & & \ddots & & 0 \\ 0 & & & h_{n-2} & 0 \end{bmatrix} \begin{bmatrix} a_1 \\ \vdots \\ a_{n-1} \end{bmatrix} + \begin{bmatrix} 1 & -1 & 0 & \dots & & 0 \\ 0 & \ddots & \ddots & & & 0 \\ \vdots & & \ddots & & & 0 \\ 0 & \dots & & 1 & -1 & 0 \\ 0 & \dots & & & 1 & -1 \end{bmatrix} \begin{bmatrix} b_1 \\ \vdots \\ b_{n-1} \end{bmatrix} + \begin{bmatrix} 0 & \frac{h_1^2 - h_2^2}{6} & 0 & 0 & \dots & 0 \\ \vdots & & \ddots & \ddots & & \vdots \\ \vdots & & & \ddots & & \vdots \\ \vdots & & & & & 0 \\ 0 & & & & 0 & \frac{h_{n-2}^2 - h_{n-1}^2}{6} \end{bmatrix} \begin{bmatrix} M_1 \\ \vdots \\ M_{n-1} \end{bmatrix} = \begin{bmatrix} 0 \\ \vdots \\ 0 \end{bmatrix} \tag{7.9.12}$$

The sizes of matrices $\boldsymbol{A}_2, \boldsymbol{B}_2, \boldsymbol{C}_2$ are $(n-2)\times(n-1)$, $(n-2)\times(n-1)$, $(n-2)\times(n-1)$, respectively.

The continuity of the derivatives of the spline functions can be expressed as the following system of $(n-2)$ linear equations

$$\boldsymbol{A}_3 \begin{bmatrix} a_1 \\ \vdots \\ a_{n-1} \end{bmatrix} + \boldsymbol{B}_3 \begin{bmatrix} b_1 \\ \vdots \\ b_{n-1} \end{bmatrix} + \boldsymbol{C}_3 \begin{bmatrix} M_1 \\ \vdots \\ M_{n-1} \end{bmatrix} = 0 \quad \text{with } \boldsymbol{B}_3 = \boldsymbol{0} \tag{7.9.13}$$

or

$$
\begin{bmatrix}
1 & -1 & 0 & \dots & & 0 \\
0 & \ddots & \ddots & & & \vdots \\
\vdots & & \ddots & & & \\
0 & \dots & & 1 & -1 & 0 \\
0 & \dots & & & 1 & -1
\end{bmatrix}
\begin{bmatrix} a_1 \\ \vdots \\ a_{n-1} \end{bmatrix}
+
\begin{bmatrix}
0 & \frac{h_1+h_2}{2} & 0 & \dots & 0 \\
0 & 0 & \ddots & 0 & 0 \\
\vdots & & \ddots & & 0 \\
0 & & & 0 & \frac{h_{n-2}+h_{n-1}}{2}
\end{bmatrix}
\begin{bmatrix} M_1 \\ \vdots \\ M_{n-1} \end{bmatrix}
=
\begin{bmatrix} 0 \\ \vdots \\ 0 \end{bmatrix}
\tag{7.9.14}
$$

The sizes of matrices $\boldsymbol{A}_3$, $\boldsymbol{C}_3$ are $(n-2)\times(n-1)$, $(n-2)\times(n-1)$, respectively.

This method with modified equations (7.9.10) has been applied to problem (7.9.1) with $\alpha = 0.1$ under the form (7.10.90) for which the theoretical solution is known. It gives Figure 7.33 which shows the evolution of the temperature profile with a time step equal to 0.1 and a final time equal to 5, and 21 equidistant points for the spatial grid. It would have been possible to use any irregular grid. The problem consisting in a linear differential–algebraic system has been solved with Dassl and Octave which gave the value $T = 0.63835$ instead of 0.63891 at $x = 0.5$ and $t = 2$, which shows a very small error, but acceptable. The results can be compared to the method of weighted residuals (Section 7.10.5) used on the same problem.

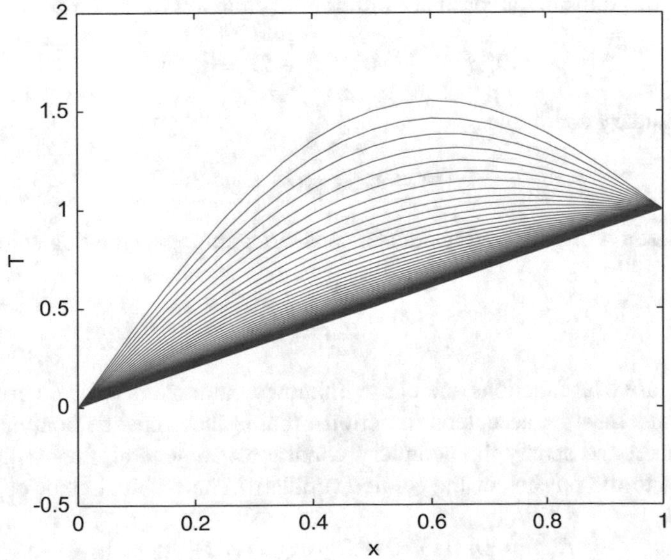

Fig. 7.33 Solution of Fourier's PDE by the method of splines

7.10 Spectral Methods

The term of spectral method is used (Canuto et al. 2006, 2007; Hesthaven et al. 2007; Karniadakis and Sherwin 2013; Trefethen 1996, 2000) to qualify methods which use an approximation of the solution by radial basis functions, also called trial functions. Among spectral methods, we find Galerkin's method, the tau method, and the collocation methods, still called pseudospectral methods. The collocation methods have known a great development since the methods of weighted residuals (Brunner 2004; Villadsen and Michelsen 1978; Villadsen and Stewart 1967). These methods are used to solve many classes of systems such as the ordinary differential equations, but also the partial differential equations (De Boor and Swartz 1973; Guillou and Soulé 1969; Jung and Don 2009; Khater et al. 2008; Peyret 2002; Russell and Shampine 1972).

7.10.1 Method of Weighted Residuals

In a general manner, the methods of weighted residuals (also called Rayleigh-Ritz methods) are based on the analytic or piecewise analytic approximation of the solution of an ordinary differential equation or a partial differential equation. The objective of these methods is to minimize the integral of the error (Schlatter 2010; Woodbury 2002). The method of weighted residuals thus allows us to solve boundary value problems by using trial functions that satisfy the boundary conditions and an integral formulation to minimize the error.

For example, consider an ordinary differential equation of the form

$$D[y(x), x] = 0\,, \quad x \in \mathcal{D} = [a, b] \tag{7.10.1}$$

with the boundary conditions

$$y(a) = y_a\,, \quad y(b) = y_b \tag{7.10.2}$$

Using the method of weighted residuals, we search an approximating solution of the form

$$\hat{y}(x) = \sum_{i=1}^{n} c_i \phi_i(x) \tag{7.10.3}$$

where $\phi_i(x)$ are trial functions and c_i are unknown parameters to be determined. The error functions must be acceptable functions, that is they must be continuous on the studied domain and satisfy the boundary conditions. In general, they will be chosen with respect to the physics of the studied problem. The residual error or residual is defined as

$$R(x) = D[\hat{y}(x), x] - D[y(x), x] = D[\hat{y}(x), x] \tag{7.10.4}$$

according to Equation (7.10.1). The residual is in general different from zero and depends on parameters c_i that are evaluated such that

$$\int_{\mathcal{D}} w_i(x)\, R(x)\, dx = 0\,, \quad i = 1, \ldots, n \tag{7.10.5}$$

where the weights $w_i(x)$ are arbitrary functions. Indeed, Equation (7.10.5) represents a system of n algebraic equations to solve.

According to the choice of the weight functions, we distinguish:

- the subdomain collocation,
- the point collocation,
- the least squares method,
- Galerkin's method,
- the method of moments.

- **Subdomain collocation**: the domain $\mathcal{D}$ is divided into N subdomains $\mathcal{D}_i$ and the weights are chosen such that

$$w_i = \begin{cases} 1 & \text{inside } \mathcal{D}_i \\ 0 & \text{outside} \end{cases} \tag{7.10.6}$$

so that we search

$$\frac{1}{\Delta x_i} \int_{\mathcal{D}_i} R(x) dx = 0 \tag{7.10.7}$$

where Δx_i represents the extent of the subdomain upon which the integration is done.

- **Point collocation method**: the weight functions are Dirac functions

$$w_i(x) = \delta(x - x_i) \tag{7.10.8}$$

so that the error is zero at nodes x_i. More generally, if the nodes are $\boldsymbol{x}_1, \ldots, \boldsymbol{x}_N$ ($\boldsymbol{x}$ is a vector of some dimension d), and if the values of some function f at the nodes are known by $f_i = f(\boldsymbol{x}_i)$, the approximation $\hat{y}(\boldsymbol{x})$ defined by

$$\hat{y}(\boldsymbol{x}) = \sum_{i=1}^{N} c_i \phi(\|\boldsymbol{x} - \boldsymbol{x}_i\|) \tag{7.10.9}$$

is such that

$$\begin{bmatrix} A_{11} \ldots A_{N1} \\ \vdots \ddots \vdots \\ A_{N1} \ldots A_{NN} \end{bmatrix} \begin{bmatrix} c_1 \\ \vdots \\ c_N \end{bmatrix} = \begin{bmatrix} f_1 \\ \vdots \\ f_N \end{bmatrix} \tag{7.10.10}$$

with $A_{ij} = \phi(\|\boldsymbol{x}_i - \boldsymbol{x}_j\|)$. Given some type of radial basis functions ϕ, Equation (7.10.10) allows us to interpolate f.

More generally, given the residual R definable by an ordinary differential equation or partial differential equation, the application of the weight function (7.10.8) implies the relation

$$\int_{\mathcal{D}} \delta(x - x_i)\, R(x)\, dx = R(x_i) = 0 \tag{7.10.11}$$

i.e. the residual is zero at any collocation point.

- **Least squares**: the weight functions are the partial derivatives of the basis functions with respect to the unknown parameters

$$w_i(x) = \frac{\partial R}{\partial c_i} \tag{7.10.12}$$

and the integral of the square of the residual is minimized

$$\int_{\mathcal{D}} R^2 \, dx \tag{7.10.13}$$

- **Galerkin's method**: the weight functions are identical to the basis functions

$$w_i(x) = \phi_i(x) \tag{7.10.14}$$

and the parameters c_i are determined so that

$$\int_{\mathcal{D}} \phi_i(x) \, R \, dx = 0 \tag{7.10.15}$$

In this case, the orthogonality of the basis functions will be often invoked.
- **Method of moments**: the weight functions are monomials

$$w_i(x) = x^i \, , \quad i = 0, \ldots, n-1 \tag{7.10.16}$$

7.10.2 Radial Basis Functions

In a general way, the approximation makes use of radial basis functions $\phi(r)$ ($r \geq 0$) which can be polynomials (Lagrange, Jacobi, Legendre, Chebyshev, Hermite, Laguerre, β-splines), often orthogonal hence the orthogonal collocation methods, or other functions (sine-cosine in Fourier series, Haar wavelets) (Buhmann 2003; Carey and Finlayson 1975; Dai and Cochran 2009; Driscoll and Fornberg 2002; Fasshauer 1997; Manni et al. 2015; Martinez and Esperanca 2007; Rocha 2009).

The choice of the radial basis functions must be done with respect to the physics of the problem. Some possible choices are described in the following.

- A Fourier series can be chosen as a radial basis function when the expected solution is periodic ($y(x) = y(x + L)$) of period L.
  The expansion of any function $f(x)$ as an infinite Fourier series in the complex domain is defined by

$$f(x) = \sum_{k=-\infty}^{k=-\infty} c_k \exp(i 2\pi k \frac{x}{L}) \tag{7.10.17}$$

where the complex Fourier coefficients c_k are equal to

$$c_k = \frac{1}{L} \int_{x_0}^{x_0+L} f(x) \exp(-i 2\pi k \frac{x}{L}) \, dx \tag{7.10.18}$$

x_0 can be arbitrarily chosen. i is the pure imaginary number.
In practice, a finite series is considered so that the approximation according the truncated Fourier series will be written as

$$\hat{y}_N(x) = \sum_{k=-K}^{K} c_k \exp(ik\alpha x) \tag{7.10.19}$$

where α is the fundamental wave number equal to $\alpha = 2\pi/L$.

- In the case of non-periodic boundary conditions, Fourier series are avoided. Lagrange polynomials can be used, but other types of polynomials are preferable. Many families of orthogonal polynomials exist (Section 2.6.1). Some important properties of Chebyshev polynomials are described in Section 2.6.1 and not recalled here. If Chebyshev polynomials $T_k(z)$ ($z \in [-1, 1]$) are used and if we make the transformation of $x \in \mathcal{D}$ into ($z \in [-1, 1]$), the approximation as a Chebyshev series expansion is written as (Boyd 2001)

$$\hat{y}_N(z) = \sum_{k=0}^{N} c_k T_k(z) \tag{7.10.20}$$

and the interpolation points or nodes are Chebyshev–Gauss–Lobatto points defined by

$$z_i = -\cos\left(\frac{\pi i}{N}\right), \quad i = 0, \ldots, N \tag{7.10.21}$$

corresponding to the positions of the extrema (located at ± 1) of Chebyshev polynomial of degree N. The Chebyshev polynomials at Chebyshev–Gauss–Lobatto nodes are

$$T_k(z_j) = -\cos\left(\frac{\pi k j}{N}\right), \quad j, k = 0, \ldots, N \tag{7.10.22}$$

In this case, the two limits of interval $[-1, +1]$ are collocation points. For example, for $N = 5$, the nodes are at $\{-1.0, -0.809, -0.309, 0.309, 0.809, 1.00\}$.

Also frequently, the nodes are chosen as the zeros of Chebyshev polynomial T_{N+1} according to Gauss quadrature i.e.

$$z_i = -\cos\left(\frac{\pi(2i+1)}{2N+2}\right), \quad i = 0, \ldots, N \tag{7.10.23}$$

In this case, the nodes are strictly inside the interval $[-1, +1]$. For example, for $N = 5$, the nodes are at $\{-0.966, -0.707, -0.259, 0.259, 0.707, 0.966\}$.

Gauss–Radau points are also used as collocation points, they are defined by

$$z_i = \cos\left(\frac{2\pi(N-i)}{2N+1}\right), \quad i = 0, \ldots, N \tag{7.10.24}$$

In this case, the node $z = 1$ is a collocation point. For example, for $N = 5$, the nodes are at $\{-0.959, -0.655, -0.142, 0.415, 0.841, 1.0\}$.

The use of irregularly spaced points such as Chebyshev–Gauss–Lobatto points is important as it avoids Runge phenomenon (Figures 1.9 and 1.11). The positions of these nodes are represented in Figure 7.34.

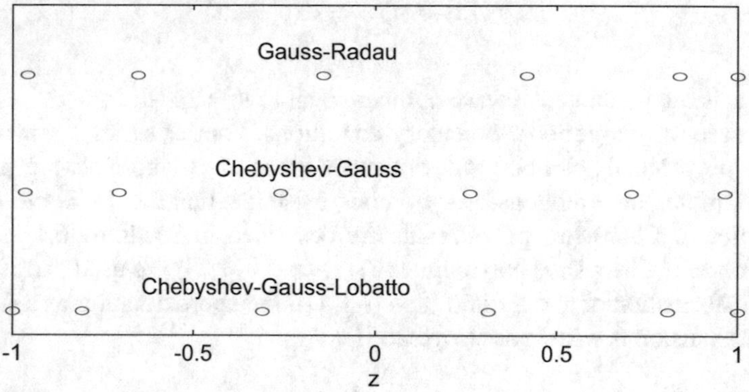

Fig. 7.34 Positions of the nodes for different collocation cases for $N = 5$. The ordinate is arbitrary

The coefficients of the Chebyshev series expansion (7.10.20) are

$$c_k = \frac{2}{N} \sum_{i=0}^{N} \frac{y(z_i)}{b_i} \cos\left(\frac{k\pi i}{N}\right), \quad k = 0, \dots, N \tag{7.10.25}$$

with $b_0 = b_N = 2$ and $b_i = 1$ $(i = 1, \dots, N-1)$. Note that, at the nodes, we have $y(z_i) = \hat{y}(z_i)$.

To evaluate the coefficients Chebyshev series expansion (7.10.20), the following recursive stable algorithm is used

$$\begin{aligned} &A_{N+1} = 0\,, \quad A_N = c_N \\ &A_k = c_k + 2\,z\,A_{k+1} - A_{k+2}\,, \quad k = N-1, \dots, 1 \\ &\hat{y}_N(z) = c_0 - A_2 + A_1\,z \end{aligned} \tag{7.10.26}$$

The transformation of the physical space $\hat{y}_N$ into the Chebyshev spectral space c_k (space of the coefficients) is called Chebyshev transformation.

Schlatter (2010) notes that the calculation of the derivative according to equation

$$\hat{y}'_N(z_i) = \sum_{i=0}^{N} D_{ij}\,\hat{y}_N(z_j) \tag{7.10.27}$$

is very sensitive to rounding errors and recommends to follow (Weideman and Reddy 2000). We can also cite Trefethen (2000).

In the case of an ordinary differential equation or a partial differential equation, Equation (7.10.20) is modified and the approximation becomes

$$\hat{y}_N(z) = \sum_{k=0}^{N} c_k(t) T_k(z) \tag{7.10.28}$$

where the coefficients $c_k(t)$ are time functions. Suppose that we desire to efficiently calculate the successive derivatives, for example, instead of writing the first order derivative as

$$\hat{y}'_N(z) = \sum_{k=0}^{N} c_k(t) T'_k(z) \tag{7.10.29}$$

indeed, it would be better to express the first order derivative under the form

$$\hat{y}'_N(z) = \sum_{k=0}^{N} c'_k(t) T_k(z) \tag{7.10.30}$$

where the coefficients c'_k can be expressed linearly (Trefethen 2000) with respect to the coefficients c_k according to the formula

$$\boldsymbol{c}' = \boldsymbol{D}_{s,N}\, \boldsymbol{c} \tag{7.10.31}$$

where $\boldsymbol{c}$ is the vector of coefficients c_k (similarly for $\boldsymbol{c}'$) and $\boldsymbol{D}_{s,N}$ is the spectral differential operator (with respect to the coefficients). The following example allows us to understand the method.

Example 7.10 :
Spectral differential operator and Chebyshev polynomials

Consider the case $N = 4$. We use Equations (7.10.28) and (7.10.30) and the expressions of Chebyshev polynomials (1.3.65) as well as the expressions of the monomials (1.3.66), we get

$$\begin{aligned} \hat{y}'_N(z) &= c_0\, T'_0(z) + c_1\, T'_1(z) + c_2\, T'_2(z) + c_3\, T'_3(z) + c_4\, T'_4(z) \\ &= c_1\, T_0(z) + c_2\, 4\, T_1(z) + c_3 (6\, T_2(z) + 3\, T_0(z)) + c_4 (8\, T_3(z) + 8\, T_1(z)) \end{aligned} \tag{7.10.32}$$

giving the relations

$$\begin{aligned} c'_0 &= c_1 + c_3 \\ c'_1 &= 4\, c_2 + 8\, c_4 \\ c'_2 &= 4\, c_3 \\ c'_3 &= 8\, c_4 \end{aligned} \tag{7.10.33}$$

hence

$$\left[\boldsymbol{c}'\right] = \begin{bmatrix} 0 & 1 & 0 & 3 & 0 \\ 0 & 0 & 4 & 0 & 8 \\ 0 & 0 & 0 & 6 & 0 \\ 0 & 0 & 0 & 0 & 8 \\ 0 & 0 & 0 & 0 & 0 \end{bmatrix} \left[\boldsymbol{c}\right] = \boldsymbol{D}_{s,4}\, \boldsymbol{c} \tag{7.10.34}$$

To obtain the second order derivative, it suffices to use twice the spectral differential (multiplication of the matrix by itself) which gives

$$\hat{y}_N^{(2)}(z) = \sum_{k=0}^{N} c''_k(t) T_k(z) \tag{7.10.35}$$

with the coefficients $\boldsymbol{c}'' = \boldsymbol{D}_{s,N}(\boldsymbol{D}_{s,N}\, \boldsymbol{c})$. The spectral differential operator is represented by a strictly upper triangular matrix whose elements can be calculated by the following relation

$$
\boldsymbol{D}_{s,N} = \begin{bmatrix} 0 & j-1 & 0 & j-1 & 0 & j-1 & 0 & \dots \\ 0 & 0 & 2(j-1) & 0 & 2(j-1) & 0 & 2(j-1) & \dots \\ 0 & 0 & 0 & 2(j-1) & 0 & 2(j-1) & 0 & \dots \\ 0 & 0 & 0 & 0 & 2(j-1) & 0 & 2(j-1) & \dots \\ \vdots & & & & \ddots & \ddots & \ddots & \\ 0 & \dots & & & & & & 0 \end{bmatrix} \tag{7.10.36}
$$

where j is the column number.

- A radial basis function which is convenient for many problems is the multiquadric function (concept of ball) (Driscoll and Fornberg 2002; Hardy 1990; Kansa 1990a,b) defined by

$$
\phi(r) = \sqrt{r^2 + \sigma^2}, \quad \sigma > 0 \tag{7.10.37}
$$

where r represents the distance with respect to a fixed center x. σ is the shape parameter. This function is also presented under the form

$$
\phi(r) = \sqrt{1 + (\epsilon r)^2} \tag{7.10.38}
$$

- Another radial basis function is the inverse multiquadric function

$$
\phi(r) = \frac{1}{\sqrt{1 + (\epsilon r)^2}} \tag{7.10.39}
$$

depending on parameter ϵ.
- Another radial basis function is the Gaussian function

$$
\phi(r) = \exp(-(\epsilon r)^2) \tag{7.10.40}
$$

depending on parameter ϵ.
- A possible general representation of radial basis functions (Larsson and Fornberg 2005; Marchi and Santin 2013; Rocha 2009), used for any type of interpolation, is

$$
\phi(r) = \phi(\|\boldsymbol{x} - \boldsymbol{x}^j\|), \quad \text{with } \|x - x_j\| = \sqrt{\sum_{i=1}^{n} |\theta_i|(x_i - x_i^j)^2} \tag{7.10.41}
$$

where θ_i are scalars. $\boldsymbol{x}$ is a vector, $\boldsymbol{x}^j$ are vectors such that the function $y(\boldsymbol{x})$ to interpolate is known at nodes $\boldsymbol{x}^j$, $j = 1, \dots, N$. The interpolation will then be given by the approximation

$$
\hat{y}(\boldsymbol{x}) = \sum_{j=1}^{N} c_j \phi(\|\boldsymbol{x} - \boldsymbol{x}^j\|) \tag{7.10.42}
$$

In this context, in addition to the quadric, inverse quadric, Gaussian functions already cited and depending on a parameter, we can cite the cubic spline $\phi(r) = r^3$ and the thin plate spline $\phi(r) = r^2 \, \ln(r)$.
- Hermite polynomials (Sections 1.3.7 and 2.6.1) are frequently used in approximation as radial basis functions. Among them, we find in particular cubic Hermite polynomials (or cubic Hermite splines).

Recall their presentation under their usual form using Hermite basis polynomials h (Section 1.3.7) defined on $[0, 1]$, i.e.

$$\begin{aligned} h_{00}(t) &= 1 - 3\,t^2 + 2\,t^3 \\ h_{01}(t) &= 3\,t^2 - 2\,t^3 \\ h_{10}(t) &= t - 2\,t^2 + t^3 \\ h_{11}(t) &= -t^2 + t^3 \end{aligned} \tag{7.10.43}$$

To perform the interpolation on the subintervals of a domain $[x_1, x_N]$ divided into N subintervals, we set

$$h(t) = h\left(\frac{x - x_i}{x_{i+1} - x_i}\right) \quad \text{with } t = \frac{x - x_i}{x_{i+1} - x_i} \tag{7.10.44}$$

hence the interpolation

$$\begin{aligned} \tilde{y}(x) = {} & y_i\, h_{00}\left(\frac{x - x_i}{x_{i+1} - x_i}\right) + y_{i+1})\, h_{01}\left(\frac{x - x_i}{x_{i+1} - x_i}\right) \\ & + y'_i\,(x_{i+1} - x_i)\, h_{10}\left(\frac{x - x_i}{x_{i+1} - x_i}\right) \\ & + y'_{i+1}\,(x_{i+1} - x_i)\, h_{11}\left(\frac{x - x_i}{x_{i+1} - x_i}\right) \end{aligned} \tag{7.10.45}$$

The following presentation (Akima 1970; Bica 2012; Fritsch and Carlson 1980; Restrepo 2003) is also based on Hermite polynomials, but is different. Consider again the domain $[x_1, x_N]$ divided into N subintervals. The cubic Hermite polynomials are defined by

$$\phi_{0,j}(x) = \begin{cases} \dfrac{(x - x_{j-1})^2[2(x_j - x) + (x_j - x_{j-1})]}{(x_j - x_{j-1})^3} & \text{if } x_{j-1} \le x \le x_j \\ \dfrac{(x - x_{j+1})^2[2(x - x_j) + (x_{j+1} - x_j)]}{(x_{j+1} - x_j)^3} & \text{if } x_j \le x < x_{j+1} \\ 0 & \text{otherwise} \end{cases} \tag{7.10.46}$$

and

$$\phi_{1,j}(x) = \begin{cases} \dfrac{(x - x_{j-1})^2(x - x_j)}{(x_j - x_{j-1})^2} & \text{if } x_{j-1} \le x \le x_j \\ \dfrac{(x - x_{j+1})^2(x - x_j)}{(x_{j+1} - x_j)^2} & \text{if } x_j \le x < x_{j+1} \\ 0 & \text{otherwise} \end{cases} \tag{7.10.47}$$

These functions possess the following properties

$$\begin{aligned} &\phi_{0,j}(x_j) = 1\,, \qquad && \phi_{1,j}(x_j) = 0 \\ &\frac{d\phi_{0,j}}{dx}(x_j) = 0\,, && \frac{d\phi_{1,j}}{dx}(x_j) = 1 \end{aligned} \tag{7.10.48}$$

The algorithm proceeds in two successive stages in the following way

- For each node x_j, sweep all the possible values of x interpolating in the domain $[x_1, x_N]$ in order to calculate $\phi_{0,j}(x)$ and $\phi_{1,j}(x)$ according to Equations (7.10.46) and (7.10.47),
- For each value of x interpolating in the domain $[x_1, x_N]$, calculate $\hat{y}(\boldsymbol{x})$ according to Equation (7.10.49).

Using the previous cubic Hermite polynomials, the following function can be taken as a differentiable and continuously differentiable interpolation function of class C^1 on the whole domain $[x_1, x_N]$

$$\hat{y}(\boldsymbol{x}) = \sum_{j=1}^{N} \left[y_j\, \phi_{0,j}(x) + \frac{dy_j}{dx}\, \phi_{1,j}(x) \right] \tag{7.10.49}$$

In this equation, y_j is the approximation of y at x_j and dy_j/dx is the approximation of dy/dx at x_j.

Example 7.11 :
Hermite polynomials: interpolation of a function

The method of Hermite polynomials has been evaluated on the following example of Restrepo (2003). The function $y(x) = \sin(\pi x)$ is considered on interval [0, 3] and the data are the ordinates y and the derivatives at [0, 0.2, 0.5, 1.2, 1.6, 2, 2.4, 2.8, 3]. The interpolation abscissas are regularly spaced in [0, 3] with a step equal to 0.02.

The approximation is represented on Figure 7.35 by two different methods. At left, the interpolation uses the original Hermite polynomials (Equation (7.10.45)) while at right it uses modified polynomials (Equations (7.10.46) and (7.10.47)). There is no noticeable difference about the quality of interpolation between both methods.

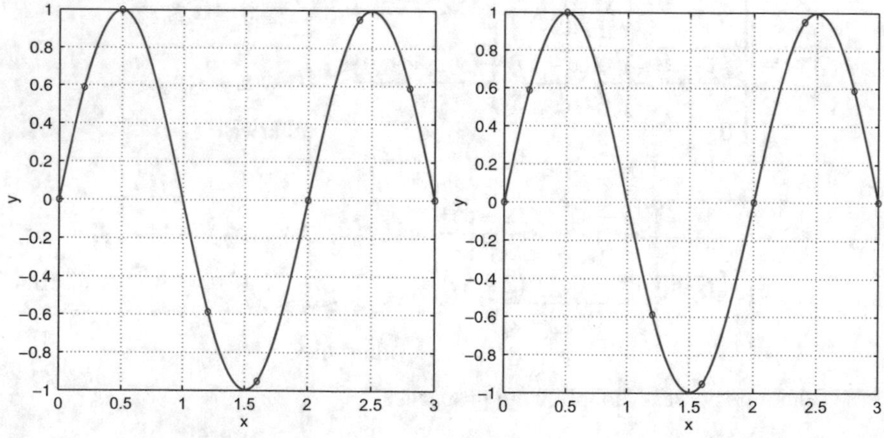

Fig. 7.35 Interpolation by Hermite polynomials. Comparison of the theoretical curve and its approximation. Left: collocation on subintervals by cubic Hermite polynomials (Equation (7.10.45)). Right: collocation on subintervals by modified cubic Hermite polynomials (Equations (7.10.46) and (7.10.47))

Recall that the interpolating polynomials h_{00}, h_{01}, h_{10}, h_{11} used in Equation (7.10.45) have been represented in Figure 1.12. The interpolating polynomials ϕ_0 have been represented on Figure 7.36. It can be noticed that, for example on interval [0.2, 0.5] located between two successive nodes 0.2 and 0.5, two interpolating polynomials are used. One of the interpolating polynomials presents its maximum 1 at 0.2 and its minimum 0 at 0.5, whereas the other interpolating polynomial presents its minimum 0 at 0.2 and its maximum 0 at 0.5. Thus both interpolating polynomials ϕ_0 take their values in [0, 1] in the same manner as h_{00} and h_{11}.

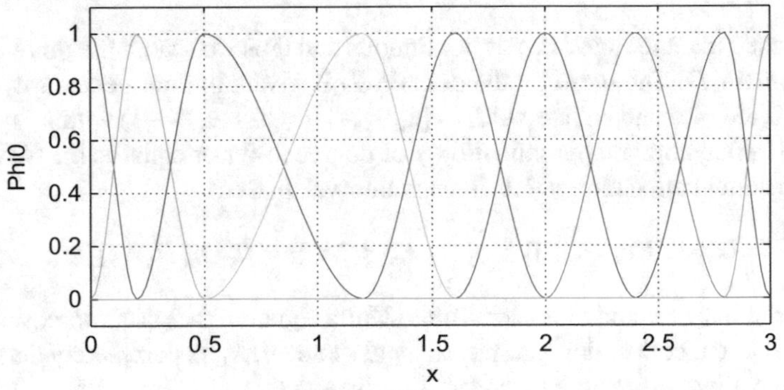

Fig. 7.36 Hermite interpolating polynomials ϕ_0

Interpolating polynomials ϕ_1 have been represented on Figure 7.37. In the same way, if we examine the interval [0.2, 0.5], we notice that two interpolating polynomials are used. However, nothing can be said about the values taken by the interpolating polynomials.

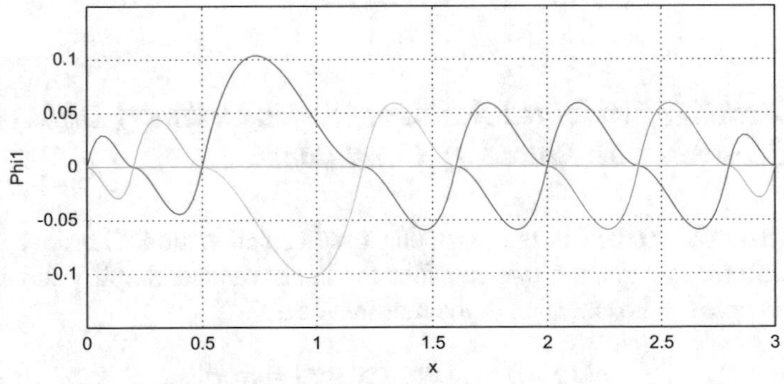

Fig. 7.37 Hermite interpolating polynomials ϕ_1

7.10.3 Polynomial Collocation for an Initial Value Ordinary Differential Equation

To introduce the collocation method, consider the case of an initial value ordinary differential equation

$$y'(t) = f(x, y), \quad x \in \mathcal{D} = [0, L], \quad y(0) = y_0 \tag{7.10.50}$$

It is supposed that the function f is continuous and thus we search the unique solution on the domain $\mathcal{D}$. The mesh on the domain $\mathcal{D}$ is given, but not necessarily uniform. Let $x_0 = 0$, $x_N = L$ and an interval $I_i = [x_i, x_{i+1}]$ $(i = 0, \dots, N-1)$ with $h_i = x_{i+1} - x_i$. We search a polynomial approximation $\hat{y}$ of degree lower or equal to m. We choose to place m distinct collocation points in each interval I_i. Set

$$X_h = \{t = t_i + c_i h_i, \quad 0 \le c_1 \dots c_m \le 1 \quad (i = 0, \dots, N-1)\} \tag{7.10.51}$$

This method must remind of m-step Runge-Kutta equation (6.2.40), moreover implicit. Provided that the collocation parameters c_i are known, X_h is completely determined.

Now, the approximation $\hat{y}$ by collocation is such that

$$\hat{y}'(x) = f(x, \hat{y}(x)), \quad x \in X_h, \quad \hat{y}(0) = y(0) = y_0 \tag{7.10.52}$$

If two successive intervals are considered, the solution $\hat{y}'$ must verify the continuity. Suppose that $c_1 = 0$ and $c_m = 1$ $(m \ge 2)$. Set $x = x_{i-1} + c_m h_{i-1} = x_i^-$ and $x = x_i + c_1 h_i = x_i^+$. The continuity imposes

$$\hat{y}'(x_i^-) = \hat{y}'(x_i^+), \quad i = 1, \dots, N-1 \tag{7.10.53}$$

7.10.4 Method of Weighted Residuals for an Ordinary Differential Equation with Boundary Conditions

Let us explain the method in the particular case of collocation. Consider the linear ordinary differential equation (we consider the linear case to simplify the following, but the principle can be extended to the nonlinear case)

$$L(y(x)) = f(x), \quad x \in \mathcal{D} = [a, b] \tag{7.10.54}$$

with the boundary conditions

$$y(a) = y_a, \quad y(b) = y_b \tag{7.10.55}$$

We use a set of radial basis functions ϕ_i $(i = 1, \dots, n)$ so that the approximation solution has the form

$$\hat{y}(x) = \sum_{i=1}^{n} c_i\, \phi_i(x) \tag{7.10.56}$$

where the coefficients c_i are unknown. As the ordinary differential equation is linear, we can write

$$L(\hat{y}(x)) = \sum_{i=1}^{n} c_i \, L(\phi_i(x)) \tag{7.10.57}$$

so that Equation (7.10.54) becomes

$$\begin{aligned} &\sum_{i=1}^{n} c_i \, L(\phi_i(x)) = f(x), \quad x \in \mathcal{D} = [a, b] \\ &\sum_{i=1}^{n} c_i \, \phi_i(a) = y_a, \quad \sum_{i=1}^{n} c_i \, \phi_i(b) = y_b \end{aligned} \tag{7.10.58}$$

Note that

$$L(\hat{y}(x)) - f(x) \tag{7.10.59}$$

is the expression of the residual. To determine the n unknown coefficients c_i, n collocation conditions are imposed to obtain a system of $n \times n$ linear equations. As 2 collocation points a and b are already used, there remain $n-2$ collocation points $x_2, \ldots, x_{n-1}$ where the ordinary differential equation is verified. Thus, we get the equations

$$\begin{aligned} &\sum_{i=1}^{n} c_i \, \phi_i(a) = y_a \\ &\sum_{i=1}^{n} c_i \, L(\phi_i(x_k)) = f(x_k), \quad k = 2, \ldots, n-1 \\ &\sum_{i=1}^{n} c_i \, \phi_i(b) = y_b \end{aligned} \tag{7.10.60}$$

that is a linear system which can be presented (by replacing a by x_1 and b by x_n) under the form

$$\begin{bmatrix} \phi_1(x_1) & \phi_2(x_1) & \ldots & \phi_n(x_1) \\ L(\phi_1(x_2)) & L(\phi_2(x_2)) & \ldots & L(\phi_n(x_2)) \\ \vdots & & & \vdots \\ L(\phi_1(x_{n-1})) & L(\phi_2(x_{n-1})) & \ldots & L(\phi_n(x_{n-1})) \\ \phi_1(x_n) & \phi_2(x_n) & \ldots & \phi_n(x_n) \end{bmatrix} \begin{bmatrix} c_1 \\ \vdots \\ c_n \end{bmatrix} = \begin{bmatrix} y_a \\ f(x_2) \\ \vdots \\ f(x_{n-1}) \\ y_b \end{bmatrix} \tag{7.10.61}$$

which can be solved provided that the square matrix in the left-hand side is not singular.

Suppose that the multiquadric function (7.10.37) is used and that the centers of each of these functions are chosen as the successive collocation points, we thus get the expression of the radial basis functions

$$\phi_i(|x - x_i|) = \sqrt{(x - x_i)^2 + \sigma^2}, \quad i = 1, \ldots, n \tag{7.10.62}$$

and the previous linear system becomes

$$\begin{bmatrix} \phi_1(|x_1 - x_1|) & \phi_2(|x_1 - x_2|) & \dots & \phi_n(|x_1 - x_n|) \\ L(\phi_1(|x_2 - x_1|)) & L(\phi_2(|x_2 - x_2|)) & \dots & L(\phi_n(|x_2 - x_n|)) \\ \vdots & & & \vdots \\ L(\phi_1(|x_{n-1} - x_1|)) & L(\phi_2(|x_{n-1} - x_2|)) & \dots & L(\phi_n(|x_{n-1} - x_n|)) \\ \phi_1(|x_n - x_1|) & \phi_2(|x_n - x_2|) & \dots & \phi_n(|x_n - x_n|) \end{bmatrix} \begin{bmatrix} c_1 \\ \vdots \\ c_n \end{bmatrix} = \begin{bmatrix} y_a \\ f(x_2) \\ \vdots \\ f(x_{n-1}) \\ y_b \end{bmatrix} \tag{7.10.63}$$

Note that $L(\sum_{i=1}^n c_i \phi_i(|x_k - x_i|)) - f(x_k)$ represents the residual $R(x_k)$.

Example 7.12 illustrates four of the methods presented in Section 7.10.1 to demonstrate the use of the residual.

Example 7.12 :

Integration of a second order ODE with boundary conditions using four residual methods

Consider the second order ordinary differential equation

$$y''(x) + y'(x) - 6y(x) = 0 \tag{7.10.64}$$

with the boundary conditions $y(0) = 1$ and $y(1) = 0.3\,\exp(2) + 0.7\,\exp(-3)$. The studied domain is thus $[0, 1]$. We note $y_1 = y(1)$. The theoretical solution is

$$y(x) = 0.3\,\exp(2x) + 0.7\,\exp(-3x) \tag{7.10.65}$$

To simplify the presentation, we decide to use as an approximation function a second degree polynomial, i.e.

$$\hat{y}(x) = a_0 + a_1\,x + a_2\,x^2 \tag{7.10.66}$$

which will be used without any subdivision on the whole interval $[0, 1]$. We impose that the approximation solution verifies the boundary conditions, i.e.

$$\hat{y}(0) = a_0 = 1\,, \quad \hat{y}(1) = a_0 + a_1 + a_2 = y_1 \tag{7.10.67}$$

It results

$$a_0 = 1\,, \quad a_2 = y_1 - 1 - a_1 \tag{7.10.68}$$

and there remains only one unknown parameter a_1.

The residual $R(x)$ is calculated

$$\begin{aligned} R(x) &= \hat{y}''(x) + \hat{y}'(x) - 6\hat{y}(x) \\ &= (2y_1 - a_1 - 8) + (2y_1 - 2 - 8a_1)x - 6(y_1 - 1 - a_1)x^2 \end{aligned} \tag{7.10.69}$$

- Collocation method
  The middle of the domain is chosen as a collocation point, i.e. $x = 0.5$. Thus, we impose $R(0.5) = 0$ which gives
  $$a_1 = \frac{3}{7}(y_1 - 5) \tag{7.10.70}$$
  On Figure 7.38, at left the theoretical solution and its approximation are represented, while the residual is represented at right. We notice on the Figure of the residual that it is zero at point $x = 0.5$.
- Collocation method on a subdomain
  We impose that

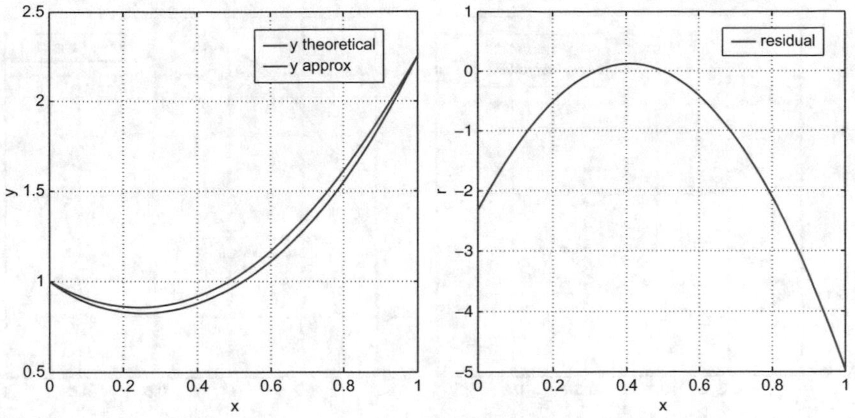

Fig. 7.38 Integration of an ordinary differential equation with boundary conditions. Collocation method. Left: theoretical solution and its approximation. Right: residual

$$\int_0^1 1\, R(x)\, dx = 0 \tag{7.10.71}$$

which gives

$$a_1 = \frac{y_1 - 7}{3} \tag{7.10.72}$$

On Figure 7.39, at left the theoretical solution and its approximation are represented, while the residual is represented at right. We must approximately notice that the surface area between the horizontal axis and the residual curve is close to 0.

Indeed, if two subdomains were considered, for example [0, 0.5] and [0.5, 1], we could determine the two missing coefficients of a third degree polynomial approximation function instead of the second degree considered here, thus four coefficients to be determined by two integrals, plus two boundary conditions that must always be considered.

More generally, if N subdomains were considered, the approximation polynomial would be of degree $(N + 1)$, corresponding to $(N + 2)$ coefficients to determine by using the N integrals of type (7.10.7) and the two boundary conditions.

- Least squares method

  The only unknown parameter is a_1 and the only necessary weight function is determined by

$$w_1(x) = \frac{\partial R(x)}{\partial a_1} = -1 - 8x + 6x^2 \tag{7.10.73}$$

Then, we impose

$$\int_0^1 w_1(x)\, R(x)\, dx = 0 \tag{7.10.74}$$

which gives

$$a_1 = \frac{38y_1 - 308}{143} \tag{7.10.75}$$

On Figure 7.40, at left the theoretical solution and its approximation are represented, while the residual is represented at right.

- Galerkin's method

  The only unknown parameter is a_1 and the only necessary weight function is determined by

$$w_1(x) = \frac{\partial \hat{y}(x)}{\partial a_1} = x - x^2 \tag{7.10.76}$$

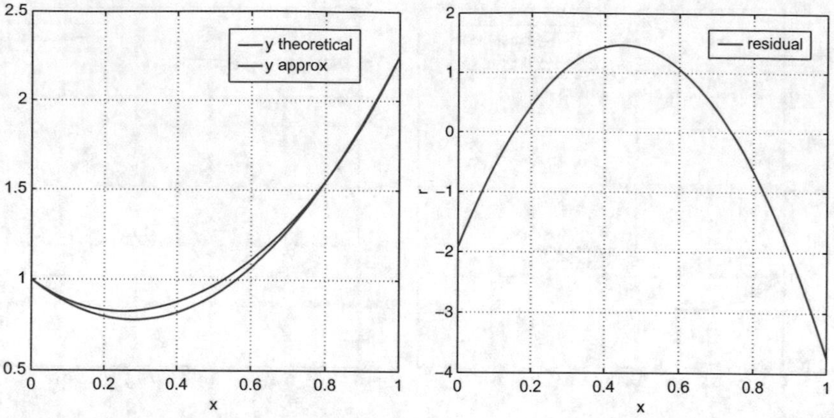

Fig. 7.39 Integration of an ordinary differential equation with boundary conditions. Collocation method on a subdomain. Left: theoretical solution and its approximation. Right: residual

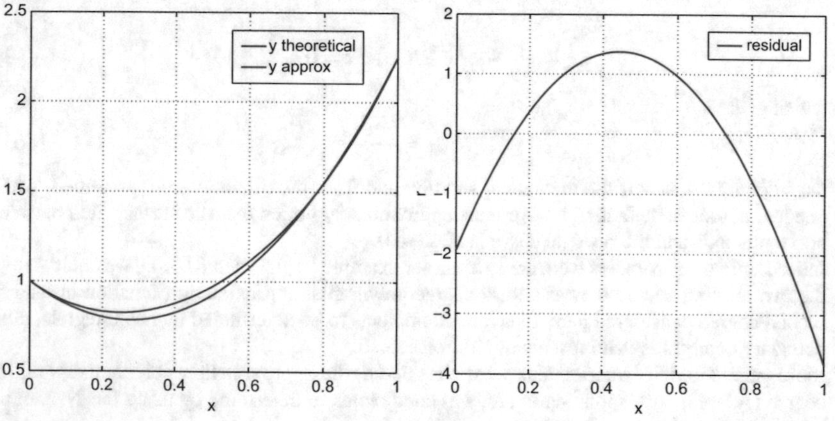

Fig. 7.40 Integration of an ordinary differential equation with boundary conditions. Least squares method. Left: theoretical solution and its approximation. Right: residual

Then, we impose

$$\int_0^1 w_1(x)\, R(x)\, dx = 0 \tag{7.10.77}$$

which gives

$$a_1 = \frac{3y_1 - 18}{8} \tag{7.10.78}$$

On Figure 7.41, at left the theoretical solution and its approximation are represented, while the residual is represented at right.

From the comparison of these methods on a single example and without subdivision of the domain into subintervals, it would be hasty to conclude about the superiority

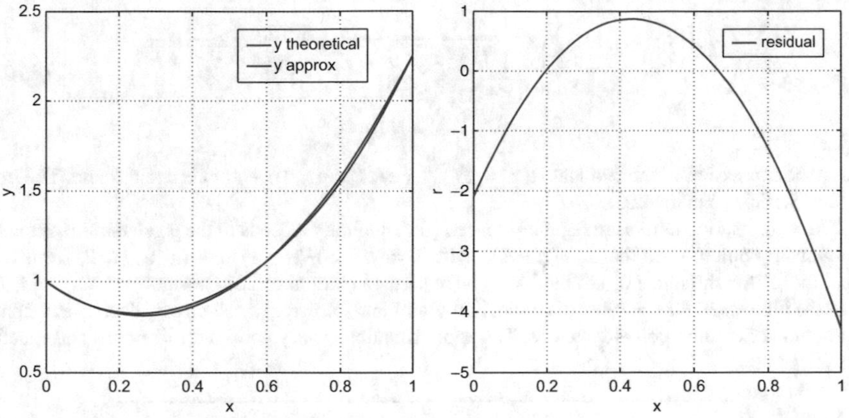

Fig. 7.41 Integration of an ordinary differential equation with boundary conditions. Galerkin's method. Left: theoretical solution and its approximation. Right: residual

of one or another method. In the following Example 7.13, more collocation points are used together with the multiquadric function.

Example 7.13 :
Integration of a second order ODE with boundary conditions using collocation and multiquadric function

The same ordinary differential equation as in previous Example 7.12 is considered, with the same domain and boundary conditions, but with collocation points in $[a, b] = [0, 1]$ and the choice of the multiquadric function as the radial basis function

$$\phi(r) = \sqrt{r^2 + \sigma^2} \tag{7.10.79}$$

thus by considering $r = |x - x_i|$

$$\phi_i(|x - x_i|) = \sqrt{(x - x_i)^2 + \sigma^2}\,, \quad i = 1, \dots, n \tag{7.10.80}$$

Equation (7.10.64) becomes

$$L(\sum_{i=1}^{n} c_i \phi_i(|x - x_i|)) = \sum_{i=1}^{n} c_i \left[\phi_i''(|x - x_i|) + \phi_i'(|x - x_i|) - 6\phi_i(|x - x_i|)\right] = f(x) = 0 \tag{7.10.81}$$

with

$$\phi'(r) = \frac{r}{\phi(r)} \Longrightarrow \phi_i'(|x - x_i|) = \frac{x - x_i}{\sqrt{(x - x_i)^2 + \sigma^2}} \tag{7.10.82}$$

and

$$\phi''(r) = \frac{1}{\phi(r)} - \frac{r}{\phi^2(r)}\phi'(r) = \frac{\sigma^2}{\phi^3(r)} \Longrightarrow \phi_i''(|x - x_i|) = \frac{\sigma^2}{((x - x_i)^2 + \sigma^2)^{3/2}} \tag{7.10.83}$$

hence

$$
\begin{aligned}
L(\sum_{i=1}^{n} c_i \phi_i(|x-x_i|)) = \sum_{i=1}^{n} c_i \left[\frac{\sigma^2}{((x-x_i)^2+\sigma^2)^{3/2}} + \frac{x-x_i}{\sqrt{(x-x_i)^2+\sigma^2}} \right. \\
\left. -6\sqrt{(x-x_i)^2+\sigma^2} \right] \\
= f(x) = 0 \qquad \forall\, x \in [a, b]
\end{aligned}
\tag{7.10.84}
$$

where the expression of the residual $R(x) = L(\sum_{i=1}^{n} c_i \phi(|x-x_i|)) - f(x)$ is again found. The linear system (7.10.63) can then be solved.

The n collocation points are regularly spaced (for simplicity reasons in the presentation), including the boundary points, such that $x_i = (i-1)h$ with $h = (b-a)/(n-1)$ and $n = 6,\ 11,\ 21$. σ is taken equal to 0.1, the influence of σ could have been studied (Mongillo 2011). Figures 7.42, 7.43, 7.44 show the influence of the number of collocation points and the error which rapidly decreases when the number of collocation points increases. The approximation is very good with 11 points and excellent with 21 points.

7.10.5 Method of Weighted Residuals for a Partial Differential Equation

The problem is set in a general manner in the framework of a partial differential equation of the form

$$
P(u(x,t)) = 0 \tag{7.10.85}
$$

valid on a domain $\mathcal{D}$ for the function $u(x,t)$ with the boundary conditions $B(u) = 0$ on the boundary δB and the initial condition $u(x,0) = u^0(x)$. The method proposes the approximation $\hat{u}_N(x,t)$ of the solution as

$$
\hat{u}_N(x,t) = u_B(x,t) + \sum_{i=1}^{N} c_i(t)\phi_i(x) \tag{7.10.86}
$$

The summation term is thus based on a separation of time and space variables. The functions ϕ are trial functions independent of time. The coefficients c_i are time-dependent. In general, the functions ϕ must verify the homogeneous boundary conditions on the boundary δB, the term $u_B(x,t)$ is used to satisfy inhomogeneous boundary conditions. In the following, the term $u_B(x,t)$ will not be considered.

Under these conditions, a kth order spatial derivative of $\hat{u}_N(x,t)$ is written as

$$
\frac{\partial^k \hat{u}_N}{\partial x^k} = \sum_{i=1}^{N} c_i(t) \frac{d^k \phi_i(x)}{dx^k} \tag{7.10.87}
$$

Using the expression of the approximation $\hat{u}_N(x,t)$ in the partial differential equation (7.10.85), the expression of the residual results

$$
R(x,t) = P(\hat{u}_N(x,t)) \tag{7.10.88}
$$

To determine the N coefficients c_i, we impose

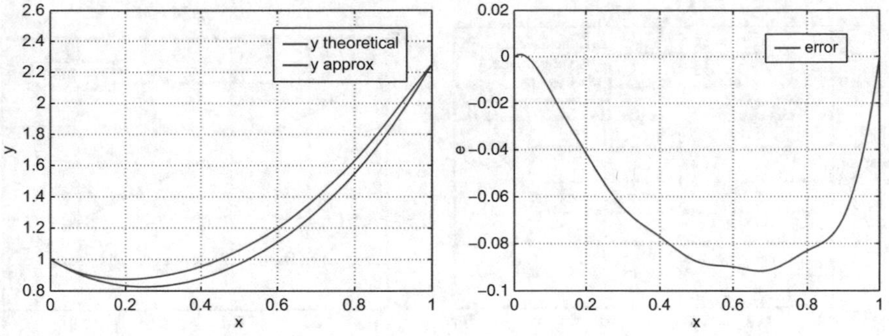

Fig. 7.42 Integration of an ordinary differential equation with boundary conditions. Collocation method with 6 points and the multiquadric function. Left: theoretical curve and its approximation. Right: error

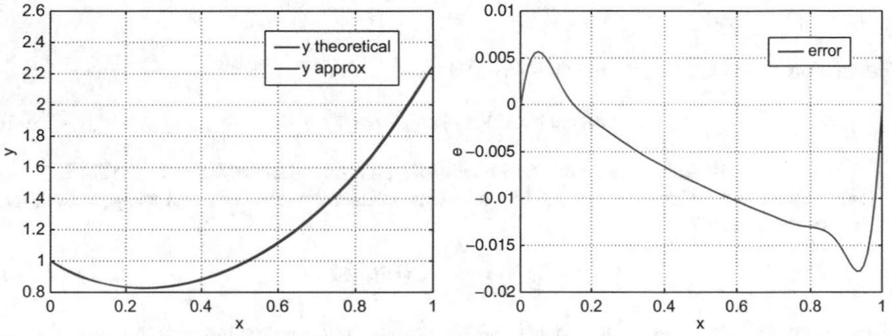

Fig. 7.43 Integration of an ordinary differential equation with boundary conditions. Collocation method with 11 points and the multiquadric function. Left: theoretical curve and its approximation. Right: error

$$\int_{\mathcal{D}} w_i(x) R(x,t) dx = 0\,, \quad i = 1, \ldots, N \tag{7.10.89}$$

which means that the residual is orthogonal to all the weight functions. Equation (7.10.89) is in fact a system of N algebraic equations verified for any t. If a collocation method is used with $w(x_i) = \delta(x - x_i)$, it implies $R(x_i, t) = 0$ at any collocation point.

Example 7.14 :
One-dimensional Fourier's law with collocation and spectral differential operator

Consider one-dimensional Fourier's law

$$\frac{\partial \theta(x,t)}{\partial t} = \alpha \frac{\partial^2 \theta(x,t)}{\partial x^2} \tag{7.10.90}$$

for $x \in [0, 1]$ with Dirichlet boundary conditions and the initial condition

$$\theta(0,t) = 0\,, \quad \theta(1,t) = 1\,, \quad \theta(x,0) = \sin(\pi x) + x \tag{7.10.91}$$

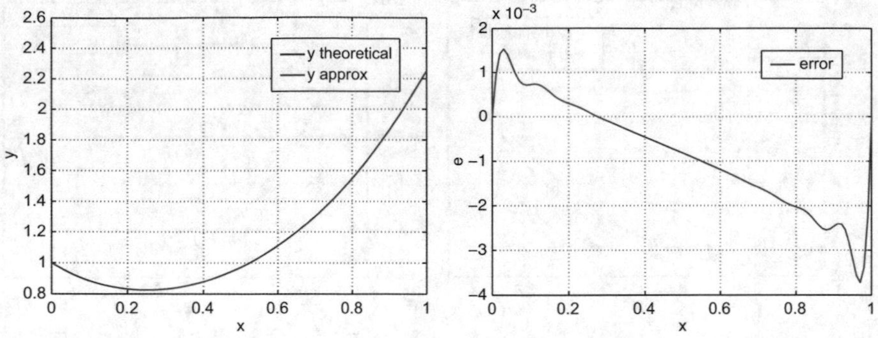

Fig. 7.44 Integration of an ordinary differential equation with boundary conditions. Collocation method with 21 points and the multiquadric function. Left: theoretical curve and its approximation. Right: error

The solution of that academic problem (Otto 2011) is

$$\theta(x, t) = \sin(\pi x) \exp(-\alpha \pi^2 t) + x \tag{7.10.92}$$

which satisfies simultaneously the boundary conditions and the initial condition of the PDE.

The numerical solution was searched by using the collocation methods and as approximation a Chebyshev series expansion as

$$\hat{\theta}(z, t) = \sum_{k=0}^{N} c_k(t) T_k(z) \tag{7.10.93}$$

where the coefficients c_k are time-dependent. The solution is thus a function with separated time and space variables. The symbol θ has been used for temperature to avoid the confusion with Chebyshev polynomials. The domain $[a, b]$ of x is transformed into $[-1, +1]$ for the variable z as

$$z = \frac{2x - a - b}{b - a} \tag{7.10.94}$$

Choosing the Chebyshev–Gauss–Lobatto collocation points, there exist $(N + 1)$ points z_i $(i = 0, \ldots, N)$ with moreover the extremities of the domain $z_0 = -1$ and $z_N = 1$. Replacing $\theta(x, t)$ according to Equation (7.10.93) in Equation (7.10.90), we obtain

$$\begin{aligned}
&\sum_{k=0}^{N} \frac{d\,c_k(t)}{dt} T_k(z) - \alpha \left(\frac{2}{b-a}\right)^2 \sum_{k=0}^{N} c_k(t) T_k^{(2)}(z) = 0 \\
&\sum_{k=0}^{N} \frac{d\,c_k(t)}{dt} T_k(z) - \alpha \left(\frac{2}{b-a}\right)^2 \sum_{k=0}^{N} c_k''(t) T_k(z) = 0 \\
&\sum_{k=0}^{N} \frac{d\,c_k(t)}{dt} T_k(z) - \alpha \left(\frac{2}{b-a}\right)^2 \sum_{k=0}^{N} D^2 c_k(t) T_k(z) = 0
\end{aligned} \tag{7.10.95}$$

where D is the spectral differential operator defined in (7.10.36) that is used twice on coefficients c_k. In this way, Equation (7.10.95) is an ordinary differential equation which is verified at each collocation point z_i in interval $[-1, 1]$. However, the boundary conditions must be respected, which can be written under the form

$$\sum_{k=0}^{N} c_k(t) T_k(z = -1) = \theta(x = 0, t)$$
$$\sum_{k=0}^{N} c_k(t) T_k(z = 1) = \theta(x = 1, t) \tag{7.10.96}$$

This induces the conditions

$$\sum_{k=0}^{N} \frac{dc_k(t)}{dt} T_k(z = -1) = 0$$
$$\sum_{k=0}^{N} \frac{dc_k(t)}{dt} T_k(z = 1) = 0 \tag{7.10.97}$$

Thus, a system of $(N + 1)$ differential equations results, taking into account the boundary conditions, with respect to the coefficients c_k that can be integrated. This system can be written under the form

$$\begin{bmatrix} T_0(z_0) & T_1(z_0) & \dots & T_N(z_0) \\ T_0(z_1) & T_1(z_1) & \dots & T_N(z_1) \\ \vdots & \vdots & \dots & \vdots \\ T_0(z_N) & T_1(z_N) & \dots & T_N(z_N) \end{bmatrix} \begin{bmatrix} \frac{dc_0(t)}{dt} \\ \frac{dc_1(t)}{dt} \\ \vdots \\ \frac{dc_N(t)}{dt} \end{bmatrix} - \alpha \left(\frac{2}{b-a}\right)^2 \begin{bmatrix} 0 \\ (\boldsymbol{D}^2\boldsymbol{c}, \boldsymbol{T}(z_1)) \\ \vdots \\ (\boldsymbol{D}^2\boldsymbol{c}, \boldsymbol{T}(z_{N-1})) \\ 0 \end{bmatrix} = 0 \tag{7.10.98}$$

where $(\boldsymbol{D}^2\boldsymbol{c}, \boldsymbol{T}(z_i))$ is the scalar product of vector $\boldsymbol{D}^2\boldsymbol{c}$ by the vector $\boldsymbol{T}$ of Chebyshev polynomials.

The matrix factor of vector $\frac{d\boldsymbol{c}}{dt}$ is fixed and invertible. This matrix can also be considered as a mass matrix according to Equation (6.5.5). This system of ordinary differential equations can be solved either approximately by a first order discretization, or more rigorously by usual integration methods. This type of last method more rigorous has been used for the present study. The chosen numerical values were $\alpha = 0.1$, a time step equal to 0.1 for the presentation of the results and a final time equal to 5. To demonstrate the feasibility with subdomains, the original Chebyshev domain $[-1, 1]$ has been divided into two adjacent subdomains of equal length and Chebyshev–Gauss–Lobatto collocation points have been chosen for each subdomain. Then all the equations have been gathered under the form (7.10.98) to be integrated. The results for two different values of N ($N = 2$ and $N = 10$, thus, respectively, 3 and 11 collocation points on each subdomain) are given in Figure 7.45. It can be verified that the boundary conditions are respected. The interpolation between the collocation points was not done according to Equation (7.10.93), but simply by joining the successive collocation points. At time $t = 5$, the steady state regime is reached. At time $t = 2$, in $x = 0.5$, we get $\hat{\theta} = 0.64025$ (with $N = 2$) and $\hat{\theta} = 0.63891$ (with $N = 10$) to compare to the theoretical solution $\theta = 0.63891$. The approximation is thus very good with $N = 2$ and excellent with $N = 10$, using subdomains. It would have been possible to work in a practically equivalent manner with a single domain and for example $N = 4$ and $N = 20$ to obtain a very close result.

7.11 Moving Grid

The problem of using an adaptive moving grid is encountered in particular for the solution of partial differential equations (PDE) when steep fronts occur (Dorfi and Drury 1987; Furzeland et al. 1990; Hyman et al. 2003; Li et al. 1998; Tang 2005; Vande Wouwer et al. 2005). This is frequently the case in partial differential equations of convection-diffusion.

The objective of a moving grid is to transform the original physical problem where a steep profile is visible into a numerical problem where the profile is smoothed.

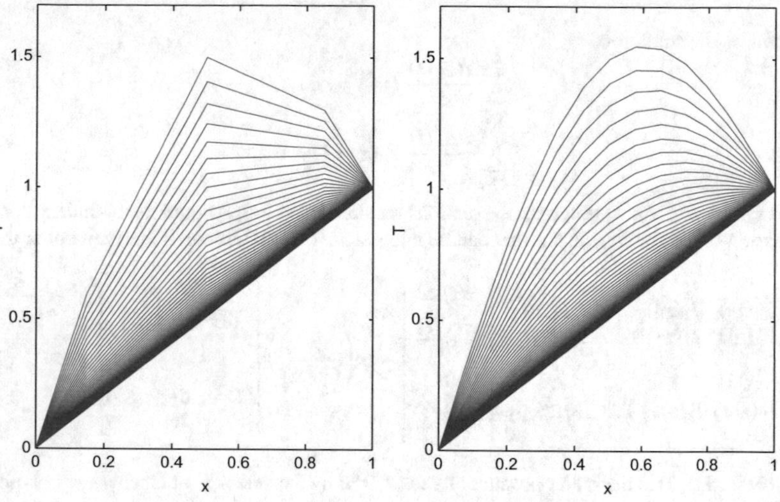

Fig. 7.45 Integration of a partial differential equation: Fourier's law, using the method of collocations and a Chebyshev series expansion with $N = 2$ (left) and $N = 10$ (right) and two subdomains

7.11.1 Theory

Consider the original PDE problem

$$\boldsymbol{u}_t = f(t, x, \boldsymbol{u}, \boldsymbol{u}_x, \boldsymbol{u}_{xx}, \dots) \tag{7.11.1}$$

with its boundary and initial conditions. A variable mesh $x(t)$, time-dependent, with N points is defined by

$$x_1 < x_2(t) < \cdots < x_i(t) < \cdots < x_N \tag{7.11.2}$$

Note that the extremities x_1 and x_N are fixed. Given this variable mesh, in the Lagrangian frame (Li et al. 1998), the PDE problem is transformed as

$$\dot{\boldsymbol{u}} = u_t + u_x\,\dot{x} = f(x, \boldsymbol{u}, \boldsymbol{u}_x, \boldsymbol{u}_{xx}, \dots) + u_x\,\dot{x} \tag{7.11.3}$$

where $\dot{\boldsymbol{u}}$ is a total derivative.

A monitor function $m(u)$ is defined to follow the spatial variations of the solution profile. Different definitions are possible (Dorfi and Drury 1987; Furzeland et al. 1990; Vande Wouwer et al. 2005). Vande Wouwer et al. (2005) use

$$m(u) = \sqrt{\alpha + \|u_x\|_2^2} \tag{7.11.4}$$

while Dorfi and Drury (1987) use

$$m(u) = \sqrt{1 + u_x^2} \tag{7.11.5}$$

or

$$M_i = \sqrt{1 + \sum_{j=1}^{k} \left(\frac{X_i}{U} \frac{u_{i+1} - u_i}{x_{i+1} - x_i} \right)^2} \tag{7.11.6}$$

where M_i is a discrete approximant of the monitor function $m(u)$ in the interval $[x_i, x_{i+1}]$. U is a natural scale associated with u and X a scale associated with the spatial length of the problem. The variable step is defined as

$$\Delta x_i = x_{i+1} - x_i \tag{7.11.7}$$

The local concentration n_i of points is defined by

$$n_i = \frac{1}{\Delta x_i} \tag{7.11.8}$$

The objective of the moving grid is to distribute the points in a balanced manner so that approximately

$$\frac{M_i}{n_i} = \text{constant} \tag{7.11.9}$$

Then, the concentration n_i would be proportional to the approximant of the monitor function. Nevertheless, it is not possible to let n vary too rapidly so that a smoothing is used at both spatial and time levels. We impose that

$$\frac{\alpha}{\alpha + 1} \leq \frac{n_{i+1}}{n_i} \leq \frac{\alpha + 1}{\alpha} \tag{7.11.10}$$

α being a measure of grid rigidity (Dorfi and Drury 1987).

The spatial smoothing is performed by replacing n_i by $\tilde{n}_i$ according to

$$\begin{aligned} &\tilde{n}_1 = n_1 - \alpha(\alpha + 1)(n_2 - n_1) \\ &\tilde{n}_i = n_i - \alpha(\alpha + 1)(n_{i+1} - 2\, n_i + n_{i-1}) \,, \quad 2 \leq i \leq N - 1 \\ &\tilde{n}_N = n_N - \alpha(\alpha + 1)(n_{N-1} - n_N) \end{aligned} \tag{7.11.11}$$

where α is a positive parameter.

Then, a time smoothing is carried out by the system of ordinary differential equations

$$\frac{\tilde{n}_{i-1} + \tau \dot{\tilde{n}}_{i-1}}{M_{i-1}} = \frac{\tilde{n}_i + \tau \dot{\tilde{n}}_i}{M_i} \,, \quad 2 \leq i \leq N - 1 \tag{7.11.12}$$

τ is a time constant to avoid rapid variations of the moving grid.

Static solving as an explicit differential system

Consider the approximant M_i of the monitor function (7.11.6). Suppose that we integrate on the total time domain $[0, t_{max}]$ and that integration intervals $[t_i, t_{i+1}]$ are considered in this domain. At any time t_{i+1}, we verify that the sum of the reciprocals of the densities n_i is equal to the length of the grid and we correct the densities so that it is verified. Then, the smoothed densities $\tilde{n}_i$ are calculated from (7.11.11) with a very small value of α. Similarly, the smoothed densities $\tilde{n}_i$ are corrected so that the sum of their reciprocals is equal to the length of the grid. Afterward, the derivatives of the smoothed densities are calculated by (7.11.12) then the integration is done on an interval $[t_i, t_{i+1}]$. This procedure qualified as static (Vande Wouwer et al. 2005) gives excellent results. However, the tuning of the parameters of the integrator and of the moving grid method is more delicate than dynamic solving as the parameters are coupled.

Dynamic solving as an implicit differential system

Indeed, Kelling et al. (2014) pose the moving grid problem under the form

$$\tau \boldsymbol{A} \frac{d\boldsymbol{x}}{dt} = \boldsymbol{s} \tag{7.11.13}$$

where they define the matrix $\boldsymbol{A}$ and the vector $\boldsymbol{s}$. The matrix $\boldsymbol{A}$ can be taken (Kelling et al. 2014; Li et al. 1998) as

$$\boldsymbol{A} = \begin{bmatrix} 1 & 0 & \dots & & & & & & 0 \\ 1 & -2 & 1 & 0 & \dots & & & & 0 \\ A_{3,1} & A_{3,2} & A_{3,3} & A_{3,4} & A_{3,5} & 0 & \dots & & 0 \\ \vdots & & & & & & & & \vdots \\ 0 & \dots & A_{i,i-2} & A_{i,i-1} & A_{i,i} & A_{i,i+1} & A_{i,i+2} & 0 & 0 \\ \vdots & & & & & & & & 0 \\ 0 & \dots & & & A_{N-2,N-4} & A_{N-2,N-3} & A_{N-2,N-2} & A_{N-2,N-1} & A_{N-2,N} \\ 0 & \dots & & & & & 1 & -2 & 1 \\ 0 & \dots & & & & & & 0 & 1 \end{bmatrix} \tag{7.11.14}$$

with its elements $(3 \leq i \leq N-2)$

$$\begin{aligned} A_{i,i-2} &= -\frac{\mu\, n_{i-2}^2}{M_{i-1}} \\ A_{i,i-1} &= \frac{\mu\, n_{i-1}^2}{M_i} + \frac{(1+2\,\mu)\, n_{i-1}^2}{M_{i-1}} + \frac{\mu\, n_{i-2}^2}{M_{i-1}} \\ A_{i,i} &= -\frac{\mu\, n_{i-1}^2}{M_i} - \frac{(1+2\,\mu)\, n_i^2}{M_i} - \frac{(1+2\,\mu)\, n_{i-1}^2}{M_{i-1}} - \frac{\mu\, n_i^2}{M_{i-1}} \\ A_{i,i+1} &= \frac{\mu\, n_{i+1}^2}{M_i} + \frac{(1+2\,\mu)\, n_i^2}{M_i} + \frac{\mu\, n_i^2}{M_{i-1}} \\ A_{i,i+2} &= -\frac{\mu\, n_{i+1}^2}{M_i} \end{aligned} \tag{7.11.15}$$

and the column vector $\boldsymbol{s}$

$$\boldsymbol{s} = \left[0\ 0\ s_3\ \dots\ s_{n-2}\ 0\ 0\right]^T \tag{7.11.16}$$

with

$$s_i = \frac{1}{M_i}\left(-\mu\, n_{i+1} + (1+2\,\mu)\, n_i - \mu\, n_{i-1}\right) - \frac{1}{M_{i-1}}\left(-\mu\, n_i + (1+2\,\mu)\, n_{i-1} - \mu\, n_{i-2}\right) \tag{7.11.17}$$

The parameter μ is equal to

$$\mu = \alpha\,(\alpha + 1) \tag{7.11.18}$$

The time constant can be taken as $\tau = 10^{-3}$ and the parameter $\alpha = 2$. The value $\alpha = 2$ corresponds to a variation of adjacent mesh points so that

$$\frac{2}{3} < \frac{\Delta x_{i+1}}{\Delta x_i} < \frac{3}{2} \tag{7.11.19}$$

The penta-diagonal matrix $\boldsymbol{A}$ is ill-conditioned and the system (7.11.13) cannot be solved like a usual system of ordinary differential equations after inversion of matrix $\boldsymbol{A}$. Instead of that, it can be solved under the form (7.11.13) as a differential–algebraic system of equations (DAE), for example with the function "dae" of Scilab, "ddassl" in Fortran.

7.11.2 Test on an Analytic Function

Dorfi and Drury (1987) consider the following analytic problem. It deals with the representation of the function

$$f(x) = \frac{1}{2}\,[1 + \tanh(1000\,(x - 0.4))]\,\exp\left(-\left(\frac{x - 0.4}{0.2}\right)^2\right) \tag{7.11.20}$$

The graph of this function (Figure 7.46) presents an extremely steep variation around $t = 0.4$ on the uniform grid. It is similar to a shock sometimes present in physical problems, such as chromatography. In Equation (7.11.6) of the monitor function, Dorfi and Drury (1987) chose $U = 1$, $X_i = 1$ and $k = 1$, When a moving grid is used according to (7.11.13), using dynamic solving as an implicit differential system, it clearly appears (Figure 7.47) that there is an important concentration of points around the abscissa 0.4. Even with less points, the curve is very well represented. If, instead of using the real abscissa, the function is represented versus the index of the abscissa (Figure 7.48), it can now be noticed that the function is smooth with respect to that index, which corresponds to the searched objective of the moving grid. The parameters (very small tolerance $rtol$, small τ and $\mu = 6$ corresponding to $\alpha = 2$) are very well adapted in this case. If we use solving as an explicit differential system by the static method (Figures 7.49 and 7.50), the results are extremely close, graphically not distinguishable. However, the value $rtol$ of the Lsode integrator must be chosen in agreement with the value α of Equation (7.11.11). The smaller the tolerance $rtol$, hence the accuracy required by the integrator is important, the smaller α must be.

7.11.3 Implementation in a Physical Problem

The chromatographic separation is in general well modelled in the scientific literature but its numerical simulation poses serious problems due to the possible presence of steep fronts (Nicoud 2015). Several models of chromatographic separation are discussed.

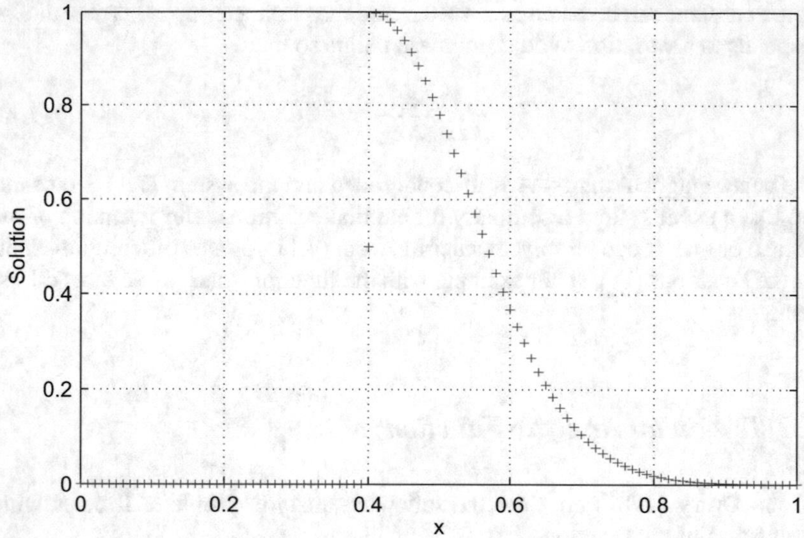

Fig. 7.46 Test function on the original regular grid with 101 points

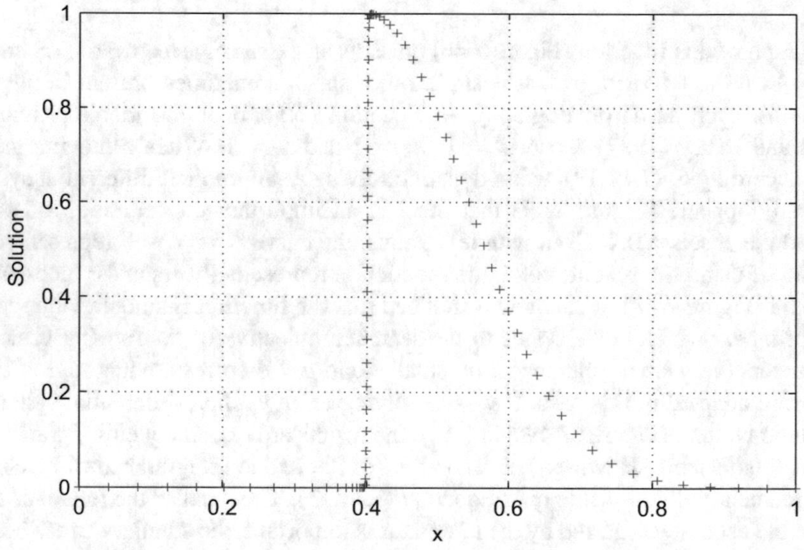

Fig. 7.47 Test function on the moving grid with 81 points with dynamic solving as an implicit differential system ($rtol = 10^{-9}, \tau = 10^{-3}, \mu = 6$)

Example 7.15 :
Moving grid for a chromatographic separation

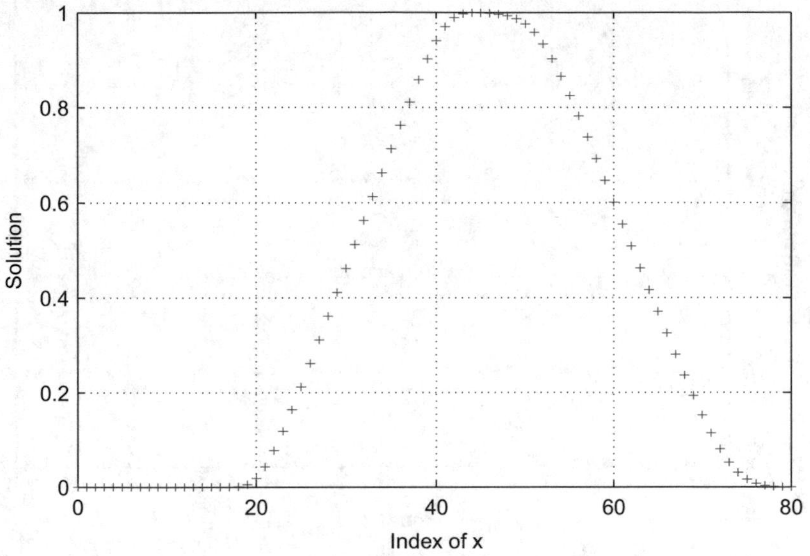

Fig. 7.48 Moving grid: Test function versus the abscissa index i with dynamic solving as an implicit differential system ($rtol = 10^{-9}$, $\tau = 10^{-3}$, $\mu = 6$)

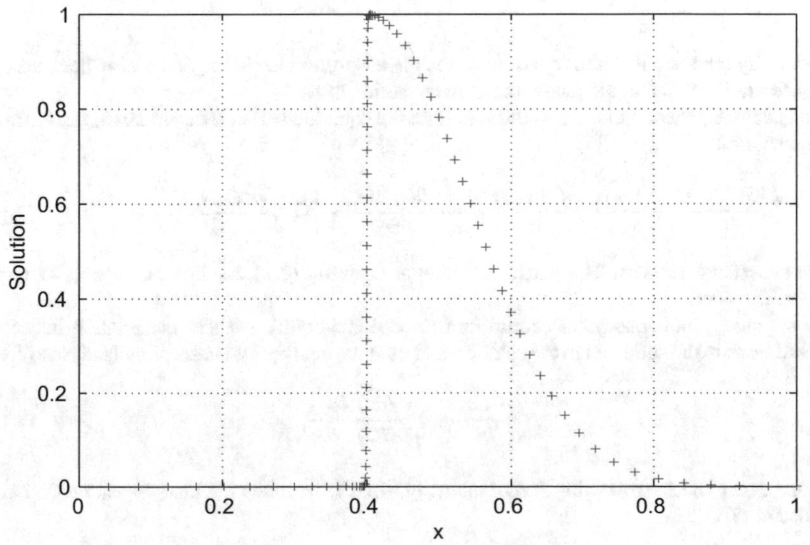

Fig. 7.49 Moving grid: Test function with 81 points with static solving as an explicit differential system ($rtol = 10^{-6}$, $\tau = 10^{-3}$, $\alpha = 1$)

The moving grid has been used to simulate the chromatographic separation of two components A and B which obey a modified Langmuir adsorption isotherm of the form

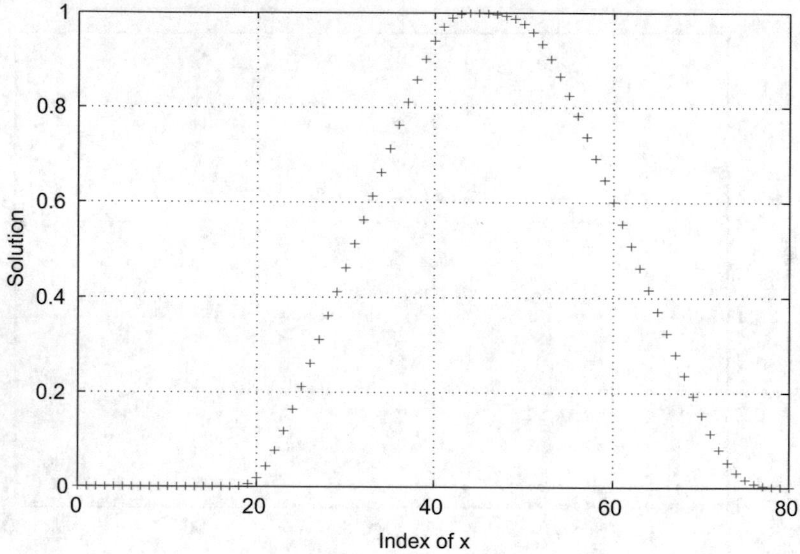

Fig. 7.50 Moving grid: Test function versus the abscissa index i with static solving as an explicit differential system ($rtol = 10^{-6}$, $\tau = 10^{-3}$, $\alpha = 1$)

$$\overline{C}_i = \lambda_i \, C_i + \frac{\overline{N}_i \, \tilde{K}_i \, C_i}{1 + \sum_j \tilde{K}_j \, C_j} \tag{7.11.21}$$

Note that any other equilibrium equation, that is the adsorption isotherm, could have been used. C is the concentration in the liquid phase and $\overline{C}$ in the solid phase.

Consider the general model of a partial differential equation (PDE) (Nicoud 2015) for a chromatographic column

$$\frac{\partial C_i(t,z)}{\partial t} + \frac{1-\epsilon_e}{\epsilon_e} \frac{\partial \overline{C}_i(t,z)}{\partial t} + \frac{v}{\epsilon_e} \frac{\partial C_i(t,z)}{\partial z} - \mathcal{D}_{ax} \frac{\partial^2 C_i(t,z)}{\partial z^2} = 0 \tag{7.11.22}$$

with v superficial velocity. The partial differential equation (7.11.22) is a remarkable example of transport-diffusion.

In the case of an adsorption equilibrium (model PD-Equil), we add the relation between the concentrations in the solid and in the solution that uses the equilibrium adsorption isotherm (7.11.21)

$$\frac{\partial \overline{C}_i}{\partial t} = \sum_j \frac{\partial \overline{C}_i}{\partial C_j} \frac{\partial C_j}{\partial t} \tag{7.11.23}$$

In the case of a Linear Driving Force (model PD-LDF), the following mass transfer equation must be added to (7.11.22)

$$\frac{\partial \overline{C}_i(t,z)}{\partial t} = \frac{1}{t^i} (\overline{C}_i^*(t,z) - \overline{C}_i(t,z)) \tag{7.11.24}$$

where t^i is a characteristic time of internal diffusion.

Consider now the case PD-Equil. Thus Equation (7.11.22) becomes

$$\frac{\partial C_i(t,z)}{\partial t} + \frac{1-\epsilon_e}{\epsilon_e}\sum_j \frac{\partial \overline{C}_i}{\partial C_j}\frac{\partial C_j(t,z)}{\partial t} + \frac{v}{\epsilon_e}\frac{\partial C_i(t,z)}{\partial z} - \mathcal{D}_{ax}\frac{\partial^2 C_i(t,z)}{\partial z^2} = 0 \tag{7.11.25}$$

or

$$\frac{\partial C_i}{\partial t} + \frac{1-\epsilon_e}{\epsilon_e}\frac{\partial \overline{C}_i}{\partial C_i}\frac{\partial C_i(t,z)}{\partial t} + \frac{1-\epsilon_e}{\epsilon_e}\sum_{j\neq i}\frac{\partial \overline{C}_i}{\partial C_j}\frac{\partial C_j(t,z)}{\partial t} + \frac{v}{\epsilon_e}\frac{\partial C_i(t,z)}{\partial z} - \mathcal{D}_{ax}\frac{\partial^2 C_i(t,z)}{\partial z^2} = 0 \tag{7.11.26}$$

or

$$\left[1 + \frac{1-\epsilon_e}{\epsilon_e}\frac{\partial \overline{C}_i}{\partial C_i}\right]\frac{\partial C_i(t,z)}{\partial t} + \frac{1-\epsilon_e}{\epsilon_e}\sum_{j\neq i}\frac{\partial \overline{C}_i}{\partial C_j}\frac{\partial C_j(t,z)}{\partial t} + \frac{v}{\epsilon_e}\frac{\partial C_i(t,z)}{\partial z} - \mathcal{D}_{ax}\frac{\partial^2 C_i(t,z)}{\partial z^2} = 0 \tag{7.11.27}$$

7.11.4 Short Presentation of the General Framework

The calculation code "spmdif.f" written by Blom and Zegeling (1994) provides Fortran subroutines designed to easily use a moving grid by means of some intermediaries. He uses the technique developed by Dorfi and Drury (1987). An additional interest of "spmdif" is to present the problems under a general form that will be followed below.

The class of allowed PDE (Blom and Zegeling 1994) by "spmdif.f" is represented by

$$\sum_{k=1}^{N_{pde}} C_{j,k}(x,t,u,u_x)\frac{\partial u^k}{\partial t} = x^{-m}\frac{\partial}{\partial x}\left(x^m \mathcal{R}_j(x,t,u,u_x)\right) - Q_j(x,t,u,u_x) \tag{7.11.28}$$

for $j = 1, \ldots, N_{pde}$ and $x \in [x_L, x_R]$, $t > t_0$, $m \in 0, 1, 2$, where N_{pde} is the number of PDEs. $u = (u^1, u^2, \ldots, u^{N_{pde}}$ is the solution vector. $\mathcal{R}_j$ and Q_j can be considered as flux and source terms, respectively. In Cartesian coordinates $m = 0$, in polar cylindrical coordinates $m = 1$, in polar spherical coordinates $m = 2$.

The boundary conditions must respect the following form

$$\beta_j(x,t)\,\mathcal{R}_j(x,t,u,u_x) = \gamma_j(x,t,u,u_x) \tag{7.11.29}$$

where it clearly appears that $\mathcal{R}$ is a flux term.

The initial conditions are defined as

$$u(x,t_0) = u^0(x) \qquad \text{for} \quad x \in [x_L, x_R] \tag{7.11.30}$$

7.11.5 Application to a Liquid Phase Chromatography, Approximation in the Equilibrium Case

In the case of PDE (7.11.22), we have $m = 0$ (Cartesian coordinates), $N_{pde} = n_c$ (n_c PDEs), so that the general equation (7.11.28) is reduced to

$$C(x,t,u,u_x)\frac{\partial u}{\partial t} = \frac{\partial}{\partial x}\left(\mathcal{R}(x,t,u,u_x)\right) - Q(x,t,u,u_x) \tag{7.11.31}$$

or

$$C(x,t,u,u_x)\frac{\partial u}{\partial t} - \frac{\partial}{\partial x}(\mathcal{R}(x,t,u,u_x)) + Q(x,t,u,u_x) = 0 \qquad (7.11.32)$$

Equation (7.11.27) is correct in the case of equilibrium. However, it presents terms $\frac{\partial C_i}{\partial t}$ and $\frac{\partial C_j}{\partial t}$ which pose a problem during programming.

Writing Equation (7.11.27) for $i = 1, \ldots, n_c$, where n_c is the number of components, a system of n_c partial differential equations results

$$A \begin{bmatrix} \frac{\partial C_1}{\partial t} \\ \vdots \\ \frac{\partial C_{n_c}}{\partial t} \end{bmatrix} + \frac{v}{\epsilon_e} \begin{bmatrix} \frac{\partial C_1}{\partial x} \\ \vdots \\ \frac{\partial C_{n_c}}{\partial x} \end{bmatrix} - \mathcal{D}_{ax} \begin{bmatrix} \frac{\partial^2 C_1}{\partial x^2} \\ \vdots \\ \frac{\partial^2 C_{n_c}}{\partial x^2} \end{bmatrix} = 0 \qquad (7.11.33)$$

with the elements of A (assuming $i \neq j$)

$$\begin{aligned} A_{i,i} &= 1 + \frac{1-\epsilon_e}{\epsilon_e}\frac{\partial \overline{C}_i}{\partial C_i} \\ A_{i,j} &= \frac{1-\epsilon_e}{\epsilon_e}\frac{\partial \overline{C}_i}{\partial C_j} \end{aligned} \qquad (7.11.34)$$

This is a system of n_c PDEs that can be solved by the moving grid code as follows.

Consider the particular case of a binary system. The PDEs of Equation (7.11.27) give two coupled PDEs

$$\begin{aligned} &\left[1 + \frac{1-\epsilon_e}{\epsilon_e}\frac{\partial \overline{C}_1}{\partial C_1}\right]\frac{\partial C_1(t,z)}{\partial t} + \frac{1-\epsilon_e}{\epsilon_e}\frac{\partial \overline{C}_1}{\partial C_2}\frac{\partial C_2(t,z)}{\partial t} + \frac{v}{\epsilon_e}\frac{\partial C_1(t,z)}{\partial z} - \mathcal{D}_{ax}\frac{\partial^2 C_1(t,z)}{\partial z^2} = 0 \\ &\left[1 + \frac{1-\epsilon_e}{\epsilon_e}\frac{\partial \overline{C}_2}{\partial C_2}\right]\frac{\partial C_2(t,z)}{\partial t} + \frac{1-\epsilon_e}{\epsilon_e}\frac{\partial \overline{C}_2}{\partial C_1}\frac{\partial C_1(t,z)}{\partial t} + \frac{v}{\epsilon_e}\frac{\partial C_2(t,z)}{\partial z} - \mathcal{D}_{ax}\frac{\partial^2 C_2(t,z)}{\partial z^2} = 0 \end{aligned} \qquad (7.11.35)$$

By referring to the notations of Equation (7.11.28) (in "spmdif"), now we have (with $N_{pde} = n_c$)

$$\begin{aligned} C_{k,k} &= 1 + \frac{1-\epsilon_e}{\epsilon_e}\frac{\partial \overline{C}_k}{\partial C_k} \\ C_{j,k} &= \frac{1-\epsilon_e}{\epsilon_e}\frac{\partial \overline{C}_j}{\partial C_k} \qquad \text{for } j \neq k \\ \mathcal{R}_k(x,t,u,u_x) &= \mathcal{D}_{ax}\frac{\partial u_k}{\partial x} \\ Q_k(x,t,u,u_x) &= \frac{v}{\epsilon_e}\frac{\partial u_k}{\partial x} \end{aligned} \qquad (7.11.36)$$

with $j = 1, \ldots, n_c$. The C_{ij} are the elements of A.

7.11.6 Application to a Liquid Phase Chromatography, Rigorous Treatment for the Case with LDF

Consider the general PDE model (7.11.22) of a liquid phase chromatography column

$$\frac{\partial C_i(t,z)}{\partial t} + \frac{1-\epsilon_e}{\epsilon_e}\frac{\partial \overline{C}_i(t,z)}{\partial t} + \frac{v}{\epsilon_e}\frac{\partial C_i(t,z)}{\partial z} - \mathcal{D}_{ax}\frac{\partial^2 C_i(t,z)}{\partial z^2} = 0, \quad i = 1, \ldots, n_c \qquad (7.11.37)$$

with the equation of Linear Driving Force

$$\frac{\partial \overline{C}_i(t,z)}{\partial t} = \frac{1}{t^i}(\overline{C}_i^* - \overline{C}_i(t,z)), \quad i = 1, \ldots, n_c \qquad (7.11.38)$$

which amounts to $2n_c$ PDEs in the frame of the code "spmdif".

To be in agreement with the standard equation (7.11.32) for the moving grid, both previous equations are written again as

$$\begin{aligned}&\frac{\partial C_i(t,z)}{\partial t}+\frac{1-\epsilon_e}{\epsilon_e}\frac{1}{t^i}(\overline{C}_i^*-\overline{C}_i(t,z))+\frac{v}{\epsilon_e}\frac{\partial C_i(t,z)}{\partial z}-\mathcal{D}_{ax}\frac{\partial^2 C_i(t,z)}{\partial z^2}=0\\&\frac{\partial \overline{C}_i(t,z)}{\partial t}-\frac{1}{t^i}(\overline{C}_i^*-\overline{C}_i(t,z))=0\end{aligned} \tag{7.11.39}$$

from which the coefficients are drawn

$$\begin{aligned}&\mathcal{C}_i(x,t,u,u_x) &&= 1 \quad \forall i=1,\ldots,2n_c\\&\mathcal{Q}_i(x,t,u,u_x) &&= \frac{1-\epsilon_e}{\epsilon_e}\frac{1}{t^i}(\overline{C}_i^*-\overline{C}_i(t,z))+\frac{v}{\epsilon_e}\frac{\partial C_i(t,z)}{\partial z} \quad \forall i=1,\ldots,n_c\\&\mathcal{R}_i(x,t,u,u_x) &&= \mathcal{D}_{ax}\frac{\partial C_i(t,z)}{\partial z} \quad \forall i=1,\ldots,n_c\\&\mathcal{Q}_{i+n_c}(x,t,u,u_x) &&= -\frac{1}{t^i}(\overline{C}_i^*-\overline{C}_i(t,z)) \quad \forall i=1,\ldots,n_c\\&\mathcal{R}_{i+n_c}(x,t,u,u_x) &&= 0 \quad \forall i=1,\ldots,n_c\end{aligned} \tag{7.11.40}$$

7.11.6.1 Boundary Conditions

Even if nowadays the Danckwerts boundary conditions are still discussed (Danckwerts 1953; Mott and Green 2015) and they are indeed questionable, they are largely used. Assimilating the chromatography column to a plug-flow reactor, the general formulation of the boundary conditions is

$$S_0^- v_0^- C_0^- = S_0^+ v_0^+ C_0^+ - S_0^+ \mathcal{D}_{ax}\left.\frac{\partial C}{\partial x}\right|_{0^+} \quad \text{and} \quad S_L^- v_L^- C_L^- - S_L^- \mathcal{D}_{ax}\left.\frac{\partial C}{\partial x}\right|_{L^-} = S_L^+ v_L^+ C_L^+ \tag{7.11.41}$$

with $S_0^+ = \epsilon_e S_0^-$ and $v_0^+ = v_0^-/\epsilon_e$. In the same manner $S_L^- = \epsilon_e S_0^+$ and $v_L^- = v_L^+/\epsilon_e$. Finally, we obtain

$$v\,C_0^- = v\,C_0^+ - \epsilon_e\,\mathcal{D}_{ax}\left.\frac{\partial C}{\partial x}\right|_{0^+} \quad \text{and} \quad v\,C_L^- - \epsilon_e\,\mathcal{D}_{ax}\left.\frac{\partial C}{\partial x}\right|_{L^-} = v\,C_L^+ \tag{7.11.42}$$

which can be compared to the previous equations.

The Danckwerts boundary conditions are

$$\frac{v}{\epsilon_e}(C_0^+ - C_0^-) - \mathcal{D}_{ax}\left.\frac{\partial C}{\partial x}\right|_{0^+} = 0 \quad \text{and} \quad \left.\frac{\partial C}{\partial x}\right|_{L^-} = 0 \tag{7.11.43}$$

with $C_0^- = C_{inj}(t)$ so that

$$\frac{v}{\epsilon_e}(C_0^+ - C_{inj}) - \mathcal{D}_{ax}\left.\frac{\partial C}{\partial x}\right|_{0^+} = 0 \quad \text{and} \quad \left.\frac{\partial C}{\partial x}\right|_{L^-} = 0 \tag{7.11.44}$$

The first boundary condition at the inlet supposes that the concentration is not the same in the injection and at $z = 0$. The second boundary condition at the outlet is also questionable as it comes from

$$\frac{v}{\epsilon_e}(C_L^- - C_L^+) - \mathcal{D}_{ax}\left.\frac{\partial C}{\partial x}\right|_{L^-} = 0 \tag{7.11.45}$$

and supposes $C_L^+ = C_L^-$.

Nevertheless, Equations (7.11.44) must be recasted in the general form for "spmdif" (7.11.29).

In the case of Example 7.11.6.4 of the chromatographic separation of two components A and B, the influence of the choice of a boundary condition at the inlet of Dirichlet type or of Neumann type, i.e. Danckwerts, has been studied (Figure 7.51). The injection is executed at $t = 1$ min. While the Dirichlet condition corresponds to a concentration step at the inlet, the Neumann condition practically acts like a first order differential system.

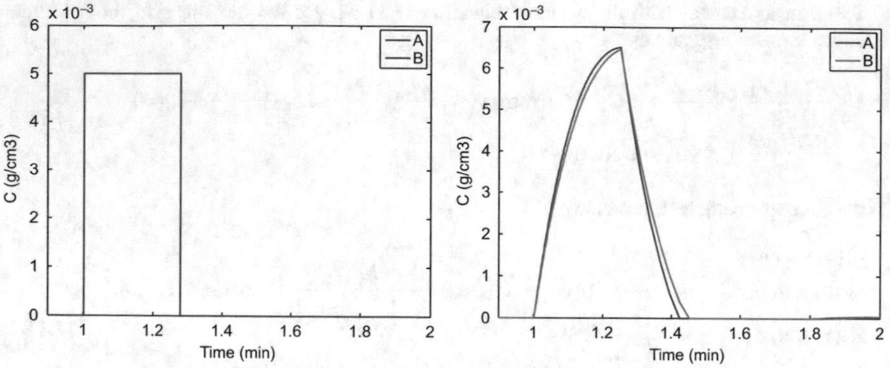

Fig. 7.51 Comparison of the concentrations at inlet ($z = 0^+$) of a column of chromatographic separation of two components A and B in the case of a Dirichlet condition (left) and a Neumann condition following Danckwerts (right)

7.11.6.2 PD-Equil Model

In this case, the general equation (7.11.22) of the model of a chromatography column coupled with the adsorption equilibrium equation (7.11.21) is considered.

The term $\mathcal{R}_j$ is imposed by the coefficients of the general form of Equation (7.11.28) and is equal to

$$\mathcal{R}_j(x, t, u, u_x) = \mathcal{D}_{ax} \frac{\partial u_j}{\partial x} \qquad j = 1, \ldots, n_c \tag{7.11.46}$$

At left ($x = 0$), it gives

$$\begin{aligned} &\beta_j = 1 \qquad j = 1, \ldots, n_c \\ &\gamma_j = \frac{v}{\epsilon_e} (C_0^+ - C_0^-) = -\frac{v}{\epsilon_e} (C_{inj} - C_0^+) \qquad j = 1, \ldots, n_c \end{aligned} \tag{7.11.47}$$

At right ($x = L$), it gives

$$\begin{aligned} &\beta_j = 1 \qquad j = 1, \ldots, n_c \\ &\gamma_j = 0 \qquad j = 1, \ldots, n_c \end{aligned} \tag{7.11.48}$$

7.11.6.3 PD-LDF Model

In this case, the general equation (7.11.22) of the model of a chromatography column coupled with the equation of LDF model (7.11.24) is considered.

The term $\mathcal{R}_j$ is imposed by the coefficients of the general form of Equation (7.11.28) and is equal to

$$\begin{aligned} &\mathcal{R}_j(x, t, u, u_x) = \mathcal{D}_{ax} \frac{\partial u_j}{\partial x} \qquad j = 1, \ldots, n_c \\ &\mathcal{R}_{j+n_c}(x, t, u, u_x) = 0 \qquad j = 1, \ldots, n_c \end{aligned} \tag{7.11.49}$$

At left ($x = 0$), it gives

$$\begin{aligned} &\beta_j = 1 \qquad j = 1, \ldots, n_c \\ &\gamma_j = \frac{v}{\epsilon_e} (C_0^+ - C_0^-) = -\frac{v}{\epsilon_e} (C_{inj} - C_0^+) \qquad j = 1, \ldots, n_c \\ &\beta_{j+n_c} = 1 \qquad j = 1, \ldots, n_c \\ &\gamma_{j+n_c} = 0 \qquad j = 1, \ldots, n_c \end{aligned} \tag{7.11.50}$$

At right ($x = L$), it gives

$$\begin{array}{ll} \beta_j = 1 & j = 1, \ldots, n_c \\ \gamma_j = 0 & j = 1, \ldots, n_c \\ \beta_{j+n_c} = 1 & j = 1, \ldots, n_c \\ \gamma_{j+n_c} = 0 & j = 1, \ldots, n_c \end{array} \tag{7.11.51}$$

Thus, in the PD-LDF model with respect to the PD-Equil model, the only difference lies in the terms indexed by $(j + n_c)$ which are very simple.

7.11.6.4 Simulation of a Chromatographic Separation

The physical simulation parameters are as follows:

$$\begin{array}{l} \overline{N}_A = 10\,\text{g}\,\text{cm}^{-3}\,, \quad \overline{N}_B = 10\,\text{g}\,\text{cm}^{-3} \\ \tilde{K}_A = 1\,\text{cm}^3\,\text{g}^{-1}\,, \quad \tilde{K}_B = 1,1\,\text{cm}^3\,\text{g}^{-1} \\ \epsilon_i = 0.5\,, \quad \epsilon_e = 0.4 \\ C_A^f = 5 \times 10^{-3}\,\text{g}\,\text{cm}^{-3}\,, \quad C_B^f = 5\,10^{-3}\,\text{g}\,\text{cm}^{-3} \\ V_{col} = 1\,\text{cm}^3\,, \quad Q = 0.4\,\text{cm}^3/\text{min} \quad \Rightarrow t_0 = 1\,\text{min} \\ V_{inj} = 0.1\,\text{cm}^3 \\ N = 200 \qquad \text{(for the PD-Equil model)} \\ Pe = 200\,, \quad J = 100 \qquad \text{(for the LDF model)} \\ t_A^i = 0.3\,\text{min}\,, \quad t_B^i = 0.3\,\text{min} \qquad \text{(for the LDF model)} \end{array} \tag{7.11.52}$$

$\overline{N}_A$, $\overline{N}_B$, $\tilde{K}_A$, $\tilde{K}_B$ are the parameters of Langmuir adsorption isotherm. ϵ_i is the intragranular porosity and ϵ_e the extragranular porosity. C_A^f and C_B^f are the concentrations of components A and B in the feed, V_{inj} the injected volume, V_{col} the column volume, Q the volume flow rate, t_0 the zero holdup time (Nicoud 2015), t_A^i and t_B^i the characteristic times of internal diffusion of components A and B in the expression of LDF law. The column diameter is about 0.23 cm and its length 25 cm. The simulation is performed with an injection occurring at time $t = 1$ min with an initially empty column to distinguish the transient regime related to the injection (cf. Figure 7.51). The simulation uses a moving grid with 101 points. Initially, the grid is uniform.

In a first stage, the simulation was done with the equilibrium assumption (PD-Equil model) which gives Figure 7.52 showing the concentrations at the outlet of the column (called breakthrough curves) both in the case of Dirichlet boundary condition or Neumann (Danckwerts) at the inlet. In the second stage, the non-equilibrium is assumed (model PD-LDF, then the concentration profiles at the outlet are very different (Figure 7.53) with smoother peaks than in the equilibrium case indicating a more difficult separation. The positions of the peak summits are close in both cases. The only practical possibility is to identify the parameters of the model by means of experimental breakthrough curves. The influence of the boundary condition type at the inlet (Dirichlet ou Neumann) is relatively low whatever in the case with equilibrium or non-equilibrium.

7.12 Finite Volume Method

7.12.1 Introduction

A first reference to discover the finite volume method is the book by Patankar (1980). Other references are Eymard et al. (2000), LeVeque and Randall (2002), Moukalled et al. (2016), Patankar et al. (1998), Versteeg and Malalasekera (1995), and Wesseling (2001).

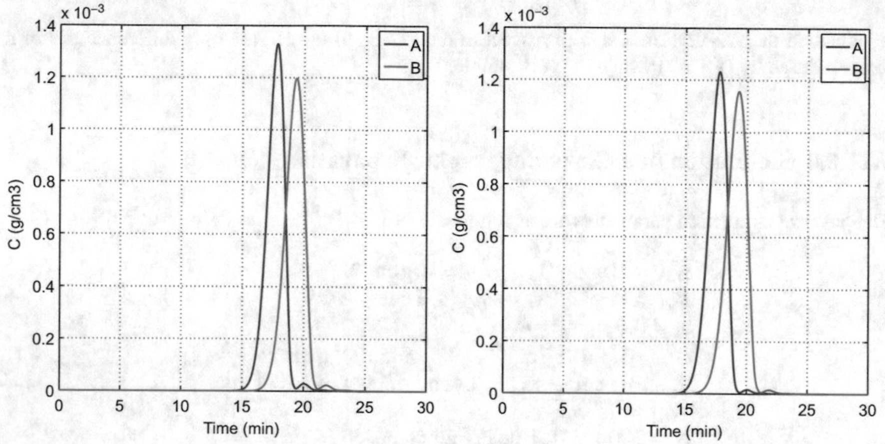

Fig. 7.52 PD-Equil model: Concentrations of components A and B at the column outlet in the case of Dirichlet boundary condition (left) or Neumann (right) at the inlet

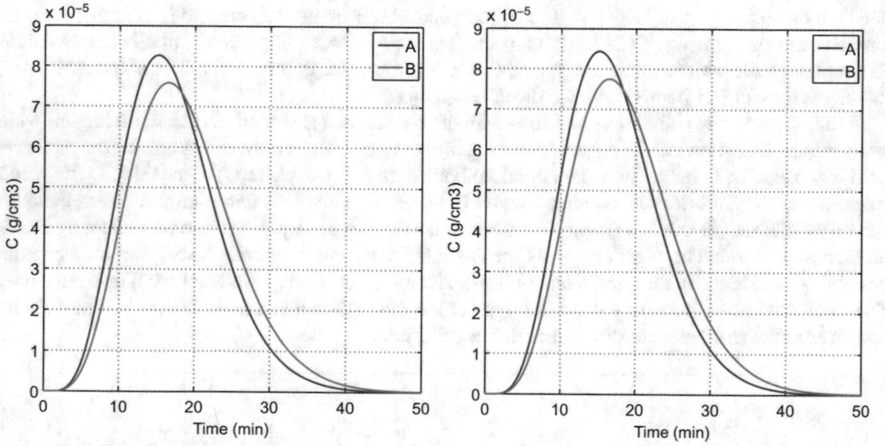

Fig. 7.53 PD-LDF model: Concentrations of components A and B at the column outlet in the case of Dirichlet boundary condition (left) or Neumann (right) at the inlet

A great interest of the finite volume method is that it is based on conservation laws (mass, energy, momentum), while with the method of finite differences, there is no insurance that the balances are verified. Thus, the finite volume method is well adapted to engineers and attractive.

To compare the techniques of finite differences and finite volumes, in a first time, we again consider the heat transfer problem in a wall illustrated by Fourier's law

$$\rho C \frac{\partial T}{\partial t} = \lambda \frac{\partial^2 T}{\partial x^2} \tag{7.12.1}$$

Here, we note C the heat capacity instead of C_p in Section 7.5 of finite differences, as the index P will be reserved for point P.

7.12.2 Mesh

The domain related to a one-dimensional calculation is divided into control volumes limited by faces (dashed lines on Figures 7.54, 7.55 and 7.56).

The points representative of control volumes are noted P_i. In a one-dimensional grid, a point P is surrounded by its neighbor W ("West") and its neighbor E ("East"). The control volume is limited by the faces w ("west") and e ("east").

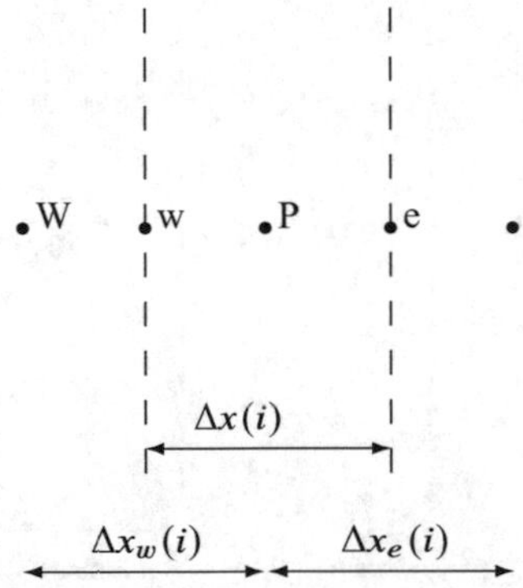

Fig. 7.54 Detail of a control volume around current point P, limited by its faces "west" and "east", and surrounded by its neighbors

The characteristic lengths are defined

$$\begin{aligned} \Delta x(i) &= x_e - x_w \\ \Delta x_e(i) &= x_E - x_P \\ \Delta x_w(i) &= x_P - x_W \end{aligned} \tag{7.12.2}$$

Two types of mesh are mentioned by Patankar (1980). Frequently, along the calculation, the hypothesis is taken that the fluid properties are constant on a control volume.

First possibility of definition of the control volume (type I) (Figure 7.55)

For any grid, e is at the middle of segment PE and w at the middle of segment WP, P is not at the middle of we. In this case, the value calculated in P is not a good representation of the control volume for the calculation of the source term, of the conductivity, …

$$\Delta x(i) = \frac{\Delta x_e(i) + \Delta x_w(i)}{2} \tag{7.12.3}$$

Second possibility of definition of the control volume (type II) (Figure 7.56)
For any grid, e is not at the middle of segment PE and w is not at the middle of segment WP, P is at the middle of we, thus at the center of the control volume. The control volume becomes the basis for the definition of the grid: we start by defining the boundaries of the control volumes, and the grid points result. It is recommended to place the faces of the control volumes where discontinuities in terms of fluid properties, source and boundary conditions occur. This is the mostly used mesh.

$$\begin{aligned} \Delta x_w(i) &= \frac{\Delta x(i-1) + \Delta x(i)}{2} \\ \Delta x_e(i) &= \frac{\Delta x(i) + \Delta x(i+1)}{2} \end{aligned} \tag{7.12.4}$$

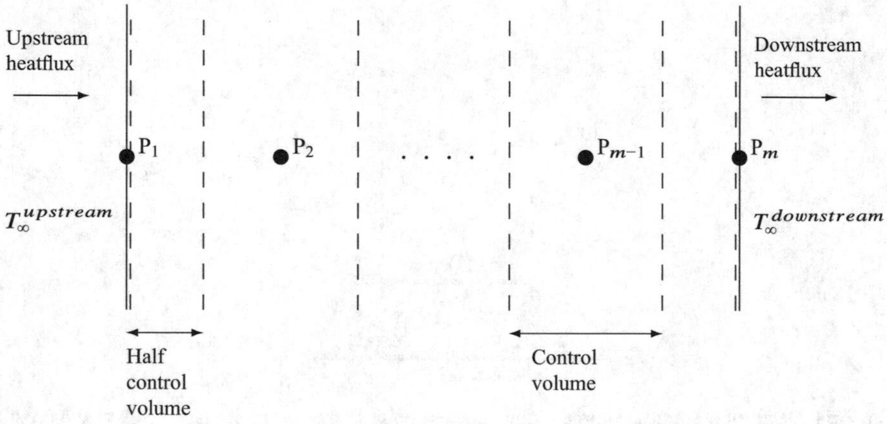

Fig. 7.55 One-dimensional grid: Type I: first choice in the definition of the control volumes and the grid

Different types of grids are possible, uniform, compressive or expansive exponential, symmetrical or dissymmetrical. This brings a considerable flexibility with respect to finite differences and allows to concentrate a larger number of volumes at the places where the gradients of variables are steeper, in general in the neighborhood of boundaries, in order to better capture the variations.

7.12.3 Integration on any Control Volume

Fourier's equation integrated on any control volume (Figure 7.54) gives

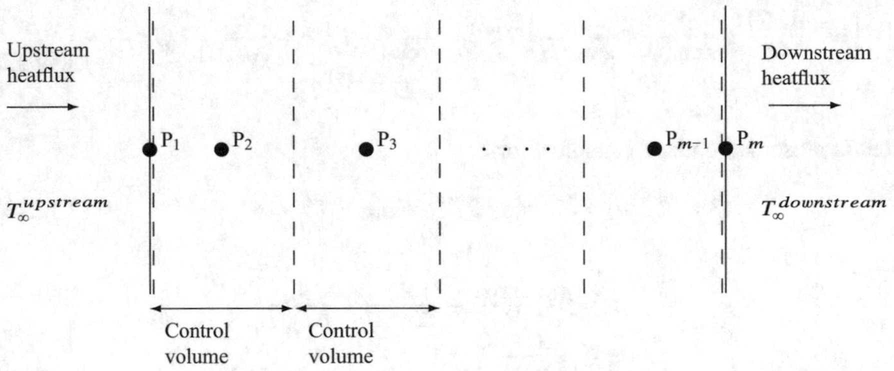

Fig. 7.56 One-dimensional grid: Type II: second choice in the definition of the control volumes and the grid

$$\rho C \underbrace{\int_V \frac{\partial T}{\partial t} dV}_{1} = \underbrace{\int_V \lambda \frac{\partial^2 T}{\partial x^2} dV}_{2} \tag{7.12.5}$$

The term 1 becomes

$$\rho C \left.\frac{\partial T}{\partial t}\right|_P V_P \tag{7.12.6}$$

Using the divergence (Green-Ostrogradsky) theorem, the term 2 becomes

$$\int_S \lambda \frac{\partial T}{\partial x} d\vec{S} = \int_S \lambda \frac{\partial T}{\partial x} dS\,\vec{n} = \lambda \left.\frac{\partial T}{\partial x}\right|_e \Delta S_e - \lambda \left.\frac{\partial T}{\partial x}\right|_w \Delta S_w \tag{7.12.7}$$

with

$$\left.\frac{\partial T}{\partial x}\right|_e = \frac{T_E - T_P}{\Delta x_e(i)} \quad \text{and} \quad \left.\frac{\partial T}{\partial x}\right|_w = \frac{T_P - T_W}{\Delta x_w(i)} \tag{7.12.8}$$

A discretization (P is the current point, n is time) can be done according to an implicit scheme

$$\rho C \frac{T_P^{n+1} - T_P^n}{\Delta t} V_P = \lambda_e \frac{T_E^{n+1} - T_P^{n+1}}{\Delta x_e} \Delta S_e - \lambda_w \frac{T_P^{n+1} - T_W^{n+1}}{\Delta x_w} \Delta S_w \tag{7.12.9}$$

As

$$V_P(i) = \Delta S_e\,\Delta x(i) = \Delta S_w\,\Delta x(i) \tag{7.12.10}$$

there comes

$$\rho C \frac{T_P^{n+1} - T_P^n}{\Delta t} \Delta x(i) = \lambda_e \frac{T_E^{n+1} - T_P^{n+1}}{\Delta x_e} - \lambda_w \frac{T_P^{n+1} - T_W^{n+1}}{\Delta x_w} \tag{7.12.11}$$

hence the writing according to finite volumes

$$\left[\rho C \frac{\Delta x(i)}{\Delta t} + \frac{\lambda_e}{\Delta x_e(i)} + \frac{\lambda_w}{\Delta x_w(i)}\right] T_P^{n+1} = \frac{\lambda_e}{\Delta x_e(i)} T_E^{n+1} + \frac{\lambda_w}{\Delta x_w(i)} T_W^{n+1} + \rho C \frac{\Delta x(i)}{\Delta t} T_P^n \tag{7.12.12}$$

that is presented under a canonical form

$$a_P T_P^{n+1} = a_E T_E^{n+1} + a_W T_W^{n+1} + b \tag{7.12.13}$$

with

$$\begin{aligned} a_P &= \rho C \frac{\Delta x(i)}{\Delta t} + \frac{\lambda_e}{\Delta x_e(i)} + \frac{\lambda_w}{\Delta x_w(i)} \\ a_E &= \frac{\lambda_e}{\Delta x_e(i)} \\ a_W &= \frac{\lambda_w}{\Delta x_w(i)} \\ b &= \rho C \frac{\Delta x(i)}{\Delta t} T_P^n \end{aligned} \tag{7.12.14}$$

The coefficients must verify

$$\begin{aligned} a_P &= a_E + a_W - S_P \Delta x + a_P^0 \\ b &= S_c \Delta x + a_P^0 T_P^0 \qquad \text{with } T_P^0 = T_P^n \end{aligned} \tag{7.12.15}$$

where S_p is a source term.

It results

$$\begin{aligned} a_P^0 &= \rho C \frac{\Delta x(i)}{\Delta t} \\ S_P &= 0 \\ S_c &= 0 \end{aligned} \quad \text{(no source term inside)} \tag{7.12.16}$$

Base rules can be prescribed:

- *Consistency at the faces of the control volumes*
  For two adjacent control volumes, on the common face, the flux crossing the face is identical in the equations for both control volumes.
- *Positive coefficients*
  All the coefficients a_E, a_W, and a_P are positive.
- *Linearization of the source term with a negative slope*
  When the source term is linearized under the form

$$S = S_c + S_P T_P \tag{7.12.17}$$

  the coefficient S_P must be negative.
- *Sum of coefficients of neighbors*
  When the function T is a solution of the ordinary differential equation and $T + c$ is also a solution (c being any constant), then necessarily

$$a_P = \sum_{neighbors} a_{neighbor} + \dots \tag{7.12.18}$$

which in one-dimension gives

$$a_P = a_E + a_W \tag{7.12.19}$$

7.12.4 Account of Boundary Conditions at Left

A type I grid has been chosen which makes use of half control volumes at the boundaries of the domain. A type II grid which is often retained could as well be chosen.

The convective flux at P_1 (Figure 7.57) is

$$\Phi = h^{upstream}(T_\infty^{upstream} - T_{P_1}) > 0 \tag{7.12.20}$$

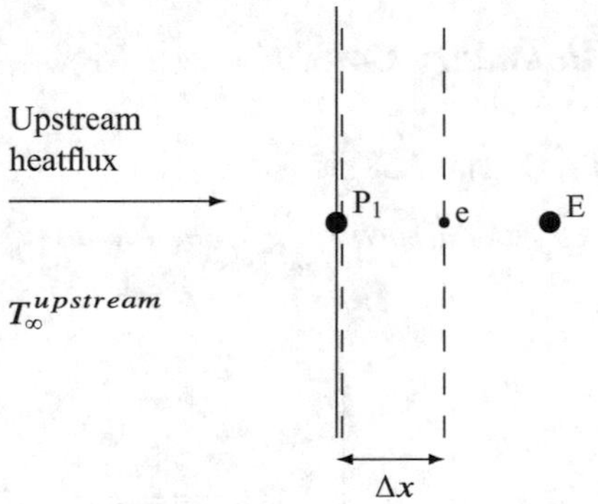

Fig. 7.57 Boundary condition at P_1

The balance discretized at P_1 is

$$\rho C \frac{T_P^{n+1} - T_P^n}{\Delta t} V_{P_1} = \lambda_e \frac{T_E^{n+1} - T_P^{n+1}}{\Delta x_e} \Delta S_e + h^{upstream} (T_\infty^{upstream} - T_{P_1}^{n+1}) \Delta S_P \tag{7.12.21}$$

According to type I grid, the half control volume at P_1 is equal to

$$V_{P_1} = \Delta S_e \, \Delta x \tag{7.12.22}$$

hence

$$\rho C \frac{T_P^{n+1} - T_P^n}{\Delta t} \Delta x = \lambda_e \frac{T_E^{n+1} - T_P^{n+1}}{\Delta x_e} + h^{upstream} (T_\infty^{upstream} - T_{P_1}^{n+1}) \tag{7.12.23}$$

that is

$$\left[\rho C \frac{\Delta x}{\Delta t} + \frac{\lambda_e}{\Delta x_e} + h^{up}\right] T_P^{n+1} = \frac{\lambda_e}{\Delta x_e} T_E^{n+1} + h^{up} T_\infty^{up} + \rho C \frac{\Delta x}{\Delta t} T_P^n \tag{7.12.24}$$

According to the previous canonical form (7.12.13), we obtain

$$
\begin{aligned}
a_P &= \rho C \frac{\Delta x(i)}{\Delta t} + \frac{\lambda_e}{\Delta x_e(i)} + h^{upstream} = a_E + a_W - S_P \Delta x + a_P^0 \\
a_E &= \frac{\lambda_e}{\Delta x_e(i)} \\
a_W &= 0 \\
S_P &= -\frac{h^{upstream}}{\Delta x} \qquad \text{(source term)} \\
b &= h^{upstream} T_\infty^{upstream} + \rho C \frac{\Delta x(i)}{\Delta t} T_P^n = S_c \Delta x + a_P^0 T_P^0 \\
a_P^0 &= \rho C \frac{\Delta x(i)}{\Delta t} \\
S_c &= \frac{h^{upstream} T_\infty^{upstream}}{\Delta x(i)} \qquad \text{(source term)}
\end{aligned}
\tag{7.12.25}
$$

7.12.5 Account of Boundary Conditions at Right

The convective flux at P_m (Figure 7.58) is

$$
\Phi = h^{downstream}(T_{P_m} - T_\infty^{downstream}) > 0 \tag{7.12.26}
$$

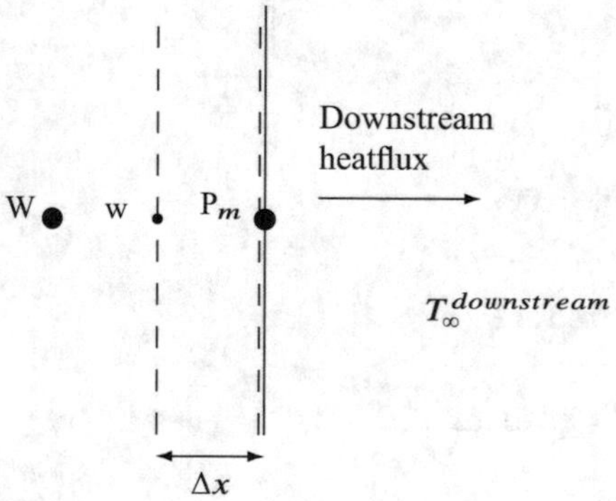

Fig. 7.58 Boundary condition at P_m

The discretized balance at P_m is

$$
\rho C \frac{T_P^{n+1} - T_P^n}{\Delta t} V_{P_m} = -h^{down} (T_{P_m}^{n+1} - T_\infty^{down}) \Delta S_P - \lambda_w \frac{T_P^{n+1} - T_W^{n+1}}{\Delta x_w} \Delta S_w \tag{7.12.27}
$$

The half control volume at P_m is equal to

$$
V_{P_m} = \Delta S_w \Delta x \tag{7.12.28}
$$

hence

$$\rho C \frac{T_P^{n+1} - T_P^n}{\Delta t} \Delta x = -h^{downstream} (T_{P_m}^{n+1} - T_\infty^{downstream}) - \lambda_w \frac{T_P^{n+1} - T_W^{n+1}}{\Delta x_w} \tag{7.12.29}$$

that is

$$\left[\rho C \frac{\Delta x}{\Delta t} + \frac{\lambda_w}{\Delta x_w} + h^{down}\right] T_P^{n+1} = \frac{\lambda_w}{\Delta x_w} T_W^{n+1} + h^{down} T_\infty^{down} + \rho C \frac{\Delta x}{\Delta t} T_P^n \tag{7.12.30}$$

According to the previous canonical form (7.12.13), we obtain

$$\begin{aligned}
a_P &= \rho C \frac{\Delta x(i)}{\Delta t} + \frac{\lambda_w}{\Delta x_w(i)} + h^{downstream} = a_E + a_W - S_P \Delta x + a_P^0 \\
a_E &= 0 \\
a_W &= \frac{\lambda_w}{\Delta x_w(i)} \\
S_P &= -\frac{h^{downstream}}{\Delta x} \qquad \text{(source term)} \\
b &= h^{downstream} T_\infty^{downstream} + \rho C \frac{\Delta x(i)}{\Delta t} T_P^n = S_c \Delta x + a_P^0 T_P^0 \\
a_P^0 &= \rho C \frac{\Delta x(i)}{\Delta t} \\
S_c &= \frac{h^{downstream} T_\infty^{downstream}}{\Delta x(i)} \qquad \text{(source term)}
\end{aligned} \tag{7.12.31}$$

7.12.6 Case of an Interface Between Two Solids of Different Conductivities

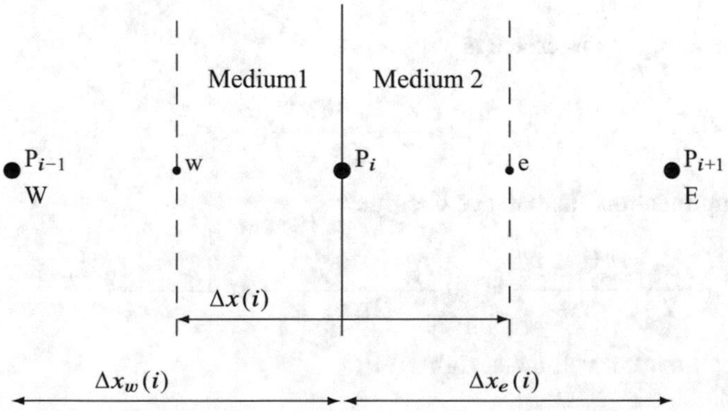

Fig. 7.59 Interface between two solids of different conductivities separated by the interface at P_i

The equality of the fluxes on both sides of the interface at P=P$_i$ (Figure 7.59) gives

$$\lambda_1^- \left.\frac{\partial T}{\partial x}\right|_{x_i^-} = \lambda_2^+ \left.\frac{\partial T}{\partial x}\right|_{x_i^+} \tag{7.12.32}$$

i.e. (in the case of a type I grid)

$$\lambda_1^- \frac{T_w - T_P}{\frac{\Delta x_w}{2}} = \lambda_2^+ \frac{T_P - T_e}{\frac{\Delta x_e}{2}} \tag{7.12.33}$$

It must be noticed that this expression makes use of temperatures T_w and T_e on the faces although we wish to only use T_W and T_E at the neighbor nodes to calculate T_P.

A composite conductivity at P is defined

$$\lambda_P = \frac{1}{\frac{\alpha}{\lambda_e} + \frac{1-\alpha}{\lambda_w}} \tag{7.12.34}$$

with

$$\alpha = \frac{\frac{\Delta x_e}{2}}{\Delta x} \quad \text{and} \quad 1 - \alpha = \frac{\frac{\Delta x_w}{2}}{\Delta x} \tag{7.12.35}$$

If $\alpha = 0.5$, the harmonic mean is

$$\lambda_P = \frac{2\,\lambda_e\,\lambda_w}{\lambda_e + \lambda_w} \tag{7.12.36}$$

For the half control volume at left of P

$$\rho_1\, C_1\, \frac{\Delta x_w}{2}\, \frac{T_P^{n+1} - T_P^n}{\Delta t} = \lambda_P\, \frac{T_e^{n+1} - T_w^{n+1}}{\Delta x} - \lambda_1\, \frac{T_P^{n+1} - T_w^{n+1}}{\frac{\Delta x_w}{2}} \tag{7.12.37}$$

as the flux crossing the interface is

$$\lambda_P\, \frac{T_e^{n+1} - T_w^{n+1}}{\Delta x} \tag{7.12.38}$$

After simplification, the balance becomes

$$\rho_1\, C_1\, \frac{\Delta x_w}{2}\, \frac{T_P^{n+1} - T_P^n}{\Delta t} = \lambda_P\, \frac{T_E^{n+1} - T_W^{n+1}}{2\Delta x} - \lambda_1\, \frac{T_P^{n+1} - T_W^{n+1}}{\Delta x_w} \tag{7.12.39}$$

For the half control volume at right of P

$$\rho_2\, C_2\, \frac{\Delta x_e}{2}\, \frac{T_P^{n+1} - T_P^n}{\Delta t} = -\lambda_P\, \frac{T_E^{n+1} - T_W^{n+1}}{2\Delta x} + \lambda_2\, \frac{T_E^{n+1} - T_P^{n+1}}{\Delta x_e} \tag{7.12.40}$$

The canonical forms according to Equation (7.12.13) can be used.

7.12.7 Numerical Solving

The solving algorithm can be summarized as:

1. Initialization of the grid and of the temperature profile.
2. Convergence search at each instant n by the sweeping method.

2.1. Noting k the iteration index during the calculation from instant n to $n+1$, we set $k = 0$ and T_P^k is equal to T_P^n.

2.2. Set $k = k + 1$. Calculation of coefficients a_P^0, a_P, a_E, a_W, and source terms S_P and S_c.

2.3. Calculate the new values of T_P^{k+1} with respect to the old values T_P^k on all points P by using equations of the form

$$T_P^{k+1} = T_P^k + r\,(\frac{a_E\,T_E^{k+1} + a_W\,T_W^{k+1} + b}{a_P} - T_P^k) \tag{7.12.41}$$

with r relaxation coefficient. The relaxation coefficient is lower than 1 in the case of under-relaxation (frequent case) and larger than 1 in the case of over-relaxation. Its most common aim is to avoid the divergence and its choice is empirical.

2.4. The residual calculated at iteration k is of the form

$$res = |a_P\,T_P^k - (a_E\,T_E^k + a_W\,T_W^k + b)| \tag{7.12.42}$$

If the residuals do not satisfy the convergence criterion, return to stage 2.2 until the convergence is verified.
When the convergence is reached, T_P^k is equal to T_P^{n+1} and we return to stage 2.1 by increasing n by one unit.

Remark: in steady state, if we note $T_{i-1} = T_W$, $T_i = T_P$, $T_{i+1} = T_E$,
the canonical equation

$$a_P\,T_P = a_E\,T_E + a_W\,T_W + b \tag{7.12.43}$$

becomes

$$a_i\,T_i = b_i\,T_{i-1} + c_i\,T_{i+1} + d_i \tag{7.12.44}$$

which allows us to represent the system under tridiagonal form, easily adapted to numerical solving.

Example 7.16 :
MT1-FVM: Transfer by diffusion of a gas though a membrane

A plane membrane is considered with transfer by diffusion of a gas through the membrane. Upstream, the pressure is assumed to be constant, while downstream the gas is accumulated in a tank noted downstream (Figure 7.60).

The hypotheses for a membrane are:

1. The resistance to mass transfer is located only in the membrane, i.e. the porous support has no influence,
2. The process is strictly diffusional in the membrane,
3. The interfacial equilibrium is instantaneous upstream

$$C = S\,P_0 \tag{7.12.45}$$

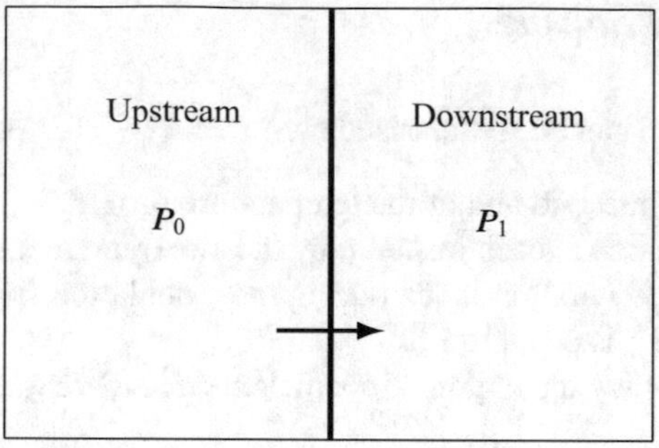

Fig. 7.60 Transfer by diffusion of a gas through a membrane separating two tanks ($P_0 > P_1$)

4. The conditions are isothermal. S is the sorption coefficient.

It results that the membrane behavior under these conditions can be modelled by Fick's law

$$\frac{\partial C}{\partial t} = \frac{\partial}{\partial x} \mathcal{D}(C) \frac{\partial C}{\partial x} \tag{7.12.46}$$

where $\mathcal{D}$ is the diffusion coefficient that may depend on concentration $C(x, t)$. In the simple case of a single membrane of thickness L subject to an upstream pressure P_0, and when the vacuum is done, initially in both upstream and downstream tanks, the boundary and initial conditions are

$$\begin{array}{ll} C(x, 0) = 0 & \forall\ 0 \leq x \leq L \\ C(0, t) = C_0 = S\,P_0 & \forall\, t \\ C(L, t) = C_1 = S\,P_1 & \forall\, t \end{array} \tag{7.12.47}$$

P_1 is the pressure in the downstream tank. Furthermore, for the downstream tank, we can write the equation

$$\frac{dP_1}{dt} = \frac{dn_{in}}{dt} \frac{RT}{V_{downstream}} \tag{7.12.48}$$

with the molar flux entering the downstream tank of volume $V_{downstream}$

$$\frac{dn_{in}}{dt} = -\mathcal{D}(C_1) \left. \frac{\partial C}{\partial x} \right|_L A \tag{7.12.49}$$

where A is the surface area offered to downstream transfer. We can also write

$$\frac{dn_{in}}{dt} = \frac{V_{downstream}}{RT} \frac{dP_{downstream}}{dt} = \frac{V_{downstream}}{SRT} \frac{dC_{downstream}}{dt} \tag{7.12.50}$$

These equations could be solved by a finite difference method. However, to improve the stability and as it ensures the conservation of masses, the finite volume method (Patankar 1980; Patankar et al. 1998; Versteeg and Malalasekera 1995) was used for all simulations in the present section.

By discretizing the membrane with m points P_i and noting the inlet by $i = 1$, the outlet by $i = m$, the equations are written as

$$\begin{aligned} & \frac{C_P^{n+1} - C_P^n}{\Delta t} V_P = 0\,, \quad i = 1 \\ & \frac{C_P^{n+1} - C_P^n}{\Delta t} V_P = -\mathcal{D} \frac{C_P^{n+1} - C_W^{n+1}}{\Delta x_w} \Delta S_w + \mathcal{D} \frac{C_E^{n+1} - C_P^{n+1}}{\Delta x_e} \Delta S_e\,, \quad 1 < i < m \\ & \frac{C_P^{n+1} - C_P^n}{\Delta t} V_P = -\mathcal{D} \frac{C_P^{n+1} - C_W^{n+1}}{\Delta x_w} \Delta S_w - \frac{V_{downstream}}{SRT\,\Delta t} (C_P^{n+1} - C_P^n)\,, \quad i = m \end{aligned} \tag{7.12.51}$$

thus, by assuming $\Delta S_e = \Delta S_w = \Delta S$

$$
\begin{aligned}
&C_P^{n+1} = \mathcal{S} P_0\,, \quad i = 1 \\
&C_P^{n+1}\left[\frac{\Delta x(i)}{\Delta t} + \frac{\mathcal{D}}{\Delta x_w} + \frac{\mathcal{D}}{\Delta x_e}\right] = \frac{\Delta x(i)}{\Delta t} C_P^n + \frac{\mathcal{D}}{\Delta x_w} C_W^{n+1} + \frac{\mathcal{D}}{\Delta x_e} C_E^{n+1}\,, \quad 1 < i < m \\
&C_P^{n+1}\left[\frac{\Delta x(i)}{\Delta t} + \frac{\mathcal{D}}{\Delta x_w} + \frac{V_{downstream}}{\mathcal{S} RT\,\Delta S\,\Delta t}\right] = \frac{\Delta x(i)}{\Delta t} C_P^n + \frac{\mathcal{D}}{\Delta x_w} C_W^{n+1} + \frac{V_{downstream}}{\mathcal{S} RT\,\Delta S\,\Delta t} C_P^n\,, \quad i = m
\end{aligned}
\tag{7.12.52}
$$

from which Patankar coefficients result (Table 7.1).

Table 7.1 Patankar coefficients

Index of point	a_P^0	a_E	a_W	S_P	S_c
$i = 1$	0	0	0	$-\frac{1}{\Delta t}$	$\frac{\mathcal{S} P_0}{\Delta t}$
$1 < i < m$	$\frac{\Delta x(i)}{\Delta t}$	$\frac{\mathcal{D}}{\Delta x_e}$	$\frac{\mathcal{D}}{\Delta x_w}$	0	0
$i = m$	$\frac{\Delta x(i)}{\Delta t} S$	0	$\frac{\mathcal{D} S}{\Delta x_w}$	$-\frac{V_{downstream}}{RT\,\Delta t\,V_P}$	$-S_P C_P^n$

An analytic solution of this problem can be found by means of some simplifying hypotheses. In the case where the downstream tank has large dimensions, the concentration $C(L, t)$ is close to zero (simplification of Equation (7.12.47)) and if, on another side, the diffusion coefficient is assumed to be constant, the solution of this problem is given by equation

$$
C(x,t) = C_0\,(1 - \frac{x}{L}) - \frac{2C_0}{\pi} \sum_{i=1}^{\infty} \left[\frac{1}{i} \sin(\frac{i\pi x}{L}) \exp(-\frac{\mathcal{D} i^2 \pi^2 t}{L^2})\right]
\tag{7.12.53}
$$

The pressure P_1 in the downstream tank can be deduced from Equation (7.12.53) by calculating the mass flux in the downstream tank

$$
P_1 = A \frac{RT\,\mathcal{D} P_0}{VL} \left(\mathcal{S} t - \frac{\mathcal{S} L^2}{6\mathcal{D}} + \frac{2\mathcal{S} L^2}{\pi^2 \mathcal{D}} \sum_{i=1}^{\infty} \left[\frac{(-1)^{i+1}}{i^2} \exp(-\frac{\mathcal{D} i^2 \pi^2 t}{L^2}) \right] \right)
\tag{7.12.54}
$$

Note that this equation is in contradiction with equation $C_1 = \mathcal{S} P_1$ that was posed in (7.12.47), as we calculate P_1 whereas we suppose that $C_1 \approx 0$. Nevertheless, it will be shown that the analytical and numerical solutions are in very good agreement as the concentration C_1 remains very low.

The simulation corresponds exactly to the described model and allows us to validate the numerical finite volume method by comparing the analytical and numerical solutions. Initially, only one component exists in the upstream tank. The diffusion coefficient is supposed to be independent of concentration C.

The integration time step is 0.1s, the number of grid points in the membrane is equal to 51. The simulation conditions are given in Table 7.2.

On Figure 7.61, the analytical and numerical concentration profiles in the membrane at final time are exactly similar. The numerical results (Table 7.3) demonstrate the excellent precision, which validates the numerical method. Even if the theoretical solution does not exactly respect the model hypotheses, it represents an excellent approximation as the downstream pressure remains very low. The theoretical concentration values have been calculated from (7.12.53) by truncating the series to $n = 10{,}000$.

The pressure evolution in the downstream tank (Figure 7.62) shows a very low pressure, related to the considerable volume of the downstream tank, chosen to validate the analytical method.

Table 7.2 Simulation conditions

Final time (s)	20
Membrane thickness (m)	10^{-3}
Diffusion coefficient (S.I.)	50×10^{-10}
Sorption coefficient (S.I.)	10^{-8}
Upstream pressure (Pa)	1.01315×10^{5}
Initial downstream pressure (Pa)	0
Temperature (K)	293.15
Surface area of the membrane (m^2)	10^{-4}
Downstream volume (m^3)	1
Number of grid points	51

Table 7.3 Example MT1-FVM: Comparison of calculated and theoretical concentrations inside the membrane at $t = 20$ s

Position (m)	Calculated concentration (mol m^{-3})	Theoretical concentration (mol m^{-3})
0.00	0.1013×10^{-2}	0.1013×10^{-2}
0.2000×10^{-3}	0.6627×10^{-3}	0.6633×10^{-3}
0.4000×10^{-3}	0.3749×10^{-3}	0.3756×10^{-3}
0.6000×10^{-3}	0.1799×10^{-3}	0.1803×10^{-3}
0.8000×10^{-3}	0.6713×10^{-4}	0.6722×10^{-4}
0.9800×10^{-3}	0.5939×10^{-5}	0.5942×10^{-5}

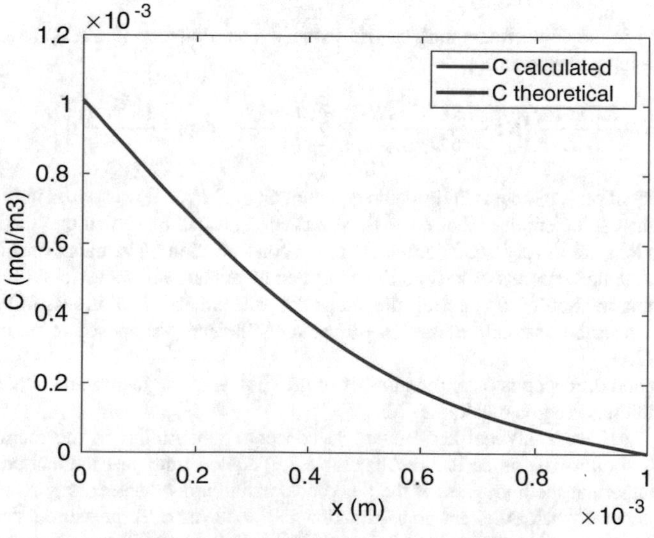

Fig. 7.61 Example MT1-FVM: Analytical and numerical concentration profiles in the membrane at $t = 20$ s

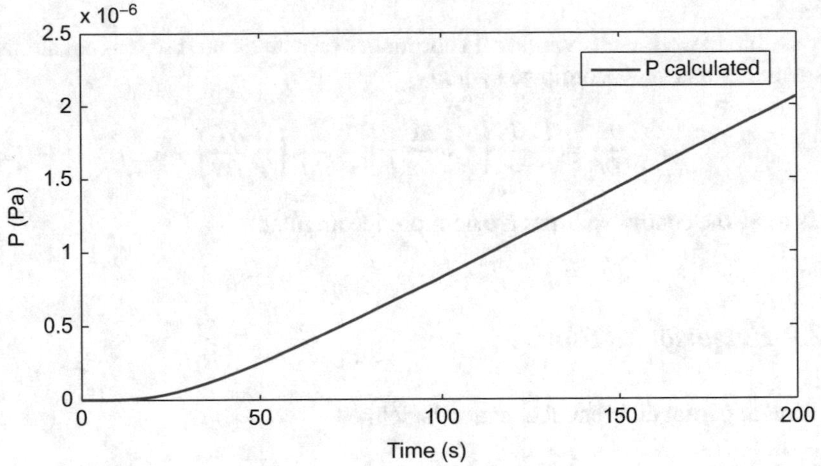

Fig. 7.62 Example MT1-FVM: Pressure evolution in the downstream tank up to $t = 200$ s

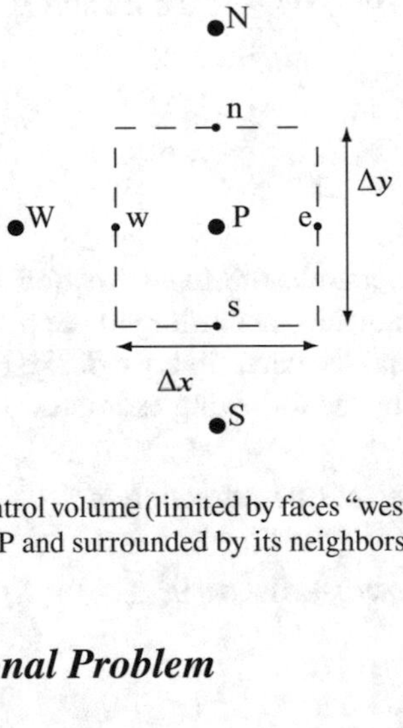

Fig. 7.63 Two-dimensional control volume (limited by faces "west", "east", "south", "north" in dashed lines) around the current point P and surrounded by its neighbors W, E, N, and S

7.12.8 Two-Dimensional Problem

Canonical equation (Figure 7.63)

$$a_P\, T_P = a_E\, T_E + a_W\, T_W + a_N\, T_N + a_S\, T_S + b \qquad (7.12.55)$$

Equations verified by the coefficients

$$\begin{aligned} a_P &= a_E + a_W + a_N + a_S - S_P\, \Delta x\, \Delta y + a_P^0 \\ b &= S_c\, \Delta x\, \Delta y + a_P^0\, T_P^0 \qquad \text{with } T_P^0 = T_P^n \end{aligned} \qquad (7.12.56)$$

Remark:
It is possible to work with cylindrical coordinates provided that the basis equations are considered in this new coordinate system

$$\rho C \frac{\partial T}{\partial t} = \frac{1}{r}\frac{\partial}{\partial r}\left(r \lambda \frac{\partial T}{\partial r}\right) + \frac{1}{r}\frac{\partial}{\partial \theta}\left(\frac{\lambda}{r}\frac{\partial T}{\partial \theta}\right) + S \tag{7.12.57}$$

In this case, the control volumes are no more rectangular.

7.12.9 Extension to Flows

The general partial differential equation is defined

$$\underbrace{\frac{\partial}{\partial t}(\rho \phi)}_{\text{transient}} + \underbrace{\frac{\partial}{\partial x_i}(\rho u_i \phi)}_{\text{convection}} = \underbrace{\frac{\partial}{\partial x_i}\left(\Gamma \frac{\partial \phi}{\partial x_i}\right)}_{\text{diffusion}} + \underbrace{S}_{\text{source}} \tag{7.12.58}$$

ϕ dependent variable,
Γ generalized diffusion coefficient,
u_i velocity,
ρ density,
S source term.
ϕ can be the enthalpy or temperature, the mass fraction of a chemical species, the velocity, or still turbulence parameters, according to the partial differential equation.

Equation (7.12.58) can be applied each time we desire to express the conservation of a variable, as demonstrated by the following examples.

Conservation of a chemical species (mass fraction m_i)

$$\frac{\partial}{\partial t}(\rho m_i) + \text{div}(\rho u m_i + J_i) = R_i \tag{7.12.59}$$

u velocity,
J_i diffusion flux de diffusion
R_i production rate of i per volume unit.

Energy balance (enthalpy h)

$$\text{div}(\rho u h) = \text{div}(k \, \text{grad} T) + S_h \tag{7.12.60}$$

T temperature,
k thermal conductivity,
S_h rate of heat production per volume unit.

Conservation of the momentum for a Newtonian fluid (velocity u)

$$\frac{\partial}{\partial t}(\rho\, u) + \operatorname{div}(\rho\, \mathbf{u}\, u) = \operatorname{div}(\mu \operatorname{grad} u) - \frac{\partial p}{\partial x} + B_x + V_x \tag{7.12.61}$$

μ viscosity,
p pressure,
B_x force in x direction per volume unit,
V_x viscous terms per volume unit.

7.12.10 Conservation Applied to a Control Volume

The convection and diffusion are combined in the expression of the total flux J_i in direction i

$$J_i = \rho\, u_i \phi - \Gamma \frac{\partial \phi}{\partial x_i} \tag{7.12.62}$$

The partial differential equation is written

$$\frac{\partial}{\partial t}(\rho\, \phi) + \frac{\partial J_i}{\partial x_i} = S \tag{7.12.63}$$

By integrating over the control volume in three dimensions, we get

$$(\rho_P\, \phi_P - \rho_P^0\, \phi_P^0) \left(\frac{\Delta V}{\Delta t}\right) + J_e\, A_e - J_w\, A_w + J_n\, A_n - J_s\, A_s + J_t\, A_t - J_b\, A_b = \bar{S}\, \Delta V \tag{7.12.64}$$

The exponent "0" means what is known at t.

For a convection-diffusion problem, the exact profile of ϕ is exponential between P and E. It follows that, for example on face e, the total flux is

$$J_e\, A_e = F_e \left(\phi_P + \frac{\phi_P - \phi_E}{\exp(Pe_e) - 1}\right) \tag{7.12.65}$$

with Pe_e Peclet number, F_e mass flow rate through the face e, D_e diffusion conductance, equal to

$$Pe_e = \frac{F_e}{D_e}\,, \quad F_e = (\rho\, u)_e\, A_e\,, \quad D_e = A_e \left[\frac{(\delta x)_e^-}{\Gamma_P} + \frac{(\delta x)_e^+}{\Gamma_E}\right]^{-1} \tag{7.12.66}$$

by considering the diffusion coefficient Γ uniform on each control volume.

If we want to avoid the expression (7.12.65) for intensive calculation reasons, Patankar (1980) recommends the following final approximation called *power-law* scheme

$$J_e\, A_e = F_e \phi_P + \{D_e\, f(|Pe_e|) + \max(-F_e, 0)\}\, (\phi_P - \phi_E) \tag{7.12.67}$$

with the power-law

$$f(|Pe|) = \max(0, (1 - 0.1|Pe|^5) \tag{7.12.68}$$

The source term may depend on ϕ, eventually in a nonlinear manner, for example for a chemical reaction. When necessary, the source term must be linearized

$$\bar{S} = S_C + S_P\, \phi_P \tag{7.12.69}$$

Under the canonical form, the discretized equation is

$$a_P\, \phi_P = a_E\, \phi_E + a_W\, \phi_W + a_N\, \phi_N + a_S\, \phi_S + a_T\, \phi_T + a_B\, \phi_B + b \tag{7.12.70}$$

(where the symbols T and B mean "Top" and "Bottom") with

$$\begin{aligned}
&a_E = D_e\, f(|Pe_e|) + \max(-F_e, 0)\,, \quad a_W = D_w\, f(|Pe_w|) + \max(F_w, 0) \ldots \\
&b = S_C\, \Delta V + a_P^0\, \phi_P^0 \\
&a_P = a_E + a_W + a_N + a_S + a_T + a_B + a_P^0 + S_P\, \Delta V \\
&a_P^0 = \rho_P^0\, \frac{\Delta V}{\Delta t}
\end{aligned} \tag{7.12.71}$$

Remark: in the case of fluid flow, in general it is necessary to use staggered grids, which avoids to find unrealistic solutions. Thus, the places of the velocity components are on the control volume faces normal to these components. The other variables, including pressure, are calculated at the grid points.

7.12.11 SIMPLER Algorithm

Many algorithms for numerical solving (SIMPLER, SIMPLE, QUICK, upwind of different orders, etc.) have been proposed. It is possible to refer to Eymard et al. (2000), LeVeque and Randall (2002), Moukalled et al. (2016), Patankar et al. (1998), Versteeg and Malalasekera (1995), and Wesseling (2001) for more details. The SIMPLER algorithm is as follows:

1. Initialize all dependent variables, in particular the velocity.
2. Calculate the coefficients of momentum equations and evaluate $\hat{u}$, $\hat{v}$, $\hat{w}$.
3. Evaluate b and obtain the pressure field p^*.
4. With p^*, solve the momentum equations and obtain u^*, v^*, w^*.
5. Calculate b and deduce the pressure correction.
6. With the new pressure field p', correct the velocities noted *.
7. Solve the equations with respect to the other variables ϕ, temperature T, concentrations, turbulence parameters.
8. Return to stage 2 with the values of the corrected velocity field and the other ϕ, until convergence.

Comparing the calculation speed and the results given by different methods for a given problem is always very informative.

7.13 Finite Element Method

The finite element method presents many common points with the spectral methods of Section 7.10; however, it is here described in a more specific frame where Galerkin's method will be favored.

The finite element method (Allaire and Alouges 2015; Ciarlet and Luneville 2009; Dhatt et al. 2014; Ern and Guermond 2004; Hutton 2004; Khennane 2013; Koutromanos 2018; Legoll 2019; Li et al. 2018; Logan 2016; Loustau 2016; Rao 2018; Spillane 2017; Zienkiewicz et al. 2013) is frequently used in the domains of solid mechanics, fluid flow, but it can also be used in heat and mass transfer (Hutton 2004; Koutromanos 2018; Logan 2016; Reddy 2015) and in process engineering (Coimbra et al. 2000, 2001, 2003, 2004, 2016; Mills and Ramachandran 1988; Sereno et al. 1991, 1992; Yu and Wang 1989). It is among the best methods when the domain is complex as it allows the calculation following a fine discretization of the domain, even in three dimensions (3D).

The finite element method can be used to find the numerical solution of ordinary differential equations or partial differential equations.

An important difference between the finite difference method and the finite element method is that this latter makes use of an approximation, for example polynomial, of the solution of the ordinary differential equations or partial differential equations whereas the finite difference method only provides numerical values at discretization points. The linear finite element method results in a system of linear equations that provides an approximate solution. However, the finite element method is much more difficult to be operated.

On another side, the finite element method is only applied to spatial variables. Consequently, when a transient problem is being solved, the finite element method is used for the spatial part and the finite difference method for the time part. Thus we will mainly consider steady-state systems in the description of the finite element method. However, the case of transient systems is also treated in detail in Section 7.13.5.

After the physical and mathematical description of the studied system, the finite element method includes the following steps:

1. Definition of elements and nodes,
2. Determination of the polynomial model,
3. Determination of linear elements,
4. Determination of the global linear system,
5. Consideration of boundary conditions,
6. Solution of the linear system,
7. Post-processing (visualization and interpretation of results).

In the case where the model of the studied system is not linear, the solution is obtained by a series of linear problems.

Suppose that a two-dimensional system (2D) is considered, for example with respect to x and y, described by a model of partial differential equations with boundary conditions of the considered volume. If that model depends on a single field variable $u(x, y)$ to be determined at points $P(x, y)$, the equation must be exactly verified at each point $P(x, y)$.

Different techniques exist to perform the finite element method:

- The direct method,
- The minimum of potential energy in mechanics,
- The Galerkin's residual method,
- The Rayleigh-Ritz method,
- The collocation method,
- The subdomain method,
- The least squares.

Several of these methods are variational methods or residual methods (Section 7.10).

The general approach for a 1D problem follows four steps:

1. Write the steady-state model of the system as fundamental spatial differential equations, to which the boundary conditions are added. Thus the *Strong Formulation* results as
$$\begin{aligned} & g(u, \frac{du}{dx}, \frac{d^2u}{dx^2}, \ldots) + a = 0, \quad 0 \leq x \leq L \\ & g_1(u(0), \frac{du}{dx}|_{x=0}) = a_1 \\ & g_2(u(L), \frac{du}{dx}|_{x=L}) = a_2 \end{aligned} \tag{7.13.1}$$
2. Transform the fundamental differential equations under the *Weak Formulation* equivalent to the strong formulation
$$\int_0^L w(x) \left[g(u, \frac{du}{dx}, \frac{d^2u}{dx^2}, \ldots) + a \right] dx = 0 \qquad \forall\, w(x) \text{ continuous} \tag{7.13.2}$$
for any continuous function $w(x)$ with the supplementary condition $w(x_i) = 0$ if at x_i a Dirichlet condition is imposed, i.e. $u(x_i) = u_i$ (Koutromanos 2018). The term of *form* is also used instead of formulation as strong and weak form. The weak formulation is often qualified as *variational formulation*. The variational formulation consists in integrating the strong formulation multiplied by a test function w. The advantage of the variational formulation with respect to the strong formulation is that, by use of Green's formula, the order of the derivative is lowered by one unit and thus the differentiability requirement for the function u is decreased.
It is possible to show that the solution of the variational formulation or weak formulation is unique and that it is also the solution of the strong formulation (equivalence theorem of Lax-Milgram) and vice versa (Allaire 2012).
The weak formulation is thus composed by integral equations on the domain. Equation (7.13.2) is in general integrated by parts so as to have the same order of differentiation for both functions u and w.
3. Introduce the approximations of the finite elements as
$$u(x) \approx \sum_i^n N_i(x) u_i\,, \quad w(x) \approx \sum_i^n N_i(x) w_i \tag{7.13.3}$$
that can be represented under matrix form
$$u(x) \approx \mathbf{Nu}\,, \quad w(x) \approx \mathbf{Nw} \tag{7.13.4}$$

where **N** is a matrix containing known functions of x, **u** is a vector of unknown parameters.

4. Use the approximations inside the weak formulation to obtain a system of linear algebraic equations with respect to **u**

$$\mathbf{K}\mathbf{u} = \mathbf{f} \tag{7.13.5}$$

where **K** and **f** can be calculated for each i if $N_i(x)$ is known.

The matrix **K** is in general called *rigidity* or *stiffness matrix* because of its origin in solid mechanics. Still in the domain of solid mechanics, **u** is the vector of nodal displacements and **f** the vector of nodal forces. The solution of this generally large size system is obtained by successive approximation (for example Gauss-Seidel). Frequently, the matrix **K** is symmetrical and Cholesky's method (Section 5.7) can be applied.

In this book, the finite element method will be briefly illustrated (Section 7.13.3.2) in the mechanical domain where it is most often treated (Hutton 2004; Khennane 2013; Koutromanos 2018; Li et al. 2018; Logan 2016; Loustau 2016; Råback et al. 2020; Rao 2018) but detailed in the heat transfer domain (Hutton 2004; Koutromanos 2018; Lewis et al. 2004; Li et al. 2018; Råback et al. 2020; Rao 2018) trying to emphasize the theoretical foundations which result in the final operating techniques. Many examples illustrating different situations will allow us to understand the operation of the method.

7.13.1 Step 1: Elements and Nodes

The studied physical domain is divided into subdomains called *finite elements*.

In dimension 1 (1D), the system that can be qualified as a bar or a rod contains two or several nodes (Figure 7.64). At left, the bar is represented by a single element with a node at each extremity. At right, the bar is divided into three successive elements each having two nodes. The elements are linear. If the bar contains one element with two nodes at the boundaries and intermediate nodes (Figure 7.65), the order of the element is larger (quadratic, cubic, . . .).

A node is a point of the finite element where the value of the field variable is calculated.

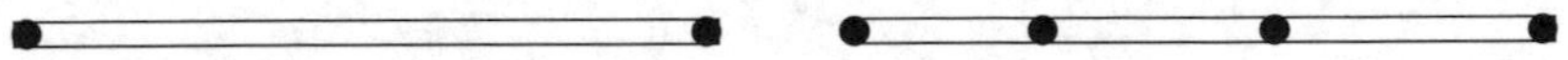

Fig. 7.64 Elements and nodes in a 1D case: one element at left, three successive elements at right

Fig. 7.65 Elements and nodes in a 1D case: one element with an internal node

In dimension 2 (2D), the elements in general are triangles or quadrilaterals, which nevertheless can be deformed to fit particular geometries. In Figure 7.66, at left, the finite element is triangular and linear with nodes at the summits. In Figure 7.66, at right, the same finite element is considered but with intermediate nodes on the edges and thus of larger order.

Fig. 7.66 Elements and nodes in a 2D case

In dimension 3 (3D), the elements in general are tetrahedrons or hexahedrons (Figure 7.67).

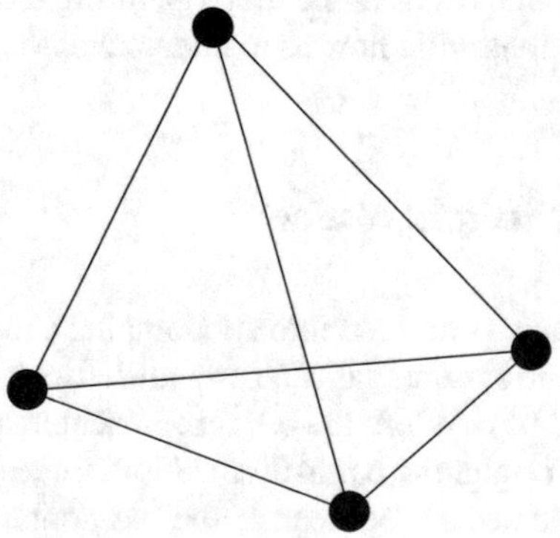

Fig. 7.67 Elements and nodes in a 3D case

A summary of some possibilities is given in Table 7.4 (Loustau 2016).

7.13.2 Step 2: Functions of Polynomial Interpolation

The interpolation functions are also called *basis or shape functions*. They are frequently polynomials, linear, quadratic, or cubic. If we consider a triangle with three nodes (Figure 7.66, left), all nodes are outside, and at each point inside the element, the field variable is expressed as

$$u(x, y) = N_1(x, y)u_1 + N_2(x, y)u_2 + N_3(x, y)u_3 \tag{7.13.6}$$

Table 7.4 Summary of some configurations of elements, nodes, and interpolation polynomials

Dimension	Geometry of the finite element	Interpolation polynomials	Degrees of freedom
1D	Segment with 2 nodes	Degree 1, Lagrange	2
1D	Segment with 2 nodes	Degree 3, Hermite	4
2D	Triangle with 3 nodes	Degree 1	3
2D	Triangle with 6 nodes	Degree 2	6
2D	Rectangle with 4 nodes	Degree 2, Lagrange	4

where u_1, u_2, u_3 are the values of the field variable at the nodes numbered from 1 to 3, while $N_1(x, y)$, N_2, N_3 are the *interpolation functions*. The values of the field variable at the nodes are constants to be determined.

The triangle element (Figure 7.66, left) has 3 degrees of freedom (Table 7.4) as 3 values at the nodes are necessary to describe the field variable in the whole element.

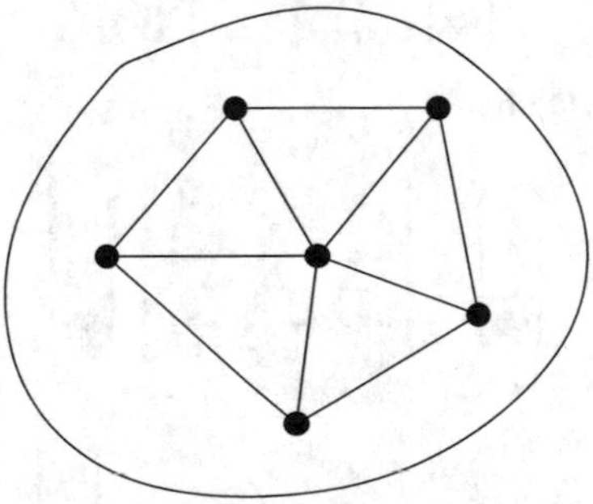

Fig. 7.68 Triangular elements and nodes for a 2D system

Figure 7.68 shows that the elements are connected between themselves by external nodes. In the finite element method, the continuity of the field variable is ensured at the nodes. Moreover, this continuity is also ensured at the frontiers between the elements. In this way, the physical continuity is guaranteed. This continuity does not lead to the continuity of the gradients.

For a finite element with two nodes placed at x_1 and x_2, the field variable is

$$u(x) = N_1(x)u_1 + N_2(x)u_2 = \frac{x - x_2}{x_1 - x_2}u_1 + \frac{x - x_1}{x_2 - x_1}u_2 \tag{7.13.7}$$

and the *shape functions* are thus linear (Figure 7.69) with $N_1(0) = 1$, $N_2(0) = 0$, $N_1(1) = 0$, $N_2(1) = 1$.

For a finite element of length L with three nodes and the inner node located at the middle of the element, a quadratic interpolation function can be defined

$$u(x) = a_0 + a_1 x + a_2 x^2 \tag{7.13.8}$$

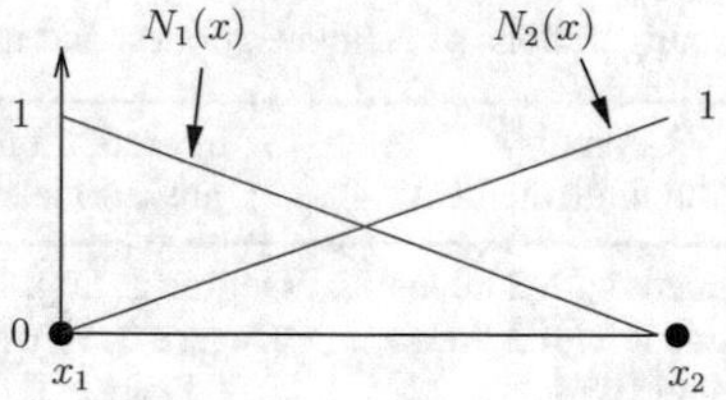

Fig. 7.69 Interpolation functions for a finite element with 2 nodes

with the conditions at the nodes (nodal conditions)

$$u(x=0)=u_1\,,\quad u(x=\frac{L}{2})=u_2\,,\quad u(x=L)=u_3 \tag{7.13.9}$$

thus

$$\begin{bmatrix} u_1\\ u_2\\ u_3 \end{bmatrix} = \begin{bmatrix} 1 & 0 & 0\\ 1 & \frac{L}{2} & \frac{L^2}{4}\\ 1 & L & L^2 \end{bmatrix} \begin{bmatrix} a_0\\ a_1\\ a_2 \end{bmatrix} \tag{7.13.10}$$

from which we get by inversion

$$\begin{bmatrix} a_0\\ a_1\\ a_2 \end{bmatrix} = \begin{bmatrix} 1 & 0 & 0\\ -\frac{3}{L} & \frac{4}{L} & -\frac{1}{L}\\ \frac{2}{L^2} & -\frac{4}{L^2} & \frac{2}{L^2} \end{bmatrix} \begin{bmatrix} u_1\\ u_2\\ u_3 \end{bmatrix} \tag{7.13.11}$$

and thus

$$\begin{aligned} u(x) &= \left[N_1(x)\; N_2(x)\; N_3(x)\right] \begin{bmatrix} u_1\\ u_2\\ u_3 \end{bmatrix}\\ &= \left[1\; x\; x^2\right] \begin{bmatrix} a_0\\ a_1\\ a_2 \end{bmatrix}\\ &\left[1\; x\; x^2\right] \begin{bmatrix} 1 & 0 & 0\\ -\frac{3}{L} & \frac{4}{L} & -\frac{1}{L}\\ \frac{2}{L^2} & -\frac{4}{L^2} & \frac{2}{L^2} \end{bmatrix} \begin{bmatrix} u_1\\ u_2\\ u_3 \end{bmatrix} \end{aligned} \tag{7.13.12}$$

The expressions of the quadratic interpolation polynomials result

$$\begin{aligned} N_1(x) &= 1 - \frac{3}{L}x + \frac{2}{L^2}x^2\\ N_2(x) &= \frac{4x}{L}\left(1-\frac{x}{L}\right)\\ N_3(x) &= \frac{x}{L}\left(\frac{2x}{L}-1\right) \end{aligned} \tag{7.13.13}$$

A similar technique can determine the interpolation polynomials in the case of a larger number of nodes for a given finite element. However, it can be noticed that

$$\begin{aligned}
N_1(0) &= 1 \quad , N_2(0) = 0 \quad , N_3(0) = 0 \\
N_1(\tfrac{L}{2}) &= 0 \quad , N_2(\tfrac{L}{2}) = 1 \quad , N_3(\tfrac{L}{2}) = 0 \\
N_1(L) &= 0 \quad , N_2(L) = 0 \quad , N_3(L) = 1
\end{aligned} \tag{7.13.14}$$

This remark can be generalized for two or more inner points irregularly placed in the finite element.

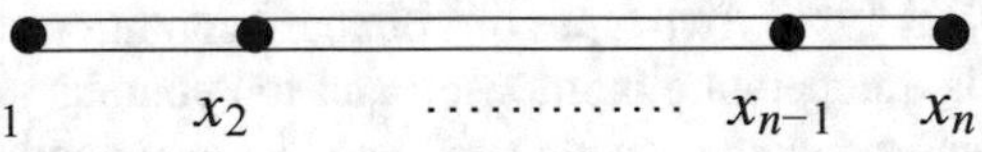

Fig. 7.70 Finite element with n nodes

Considering the general case of a finite element with n successive nodes located at $\{x_1, x_2, \ldots, x_n\}$ (Figure 7.70), the interpolation polynomials are determined as

$$N_k(x) = \prod_{\substack{j=1 \\ j \neq k}}^{n} \frac{x - x_j}{x_k - x_j} \tag{7.13.15}$$

Such polynomials are Lagrange polynomials (Section 1.3.3) verifying Kronecker property $N_k(x_k) = \delta_{kk} = 1, N_k(x_j) = \delta_{kj} = 0 \quad \forall j \neq k$.

Moreover, the shape functions verify the unit value property

$$\sum_{i=1}^{n} N_i(x) = 1 \qquad \forall x \in [x_1, x_n] \tag{7.13.16}$$

for any x belonging to the considered finite element.

The interpolation functions must respect the continuity between intermediate nodes.

Other interpolation functions are possible (Hermite polynomials, splines, trigonometric series), however, in the codes, the most currently used interpolation functions are linear and quadratic polynomials.

7.13.3 Steps 3–4: Determination of the Conductance Matrices and Nodal Flux, Determination by Assembling of the Global Conductance Matrix and the Global Equivalent Nodal Flux Vector

The equations in heat transfer and linear elasticity are very close. The case of heat transfer will be examined in detail whereas the similarity between heat transfer and linear elasticity is shown in the following.

7.13.3.1 Heat Transfer

A thermal system will be solved by different techniques of finite elements. It must be recalled that the finite element method takes into account the steady-state problems, as the potential dynamic aspect will be treated by explicit or implicit time discretization. The fundamental equations are considered under a rate form, that is per time unit.

The main boundary conditions met in heat transfer are summarized in Table 7.5 (Incropera and DeWitt 1996). Two types of boundary conditions are present, Dirichlet conditions where the temperature is imposed and the boundary surface is then symbolized by Γ_D (*essential boundary condition*) and Neumann conditions where the flux is specified for which the boundary surface is symbolized by Γ_N (*natural boundary condition*). The essential boundary conditions (of Dirichlet type) are not part of the variational formulation and are present under the form of a separate equation whereas the natural boundary conditions (of Neumann type) are taken into account in the variational formulation (Koutromanos 2018).

The dimensions of the considered variables are the following: temperature T (K), position x (m), heat power (or rate of generation of heat) $\dot{q}$ (W), heat flux $\dot{q}''$ (W m^{-2}), thermal conductivity λ (W m^{-1} K^{-1}). In many books, λ is noted as k, but we preferred the notation λ to avoid the confusion with a possible index k. In the flux direction, the temperature gradient is negative, the heat being transferred in the direction of decreasing temperature.

The fundamental equation governing heat conduction in 1D (Koutromanos 2018) is

$$\frac{d}{dx}\left(\lambda A \frac{dT}{dx}\right) + S(x) = 0 \tag{7.13.17}$$

where $S(x)$ is the source term which is the rate of heat diffusion (or energy) added to the system per unit of length or time (units W m^{-1}) and coming from a source. S is positive if it is added to the system. The thermal conductivity λ is a function of the temperature and thus generally depends on x. The surface area A of the cross section can also depend on the position x. Equation (7.13.17) accompanied by the boundary conditions represents the strong formulation.

Frequently, Equation (7.13.17) is presented under a slightly different form

$$\rho C_p \frac{\partial T}{\partial t} = \frac{\partial}{\partial x}\left(\lambda \frac{\partial T}{\partial x}\right) + \dot{q}''' \tag{7.13.18}$$

which takes into account the dynamic aspect, but where especially the source term $\dot{q}'''$ is expressed in W m^{-3}.

7.13.3.2 Linear Elasticity

Let an homogeneous bar that can be considered as one-dimensional (Figure 7.71). The transverse displacements are ignored. Suppose that the bar is fixed at the extremity $x = 0$ and that a tensile force F is applied at the extremity $x = L$. The bar is subjected to a distributed load $b(x)$ on all its length. Let $u(x)$ be the axial displacement. Furthermore, Young modulus $E(x)$ and the cross section area $A(x)$ are assumed to be depending on the position.

Table 7.5 Main boundary conditions encountered in heat transfer (schematic transient representation)

Boundary condition	Illustration
Temperature imposed at the wall Dirichlet boundary condition (first kind condition) $T(x = 0, t) = T_0$	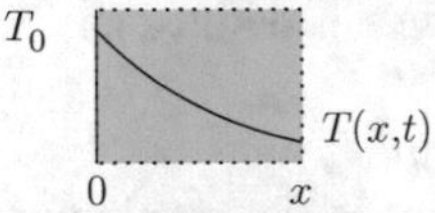
Insulated wall Neumann boundary condition (second kind condition) $\left.\frac{\partial T}{\partial x}\right\|_{x=0} = 0$	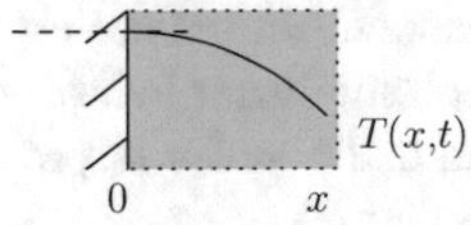
Finite flux imposed at the wall Neumann boundary condition (second kind condition) $-\lambda \left.\frac{\partial T}{\partial x}\right\|_{x=0} = \dot{q}''_s$	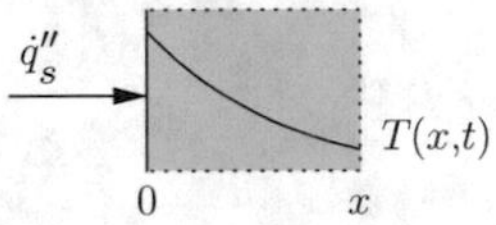
Convective flux at the wall Neumann boundary condition (third kind condition) $-\lambda \left.\frac{\partial T}{\partial x}\right\|_{x=0} = \dot{q}''$ with $\dot{q}'' = h[T_\infty - T(x = 0, t)]$	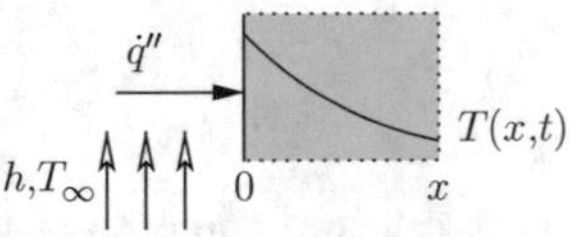
Radiative flux at the wall Neumann boundary condition (third kind condition) $-\lambda \left.\frac{\partial T}{\partial x}\right\|_{x=0} = \dot{q}''_{\text{rad}}$ with $\dot{q}''_{\text{rad}} = \sigma[\epsilon T_g^4 - \alpha T^4(x = 0, t)]$	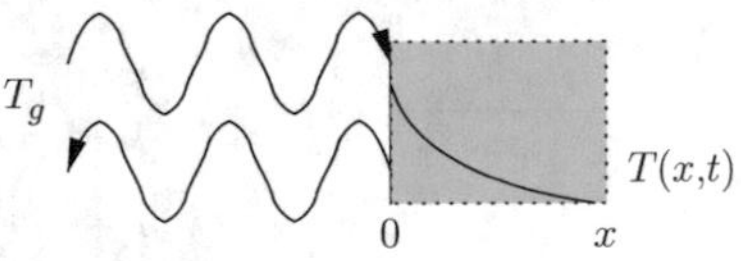

The following equations are written in a 1D framework and avoid the tensor notations necessary in 2D and 3D (Allaire and Alouges 2015; Ern and Guermond 2004; Reddy 2015).

Let $N(x)$ be the axial force and $\sigma(x)$ the axial stress at any point x. The axial force is equal to

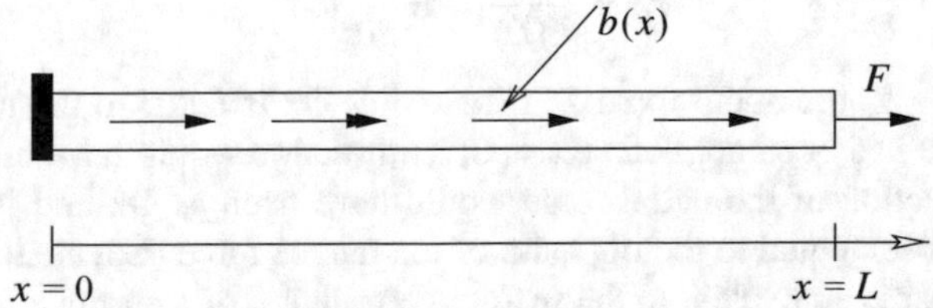

Fig. 7.71 One-dimensional bar, fixed at $x = 0$, subjected to a tensile force F at $x = L$ and a distributed load $b(x)$

$$N(x) = A(x)\sigma(x) \tag{7.13.19}$$

In 1D, the strain ϵ is defined as

$$\epsilon = \frac{du}{dx} \tag{7.13.20}$$

The elastic limit is the maximum stress supportable by a material before the strain remains permanent. Above the elasticity limit, the plastic deformation occurs. Presently supposing an elastic linear behavior of the material, i.e. when the stress is lower than the elastic limit, the stress is proportional to the strain according to Hooke's law (stress-strain relation)

$$\sigma(x) = E(x)\epsilon(x) \tag{7.13.21}$$

hence the axial force

$$N(x) = E(x)A(x)\frac{du}{dx} \tag{7.13.22}$$

If A and E were constant, and the deformation also constant, we would have

$$F = EA\frac{u(L) - 0}{L} \tag{7.13.23}$$

The force equilibrium equation (Ferreira 2009; Koutromanos 2018) is

$$\frac{dN}{dx} + b(x) = 0 \tag{7.13.24}$$

thus

$$\frac{d}{dx}\left(E(x)A(x)\frac{du}{dx}\right) + b(x) = 0 \tag{7.13.25}$$

with the boundary conditions

$$\begin{aligned} \sigma_L &= \frac{F}{A(L)} = nE\left.\frac{du}{dx}\right|_{x=L} = E\left.\frac{du}{dx}\right|_{x=L} \\ u(x = 0) &= 0 \end{aligned} \tag{7.13.26}$$

n is the intensity of the outward unit normal vector $\vec{n}$ at a boundary, with $n = -1$ at $x = 0$ as $\vec{n}$ is oriented in opposite direction to Ox, and thus $n = 1$ at $x = L$. The tensile stress σ_L is

$$\sigma_L = \frac{F}{A(L)} = E\left.\frac{du}{dx}\right|_{x=L} \tag{7.13.27}$$

At the node $x = L$, the nodal force f_L oriented as Ox is equal to the tensile force. If the tensile force had been applied at $x = 0$, its direction would have been opposed to Ox but the direction of the nodal force would have been as Ox and the nodal force f_0 would have been equal to the opposite of the tensile force (Logan 2016). The ratio AE/L for the bar is equivalent to the stiffness constant k of a spring.

Formally, Equation (7.13.25) is similar to Equation (7.13.17) of heat transfer. The fundamental equation (7.13.25) in linear elasticity is thus the strong formulation. The boundary conditions are identified as essential boundary condition at $x = 0$ (Γ_D:

Dirichlet condition) and natural boundary condition at $x = L$ (Γ_N: Neumann condition) in the same manner as in heat transfer.

In a similar way to heat transfer, the weak formulation is written as

$$\int_0^L w(x)\left[\frac{d}{dx}\left(E(x)A(x)\frac{du}{dx}\right) + b(x)\right] dx = 0 \qquad \forall w(x) \text{ continuous} \tag{7.13.28}$$

After integration by parts, Equation (7.13.28) becomes

$$\left[w(x)E(x)A(x)\frac{du}{dx}\right]_0^L - \int_0^L \frac{dw}{dx}E(x)A(x)\frac{du}{dx}dx + \int_0^L w(x)b(x)dx = 0 \tag{7.13.29}$$

hence the weak formulation

$$\begin{aligned} & -\int_0^L \frac{dw}{dx}E(x)A(x)\frac{du}{dx}dx + \int_0^L w(x)b(x)dx + w(L)A(L)\sigma_L = 0 \quad \forall w(x) \\ & u(0) = 0 \end{aligned} \tag{7.13.30}$$

by using the influence of the essential boundary condition

$$w(0) = 0 \tag{7.13.31}$$

Equation (7.13.30) is formally similar to Equation (7.13.39), as confirmed by the correspondence of Table 7.6.

Table 7.6 Correspondence between the variables and parameters of mechanics and heat transfer

Mechanics	Heat transfer
Displacement u	Temperature T
Strain $\epsilon = \frac{du}{dx}$	Temperature gradient $\frac{dT}{dx}$
Young modulus E	Heat conductivity λ
Stress $n\,E\frac{du}{dx}$	Heat flux $-n\,\lambda\frac{dT}{dx}$
Axial force $N = EA\frac{du}{dx}$	Heat power $\lambda A\frac{dT}{dx}$
Internal force b	Heat source S

7.13.3.3 Heat Transfer (Following)

In the following, finite elements are mainly illustrated in the heat transfer domain by means of examples and theoretical developments.

Example 7.17 :
HT1-FEM-1D: 1D metal rod insulated on its periphery

We consider the 1D physical example of a cylindrical metal rod, treated in a one-dimensional way, of constant section, constant conductivity λ, insulated on its periphery and subjected to a heat flux $\dot{q}''_0$ at left (Neumann boundary condition) and a constant temperature T_L at right (Dirichlet boundary

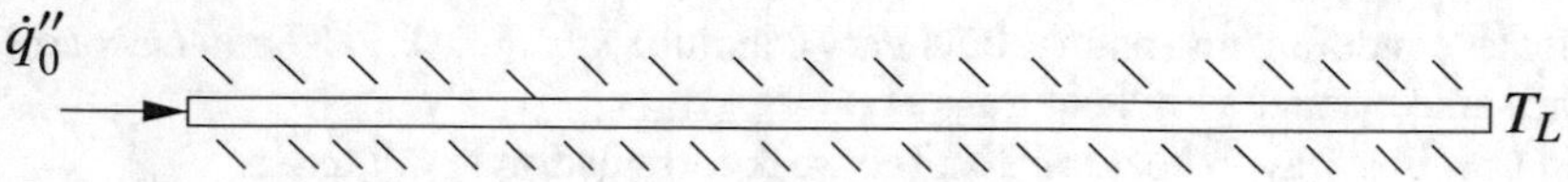

Fig. 7.72 Example HT1-FEM-1D: 1D insulated rod with imposed heat flux at left and imposed temperature at right

condition) according to Figure 7.72 (Hutton 2004). Assume that an internal power $\dot{q}'''_g$ per unit of volume (units W m^{-3}) is generated (for example, due to a reaction that can be nuclear) inside the rod. The steady-state model or fundamental equation of the rod can be written under the following strong formulation with the associated boundary conditions

$$\begin{aligned} &\lambda \frac{d^2T}{dx^2} + \dot{q}'''_g = 0 \\ &-\lambda \left.\frac{dT}{dx}\right|_{x=0} = \dot{q}''_0 \\ &T(x = L) = T_L \end{aligned} \tag{7.13.32}$$

where T is the function or field variable. In this case, the source term $S(x)$ is equal to $A\dot{q}'''_g$ and we find again the strong formulation (7.13.17).

Example 7.18 :
HT2-FEM-1D: 1D metal rod receiving a constant thermal power on its periphery

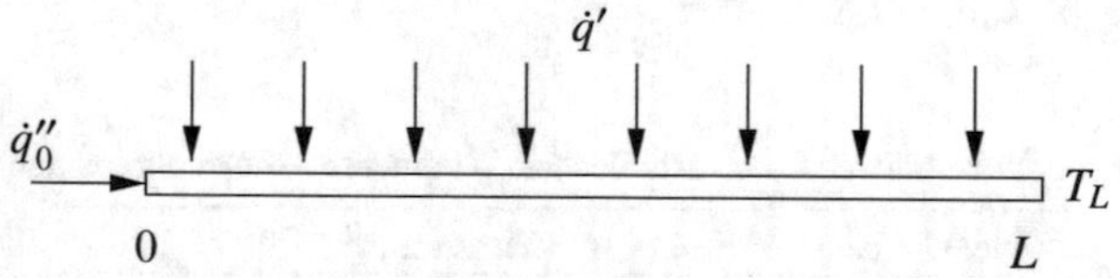

Fig. 7.73 Example HT2-FEM-1D: 1D rod with thermal power $\dot{q}'$ received laterally and imposed temperature at right

In this case, the rod is not insulated on its periphery but a thermal power per unit of length $\dot{q}'$ is applied on the side uniformly to equivalently simulate an internal heat generation. The same boundary conditions at the extremities are considered as in the case HT1-FEM-1D (Example 7.17), i.e. a heat flux at left (Neumann boundary condition) and a constant temperature at right (Dirichlet boundary condition) (Figure 7.73). The strong formulation of case HT2-FEM-1D (Example 7.18) is very close to that of case HT1-FEM-1D (Example 7.17) (Figure 7.72)

$$\begin{aligned} &A\lambda \frac{d^2T}{dx^2} + \dot{q}' = 0 \\ &-\lambda \left.\frac{dT}{dx}\right|_{x=0} = \dot{q}''_0 \\ &T(x = L) = T_L \end{aligned} \tag{7.13.33}$$

The source term $S(x)$ is equal to $\dot{q}'$ and Equation (7.13.33) can be written under the strong formulation (7.13.17).

7.13.3.4 Treatment of 1D Heat Transfer by Finite Elements

Rod represented by a single finite element having two nodes

A single finite element of extremities $x_1 = 0$ and $x_2 = L$ (Figure 7.74) is considered in this rod with a node located at each extremity. Inside the finite element, the approximation of the temperature distribution using linear interpolation functions is written as

$$\tilde{T}(x) = N_1(x)T_1 + N_2(x)T_2 \tag{7.13.34}$$

where T_1 and T_2 are the temperatures at nodes 1 and 2.

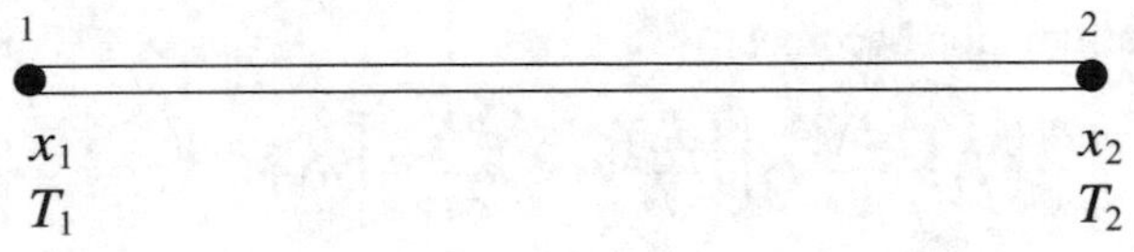

Fig. 7.74 One element with two nodes

The strong formulation (7.13.17) can be written according to the general relation

$$w(x)\left[\frac{d}{dx}\left(\lambda A \frac{dT}{dx}\right) + S(x)\right] = 0 \tag{7.13.35}$$

where $w(x)$ is any continuous function such that $w(x) = 0$ on an essential boundary Γ_D (Dirichlet condition). $w(x)$ is a weight function.

In the method of weighted residuals, Section 7.10.1, Equation (7.10.5), the volume integration of the fundamental equation multiplied by the interpolation function (shape function) is performed. Suppose a constant thermal conductivity λ. In the present case, as the geometry is 1D, the integration deals only with the variable x on the domain $[x_1, x_2]$. We thus obtain the following equations

$$\int_{x_1}^{x_2} w(x)\left[\lambda A \frac{d^2T}{dx^2} + S(x)\right] dx = 0 \tag{7.13.36}$$

Performing an integration by parts acting on the term d^2T/dx^2, we deduce

$$\lambda A \left[w(x)\frac{dT}{dx}\right]_{x_1}^{x_2} - \lambda A \int_{x_1}^{x_2} \frac{dw}{dx}\frac{dT}{dx} dx + \int_{x_1}^{x_2} w(x)S(x)dx = 0 \tag{7.13.37}$$

The Dirichlet condition at $x = x_2$ implies that $w(x_2) = 0$ thus

$$\left[w(x)\frac{dT}{dx}\right]_{x_1}^{x_2} = -w(x_1) \left.\frac{dT}{dx}\right|_{x=x_1} \tag{7.13.38}$$

with $-\lambda \left.\frac{dT}{dx}\right|_{x=x_1} = \dot{q}_0''$.

The weak formulation results, by adding the Dirichlet boundary condition at the essential boundary Γ_D at right

$$\begin{aligned} & -\lambda A \int_{x_1}^{x_2} \frac{dw}{dx}\frac{dT}{dx} dx + \int_{x_1}^{x_2} w(x) S(x) dx + A w(x_1) \dot{q}_0'' = 0 \\ & T(x = L) = T_L \end{aligned} \tag{7.13.39}$$

Galerkin's formulation

In Galerkin's method (Section 7.10.1), the weight functions are equal to the shape functions, thus

$$w(x) = \tilde{T}(x) = N_1(x)T_1 + N_2(x)T_2 \tag{7.13.40}$$

The weak formulation becomes

$$\begin{aligned} & -\lambda A \int_{x_1}^{x_2} \left[\frac{dN_1}{dx}T_1 + \frac{dN_2}{dx}T_2\right]\left[\frac{dN_1}{dx}T_1 + \frac{dN_2}{dx}T_2\right] dx + \\ & \int_{x_1}^{x_2} \left[\frac{dN_1}{dx}T_1 + \frac{dN_2}{dx}T_2\right] S(x) dx + A[N_1(x_1)T_1 + N_2(x_1)T_2]\, \dot{q}_0'' = 0 \\ & T(x = L) = N_1(L)T_1 + N_2(L)T_2 = T_L \end{aligned} \tag{7.13.41}$$

Note that, as it is a single element, $N_1(L) = 0$, $N_2(L) = 1$, thus the Dirichlet condition at $x = L$ is automatically verified due to the wise choice of the shape functions $N_1(x)$ and $N_2(x)$ (verifying Kronecker condition).

Indeed, from the general hypotheses on $w(x)$, this latter can be any function, thus the equation can be written separately for $w(x) = N_1(x)$ or $w(x) = N_2(x)$, which gives

$$\begin{aligned} & -\lambda A \int_{x_1}^{x_2} \frac{dN_i}{dx}T_i \left[\frac{dN_1}{dx}T_1 + \frac{dN_2}{dx}T_2\right] dx + \\ & \int_{x_1}^{x_2} \frac{dN_i}{dx}T_i S(x) dx + A N_i(x_1) T_i\, \dot{q}_0'' = 0\,, \quad i = 1, 2 \\ & T(x = L) = N_1(L)T_1 + N_2(L)T_2 = T_L \end{aligned} \tag{7.13.42}$$

Notice that the derivative of the approximation function of temperature can be written as

$$\frac{d\tilde{T}}{dx} = \frac{dN_1}{dx}T_1 + \frac{dN_2}{dx}T_2 = \begin{bmatrix} \frac{dN_1}{dx} & \frac{dN_2}{dx} \end{bmatrix} \begin{bmatrix} T_1 \\ T_2 \end{bmatrix} \tag{7.13.43}$$

using the expression of the linear shape functions (7.13.7), it results

$$\begin{bmatrix} \frac{dN_1}{dx} & \frac{dN_2}{dx} \end{bmatrix} = \begin{bmatrix} \frac{1}{x_1 - x_2} & \frac{1}{x_2 - x_1} \end{bmatrix} = \begin{bmatrix} -\frac{1}{L} & \frac{1}{L} \end{bmatrix} \tag{7.13.44}$$

thus

$$\frac{d\tilde{T}}{dx} = \begin{bmatrix} -\frac{1}{L} & \frac{1}{L} \end{bmatrix} \begin{bmatrix} T_1 \\ T_2 \end{bmatrix} = \mathbf{B}^{(e)}\mathbf{T}^{(e)} \tag{7.13.45}$$

Rod represented by n finite elements

We consider that the rod is represented by n_e finite elements noted $^{(e)}$ (Figure 7.75), each of them having the extremities $x_1^{(e)}$ and $x_2^{(e)}$. Each element (Figure 7.76) thus constitutes a subdomain of the complete domain $[0, L]$.

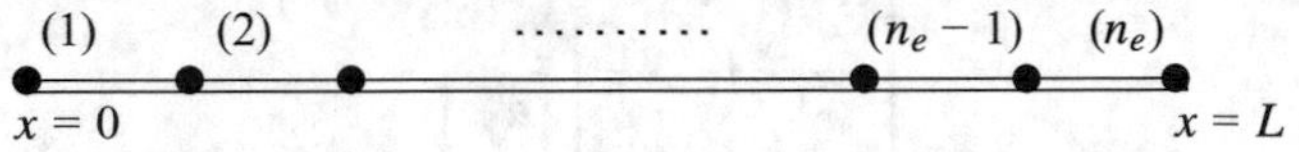

Fig. 7.75 Complete domain with $n_e + 1$ nodes

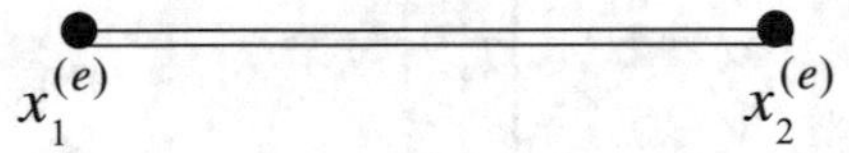

Fig. 7.76 One subdomain of the complete domain with 2 nodes

On each finite element $[x_1^{(e)}, x_n^{(e)}]$, the temperature is approximated as

$$T(x) \approx T^{(e)}(x) = \sum_{j=1}^{n} N_j^{(e)}(x) T_j^{(e)} = \mathbf{N}^{(e)}\,\mathbf{T}^{(e)} \tag{7.13.46}$$

where $\mathbf{N}^{(e)}$ is a row vector, while $\mathbf{T}^{(e)}$ is the column vector of temperatures at the nodes of the considered element.

The weight or approximation function $w(x)$ is written in the same way with the same shape functions

$$w(x) \approx w^{(e)} = \sum_{j=1}^{n} N_j^{(e)}(x) w_j^{(e)} = \mathbf{N}^{(e)}\,\mathbf{w}^{(e)} \tag{7.13.47}$$

In the case of a finite element with two nodes, Equation (7.13.45) can be generalized for any finite element $^{(e)}$

$$\frac{d\tilde{T}}{dx} = \begin{bmatrix} -\frac{1}{l^{(e)}} & \frac{1}{l^{(e)}} \end{bmatrix} \begin{bmatrix} T_1^{(e)} \\ T_2^{(e)} \end{bmatrix} = \mathbf{B}^{(e)}\mathbf{T}^{(e)} \tag{7.13.48}$$

In the case of a finite element with three nodes, where the intermediate node is placed anywhere between the two external nodes, we suppose that the approximation function is quadratic

$$\tilde{T} = a_0^{(e)} + a_1^{(e)} x + a_2^{(e)} x^2 \tag{7.13.49}$$

We deduce

$$\begin{bmatrix} T_1^{(e)} \\ T_2^{(e)} \\ T_3^{(e)} \end{bmatrix} = \begin{bmatrix} a_0^{(e)} + a_1^{(e)} x_1^{(e)} + a_2^{(e)} \left(x_1^{(e)}\right)^2 \\ a_0^{(e)} + a_1^{(e)} x_2^{(e)} + a_2^{(e)} \left(x_2^{(e)}\right)^2 \\ a_0^{(e)} + a_1^{(e)} x_3^{(e)} + a_2^{(e)} \left(x_3^{(e)}\right)^2 \end{bmatrix} = \begin{bmatrix} 1 \; x_1^{(e)} \; \left(x_1^{(e)}\right)^2 \\ 1 \; x_2^{(e)} \; \left(x_2^{(e)}\right)^2 \\ 1 \; x_3^{(e)} \; \left(x_3^{(e)}\right)^2 \end{bmatrix} \begin{bmatrix} a_0^{(e)} \\ a_1^{(e)} \\ a_2^{(e)} \end{bmatrix} = \mathbf{M^{(e)} a^{(e)}} \tag{7.13.50}$$

and reciprocally

$$\mathbf{a}^{(\mathbf{e})}\mathbf{M}^{(\mathbf{e})^{-1}} \begin{bmatrix} T_1^{(e)} \\ T_2^{(e)} \\ T_3^{(e)} \end{bmatrix} = \mathbf{M}^{(\mathbf{e})^{-1}} \mathbf{T}^{(e)} \tag{7.13.51}$$

with the notation

$$\mathbf{T}^{(e)} = \begin{bmatrix} T_1^{(e)} \\ T_2^{(e)} \\ T_3^{(e)} \end{bmatrix} \tag{7.13.52}$$

Noting $\mathbf{p}(x)$ the vector representing the monomials of the polynomial, we have

$$\mathbf{p}(x) = \left[1 \; x \; x^2\right] \tag{7.13.53}$$

hence the general expression of the vector of interpolation functions. In this case of a finite element with three nodes, at any point x of the finite element

$$\mathbf{N}^{(e)}(x) = \left[N_1^{(e)}(x) \; N_2^{(e)}(x) \; N_3^{(e)}(x)\right] = \mathbf{p}(x)\mathbf{M}^{(\mathbf{e})^{-1}} \mathbf{T}^{(e)} \tag{7.13.54}$$

Still in the case of this finite element with three nodes, we have

$$\left[\frac{d\mathbf{N}^{(e)}(x)}{dx}\right] = \left[\frac{dN_1^{(e)}(x)}{dx} \; \frac{dN_2^{(e)}(x)}{dx} \; \frac{dN_3^{(e)}(x)}{dx}\right] \tag{7.13.55}$$

with the shape functions which are Lagrange polynomials (with $n = 3$)

$$N_k(x) = \prod_{\substack{j=1 \\ j \neq k}}^{n} \frac{x - x_j}{x_k - x_j} \tag{7.13.56}$$

thus

$$\begin{aligned} \mathbf{B}^{(e)} &= \left[\frac{d\mathbf{N}^{(e)}(x)}{dx}\right] \\ &= \left[\frac{2x - x_2 - x_3}{(x_1 - x_2)(x_1 - x_3)} \; \frac{2x - x_1 - x_3}{(x_2 - x_1)(x_2 - x_3)} \; \frac{2x - x_1 - x_2}{(x_3 - x_1)(x_3 - x_2)}\right] \end{aligned} \tag{7.13.57}$$

While $\mathbf{B}^{(e)}$ was constant for a finite element with two nodes, this is no more true for a finite element with three nodes.

7.13.3.5 Application to the Entire Domain

We consider the weak formulation (7.13.37) that makes use of the same partial derivatives under the same first order only, but for any finite element

$$\left[w(x)\lambda A\frac{dT}{dx}\right]_{x_1^{(e)}}^{x_2^{(e)}} - \int_{x_1^{(e)}}^{x_2^{(e)}} \frac{dw}{dx}\lambda A\frac{dT}{dx}dx + \int_{x_1^{(e)}}^{x_2^{(e)}} w(x)S(x)dx = 0 \tag{7.13.58}$$

even assuming that λ and A depend on the position for more generality. The boundary conditions on the rod intervene at $x_1^{(1)} = 0$ (Neumann condition) and at $x_2^{(n_e)} = 0$ (Dirichlet condition). This equation is again written as

$$\int_{x_1^{(e)}}^{x_2^{(e)}} \frac{dw}{dx}\lambda A\frac{dT}{dx}dx = \int_{x_1^{(e)}}^{x_2^{(e)}} w(x)S(x)dx + \left[w(x)\lambda A\frac{dT}{dx}\right]_{x_1^{(e)}}^{x_2^{(e)}} \tag{7.13.59}$$

We make the sum on all the elements

$$\sum_{e=1}^{n_e}\int_{x_1^{(e)}}^{x_2^{(e)}} \frac{dw}{dx}\lambda A\frac{dT}{dx}dx = \sum_{e=1}^{n_e}\int_{x_1^{(e)}}^{x_2^{(e)}} w(x)S(x)dx + \sum_{e=1}^{n_e}\left[w(x)\lambda A\frac{dT}{dx}\right]_{x_1^{(e)}}^{x_2^{(e)}} \tag{7.13.60}$$

First consider the left-hand side of (7.13.60) and take any element of the sum

$$\int_{x_1^{(e)}}^{x_2^{(e)}} \frac{dw}{dx}\lambda A\frac{dT}{dx}dx \tag{7.13.61}$$

The approximation of the temperature on the finite element $^{(e)}$ is

$$T(x) \approx \tilde{T}^{(e)}(x) = \sum_{j=1}^{n} N_j^{(e)}(x)T_j^{(e)} = \mathbf{N}^{(e)}\mathbf{T}^{(e)} \tag{7.13.62}$$

with

$$\mathbf{N}^{(e)} = \left[N_1^{(e)}\ N_2^{(e)}\ \dots\ N_n^{(e)}\right], \quad \mathbf{T}^{(e)} = \begin{bmatrix} T_1^{(e)} \\ T_2^{(e)} \\ \vdots \\ T_n^{(e)} \end{bmatrix} \tag{7.13.63}$$

Similarly, for the temperature gradient

$$\frac{dT}{dx} \approx \frac{d\tilde{T}^{(e)}}{dx} = \sum_{j=1}^{n} \frac{dN_j^{(e)}(x)}{dx}T_j^{(e)} = \sum_{j=1}^{n} B_j^{(e)}(x)T_j^{(e)} = \mathbf{B}^{(e)}\mathbf{T}^{(e)} \tag{7.13.64}$$

with

$$\mathbf{B}^{(e)} = \left[\frac{dN_1^{(e)}(x)}{dx} \; \frac{dN_2^{(e)}(x)}{dx} \; \dots \; \frac{dN_n^{(e)}(x)}{dx}^{(e)} \right] \tag{7.13.65}$$

For the weight functions, we have

$$w(x) \approx \tilde{w}^{(e)}(x) = \sum_{j=1}^{n} N_j^{(e)} w_j^{(e)} = \mathbf{N}^{(e)} \mathbf{W}^{(e)} \tag{7.13.66}$$

and for the derivative of the weight functions

$$\frac{dw(x)}{dx} \approx \frac{d\tilde{w}^{(e)}(x)}{dx} = \sum_{j=1}^{n} \frac{dN_j^{(e)}}{dx} w_j^{(e)} = \mathbf{B}^{(e)} \mathbf{W}^{(e)} \tag{7.13.67}$$

with

$$\mathbf{W}^{(e)} = \begin{bmatrix} w_1^{(e)} \\ w_2^{(e)} \\ \vdots \\ w_n^{(e)} \end{bmatrix} \tag{7.13.68}$$

In the 1D case, we can also write

$$w^{(e)}(x) \approx \mathbf{W}^{(e)T} \mathbf{N}^{(e)T} \tag{7.13.69}$$

and

$$\frac{dw^{(e)}(x)}{dx} \approx \mathbf{W}^{(e)T} \mathbf{B}^{(e)T} \tag{7.13.70}$$

Each contribution to the left-hand side of Equation (7.13.60) can now be written

$$\begin{aligned}
\int_{x_1^{(e)}}^{x_2^{(e)}} \frac{dw}{dx} \lambda A \frac{dT}{dx} dx &\approx \int_{x_1^{(e)}}^{x_2^{(e)}} \frac{d\tilde{w}^{(e)}}{dx} \lambda^{(e)} A^{(e)} \frac{d\tilde{T}^{(e)}}{dx} dx \\
&= \int_{x_1^{(e)}}^{x_2^{(e)}} \mathbf{W}^{(e)T} \mathbf{B}^{(e)T} \lambda^{(e)} A^{(e)} \mathbf{B}^{(e)} \mathbf{T}^{(e)} dx \\
&= \mathbf{W}^{(e)T} \left(\int_{x_1^{(e)}}^{x_2^{(e)}} \mathbf{B}^{(e)T} \lambda^{(e)} A^{(e)} \mathbf{B}^{(e)} dx \right) \mathbf{T}^{(e)} \\
&= \mathbf{W}^{(e)T} \mathbf{K}^{(e)} \mathbf{T}^{(e)}
\end{aligned} \tag{7.13.71}$$

using the fact that $\mathbf{W}^{(e)}$ and $\mathbf{T}^{(e)}$ are constant. In heat transfer domain, $\mathbf{K}^{(e)}$ is called *conductance matrix* and is equal to

$$\mathbf{K}^{(e)} = \left[\int_{x_1^{(e)}}^{x_2^{(e)}} \mathbf{B}^{(e)T} \lambda^{(e)} A^{(e)} \mathbf{B}^{(e)} dx \right] \tag{7.13.72}$$

The conductance matrix is symmetrical, but it is also singular, that is non invertible (Koutromanos 2018). In solid mechanics, there exists a matrix totally equivalent called stiffness matrix (Ferreira 2009; Koutromanos 2018; Pozrikidis 2014; Rao 2018).

To avoid the singularity problems, constraints on temperature need to be specified.

Now, consider the first term of the right-hand side of Equation (7.13.60)

$$\begin{aligned}\sum_{e=1}^{n_e}\int_{x_1^{(e)}}^{x_2^{(e)}} w(x)S(x)dx &\approx \sum_{e=1}^{n_e}\int_{x_1^{(e)}}^{x_2^{(e)}} \tilde{w}^{(e)}(x)S(x)dx\\ &= \sum_{e=1}^{n_e}\int_{x_1^{(e)}}^{x_2^{(e)}} \mathbf{W}^{(e)^T}\mathbf{N}^{(e)^T}S(x)dx\\ &= \sum_{e=1}^{n_e}\left(\mathbf{W}^{(e)^T}\int_{x_1^{(e)}}^{x_2^{(e)}}\mathbf{N}^{(e)^T}S(x)dx\right)\end{aligned} \tag{7.13.73}$$

Finally, consider the second term of the right-hand side of Equation (7.13.60)

$$\begin{aligned}\sum_{e=1}^{n_e}\left[w(x)\lambda A\frac{dT}{dx}\right]_{x_1^{(e)}}^{x_2^{(e)}} \approx\; &\cancel{\left[\tilde{w}(x)\lambda(x)A(x)\frac{dT}{dx}\right]_{x_2^{(1)}}} - \left[\tilde{w}(x)\lambda(x)A(x)\frac{dT}{dx}\right]_{x_1^{(1)}}\\ &+ \cancel{\left[\tilde{w}(x)\lambda(x)A(x)\frac{dT}{dx}\right]_{x_2^{(2)}}} - \cancel{\left[\tilde{w}(x)\lambda(x)A(x)\frac{dT}{dx}\right]_{x_1^{(2)}}}\\ &\vdots\\ &+ \cancel{\left[\tilde{w}(x)\lambda(x)A(x)\frac{dT}{dx}\right]_{x_2^{(ne-1)}}} - \cancel{\left[\tilde{w}(x)\lambda(x)A(x)\frac{dT}{dx}\right]_{x_1^{(ne-1)}}}\\ &+ \left[\tilde{w}(x)\lambda(x)A(x)\frac{dT}{dx}\right]_{x_2^{(ne)}} - \cancel{\left[\tilde{w}(x)\lambda(x)A(x)\frac{dT}{dx}\right]_{x_1^{(ne)}}}\\ =\; &\left[\tilde{w}(x)\lambda(x)A(x)\frac{dT}{dx}\right]_{x_2^{(ne)}} - \left[\tilde{w}(x)\lambda(x)A(x)\frac{dT}{dx}\right]_{x_1^{(1)}}\end{aligned} \tag{7.13.74}$$

In Example 7.18 of the rod (Figure 7.73), the previous equation yields

$$\begin{aligned}\sum_{e=1}^{n_e}\left[w(x)\lambda A\frac{dT}{dx}\right]_{x_1^{(e)}}^{x_2^{(e)}} &\approx \left[\tilde{w}(x)\lambda(x)A(x)\frac{dT}{dx}\right]_L - \left[\tilde{w}(x)\lambda(x)A(x)\frac{dT}{dx}\right]_0\\ &= \tilde{w}(0)A(0)\dot{q}_0''\\ &= [\tilde{w}(x)A(x)\dot{q}'']_{\Gamma_N}\end{aligned} \tag{7.13.75}$$

by using $x_1^{(i)} = x_2^{(i-1)}$ (continuity of the finite elements, $2 \le i \le n_e$) and $w(L) = 0$ (essential boundary condition at Γ_D). Γ_N symbolizes the natural boundaries (flux or Neumann condition). Equation (7.13.75) can be modified as

$$\begin{aligned}[\tilde{w}(x)A(x)q'']_{\Gamma_N} &= \sum_{e=1}^{n_e}[\tilde{w}(x)A(x)\dot{q}'']_{\Gamma_N}\\ &= \sum_{e=1}^{n_e}\left(\mathbf{W}^{(e)^T}\left[\mathbf{N}^{(e)^T}A(x)\dot{q}''\right]_{\Gamma_N}\right)\end{aligned} \tag{7.13.76}$$

This writing allows us to obtain the different terms of Equation (7.13.60) under a similar form.

By using the previous developments, each contribution on the finite element $^{(e)}$ of the right-hand side Equation (7.13.60) can be expressed as

$$\begin{aligned}
&\int_{x_1^{(e)}}^{x_2^{(e)}} \tilde{w}(x)S(x)dx + [\tilde{w}(x)A(x)\dot{q}'']_{\Gamma_N} \\
&= \mathbf{W}^{(e)^T} \int_{x_1^{(e)}}^{x_2^{(e)}} \mathbf{N}^{(e)^T} S(x)dx + \mathbf{W}^{(e)^T} \left[\mathbf{N}^{(e)^T} A(x)\dot{q}''\right]_{\Gamma_N} \\
&= \mathbf{W}^{(e)^T} \left\{\int_{x_1^{(e)}}^{x_2^{(e)}} \mathbf{N}^{(e)^T} S(x)dx + \left[\mathbf{N}^{(e)^T} A(x)\dot{q}''\right]_{\Gamma_N}\right\} \\
&= \mathbf{W}^{(e)^T} \mathbf{q}^{(e)}
\end{aligned} \tag{7.13.77}$$

by setting

$$\mathbf{q}^{(e)} = \int_{x_1^{(e)}}^{x_2^{(e)}} \mathbf{N}^{(e)^T} S(x)dx + \left[\mathbf{N}^{(e)^T} A(x)q''\right]_{\Gamma_N} \tag{7.13.78}$$

In heat transfer, $\mathbf{q}^{(e)}$ is the *equivalent nodal flux vector* of the finite element $^{(e)}$. In elasticity, we get $\mathbf{f}^{(e)}$ the *equivalent nodal force vector* of the finite element $^{(e)}$.

Then, we gather the previous results (7.13.71), (7.13.77) for the whole set of finite elements according to the so-called *assembling* technique for matrices (or vectors)

$$\sum_{e=1}^{n_e} \left(\mathbf{W}^{(e)^T} \mathbf{K}^{(e)} \mathbf{T}^{(e)}\right) = \sum_{e=1}^{n_e} \left(\mathbf{W}^{(e)^T} \mathbf{q}^{(e)}\right) \tag{7.13.79}$$

The global temperature vector for the set of finite elements can be noted

$$\mathbf{T} = \begin{bmatrix} T_1^{(1)} \\ T_1^{(2)} \\ \vdots \\ T_1^{(n_e)} \\ T_2^{(n_e)} \end{bmatrix} \tag{7.13.80}$$

which gathers all the temperatures at the nodes by using the continuity $T_1^{(i)} = T_2^{(i-1)}$ for $(2 \leq i \leq n_e)$. Then, we can set

$$\mathbf{T}^{(e)} = \mathbf{L}^{(e)} \mathbf{T} \tag{7.13.81}$$

where $\mathbf{L}^{(e)}$ is simply a matrix (formed by 0 and 1) which expresses the way according to which the vector $\mathbf{T}^{(e)}$ for an element depends on the global vector $\mathbf{T}$.

Similarly, it can be written

$$\mathbf{W}^{(e)} = \mathbf{L}^{(e)} \mathbf{W} \quad \text{hence} \quad \mathbf{W}^{(e)^T} = \mathbf{W}^T \mathbf{L}^{(e)^T} \tag{7.13.82}$$

Equation (7.13.79) becomes

$$\sum_{e=1}^{n_e}\left(\mathbf{W}^T\mathbf{L}^{(e)^T}\mathbf{K}^{(e)}\mathbf{L}^{(e)}\mathbf{T}\right)=\sum_{e=1}^{n_e}\left(\mathbf{W}^T\mathbf{L}^{(e)^T}\mathbf{q}^{(e)}\right) \tag{7.13.83}$$

As $\mathbf{W}^T$ and $\mathbf{T}$ do not depend on the summation aspect, it can be written

$$\mathbf{W}^T\left\{\sum_{e=1}^{n_e}\left(\mathbf{L}^{(e)^T}\mathbf{K}^{(e)}\mathbf{L}^{(e)}\right)\right\}\mathbf{T}=\mathbf{W}^T\left\{\sum_{e=1}^{n_e}\left(\mathbf{L}^{(e)^T}\mathbf{q}^{(e)}\right)\right\} \tag{7.13.84}$$

or

$$\mathbf{W}^T\left[\left\{\sum_{e=1}^{n_e}\left(\mathbf{L}^{(e)^T}\mathbf{K}^{(e)}\mathbf{L}^{(e)}\right)\right\}\mathbf{T}-\sum_{e=1}^{n_e}\left(\mathbf{L}^{(e)^T}\mathbf{q}^{(e)}\right)\right]=0 \tag{7.13.85}$$

and finally

$$\mathbf{W}^T\left(\mathbf{KT}-\mathbf{q}\right)=0 \qquad \forall\,\mathbf{W} \tag{7.13.86}$$

In heat transfer, $\mathbf{K}$ is the *global (or structural) conductance matrix* equal to

$$\mathbf{K}=\sum_{e=1}^{n_e}\left(\mathbf{L}^{(e)^T}\mathbf{K}^{(e)}\mathbf{L}^{(e)}\right) \tag{7.13.87}$$

$\mathbf{q}$ is the *global equivalent nodal flux vector* equal to

$$\mathbf{q}=\sum_{e=1}^{n_e}\left(\mathbf{L}^{(e)^T}\mathbf{q}^{(e)}\right) \tag{7.13.88}$$

In solid mechanics, $\mathbf{K}$ is the *global stiffness matrix*. In the same way as $\mathbf{K}^{(e)}$ was symmetrical and singular, the matrix $\mathbf{K}^{(e)}$ is symmetrical and singular. To solve the problem, constraints on temperature must be used.

Case of a fixed temperature at one node *m* (essential boundary)

In this case, the temperature T_m is no more an unknown and it should be removed from the vector $\mathbf{T}$. Indeed, this implies that the corresponding weight function W_m does not exist any more at that node $W_m = 0$. We thus suppress the equation of rank m from the global equations

$$W_m\left[\sum_{j=1}^{n_e}K_{mj}T_j-q_m\right]=0\,,\quad \text{with } W_m=0 \tag{7.13.89}$$

thus the term between brackets does not need to be zero so that this equation be satisfied and we set

$$\sum_{j=1}^{n_e}K_{mj}T_j-q_m=r \tag{7.13.90}$$

where r is a residual, a flux term which allows us to impose the temperature at node m. The same concepts exist in solid mechanics for the essential boundaries. The number of equations to be solved is then decreased by one unit.

Example 7.19 :
Obtaining the matrices $\mathbf{L}^{(e)}$.

Let a system have 5 nodes and 4 elements of 2 nodes each (similar to Figure 7.75). The temperature vector at the nodes is

$$\mathbf{T} = \begin{bmatrix} T_1 & T_2 & T_3 & T_4 & T_5 \end{bmatrix}^T \tag{7.13.91}$$

Consider the second finite element $^{(2)}$ whose extremities are $x_1^{(2)}$ and $x_2^{(2)}$ and the corresponding temperatures T_2 and T_3. Thus we get

$$\mathbf{T}^{(2)} = \begin{bmatrix} T_2 \\ T_3 \end{bmatrix} = \begin{bmatrix} 0 & 1 & 0 & 0 & 0 \\ 0 & 0 & 1 & 0 & 0 \end{bmatrix} \begin{bmatrix} T_1 \\ T_2 \\ T_3 \\ T_4 \\ T_5 \end{bmatrix} = \mathbf{L}^{(2)}\mathbf{T} \tag{7.13.92}$$

hence

$$\mathbf{L}^{(2)} = \begin{bmatrix} 0 & 1 & 0 & 0 & 0 \\ 0 & 0 & 1 & 0 & 0 \end{bmatrix} \tag{7.13.93}$$

The number of rows of matrix $\mathbf{L}^{(e)}$ is equal to the number of nodes of element $^{(e)}$ and the number of columns is equal to the total number of nodes.

Examples of application of the finite elements in heat transfer

The following examples are completely developed in the following aim of highlighting various issues occurring during solving a heat transfer problem by means of finite elements:

- HT3-FEM-1D: Study of a 1D metal rod receiving a constant thermal power on its periphery (source term), an imposed flux at $x = 0$ (Neumann boundary condition), providing a convective flux at $x = L$ (Neumann boundary condition). Expression of the weak formulation. Finite elements with 2 nodes. Linear shape functions. Assembling. Solution.
- HT4-FEM-1D: Similar to HT3-FEM-1D, but insulated rod on the periphery (Neumann boundary condition).
- HT5-FEM-1D: Insulated rod with an imposed flux at $x = 0$ (Neumann boundary condition) and an imposed temperature at $x = L$ (Dirichlet boundary condition). Consequences related to that latter condition.
- HT6-FEM-1D: Similar to HT5-FEM-1D but finite elements with 3 nodes. Quadratic shape functions. Consequences on the shape functions and their derivatives.
- HT7-FEM-1D: Rod receiving an imposed flux at $x = 0$, exchanging a convective flux on the periphery (Neumann boundary condition depending on the position) and a convective flux at $x = L$ (Neumann boundary condition). Influence on the global conductance matrix and the global equivalent nodal flux vector.
- HT8-FEM-1D: Transient regime of the rod considered in HT3-FEM-1D case.

Example 7.20 :
HT3-FEM-1D: 1D metal rod receiving a constant thermal power on its periphery, an imposed flux at $x = 0$, providing a convective flux at $x = L$.

We consider Example HT3-FEM-1D close to example HT2-FEM-1D (Example 7.18) but with a convective flux at $x = L$ (Figure 7.77).

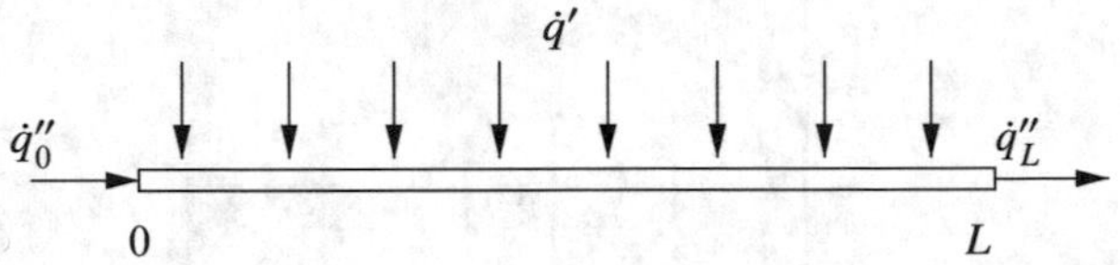

Fig. 7.77 Example HT3-FEM-1D: 1D rod with thermal power $\dot{q}'$ received laterally

The transversal influence is neglected and the rod is modelled as one-dimensional following its axis.

The rod can be modelled by general Equation (7.13.60) of the weak formulation

$$\sum_{e=1}^{n_e} \int_{x_1^{(e)}}^{x_2^{(e)}} \frac{dw}{dx} \lambda A \frac{dT}{dx} dx = \sum_{e=1}^{n_e} \int_{x_1^{(e)}}^{x_2^{(e)}} w(x) S(x) dx + \sum_{e=1}^{n_e} \left[w(x) \lambda A \frac{dT}{dx} \right]_{x_1^{(e)}}^{x_2^{(e)}} \tag{7.13.94}$$

To simplify, it is assumed that the cross section of the rod and the steel conductivity are constant. The source term is $S(x) = \dot{q}'$ (thermal power per length unit). At left, a Neumann condition with imposed constant flux $\dot{q}_0''$ prevails and at right a Neumann condition with convective flux $\dot{q}_L'' = h(T_L - T_\infty)$.

Equation (7.13.94) can be simplified as

$$\lambda A \sum_{e=1}^{n_e} \int_{x_1^{(e)}}^{x_2^{(e)}} \frac{dw}{dx} \frac{dT}{dx} dx = \sum_{e=1}^{n_e} \int_{x_1^{(e)}}^{x_2^{(e)}} w(x) q' dx + \lambda A \sum_{e=1}^{n_e} \left[w(x) \frac{dT}{dx} \right]_{x_1^{(e)}}^{x_2^{(e)}} \tag{7.13.95}$$

with the natural boundary conditions Γ_N

$$\begin{aligned} -\lambda \left. \frac{dT}{dx} \right|_{x_1^{(1)}} &= \dot{q}_0'' \\ -\lambda \left. \frac{dT}{dx} \right|_{x_n^{(ne)}} &= \dot{q}_L'' = h(T_L - T_\infty) \end{aligned} \tag{7.13.96}$$

supposing that each finite element $^{(e)}$ is limited by $x_1^{(e)}$ and $x_n^{(e)}$. In the following, finite elements with two nodes and $n_e = 3$ finite elements are considered, thus resulting in a total of 4 nodes.

The approximation of temperature is

$$T(x) \approx T^{(e)}(x) = N_1^{(e)}(x) T_1^{(e)} + N_2^{(e)}(x) T_2^{(e)} \tag{7.13.97}$$

The interpolation functions for finite elements with two nodes are linear, according to Equation (7.13.7), i.e.

$$N_1^{(e)}(x) = \frac{x - x_2^{(e)}}{x_1^{(e)} - x_2^{(e)}}, \quad N_2^{(e)}(x) = \frac{x - x_1^{(e)}}{x_2^{(e)} - x_1^{(e)}} \tag{7.13.98}$$

hence

$$\frac{dN_1^{(e)}(x)}{dx} = \frac{1}{x_1^{(e)} - x_2^{(e)}} = -\frac{1}{l^{(e)}}, \quad \frac{dN_2^{(e)}(x)}{dx} = \frac{1}{x_2^{(e)} - x_1^{(e)}} = \frac{1}{l^{(e)}} \tag{7.13.99}$$

The following matrices result (indeed, in the 1D case, row vectors)

$$\mathbf{N}^{(e)} = \begin{bmatrix} \frac{x - x_2^{(e)}}{x_1^{(e)} - x_2^{(e)}} & \frac{x - x_1^{(e)}}{x_2^{(e)} - x_1^{(e)}} \end{bmatrix}, \quad \mathbf{B}^{(e)} = \begin{bmatrix} -\frac{1}{l^{(e)}} & \frac{1}{l^{(e)}} \end{bmatrix} \tag{7.13.100}$$

Furthermore, the conductance matrix $\mathbf{K}^{(e)}$ is

$$\begin{aligned} \mathbf{K}^{(e)} &= \left[\int_{x_1^{(e)}}^{x_2^{(e)}} \mathbf{B}^{(e)T} \lambda^{(e)} A^{(e)} \mathbf{B}^{(e)} dx \right] \\ &= \left[\int_{x_1^{(e)}}^{x_2^{(e)}} \begin{bmatrix} -\frac{1}{l^{(e)}} \\ \frac{1}{l^{(e)}} \end{bmatrix} \lambda A \begin{bmatrix} -\frac{1}{l^{(e)}} & \frac{1}{l^{(e)}} \end{bmatrix} dx \right] \\ &= \lambda A \left[\int_{x_1^{(e)}}^{x_2^{(e)}} \begin{bmatrix} (\frac{1}{l^{(e)}})^2 & -(\frac{1}{l^{(e)}})^2 \\ -(\frac{1}{l^{(e)}})^2 & (\frac{1}{l^{(e)}})^2 \end{bmatrix} dx \right] \\ &= \lambda A \begin{bmatrix} \frac{1}{l^{(e)}} & -\frac{1}{l^{(e)}} \\ -\frac{1}{l^{(e)}} & \frac{1}{l^{(e)}} \end{bmatrix} \end{aligned} \tag{7.13.101}$$

The matrices $\mathbf{L}^{(e)}$ are equal to

$$\left[\boldsymbol{L}^{(1)}\right] = \begin{bmatrix} 1 & 0 & 0 & 0 \\ 0 & 1 & 0 & 0 \end{bmatrix}, \quad \left[\boldsymbol{L}^{(2)}\right] = \begin{bmatrix} 0 & 1 & 0 & 0 \\ 0 & 0 & 1 & 0 \end{bmatrix}, \quad \left[\boldsymbol{L}^{(3)}\right] = \begin{bmatrix} 0 & 0 & 1 & 0 \\ 0 & 0 & 0 & 1 \end{bmatrix} \tag{7.13.102}$$

The global conductance matrix results

$$\begin{aligned} \mathbf{K} &= \sum_{e=1}^{n_e} \left(\mathbf{L}^{(e)T} \mathbf{K}^{(e)} \mathbf{L}^{(e)} \right) \\ &= \lambda A \left\{ \begin{bmatrix} \frac{1}{l^{(1)}} & -\frac{1}{l^{(1)}} & 0 & 0 \\ -\frac{1}{l^{(1)}} & \frac{1}{l^{(1)}} & 0 & 0 \\ 0 & 0 & 0 & 0 \\ 0 & 0 & 0 & 0 \end{bmatrix} + \begin{bmatrix} 0 & 0 & 0 & 0 \\ 0 & \frac{1}{l^{(2)}} & -\frac{1}{l^{(2)}} & 0 \\ 0 & -\frac{1}{l^{(2)}} & \frac{1}{l^{(2)}} & 0 \\ 0 & 0 & 0 & 0 \end{bmatrix} + \begin{bmatrix} 0 & 0 & 0 & 0 \\ 0 & 0 & 0 & 0 \\ 0 & 0 & \frac{1}{l^{(3)}} & -\frac{1}{l^{(3)}} \\ 0 & 0 & -\frac{1}{l^{(3)}} & \frac{1}{l^{(3)}} \end{bmatrix} \right\} \\ &= \lambda A \begin{bmatrix} \frac{1}{l^{(1)}} & -\frac{1}{l^{(1)}} & 0 & 0 \\ -\frac{1}{l^{(1)}} & \frac{1}{l^{(1)}} + \frac{1}{l^{(2)}} & -\frac{1}{l^{(2)}} & 0 \\ 0 & -\frac{1}{l^{(2)}} & \frac{1}{l^{(2)}} + \frac{1}{l^{(3)}} & -\frac{1}{l^{(3)}} \\ 0 & 0 & -\frac{1}{l^{(3)}} & \frac{1}{l^{(3)}} \end{bmatrix} \end{aligned} \tag{7.13.103}$$

The equivalent nodal flux vector is

$$\begin{aligned} \mathbf{q}^{(e)} &= \int_{x_1^{(e)}}^{x_2^{(e)}} \mathbf{N}^{(e)T} S(x) dx + \left[\mathbf{N}^{(e)T} A(x) \dot{q}'' \right]_{\Gamma_N} \\ &= \dot{q}' \int_{x_1^{(e)}}^{x_2^{(e)}} \mathbf{N}^{(e)T} dx + A \left[\mathbf{N}^{(e)T} \dot{q}'' \right]_{\Gamma_N} \\ &= \dot{q}' \begin{bmatrix} \frac{1}{2}(x_2^{(e)} - x_1^{(e)}) \\ \frac{1}{2}(x_2^{(e)} - x_1^{(e)}) \end{bmatrix} + A \left[\begin{bmatrix} \frac{x - x_2^{(e)}}{x_1^{(e)} - x_2^{(e)}} \\ \frac{x - x_1^{(e)}}{x_2^{(e)} - x_1^{(e)}} \end{bmatrix} \dot{q}'' \right]_{\Gamma_N} \end{aligned} \tag{7.13.104}$$

yielding

$$\left[q^{(1)}\right] = \frac{\dot{q}'}{2}\begin{bmatrix} x_2^{(1)} - x_1^{(1)} \\ x_2^{(1)} - x_1^{(1)} \end{bmatrix} + A\begin{bmatrix} \dfrac{x_1^{(1)} - x_2^{(1)}}{x_1^{(1)} - x_2^{(1)}} \\ \dfrac{x_1^{(1)} - x_1^{(1)}}{x_2^{(1)} - x_1^{(1)}} \end{bmatrix} \dot{q}_0'' = \frac{\dot{q}'}{2}\begin{bmatrix} x_2^{(1)} \\ x_2^{(1)} \end{bmatrix} + A\begin{bmatrix} 1 \\ 0 \end{bmatrix} \dot{q}_0''$$

$$\left[q^{(2)}\right] = \frac{\dot{q}'}{2}\begin{bmatrix} x_2^{(2)} - x_1^{(2)} \\ x_2^{(2)} - x_1^{(2)} \end{bmatrix} \tag{7.13.105}$$

$$\left[q^{(3)}\right] = \frac{\dot{q}'}{2}\begin{bmatrix} x_2^{(3)} - x_1^{(3)} \\ x_2^{(3)} - x_1^{(3)} \end{bmatrix} - A\begin{bmatrix} \dfrac{x_2^{(3)} - x_2^{(3)}}{x_1^{(3)} - x_2^{(3)}} \\ \dfrac{x_2^{(3)} - x_1^{(3)}}{x_2^{(3)} - x_1^{(3)}} \end{bmatrix} \dot{q}_L'' = \frac{\dot{q}'}{2}\begin{bmatrix} L - x_1^{(3)} \\ L - x_1^{(3)} \end{bmatrix} - A\begin{bmatrix} 0 \\ 1 \end{bmatrix} \dot{q}_L''$$

The global equivalent nodal flux vector $\mathbf{q}$ is equal to

$$\begin{aligned} \mathbf{q} &= \sum_{e=1}^{n_e} \left(\mathbf{L}^{(e)T}\mathbf{q}^{(e)}\right) \\ &= \frac{\dot{q}'}{2}\left\{\begin{bmatrix} x_2^{(1)} \\ x_2^{(1)} \\ 0 \\ 0 \end{bmatrix} + \begin{bmatrix} 0 \\ x_2^{(2)} - x_1^{(2)} \\ x_2^{(2)} - x_1^{(2)} \\ 0 \end{bmatrix} + \begin{bmatrix} 0 \\ 0 \\ L - x_1^{(3)} \\ L - x_1^{(3)} \end{bmatrix}\right\} + A\begin{bmatrix} 1 \\ 0 \\ 0 \\ 0 \end{bmatrix}\dot{q}_0'' - A\begin{bmatrix} 0 \\ 0 \\ 0 \\ 1 \end{bmatrix}\dot{q}_L'' \\ &= \frac{\dot{q}'}{2}\begin{bmatrix} x_2^{(1)} \\ x_2^{(2)} \\ L - x_1^{(2)} \\ L - x_1^{(3)} \end{bmatrix} + A\begin{bmatrix} 1 \\ 0 \\ 0 \\ 0 \end{bmatrix}\dot{q}_0'' - A\begin{bmatrix} 0 \\ 0 \\ 0 \\ 1 \end{bmatrix}\dot{q}_L'' \end{aligned} \tag{7.13.106}$$

using $x_1^{(1)} = 0$, $x_1^{(2)} = x_2^{(1)}$, $x_1^{(3)} = x_2^{(3)}$, $x_2^{(3)} = L$.

From Equation (7.13.86)

$$\mathbf{W}^T \left\{\mathbf{KT} - \mathbf{q}\right\} = 0 \qquad \forall\,\mathbf{W} \tag{7.13.107}$$

true whatever $\mathbf{W}$, it results that

$$\mathbf{KT} - \mathbf{q} = 0 \tag{7.13.108}$$

The equivalent of this relation in solid mechanics is called *equilibrium relation.*

Numerical application:

Suppose $L = 1$ m, the nodes placed at 0, 0.1, 0.3, 1, transfer surface at extremities equal to $A = 0.01\,\text{m}^2$, conductivity $\lambda = 26\,\text{W}\,\text{m}^{-1}\,\text{K}^{-1}$, flux $\dot{q}_0'' = 120\,\text{W}\,\text{m}^{-2}$, heat transfer coefficient $h = 50\,\text{W}\,\text{m}^{-2}\,\text{K}^{-1}$, thermal power $\dot{q}' = 10\,\text{W}\,\text{m}^{-1}$, environment temperature $T_\infty = 293\,\text{K}$.

It is necessary to modify the global conductance matrix (7.13.103) to take into account the boundary condition of convective flux at $x_4 = L$, thus $\dot{q}_L'' = h(T_4 - T_\infty)$. For that purpose, it suffices to explain the term $\dot{q}_L''$ in (7.13.106) and to transfer the term including the unknown temperature T_4 in the part $\mathbf{KT}$ so that Equation (7.13.108) becomes

$$\left\{\lambda A \begin{bmatrix} \frac{1}{l^{(1)}} & -\frac{1}{l^{(1)}} & 0 & 0 \\ -\frac{1}{l^{(1)}} & \frac{1}{l^{(1)}}+\frac{1}{l^{(2)}} & -\frac{1}{l^{(2)}} & 0 \\ 0 & -\frac{1}{l^{(2)}} & \frac{1}{l^{(2)}}+\frac{1}{l^{(3)}} & -\frac{1}{l^{(3)}} \\ 0 & 0 & -\frac{1}{l^{(3)}} & \frac{1}{l^{(3)}} \end{bmatrix} + \begin{bmatrix} 0&0&0&0\\0&0&0&0\\0&0&0&0\\0&0&0&Ah \end{bmatrix}\right\} \begin{bmatrix} T_1\\T_2\\T_3\\T_4 \end{bmatrix} = \frac{\dot{q}'}{2}\begin{bmatrix} x_2^{(1)} \\ x_2^{(2)} \\ L-x_1^{(2)} \\ L-x_1^{(3)} \end{bmatrix} + \begin{bmatrix} A\dot{q}_0'' \\ 0 \\ 0 \\ AhT_\infty \end{bmatrix} \tag{7.13.109}$$

With $l^{(1)} = 0.1$, $l^{(2)} = 0.2$, $l^{(3)} = 0.5$, numerically we get

$$\begin{aligned} \mathbf{K} &= \begin{bmatrix} 2.60 & -2.60 & 0 & 0 \\ -2.60 & 3.90 & -1.30 & 0 \\ 0 & -1.30 & 1.67 & -0.37 \\ 0 & 0 & -0.37 & 0.37 \end{bmatrix} + \begin{bmatrix} 0&0&0&0\\0&0&0&0\\0&0&0&0\\0&0&0&Ah \end{bmatrix} \\ &= \begin{bmatrix} 2.60 & -2.60 & 0 & 0 \\ -2.60 & 3.90 & -1.30 & 0 \\ 0 & -1.30 & 1.67 & -0.37 \\ 0 & 0 & -0.37 & 0.87 \end{bmatrix} \end{aligned} \tag{7.13.110}$$

This matrix is tridiagonal and, even if the number of nodes was very high, the solution of the linear problem (7.13.108) can be obtained directly according to algorithm of Section 5.11 without inverting matrix **K**. Other algorithms such as Cholesky direct method when matrix **K** is symmetrical, or Gauss-Seidel iterative method (Equation (5.10.19)) are frequently used, this latter in particular in nonlinear cases (Braess 2010; Malinen and Råback 2013; Råback et al. 2020; Ruokolainen et al. 2020).

The global equivalent nodal flux vector **q** is equal to

$$\mathbf{q} = \left[0.5 + A\dot{q}_0'' \ 1.5 \ 4.5 \ 3.5 + AhT_\infty\right]^T = \left[1.7 \ 1.5 \ 4.5 \ 150\right]^T \tag{7.13.111}$$

The solution is $\boldsymbol{T} = \left[339.25 \ 338.59 \ 336.13 \ 315.40\right]^T$ (Figure 7.78). The interpolation between the nodes is linear, according to the hypothesis on polynomials $N_j^{(e)}$.

Example 7.21 :

HT4-FEM-1D: 1D insulated metal rod, receiving an imposed flux at $x = 0$, yielding a convective flux at $x = L$.

This example is the same as Example 7.20 except that it is assumed that the thermal power $\dot{q}' = 0$, which amounts to consider that the rod is insulated along its length except at extremities. Thus, there is a specified flux $\dot{q}_0''$ at $x = 0$ and a convective flux at $x = L$ (Figure 7.79).

The theoretical calculations are identical to those of Example 7.20. Finally, it suffices to substitute $\dot{q}'$ by zero. The global conductance matrix is identical to that of Example 7.20.

The global equivalent nodal flux vector is modified with respect to Example 7.20 by imposing $\dot{q}' = 0$ in (7.13.109) hence

$$\mathbf{q} = \begin{bmatrix} 1.2 \\ 0 \\ 0 \\ 146.5 \end{bmatrix} \tag{7.13.112}$$

The nodal temperatures are $\boldsymbol{T} = \left[300.01 \ 299.55 \ 298.63 \ 295.40\right]^T$. In this case, the temperature profile is linear (Figure 7.80).

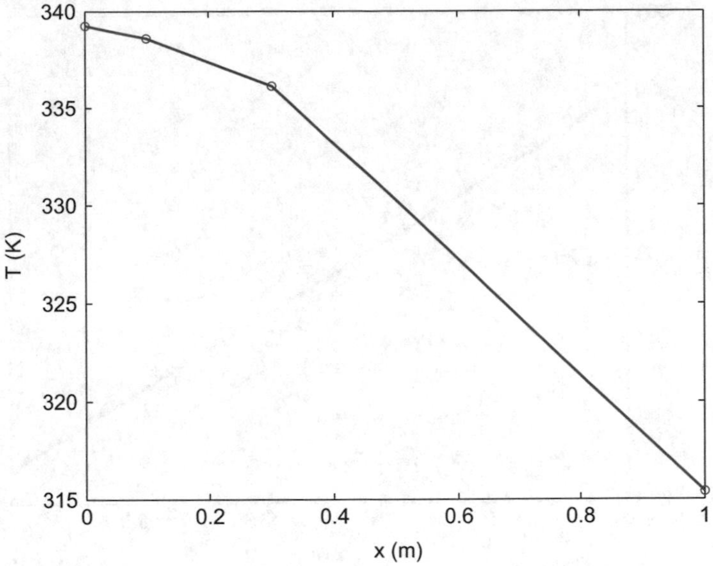

Fig. 7.78 Example HT3-FEM-1D: Temperature profile

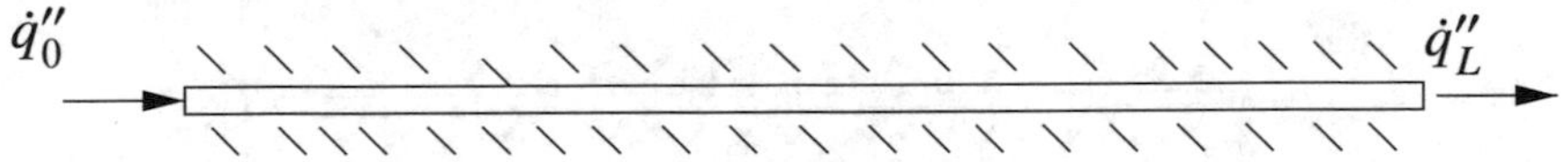

Fig. 7.79 Example HT4-FEM-1D: 1D insulated rod, receiving an imposed flux at $x = 0$, yielding a convective flux at $x = L$

Example 7.22 :

HT5-FEM-1D: 1D insulated metal rod, receiving an imposed flux at $x = 0$, with imposed temperature at $x = L$.

The same nodes and physical properties as in Example 7.20 are taken. At $x = 0$, the Neumann condition reigns with $\dot{q}''_0$ and at $x = L$ a Dirichlet condition with imposed $T_L = 293$ K. The temperature T_4 is no more an unknown (Figure 7.81).

The global conductance matrix is unchanged and given by (7.13.103). The global equivalent nodal flux vector is modified according to (7.13.106) and equal to

$$\mathbf{q} = A \begin{bmatrix} 1 \\ 0 \\ 0 \\ 0 \end{bmatrix} \dot{q}''_0 \tag{7.13.113}$$

If Equation (7.13.86) remains valid, it means that, by developing and transferring T_4 in the known right-hand side, it can be written

$$\begin{bmatrix} K_{11} & K_{12} & K_{13} \\ K_{21} & K_{22} & K_{23} \\ K_{31} & K_{32} & K_{33} \\ K_{41} & K_{42} & K_{43} \end{bmatrix} \begin{bmatrix} T_1 \\ T_2 \\ T_3 \end{bmatrix} = \begin{bmatrix} q_1 - K_{14}T_4 \\ q_2 - K_{24}T_4 \\ q_3 - K_{34}T_4 \\ q_4 - K_{44}T_4 \end{bmatrix} \tag{7.13.114}$$

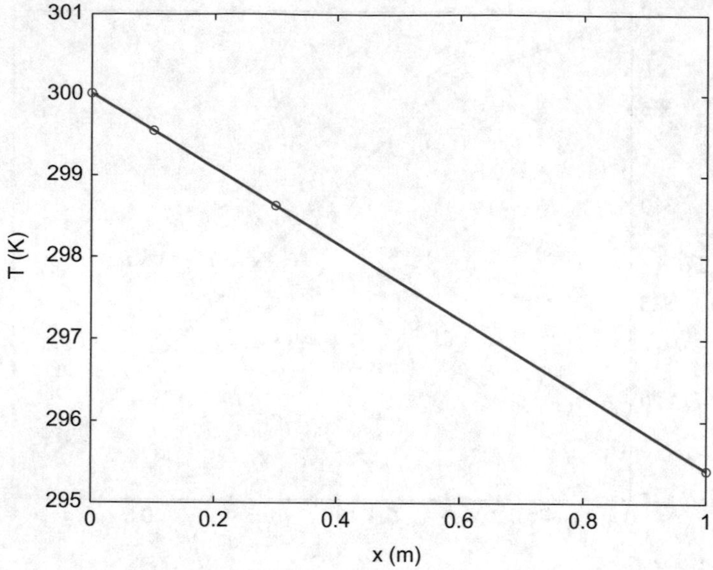

Fig. 7.80 Example HT4-FEM-1D: Temperature profile

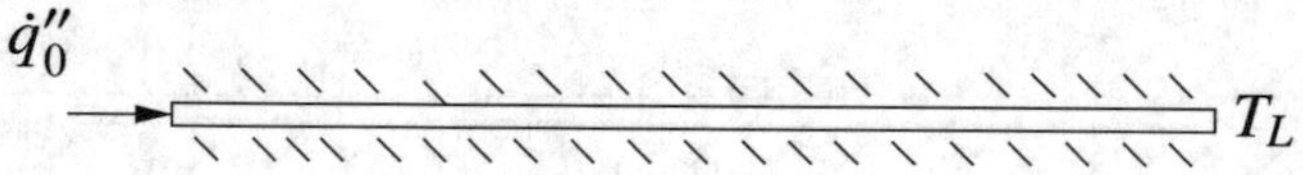

Fig. 7.81 Example HT5-FEM-1D: 1D insulated rod, receiving an imposed flux at $x = 0$, with imposed temperature at $x = L$

which poses a problem, as there are more equations than unknowns. That problem is solved by considering equations (7.13.89)–(7.13.90).

Equation (7.13.108) originally of the form

$$\mathbf{KT} - \mathbf{q} = 0 \tag{7.13.115}$$

is written under the modified form

$$\mathbf{K} \begin{bmatrix} T_1 \\ T_2 \\ T_3 \\ T_4 \end{bmatrix} - \mathbf{q} = \begin{bmatrix} 0 \\ 0 \\ 0 \\ r_4 \end{bmatrix} \tag{7.13.116}$$

to signify that r_4 takes into account the fact that T_4 is known.

Numerically, that gives

$$\begin{bmatrix} 2.60 & -2.60 & 0 & 0 \\ -2.60 & 3.90 & -1.30 & 0 \\ 0 & -1.30 & 1.67 & -0.37 \\ 0 & 0 & -0.37 & 0.37 \end{bmatrix} \begin{bmatrix} T_1 \\ T_2 \\ T_3 \\ 293 \end{bmatrix} - \begin{bmatrix} 1.2 \\ 0 \\ 0 \\ 0 \end{bmatrix} = \begin{bmatrix} 0 \\ 0 \\ 0 \\ r_4 \end{bmatrix} \tag{7.13.117}$$

thus

$$\left[\begin{array}{ccc|c} 2.60 & -2.60 & 0 & 0 \\ -2.60 & 3.90 & -1.30 & 0 \\ 0 & -1.30 & 1.67 & -0.37 \\ \hline 0 & 0 & -0.37 & 0.37 \end{array}\right] \left[\begin{array}{c} T_1 \\ T_2 \\ T_3 \\ \hline 293 \end{array}\right] - \left[\begin{array}{c} 1.2 \\ 0 \\ 0 \\ \hline 0 \end{array}\right] = \left[\begin{array}{c} 0 \\ 0 \\ 0 \\ \hline r_4 \end{array}\right] \tag{7.13.118}$$

The three first equations give

$$\begin{bmatrix} 2.60 & -2.60 & 0 \\ -2.60 & 3.90 & -1.30 \\ 0 & -1.30 & 1.67 \end{bmatrix} \begin{bmatrix} T_1 \\ T_2 \\ T_3 \end{bmatrix} + \begin{bmatrix} 0 \\ 0 \\ -0.37 \end{bmatrix} 293 - \begin{bmatrix} 1.2 \\ 0 \\ 0 \end{bmatrix} = \begin{bmatrix} 0 \\ 0 \\ 0 \end{bmatrix} \tag{7.13.119}$$

that can be solved. We thus get $\mathbf{T} = \left[297.61\ 297.15\ 296.23\ 293.00\right]^T$. The profile is linear as in Example 7.22 (Figure 7.82). We could deduce $r_4 = -1.2$ which is a flux term.

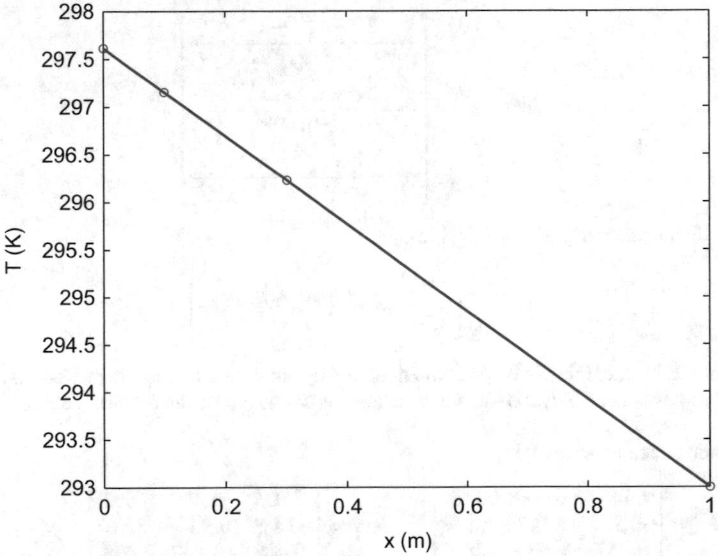

Fig. 7.82 Example HT5-FEM-1D: Temperature profile

Example 7.23 :

HT6-FEM-1D: 1D metal rod receiving a constant thermal power on its periphery, an imposed flux at $x = 0$, yielding a convective flux at $x = L$.

Example 7.20 is again considered by choosing finite elements with three nodes and quadratic Lagrange polynomials for the interpolation functions. The nodes are located at $\{0\ 0.05\ 0.1\ 0.2\ 0.3\ 0.65\ 1\}$ thus the rod is modelled by three finite elements.

The approximation of temperature is

$$\tilde{T}^{(e)}(x) = \sum_{j=1}^{n} N_j^{(e)}(x) T_j^{(e)} = \mathbf{N}^{(e)}(x)\mathbf{T}^{(e)} = \left[N_1^{(e)}\ N_2^{(e)}\ N_3^{(e)}\right] \begin{bmatrix} T_1^{(e)} \\ T_2^{(e)} \\ T_3^{(e)} \end{bmatrix} \tag{7.13.120}$$

with

$$[N^{(e)}(x)]^T = \begin{bmatrix} \dfrac{(x - x_2^{(e)})(x - x_3^{(e)})}{(x_1^{(e)} - x_2^{(e)})(x_1^{(e)} - x_3^{(e)})} \\ \dfrac{(x - x_1^{(e)})(x - x_3^{(e)})}{(x_2^{(e)} - x_1^{(e)})(x_2^{(e)} - x_3^{(e)})} \\ \dfrac{(x - x_1^{(e)})(x - x_2^{(e)})}{(x_3^{(e)} - x_1^{(e)})(x_3^{(e)} - x_2^{(e)})} \end{bmatrix} \tag{7.13.121}$$

Moreover

$$\frac{d\tilde{T}^{(e)}(x)}{dx} = \sum_{j=1}^{n} \frac{dN_j^{(e)}}{dx} T_j^{(e)} = \sum_{j=1}^{n} B_j^{(e)} T_j^{(e)} = \mathbf{B}^{(e)}(x)\mathbf{T}^{(e)} = \begin{bmatrix} B_1^{(e)} & B_2^{(e)} & B_3^{(e)} \end{bmatrix} \begin{bmatrix} T_1^{(e)} \\ T_2^{(e)} \\ T_3^{(e)} \end{bmatrix} \tag{7.13.122}$$

with

$$\mathbf{B}^{(e)}(x)^T = \begin{bmatrix} \dfrac{2x - x_2^{(e)} - x_3^{(e)}}{(x_1^{(e)} - x_2^{(e)})(x_1^{(e)} - x_3^{(e)})} \\ \dfrac{2x - x_1^{(e)} - x_3^{(e)}}{(x_2^{(e)} - x_1^{(e)})(x_2^{(e)} - x_3^{(e)})} \\ \dfrac{2x - x_1^{(e)} - x_2^{(e)}}{(x_3^{(e)} - x_1^{(e)})(x_3^{(e)} - x_2^{(e)})} \end{bmatrix} \tag{7.13.123}$$

The conductance matrix $\mathbf{K}^{(e)}$ is equal to

$$\mathbf{K}^{(e)} = \left[\int_{x_1^{(e)}}^{x_2^{(e)}} \mathbf{B}^{(e)T} \lambda^{(e)} A^{(e)} \mathbf{B}^{(e)} dx \right] \tag{7.13.124}$$

The calculations of integrals can be performed exactly by noticing that they are polynomial functions, except if A or k was not a polynomial, and by using Gauss–Legendre quadrature (Section 2.6.2) in all cases.

The conductance matrices are

$$\mathbf{K}^{(1)} = \begin{bmatrix} 6.067 & -6.933 & 0.867 \\ -6.933 & 13.867 & -6.933 \\ 0.867 & -6.933 & 6.067 \end{bmatrix}, \quad \mathbf{K}^{(2)} = \begin{bmatrix} 3.033 & -3.467 & 0.433 \\ -3.467 & 6.933 & -3.467 \\ 0.433 & -3.467 & 3.033 \end{bmatrix}$$
$$\mathbf{K}^{(3)} = \begin{bmatrix} 0.867 & -0.990 & 0.124 \\ -0.990 & 1.981 & -0.990 \\ 0.124 & -0.990 & 0.867 \end{bmatrix} \tag{7.13.125}$$

hence the global conductance matrix equals to

$$\mathbf{K} = \begin{bmatrix} 6.067 & -6.933 & 0.867 & 0 & 0 & 0 & 0 \\ -6.933 & 13.867 & -6.933 & 0 & 0 & 0 & 0 \\ 0.867 - 6.933 & 9.100 & -3.467 & 0.433 & 0 & 0 \\ 0 & 0 & -3.467 & 6.933 & -3.467 & 0 & 0 \\ 0 & 0 & 0.433 & -3.467 & 3.900 & -0.990 & 0.124 \\ 0 & 0 & 0 & 0 & -0.990 & 1.981 & -0.990 \\ 0 & 0 & 0 & 0 & 0.124 & -0.990 & 0.867 + Ah \end{bmatrix} \tag{7.13.126}$$

with $Ah = 0.500$.

The global equivalent nodal flux vector is equal to

$$\mathbf{q} = \begin{bmatrix} 0.167 + A\dot{q}_0'' \\ 0.667 \\ 0.500 \\ 1.333 \\ 1.500 \\ 4.667 \\ 1.167 + AhT_\infty \end{bmatrix} = \begin{bmatrix} 1.3667 \\ 0.6667 \\ 0.5000 \\ 1.3333 \\ 1.5000 \\ 4.6667 \\ 147.6667 \end{bmatrix} \tag{7.13.127}$$

The nodal temperatures are calculated

$$\mathbf{T} = \begin{bmatrix} 339.25 \\ 338.97 \\ 338.59 \\ 337.55 \\ 336.13 \\ 328.12 \\ 315.40 \end{bmatrix} \tag{7.13.128}$$

and it can be noticed that we find the same values at the less numerous nodes which had been considered in Example 7.20, with now intermediary values and moreover quadratic interpolation polynomials (Figure 7.83). Clearly, the quadratic interpolation gives a more realistic profile than Figure 7.78 that was obtained with the same number of finite elements, but two nodes per element hence a linear interpolation, instead of three nodes in the present case Example 7.23.

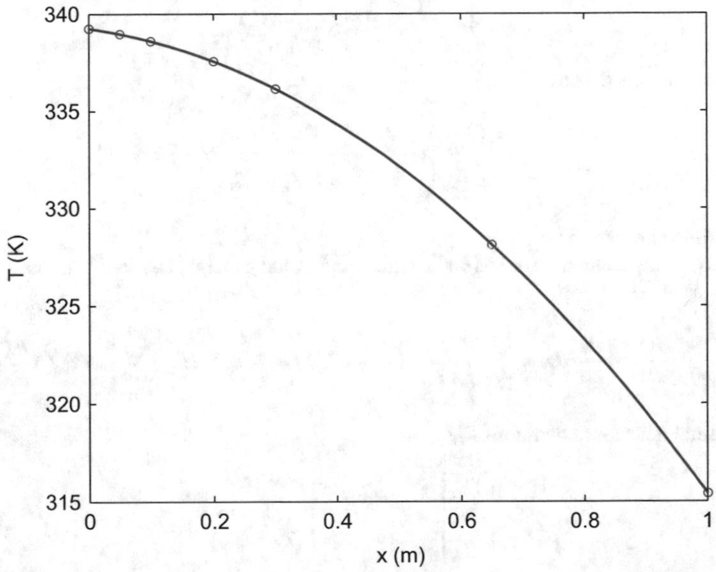

Fig. 7.83 Example HT6-FEM-1D: Temperature profile obtained with three nodes per element to compare with Figure 7.78 with two nodes per element

---

Example 7.24 :

HT7-FEM-1D: 1D metal rod receiving an imposed flux at $x = 0$, exchanging laterally a convective flux $\dot{q}_s''$ and yielding a convective flux at $x = L$.

Example HT7-FEM-1D represents a real situation often encountered for a 1D rod receiving an imposed flux at $x = 0$, exchanging laterally a convective flux $\dot{q}''_s$ and yielding a convective flux at $x = L$ (Figure 7.84).

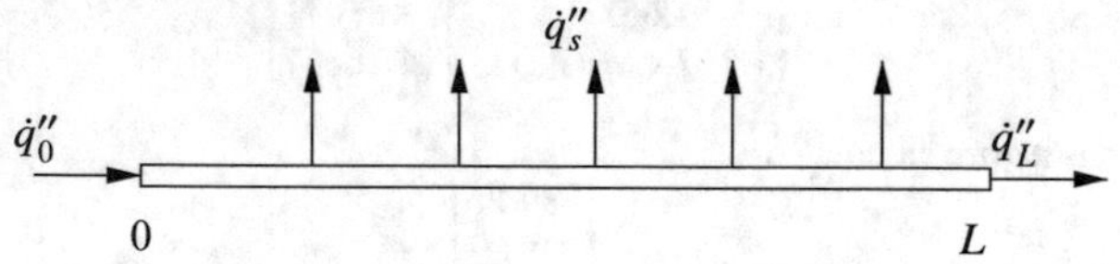

Fig. 7.84 Example HT7-FEM-1D: 1D rod with imposed flux received at $x = 0$, convective flux $\dot{q}''_s$ exchanged laterally and convective flux $\dot{q}''_s$ yielded at $x = L$

The lateral convective flux is equal to

$$\dot{q}''_s = h(T(x) - T_\infty) \tag{7.13.129}$$

Assuming the thermal conductivity λ and the surface area A of the cross section constant, the rod is modelled by the following one-dimensional physical model

$$A\lambda \frac{d^2T}{dx^2} = Ph(T(x) - T_\infty) \tag{7.13.130}$$

with the boundary conditions

$$\begin{aligned} -\lambda \left.\frac{dT}{dx}\right|_{x_0} &= \dot{q}''_0 \\ -\lambda \left.\frac{dT}{dx}\right|_{x_L} &= \dot{q}''_L = h(T_L - T_\infty) \end{aligned} \tag{7.13.131}$$

P is the perimeter of the rod.

In terms of finite elements, the rod can be modelled by the general equation (7.13.60) of the weak formulation as

$$\sum_{e=1}^{n_e} \int_{x_1^{(e)}}^{x_2^{(e)}} \frac{dw}{dx} \lambda A \frac{dT}{dx} dx = \sum_{e=1}^{n_e} \int_{x_1^{(e)}}^{x_2^{(e)}} w(x)[-Ph(T(x) - T_\infty)]dx + \sum_{e=1}^{n_e} \left[w(x)\lambda A \frac{dT}{dx} \right]_{x_1^{(e)}}^{x_2^{(e)}} \tag{7.13.132}$$

with the natural boundary conditions Γ_N

$$\begin{aligned} -\lambda \left.\frac{dT}{dx}\right|_{x_1^{(1)}} &= \dot{q}''_0 \\ -\lambda \left.\frac{dT}{dx}\right|_{x_n^{(ne)}} &= \dot{q}''_L = h(T_L - T_\infty) \end{aligned} \tag{7.13.133}$$

In expression (7.13.132), the dependency with respect to the position of the source term $S(x) = -Ph(T(x) - T_\infty)$ can be noticed.

To simplify the following presentation, we consider that the rod is discretized like in Example 7.20 by $n_e = 3$ finite elements with two nodes each, thus a total of four nodes.

$$T(x) \approx T^{(e)}(x) = N_1^{(e)}(x)T_1^{(e)} + N_2^{(e)}(x)T_2^{(e)} \tag{7.13.134}$$

The interpolation functions are linear and identical to those of Example 7.20. The matrices $\mathbf{N}^{(e)}$ and $\mathbf{B}^{(e)}$ are

$$\mathbf{N}^{(e)} = \left[\frac{x - x_2^{(e)}}{x_1^{(e)} - x_2^{(e)}} \quad \frac{x - x_1^{(e)}}{x_2^{(e)} - x_1^{(e)}} \right], \quad \mathbf{B}^{(e)} = \left[\frac{1}{x_1^{(e)} - x_2^{(e)}} \quad \frac{1}{x_2^{(e)} - x_1^{(e)}} \right] \tag{7.13.135}$$

Furthermore, the conductance matrix $\mathbf{K}^{(e)}$ is identical to that of Example 7.20

$$\begin{aligned}
\mathbf{K}^{(e)} &= \left[\int_{x_1^{(e)}}^{x_2^{(e)}} \mathbf{B}^{(e)^T} \lambda^{(e)} A^{(e)} \mathbf{B}^{(e)} dx \right] \\
&= \lambda A \begin{bmatrix} \frac{1}{x_2^{(e)} - x_1^{(e)}} & \frac{1}{x_1^{(e)} - x_2^{(e)}} \\ \frac{1}{x_1^{(e)} - x_2^{(e)}} & \frac{1}{x_2^{(e)} - x_1^{(e)}} \end{bmatrix} = \lambda A \begin{bmatrix} \frac{1}{l^{(e)}} & -\frac{1}{l^{(e)}} \\ -\frac{1}{l^{(e)}} & \frac{1}{l^{(e)}} \end{bmatrix}
\end{aligned} \tag{7.13.136}$$

The matrices $\mathbf{L}^{(e)}$ are identical to those of Example 7.20 as the discretization is unchanged.

The global conductance matrix is identical

$$\mathbf{K} = \sum_{e=1}^{n_e} \left(\mathbf{L}^{(e)^T} \mathbf{K}^{(e)} \mathbf{L}^{(e)} \right) = \lambda A \begin{bmatrix} \frac{1}{l^{(1)}} & -\frac{1}{l^{(1)}} & 0 & 0 \\ -\frac{1}{l^{(1)}} & \frac{1}{l^{(1)}} + \frac{1}{l^{(2)}} & -\frac{1}{l^{(2)}} & 0 \\ 0 & -\frac{1}{l^{(2)}} & \frac{1}{l^{(2)}} + \frac{1}{l^{(3)}} & -\frac{1}{l^{(3)}} \\ 0 & 0 & -\frac{1}{l^{(3)}} & \frac{1}{l^{(3)}} \end{bmatrix} \tag{7.13.137}$$

Considering the right-hand side of Equation (7.13.132)

$$-\sum_{e=1}^{n_e} \int_{x_1^{(e)}}^{x_2^{(e)}} w(x) Ph(T(x) - T_\infty) dx + \sum_{e=1}^{n_e} \left[w(x) \lambda A \frac{dT}{dx} \right]_{x_1^{(e)}}^{x_2^{(e)}} \tag{7.13.138}$$

with (noting $S(x) = -Ph(T(x) - T_\infty)$)

$$\begin{aligned}
&\int_{x_1^{(e)}}^{x_2^{(e)}} \tilde{w}(x) S(x) dx + [\tilde{w}(x) A(x) \dot{q}'']_{\Gamma_N} \\
&= \mathbf{W}^{(e)^T} \left\{ \int_{x_1^{(e)}}^{x_2^{(e)}} \mathbf{N}^{(e)^T} S(x) dx + \left[\mathbf{N}^{(e)^T} A \dot{q}'' \right]_{\Gamma_N} \right\} \\
&= \mathbf{W}^{(e)^T} \mathbf{q}^{(e)}
\end{aligned} \tag{7.13.139}$$

the equivalent nodal flux vector is expressed for a finite element

$$\begin{aligned}
\mathbf{q}^{(e)} &= -\int_{x_1^{(e)}}^{x_2^{(e)}} \mathbf{N}^{(e)^T} Ph(N_1^{(e)}(x) T_1^{(e)} + N_2^{(e)}(x) T_2^{(e)} - T_\infty) dx + \left[\mathbf{N}^{(e)^T} A q'' \right]_{\Gamma_N} \\
&= -Ph \int_{x_1^{(e)}}^{x_2^{(e)}} \mathbf{N}^{(e)^T} (\mathbf{N}^{(e)} \mathbf{T}^{(e)} - T_\infty) dx + A \left[\mathbf{N}^{(e)^T} q'' \right]_{\Gamma_N} \\
&= -Ph \left(\int_{x_1^{(e)}}^{x_2^{(e)}} \mathbf{N}^{(e)^T} \mathbf{N}^{(e)} dx \right) \mathbf{T}^{(e)} + Ph \left(\int_{x_1^{(e)}}^{x_2^{(e)}} \mathbf{N}^{(e)^T} dx \right) T_\infty \\
&\quad + A \left[\mathbf{N}^{(e)^T} q'' \right]_{\Gamma_N}
\end{aligned} \tag{7.13.140}$$

As $\mathbf{q}^{(e)}$ depends on $\mathbf{T}^{(e)}$, it is necessary to take into account that characteristics in order to add it to the existing matrix $\mathbf{K}$.

Knowing the interpolation polynomials, after analytical integration and simplification, we get

$$
\begin{aligned}
\mathbf{q}^{(e)} &= -Ph \begin{bmatrix} \frac{1}{3}(x_2^{(e)} - x_1^{(e)}) & \frac{1}{6}(x_2^{(e)} - x_1^{(e)}) \\ \frac{1}{6}(x_2^{(e)} - x_1^{(e)}) & \frac{1}{3}(x_2^{(e)} - x_1^{(e)}) \end{bmatrix} \mathbf{T}^{(e)} + Ph \begin{bmatrix} \frac{1}{2}(x_2^{(e)} - x_1^{(e)}) \\ \frac{1}{2}(x_2^{(e)} - x_1^{(e)}) \end{bmatrix} T_\infty \\
&\quad + A \left[\mathbf{N}^{(e)T} q'' \right]_{\Gamma_N} \\
&= -Ph \frac{1}{6}(x_2^{(e)} - x_1^{(e)}) \begin{bmatrix} 2T_1^{(e)} + T_2^{(e)} - 3T_\infty \\ T_1^{(e)} + 2T_2^{(e)} - 3T_\infty \end{bmatrix} + A \left[\mathbf{N}^{(e)T} q'' \right]_{\Gamma_N}
\end{aligned}
\tag{7.13.141}
$$

hence

$$
\begin{aligned}
\left[\boldsymbol{q}^{(1)}\right] &= -Ph \frac{1}{6}(x_2^{(1)} - x_1^{(1)}) \begin{bmatrix} 2T_1^{(1)} + T_2^{(1)} - 3T_\infty \\ T_1^{(1)} + 2T_2^{(1)} - 3T_\infty \end{bmatrix} + A \begin{bmatrix} 1 \\ 0 \end{bmatrix} \dot{q}_0'' \\
\left[\boldsymbol{q}^{(2)}\right] &= -Ph \frac{1}{6}(x_2^{(2)} - x_1^{(2)}) \begin{bmatrix} 2T_1^{(2)} + T_2^{(2)} - 3T_\infty \\ T_1^{(2)} + 2T_2^{(2)} - 3T_\infty \end{bmatrix} \\
\left[\boldsymbol{q}^{(3)}\right] &= -Ph \frac{1}{6}(x_2^{(3)} - x_1^{(3)}) \begin{bmatrix} 2T_1^{(3)} + T_2^{(3)} - 3T_\infty \\ T_1^{(3)} + 2T_2^{(3)} - 3T_\infty \end{bmatrix} - A \begin{bmatrix} 0 \\ 1 \end{bmatrix} \dot{q}_L''
\end{aligned}
\tag{7.13.142}
$$

The global equivalent nodal flux vector $\mathbf{q}$ is equal to

$$
\mathbf{q} = \sum_{e=1}^{n_e} \left(\mathbf{L}^{(e)T} \mathbf{q}^{(e)} \right)
\tag{7.13.143}
$$

A last difficulty resides in the expression of the flux $\dot{q}_L'' = h(T_L - T_\infty)$ which must be considered in the same way as in Example 7.20.

The global conductance matrix $\mathbf{K}$ becomes

$$
\mathbf{K} = \lambda A \begin{bmatrix} \frac{1}{l^{(1)}} & -\frac{1}{l^{(1)}} & 0 & 0 \\ -\frac{1}{l^{(1)}} & \frac{1}{l^{(1)}} + \frac{1}{l^{(2)}} & -\frac{1}{l^{(2)}} & 0 \\ 0 & -\frac{1}{l^{(2)}} & \frac{1}{l^{(2)}} + \frac{1}{l^{(3)}} & -\frac{1}{l^{(3)}} \\ 0 & 0 & -\frac{1}{l^{(3)}} & \frac{1}{l^{(3)}} \end{bmatrix} + A \begin{bmatrix} 0 & 0 & 0 & 0 \\ 0 & 0 & 0 & 0 \\ 0 & 0 & 0 & 0 \\ 0 & 0 & 0 & h \end{bmatrix}
\tag{7.13.144}
$$

The global equivalent nodal flux vector $\mathbf{q}$ is equal to

$$
\mathbf{q} = -Ph \frac{1}{6} \begin{bmatrix} l^{(1)}(2T_1^{(1)} + T_2^{(1)} - 3T_\infty) \\ l^{(1)}(T_1^{(1)} + 2T_2^{(1)} - 3T_\infty) + l^{(2)}(2T_1^{(2)} + T_2^{(2)} - 3T_\infty) \\ l^{(2)}(T_1^{(2)} + 2T_2^{(2)} - 3T_\infty) + l^{(3)}(2T_1^{(3)} + T_2^{(3)} - 3T_\infty) \\ l^{(3)}(T_1^{(3)} + 2T_2^{(3)} - 3T_\infty) \end{bmatrix} + \begin{bmatrix} A\dot{q}_0'' \\ 0 \\ 0 \\ AhT_\infty \end{bmatrix}
\tag{7.13.145}
$$

Indeed, in this writing, $\mathbf{q}$ depends on $\mathbf{T}$ because of the natural convection varying along the rod, thus a transformation is necessary to obtain an acceptable form

$$
\begin{aligned}
\mathbf{q} &= -Ph \frac{1}{6} \begin{bmatrix} 2l^{(1)} & l^{(1)} & 0 & 0 \\ l^{(1)} & 2l^{(1)} + 2l^{(2)} & l^{(2)} & 0 \\ 0 & l^{(2)} & 2l^{(2)} + 2l^{(3)} & l^{(3)} \\ 0 & 0 & l^{(3)} & 2l^{(3)} \end{bmatrix} \begin{bmatrix} T_1 \\ T_2 \\ T_3 \\ T_4 \end{bmatrix} \\
&\quad + Ph \frac{1}{2} \begin{bmatrix} l^{(1)} \\ l^{(1)} + l^{(2)} \\ l^{(2)} + l^{(3)} \\ l^{(3)} \end{bmatrix} T_\infty + \begin{bmatrix} A\dot{q}_0'' \\ 0 \\ 0 \\ AhT_\infty \end{bmatrix}
\end{aligned}
\tag{7.13.146}
$$

The first term of that equation must then be taken into account in the global conductance matrix which finally becomes

$$\mathbf{K} = \lambda A \begin{bmatrix} \frac{1}{l^{(1)}} & -\frac{1}{l^{(1)}} & 0 & 0 \\ -\frac{1}{l^{(1)}} & \frac{1}{l^{(1)}} + \frac{1}{l^{(2)}} & -\frac{1}{l^{(2)}} & 0 \\ 0 & -\frac{1}{l^{(2)}} & \frac{1}{l^{(2)}} + \frac{1}{l^{(3)}} & -\frac{1}{l^{(3)}} \\ 0 & 0 & -\frac{1}{l^{(3)}} & \frac{1}{l^{(3)}} \end{bmatrix} + Ph\frac{1}{6}\begin{bmatrix} 2l^{(1)} & l^{(1)} & 0 & 0 \\ l^{(1)} & 2l^{(1)} + 2l^{(2)} & l^{(2)} & 0 \\ 0 & l^{(2)} & 2l^{(2)} + 2l^{(3)} & l^{(3)} \\ 0 & 0 & l^{(3)} & 2l^{(3)} \end{bmatrix} + A\begin{bmatrix} 0 & 0 & 0 & 0 \\ 0 & 0 & 0 & 0 \\ 0 & 0 & 0 & 0 \\ 0 & 0 & 0 & h \end{bmatrix} \tag{7.13.147}$$

while the global equivalent nodal flux vector $\mathbf{q}$ is reduced to

$$\mathbf{q} = Ph\frac{1}{2}\begin{bmatrix} l^{(1)} \\ l^{(1)} + l^{(2)} \\ l^{(2)} + l^{(3)} \\ l^{(3)} \end{bmatrix} T_\infty + \begin{bmatrix} A\dot{q}_0'' \\ 0 \\ 0 \\ AhT_\infty \end{bmatrix} \tag{7.13.148}$$

It can be noticed in the global conductance matrix given by Equation (7.13.147) that the first term corresponds to the conduction in the rod, the second one to the convective flux along the rod, the third one to the convective flux at $x = L$. In the global equivalent nodal flux vector, the first term comes from the convective flux along the rod and the second term from the boundary conditions.

The system linear can now be solved

$$\mathbf{KT} = \mathbf{q} \tag{7.13.149}$$

to obtain the temperatures at the four nodes.

Numerical application:

Suppose $L = 1$ m, the nodes placed at 0, 0.1, 0.3, 1, the transfer surface at extremities equal to $A = 0.01$ m^2, the conductivity $\lambda = 26\,\text{W m}^{-1}\,\text{K}^{-1}$, the flux $\dot{q}_0'' = 120\,\text{W m}^{-2}$, the heat transfer coefficient $h = 50\,\text{W m}^{-2}\,\text{K}^{-1}$, the environment temperature $T_\infty = 293$ K. The perimeter is $P = 0.355$ m.

The global conductance matrix is equal to

$$\mathbf{K} = \begin{bmatrix} 2.60 & -2.60 & 0 & 0 \\ -2.60 & 3.90 & -1.30 & 0 \\ 0 & -1.30 & 1.671 & -0.371 \\ 0 & 0 & -0.371 & 0.371 \end{bmatrix} + \begin{bmatrix} 0.591 & 0.295 & 0 & 0 \\ 0.295 & 1.772 & 0.591 & 0 \\ 0 & 0.591 & 5.317 & 2.068 \\ 0 & 0 & 2.068 & 4.136 \end{bmatrix} + \begin{bmatrix} 0 & 0 & 0 & 0 \\ 0 & 0 & 0 & 0 \\ 0 & 0 & 0 & 0 \\ 0 & 0 & 0 & 0.50 \end{bmatrix} = \begin{bmatrix} 3.191 & -2.305 & 0 & 0 \\ -2.305 & 5.672 & -0.709 & 0 \\ 0 & -0.709 & 6.989 & 1.700 \\ 0 & 0 & 1.696 & 5.007 \end{bmatrix} \tag{7.13.150}$$

The global equivalent nodal flux vector is equal to

$$\mathbf{q} = \begin{bmatrix} 259.66 \\ 778.99 \\ 2336.98 \\ 1817.65 \end{bmatrix} + \begin{bmatrix} 1.20 \\ 0 \\ 0 \\ 0 \end{bmatrix} + \begin{bmatrix} 0 \\ 0 \\ 0 \\ 146.5 \end{bmatrix} = \begin{bmatrix} 260.86 \\ 778.99 \\ 2336.98 \\ 1964.15 \end{bmatrix} \tag{7.13.151}$$

By solving (7.13.149), the temperatures at nodes are obtained. They are given in Table 7.7.

The temperatures at $T = L$ are close to T_∞ in particular for $h = 50$, (Figure 7.85) as the heat loss per convection is very important with respect to the provided flux $\dot{q}_0''$. When h decreases, the deviation with respect to T_∞ increases as expected.

Table 7.7 Comparison of the nodal temperatures calculated for different heat transfer coefficients h

h	$\mathbf{T}$ (at $x = 0, 0.1, 0.3, 1$)
50	$[293.54\ 293.22\ 293.02\ 292.99]^T$
5	$[294.72\ 294.31\ 293.75\ 293.15]^T$
1	$[297.63\ 297.20\ 296.51\ 295.50]^T$

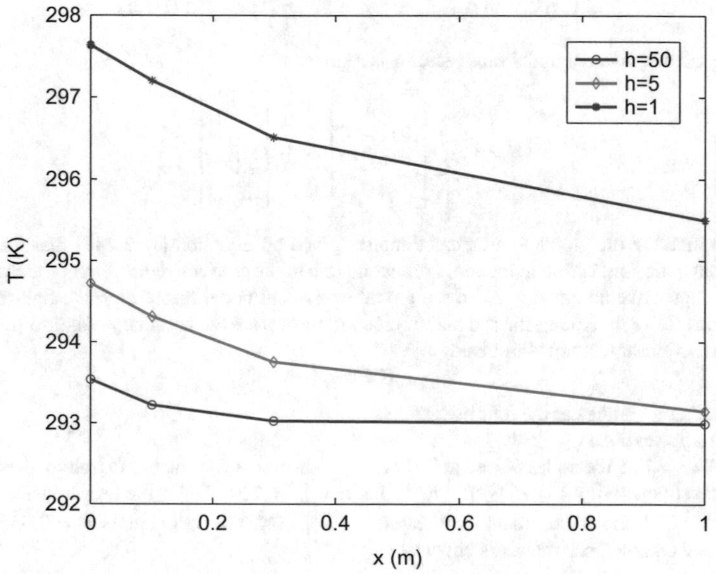

Fig. 7.85 Example HT7-FEM-1D: Temperature profiles obtained with two nodes per element for different heat transfer coefficients h

7.13.4 Convergence, Compatibility, Completeness

When a complex system is studied and simulated by the finite element method, the question of the quality of the approximation of the obtained solution arises, i.e. the *convergence* (Oden et al. 2005; Zienkiewicz and Zhu 1987; Zienkiewicz et al. 2013). It can be judged by considering a refined mesh. In fact, this refinement can be obtained in two ways, the first one called h-convergence or h-refinement by increasing the number of elements in the domain, the second one called p-convergence or p-refinement by keeping the same number of elements but by increasing the degree of the interpolation polynomials. The mathematical demonstrations deal with structured meshes, but the same tendencies are obtained with unstructured meshes. According to Lax equivalence theorem, Butzer and Weis (1975) and Lax and Richtmyer (1956) the stability of a finite difference scheme applied to a linear continuous problem properly posed with initial values is a necessary and sufficient condition for convergence provided that it is consistent. This theorem is often invoked in the theory of finite elements.

Another important notion is *compatibility*. Through the frontiers of the elements and inside the elements, the field variable and its derivatives up to the value minus one of the highest order of derivative appearing in the weak formulation must be continuous (Hutton 2004; Koutromanos 2018; Zienkiewicz et al. 2013). If the highest order of derivative in the weak formulation is k, it requires a continuity of type C^{k-1}. In Example 7.20, the highest order in the weak formulation is 1, thus the required continuity for temperature is C^0.

The third notion is *completeness*. Inside the elements, the trial function and the weight functions must perfectly describe a given function. A function is k-complete if it can represent any polynomial of degree k. Thus, the function $u(x) = a_0 + a_1 x$ is 1-complete. The function $u(x) = a_0 + a_1 x + a_2 x^2$ is 2-complete but the function $u(x) = a_0 + a_1 x + a_3 x^3$ is only 1-complete.

To ensure the convergence, the approximation and weight functions and their derivatives until the highest order of differentiation appearing in the weak formulation must be able to represent constant values (Hutton 2004; Koutromanos 2018; Zienkiewicz et al. 2013). This amounts to require that the approximated solution has a degree of completeness at least equal to 1. Thus, the approximation $u(x) = a_0 + a_2 x^2 + a_3 x^3$ cannot converge as it cannot represent a function with a constant derivative (Koutromanos 2018).

In Example 7.20, the temperature is approximated by means of linear functions, i.e. first degree polynomials. The temperature is continuous between the elements, but the flux which is proportional to the first derivative of temperature is not continuous, hence Figure 7.78 which shows the discontinuities of the derivatives at the nodes. in Example 7.23, the temperature is approximated by second degree polynomials. In this case, not only the temperature is continuous between elements but also the flux proportional to the first derivative, inducing a greater smoothness of the profile (Figure 7.83).

7.13.5 Case of Transient Systems

In the problems of heat transfer, fluid flow, chemical reaction, the user frequently desires to solve a transient problem, that is obtain a time-dependent solution.

7.13.5.1 Development of the Method

If the general steady-state equation is symbolized under the usual form in solid mechanics or similar in other domains as

$$\mathbf{K}\mathbf{X} = \mathbf{f} \tag{7.13.152}$$

the corresponding differential equation is written as

$$\mathbf{C}\dot{\mathbf{X}}(t) + \mathbf{K}\mathbf{X}(t) = \mathbf{f}(t) \tag{7.13.153}$$

The exact determination of $\mathbf{X}(t)$ as a time function is very difficult, not to say impossible. Only step-by-step solutions are possible (Koutromanos 2018; Lewis et al. 2004; Pozrikidis 2014; Tham and Cheung 1982; Yu and Hsu 1985). Thus we will only obtain approximations of the exact solution, called closed-form which is an analytical solution.

A time discretization must be performed (cf. Chapter 6 for ordinary differential equations and Chapter 7, Section 7.5, for finite differences applied to partial differential equations).

The dynamic equation at times t_n and t_{n+1} gives

$$\begin{aligned} C\dot{\mathbf{X}}_n + \mathbf{K}\mathbf{X}_n &= \mathbf{f}_n \\ C\dot{\mathbf{X}}_{n+1} + \mathbf{K}\mathbf{X}_{n+1} &= \mathbf{f}_{n+1} \end{aligned} \tag{7.13.154}$$

We can write for an intermediary time indexed $n+\theta$ included between n and $n+1$ (Hughes 2000; Koutromanos 2018)

$$\dot{\mathbf{X}}_{n+\theta} = (1-\theta)\dot{\mathbf{X}}_n + \theta\dot{\mathbf{X}}_{n+1}\,, \quad 0 \le \theta \le 1\,, \quad n \le n+\theta \le n+1 \tag{7.13.155}$$

from which we get (if $\theta \neq 0$)

$$\dot{\mathbf{X}}_{n+1} = \frac{1}{\theta}\dot{\mathbf{X}}_{n+\theta} - \frac{1-\theta}{\theta}\dot{\mathbf{X}}_n \tag{7.13.156}$$

and

$$C\left(\frac{1}{\theta}\dot{\mathbf{X}}_{n+\theta} - \frac{1-\theta}{\theta}\dot{\mathbf{X}}_n\right) + \mathbf{K}\mathbf{X}_{n+1} = \mathbf{f}_{n+1} \tag{7.13.157}$$

Setting

$$\dot{\mathbf{X}}_{n+\theta} = \frac{1}{\Delta t}\left(\mathbf{X}_{n+1} - \mathbf{X}_n\right) \tag{7.13.158}$$

we obtain

$$C\left(\frac{1}{\theta\Delta t}\left(\mathbf{X}_{n+1} - \mathbf{X}_n\right)\right) - \frac{1-\theta}{\theta}\dot{\mathbf{X}}_n + \mathbf{K}\mathbf{X}_{n+1} = \mathbf{f}_{n+1} \tag{7.13.159}$$

or in an equivalent way

$$\underbrace{\left[\frac{1}{\theta\Delta t}C + \mathbf{K}\right]}_{\tilde{\mathbf{K}}}\mathbf{X}_{n+1} = \underbrace{\frac{1}{\theta\Delta t}C\mathbf{X}_n + \frac{1-\theta}{\theta}C\dot{\mathbf{X}}_n + \mathbf{f}_{n+1}}_{\tilde{\mathbf{f}}_{n+1}} \tag{7.13.160}$$

Equation (7.13.160) can also be written under the form (Lewis et al. 2004)

$$\begin{aligned} (C + \theta\Delta t\mathbf{K})\,\mathbf{X}_{n+1} = {} & (C - (1-\theta)\Delta t\mathbf{K})\,\mathbf{X}_n \\ & + \Delta t\left((1-\theta)\mathbf{f}_n + \theta\mathbf{f}_{n+1}\right) \end{aligned} \tag{7.13.161}$$

However, it can be noticed in Equation (7.13.161) that it is necessary to know $\mathbf{f}_{n+1}$ to deduce $\mathbf{X}_{n+1}$. This can be performed by an iterative scheme, first a prediction, then a correction.

Equations (7.13.161) and (7.13.160) could have been obtained by setting, since the start, the generalized trapezoidal rule (Logan 2016; Reddy 2015)

$$\mathbf{X}_{n+1} = \mathbf{X}_n + \Delta t \left((1-\theta)\dot{\mathbf{X}}_n + \theta \dot{\mathbf{X}}_{n+1}\right) \tag{7.13.162}$$

Indeed, multiplying by $\boldsymbol{C}$, then replacing $\boldsymbol{C}\dot{\mathbf{X}}_n$ and $\boldsymbol{C}\dot{\mathbf{X}}_{n+1}$ by their expressions, we get

$$\begin{aligned} \boldsymbol{C}\mathbf{X}_{n+1} = \boldsymbol{C}\mathbf{X}_n + \Delta t\,((1-\theta)\,(\mathbf{f}_n - \mathbf{K}\mathbf{X}_n) \\ +\theta\,(\mathbf{f}_{n+1} - \mathbf{K}\mathbf{X}_{n+1})) \end{aligned} \tag{7.13.163}$$

thus

$$\begin{aligned} (\boldsymbol{C} + \theta\Delta t\mathbf{K})\,\mathbf{X}_{n+1} &= (\boldsymbol{C} - (1-\theta)\Delta t\mathbf{K})\,\mathbf{X}_n \\ &+\Delta t\,((1-\theta)\mathbf{f}_n + \theta\mathbf{f}_{n+1}) \end{aligned} \tag{7.13.164}$$

which is identical to Equation (7.13.161) (Lewis et al. 2004).

If $\theta = 0$, we obtain

$$\begin{aligned} \boldsymbol{C}\mathbf{X}_{n+1} &= \boldsymbol{C}\mathbf{X}_n + \Delta t\,\boldsymbol{C}\dot{\mathbf{X}}_{n+1} \\ &= \boldsymbol{C}\mathbf{X}_n + \Delta t\left(\tilde{\mathbf{f}}_n - \mathbf{K}\mathbf{X}_n\right) \end{aligned} \tag{7.13.165}$$

If $\theta \geq 1/2$, the scheme is unconditionnally stable.

If $\theta < 1/2$, the upper stability limit (Hughes 2000; Reddy 2015) is

$$\Delta t = \frac{2}{(1-2\theta)\lambda_{max}} \tag{7.13.166}$$

where λ_{max} is the largest eigenvalue of equations

$$(\mathbf{K} - \lambda \boldsymbol{C})\,\mathbf{X} = 0 \tag{7.13.167}$$

where the formulation $\{\boldsymbol{C}, \mathbf{K}\}$ refers to finite element equations after assembling, taking into account the boundary conditions (Reddy 2015). Indeed, if the time approximation is equivalent to a scheme of the form

$$\hat{\mathbf{K}}\mathbf{X}_{n+1} = \hat{\mathbf{K}}\mathbf{X}_n \Longrightarrow \mathbf{X}_{n+1} = \hat{\mathbf{K}}^{-1}\hat{\mathbf{K}}\mathbf{X}_n = \hat{\mathbf{A}}\mathbf{X}_n \tag{7.13.168}$$

this amounts to study the eigenvalues of the amplification matrix $\hat{\mathbf{A}}$.

To present the algorithms under a general form, it is practical to use the expression (Koutromanos 2018) with the residual

$$\begin{aligned} \mathbf{R}(t) &= \boldsymbol{C}\dot{\mathbf{X}}(t) \\ &= \mathbf{f}(t) - \mathbf{K}\mathbf{X}(t) \end{aligned} \tag{7.13.169}$$

hence the generalized trapezoidal rule as

$$\mathbf{R}_{n+\theta} = (1-\theta)\mathbf{R}_n + \theta\mathbf{R}_{n+1}\,, \quad 0 \leq \theta \leq 1\,, \quad n \leq n+\theta \leq n+1 \tag{7.13.170}$$

7.13.5.2 One-Step Methods

Among one-step methods, using the generalized trapezoidal rule, we find:

- the explicit forward Euler method with $\theta = 0$,

$$\mathbf{R}_{n+\theta} = \mathbf{R}_n = \mathbf{f}_n - \mathbf{K}\mathbf{X}_n$$

This method is stable only if Δt is lower than (7.13.166).

- the unconditionally stable Crank-Nicolson method with $\theta = 1/2$. It is the trapezoidal rule,

$$\mathbf{R}_{n+\theta} = \frac{1}{2}(\mathbf{R}_n + \mathbf{R}_{n+1})$$

Oscillations can occur in the case of strong transients.

- the unconditionally stable Galerkin's method, with $\theta = 2/3$

$$\mathbf{R}_{n+\theta} = \frac{1}{3}\mathbf{R}_n + \frac{2}{3}\mathbf{R}_{n+1}$$

- the implicit backward Euler method with $\theta = 1$. This method is unconditionally stable,

$$\mathbf{R}_{n+\theta} = \mathbf{R}_{n+1} = \mathbf{f}_{n+1} - \mathbf{K}\mathbf{X}_{n+1}$$

The methods with $\theta > 0$ are either semi-implicit or implicit.

These methods for $0 < \theta \leq 1$ will be performed according to the following predictor–corrector algorithm:

Stage 1: Initial conditions. Given an initial state $\mathbf{X}_0$, $\dot{\mathbf{X}}_0$ is calculated from

$$C\dot{\mathbf{X}}_0 = \mathbf{f}_0 - \mathbf{K}\mathbf{X}_0$$

Then, stages 2 to 5 are performed at each step n.

Stage 2: Given $\mathbf{X}_n$ and $\dot{\mathbf{X}}_n$, for each element $^{(e)}$, calculate the nodal field and the velocity of the nodal field

$$\mathbf{X}_n^{(e)} = \mathbf{L}^{(e)}\mathbf{X}_n \,, \quad \dot{\mathbf{X}}_n^{(e)} = \mathbf{L}^{(e)}\dot{\mathbf{X}}_n$$

Calculate the prediction of the nodal flux

$$\tilde{\mathbf{f}}_{n+1}^{(e)} = \mathbf{f}_{n+1}^{(e)} + \frac{1}{\theta\Delta t}C^{(e)}\mathbf{X}_n^{(e)} + \frac{1-\theta}{\theta}C^{(e)}\dot{\mathbf{X}}_n^{(e)}$$

Calculate the prediction of the conductance matrix

$$\tilde{\mathbf{K}}^{(e)} = \frac{1}{\theta\Delta t}C^{(e)} + \mathbf{K}^{(e)}$$

Stage 3: Assemble the vectors and matrices

$$\tilde{\mathbf{f}}_{n+1} = \sum_{e=1}^{n_e}\left(\mathbf{L}^{(e)^T}\tilde{\mathbf{f}}_{n+1}^{(e)}\right)$$

$$\tilde{\mathbf{K}} = \sum_{e=1}^{n_e}\left(\mathbf{L}^{(e)^T}\tilde{\mathbf{K}}^{(e)}\mathbf{L}^{(e)}\right)$$

If the time step Δt is constant, $\tilde{\mathbf{K}}$ can be calculated only once.

Stage 4: Calculate the corrected solution vector as

$$\tilde{\mathbf{K}}\mathbf{X}_{n+1} = \tilde{\mathbf{f}}_{n+1}$$

$$\dot{\mathbf{X}}_{n+1} = \frac{1}{\theta \Delta t}(\mathbf{X}_{n+1} - \mathbf{X}_n) - \frac{1-\theta}{\theta}\dot{\mathbf{X}}_n$$

Stage 5: If the convergence to the steady-state solution is reached, stop, otherwise return to stage 2.

Example 7.25 :

Transient heat transfer

In the case of heat transfer, the general model follows Fourier's law

$$\rho c_p \frac{\partial T}{\partial t} = \lambda \frac{\partial^2 T}{\partial x^2} + \dot{q}''' \tag{7.13.171}$$

to which the boundary conditions must be added. After transformation of the integral formulation, taking into account the source term $S(x)$ and the natural boundary conditions, the weak formulation of the transient general equation is written as

$$\begin{aligned} &\sum_{e=1}^{n_e} \int_{x_1^{(e)}}^{x_2^{(e)}} w(x) \rho c_p A \frac{\partial T}{\partial t} dx + \sum_{e=1}^{n_e} \int_{x_1^{(e)}}^{x_2^{(e)}} \frac{dw}{dx} \lambda A \frac{\partial T}{\partial x} dx = \\ &\sum_{e=1}^{n_e} \int_{x_1^{(e)}}^{x_2^{(e)}} w(x) S(x) dx + \sum_{e=1}^{n_e} \left[w(x) \lambda A \frac{\partial T}{\partial x} \right]_{x_1^{(e)}}^{x_2^{(e)}} \end{aligned} \tag{7.13.172}$$

where ρ is the density (kg m^{-3}) and c_p the heat capacity (J kg^{-1} K^{-1}). In the expression (7.13.172), the physical parameters ρ, c_p and λ can depend on temperature, thus on position, and the surface area A of the cross section can also depend on position. The writing is simplified when these parameters are independent of the position.

Using the temperature approximation

$$T(x) \approx \tilde{T}^{(e)}(x) = \sum_{j=1}^{n} N_j^{(e)}(x) T_j^{(e)} = \mathbf{N}^{(e)}(x)\mathbf{T}^{(e)} \tag{7.13.173}$$

and for the interpolation functions

$$w(x) \approx \tilde{w}^{(e)}(x) = \sum_{j=1}^{n} N_j^{(e)} w_j^{(e)} = \mathbf{N}^{(e)}\mathbf{W}^{(e)} = \mathbf{W}^{(e)^T} \mathbf{N}^{(e)^T} \tag{7.13.174}$$

we obtain for the term of the time derivative of temperature

$$\begin{aligned} \int_{x_1^{(e)}}^{x_2^{(e)}} w(x) \rho c_p A \frac{\partial T}{\partial t} dx &\approx \int_{x_1^{(e)}}^{x_2^{(e)}} \tilde{w}^{(e)}(x) \rho c_p A \frac{\partial \tilde{T}^{(e)}}{\partial t} dx \\ &= \int_{x_1^{(e)}}^{x_2^{(e)}} \mathbf{W}^{(e)^T} \mathbf{N}^{(e)^T} \rho^{(e)} c_p^{(e)} A^{(e)} \mathbf{N}^{(e)} \frac{\partial \mathbf{T}^{(e)}}{\partial t} dx \\ &= \mathbf{W}^{(e)^T} \left(\int_{x_1^{(e)}}^{x_2^{(e)}} \mathbf{N}^{(e)^T} \rho^{(e)} c_p^{(e)} A^{(e)} \mathbf{N}^{(e)} dx \right) \frac{\partial \mathbf{T}^{(e)}}{\partial t} \\ &= \mathbf{W}^{(e)^T} \mathbf{C}^{(e)} \left[\dot{\boldsymbol{T}}^{(e)} \right] \end{aligned} \tag{7.13.175}$$

where $C^{(e)}$ is the *capacitance matrix* equal to

$$C^{(e)} = \int_{x_1^{(e)}}^{x_2^{(e)}} \mathbf{N}^{(e)T} \rho^{(e)} c_p^{(e)} A^{(e)} \mathbf{N}^{(e)} dx \qquad (7.13.176)$$

Then, the global capacitance matrix is obtained by assembling

$$C = \sum_{e=1}^{n_e} \left(\mathbf{L}^{(e)T} C^{(e)} \mathbf{L}^{(e)} \right) \qquad (7.13.177)$$

The differential equation to be solved becomes

$$C \left[\dot{T} \right] + \mathbf{KT} = \mathbf{q} \qquad (7.13.178)$$

Example 7.26 :

HT8-FEM-1D: Transient regime for a 1D metal rod receiving a constant thermal power on its periphery, an imposed flux at $x = 0$, yielding a convective flux at $x = L$.

The conditions are the same as in Example 7.20 except that this latter was only modelled from a steady-state point of view, whereas in the case 7.26, the transient regime will be determined.

Numerical application:

The physical data are totally identical to those of Example 7.20 except that here we specify the density $\rho = 7800\,\text{kg m}^{-3}$, the heat capacity $C_p = 500\,\text{J kg}^{-1}\,\text{K}^{-1}$, and the initial conditions $T = 293$ K in the whole rod. The number of nodes is 4 like in Example 7.20 and each element has two nodes. The interpolation functions are the same linear polynomials as for Example 7.20.

The capacitance matrix $C^{(e)}$ for one element is

$$C^{(e)} = A\rho c_p \begin{bmatrix} \frac{1}{3}(x_2^{(e)} - x_1^{(e)}) & \frac{1}{6}(x_2^{(e)} - x_1^{(e)}) \\ \frac{1}{6}(x_2^{(e)} - x_1^{(e)}) & \frac{1}{3}(x_2^{(e)} - x_1^{(e)}) \end{bmatrix} \qquad (7.13.179)$$

Numerically, the global capacitance matrix is equal to

$$10^4 \begin{bmatrix} 0.130 & 0.065 & 0 & 0 \\ 0.065 & 0.390 & 0.130 & 0 \\ 0 & 0.130 & 1.170 & 0.455 \\ 0 & 0 & 0.455 & 0.910 \end{bmatrix} \qquad (7.13.180)$$

If the term $\mathbf{q}$ is constant, the calculation of the temperature vector can be performed according to the relation

$$(C + \theta \Delta t \mathbf{K})\,\mathbf{X}_{n+1} = (C - (1-\theta)\Delta t \mathbf{K})\,\mathbf{X}_n + \Delta t \mathbf{f}_n \qquad (7.13.181)$$

obtained by setting $\mathbf{f}_{n+1} = \mathbf{f}_n$ in Equation (7.13.164).

The chosen time step is $\Delta t = 10$ s and we execute 100,000 time steps. The one-step Galerkin's method will be used, corresponding to parameter θ equal to 2/3. The profile is plotted every 2000 steps, thus 51 profiles are represented in Figure 7.86. The final profile gives the nodal temperatures $\boldsymbol{T} = [339.227\ 338.573\ 336.112\ 315.391]^T$ to be compared to the values of Example 7.20 for the steady state, i.e. $\boldsymbol{T} = [339.25\ 338.59\ 336.13\ 315.40]^T$. It shows that the last temperature vector obtained by the transient calculation is very close to the steady state previously calculated in Example 7.20.

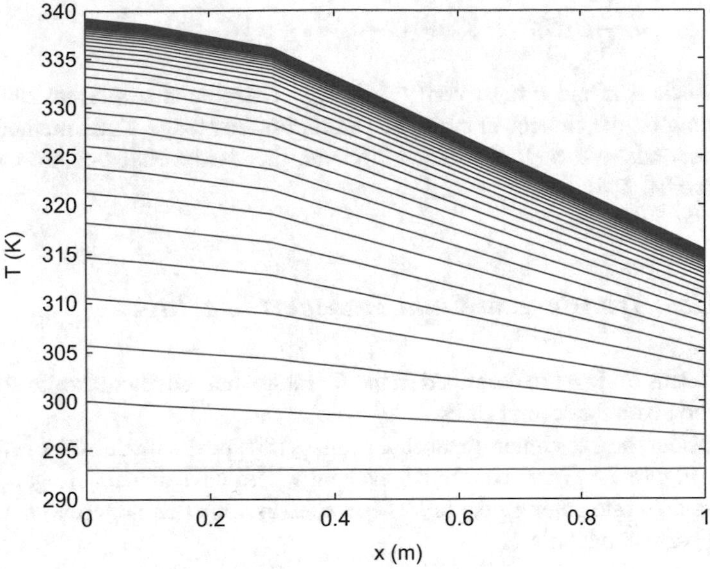

Fig. 7.86 Example HT8-FEM-1D: Temperature profiles obtained with respect to time. Two successive profiles are separated by 20,000s and 51 profiles are displayed

7.13.5.3 Multiple Step Methods

The multiple step methods of Chapter 6, Section 6.2.2, are again found. Only a few examples will be cited here.

- Explicit Adams–Bashforth method of order 2

$$\frac{1}{\Delta t} C\,(\mathbf{X}_{n+1} - \mathbf{X}_n) = \frac{1}{2}\left(-\left[\boldsymbol{R}_{n-1}\right] + 3\mathbf{R}_n\right) \tag{7.13.182}$$

- Implicit Adams-Moulton method of order 3

$$\frac{1}{\Delta t} C\,(\mathbf{X}_{n+1} - \mathbf{X}_n) = \frac{1}{12}\left(-\left[\boldsymbol{R}_{n-1}\right] + 8\mathbf{R}_n + 5\mathbf{R}_{n+1}\right) \tag{7.13.183}$$

- Predictor–corrector Runge-Kutta method of order 2

Predictor

$$\left[\boldsymbol{R}_{pred}\right] = \left[f(t_n + b\Delta t)\right] - \mathbf{K}\left[\boldsymbol{X}_{pred}\right] \tag{7.13.184}$$

with $\left[\boldsymbol{X}_{pred}\right]$ solution of following equation

$$\frac{1}{b\Delta t} C\left(\left[\boldsymbol{X}_{pred}\right] - \mathbf{X}_n\right) = \mathbf{R}_n \tag{7.13.185}$$

Corrector

$$\frac{1}{\Delta t}\boldsymbol{C}\left(\mathbf{X}_{n+1}-\mathbf{X}_n\right)=(1-a)\mathbf{R}_n+a\left[\boldsymbol{R}_{pred}\right] \tag{7.13.186}$$

The constants a and b must verify $2ab = 1$. There exist several known methods depending on the choices of a and b. The mid-point Runge-Kutta method of order 2 corresponds to $a = 1$, $b = 0.5$. Moreover, there exist Runge-Kutta methods of higher order, 3, 4, ...

7.13.6 Heat Transfer and Fluid Transport in a Tube

Different forms of flow are described in the literature and the references to specialized texts will be given for several cases.

We consider the one-dimensional case of mass transport with diffusive heat transfer in a tube (Figure 7.87) and convective exchange with environment (Logan 2016). $\dot{m}$ is the mass flow rate entering the tube. Moreover, assume that a thermal power $\dot{q}_g'''$ is generated per volume unit.

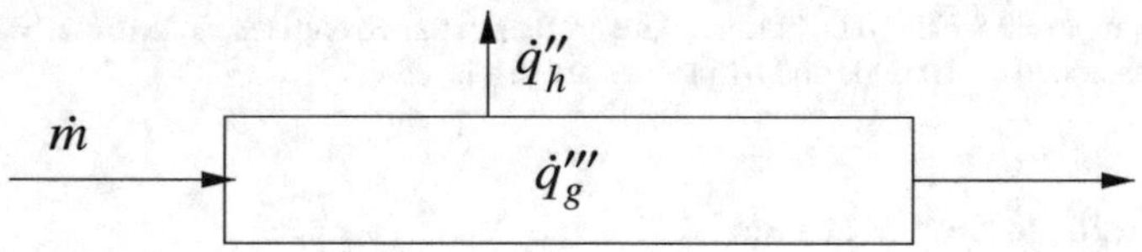

Fig. 7.87 Heat transfer and mass transport coupled in the one-dimensional case

In transient regime, the fundamental equation ruling that phenomenon is written as

$$\rho C_p \frac{\partial T}{\partial t} = -\dot{m}\frac{C_p}{A}\frac{\partial T}{\partial x} + \frac{\partial}{\partial x}\left(\lambda \frac{\partial T}{\partial x}\right) + \dot{q}_g''' + h\frac{P}{A}(T_\infty - T) \tag{7.13.187}$$

where h is the convective heat transfer coefficient with environment, P the perimeter and T_∞ the environment temperature.

Consider this steady state equation which constitutes the strong formulation and assume that the source term is zero. The strong formulation can be written differently as

$$w(x)\left[\dot{m}\frac{C_p}{A}\frac{dT}{dx} - \frac{d}{dx}\left(\lambda\frac{dT}{dx}\right) + h\frac{P}{A}(T - T_\infty)\right] = 0 \qquad \forall\, w \tag{7.13.188}$$

with the condition that $w(x) = 0$ on an essential boundary of Γ_D type (Dirichlet condition). We integrate this equation on the domain $[x_1, x_2]$ i.e.

$$\int_{x_1}^{x_2} w(x)\left[\dot{m}\frac{C_p}{A}\frac{dT}{dx} - \frac{d}{dx}\left(\lambda\frac{dT}{dx}\right) + h\frac{P}{A}(T - T_\infty)\right] = 0 \tag{7.13.189}$$

We perform an integration by parts of the diffusion term hence

$$\int_{x_1}^{x_2} w(x)\left[\dot{m}\frac{C_p}{A}\frac{dT}{dx}+h\frac{P}{A}(T-T_\infty)\right]dx+\int_{x_1}^{x_2}\left[\lambda\frac{dw}{dx}\frac{dT}{dx}\right]dx=\left[w(x)\lambda\frac{dT}{dx}\right]_{x_1}^{x_2} \tag{7.13.190}$$

This equation constitutes the weak formulation provided the boundary conditions are added which allows us to completely define the problem.

To apply Galerkin's method, we consider the weight functions equal to the interpolation functions

$$w(x)=\tilde{T}(x)=N_1(x)T_1+N_2(x)T_2 \tag{7.13.191}$$

hence the weak formulation

$$\begin{aligned}&\int_{x_1}^{x_2}(N_1T_1+N_2T_2)\left[\dot{m}\frac{C_p}{A}(\frac{dN_1}{dx}T_1+\frac{dN_2}{dx}T_2)+h\frac{P}{A}(N_1T_1+N_2T_2-T_\infty)\right]dx\\&+\int_{x_1}^{x_2}\left[\lambda(\frac{dN_1}{dx}T_1+\frac{dN_2}{dx}T_2)(\frac{dN_1}{dx}T_1+\frac{dN_2}{dx}T_2)\right]dx=\\&\left[(N_1T_1+N_2T_2)\lambda(\frac{dN_1}{dx}T_1+\frac{dN_2}{dx}T_2)\right]_{x_1}^{x_2}\end{aligned} \tag{7.13.192}$$

Of course, the boundary conditions must be added. Indeed, as $w(x)$ can be any, the previous equation is true separately for $w(x)=N_1(x)$ or $w(x)=N_2(x)$, thus

$$\begin{aligned}&\int_{x_1}^{x_2}N_iT_i\left[\dot{m}\frac{C_p}{A}(\frac{dN_1}{dx}T_1+\frac{dN_2}{dx}T_2)+h\frac{P}{A}(N_1T_1+N_2T_2-T_\infty)\right]dx\\&+\int_{x_1}^{x_2}\left[\lambda\frac{dN_i}{dx}T_i(\frac{dN_1}{dx}T_1+\frac{dN_2}{dx}T_2)\right]dx=\\&\left[N_iT_i\lambda(\frac{dN_1}{dx}T_1+\frac{dN_2}{dx}T_2)\right]_{x_1}^{x_2},\quad i=1,2\end{aligned} \tag{7.13.193}$$

Consider an element $[x_1^{(e)},x_2^{(e)}]$. Taking again the already defined notations (Section 7.13.3.5)

$$\begin{aligned}&T(x)\approx\tilde{T}^{(e)}(x)=\sum_{j=1}^{n}N_j^{(e)}(x)T_j^{(e)}=\mathbf{N}^{(e)}\mathbf{T}^{(e)}\\&\frac{dT}{dx}\approx\frac{d\tilde{T}^{(e)}}{dx}=\sum_{j=1}^{n}\frac{dN_j^{(e)}(x)}{dx}T_j^{(e)}=\sum_{j=1}^{n}B_j^{(e)}(x)T_j^{(e)}=\mathbf{B}^{(e)}\mathbf{T}^{(e)}\\&w^{(e)}(x)\approx\mathbf{W}^{(e)^T}\mathbf{N}^{(e)T}\\&\frac{dw^{(e)}(x)}{dx}\approx\mathbf{W}^{(e)T}\mathbf{B}^{(e)T}\end{aligned} \tag{7.13.194}$$

we can write on an element $[x_1^{(e)},x_2^{(e)}]$ the integral terms under the form

$$\begin{aligned}
&\int_{x_1^{(e)}}^{x_2^{(e)}} \mathbf{W}^{(e)T}\mathbf{N}^{(e)T}\left[\dot{m}\frac{C_p}{A}\mathbf{B}^{(e)}\mathbf{T}^{(e)} + h\frac{P}{A}(\mathbf{N}^{(e)}\mathbf{T}^{(e)} - T_\infty)\right] dx \\
&+ \int_{x_1}^{x_2}\left[\lambda \mathbf{W}^{(e)T}\mathbf{B}^{(e)T}\mathbf{B}^{(e)}\mathbf{T}^{(e)}\right] dx \\
&= \mathbf{W}^{(e)T}\left\{\int_{x_1^{(e)}}^{x_2^{(e)}}\left(\mathbf{N}^{(e)T}\left[\dot{m}\frac{C_p}{A}\mathbf{B}^{(e)} + h\frac{P}{A}\mathbf{N}^{(e)}\right]\right.\right. \\
&\left.\left.+\lambda\mathbf{B}^{(e)T}\mathbf{B}^{(e)}\right) dx\right\}\mathbf{T}^{(e)} - \mathbf{W}^{(e)T}\left(\int_{x_1^{(e)}}^{x_2^{(e)}}\mathbf{N}^{(e)T} dx\right) h\frac{P}{A}T_\infty
\end{aligned} \tag{7.13.195}$$

hence

$$\begin{aligned}
&\mathbf{W}^{(e)T}\left\{\int_{x_1^{(e)}}^{x_2^{(e)}}\left(\mathbf{N}^{(e)T}\left[\dot{m}\frac{C_p}{A}\mathbf{B}^{(e)} + h\frac{P}{A}\mathbf{N}^{(e)}\right]\right.\right. \\
&\left.\left.+\lambda\mathbf{B}^{(e)T}\mathbf{B}^{(e)}\right) dx\right\}\mathbf{T}^{(e)} \\
&= \mathbf{W}^{(e)T}\left(\int_{x_1^{(e)}}^{x_2^{(e)}}\mathbf{N}^{(e)T} dx\right) h\frac{P}{A}T_\infty + \mathbf{W}^{(e)T}\left[\mathbf{N}^{(e)T}\lambda\mathbf{B}^{(e)}\mathbf{T}^{(e)}\right]_{x_1^{(e)}}^{x_2^{(e)}}
\end{aligned} \tag{7.13.196}$$

By assembling the elements on the entire domain $[0, L]$, we obtain

$$\begin{aligned}
&\sum_{e=1}^{n_e}\mathbf{W}^{(e)T}\left\{\int_{x_1^{(e)}}^{x_2^{(e)}}\left(\mathbf{N}^{(e)T}\left[\dot{m}\frac{C_p}{A}\mathbf{B}^{(e)} + h\frac{P}{A}\mathbf{N}^{(e)}\right] + \lambda\mathbf{B}^{(e)T}\mathbf{B}^{(e)}\right) dx\right\}\mathbf{T}^{(e)} \\
&= \sum_{e=1}^{n_e}\mathbf{W}^{(e)T}\left(\int_{x_1^{(e)}}^{x_2^{(e)}}\mathbf{N}^{(e)T} dx\right) h\frac{P}{A}T_\infty + \sum_{e=1}^{n_e}\mathbf{W}^{(e)T}\left[\mathbf{N}^{(e)T}\lambda\mathbf{B}^{(e)}\mathbf{T}^{(e)}\right]_{x_1^{(e)}}^{x_2^{(e)}}
\end{aligned} \tag{7.13.197}$$

First consider the second term between brackets of the right-hand side. As in Equation (7.13.74), except a discontinuity between 0 and L, the terms are simplified between themselves except those of inlet and outlet so that this sum is reduced to

$$\begin{aligned}
&\sum_{e=1}^{n_e}\mathbf{W}^{(e)T}\left[\mathbf{N}^{(e)T}\lambda\mathbf{B}^{(e)}\mathbf{T}^{(e)}\right]_{x_1^{(e)}}^{x_2^{(e)}} = \\
&\left[\mathbf{W}^{(n_e)}\right]^T\left[\mathbf{N}^{(e)T}\lambda\mathbf{B}^{(e)}\mathbf{T}^{(e)}\right]_{x_2^{(n_e)}} - \left[\mathbf{W}^{(1)}\right]^T\left[\mathbf{N}^{(e)T}\lambda\mathbf{B}^{(e)}\mathbf{T}^{(e)}\right]_{x_1^{(1)}}
\end{aligned} \tag{7.13.198}$$

Note that $x_1^{(1)} = 0$ and $x_2^{(n_e)} = L$.

The equation then takes the form

$$\begin{aligned}
&\sum_{e=1}^{n_e}\mathbf{W}^{(e)T}\left\{\int_{x_1^{(e)}}^{x_2^{(e)}}\left(\mathbf{N}^{(e)T}\left[\dot{m}\frac{C_p}{A}\mathbf{B}^{(e)} + h\frac{P}{A}\mathbf{N}^{(e)}\right] + \lambda\mathbf{B}^{(e)T}\mathbf{B}^{(e)}\right) dx\right\}\mathbf{T}^{(e)} \\
&- \left[\mathbf{W}^{(n_e)}\right]^T\left[\mathbf{N}^{(e)T}\lambda\mathbf{B}^{(e)}\mathbf{T}^{(e)}\right]_{x_2^{(n_e)}} \\
&= \sum_{e=1}^{n_e}\mathbf{W}^{(e)T}\left(\int_{x_1^{(e)}}^{x_2^{(e)}}\mathbf{N}^{(e)T} dx\right) h\frac{P}{A}T_\infty - \left[\mathbf{W}^{(1)}\right]^T\left[\mathbf{N}^{(e)T}\lambda\mathbf{B}^{(e)}\mathbf{T}^{(e)}\right]_{x_1^{(1)}}
\end{aligned} \tag{7.13.199}$$

where the unknown temperatures are in the left-hand side and those known in the right-hand side, except the inlet temperature which intervenes in $\left[\boldsymbol{T}^{(1)}\right]$.

Setting the transport-convection-conduction matrix equal to

$$\mathbf{K}^{(e)} = \left[\int_{x_1^{(e)}}^{x_2^{(e)}} \left(\mathbf{N}^{(e)^T}\left[\dot{m}\frac{C_p}{A}\mathbf{B}^{(e)} + h\frac{P}{A}\mathbf{N}^{(e)}\right] + \lambda\mathbf{B}^{(e)^T}\left[\boldsymbol{B}^{(e)}\right]\right)dx\right] \tag{7.13.200}$$

and the convective heat transfer flux

$$\mathbf{q}^{(e)} = \left(\int_{x_1^{(e)}}^{x_2^{(e)}} \mathbf{N}^{(e)^T}\,dx\right) h\frac{P}{A}T_\infty \tag{7.13.201}$$

it results

$$\begin{aligned} &\sum_{e=1}^{n_e}\left(\mathbf{W}^{(e)^T}\mathbf{K}^{(e)}\mathbf{T}^{(e)}\right) - \left[W^{(n_e)}\right]^T\left[\mathbf{N}^{(e)^T}\lambda\mathbf{B}^{(e)}\mathbf{T}^{(e)}\right]_{x_2^{(ne)}} \\ &= \sum_{e=1}^{n_e}\left(\mathbf{W}^{(e)^T}\mathbf{q}^{(e)}\right) - \left[W^{(1)}\right]^T\left[\mathbf{N}^{(e)^T}\lambda\mathbf{B}^{(e)}\mathbf{T}^{(e)}\right]_{x_1^{(1)}} \end{aligned} \tag{7.13.202}$$

The expression (7.13.200) of the transport-convection-conduction matrix shows that it is composed by three terms, one for fluid transport, one for convective heat transfer with environment and one for heat diffusion. In most cases, the contribution of the diffusion term will be negligible with respect to the transport term.

Using the matrices $\mathbf{L}^{(e)}$ such that

$$\mathbf{T}^{(e)} = \mathbf{L}^{(e)}\mathbf{T}\,, \quad \mathbf{W}^{(e)} = \mathbf{L}^{(e)}\mathbf{W} \tag{7.13.203}$$

we obtain

$$\begin{aligned} &\sum_{e=1}^{n_e}\left(\mathbf{W}^T\mathbf{L}^{(e)^T}\mathbf{K}^{(e)}\mathbf{L}^{(e)}\mathbf{T}\right) \\ &-\left[\mathbf{W}^T\mathbf{L}^{(e)^T}\mathbf{N}^{(e)^T}\lambda\mathbf{B}^{(e)}\mathbf{L}^{(e)}\mathbf{T}\right]_{x_2^{(ne)}} \\ &= \sum_{e=1}^{n_e}\left(\mathbf{W}^T\mathbf{L}^{(e)^T}\mathbf{q}^{(e)}\right) - \left[\mathbf{W}^T\mathbf{L}^{(e)^T}\mathbf{N}^{(e)^T}\lambda\mathbf{B}^{(e)}\mathbf{L}^{(e)}\mathbf{T}\right]_{x_1^{(1)}} \end{aligned} \tag{7.13.204}$$

Setting the global transport-convection-conduction matrix $\mathbf{K}$

$$\mathbf{K} = \sum_{e=1}^{n_e}\left(\mathbf{L}^{(e)^T}\mathbf{K}^{(e)}\mathbf{L}^{(e)}\right) \tag{7.13.205}$$

and the global equivalent nodal flux vector $\mathbf{q}$ equal to

$$\mathbf{q} = \sum_{e=1}^{n_e}\left(\mathbf{L}^{(e)^T}\mathbf{q}^{(e)}\right) \tag{7.13.206}$$

we finally obtain

$$\begin{aligned}\mathbf{W}^T\mathbf{K}\mathbf{T} - \left[\mathbf{W}^T\mathbf{L}^{(e)^T}\mathbf{N}^{(e)^T}\lambda\mathbf{B}^{(e)}\mathbf{L}^{(e)}\mathbf{T}\right]_{x_2^{(ne)}} \\ = \mathbf{W}^T\mathbf{q} - \left[\mathbf{W}^T\mathbf{L}^{(e)^T}\mathbf{N}^{(e)^T}\lambda\mathbf{B}^{(e)}\mathbf{L}^{(e)}\mathbf{T}\right]_{x_1^{(1)}} \qquad \forall\, \mathbf{W}\end{aligned} \tag{7.13.207}$$

It must be noticed that in the vector $\mathbf{T}$, the inlet temperature $T_1^{(1)} = T_0$ is known.

Example 7.27 :

HTC1-FEM-1D: Fluid transport in a tube with convective exchange with the environment.

We consider the example from Bergman et al. (2011, page 550), of a tube in which air circulates (Figure 7.87).

Numerical application:

Mass flow rate $\dot{m} = 0.050\,\text{kg}\,\text{s}^{-1}$, heat transfer coefficient $h = 4.12\,\text{W}\,\text{m}^{-2}\,\text{K}^{-1}$, heat capacity $C_p = 1011\,\text{J}\,\text{kg}^{-1}\,\text{K}^{-1}$, thermal conductivity $\lambda = 0.0306\,\text{W}\,\text{m}^{-1}\,\text{K}^{-1}$, tube length $L = 5\,\text{m}$, tube diameter $D = 0.15\,\text{m}$, inlet temperature $T_{in} = 376.15\,\text{K}$, environment temperature $T_\infty = 273.15\,\text{K}$.

We choose linear polynomials as interpolation functions

$$\mathbf{N}^{(e)} = \left[\frac{x - x_2^{(e)}}{x_1^{(e)} - x_2^{(e)}} \quad \frac{x - x_1^{(e)}}{x_2^{(e)} - x_1^{(e)}}\right], \quad \mathbf{B}^{(e)} = \left[-\frac{1}{l^{(e)}} \ \frac{1}{l^{(e)}}\right] \tag{7.13.208}$$

and we consider three finite elements such that the coordinates of the nodes are $\{0, 0.1, 0.3, 1\}L$.

We separately calculate the terms of the matrix $\boldsymbol{K}^{(e)}$ (Equation (7.13.200)) by assuming constants physical parameters

The transport term

$$\mathbf{K}_t^{(e)} = \int_{x_1^{(e)}}^{x_2^{(e)}} \mathbf{N}^{(e)^T} \dot{m}\frac{C_P}{A}\mathbf{B}^{(e)} dx = \dot{m}\frac{C_P}{2A}\begin{bmatrix}-1 & 1\\ -1 & 1\end{bmatrix} \tag{7.13.209}$$

The convective transfer term

$$\mathbf{K}_{cv}^{(e)} = \int_{x_1^{(e)}}^{x_2^{(e)}} \mathbf{N}^{(e)^T} h\frac{P}{A}\mathbf{N}^{(e)} dx = h\frac{P l^{(e)}}{6A}\begin{bmatrix}2 & 1\\ 1 & 2\end{bmatrix} \tag{7.13.210}$$

The conduction term

$$\mathbf{K}_{cd}^{(e)} = \int_{x_1^{(e)}}^{x_2^{(e)}} \lambda\mathbf{B}^{(e)^T}\left[\boldsymbol{B}^{(e)}\right] dx = \frac{\lambda}{l^{(e)}}\begin{bmatrix}1 & -1\\ -1 & 1\end{bmatrix} \tag{7.13.211}$$

It must be noticed that the matrix $\mathbf{K}_t^{(e)}$ related to transport is not symmetrical opposite to the two other matrices.

The flux vector is

$$\mathbf{q}^{(e)} = h\frac{P}{A}T_\infty \int_{x_1^{(e)}}^{x_2^{(e)}} \mathbf{N}^{(e)^T} dx = h\frac{P}{2A}T_\infty \begin{bmatrix}1\\ 1\end{bmatrix} \tag{7.13.212}$$

Let us explain the calculation of the term

$$\left[\mathbf{N}^{(e)T}\lambda\mathbf{B}^{(e)}\mathbf{T}^{(e)}\right]_{x_1^{(e)}}^{x_2^{(e)}} = \lambda\left[\begin{bmatrix}\dfrac{x-x_2^{(e)}}{x_1^{(e)}-x_2^{(e)}}\\ \dfrac{x-x_1^{(e)}}{x_2^{(e)}-x_1^{(e)}}\end{bmatrix}\begin{bmatrix}-\dfrac{1}{l^{(e)}} & \dfrac{1}{l^{(e)}}\end{bmatrix}\mathbf{T}^{(e)}\right]_{x_1^{(e)}}^{x_2^{(e)}}$$

$$= \lambda\left[\begin{bmatrix}\dfrac{x-x_2^{(e)}}{(l^{(e)})^2} & -\dfrac{x-x_2^{(e)}}{(l^{(e)})^2}\\ -\dfrac{x-x_1^{(e)}}{(l^{(e)})^2} & \dfrac{x-x_1^{(e)}}{(l^{(e)})^2}\end{bmatrix}\mathbf{T}^{(e)}\right]_{x_1^{(e)}}^{x_2^{(e)}} = \lambda\left[\begin{bmatrix}\dfrac{x-x_2^{(e)}}{(l^{(e)})^2}T_1^{(e)} - \dfrac{x-x_2^{(e)}}{(l^{(e)})^2}T_2^{(e)}\\ -\dfrac{x-x_1^{(e)}}{(l^{(e)})^2}T_1^{(e)} + \dfrac{x-x_1^{(e)}}{(l^{(e)})^2}T_2^{(e)}\end{bmatrix}\right]_{x_1^{(e)}}^{x_2^{(e)}} \quad (7.13.213)$$

$$= \frac{\lambda}{l^{(e)}}(T_1^{(e)} - T_2^{(e)})\begin{bmatrix}1\\-1\end{bmatrix} = \frac{\lambda}{l^{(e)}}\begin{bmatrix}1 & -1\\ -1 & 1\end{bmatrix}\begin{bmatrix}T_1^{(e)}\\ T_2^{(e)}\end{bmatrix}$$

Finally, Equation (7.13.196) for any finite element is written by omitting the factor $\mathbf{W}^{(e)}$

$$\left(\dot{m}\frac{C_P}{2A}\begin{bmatrix}-1 & 1\\ -1 & 1\end{bmatrix} + h\frac{Pl^{(e)}}{6A}\begin{bmatrix}2 & 1\\ 1 & 2\end{bmatrix} + \frac{\lambda}{l^{(e)}}\begin{bmatrix}1 & -1\\ -1 & 1\end{bmatrix}\right)\mathbf{T}^{(e)} - \frac{\lambda}{l^{(e)}}(T_1^{(e)} - T_2^{(e)})\begin{bmatrix}1\\ -1\end{bmatrix} = h\frac{P}{2A}T_\infty\begin{bmatrix}1\\ 1\end{bmatrix} \quad (7.13.214)$$

or still

$$\left(\dot{m}\frac{C_P}{2A}\begin{bmatrix}-1 & 1\\ -1 & 1\end{bmatrix} + h\frac{Pl^{(e)}}{6A}\begin{bmatrix}2 & 1\\ 1 & 2\end{bmatrix} + \frac{\lambda}{l^{(e)}}\begin{bmatrix}1 & -1\\ -1 & 1\end{bmatrix}\right)\mathbf{T}^{(e)} - \frac{\lambda}{l^{(e)}}\begin{bmatrix}1 & -1\\ -1 & 1\end{bmatrix}\begin{bmatrix}T_1^{(e)}\\ T_2^{(e)}\end{bmatrix} = h\frac{P}{2A}T_\infty\begin{bmatrix}1\\ 1\end{bmatrix} \quad (7.13.215)$$

In a general manner, the temperature vector for a finite element with n nodes is

$$\mathbf{T}^{(e)} = \left[T_1^{(e)}\ T_2^{(e)} \ldots\ T_n^{(e)}\right]^T \quad (7.13.216)$$

and is equal to the vector $\left[T_1^{(e)}\ 0 \ldots\ T_n^{(e)}\right]^T$ only if the number of nodes of the element is reduced to 2 (linear interpolation). In the present case, the conduction terms cancel themselves. It would not be the case if there were more than two nodes. In Equation (7.13.215), the conduction term with sign + refers to full vector $\mathbf{T}^{(e)}$ while the conduction term with sign – refers to vector $\mathbf{T}^{(e)}$ with only the vector extremities and zeros inside. In all cases, Equation (7.13.207) is the reference.

The matrices $\mathbf{L}^{(e)}$ are given by Equation (7.13.102) as we have the same number of elements and nodes per element.

Then, we numerically obtain for the whole domain

$$\begin{bmatrix}-1411.96 & 1439.43 & 0 & 0\\ -1421.11 & 54.93 & 1448.58 & 0\\ 0 & -1411.96 & 1648.00 & 1494.36\\ 0 & 0 & -1366.18 & 1558.45\end{bmatrix}\mathbf{T} = \begin{bmatrix}7502.52\\ 22507.56\\ 67522.68\\ 52517.64\end{bmatrix} \quad (7.13.217)$$

that can be set under the form

$$\mathbf{K}_g\mathbf{T} = \mathbf{q}_g \quad (7.13.218)$$

The temperature T_1 being known at the inlet of the tube $T_1 = T_{in}$, the solving method similar to that of Example 7.22 can be used. Equation (7.13.218) is now written as

$$\mathbf{K}_g \begin{bmatrix} T_{in} \\ T_2 \\ T_3 \\ T_4 \end{bmatrix} - \mathbf{q}_g = \begin{bmatrix} r_1 \\ 0 \\ 0 \\ 0 \end{bmatrix} \tag{7.13.219}$$

where r_1 is an unknown flux. The matrices and vectors are partitioned under the numerical form

$$\left[\begin{array}{c|ccc} -1411.96 & 1439.43 & 0 & 0 \\ \hline -1421.12 & 54.93 & 1448.58 & 0 \\ 0 & -1411.96 & 1648.01 & 1494.36 \\ 0 & 0 & -1366.18 & 1558.45 \end{array}\right] \left[\begin{array}{c} T_{in} \\ \hline T_2 \\ T_3 \\ T_4 \end{array}\right] - \left[\begin{array}{c} 7502.52 \\ \hline 22507.56 \\ 67522.68 \\ 52517.64 \end{array}\right] = \left[\begin{array}{c} r_1 \\ \hline 0 \\ 0 \\ 0 \end{array}\right] \tag{7.13.220}$$

The three last equations give

$$\begin{bmatrix} 54.93 & 1448.58 & 0 \\ -1411.96 & 1648.01 & 1494.36 \\ 0 & -1366.18 & 1558.45 \end{bmatrix} \begin{bmatrix} T_2 \\ T_3 \\ T_4 \end{bmatrix} = \begin{bmatrix} 22507.56 \\ 67522.68 \\ 52517.64 \end{bmatrix} - \begin{bmatrix} -1421.12 \\ 0 \\ 0 \end{bmatrix} T_{in} \tag{7.13.221}$$

Thus, the nodal temperature vector is obtained

$$\mathbf{T} = \begin{bmatrix} 376.150 & 374.673 & 370.347 & 358.3558 \end{bmatrix} \tag{7.13.222}$$

Bergman et al. (2011) gives 358.15 K as the temperature at the tube outlet thus the agreement is very good (Figure 7.88). With eleven nodes regularly spaced, the terminal temperature is 358.1540 K, thus the agreement is nearly perfect and the profile more smooth (Figure 7.88)

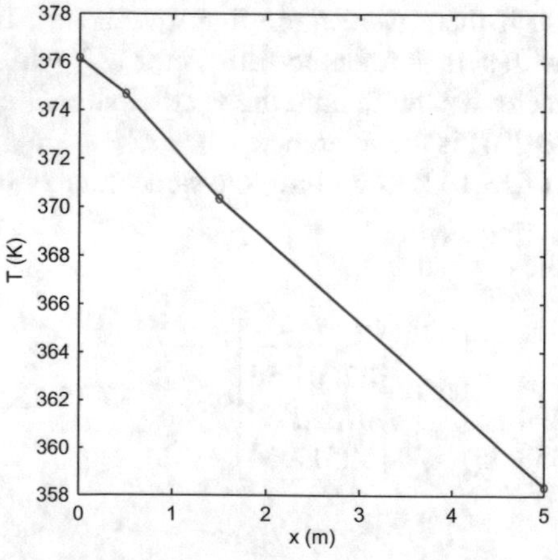

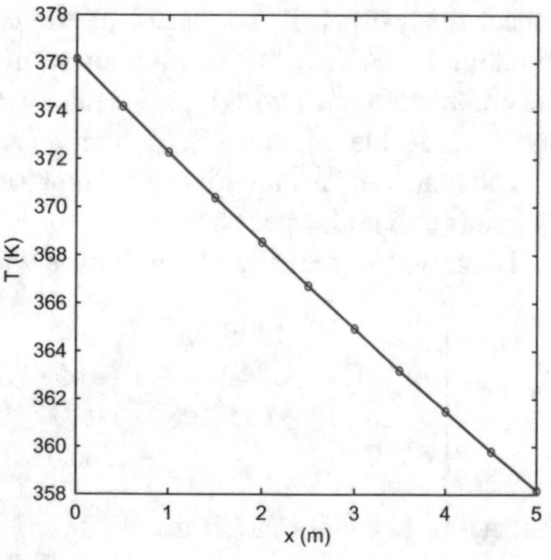

Fig. 7.88 Example HTC1-FEM-1D: Tube with air flow. Left: Temperature profile in the tube with 4 nodes irregularly spaced. Right: Temperature profile in the tube with 11 nodes regularly spaced

Lewis et al. (2004) describes in detail the solution of the transient convection-diffusion problem which is modelled in the multi-dimensional case as

$$\frac{\partial \phi}{\partial t} + u_i \frac{\partial \phi}{\partial x_i} + \phi \frac{\partial u_i}{\partial x_i} - \frac{\partial}{\partial t}\left(\kappa \frac{\partial \phi}{\partial x_i}\right) + Q = 0 \tag{7.13.223}$$

where $\kappa = \lambda$ and $\phi = T$ in the case of thermal diffusion. u_i are the velocity components and Q a source term. The consideration of convection must be carefully done to avoid oscillations beyond a given Peclet number. Then, for the finite elements, upwind schemes comparable to those existing in finite volumes must be used. In the case of time-dependent equations, there exist Taylor-Galerkin or the characteristic Galerkin's scheme (Zienkiewicz et al. 2013). The treatment of momentum equations is done in two main parts, first neglecting the influence of pressure and using intermediary velocities (stage 1), then, using the pressure calculation (stage 2), the velocities are corrected (stage 3), finally the temperatures are calculated (stage 4).

7.13.7 Flow in a Porous Medium

The flow of an incompressible fluid in a porous medium is described by Bear (1988), Guyon et al. (2015), Koutromanos (2018), Lewis et al. (2004), and Logan (2016). In the one-dimensional case, the fundamental equation is Darcy's law that can be written as

$$\vec{q} = -\frac{k}{\mu}\vec{\nabla} P \tag{7.13.224}$$

where q if the volume flux (volume flow rate per surface unit) k the permeability of the porous medium, μ the fluid viscosity, P the pressure. Lewis et al. (2004) applies to the finite elements the generalized Darcy's law modified by Forchheimer or Brinkman or Ergun, with moreover the energy balance.

The strong formulation of Darcy's equation is written as

$$-\langle \vec{\nabla}, \left(\frac{k}{\mu}\vec{\nabla} P\right)\rangle = 0 \tag{7.13.225}$$

with the boundary conditions

$$\begin{aligned} &P = \bar{P} \text{ on } \Gamma_D \quad \text{(essential boundary condition)} \\ &\langle \vec{q}, \vec{n} \rangle = -\langle \left(\frac{k}{\mu}\vec{\nabla} P\right), \vec{n} \rangle = 0 \text{ on } \Gamma_N \quad \text{(natural boundary condition)} \end{aligned} \tag{7.13.226}$$

where $\bar{P}$ is a given pressure.

7.13.8 Diffusion—Chemical Reaction

The diffusion of a chemical species from a medium of high concentration towards a medium of lower concentration is ruled by Fick's law

$$\vec{q} = -\mathcal{D}\vec{\nabla}C \tag{7.13.227}$$

where C is the species concentration, $\mathcal{D}$ the diffusion coefficient and $\vec{q}$ the mass flux. If a chemical reaction is added to diffusion, the strong formulation is written as

$$-\langle\vec{\nabla},\left(\mathcal{D}\vec{\nabla}C\right)\rangle = \mathcal{R}(x,y) \tag{7.13.228}$$

where $\mathcal{R}(x,y)$ is the rate of reaction, expressed as a mass par volume and time units. This rate is null in the case of pure diffusion. The boundary conditions are

$$\begin{aligned} &C = \bar{C} \text{ on } \Gamma_D \quad \text{(essential boundary condition)} \\ &\langle\vec{q},\vec{n}\rangle = -\langle\left(\mathcal{D}\vec{\nabla}C\right),\vec{n}\rangle = 0 \text{ on } \Gamma_N \quad \text{(natural boundary condition)} \end{aligned} \tag{7.13.229}$$

where $\bar{C}$ is a given concentration.

Fick's law is similar to Fourier's law and the treatment of diffusion will be done in the same way as for heat transfer.

Example 7.28 :

MT1-FEM-1D: Transient regime of gas transfer through a membrane

We consider Example 7.16 already treated by finite volumes accompanied by Table 7.2 summarizing the simulation conditions.

The strong formulation is that of Fick's law

$$\frac{\partial C}{\partial t} = \mathcal{D}\frac{\partial^2 C}{\partial x^2} \tag{7.13.230}$$

To simplify, the diffusion coefficient $\mathcal{D}$ is assumed to be constant. The boundary conditions and the initial condition are

$$C(x=0,t) = C_0 = SP_0\,, \quad C(x=L,t) = C_1 = SP_1\,, \quad C(x,t=0) = 0 \tag{7.13.231}$$

where S is the sorption coefficient, P_0 the upstream pressure assumed to be fixed, P_1 the variable downstream pressure following the equation

$$\frac{dP_1}{dt} = \frac{dn_{in}}{dt}\frac{RT}{V_{downstream}} \quad \text{with} \quad \frac{dn_{in}}{dt} = -A\mathcal{D}\left.\frac{\partial C}{\partial x}\right|_L = \frac{V_{downstream}}{RT}\frac{1}{S}\frac{dC(x=L,t)}{dt} \tag{7.13.232}$$

The weak formulation is written as

$$\sum_{e=1}^{n_e}\int_{x_1^{(e)}}^{x_2^{(e)}} w(x)\frac{\partial C}{\partial t}dx + \sum_{e=1}^{n_e}\int_{x_1^{(e)}}^{x_2^{(e)}} \frac{dw}{dx}\mathcal{D}\frac{\partial C}{\partial x}dx = \sum_{e=1}^{n_e}\left[w(x)\mathcal{D}\frac{\partial C}{\partial x}\right]_{x_1^{(e)}}^{x_2^{(e)}} \tag{7.13.233}$$

The concentration approximation is

$$C(x) \approx \tilde{C}^{(e)}(x) = \sum_{j=1}^{n} N_j^{(e)}(x) C_j^{(e)} = \left[N^{(e)}(x)\right] \left[\mathbf{C}^{(e)}\right] \tag{7.13.234}$$

and the interpolation functions

$$w(x) \approx \tilde{w}^{(e)}(x) = \sum_{j=1}^{n} N_j^{(e)}(x) w_j^{(e)} = \mathbf{N}^{(e)}(x)\mathbf{W}^{(e)} = \mathbf{W}^{(e)^T} \mathbf{N}^{(e)^T}(x) \tag{7.13.235}$$

where n is the number of nodes of element $^{(e)}$. The time derivative term of concentration is expressed as

$$\begin{aligned} \int_{x_1^{(e)}}^{x_2^{(e)}} w(x) \frac{\partial C}{\partial t} dx &\approx \mathbf{W}^{(e)^T} \int_{x_1^{(e)}}^{x_2^{(e)}} \left(\mathbf{N}^{(e)^T} \mathbf{N}^{(e)} dx\right) \frac{\partial \mathbf{C}^{(e)}}{\partial t} \\ &= \mathbf{W}^{(e)^T} \mathcal{C}^{(e)} \dot{\mathbf{C}}^{(e)} \end{aligned} \tag{7.13.236}$$

where $\mathcal{C}^{(e)}$ is the capacitance matrix equal to

$$\mathcal{C}^{(e)} = \int_{x_1^{(e)}}^{x_2^{(e)}} \mathbf{N}^{(e)^T} \mathbf{N}^{(e)} dx = \frac{1}{6} \begin{bmatrix} 2l^{(e)} & l^{(e)} \\ l^{(e)} & 2l^{(e)} \end{bmatrix} \tag{7.13.237}$$

By assembling, we obtain the global capacitance matrix equal to

$$\mathcal{C} = \sum_{e=1}^{n_e} \left(\mathbf{L}^{(e)^T} \mathcal{C}^{(e)} \mathbf{L}^{(e)}\right) \tag{7.13.238}$$

and the differential equation to be solved is

$$\mathcal{C}\dot{\mathbf{C}} + \mathbf{K}\mathbf{C} = \mathbf{q} \tag{7.13.239}$$

with the conductance matrix

$$\mathbf{K}^{(e)} = \left[\int_{x_1^{(e)}}^{x_2^{(e)}} \mathbf{B}^{(e)^T} \mathcal{D} \mathbf{B}^{(e)} dx\right] \tag{7.13.240}$$

and the global conductance matrix

$$\mathbf{K} = \sum_{e=1}^{n_e} \left(\mathbf{L}^{(e)^T} \mathbf{K}^{(e)} \mathbf{L}^{(e)}\right) \tag{7.13.241}$$

For the needs of the following demonstration, so as to simplify the notations, the membrane is discretized into three finite elements with two nodes each and linear approximation functions are taken, identical to those of Example 7.20.

We thus obtain

$$\mathbf{K}^{(e)} = \mathcal{D} \begin{bmatrix} \frac{1}{l^{(e)}} & -\frac{1}{l^{(e)}} \\ -\frac{1}{l^{(e)}} & \frac{1}{l^{(e)}} \end{bmatrix} \tag{7.13.242}$$

and

$$\mathbf{K} = \mathcal{D} \begin{bmatrix} \frac{1}{l^{(1)}} & -\frac{1}{l^{(1)}} & 0 & 0 \\ -\frac{1}{l^{(1)}} & \frac{1}{l^{(1)}} + \frac{1}{l^{(2)}} & -\frac{1}{l^{(2)}} & 0 \\ 0 & -\frac{1}{l^{(2)}} & \frac{1}{l^{(2)}} + \frac{1}{l^{(3)}} & -\frac{1}{l^{(3)}} \\ 0 & 0 & -\frac{1}{l^{(3)}} & \frac{1}{l^{(3)}} \end{bmatrix} \tag{7.13.243}$$

The global flux vector is

$$\mathbf{q} = \left[r_1 \; 0 \ldots 0 \; q_L''\right]^T \tag{7.13.244}$$

with the variable flux q''_L depending on pressure P_1, thus on $C_2^{(3)}$. r_1 is an unknown flux to take into account the constraint $C_1^{(1)} = \mathcal{S} P_0$. Moreover

$$q''_L = -\frac{V_{downstream}}{ART\mathcal{S}} \frac{dC_2^{(3)}}{dt} \tag{7.13.245}$$

Thus, we can write

$$\mathbf{C}\dot{\mathbf{C}} + \mathbf{K}\mathbf{C} = \begin{bmatrix} r_1 \\ 0 \\ \vdots \\ 0 \end{bmatrix} + \begin{bmatrix} 0 \ldots & & 0 \\ \vdots & & \vdots \\ 0 \ldots & & 0 \\ 0 \ldots & 0 & -\frac{V_{downstream}}{ART\mathcal{S}} \end{bmatrix} \dot{\mathbf{C}} \tag{7.13.246}$$

hence

$$\left(\mathbf{C} + \begin{bmatrix} 0 \ldots & & 0 \\ \vdots & & \vdots \\ 0 \ldots & & 0 \\ 0 \ldots & 0 & \frac{V_{downstream}}{ART\mathcal{S}} \end{bmatrix} \right) \dot{\mathbf{C}} + \mathbf{K}\mathbf{C} = \begin{bmatrix} r_1 \\ 0 \\ \vdots \\ 0 \end{bmatrix} \tag{7.13.247}$$

As $C_1^{(1)}$ is known and fixed, solving is resumed as

$$\left(\begin{bmatrix} C_{22} & \ldots & C_{24} \\ \vdots & & \vdots \\ C_{42} & \ldots & C_{44} \end{bmatrix} + \begin{bmatrix} 0\;0 & & 0 \\ 0\;0 & & 0 \\ 0\;0 & & \frac{V_{downstream}}{ART\mathcal{S}} \end{bmatrix} \right) \begin{bmatrix} \dot{C}_2^{(1)} \\ \dot{C}_2^{(2)} \\ \dot{C}_2^{(3)} \end{bmatrix} + \begin{bmatrix} K_{22} & \ldots & K_{24} \\ \vdots & & \vdots \\ K_{42} & \ldots & K_{44} \end{bmatrix} \begin{bmatrix} C_2^{(1)} \\ C_2^{(2)} \\ C_2^{(3)} \end{bmatrix} = \begin{bmatrix} 0 \\ \vdots \\ 0 \end{bmatrix} - \begin{bmatrix} K_{21} \\ \ldots \\ K_{41} \end{bmatrix} C_1^{(1)} \tag{7.13.248}$$

For the numerical simulation, the thickness of the membrane has been divided into 20 identical finite elements of 2 nodes each. Different transient simulations have been performed, either with explicit Euler method or with semi-implicit Galerkin's method ($\theta = 2/3$) under predictor–corrector form, moreover with different time steps and a final equal to 20 s. The explicit Euler method becomes unstable for a time step of 10^{-1} s whereas the semi-implicit Galerkin's method still gives extremely close results with this step. When Euler method is stable, both methods give very close results. The theoretical values of concentrations have been calculated from Equation (7.12.53) by truncating the series at $n = 10{,}000$. However, this theoretical concentration is very slightly erroneous as it is based on the hypothesis that the downstream pressure P_1 is zero whereas its increase is taken into account in the calculation by finite elements. The pressure P_1 is equal to $P_1 = \mathcal{S}C_2^{(3)}$. A few results are shown in Tables 7.8, 7.9, and Figure 7.89. The agreement is very good.

7.13.9 Fluid Mechanics

The domain is extremely vast (Batchelor 2000; Guyon et al. 2015; Lewis et al. 2004; Pletcher et al. 2013; Reddy and Gartling 2010; White 2016; Zienkiewicz et al. 2014) and here a few main characteristics will be outlined. Only incompressible fluids, Newtonian (nonviscous), will be concerned. The presented approach is simplified as Navier–Stokes equations are not completely considered and as mainly the steady-state

Table 7.8 Example MT1-FEM-1D: Comparison of calculated and theoretical concentrations inside the membrane at $t = 20$ s with a time step of 10^{-4}. In each case, twenty finite elements with two nodes each

Position (m)	Calculated concentration (mol m^{-3}) Explicit Euler	Theoretical concentration (mol m^{-3})
0.00	0.1013×10^{-2}	0.1013×10^{-2}
0.20×10^{-3}	0.6641×10^{-3}	0.6633×10^{-3}
0.40×10^{-3}	0.3767×10^{-3}	0.3756×10^{-3}
0.60×10^{-3}	0.1811×10^{-3}	0.1803×10^{-3}
0.80×10^{-3}	0.6756×10^{-4}	0.6722×10^{-4}

Table 7.9 Example MT1-FEM-1D: Comparison of calculated and theoretical concentrations at $x = 0.4 \times 10^{-3}$ m, at $t = 20$ s with different time steps and different methods. In each case, twenty finite elements with two nodes each

Method of integration	Time step (s)	Calculated concentration (mol m^{-3})	Theoretical concentration (mol m^{-3})
Explicit Euler	10^{-1}	Unstable	0.375622×10^{-3}
Explicit Euler	10^{-2}	0.376768×10^{-3}	idem
Explicit Euler	10^{-3}	0.376706×10^{-3}	
Explicit Euler	10^{-4}	0.376700×10^{-3}	
Galerkin	10^{-1}	0.376069×10^{-3}	
Galerkin	10^{-2}	0.376636×10^{-3}	
Galerkin	10^{-3}	0.376693×10^{-3}	
Galerkin	10^{-4}	0.376698×10^{-3}	

regime is considered. For a more general treatment of Navier–Stokes equations in the frame of finite elements, among numerous books, the reader can refer to Pozrikidis (2014) and Rao (2018).

In transient regime, in Cartesian coordinates and in 3D, the continuity equation is

$$\frac{\partial \rho}{\partial t} + \text{div}(\rho \vec{V}) = \frac{\partial \rho}{\partial t} + u\frac{\partial \rho}{\partial x} + v\frac{\partial \rho}{\partial y} + w\frac{\partial \rho}{\partial z} + \rho\left(\frac{\partial u}{\partial x} + \frac{\partial v}{\partial y} + \frac{\partial w}{\partial z}\right) = 0 \quad (7.13.249)$$

where u, v, w are the velocity components, ρ the density. In the case of a steady incompressible flow, this equation is reduced as

$$\text{div}\vec{V} = \frac{\partial u}{\partial x} + \frac{\partial v}{\partial y} + \frac{\partial w}{\partial z} = 0 \quad (7.13.250)$$

When the flow is irrotational, the following equations are used

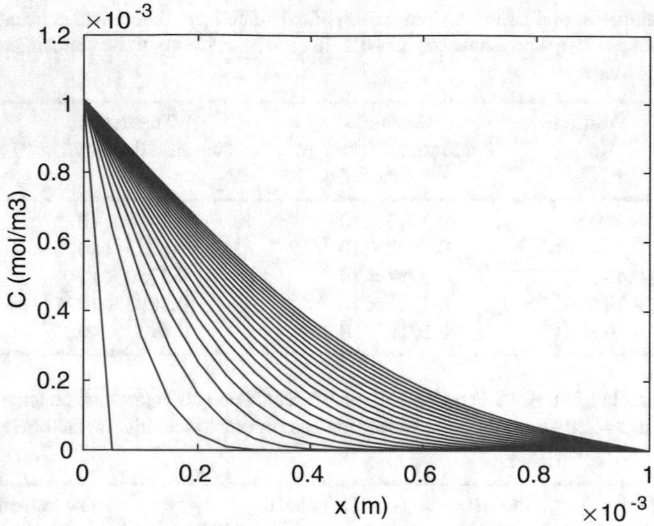

Fig. 7.89 Example MT1-FEM-1D: Evolution of the numerical concentration profile in the membrane until $t = 20$ s every second

$$\vec{\text{rot}}(\vec{V}) = 0 \Longrightarrow \begin{cases} \dfrac{\partial w}{\partial y} - \dfrac{\partial v}{\partial z} = 0 \\ \dfrac{\partial u}{\partial z} - \dfrac{\partial w}{\partial x} = 0 \\ \dfrac{\partial v}{\partial x} - \dfrac{\partial u}{\partial y} = 0 \end{cases} \tag{7.13.251}$$

For simplicity reasons, a one-dimensional flow is described. The stream function $\psi(x, y)$ is defined as

$$u = \frac{\partial \psi}{\partial y}, \quad v = -\frac{\partial \psi}{\partial x} \tag{7.13.252}$$

so that Laplace's equation results which constitutes the strong formulation

$$\frac{\partial^2 \psi}{\partial x^2} + \frac{\partial^2 \psi}{\partial y^2} = \Delta \psi = \nabla^2 \psi = 0 \tag{7.13.253}$$

by adding the boundary conditions. Frequently, the streamlines are considered along which the stream function is constant.

The simplest introduction to flows by means of finite elements consists in considering the stream function and discretizing it into finite elements on the studied domain

$$\psi(x, y) = \sum_{i=1}^{n} N_i(x, y)\psi_i = \mathbf{N}\boldsymbol{\psi} \tag{7.13.254}$$

To solve the transient Navier–Stokes equations in the framework of finite elements, Pozrikidis (2014) discretizes the velocity fields and the pressure fields with different sets of interpolation functions. He thus obtains a system of differential equations for the nodal velocities where the nodal pressures are also present. Moreover, there exist a system of algebraic equations for the nodal velocities associated with the continuity equation which acts as a constraint.

The application of Galerkin's method on a domain $\Omega^{(e)}$ gives

$$\int_{\Omega^{(e)}} N_i(x,y)\left(\frac{\partial^2\psi}{\partial x^2}+\frac{\partial^2\psi}{\partial y^2}\right)dx\,dy=0\,,\quad i=1,\ldots,n \tag{7.13.255}$$

or still

$$\int_{\Omega^{(e)}} \mathbf{N}^T\left(\frac{\partial^2\psi}{\partial x^2}+\frac{\partial^2\psi}{\partial y^2}\right)dx\,dy=0 \tag{7.13.256}$$

By applying Green-Ostrogradsky theorem, we obtain

$$\begin{aligned}&\int_{S^{(e)}} \mathbf{N}^T\frac{\partial\psi}{\partial x}n_x dS-\int_{\Omega^{(e)}}\frac{\partial\mathbf{N}^T}{\partial x}\frac{\partial\psi}{\partial x}dx\,dy+\\&\int_{S^{(e)}} \mathbf{N}^T\frac{\partial\psi}{\partial y}n_y dS-\int_{\Omega^{(e)}}\frac{\partial\mathbf{N}^T}{\partial y}\frac{\partial\psi}{\partial y}dx\,dy=0\end{aligned} \tag{7.13.257}$$

where $S^{(e)}$ is the surface surrounding the volume $\Omega^{(e)}$. n_x and n_y are the components of the vector normal to the surface and directed towards the outside.

It results

$$\int_{\Omega^{(e)}}\left(\frac{\partial\mathbf{N}^T}{\partial x}\frac{\partial\mathbf{N}}{\partial x}+\frac{\partial\mathbf{N}^T}{\partial y}\frac{\partial\mathbf{N}}{\partial y}\right)dx\,dy\,\boldsymbol{\psi}=\int_{S^{(e)}}\mathbf{N}^T(un_y-vn_x)dS \tag{7.13.258}$$

This equation is of the form

$$\mathbf{K}^{(e)}\psi=\mathbf{f}^{(e)} \tag{7.13.259}$$

similar to Equation (7.13.86) in heat transfer. $\mathbf{K}^{(e)}$ is the stiffness matrix and $\mathbf{f}^{(e)}$ the nodal force vector.

The problem is only completely defined when the boundary conditions are specified. The partial differential equation of the two-dimensional stream function being of order 2, four boundary conditions are required that may deal with the velocity or the stream function or the nodal forces. For example, null nodal forces mean that the streamlines are normal to the boundaries. If the horizontal axis is a symmetry axis, the velocity component v is zero and that axis is a streamline, thus the stream function ψ is constant along this axis. If a surface limiting the system is impenetrable, the velocity component perpendicular to this surface is zero and the streamline along this surface is constant.

Another manner to solve the two-dimensional incompressible flow problem is to use the potential function of velocities such that

$$u=-\frac{\partial\phi}{\partial x}\,,\quad v=-\frac{\partial\phi}{\partial x} \tag{7.13.260}$$

the condition of irrotationality (7.13.251) is automatically verified. The continuity equation (7.13.250) becomes Laplace's equation which constitutes the strong formulation

$$\frac{\partial^2 \phi}{\partial x^2} + \frac{\partial^2 \phi}{\partial y^2} = \Delta\phi = \nabla^2\phi = 0 \tag{7.13.261}$$

In 2D, we deduce

$$d\phi = \frac{\partial \phi}{\partial x}dx + \frac{\partial \phi}{\partial y}dy = -(udx+vdy) = -\langle \begin{bmatrix} u \\ v \end{bmatrix}, \begin{bmatrix} dx \\ dy \end{bmatrix} \rangle = -\begin{bmatrix} u \\ v \end{bmatrix} \bullet \begin{bmatrix} dx \\ dy \end{bmatrix} = 0 \tag{7.13.262}$$

The notation $\langle \vec{V}_1, \vec{V}_2 \rangle$ or $\bullet$ stands for the scalar product. As this scalar product is zero, the streamlines and the equipotential lines are orthogonal.

The formulation in finite elements based on the potential function is similar to that based on the stream function as both functions verify Laplace's equation.

We thus get

$$\int_{\Omega^{(e)}} \left(\frac{\partial \mathbf{N}^T}{\partial x}\frac{\partial \mathbf{N}}{\partial x} + \frac{\partial \mathbf{N}^T}{\partial y}\frac{\partial \mathbf{N}}{\partial y} \right) dx\, dy\, \boldsymbol{\phi} = -\int_{S^{(e)}} \mathbf{N}^T (un_x + vn_y) dS \tag{7.13.263}$$

This equation is of the form

$$\mathbf{K}^{(e)}\boldsymbol{\phi} = \mathbf{f}^{(e)} \tag{7.13.264}$$

The stiffness matrix is the same as in Equation (7.13.258), but the nodal force vector is different.

The mathematical consequences of the boundary conditions in terms of finite elements are rather complex (Hutton 2004; Koutromanos 2018) to be developed and can be summarized as

$$\begin{aligned} &\phi = \text{constant on } \Gamma_D \quad \text{(essential boundary condition)} \\ &\langle \vec{V}, \vec{n} \rangle = \langle \vec{\nabla}\phi, \vec{n} \rangle = V \ \text{ on } \Gamma_N \quad \text{(natural boundary condition)} \end{aligned} \tag{7.13.265}$$

where $\vec{n}$ is the outward normal vector normal at boundary Γ_N and V the value of the considered velocity.

For the treatment of coupled general flows and heat transfer, the reader can consult (Lewis et al. 2004).

7.13.10 Two-Dimensional Formulation

In two-dimensional formulation (2D), different types of elements can be considered, triangles, quadrilaterals (Rao 2018). To simplify the presentation, a triangular element $^{(e)}$ is considered (Figure 7.90).

The two-dimensional heat transfer closely follows the one-dimensional (1D) formulation already presented (Koutromanos 2018; Lewis et al. 2004; Logan 2016). The method can be described by the following stages

1. Determine the strong formulation constituted by the model that describes the physics of the studied heat transfer problem, including the conduction inside the system, the heat generation, the boundary conditions (fixed temperature: essential condition or

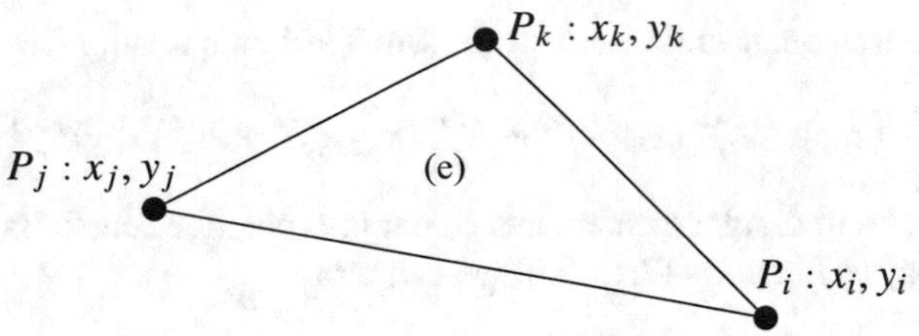

Fig. 7.90 Triangular finite element

Dirichlet, fixed flux: natural condition or Neumann). If the problem is transient, add the initial conditions. The general equation of the form inside the system is written as

$$\langle \vec{\nabla}, - \begin{bmatrix} \lambda_{xx} & \lambda_{xy} \\ \lambda_{yx} & \lambda_{yy} \end{bmatrix} \begin{bmatrix} \frac{\partial T}{\partial x} \\ \frac{\partial T}{\partial y} \end{bmatrix} \rangle - s(x, y) = 0 \tag{7.13.266}$$

or

$$\langle \vec{\nabla}, -\mathbf{K}\vec{\nabla}T \rangle - s(x, y) = -\vec{\nabla} \cdot \mathbf{K}\vec{\nabla}T - s(x, y) = 0 \tag{7.13.267}$$

where $s(x, y)$ is a source term and $\vec{\nabla} = \frac{\partial}{\partial x}\vec{i} + \frac{\partial}{\partial y}\vec{j}$. To this equation, the boundary conditions must be added to obtain the complete strong formulation.
The developed strong formulation is written as

$$\frac{\partial}{\partial x}\left[\lambda_{xx}\frac{\partial T}{\partial x} + \lambda_{xy}\frac{\partial T}{\partial y}\right] + \frac{\partial}{\partial y}\left[\lambda_{yx}\frac{\partial T}{\partial x} + \lambda_{yy}\frac{\partial T}{\partial y}\right] + s(x, y) = 0 \tag{7.13.268}$$

2. Deduce the weak formulation by integration on the finite element and use of Green-Ostrogradsky formula

$$\iint_{\Omega} \vec{\nabla}w \cdot \mathbf{K}\vec{\nabla}T \, dV = \iint_{\Omega} w \, s \, dV - \int_{\Gamma_N} w \, \bar{q} \, dS \tag{7.13.269}$$

where $w(x, y)$ is any function such that $w = 0$ at the essential boundaries Γ_D. The flux value is $\bar{q} = \vec{q} \cdot \vec{n}$ where $\vec{n}$ is the vector normal to Γ_N. The weak formulation is also written under the form

$$\iint_{\Omega} [\nabla w]^T \, \mathbf{K}\nabla T \, dV = \iint_{\Omega} w \, s \, dV - \int_{\Gamma_N} w \, \bar{q} \, dS \tag{7.13.270}$$

After having determined this weak formulation, the solving stages are

3. Selection of the type of 2D finite element (triangle, quadrilateral).
4. Choice of the approximation function for temperature (linear, quadratic). The temperature is approximated as

$$T \approx \tilde{T}^{(e)}(x, y) = \sum_{\alpha=1}^{n} N_{\alpha}^{(e)}(x, y) T_{\alpha}^{(e)} = \mathbf{N}^{(e)}\mathbf{T}^{(e)} \tag{7.13.271}$$

Assuming a triangular finite element (Figure 7.90), this relation becomes

$$T \approx \tilde{T}^{(e)}(x, y) = N_i^{(e)}(x, y)T_i^{(e)} + N_j^{(e)}(x, y)T_j^{(e)} + N_k^{(e)}(x, y)T_k^{(e)} \quad (7.13.272)$$

In the case of a triangular element and linear interpolation functions, proceeding in the same way as Equation (7.13.50), we can write

$$\tilde{T}_\alpha^{(e)} = a_0 + a_1 x_\alpha + a_2 y_\alpha \quad (7.13.273)$$

hence

$$\mathbf{T}^{(e)} = \begin{bmatrix} \tilde{T}_i^{(e)} \\ \tilde{T}_j^{(e)} \\ \tilde{T}_k^{(e)} \end{bmatrix} = \begin{bmatrix} 1 & x_i^{(e)} & y_i^{(e)} \\ 1 & x_j^{(e)} & y_j^{(e)} \\ 1 & x_k^{(e)} & y_k^{(e)} \end{bmatrix} \begin{bmatrix} a_0 \\ a_1 \\ a_2 \end{bmatrix} = \mathbf{M}^{(e)} \mathbf{a}^{(e)} \quad (7.13.274)$$

We set

$$\mathbf{p}(x, y) = \begin{bmatrix} 1 & x & y \end{bmatrix} \quad (7.13.275)$$

and we obtain the interpolation functions

$$\mathbf{N}^{(e)}(x, y) = \mathbf{p}(x, y) {\mathbf{M}^{(e)}}^{-1} \mathbf{T}^{(e)} \quad (7.13.276)$$

5. Calculation of the temperature gradient and flux
The temperature gradient is

$$\mathbf{g} = \begin{bmatrix} \frac{\partial T}{\partial x} \\ \frac{\partial T}{\partial y} \end{bmatrix} = \begin{bmatrix} \frac{\partial N_i}{\partial x} & \frac{\partial N_j}{\partial x} & \frac{\partial N_k}{\partial x} \\ \frac{\partial N_i}{\partial y} & \frac{\partial N_j}{\partial y} & \frac{\partial N_k}{\partial y} \end{bmatrix} \begin{bmatrix} T_i \\ T_j \\ T_k \end{bmatrix} = \mathbf{B}^{(e)} \mathbf{T}^{(e)} \quad (7.13.277)$$

and the flux

$$\mathbf{q}^{(e)} = \begin{bmatrix} q_x \\ q_y \end{bmatrix} \quad (7.13.278)$$

hence for the finite element $^{(e)}$

$$ {\mathbf{W}^{(e)}}^T \mathbf{K}^{(e)} \mathbf{T}^{(e)} = {\mathbf{W}^{(e)}}^T \mathbf{q}^{(e)} \quad (7.13.279)$$

with the conductance matrix

$$\mathbf{K}^{(e)} = \begin{bmatrix} \lambda_{xx} & \lambda_{xy} \\ \lambda_{yx} & \lambda_{yy} \end{bmatrix} \quad (7.13.280)$$

In the case of isotropic materials, the matrix $\mathbf{K}^{(e)}$ is diagonal. In Galerkin's formulation, the weight functions w are equal to the interpolation functions $\tilde{T}$.

6. Assemble matrices and vectors. The influence of the system boundary conditions clearly appears at this level. Finally, we must obtain a general equation of the form

$$\mathbf{W}^T (\mathbf{K}_g \mathbf{T} - \mathbf{q}_g) = 0 \qquad \forall \mathbf{W} \quad (7.13.281)$$

where the notations $\mathbf{K}_g$ and $\mathbf{q}_g$ refer to final global equations.

7.13.11 Examples of 2D and 3D Simulations

The 2D and 3D simulations were performed by means of the free code Gmsh (Geuzaine and Remacle 2009) for the mesh, the free code Elmer (Malinen and Råback 2013) for the finite element simulation and the free code Paraview for postprocessing.

7.13.11.1 2D Simulation

The 2D simulation is demonstrated by the single Example 7.29.

Example 7.29 :
HT1-FEM-2D: Rectangular plate subjected to Dirichlet conditions

This example has first been treated by the finite difference method and alternate directions (cf. Example 7.5).

In steady-state regime, in two dimensions, Fourier's equation becomes Laplace's equation

$$\frac{\partial^2 T}{\partial x^2} + \frac{\partial^2 T}{\partial y^2} = 0 \tag{7.13.282}$$

The example deals with a metal plate subjected to different Dirichlet conditions on four sides

$$\begin{array}{l} T(x=0,y) = 450\,\mathrm{K}\,,\ T(x=L_x,y) = 550\,\mathrm{K} \\ T(x,y=0) = 400\,\mathrm{K}\,,\ T(x,y=L_y) = 500\,\mathrm{K} \end{array} \tag{7.13.283}$$

Numerical application:

Dimensions $L_x = 0.02\,\mathrm{m}$, $L_y = 0.05\,\mathrm{m}$, density $\rho = 7800\,\mathrm{kg\,m^{-3}}$, heat capacity $C_p = 500\,\mathrm{J\,kg^{-1}\,K^{-1}}$, thermal conductivity $\lambda = 26\,\mathrm{W\,m^{-1}\,K^{-1}}$.

Within the framework of the finite element method, first a mesh has been performed with the open source code Gmsh (Figure 7.91) by using 488 triangular elements. Then, the simulation in order to obtain the steady-state regime was done with the open source simulation code Elmer by finite elements (Figure 7.91). The results are similar to those obtained by finite differences (Figure 7.27 of Example 7.5).

7.13.11.2 3D Simulations

Different configurations dealing with the rod previously studied as a 1D object (Examples 7.20 to 7.26) are worked again in order to take into account the transversal dimension of the rod considered as a cylinder. Only the steady state case is considered. In these 3D simulations, the radius influence will be taken into account. In the following examples (7.30 to 7.33), the rod will be treated as a 3D object and the simulation results will be compared between the 1D and 3D simulations.

The 3D studied cases are:

- HT3-FEM-3D: Transposition of HT3-FEM-1D. Study of a 3D metal rod receiving a constant thermal power on its periphery (Neumann boundary condition), an imposed flux at $x = 0$ (Neumann boundary condition), providing a convective flux at $x = L$ (Neumann boundary condition).

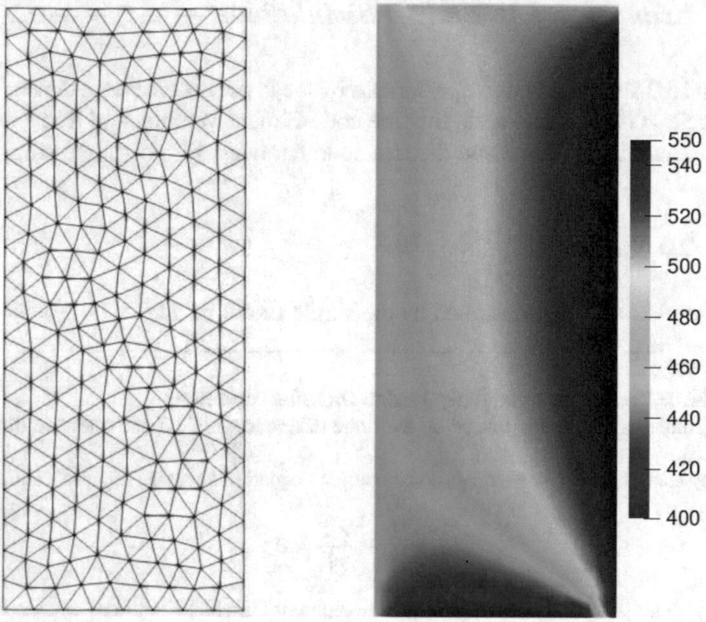

Fig. 7.91 Example HT1-FEM-2D: Plane wall subjected to heat transfer with Dirichlet conditions on four sides. Left: 2D mesh of the plate by the code Gmsh. Right: Temperature surface of 2D simulation by the code Elmer

- HT4-FEM-3D: Transposition of HT4-FEM-1D. Similar to HT3-FEM-3D, but insulated rod on the periphery (Neumann boundary condition).
- HT5-FEM-3D: Transposition of HT5-FEM-1D. Insulated rod with an imposed flux at $x = 0$ (Neumann boundary condition) and an imposed temperature at $x = L$ (Dirichlet boundary condition).
- HT7-FEM-3D: Transposition of HT7-FEM-1D. Rod receiving an imposed flux at $x = 0$, exchanging a convective flux on the periphery (Neumann boundary condition depending on the position) and a convective flux at $x = L$ (Neumann boundary condition).

Example 7.30 :

HT3-FEM-3D: 3D metal rod receiving a constant thermal power on its periphery, an imposed flux at $x = 0$, yielding a convective flux at $x = L$.

That example makes use of the data of Example HT3-FEM-1D 7.20 but treats it as a three-dimensional problem (Figure 7.92) instead of a one-dimensional problem.

Numerical application:

Dimensions: $L = 1$ m, $r = 0.056419$ m. Transfer surface area at extremities $A = 0.01\,\text{m}^2$. Density $\rho = 7800\,\text{kg m}^{-3}$, heat capacity $C_p = 500\,\text{J kg}^{-1}\,\text{K}^{-1}$, thermal conductivity $\lambda = 26\,\text{W m}^{-1}\,\text{K}^{-1}$, flux $\dot{q}_0'' = 120\,\text{W m}^{-2}$, heat transfer coefficient $h = 50\,\text{W m}^{-2}\,\text{K}^{-1}$, environment temperature $T_\infty = 293$ K.

In the Elmer code of 3D simulation by finite elements, the lateral thermal power $\dot{q}'$ which was defined in the 1D data as equal to $10\,\text{W m}^{-1}$ (cf. Example 7.20) has been replaced by a flux $\dot{q}''$ equal to $28.209\,\text{W m}^{-2}$ in order to respect the energy of 10 W provided (the rod having a length of 1 m) taking into account the lateral surface area of the rod equal to $0.3545\,\text{m}^2$.

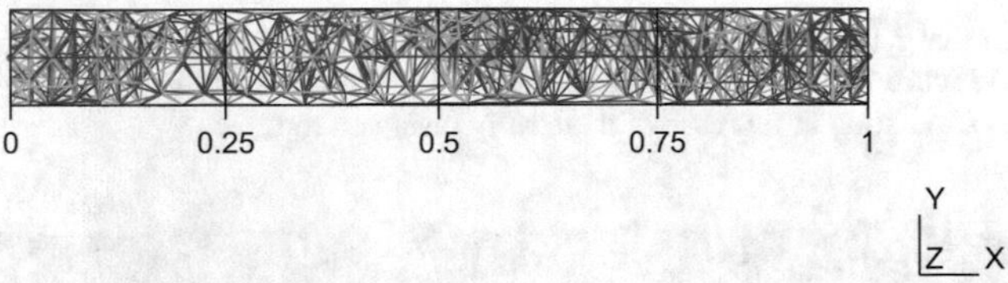

Fig. 7.92 Example HT3-FEM-3D: 3D mesh of the rod as a cylinder by the code Gmsh

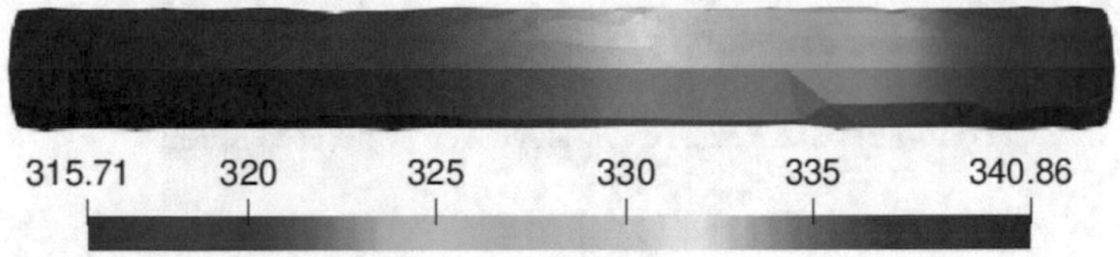

Fig. 7.93 Example HT3-FEM-3D: Results of 3D simulation by the code Elmer

The 1D (Example 7.20) and 3D (Example 7.30) simulations differ by the consideration of the source term $\dot{q}'$. In the 1D simulation, this term was perfectly taken into account, which was not exactly the case in the 3D simulation. However, both simulations give very close results (Table 7.10 and Figure 7.93) as the rod had a stretched form.

Table 7.10 Example HT3-FEM-3D: Comparison of the minimum and maximum temperatures calculated in 1D and 3D simulations

	T_{min} (K)	T_{max} (K)
1D Simulation	315.40	339.25
3D Simulation	315.71	340.86

Example 7.31 :

HT4-FEM-3D: 3D insulated metal rod, receiving an imposed flux at $x = 0$, yielding a convective flux at $x = L$.

The rod considered in Example 7.21 (Figure 7.79) is again studied as a 3D object with the same mesh (Figure 7.92) as in Example 7.30. The numerical data are identical to those of Example 7.30 (except $\dot{q}' = 0$).

Table 7.11 Example HT4-FEM-3D: Comparison of the minimum and maximum temperatures calculated in 1D and 3D simulations

	T_{min} (K)	T_{max} (K)
1D Simulation	295.40	300.01
3D simulation	295.39	300.20

The problem that has been treated in 3D by the code Elmer has given nearly identical results (Table 7.11 and Figure 7.94) to those of Example 7.21. Thus, the 3D simulation gives very close results to the 1D simulation in this case of a laterally insulated rod.

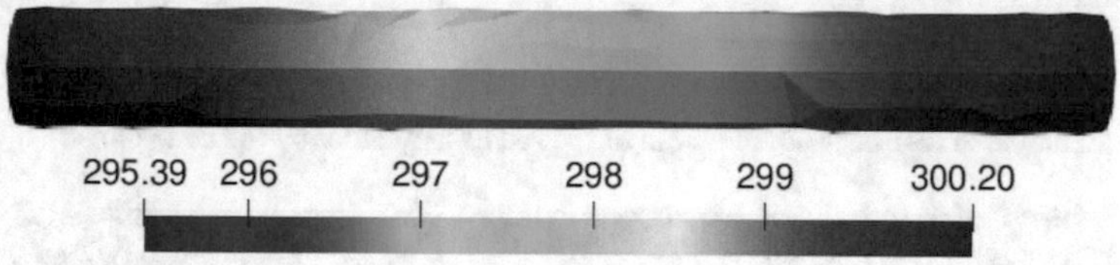

Fig. 7.94 Example HT4-FEM-3D: Results of 3D simulation by the code Elmer

---

Example 7.32 :

HT5-FEM-3D: 3D insulated metal rod, receiving an imposed flux at $x = 0$, with imposed temperature at $x = L$.

The system and the numerical data are identical to those of Example 7.22 (Figure 7.81). The mesh is 3D (Figure 7.92) and the problem has been treated in 3D by the code Elmer which has given close results (Table 7.12 and Figure 7.95) to those of Example 7.22. The results of the 3D simulation are close to those of the 1D simulation.

Table 7.12 Example HT5-FEM-3D: Comparison of minimum and maximum temperatures calculated in 1D and 3D simulations

	T_{min} (K)	T_{max} (K)
1D Simulation	293.00	297.61
3D Simulation	293.00	297.80

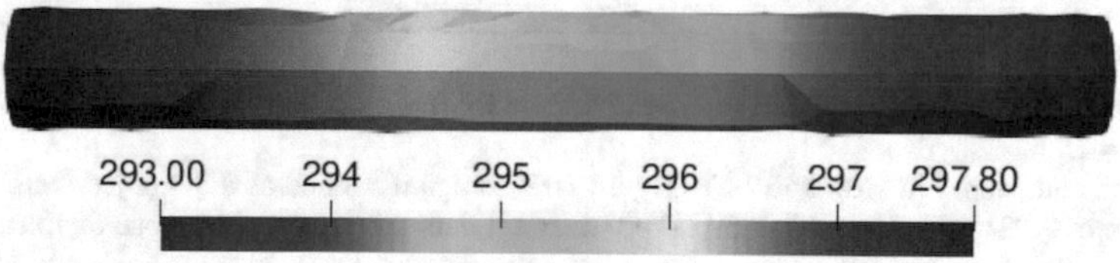

Fig. 7.95 Example HT5-FEM-3D: results of 3D simulation by the code Elmer

---

Example 7.33 :

HT7-FEM-3D: 3D metal rod receiving an imposed flux at $x = 0$, exchanging laterally a convective flux $\dot{q}_s''$ and yielding a convective flux at $x = L$.

The system and the numerical data are identical to those of Example 7.24. A 3D simulation has then been performed using the mesh of the cylinder (Figure 7.92) Thus, the nearly perfect agreement with the 1D simulation can be observed (Table 7.13 and Figure 7.96).

Table 7.13 Example HT7-FEM-3D: Comparison of minimum and maximum temperatures calculated in 1D and 3D simulations

	T_{min} (K)	T_{max} (K)
1D Simulation	292.99	293.54
3D Simulation	293.00	293.57

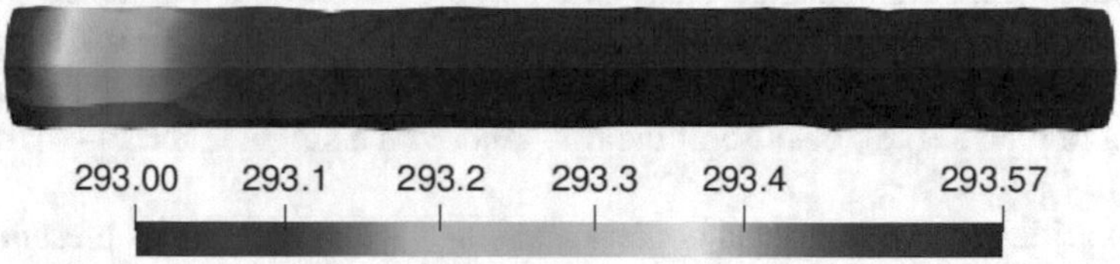

Fig. 7.96 Example HT7-FEM-3D: Results of the 3D simulation 3D by the code Elmer

In all studied cases dealing with the rod (Examples 7.30 to 7.33), the 1D and 3D simulations gave very close results. This is strongly due to the stretched shape of the rod which makes this 3D object close to a 1D model.

7.14 Boundary Element Method

The boundary element method (Beer et al. 2008; Brebbia et al. 1984; Chengy and Cheng 2005; Gaul et al. 2003; Katsikadelis 2016; Pepper et al. 2014; Sauter and Schwab 2011) is based on boundary integral equations to yield an approximate solution that uses an exact solution of the integral equation supported on the boundaries issued from the partial differential equation supported in the inner domain. Green's function method is then used. However, hybrid analytical-numerical methods exist. Variants of the method exist such as methods combining boundary integral equations with the variational formulation (Lachat and Watson 1976), also the fast multipole acceleration method (Liu 2009) or the scaled boundary finite element method (Dsouza et al. 2021) which uses features of domain based and boundary based approaches.

This method presents several advantages (Costabel 1986) over the finite element method:

- It requires only the discretization of the boundaries, not the inner domain limited by the boundaries,
- It is particularly well adapted to physical problems which present singularities or rapidly changing variables, very complicated geometries like cavities, nearly punctual heat sources, complex nonhomogeneous media (Dsouza et al. 2021), fluid-structure interaction (Dargush and Banerjee 1991), free surfaces. Then, the solution is very much influenced by some boundary values. In this case, compared to the finite element method, it provides accurate results with a reduced mesh and thus

a lower computational load (Smigaj et al. 2015). It can also solve unbounded domains occurring in two-phase flows (Lemonnier 1996), electric or magnetic fields, acoustics (Kirkup 2007), wave propagation (Chaillat and Bonnet 2013), free surface flow.

The boundary element method also present difficulties:

- It requires the analytical solution of the partial differential equation in the domain. The subsequent mathematical analysis is complex in particular in the case of equations other than linear with constant coefficients (Pozrikidis 2014). The boundary integral equations may be ordinary Fredholm integral equations of the second kind with a regular kernel, but also of the first kind with a singular kernel (Arfken et al. 2013).
- Singularities can be present at corners and edges when boundaries present discontinuities.

It must be noted that the finite element method and the boundary element method can be coupled (Gwinner and Stephan 2018), which is the case even in some commercial codes such as Comsol.

The mathematical basis of the boundary element method will be presented in the following without entering too much into the mathematical details and demonstrations. We will concentrate on steady-state conductive heat transfer (Wrobel and Kassab 2000) examples similar to Example HT1-FEM-2D 7.29 of conductive heat transfer in a plane plate where Dirichlet boundary conditions were present. We will consider both cases of Dirichlet boundary conditions and Neumann boundary conditions. The transient conductive and convective heat transfer is studied by Erhart et al. (2006), Lewandowski (2013), Shi and Banerjee (1993), Sutradhar and Paulino (2004), and Zang et al. (2021), with possible use of the Laplace transformation to cope with the time problem.

The 2D governing steady-state equation in the domain Ω is a potential problem written as

$$\nabla^2 T(x, y) = -u_g(x, y)\,, \quad (x, y) \in \Omega \tag{7.14.1}$$

with boundary conditions (Dirichlet or Neumann or Robin, this latter being a weighted sum of Dirichlet and Neumann conditions). The boundary noted Γ encloses the domain Ω. u_g is a generation term. The sign minus in the right-hand term comes from the notation mostly encountered in physics.

7.14.1 Mathematical Preliminaries

The role of Green's function in the boundary element method is essential so that some mathematical preliminaries are introduced (Arfken et al. 2013; Challis and Sheard 2003; Stakgold and Holst 2011).

Consider L a linear differential or partial differential operator such that

$$Lu(\boldsymbol{x}) = f(\boldsymbol{x}) \quad , \quad \boldsymbol{x} \text{ in } \Omega \tag{7.14.2}$$

where u is a distribution and the equation is defined without boundary conditions.

A *fundamental solution* is a solution of the corresponding inhomogeneous equation

$$\mathcal{L}u(\boldsymbol{x}) = \delta(\boldsymbol{x}) \tag{7.14.3}$$

where δ is Dirac delta function, indeed a distribution.

A *Green's function* is a fundamental solution that satisfies given conditions at the boundaries Γ that completely define the problem (7.14.2). Often as encountered in many cited examples, but *not necessarily*, the boundary conditions are homogeneous, i.e. they are of the form

$$\begin{array}{lll} u = 0 & \text{on } \Gamma_1 & \text{and not } u = c_1 \\ \dfrac{\partial u}{\partial \boldsymbol{n}} = 0 & \text{on } \Gamma_2 & \text{and not } \dfrac{\partial u}{\partial \boldsymbol{n}} = c_2 \end{array} \tag{7.14.4}$$

In this case, the operator L is Hermitian. When the boundary conditions are inhomogeneous, determining Green's function is a case by case problem, as Green's function is dependent on the boundary conditions.

In mathematics, Green's function is defined as $G(x, s)$ such that

$$\mathcal{L}G(\boldsymbol{x}, \boldsymbol{s}) = \delta(\boldsymbol{s} - \boldsymbol{x}) \tag{7.14.5}$$

In physics, Green's function is often defined with the opposite sign (Arfken et al. 2013) as the sign of f in Equation (7.14.2) is negative like in electrostatics and the Laplacian refers to $-|f|$

$$\mathcal{L}G(\boldsymbol{x}, \boldsymbol{s}) = -\delta(\boldsymbol{x} - \boldsymbol{s}) \tag{7.14.6}$$

but the function is symmetric

$$G(\boldsymbol{x}, \boldsymbol{s}) = G(\boldsymbol{s}, \boldsymbol{x})^* \tag{7.14.7}$$

where G^* is the conjugate of G.

The solution of the initial value problem (7.14.2) is the convolution product

$$u(\boldsymbol{x}) = G * f = \int_\Omega G(\boldsymbol{x}, \boldsymbol{y}) f(\boldsymbol{y}) d\boldsymbol{y} \tag{7.14.8}$$

since

$$\mathcal{L}u(\boldsymbol{x}) = \mathcal{L}\int_\Omega G(\boldsymbol{x}, \boldsymbol{y}) f(\boldsymbol{y}) d\boldsymbol{y} = \int_\Omega \mathcal{L}G(\boldsymbol{x}, \boldsymbol{y}) f(\boldsymbol{y}) d\boldsymbol{y} = \int_\Omega \delta(\boldsymbol{x} - \boldsymbol{y}) f(\boldsymbol{y}) d\boldsymbol{y} = f(\boldsymbol{x}) \tag{7.14.9}$$

Defining the orthonormal eigenfunctions ϕ and corresponding eigenvalues λ of the operator $\mathcal{L}$ such that $\mathcal{L}\phi = \lambda\phi$, Green's function is given (Arfken et al. 2013) by

$$G(\boldsymbol{x}, \boldsymbol{s}) = \sum_{i=0}^{\infty} \frac{1}{\lambda_i} \phi_i^*(\boldsymbol{s}) \phi_i(\boldsymbol{x}) \tag{7.14.10}$$

7.14.2 Potential Problems

In this section, on the contrary of the following Section 7.14.3, the signs are those adopted in mathematics. They differ from those used in physics.

The general potential problem can be written as *Poisson's equation*

$$\begin{array}{ll} \nabla^2 u(\boldsymbol{x}) = f\,, & \boldsymbol{x} \in \Omega \\ u = 0\,, & \boldsymbol{x} \text{ on } \Gamma \end{array} \tag{7.14.11}$$

where, in physics, u is a scalar potential, f is a source term or a force that derives from a potential. When $f = 0$, this is Laplace equation. In electrostatics, u is the electric potential and $f = -\rho/\epsilon_0$ is the charge density. In classical mechanics, given the potential ϕ, the gravitational field equation is $g = F/m = -\nabla\phi$ and the field equation is $\nabla g = -\nabla^2\phi = -4\pi\mathcal{G}\rho$ where $\mathcal{G}$ is the universal gravitational constant.

Using Green's function G, the solution to Equation (7.14.11) can be written as

$$u(\boldsymbol{x}) = \int_\Omega G(\boldsymbol{y}, \boldsymbol{x}) f(\boldsymbol{y}) d\Omega_{\boldsymbol{y}} \tag{7.14.12}$$

In mathematics, with respect to the potential problem, *Green's function G* obeys the particular Poisson equation

$$\begin{array}{ll} \nabla^2 G(\boldsymbol{y}, \boldsymbol{x}) = \Delta G(\boldsymbol{y}, \boldsymbol{x}) = \delta(\boldsymbol{x} - \boldsymbol{y})\,, & \boldsymbol{x} \in \Omega \\ G = 0\,, & \boldsymbol{x} \text{ on } \Gamma \end{array} \tag{7.14.13}$$

Green's function is known as *Newton's kernel* or *Newton's potential*, i.e. the potential from a point source at $\boldsymbol{x}$ with the boundary maintained at zero potential.

For Equation (7.14.13), Green's function (the signs are those used in mathematics Arfken et al. 2013) is equal to

$$\begin{array}{ll} G(\boldsymbol{y}, \boldsymbol{x}) = \frac{1}{2} r(\boldsymbol{x}, \boldsymbol{y}) & \text{in 1D} \\ G(\boldsymbol{y}, \boldsymbol{x}) = \frac{1}{2\pi} \ln(r(\boldsymbol{x}, \boldsymbol{y})) & \text{in 2D} \\ G(\boldsymbol{y}, \boldsymbol{x}) = -\frac{1}{4\pi r(\boldsymbol{x}, \boldsymbol{y})} & \text{in 3D} \end{array} \tag{7.14.14}$$

where r is the distance between $\boldsymbol{x}$ and $\boldsymbol{y}$. Thus, in 2D

$$r^2 = (x_1 - y_1)^2 + (x_2 - y_2)^2 \tag{7.14.15}$$

The boundary conditions are arbitrary for 1D and 2D, and $G \to 0$, when $\boldsymbol{x} \to \infty$ in 3D.

Green's function is illustrated in Figure 7.97.

In the case of a more general potential problem with Dirichlet or Neumann boundary conditions such as

$$\begin{array}{ll} \nabla^2 u(\boldsymbol{x}) = f\,, & \boldsymbol{x} \in \Omega \\ u = \overline{u}\,, & \text{on } \Gamma_1 \\ \frac{\partial u}{\partial \boldsymbol{n}} = \overline{\partial u_n}\,, & \text{on } \Gamma_2 \end{array} \tag{7.14.16}$$

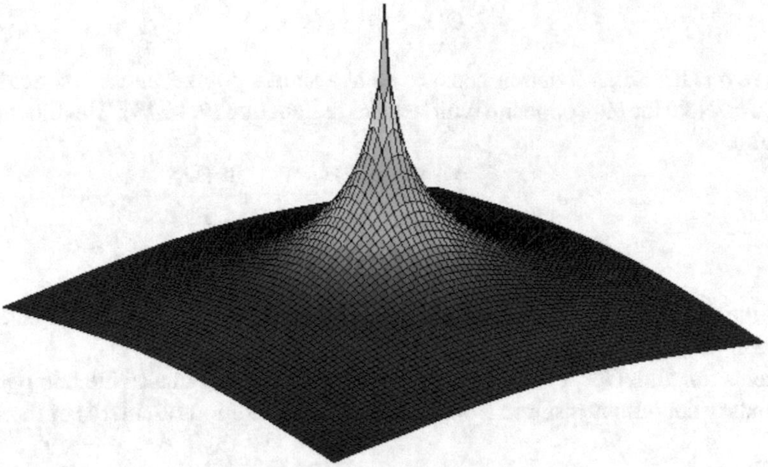

Fig. 7.97 Green's function in 2D at (0, 0)

the potential $u(\xi)$, for a point ξ inside the domain Ω, is equal to

$$u(\boldsymbol{\xi}) = \oint_{\Gamma} \left(u \frac{\partial G}{\partial \boldsymbol{n}} - G \frac{\partial f}{\partial \boldsymbol{n}} \right) d\Gamma(\boldsymbol{x}) + \int_{\Omega} G \; f \; d\Omega(\boldsymbol{x}) \tag{7.14.17}$$

In 2D, using Green's fundamental solution, it yields

$$\begin{aligned} u(\boldsymbol{\xi}) = \frac{1}{2\pi} \oint_{\Gamma} \left(u \frac{\partial \ln r(\boldsymbol{x}, \boldsymbol{\xi})}{\partial \boldsymbol{n}} - \ln r(\boldsymbol{x}, \boldsymbol{\xi}) \frac{\partial u}{\partial \boldsymbol{n}} \right) d\Gamma(\boldsymbol{x}) \\ + \frac{1}{2\pi} \int_{\Omega} \ln r(\boldsymbol{x}, \boldsymbol{\xi}) \; f \; d\Omega(\boldsymbol{x}) \end{aligned} \tag{7.14.18}$$

7.14.3 Green's Function Method

In this part and in the following, the sign convention is taken with respect to physics.

The methodology (Ang 2007; Antes 2010; Banerjee 1994; Becker 1992; Kassab 2018) is illustrated in the case of the following convective heat transfer problem (7.14.1)

$$\nabla^2 T(\boldsymbol{x}) = -u_g(\boldsymbol{x}) \, , \quad \boldsymbol{x} \in \Omega \tag{7.14.19}$$

as follows:

- A test function Θ is chosen as the fundamental solution or Green's free space solution.
- In the case of Poisson equation (7.14.19) or Laplace equation if $u_g = 0$, the fundamental solution $\Theta(\boldsymbol{x}, \boldsymbol{\xi})$ is the solution of

$$\nabla^2\Theta(\boldsymbol{x},\boldsymbol{\xi}) = -\delta(\boldsymbol{x}-\boldsymbol{\xi}) \tag{7.14.20}$$

where δ is the Dirac function considered at a source point $\boldsymbol{\xi}$ and $\boldsymbol{x}$ is a field point, $\boldsymbol{x} \in \mathbb{R}^2$. Note the sign opposite with respect to Equation (7.14.13). The fundamental solution is

$$\begin{aligned} \Theta(\boldsymbol{x},\boldsymbol{\xi}) &= -\frac{1}{2} r(\boldsymbol{x},\boldsymbol{\xi}) && \text{in 1D} \\ \Theta(\boldsymbol{x},\boldsymbol{\xi}) &= -\frac{1}{2\pi}\ln(r(\boldsymbol{x},\boldsymbol{\xi})) && \text{in 2D} \\ \Theta(\boldsymbol{x},\boldsymbol{\xi}) &= \frac{1}{4\pi r(\boldsymbol{x},\boldsymbol{\xi})} && \text{in 3D} \end{aligned} \tag{7.14.21}$$

r is the distance between $\boldsymbol{x}$ and $\boldsymbol{\xi}$. The sign notations used in (7.14.21) are those used in physics.

- Green's function $G(\boldsymbol{x},\boldsymbol{\xi})$ must verify Equation (7.14.20) but also the homogeneous boundary conditions inspired from the boundary conditions (7.14.16) of the system

$$\begin{aligned} G &= 0\,, && \boldsymbol{x} \text{ on } \Gamma_1 \\ \frac{\partial G}{\partial \boldsymbol{n}} &= 0\,, && \boldsymbol{x} \text{ on } \Gamma_2 \end{aligned} \tag{7.14.22}$$

- The governing equation (7.14.19) multiplied by the test function is integrated over space and time (if the problem is time-dependent)

$$\int_\Omega \Theta\, \nabla^2 T\, d\Omega(\boldsymbol{x}) = -\int_\Omega \Theta\, u_g\, d\Omega(\boldsymbol{x}) \tag{7.14.23}$$

- Green's second identity (reciprocal theorem) stipulates that

$$\int_\Omega \left(\phi\, \nabla^2\psi - \psi\, \nabla^2\phi\right)\, d\Omega(\boldsymbol{x}) = \oint_\Gamma \left(\phi\frac{\partial\psi}{\partial\boldsymbol{n}} - \psi\frac{\partial\phi}{\partial\boldsymbol{n}}\right) d\Gamma(\boldsymbol{x}) \tag{7.14.24}$$

where $\boldsymbol{n}$ is the outward pointing normal vector at the boundary Γ and the contour integral on Γ is integrated anticlockwise. ϕ and ψ are two twice continuously differentiable functions on Ω. Setting $\phi = \Theta$ and $\psi = T$, Green's Equation (7.14.24) can be written either as the first form

$$\int_\Omega \left(\Theta\, \nabla^2 T - T\, \nabla^2\Theta\right)\, d\Omega(\boldsymbol{x}) = \oint_\Gamma \left(\Theta\frac{\partial T}{\partial\boldsymbol{n}} - T\frac{\partial\Theta}{\partial\boldsymbol{n}}\right) d\Gamma(\boldsymbol{x}) \tag{7.14.25}$$

or, using Equations (7.14.20) and (7.14.23), as the second form

$$\int_\Omega \left(\Theta\, \nabla^2 T - T\, \nabla^2\Theta\right)\, d\Omega(\boldsymbol{x}) = -\int_\Omega \Theta\, u_g\, d\Omega(\boldsymbol{x}) + \int_\Omega T\, \delta(\boldsymbol{\xi}-\boldsymbol{x})\, d\Omega(\boldsymbol{x}) \tag{7.14.26}$$

so that it results

$$\oint_\Gamma \left(\Theta\frac{\partial T}{\partial\boldsymbol{n}} - T\frac{\partial\Theta}{\partial\boldsymbol{n}}\right) d\Gamma(\boldsymbol{x}) + \int_\Omega \Theta\, u_g\, d\Omega(\boldsymbol{x}) = \int_\Omega T\, \delta(\boldsymbol{\xi}-\boldsymbol{x})\, d\Omega(\boldsymbol{x}) \tag{7.14.27}$$

- The choice of Θ in Equation (7.14.20) used in (7.14.26) was made wisely so that the right-hand side can be simplified as

$$\int_\Omega T\,\delta(\boldsymbol{\xi}-\boldsymbol{x})\,d\Omega(\boldsymbol{x}) = T(\boldsymbol{\xi}) \tag{7.14.28}$$

- Finally, the following integral equation results

$$T(\boldsymbol{\xi}) = \oint_\Gamma \left(\Theta\frac{\partial T}{\partial \boldsymbol{n}} - T\frac{\partial \Theta}{\partial \boldsymbol{n}}\right) d\Gamma(\boldsymbol{x}) + \int_\Omega \Theta\, u_g\, d\Omega(\boldsymbol{x}) \tag{7.14.29}$$

We can set $\boldsymbol{\xi} = (x_i, y_i)$ as any point in the domain.

Remark: *all equations from (7.14.23) to (7.14.29) are exactly the same if the test function is taken as $\Theta(\boldsymbol{x}, \boldsymbol{\xi})$ that is the fundamental solution or if it is replaced by Green's function $G(\boldsymbol{x}, \boldsymbol{\xi})$.*

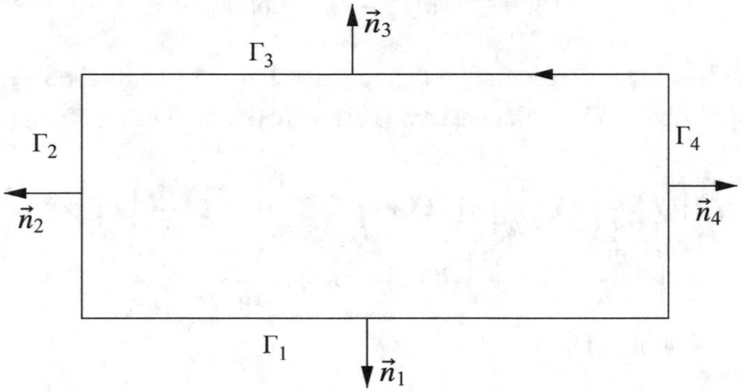

Fig. 7.98 Boundary element method: Rectangular plane plate

To fix a few ideas, consider the example of steady-state conductive heat transfer for a rectangular plane plate as in Figure 7.98. In the absence of internal heat generation, the system is modelled by Laplace equation

$$\begin{aligned} &\nabla.(\lambda\nabla T(\boldsymbol{x})) = 0 && \text{or} \\ &\nabla^2 T(\boldsymbol{x}) = 0 && \text{with constant } \lambda \end{aligned} \tag{7.14.30}$$

On each boundary Γ_i, different types of boundary conditions can occur which must be satisfied by T:

- Dirichlet: T_i fixed,
- Neumann with zero heat flux for an isolated wall: $\dfrac{\partial T_i}{\partial n_i} = 0$,
- Neumann with fixed heat flux: $\lambda\dfrac{\partial T_i}{\partial n_i} = \dot{q}_i''$,
- Neumann with convective heat flux: $\lambda\dfrac{\partial T_i}{\partial n_i} = h(T_i - T_\infty)$.

Equation (7.14.29) is developed as

$$\begin{aligned} T(\boldsymbol{\xi}) = & \oint_{\Gamma_1} \left(\Theta \frac{\partial T}{\partial \boldsymbol{n}} - T \frac{\partial \Theta}{\partial \boldsymbol{n}} \right) d\Gamma(\boldsymbol{x}) + \oint_{\Gamma_2} \left(\Theta \frac{\partial T}{\partial \boldsymbol{n}} - T \frac{\partial \Theta}{\partial \boldsymbol{n}} \right) d\Gamma(\boldsymbol{x}) \\ & + \oint_{\Gamma_3} \left(\Theta \frac{\partial T}{\partial \boldsymbol{n}} - T \frac{\partial \Theta}{\partial \boldsymbol{n}} \right) d\Gamma(\boldsymbol{x}) + \oint_{\Gamma_4} \left(\Theta \frac{\partial T}{\partial \boldsymbol{n}} - T \frac{\partial \Theta}{\partial \boldsymbol{n}} \right) d\Gamma(\boldsymbol{x}) \\ & + \int_{\Omega} \Theta \, u_g \, d\Omega(\boldsymbol{x}) \end{aligned} \tag{7.14.31}$$

Using the boundary conditions, Equation (7.14.31) is simplified.

Imagine the following boundary conditions

$$\begin{aligned} & T = T_1 && \text{on } \Gamma_1 \\ & \frac{\partial T_2}{\partial n_2} = 0 && \text{on } \Gamma_2 \\ & \lambda \frac{\partial T_3}{\partial n_3} = \dot{q}_3'' && \text{on } \Gamma_3 \\ & \lambda \frac{\partial T_4}{\partial n_4} = h(T_4 - T_\infty) && \text{on } \Gamma_4 \end{aligned} \tag{7.14.32}$$

If Equation (7.14.31) is considered, it appears that T is not specified on Γ_2, Γ_3, Γ_4 and $\frac{\partial T}{\partial n}$ is not specified on Γ_1 as emphasized (surrounded terms) in Equation (7.14.33)

$$\begin{aligned} T(\boldsymbol{\xi}) = & \oint_{\Gamma_1} \left(\Theta \boxed{\frac{\partial T}{\partial \boldsymbol{n}}} - T \frac{\partial \Theta}{\partial \boldsymbol{n}} \right) d\Gamma(\boldsymbol{x}) + \oint_{\Gamma_2} \left(\Theta \frac{\partial T}{\partial \boldsymbol{n}} - \boxed{T} \frac{\partial \Theta}{\partial \boldsymbol{n}} \right) d\Gamma(\boldsymbol{x}) \\ & + \oint_{\Gamma_3} \left(\Theta \frac{\partial T}{\partial \boldsymbol{n}} - \boxed{T} \frac{\partial \Theta}{\partial \boldsymbol{n}} \right) d\Gamma(\boldsymbol{x}) + \oint_{\Gamma_4} \left(\Theta \frac{\partial T}{\partial \boldsymbol{n}} - \boxed{T} \frac{\partial \Theta}{\partial \boldsymbol{n}} \right) d\Gamma(\boldsymbol{x}) \\ & + \int_{\Omega} \Theta \, u_g \, d\Omega(\boldsymbol{x}) \end{aligned} \tag{7.14.33}$$

Thus, when the terms T or $\partial T / \partial \boldsymbol{n}$ are not specified by the boundary conditions (the surrounded terms in (7.14.33)), they must vanish. Θ *is introduced as* Green's function *with homogeneous boundary conditions corresponding to* (7.14.32) such that

$$\begin{aligned} & \nabla^2 \Theta = \delta(\boldsymbol{x} - \boldsymbol{\xi}) && \text{on } \Omega \\ & \Theta = 0 && \text{on } \Gamma_1 \\ & \frac{\partial \Theta}{\partial \boldsymbol{n}} = 0 && \text{on } \Gamma_2 \\ & \frac{\partial \Theta}{\partial \boldsymbol{n}} = 0 && \text{on } \Gamma_3 \\ & \frac{\partial \Theta}{\partial \boldsymbol{n}} = 0 && \text{on } \Gamma_4 \end{aligned} \tag{7.14.34}$$

Equation (7.14.33) then becomes

$$\begin{aligned} T(\boldsymbol{\xi}) = & - \oint_{\Gamma_1} T \frac{\partial \Theta}{\partial \boldsymbol{n}} d\Gamma(\boldsymbol{x}) + \oint_{\Gamma_2} \Theta \frac{\partial T}{\partial \boldsymbol{n}} d\Gamma(\boldsymbol{x}) + \oint_{\Gamma_3} \Theta \frac{\partial T}{\partial \boldsymbol{n}} d\Gamma(\boldsymbol{x}) + \oint_{\Gamma_4} \Theta \frac{\partial T}{\partial \boldsymbol{n}} d\Gamma(\boldsymbol{x}) \\ & + \int_{\Omega} \Theta \, u_g \, d\Omega(\boldsymbol{x}) \end{aligned} \tag{7.14.35}$$

which yields the temperature at any point ξ inside the domain Ω. This is called *Green's function method.*

The method of images is also used to obtain Green's functions in some particular cases, for example a half plane (in 2D) (Ang 2007) with boundary conditions such as

$$G_1(x, y; \xi, \eta) = 0 \quad \text{for } y = 0 \quad \text{and } -\infty < x < \infty \tag{7.14.36}$$

yielding

$$G_1(x, y; \xi, \eta) = \frac{1}{4\pi} \ln([x-\xi]^2 + [y-\eta]^2 - \frac{1}{4\pi} \ln([x-\xi]^2 + [y+\eta]^2 \tag{7.14.37}$$

or

$$\frac{\partial G_2(x, y; \xi, \eta)}{\partial y} = 0 \quad \text{for } y = 0 \quad \text{and } -\infty < x < \infty \tag{7.14.38}$$

yielding

$$G_1(x, y; \xi, \eta) = \frac{1}{4\pi} \ln([x-\xi]^2 + [y-\eta]^2 + \frac{1}{4\pi} \ln([x-\xi]^2 + [y+\eta]^2 \tag{7.14.39}$$

Other expressions of Green's functions can be found for the exterior region of a circle or different geometries.

Thus, Green's functions can be calculated analytically in general by means of complex expressions and only in a limited number of cases, but they can also be calculated numerically (Duhamel 2007). Nowadays, it appears that an automatic calculation of Green's functions is much preferred in most cases (Beer et al. 2008).

Let us now introduce *the boundary element method.* In Equation (7.14.35), when the temperature is imposed by a boundary condition, the heat flux is unknown and vice versa. In the boundary element method, to make the equations only dependent on the boundary values, *the interior point ξ is moved to the boundary* Γ. However, due to this, Θ and $\frac{\partial \Theta}{\partial \boldsymbol{n}}$ become weakly and strongly singular, respectively (Antes 2010). A new form of the integral equation called *boundary integral equation* results *for a point ξ on the boundary Γ or inside the domain Ω*

$$\frac{\Delta\Omega(\xi)}{2\pi} T(\boldsymbol{\xi}) = \oint_\Gamma \Theta \frac{\partial T}{\partial \boldsymbol{n}} d\Gamma(\boldsymbol{x}) - \oint_\Gamma T \frac{\partial \Theta}{\partial \boldsymbol{n}} d\Gamma(\boldsymbol{x}) + \int_\Omega \Theta \, u_g \, d\Omega(\boldsymbol{x}) \tag{7.14.40}$$

where $\Delta\Omega(\xi)$ is the internal angle in 2D, or the inner solid angle in 3D, of the boundary Γ at point ξ. Thus in 2D, $\Delta\Omega(\xi) = \pi$ except at corners, and in 3D, $\Delta\Omega(\xi) = 2\pi$ except at corners and edges. Consider the coefficient

$$c_i = \frac{\Delta\Omega(\xi)}{2\pi} \tag{7.14.41}$$

with $c_i = 1$ inside the domain Ω (ξ in Ω) and $c_i = 0$ outside the domain Ω. In general, $c_i = 1/2$ at the boundary Γ.

Generally, constructing Green's functions is difficult and the only remaining possibility is often the analytical-numerical boundary element method described in Section 7.14.4.

Example 7.34 :
HT1-BEM-1D: 1D insulated metal rod, receiving an imposed flux at $x = 0$, with imposed temperature at $x = L$.

This is the same steady-state example as 7.22 (Figure 7.81) but it will be solved by Green's function method. The imposed flux at $x = 0$ is $\dot{q}''_0 = 120\,\mathrm{W\,m^{-2}}$ and the imposed temperature at $x = L$ is $T_L = 293$ K. The thermal conductivity is $\lambda = 26\,\mathrm{W\,m^{-1}\,K^{-1}}$.

The boundary conditions for the system are

$$\begin{aligned} -\lambda \left.\frac{\partial T}{\partial x}\right|_{x=0} &= \dot{q}''_0 \\ T(x = L) &= T_L \end{aligned} \tag{7.14.42}$$

As the problem is one-dimensional, according to Equation (7.14.14), the fundamental solution Θ is

$$\Theta(x, \xi) = \frac{1}{2}|x - \xi| \tag{7.14.43}$$

corresponding to Equation (7.14.20).

Green's function $G(x, \xi)$ must obey the following homogeneous boundary conditions

$$\begin{aligned} -\lambda \left.\frac{\partial G(x, \xi)}{\partial x}\right|_{x=0} &= 0 \\ G(x = L, \xi) &= 0 \end{aligned} \tag{7.14.44}$$

Derived from the fundamental solution Θ, the solution for Green's function ensuring these boundary conditions is

$$G(x, \xi) = \underbrace{\Theta(x, \xi) - \Theta(L, \xi)}_{\text{to ensure } G=0 \text{ at } x=L} - \underbrace{\left.\frac{\partial \Theta}{\partial x}\right|_{x=0} (x - L)}_{\text{to ensure } \partial G/\partial x \text{ at } x=0} \tag{7.14.45}$$

This can be verified as

$$\begin{aligned} &\text{At } x = 0, \ \frac{\partial G}{\partial x} = \frac{\partial \Theta}{\partial x} - \left.\frac{\partial \Theta}{\partial x}\right|_{x=0} = 0 \\ &\text{At } x = L, \ G(L, \xi) = 0 \end{aligned} \tag{7.14.46}$$

Now, according to Equation (7.14.29), the temperature at any point in the rod is

$$\begin{aligned} T(\xi) &= \oint_\Gamma \left(G\frac{\partial T}{\partial x} - T\frac{\partial G}{\partial x} \right) d\Gamma(x) + \int_\Omega G u_g \, d\Omega(x) \\ &= \left[G\frac{\partial T}{\partial x} - T\frac{\partial G}{\partial x} \right]_0^L + 0 \\ &= G_L \left.\frac{\partial T}{\partial x}\right|_L - T_L \left.\frac{\partial G}{\partial x}\right|_L - G_0 \left.\frac{\partial T}{\partial x}\right|_0 + T_0 \left.\frac{\partial G}{\partial x}\right|_0 \\ &= -T_L \left.\frac{\partial G}{\partial x}\right|_L - G_0 \left.\frac{\partial T}{\partial x}\right|_0 \\ &= -T_L \left.\frac{\partial G}{\partial x}\right|_L + G_0 \frac{1}{\lambda} \dot{q}''_0 \end{aligned} \tag{7.14.47}$$

using $u_g = 0$, boundary conditions (7.14.44) and boundary conditions (7.14.42).

Using the expression of the fundamental solution (7.14.43) in (7.14.45), it results

$$\begin{aligned} G(0, \xi) &= -\frac{1}{2}|0 - \xi| + \frac{1}{2}|L - \xi| - \left(\frac{1}{2}\right)(0 - L) = -\frac{1}{2}\xi + \frac{1}{2}L - \frac{1}{2}\xi + \frac{1}{2}L = L - \xi \\ &\text{with } \left.\frac{\partial \Theta}{\partial x}\right|_{x=0} = \frac{1}{2} \end{aligned} \tag{7.14.48}$$

and

$$\begin{aligned}
\left.\frac{\partial G(x,\xi)}{\partial x}\right|_{x=L} &= \left.\frac{\partial \Theta}{\partial x}\right|_{x=L} - \left.\frac{\partial \Theta}{\partial x}\right|_{x=0} \\
&= \left.\frac{\partial\left(-\frac{1}{2}(x-\xi)\right)}{\partial x}\right|_{x=L} - \left.\frac{\partial\left(-\frac{1}{2}(\xi-x)\right)}{\partial x}\right|_{x=0} \\
&= -\frac{1}{2} - \left(\frac{1}{2}\right) = -1
\end{aligned} \tag{7.14.49}$$

Finally, Equation (7.14.47) gives the analytical solution

$$\begin{aligned}
T(\xi) &= T_L(1) + \left(\frac{1}{\lambda}\dot{q}_0''\right)(L-\xi) \\
&= T_L + \frac{\dot{q}_0''}{\lambda}(L-\xi)
\end{aligned} \tag{7.14.50}$$

The steady-state temperature profile in the rod is linear and, numerically, we find $T(0) = 297.615$ K in perfect agreement with Table 7.12 resulting from the finite element method.

Example 7.35 :

HT2-BEM-1D: 1D metal rod, receiving an imposed flux at $x = 0$, receiving a thermal power on its periphery, with imposed temperature at $x = L$.

In this case (Figure 7.18), this is again the same rod, but compared to previous Example 7.34, the lateral thermal power is $\dot{q}' = 10\,\mathrm{W\,m^{-1}}$ instead of being zero. The imposed flux at $x = 0$ is $\dot{q}_0'' = 120\,\mathrm{W\,m^{-2}}$ and the imposed temperature at $x = L$ is $T_L = 293$ K. The thermal conductivity is $\lambda = 26\,\mathrm{W\,m^{-1}\,K^{-1}}$. The transfer surface at extremities is equal to $A = 0.01\,\mathrm{m^2}$.

The governing equation is no more Laplace's equation, but Poisson's equation

$$A\lambda\frac{d^2T}{dx^2} + \dot{q}' = 0 \tag{7.14.51}$$

Compared to Equation (7.14.47), the source term must be added. Thus, in the present case

$$u_g = \frac{\dot{q}'}{A\lambda} \tag{7.14.52}$$

With regard to Equation (7.14.47), the additional contribution to temperature $T(\xi)$ is

$$\int_\Omega G u_g \, d\Omega(\boldsymbol{x}) = \int_0^L \left[\Theta(x,\xi) - \Theta(L,\xi) - \left.\frac{\partial T}{\partial x}\right|_0 (x-L)\right]\frac{\dot{q}'}{A\lambda}dx \tag{7.14.53}$$

with

$$\begin{aligned}
G(x,\xi) &= -\frac{1}{2}(\xi - x) + L - \frac{1}{2}\xi - \frac{1}{2}x \quad &\text{for } x \in [0,\xi] \\
G(x,\xi) &= -\frac{1}{2}(x - \xi) + L - \frac{1}{2}\xi - \frac{1}{2}x \quad &\text{for } x \in [\xi,L]
\end{aligned} \tag{7.14.54}$$

thus

$$\begin{aligned}
\int_\Omega G u_g \, d\Omega(\boldsymbol{x}) &= \left[\int_0^\xi (L-\xi)dx + \int_\xi^L (L-x)dx\right]\frac{\dot{q}'}{A\lambda} \\
&= \left[(L-\xi)\xi + \frac{1}{2}(L-\xi)^2\right]\frac{\dot{q}'}{A\lambda} \\
&= \left[(L-\xi)(\frac{3}{2}L - \frac{1}{2}\xi\right]\frac{\dot{q}'}{A\lambda}
\end{aligned} \tag{7.14.55}$$

so that the temperature at any point ξ in the rod is equal to

$$T(\xi) = T_L + \frac{\dot{q}_0''}{\lambda}(L-\xi) + \frac{\dot{q}'}{A\lambda}(L-\xi)(\frac{3}{2}L - \frac{1}{2}\xi) \tag{7.14.56}$$

This gives the analytical expression of the temperature profile which is shown in Figure 7.99.

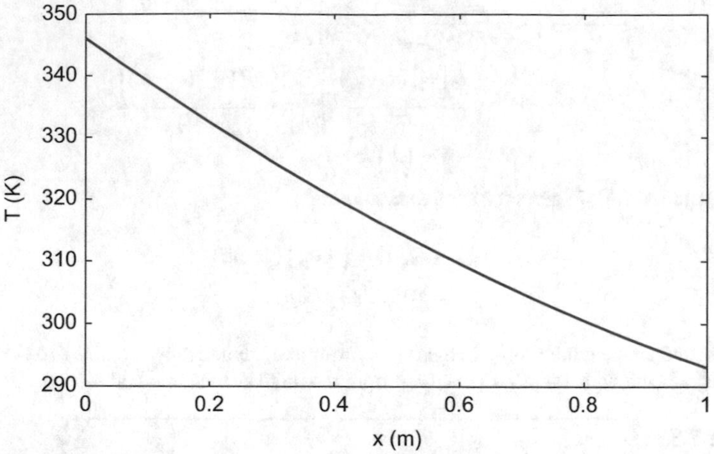

Fig. 7.99 Example HT2-BEM-1D: Temperature profile by the boundary element method

7.14.4 Analytical-Numerical Boundary Element Method

In the analytical-numerical boundary element method, several differences exist with respect to Green's function method:

- The source point is taken at the boundary,
- An exact boundary integral equation is used,
- The boundaries are divided into elements and the integral equation is discretized at the collocation points,
- A system of linear equations with respect to T and $\frac{\partial T}{\partial \boldsymbol{n}}$ results which takes into account the boundary conditions. Thus, in the boundary element method, with respect to Green's function method, the boundary conditions are not imposed at the same time.

In the boundary element method, using Equation (7.14.12), Green's function is replaced by the fundamental solution $g(\boldsymbol{y}, \boldsymbol{x})$ to yield the approximate solution (Antes 2010; Hartman and Katz 2008)

$$cu_h(\boldsymbol{\xi}) = \oint_\Gamma \left[g(\boldsymbol{x}, \boldsymbol{\xi}) \frac{\partial u_h}{\partial \boldsymbol{n}}(\boldsymbol{x}) - u_h(\boldsymbol{x}) \frac{\partial g(\boldsymbol{x}, \boldsymbol{\xi})}{\partial \boldsymbol{n}} \right] d\Gamma(\boldsymbol{x}) + \int_\Omega g(\boldsymbol{x}, \boldsymbol{\xi}) p(\boldsymbol{x}) d\Omega(\boldsymbol{x}) \tag{7.14.57}$$

where $\boldsymbol{\xi}$ is on any boundary Γ or inside the domain Ω. This is the analog to Equation (7.14.29). c is the coefficient defined in Equation (7.14.40) depending on the position of $\boldsymbol{\xi}$ with respect to the boundaries and the domain.

The fundamental solution g is given by Equation (7.14.21) in 2D and 3D.

The main characteristics of the boundary element method (Banerjee 1994; Becker 1992; Brebbia 2017; Brebbia and Dominguez 1998) are as follows:

- To simplify the notations, in the following, the partial derivative $\partial u/\partial \boldsymbol{n}$ of the potential u which is a flux is noted q so that Equation (7.14.57) is written as

$$cu_h(\boldsymbol{\xi}) = \oint_\Gamma \left[g(\boldsymbol{x}, \boldsymbol{\xi}) q_h(\boldsymbol{x}) - u_h(\boldsymbol{x}) \frac{\partial g(\boldsymbol{x}, \boldsymbol{\xi})}{\partial \boldsymbol{n}} \right] d\Gamma(\boldsymbol{x}) + \int_\Omega g(\boldsymbol{x}, \boldsymbol{\xi}) p(\boldsymbol{x}) d\Omega(\boldsymbol{x}) \tag{7.14.58}$$

- The boundary curve Γ is first partitioned in non-intersecting boundary elements Γ^e by introducing nodal points $\boldsymbol{x}^{ie}$ $(i = 1, \ldots, N)$. At the interface between two adjacent elements Γ^e and Γ^{e+1}, the nodal points are identical $\boldsymbol{x}^{Ne} = \boldsymbol{x}^{1,e+1}$. In each element Γ^e, the boundary curve and the states are approximated by polynomial functions as it was the case for finite elements. The respective polynomials for the curve and the states can be of different degrees. If they have the same degree, the problem is isoparametric. In 3D, the boundary surface is often composed of triangular or quadrilateral elements exactly as in the finite element method. The boundary states are approximated by shape functions in the same way as for the finite element method. The shape functions can be linear, quadratic, or cubic splines.
- If the field variables are not considered constant on the discretized boundary elements, the functions u and q are expressed on each element by use of shape functions as

$$u(x) = \sum_k u_k N_k(x)\,, \quad q(x) = \sum_k q_k N_k(x) \tag{7.14.59}$$

where k is the index of any node of an element having K nodes $(k = 1, \ldots, K)$. u_k and q_k are the nodal values of u and q, respectively.

In 2D, assuming that the normalized coordinate $\eta = \zeta/l$ is used where η describes the variation along the boundary element, the shape functions are
For a linear polynomial

$$N_1(\eta) = \frac{1}{2}(1+\eta)\,, \quad N_2(\eta) = \frac{1}{2}(1-\eta) \tag{7.14.60}$$

For a quadratic polynomial

$$N_1(\eta) = -\frac{1}{2}\eta(1-\eta)\,, \quad N_2(\eta) = 1-\eta^2\,, \quad N_3(\eta) = \frac{1}{2}\eta(1+\eta) \tag{7.14.61}$$

For a cubic polynomial

$$\begin{aligned} N_1(\eta) &= \frac{1}{16}(1-\eta)(1+3\eta)(3\eta-1)\,, \quad N_2(\eta) = \frac{9}{16}(1+\eta)(1-\eta)(1-3\eta) \\ N_3(\eta) &= \frac{9}{16}(1+\eta)(1-\eta)(1+3\eta)\,, \quad N_4(\eta) = \frac{1}{16}(1+\eta)(1+3\eta)(3\eta-1) \end{aligned} \tag{7.14.62}$$

In the case of a quadratic shape function, the model geometry (Figure 7.100) is described by

$$x(\eta) = \sum_k x_k N_k(\eta)\,, \quad y(\eta) = \sum_k y_k N_k(\eta) \tag{7.14.63}$$

For a continuous element or constant element, a field variable u is constant on the element Γ, similarly for the gradient.

If the elements are *isoparametric*, the same shape functions are used for the potential functions u and their gradients as for the geometries

$$u(\eta) = \sum_k u_k N_k(\eta), \quad q(\eta) = \sum_k q_k N_k(\eta) \tag{7.14.64}$$

The use of isoparametric elements provides a good compromise between accuracy and efficiency (Becker 1992).

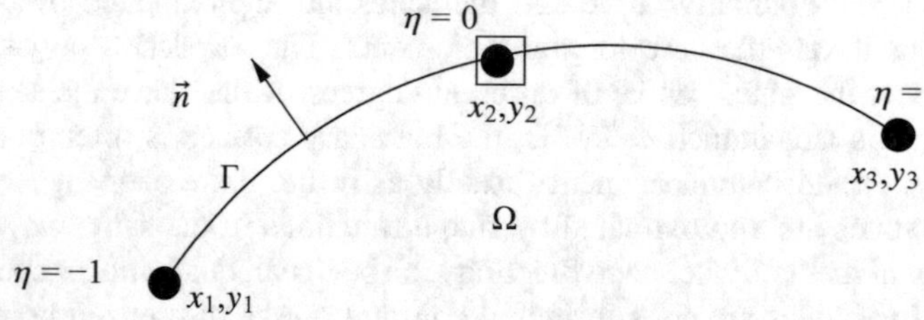

Fig. 7.100 Geometry and notations for a continuous 2D boundary element for a quadratic shape function. T and q are evaluated at the point (x_2, y_2) while the three nodes of the boundary element are (x_i, y_i), $(i = 1, 2, 3)$

A typical integral over the boundary element Γ of Figure 7.100 is given as

$$\oint_\Gamma u(x)q(x,\xi)d\Gamma(x) = \int_{-1}^{1} u(\eta)q(\eta,\xi)J(\eta)d\eta \tag{7.14.65}$$

with the Jacobian J equal to

$$J(\eta) = \sqrt{\left(\frac{dx(\eta)}{d\eta}\right)^2 + \left(\frac{dy(\eta)}{d\eta}\right)^2} \tag{7.14.66}$$

This integral can be calculated by Gauss–Legendre quadrature. This type of *continuous element* requires the connectivity at the extremities of the element and poses more difficulties for coding in 3D, especially at corners.

Another type of element called *discontinuous* is given in Figure 7.101 (Ang 2007; Kassab 2018). In this case, the field variables are discontinuous, they are interpolated on $\eta = [-3/4, 3/4]$ and extrapolated on $\eta = [-1, -3/4[$ and $\eta =]3/4, 1]$, the nodes being placed at $\eta = \{-3/4, 0, 3/4\}$. This technique poses less difficulties at the corners, is easier for coding in 3D, but requires more nodes than the continuous elements. The shape functions are different. For a quadratic polynomial, they are

$$M_1(\eta) = \frac{2}{9}\eta(4\eta-3), \quad M_2(\eta) = \frac{1}{9}(3-4\eta)(3+4\eta), \quad M_3(\eta) = \frac{2}{3}\eta(4\eta+3) \tag{7.14.67}$$

and a field variable u (similarly for q) is estimated as

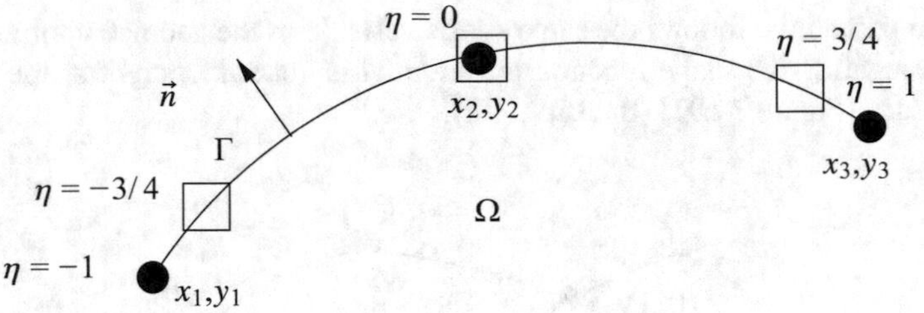

Fig. 7.101 Geometry and notations for a discontinuous 2D boundary element for a quadratic shape function

$$u(\eta) = \sum_k u_k M_k(\eta) \tag{7.14.68}$$

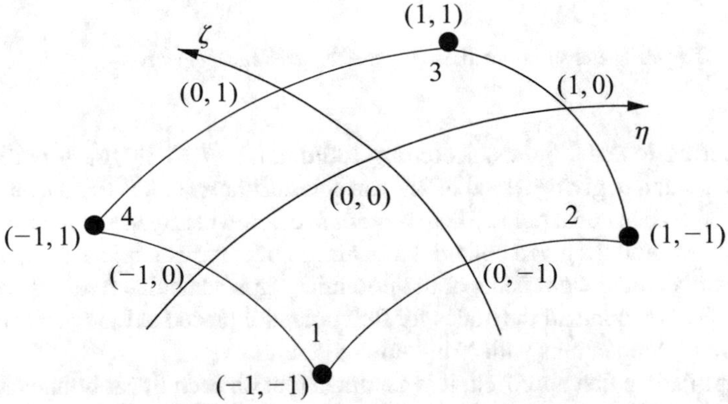

Fig. 7.102 3D quadrilateral surface with isoparametric linear elements

In 3D, for an isoparametric linear quadrilateral element (Figure 7.102), the shape functions are

$$\begin{aligned} N_1(\eta,\zeta) &= \tfrac{1}{4}(1-\eta)(1-\zeta)\,, \quad N_2(\eta,\zeta) = \tfrac{1}{4}(1+\eta)(1-\zeta) \\ N_3(\eta,\zeta) &= \tfrac{1}{4}(1+\eta)(1+\zeta)\,, \quad N_4(\eta,\zeta) = \tfrac{1}{4}(1-\eta)(1+\zeta) \end{aligned} \tag{7.14.69}$$

and for a bilinear discontinuous element (nodes at $\eta = [-3/4, 3/4]$ and $\zeta = [-3/4, 3/4]$) (Kassab 2018)

$$\begin{aligned} M_1(\eta,\zeta) &= \tfrac{1}{36}(3-4\eta)(3-4\zeta)\,, \quad M_2(\eta,\zeta) = \tfrac{1}{36}(3+4\eta)(3-4\zeta) \\ M_3(\eta,\zeta) &= \tfrac{1}{36}(3+4\eta)(3+4\zeta)\,, \quad M_4(\eta,\zeta) = \tfrac{1}{36}(3-4\eta)(3+4\zeta) \end{aligned} \tag{7.14.70}$$

Again, the potential u follows the same shape variable as the geometry for an isoparametric element. If the shape function polynomial is quadratic, eight shape functions are necessary (Figure 7.103) (Becker 1992).

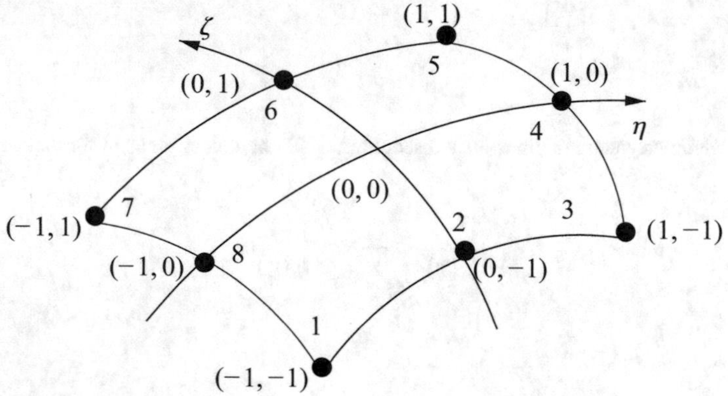

Fig. 7.103 3D quadrilateral surface with isoparametric quadratic elements

- Considering N nodal points, according to Equation (7.14.58), each point has two variables u and q, giving a total of $2N$ variables. Either u or q is fixed by a boundary condition at each nodal point. There remain N unknowns. By use of the fundamental solution, assuming a given potential at a given node, for example starting at node 1, the potential and the gradients of the potential can be calculated for all nodes from 1 to N. By sweeping all the nodes for that potential placed at the successive nodes, a system of N equations with N unknowns results.
- To simplify, in a first approach, it is assumed that on each linear boundary element Γ_e, the variable u and its gradient q (the subscript "h" for the approximation has been dropped) are constant so that we can write (7.14.58) at each boundary node ξ_i of a boundary element Γ_j ($j = 1, \ldots, N_e$)

$$c_i u_i = \sum_{j=1}^{N_e} q_j \oint_{\Gamma_j} g(\boldsymbol{x}, \boldsymbol{\xi}_i) d\Gamma_j(\boldsymbol{x}) - \sum_{j=1}^{N_e} u_j \oint_{\Gamma_j} \frac{\partial g(\boldsymbol{x}, \boldsymbol{\xi}_i)}{\partial \boldsymbol{n}} d\Gamma_j(\boldsymbol{x}) + \int_{\Omega} g(\boldsymbol{x}, \boldsymbol{\xi}_i) p(\boldsymbol{x}) d\Omega_{\boldsymbol{x}} \tag{7.14.71}$$

We adopt the usual definition

$$\begin{aligned} G_{ij} &= \oint_{\Gamma_j} g(\boldsymbol{x}, \boldsymbol{\xi}) d\Gamma_j(\boldsymbol{x}) \\ H_{ij} &= \oint_{\Gamma_j} \frac{\partial g(\boldsymbol{x}, \boldsymbol{\xi})}{\partial \boldsymbol{n}} d\Gamma_j(\boldsymbol{x}) + c_i \delta_{ij} = \hat{H}_{ij} + c_i \delta_{ij} \end{aligned} \tag{7.14.72}$$

where δ_{ij} is the Kronecker notation ($\delta_{ii} = 1, \delta_{ij} = 0$ if $i \neq j$), thus c_i influences only the diagonal element. H_{ij} and G_{ij} are called *influence coefficients*. The functions $g(\boldsymbol{x}, \boldsymbol{\xi})$ and $\dfrac{\partial g(\boldsymbol{x}, \boldsymbol{\xi})}{\partial \boldsymbol{n}}$ are called the *kernels* which *depend only on the geometry*.

Thus Equation (7.14.71) becomes

$$\sum_{j=1}^{N_e} u_j H_{ij} = \sum_{j=1}^{N_e} q_j G_{ij} + \int_{\Omega} g(\boldsymbol{x}, \boldsymbol{\xi}) p(\boldsymbol{x}) d\Omega_{\boldsymbol{x}}\,, \quad \text{for } i = 1, \dots, N_e \tag{7.14.73}$$

which gives the following system

$$[\boldsymbol{H}]\ \{\boldsymbol{u}\} = [\boldsymbol{G}]\ \{\boldsymbol{q}\} + \{\boldsymbol{S}\} \tag{7.14.74}$$

where $[\boldsymbol{H}]$ and $[\boldsymbol{G}]$ are square matrices of dimension N_e, $\{\boldsymbol{u}\}$ is the vector of field variables and $\{\boldsymbol{q}\}$ the vector of gradients of field variables. $\boldsymbol{S}$ stands for the source term.
Then, the known boundary conditions at each nodal point (all nodal points are considered) are imposed giving some $\boldsymbol{u}$ or $\boldsymbol{q}$, thus the known values of $\{\boldsymbol{u}\}$ are transferred in the right-hand side of the matrix form while the unknown values of $\{\boldsymbol{q}\}$ are transferred in the left-hand side. The matrices $[\boldsymbol{H}]$ and $[\boldsymbol{G}]$ are partitioned (Kakuba 2011) so that

$$\begin{bmatrix} \boldsymbol{H}_1 & \boldsymbol{H}_2 \end{bmatrix} \begin{bmatrix} \tilde{\boldsymbol{u}}_1 \\ \boldsymbol{u}_2 \end{bmatrix} = \begin{bmatrix} \boldsymbol{G}_1 & \boldsymbol{G}_2 \end{bmatrix} \begin{bmatrix} \boldsymbol{q}_1 \\ \tilde{\boldsymbol{q}}_2 \end{bmatrix} \tag{7.14.75}$$

where $\tilde{\boldsymbol{u}}_1$ and $\tilde{\boldsymbol{q}}_2$ are the known quantities. It results

$$\begin{bmatrix} \boldsymbol{H}_2 & -\boldsymbol{G}_1 \end{bmatrix} \begin{bmatrix} \boldsymbol{u}_2 \\ \boldsymbol{q}_1 \end{bmatrix} = \begin{bmatrix} -\boldsymbol{H}_1 & \boldsymbol{G}_2 \end{bmatrix} \begin{bmatrix} \tilde{\boldsymbol{u}}_1 \\ \tilde{\boldsymbol{q}}_2 \end{bmatrix} \tag{7.14.76}$$

Setting

$$\boldsymbol{A} = \begin{bmatrix} \boldsymbol{H}_2 & -\boldsymbol{G}_1 \end{bmatrix}, \quad \boldsymbol{Z} = \begin{bmatrix} \boldsymbol{u}_2 \\ \boldsymbol{q}_1 \end{bmatrix}, \quad \boldsymbol{B} = -\boldsymbol{H}_1 \tilde{\boldsymbol{u}}_1 + \boldsymbol{G}_2 \tilde{\boldsymbol{q}}_2 \tag{7.14.77}$$

the new system results, in the absence of the source term,

$$\boldsymbol{A}\,\boldsymbol{Z} = \boldsymbol{B} \tag{7.14.78}$$

where $\boldsymbol{Z}$ is the vector of unknown variables.
The source term $\boldsymbol{S}$ requires a special treatment as it can be the consequence of point sources or distributed sources. If they are distributed, the calculation must take into account the boundary sources and the internal sources.
The elements of the matrices $[\boldsymbol{H}]$ and $[\boldsymbol{G}]$ depend only on the geometry and are generally calculated by Gauss–Legendre quadratures.
If subdomains are considered, then the systems of equations are gathered by an assembling procedure.
The system (7.14.78) can typically be solved by quasi-Newton or Levenberg–Marquardt methods.

- Not only the steady-state problems can be treated by the boundary element method, but also the transient problems (Banerjee 1994) which can be solved by usual integration methods such as Crank-Nicholson. The technique of Laplace transforms is also used to solve the transient boundary element method (Erhart et al. 2006).

7.14.5 Boundary Element Method in 2D Heat Transfer

Equation (7.14.71) must be slightly modified in the case of heat transfer (Antes 2010; Divo and Kassab 2003; Kassab 2018; Kassab and Wrobel 2000).

Starting from the fundamental solution θ for the two-dimensional Laplace equation

$$\nabla^2 \theta = -\delta(\boldsymbol{x}, \boldsymbol{\xi}) \tag{7.14.79}$$

yielding the solution (7.14.21), the flux corresponding to the fundamental solution is given by

$$q = \lambda \frac{\partial \theta}{\partial \boldsymbol{n}} \tag{7.14.80}$$

The usual heat flux is given by

$$\dot{q}'' = \lambda \frac{\partial T}{\partial \boldsymbol{n}} \tag{7.14.81}$$

Equation (7.14.71) is modified as

$$\begin{aligned} c_i u_i(\xi, \eta) = \frac{1}{\lambda} & \left[\sum_{j=1}^{N_e} q_j \oint_{\Gamma_j} \theta(\boldsymbol{x}, \boldsymbol{\xi}_i) d\Gamma_j(\boldsymbol{x}) - \sum_{j=1}^{N_e} u_j \oint_{\Gamma_j} \frac{\partial \theta(\boldsymbol{x}, \boldsymbol{\xi}_i)}{\partial \boldsymbol{n}} d\Gamma_j(\boldsymbol{x}) \right] \\ & + \int_{\Omega} \theta(\boldsymbol{x}, \boldsymbol{\xi}_i) u_g(\boldsymbol{x}) d\Omega_{\boldsymbol{x}} \end{aligned} \tag{7.14.82}$$

taking into account the thermal conductivity λ and $u = T$. In the following, the term u_g of heat generation is not considered.

Some details of calculation are given below in the case of constant elements.

The boundary Γ is approximated by N linear discrete boundary elements $\Gamma^{(1)}, \ldots, \Gamma^{(N)}$ in the counterclockwise direction (Figure 7.104).

It is assumed that the functions $u = T$ and $\lambda \partial u / \partial \boldsymbol{n} = q$ are constant on the discrete boundary elements and their value is considered at the midpoint of the given element

$$T = \bar{T}^{(k)} \qquad \text{and} \quad \lambda \frac{\partial T}{\partial \boldsymbol{n}} = \bar{q}^{(k)} \quad \text{for } (x, y) \in \Gamma^{(k)} \tag{7.14.83}$$

Consider a point (x, y) on a discrete linear boundary element $\Gamma^{(k)}$. Its position can be parameterized by t such that

$$\begin{aligned} x &= x^{(k)} - t\, l^{(k)}\, n_y^{(k)} \\ y &= y^{(k)} + t\, l^{(k)}\, n_x^{(k)}, \quad t \in [0, 1] \end{aligned} \tag{7.14.84}$$

where the unit outward normal vector $\boldsymbol{n}$ at (x, y) is given by

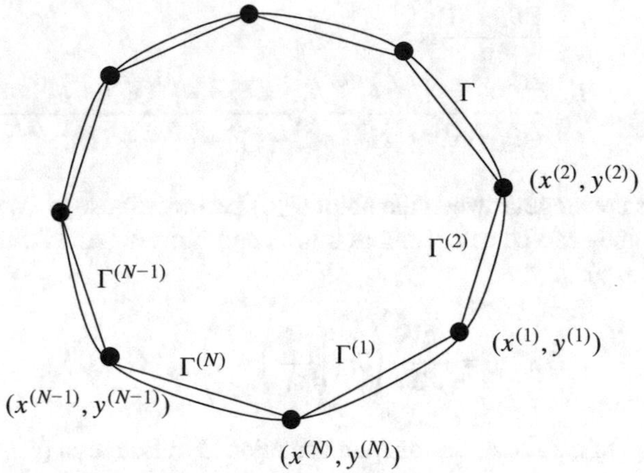

Fig. 7.104 Approximation of the boundary Γ by discrete linear elements $\Gamma^{(i)}$

$$\begin{aligned} n_x &= \frac{y^{(k+1)} - y^{(k)}}{l^{(k)}} \\ n_y &= -\frac{x^{(k+1)} - x^{(k)}}{l^{(k)}} \end{aligned} \tag{7.14.85}$$

where $l^{(k)}$ is the length of the boundary element $\Gamma^{(k)}$. Thus, the curvilinear variable s for integration along Γ is given by

$$ds^2 = dx^2 + dy^2 = l^2\, dt \tag{7.14.86}$$

and

$$r^2 = (x - \xi)^2 + (y - \eta)^2 \tag{7.14.87}$$

becomes a second degree polynomial with respect to the parameter t. The point (ξ, η) is anywhere, either on the boundary Γ or in the domain Ω. Thus, the influence coefficients $G^{(k)}$ and $H^{(k)}$ can be even calculated analytically in the case of constant elements (Ang 2007). However, it is more general to calculate them by means of Gauss–Legendre quadrature (cf. Section 2.6.2)

$$\begin{aligned} H(\xi, \eta) &= \oint_{\Gamma^{(k)}} \theta(\xi, \eta) ds = -\oint_{\Gamma^{(k)}} \frac{1}{2\pi} \log(r) ds \\ &= -\oint_0^1 \frac{l^{(k)}}{4\pi} \log\left((x^{(k)} - t\, l^{(k)}\, n_y^{(k)} - \xi)^2 + (y^{(k)} + t\, l^{(k)}\, n_x^{(k)} - \eta)^2 \right) dt \end{aligned} \tag{7.14.88}$$

where r is given by (7.14.87).

$$
\begin{aligned}
G(\xi,\eta) &= -\lambda \oint_{\Gamma^{(k)}} \frac{\partial\theta(\xi,\eta)}{\partial \boldsymbol{n}} ds \\
&= -\lambda \oint_0^1 \frac{l^{(k)}}{2\pi} \frac{n_x^{(k)}(x^{(k)} - t\, l^{(k)}\, n_y^{(k)} - \xi) + n_y^{(k)}(y^{(k)} + t\, l^{(k)}\, n_x^{(k)} - \eta)}{(x^{(k)} - t\, l^{(k)}\, n_y^{(k)} - \xi)^2 + (y^{(k)} + t\, l^{(k)}\, n_x^{(k)} - \eta)^2} dt
\end{aligned}
\tag{7.14.89}
$$

These integrals are singular when the point (ξ, η) becomes close to (x, y) or r is close to 0. However, they can be calculated as a limit and it gives (Ang 2007; Antes 2010) when $(\xi, \eta) = (x, y)$

$$
H(\xi,\eta) = -\frac{l^{(k)}}{2\pi}\left(\log\left(\frac{l^{(k)}}{2}\right) - 1\right), \quad G = 0 \tag{7.14.90}
$$

The elements of matrix $\boldsymbol{A}$, vector $\boldsymbol{B}$ and solution $\boldsymbol{Z}$ in Equation (7.14.77) are given by

$$
\begin{aligned}
A_{ij} &= \begin{cases} H^{(j)}(x^{(i)}, y^{(i)}) & \text{if } T \text{ specified on } \Gamma^{(j)}(\text{Dirichlet}) \\ -G^{(j)}(x^{(i)}, y^{(i)}) - \frac{1}{2}\lambda\delta_{ij} & \text{if } q \text{ specified on } \Gamma^{(j)}(\text{Neumann}) \end{cases} \\
B_i &= \begin{cases} \sum_j \bar{T}^{(j)}(G^{(j)}(x^{(i)}, y^{(i)}) + \frac{1}{2}\lambda\delta_{ij}) & \text{if } T \text{ specified on } \Gamma^{(j)}(\text{Dirichlet}) \\ \sum_j \bar{q}^{(j)} H^{(j)} & \text{if } q \text{ specified on } \Gamma^{(j)}(\text{Neumann}) \end{cases} \\
Z_j &= \begin{cases} \bar{q}^{(j)} & \text{if } \bar{T} \text{ specified on } \Gamma^{(j)}(\text{Dirichlet}) \\ \bar{T}^{(j)} & \text{if } \bar{q} \text{ specified on } \Gamma^{(j)}(\text{Neumann}) \end{cases}
\end{aligned}
\tag{7.14.91}
$$

where δ_{ij} is the Kronecker symbol ($\delta_{ii} = 1$, $\delta_{ij} = 0$ if $i \neq j$). A_{ii} is given by Equation (7.14.90).

Thus, solving (7.14.78) gives the temperatures and the fluxes at the midpoints of the boundary elements.

To obtain the temperature field in the domain for observation points (ξ, η), Equation (7.14.82) is used with $c = 1$. It requires to calculate the influence coefficients for each domain point, but at this stage, all the temperatures and fluxes on the boundary are known.

Example 7.36 :

HT3-BEM-2D: 2D plane plate, subjected to Dirichlet boundary conditions on all sides.

This example solved by the boundary element method (Ang 2007, 2008; Antes 2010; Kirkup and Yazdani 2008) considers the same problem as Example 7.5 which was solved by finite differences and Example 7.29 which was solved by finite elements. A plane plate of dimensions 0.02×0.05 m is subjected to Dirichlet boundary conditions on all sides (Figure 7.105)

$$
\begin{aligned}
&T(x = 0, y) = 450\,\text{K}\,, \; T(x = L_x, y) = 550\,\text{K} \\
&T(x, y = 0) = 400\,\text{K}\,, \; T(x, y = L_y) = 500\,\text{K}
\end{aligned}
\tag{7.14.92}
$$

We desire to find the steady state temperature field.

- The mesh was done by selecting twenty boundary elements for each side of the rectangle.
- After defining the geometrical characteristics and the boundary conditions, a simplification was operated by approximating the temperatures and fluxes on the boundaries as constant values over

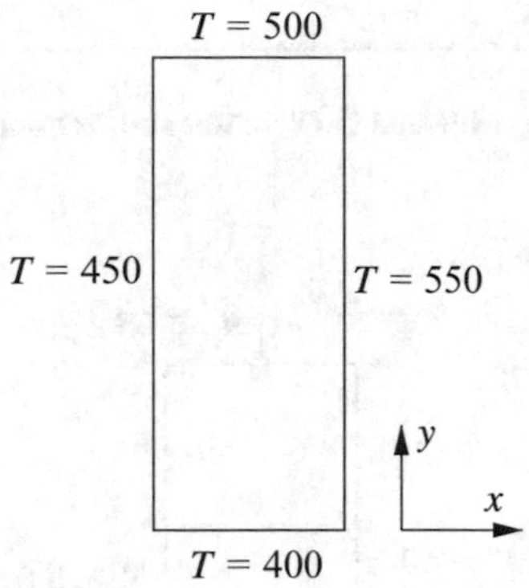

Fig. 7.105 Example HT3-BEM-2D: boundary conditions

each element and equal to their respective values at the midpoint of each boundary element k. The influence coefficients, constant per boundary element, are given by Equations (7.14.89) and (7.14.89). The integrals were calculated by Gauss–Legendre quadrature as described previously. Equation (7.14.82) is first used on the boundaries (with $c = 1/2$).

- Then, the system (7.14.78) was defined and solved. Thus, the unknown temperatures and fluxes on the boundaries were found. Indeed, in this particular case, in the absence of any Neumann boundary condition, all the temperatures were already known as resulting from the Dirichlet boundary conditions.
- Finally, knowing the boundary temperatures and fluxes, the temperatures inside the domain were calculated by Equation (7.14.82) (with $c = 1$). To give a fine graphical representation, a grid of 100×100 points was selected.
- The temperature field is represented in Figure 7.106 which is in good agreement with Figures 7.27 and 7.91 issued from the two Examples 7.5 and 7.29.

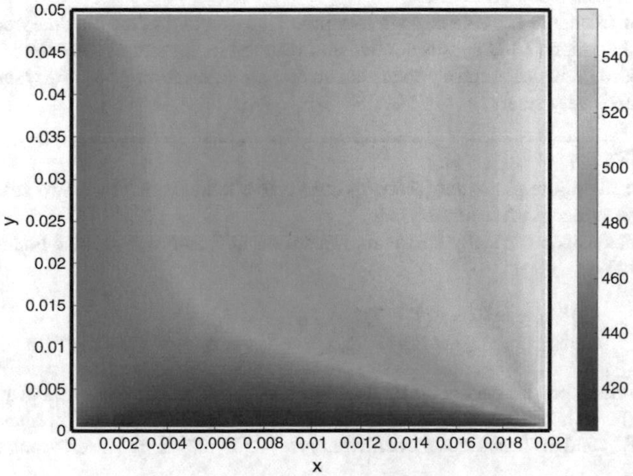

Fig. 7.106 Example HT3-BEM-2D: Temperature surface in 2D plate by the boundary element method

Example 7.37 :

HT4-BEM-2D: 2D plane plate, subjected to Dirichlet and Neumann boundary conditions on all sides.

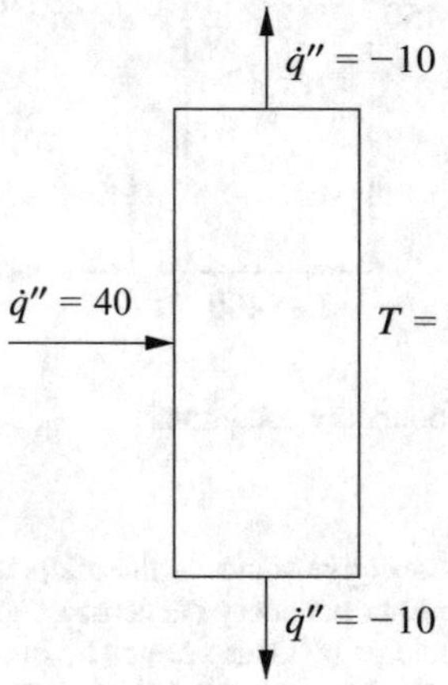

Fig. 7.107 Example HT4-BEM-2D: boundary conditions

The plane plate is geometrically identical to that used in Example 7.36. The boundary conditions (Figure 7.107) are:

$$\begin{array}{c}\dot{q}''(x, y = 0) = -10\,\mathrm{W\,m^{-2}}\,,\ T(x = L_x, y) = 300\,\mathrm{K}\\ \dot{q}''(x, y = L_y) = -10\,,\ \dot{q}''(x = 0, y) = 40\,\mathrm{W\,m^{-2}}\end{array} \tag{7.14.93}$$

thus the two horizontal sides and the left vertical side are subjected to fixed heat fluxes whereas the right vertical side is subjected to a fixed temperature. This results in mixed Dirichlet and Neumann boundary conditions. The thermal conductivity is $\lambda = 26\,\mathrm{W\,m^{-1}\,K^{-1}}$. Twenty boundary elements were selected for each side of the rectangle.

The solution procedure is the same as in Example 7.36 except that the boundary conditions differ. The method of Equation (7.14.78) was applied and resulted in the temperature field of Figure 7.108. Due to the imposed Dirichlet boundary condition on a single side, the range of steady state temperatures in the plane plate is very small.

Example 7.38 :

HT5-BEM-2D: 2D plane plate, subjected to convective heat transfer on two sides, an imposed heat flux on one side and an insulated side.

The plane plate is geometrically identical to that used in Example 7.36. The boundary conditions (Figure 7.109) are

$$\begin{array}{c}\dot{q}''(x, y = 0) = h(T - T_{env})\,,\ \dot{q}''(x = L_x, y) = 0\\ \dot{q}''(x, y = L_y) = h(T - T_{env})\,,\ \dot{q}''(x = 0, y) = 120\,\mathrm{W\,m^{-2}}\end{array} \tag{7.14.94}$$

thus convective heat transfer occurs on the two horizontal sides. The left vertical side is subjected to a fixed heat flux. The right vertical side is insulated. In Figure 7.109, the thermal conductivity is $\lambda = 26\,\mathrm{W\,m^{-1}\,K^{-1}}$ and the heat transfer coefficient is $h = 5\,\mathrm{W\,m^{-2}\,K^{-1}}$. The environment temperature is $T_{env} = 293\,\mathrm{K}$.

The boundary condition corresponding to the convective heat transfer is not a Dirichlet boundary condition, nor a boundary condition Neumann. In the convective boundary condition, both temperature and flux are unknown at the boundary. Consequently, the previous method symbolized by

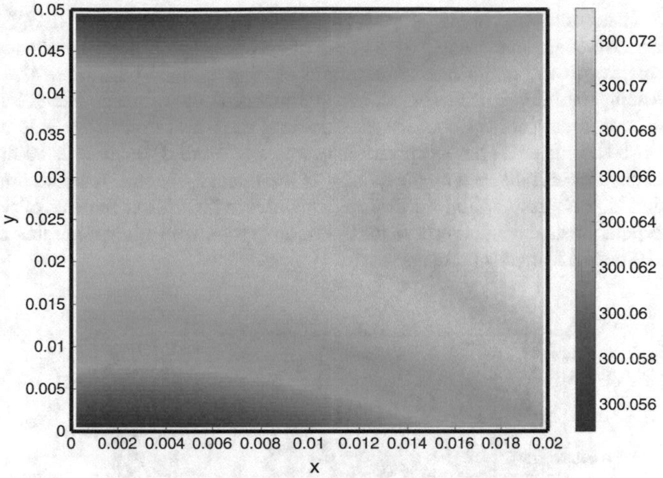

Fig. 7.108 Example HT4-BEM-2D: Temperature surface in 2D plate by the boundary element method

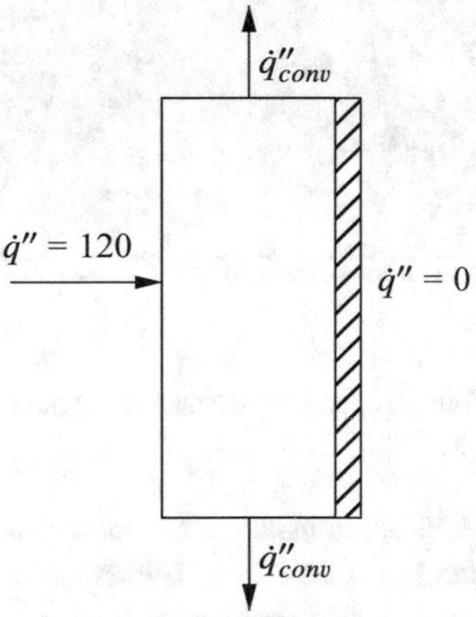

Fig. 7.109 Example HT5-BEM-2D: boundary conditions

Equation (7.14.78) and used in Examples 7.36 and 7.37 is not directly applicable. We have chosen the following optimization method. At the midpoints of the boundary elements where the convective heat transfer takes place, the temperature is initially assumed to be equal to $T = 300$ K. Thus, the method of Equation (7.14.78) can be executed. It yields values of the conductive heat flux at the same midpoints of the boundary elements. Obviously, the conductive heat flux thus calculated and the effective convective heat transfer differs. An optimization criterion dealing only with the convective heat transfer boundary elements can be designed as

$$C = \sum(\dot{q}''_{conductive} - \dot{q}''_{convective})^2 \tag{7.14.95}$$

By minimizing C with respect to the vector of boundary temperatures at midpoints of the conductive heat transfer boundary elements, provided that the optimization succeeds, the problem can be solved. The optimization was performed by function Fmincon which is able to handle nonlinear optimization in the presence of any constraint in Matlab™. It could have been done by using NLPQL (Schittkowski 1981) in Fortran. No constraint was included in the optimization process except the lower and upper bounds with respect to the optimized temperatures. It was noticed that, when the dimension of the optimized temperature vector increased, the optimization convergence became more difficult. The optimization was easy for 5 boundary elements on each heat convective side resulting in 10 temperatures to be optimized. The optimization was neatly more difficult with 10 or 20 boundary elements. This has a small influence on the quality of the final steady-state temperature field. To help the optimization, a previously optimized profile calculated with a lower number of points could be used after interpolation as the initial profile for the optimization with a larger number of points.

Figure 7.110 results from this procedure.

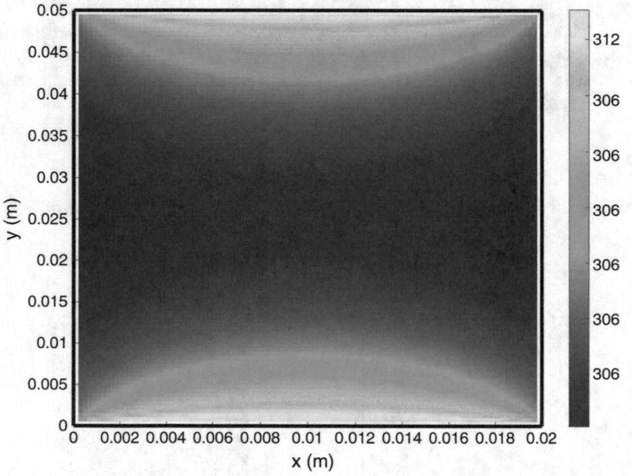

Fig. 7.110 Example HT5-BEM-2D: Temperature surface in 2D plate by the boundary element method

The influence of the main physical parameters, the convective heat transfer coefficient h and heat conductivity λ, is highlighted in Figure 7.111. The temperature profiles at mid-height in the plate are flattened when the convective heat transfer coefficient or the heat conductivity increase.

7.15 Discussion and Conclusion

It is assumed that the modeler has correctly set the partial differential equations which describe his system, with the right initial conditions and especially the right boundary conditions. As a general rule, these latter are the cause of the difficulties and they determine the solution.

The finite difference method is always attractive because of its relatively easy approach and its apparent operating simplicity. It hides numerous traps and is not conservative opposite to the finite volume method. It is no more used in commercial codes where finite volumes and finite elements are generally used. When the finite

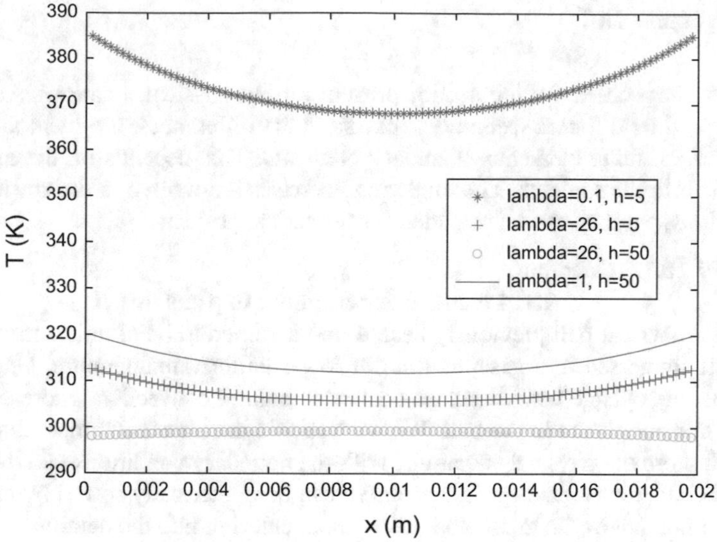

Fig. 7.111 Example HT5-BEM-2D: Influence of convective heat transfer coefficient h and heat conductivity λ on the temperature profiles at mid-height of the plate

difference method is used, it is frequent to observe instabilities at the limits of the domain.

The method of characteristics allows to explain in an elegant manner a certain number of physical phenomena for some classes of equations. It avoids the instabilities often met when using the finite difference method in the neighborhood of boundaries. It seems to be more used by researchers than by engineers but is little encountered nowadays.

The finite volume method respects the conservation equations, is very efficient and much used for fluid mechanics, heat transfer and chemical engineering (ANSYS FLUENT). It is perfectly adapted to someone who is willing to use it, provided he has enough time ahead of him.

The finite element method is more difficult theoretically, but very powerful by allowing to take into account the more complex geometries. Many codes are based on this method, commercial codes (COMSOL, ANSYS FLUENT, ABAQUS) as well as open software (FreeFem, OpenFEM, MODULEF). It is the reference method in the domain of solid mechanics, but it is also very much used in fluid mechanics and heat transfer, neatly less in chemical engineering.

The boundary element method is well adapted to systems presenting complex geometries. It is less used than the finite element method (COMSOL) but it can alleviate the calculation burden.

7.16 Exercise Set

Preamble: The problems of this section present various physical situations. All of them can be solved by different techniques, classic finite differences, the method of lines, finite volumes, finite elements, boundary elements. This depends on the amount of time allotted to the students. The following exercises are written in the spirit of short examinations, but they can be extended for numerical projects.

Exercise 7.16.1 (Medium)

Heat transfer for a sphere in a gas

A steel spherical ball previously heated and assumed to be at uniform inner temperature is immersed in a gaseous fluid at lower uniform temperature. Under these conditions, the outside heat exchange occurs by natural convection whereas conductive heat transfer occurs in the ball. We wish to find the transient temperature profile in the ball. Pose the physical equations, with the boundary conditions and initial condition. Then, demonstrate how this problem can be numerically solved by presenting a flow diagram (show the main modules without entering into the details).
Physical data: Initial temperature of the ball $T = 800\,\text{K}$, Initial temperature of air $T = 300\,\text{K}$. Ball radius $R = 0.02\,\text{m}$. Global heat transfer coefficient $h = 30\,\text{W}\,\text{m}^{-2}\,\text{K}^{-1}$. Thermal conductivity of steel is $\lambda = 26\,\text{W}\,\text{m}^{-1}\,\text{K}^{-1}$, density $\rho = 7800\,\text{kg}\,\text{m}^{-3}$, heat capacity $C_p = 500\,\text{J}\,\text{kg}^{-1}\,\text{K}^{-1}$.

Exercise 7.16.2 (Medium)

Heat transfer in a one-dimensional plane plate

A steel plane plate ruled by conductive heat transfer is represented in Figure 7.112. On the left wall, the plate is subjected to a fire. On the right wall, the plate is perfectly isolated. The thickness of the plate is $l = 0.02\,\text{m}$. 5 regularly spaced nodes are considered including the wall, noted $x(i)$, $i = 1, \ldots, 5$. The temperatures are noted T_i.

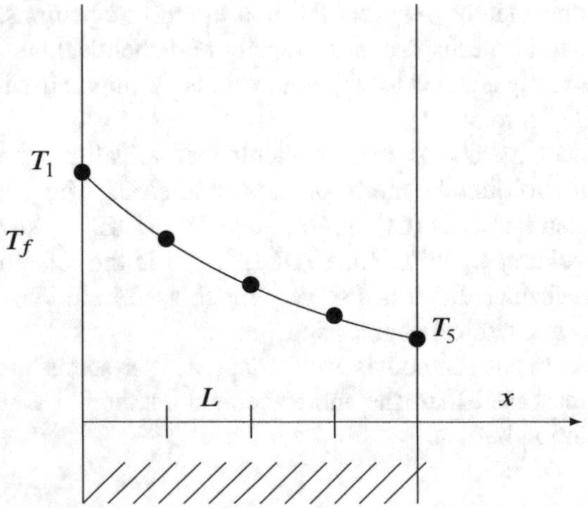

Fig. 7.112 Steel plane plate subjected to a fire

At initial time, the plate is at uniform temperature 300 K. The physical characteristics of steel are: thermal conductivity $\lambda = 16\ \mathrm{W\,m^{-1}\,K^{-1}}$, density $\rho = 8000\,\mathrm{kg\,m^{-3}}$, heat capacity $C_p = 480\,\mathrm{J\,kg^{-1}\,K^{-1}}$.

To represent the radiative transfer on the left wall, an equivalent flame temperature equal to $T_f = 1246\,\mathrm{K}$ is calculated (temperature at infinite distance with respect to the plate). The radiative flux equal to $\phi = \sigma\epsilon(T_f^4 - T_1^4)$ is calculated as an equivalent flux at the wall equal to: $\phi = h\,(T_f - T_1)$ with the heat transfer coefficient $h = 95\,\mathrm{W\,m^{-2}\,K^{-1}}$. σ is Stefan-Boltzmann constant equal to $\sigma = 5.67 \times 10^{-8}\,\mathrm{W\,m^{-2}\,K^{-4}}$, and ϵ is the emissivity equal to $\epsilon = 0.66$.

To express the fact that the plate is isolated on the right wall, it is supposed that the flux is zero at the wall.

As a first estimation, we desire to calculate the transient temperature profile in the plate on four time steps with $\Delta t = 1\,\mathrm{s}$. An implicit finite difference method will be used.

1. Clearly summarize the fundamental equations of the physical problem to solve.
2. Explain the finite difference in the studied case while keeping the dimensional variables.
3. Give the numerical results in a table while explaining how the problem was solved numerically (in dimensional variables).

Exercise 7.16.3 (Medium)

Heat transfer in a two-dimensional plane plate

The heat transfer in the two-dimensional plane plate represented in Figure 7.113 is due to conduction. The plate is subjected to the following boundary conditions

$$\begin{array}{llll}\dot{q}_{x,0} = 25\ \mathrm{W.m^{-2}} & \text{at} & y = 0, & 0 \le x \le L_x \\ \dot{q}_{L_x,y} = h(T(L_x, y) - T_{env}) & \text{at} & x = L_x, & 0 \le y \le L_y \\ \dot{q}_{0,y} = h(T(0, y) - T_{env}) & \text{at} & x = 0, & 0 \le y \le L_y \\ \dot{q}_{x,L_y} = h(T(x, L_y) - T_{env}) & \text{at} & y = L_y, & 0 \le x \le L_x\end{array} \tag{7.16.1}$$

Thus, the plate is heated on one side and surrounded on other sides by air at environment temperature with which convective heat transfer occurs. The initial temperature in the plate is $T = 300\,\mathrm{K}$.

Study the transient regime and determine the steady-state temperatures in the plate.

Physical data: thermal conductivity $\lambda = 16\,\mathrm{W\,m^{-1}\,K^{-1}}$, density $\rho = 8000\,\mathrm{kg\,m^{-3}}$, heat capacity $C_p = 480\,\mathrm{J\,kg^{-1}\,K^{-1}}$. Global heat transfer coefficient $h = 10\,\mathrm{W\,m^{-2}\,K^{-1}}$. Environment temperature $T_{env} = 300\,\mathrm{K}$. Dimensions $L_x = 0.1\,\mathrm{m}$, $L_y = 0.05\,\mathrm{m}$.

Exercise 7.16.4 (Difficult)

Mass transfer in a membrane

A polymer membrane (Figure 7.114) permeable to a gas is considered to obey Fick's law

$$\frac{\partial C}{\partial t} = \mathcal{D}\frac{\partial^2 C}{\partial x^2} \tag{7.16.2}$$

where x is the abscissa in the membrane of thickness L.

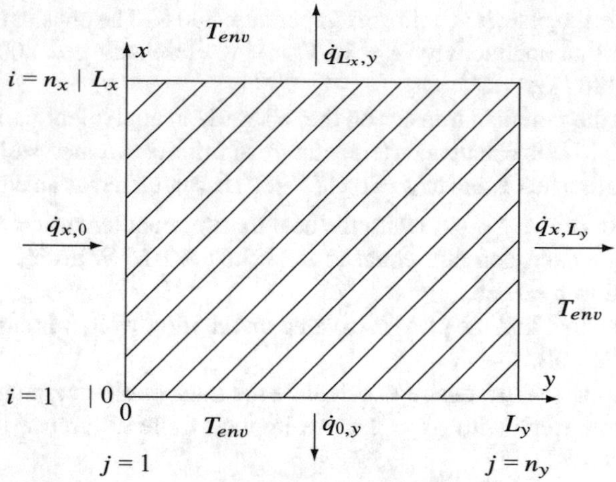

Fig. 7.113 Two-dimensional plane plate ruled by conductive heat transfer with various boundary conditions on the four sides

On the left wall, the membrane is in contact with a gas at higher pressure P_A that is assumed to be constant, as the upstream tank has a large volume, considered as infinite. On the right wall, the gas leaving the membrane flows into a tank R_B of low finite volume V_B. Initially, the tank R_B is empty. The pressure P_B in the tank R_B thus increases with time. The temperature T_B is supposed to be constant.

The experimenter would like to model the transient pressure P_B to deduce the diffusion coefficient $\mathcal{D}$ in the membrane. For that purpose, he/she will perform experiments with a given membrane, assume a given diffusion coefficient and compare the experimental and theoretical profiles of pressure $P_B(t)$.

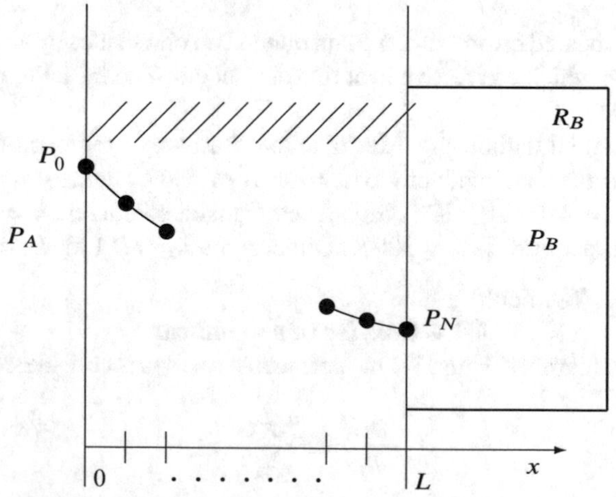

Fig. 7.114 Polymer membrane permeable to a gas

We use the relation

$$C(x) = \frac{P(x)}{RT} \tag{7.16.3}$$

so that Fick's law (7.16.2) becomes equivalent to

$$\frac{\partial P}{\partial t} = \mathcal{D}\frac{\partial^2 P}{\partial x^2} \tag{7.16.4}$$

In the present case, the boundary left condition is

$$P(x=0) = P_A = \text{constant} \tag{7.16.5}$$

whereas the right boundary condition is obtained by writing that the gas crosses the contact surface area S between the membrane and the tank R_B

$$-S\,\mathcal{D}\left(\frac{\partial C}{\partial x}\right)_{x=L} = \frac{dn_B}{dt} = \frac{V_B}{RT_B}\frac{dP_B}{dt} \tag{7.16.6}$$

Note that: $P_N = P_B$.

An implicit finite difference method is used to solve this problem.

Explain the finite difference method in the studied case while keeping the dimensional variables. Explain how it would be solved and comment how we get the profile $P_B(t)$.

Numerical data: $\mathcal{D} = 5{\times}10^{-9}\,\text{m}^2\,\text{s}^{-1}$, $P_A = 10^5\,\text{Pa}$, $L = 10^{-5}\,\text{m}$, $R = 8.314\,\text{J mol}^{-1}\,\text{K}^{-1}$, $S = 10^{-4}\,\text{m}^2$, $V_B = 10^{-6}\,\text{m}^2$, $T_B = 300\,\text{K}$.

Exercise 7.16.5 (Medium)

Wave equation for a string

Consider the one-dimensional wave equation for the displacement $u(x,t)$ of a string

$$\frac{\partial^2 u}{\partial t^2} = \frac{\partial^2 u}{\partial x^2}, \quad 0 \le x \le L \tag{7.16.7}$$

with the boundary conditions

$$u(0,t) = u(L,t) = 0 \qquad \forall t \ge 0 \tag{7.16.8}$$

and the initial conditions

$$u(x,0) = \sin\left(\frac{\pi x}{L}\right), \quad \left.\frac{\partial u}{\partial t}\right|_{(x,0)} = 0 \qquad \forall\, 0 \le x \le L \tag{7.16.9}$$

Discretize this equation by noting Δt the time step and Δx the space step. Give the resulting discrete system while indicating how you determine the solutions at instants $k\Delta t$.

Note $x_0 = 0$ and $x_N = L$.

References

H. Akima. A new method for interpolation and smooth-curve fitting based on local procedures. *J. Assoc. Comp. Machinery*, 4:589–602, 1970.

G. Allaire. *Numerical Analysis and optimization*. Oxford University Press, Oxford, 2007.

G. Allaire. *Analyse numérique et optimisation*. Editions de l'Ecole Polytechnique, Palaiseau, 2012.

G. Allaire and F. Alouges. *Analyse variationnelle des équations aux dérivées partielles*. Ecole Polytechnique, Palaiseau, 2015.

W. F. Ames. *Numerical Methods for Partial Differential Equations*. Academic Press, London, 3rd edition, 1992.

J. D. Anderson. *Modern Compressible Flow with Historical Perspective*. McGraw Hill, Boston, 3rd edition, 2003.

W. T. Ang. *A Beginner's Course in Boundary Element Methods*. Universal Publishers, Boca Raton, 2007.

W. T. Ang. Introducing the boundary element method with MATLAB. *International Journal of Mathematical Education in Science and Technology*, 39(4):505–519, 2008.

H. Antes. A short course on boundary element methods. Technische Universität Braunschweig, 2010.

G. B. Arfken, H. J. Weber, and F. E. Harris. *Mathematical Methods for Physicists*. Elsevier, Amsterdam, 7th edition, 2013.

R. Aris and N. R. Amundson. *Mathematical Methods in Chemical Engineering*. Prentice-Hall, Englewood Cliffs, N.J., 1973.

P. K. Banerjee. *The Boundary Element Methods in Engineering*. London, McGraw-Hill, 2nd edition, 1994.

G. K. Batchelor. *An Introduction to Fluid Dynamics*. Cambridge University Press, Cambridge, 3rd edition, 2000.

J. Bear. *Dynamics of Fluids in Porous Media*. Dover Publications, 1988.

A. A. Becker. *The Boundary Element Method in Engineering - A Complete Course*. McGraw-Hill, New York, 1992.

G. Beer, I. Smith, and C. Duenser. *The Boundary Element Method with Programming*. Springer, Wien, 2008.

T. L. Bergman, A. S. Lavine, F. P. Incropera, and D. P. DeWitt. *Fundamentals of Heat and Mass Transfer*. John Wiley, New York, 7th edition, 2011.

A. M. Bica. Fitting data using optimal Hermite type cubic interpolating splines. *Applied Mathematics Letters*, 25:2047–2051, 2012.

J. G. Blom and P. A. Zegeling. Algorithm 731: A moving-grid interface for systems of one-dimensional time-dependent partial differential equations. *ACM Transactions on Mathematical Software*, 20(2):194–214, 1994.

J. P. Boyd. *Chebyshev and Fourier Spectral Methods*. Dover, New York, second edition, 2001.

D. Braess. *Finite Elements - Theory, Fast Solvers, and Applications in Solid Mechanics*. Cambridge University Press, 2010.

C. A. Brebbia. The birth of the boundary element method from conception to application. *Engineering Analysis with Boundary Elements*, 77:iii–x, 2017.

C. A. Brebbia and I. Dominguez. *Boundary Element Techniques*. WIT Press, Southampton, 2nd edition, 1998.

C. A. Brebbia, J. C. F. Telles, and L. C. Wrobel. *Boundary Element Techniques*. Springer-Verlag, Berlin, 1984.

H. Brunner. *Collocation Methods for Volterra Integral and Related Functional Differential Equations*. Cambridge University Press, Cambridge, 2004.

M. Buhmann. *Radial Basis Functions: Theory and Implementations*. Cambridge University Press, Cambridge, UK, 2003.

P. L. Butzer and R. Weis. On the Lax equivalence theorem equipped with orders. *Journal of Approximation Theory*, 19:239–252, 1975.

C. Canuto, M. Hussaini, A. Quarteroni, and T. Zang. *Spectral Methods. Fundamentals in Single Domains*. Springer-Verlag, Berlin, 2006.

C. Canuto, M. Hussaini, A. Quarteroni, and T. Zang. *Spectral Methods. Evolution to Complex Geometries and Applications to Fluid Dynamics*. Springer-Verlag, Berlin, 2007.

G. F. Carey and B. A. Finlayson. Orthogonal collocation on finite elements. *Chem. Eng. Sci.*, 30:587–596, 1975.

S. Chaillat and M. Bonnet. Recent advances on the fast multipole accelerated boundary element method for 3D time-harmonic elastodynamics. *Wave Motion*, 50(7):1090–1104, 2013.

L. Challis and F. Sheard. The Green of Green's functions. *Physics Today*, 56(12): 41–46, 2003.

J. G. Charnay, R. Fjørtoft, and J. Von Neumann. Numerical integration of the barotropic vorticity equation. *Tellus*, 2:237–254, 1950.

A. H. D. Chengy and D. T. Cheng. Heritage and early history of the boundary element method. *Engineering Analysis with Boundary Elements*, 29:268–302, 2005.

P. Ciarlet and E. Luneville. *La méthode des éléments finis*. Les Presses de l'ENSTA, 2009.

M. D. C. Coimbra, C. Sereno, and A. Rodrigues. Modelling multicomponent adsorption process by a moving finite element method. *Journal of Computational and Applied Mathematics*, 115:169–179, 2000.

M. D. C. Coimbra, C. Sereno, and A. Rodrigues. Applications of a moving finite element method. *Chem. Eng. J.*, 84:23–29, 2001.

M. D. C. Coimbra, C. Sereno, and A. Rodrigues. A moving finite element method for the solution of two-dimensional time-dependent models. *Applied Numerical Mathematics*, 44:449–469, 2003.

M. D. C. Coimbra, C. Sereno, and A. Rodrigues. Moving finite element method: applications to science and engineering problems. *Comp. Chem. Engn.*, 28:597–603, 2004.

M. D. C. Coimbra, A. E. Rodrigues, J. D. Rodrigues, R. J. M. Robalo, and R. M. P. Almeida. *Moving Finite Element Method: Fundamentals and Applications in Chemical Engineering*. CRC Press, Boca Raton, 2016.

M. Costabel. Principles of boundary element methods. In *Finite Elements in Physics*, First Graduate Course in Computational Physics, Lausanne, 1986.

R. Courant, K. Friedrichs, and H. Lewy. Über die partiellen Differenzengleichungen der mathematischen Physik. *Mathematische Annalen,*, 100(1):32–74, 1928.

R. Dai and J. E. Cochran. Wavelet collocation method for optimal control problems. *J. Optim. Theory Appl.*, 143:265–278, 2009.

P. V. Danckwerts. Continuous flow systems: Distribution of residence times. *Chem. Eng. Sci.*, 2(1):1–13, 1953.

G. F. Dargush and P. K. Banerjee. A boundary element method for steady incompressible thermoviscous flow. *International Journal for Numerical Methods in Engineering*, 31:1605–1626, 1991.

C. De Boor and B. Swartz. Collocation at Gaussian points. *SIAM Journal on Numerical Analysis*, 10(4):582–606, 1973.

G. Dhatt, G. Touzot, and E. Lefrançois. *Méthode des éléments finis*. Hermès, 2014.

E. Divo and A. J. Kassab. *Boundary Element Method for Heat Conduction: With Applications in Non-homogenous Media.* WIT Press, 2003.

E. A. Dorfi and L. O'C. Drury. Simple adaptive grids for 1-D initial value problems. *Journal of Computational Physics*, 69:175–195, 1987.

T. A. Driscoll and B. Fornberg. Interpolation in the limit of increasingly flat radial basis functions. *Computers & Mathematics with Applications*, 43:413–422, 2002.

S. M. Dsouza, A. L. N. Pramod, E. T. Ooi, C. Song, and S. Natarajan. Robust modelling of implicit interfaces by the scaled boundary finite element method. *Engineering Analysis with Boundary Elements*, 124:266–286, 2021.

D. Duhamel. Finite element computation of Green's functions. *Engineering Analysis with Boundary Elements*, 31:919–930, 2007.

G. Dunnebier, I. Weirich, and K. U. Klatt. Computationally efficient dynamic modeling and simulation of simulated moving-bed chromatographic processes with linear isotherms. *Chem. Eng. Sci.*, 53(14):2537–2546, 1998.

J. F. Epperson. *An Introduction to Numerical Methods and Analysis*. Wiley, Hoboken, 2nd edition, 2013.

K. Erhart, E. Divo, and A. J. Kassab. A parallel domain decomposition boundary element method approach for the solution of large-scale transient heat conduction problems. *Engineering Analysis with Boundary Elements*, 30:553–563, 2006.

A. Ern and J. L. Guermond. *Theory and practice of finite elements*, volume 159 of *Applied Mathematical Series*. Springer, New York, 2004.

D. Euvrard. *Résolution numérique des équations aux dérivées partielles*. Masson, Paris, 1988.

G. Evans, J. Blackledge, and P. Yardley. *Numerical methods for partial differential equations*. Springer, London, 2000.

R. Eymard, T. Gallouët, and R. Herbin. *Handbook of Numerical Analysis*, volume VII, chapter The finite volume method, pages 713–1020. Elsevier, 2000.

G. E. Fasshauer. *Surface Fitting and Multiresolution Methods*, chapter Solving partial differential equations by collocation with radial basis functions, pages 131–138. A. Le Méhauté and C. Rabut and L. L. Schumaker, Vanderbilt University Press, Nashville, TN, 1997.

A. J. M. Ferreira. *Matlab Codes for Finite Element Analysis - Solids and Structures*. Springer, New York, 2009.

B. Fornberg. Generation of finite difference formulas on arbitrarily spaces grids. *Mathematics of Computation*, 51(184):699–706, 1988.

B. Fornberg. Calculation of weights in finite difference formulas. *SIAM Rev.*, 40(3): 685–691, 1998.

F. Fritsch and R. Carlson. Monotone piecewise cubic interpolation. *SIAM Journal on Numerical Analysis*, 17(2):238–246, 1980.

R. M. Furzeland, J. G. Verwer, and P. A. Zegeling. A numerical study of three moving-grid methods for one-dimensional partial differential equations which are based on the method of lines. *Journal of Computational Physics*, 89:349–388, 1990.

L. Gaul, M. Kögl, and M. Wagner. *Boundary Element Methods for Engineers and Scientists*. Springer, Berlin, 2003.

C. Geuzaine and J.-F. Remacle. Gmsh: a three-dimensional finite element mesh generator with built-in pre- and post-processing facilities. *International Journal for Numerical Methods in Engineering*, 79(11):1309–1331, 2009.

A. Guillou and J. L. Soulé. La résolution numérique des problèmes différentiels aux conditions initiales par des méthodes de collocation. *R.I.R.O.*, 3:17–44, 1969.

E. Guyon, J. P. Hulin, L. Petit, and C. D. Mitescu. *Physical Hydrodynamics*. Oxford University Press, New York, 2nd edition, 2015.

R. L. Hardy. Theory and applications of the multiquadric-biharmonic method: 20 years of discovery. *Computers and Mathematics with Applications*, 19(8):163–208, 1990.

F. Hartman and C. Katz. *Structural Analysis with Finite Elements*. Springer, Berlin, 2nd edition, 2008.

J. L. M. Hensen and A. Nakhi. Fourier and Biot numbers and the accuracy of conduction modelling. In *BEP'94 Conference "Facing the Future"*, pages 247–256, York, 1994. Building Environmental Performance Analysis Club (BEPAC).

J. Hesthaven, S. Gottlied, and D. Gottlieb. *Spectral Methods for Time-Dependent Problems*. Cambridge University Press, Cambridge, 2007.

T. J. R. Hughes. *The Finite Element Method - Linear Static and Dynamic Finite Element Analysis*. Dover, New York, second edition, 2000.

D. V. Hutton. *Fundamentals of Finite Element Analysis*. Mc Graw Hill, New York, 2004.

J. M. Hyman, S. Li, and L. R. Petzold. An adaptive moving mesh method with static rezoning for partial differential equations. *Computers and Mathematics with Applications*, 46:1511–1524, 2003.

F. P. Incropera and D. P. DeWitt. *Fundamentals of Heat and Mass Transfer*. John Wiley, New York, 4th edition, 1996.

A. Iserles. *A First Course in the Numerical Analysis of Differential Equations*. Cambridge University Press, Cambridge, 2nd edition, 2014.

J.-H. Jung and Wai Sun Don. Collocation methods for hyperbolic partial differential equations with singular sources. *Advances in Applied Mathematics and Mechanics*, 1(6):769–780, 2009.

G. Kakuba. *The boundary element method: errors and gridding for problems with hot spots*. PhD thesis, Technische Universiteit Eindhoven, Netherlands, 2011.

E. J. Kansa. Multiquadrics - a scattered data approximation scheme with applications to computational fluid dynamics - I surface approximations and partial derivative estimates. *Computers and Mathematics with Applications*, 19(8):127–145, 1990a.

E. J. Kansa. Multiquadrics - a scattered data approximation scheme with applications to computational fluid-dynamics - II solutions to parabolic, hyperbolic and elliptic partial differential equations. *Computers and Mathematics with Applications*, 8: 147–161, 1990b.

G. Karniadakis and S. Sherwin. *Spectral/hp Element Methods for Computational Fluid Dynamics*. Oxford University Press, Oxford, second edition, 2013.

A. J. Kassab. Boundary elements in heat transfer. In *HPC for Engineering and Chemistry*, Ljubljana, Slovenia, 2018. PRACE Autumn School 2018 -.

A. J. Kassab and L. C. Wrobel. *Advances in Numerical Heat Transfer*, chapter Boundary Element Methods for Heat Conduction. CRC Press, 2000.

J. T. Katsikadelis. *Boundary elements theory and applications*. Academic Press, London, 2nd edition, 2016.

R. Kelling, J. Bickel, U. Nieken, and P. A. Zegeling. An adaptive moving grid method for solving convection dominated transport equations in chemical engineering. *Comp. Chem. Engng.*, 71:467–477, 2014.

A. H. Khater, R. S. Temsah, and M. M. Hassan. A Chebyshev spectral collocation method for solving Burgers-type equations. *Journal of Computational and Applied Mathematics*, 222:333–350, 2008.

A. Khennane. *Introduction to Finite Element Analysis Using MATLAB and Abaqus*. CRC Press, Boca Raton, 2013.

S. Kirkup. *The Boundary Element Method in Acoustics*. Integrated Sound Software, 2007.

S. Kirkup and J. Yazdani. A gentle introduction to the boundary element method in Matlab/Freemat. In *MAMECTIS'08: Proceedings of the 10th WSEAS international conference on Mathematical methods, computational techniques and intelligent systems*, pages 46–52, 2008.

I. Koutromanos. *Fundamentals of Finite Element Analysis - Linear Finite Element Analysis*. Wiley, Croydon, 2018.

J. C. Lachat and J. O. Watson. Effective numerical treatment of boundary integral equations: A formulation for three-dimensional elastostatics. *International Journal for Numerical Methods in Engineering*, 10:991–1005, 1976.

L. Lapidus and G. F. Pinder. *Numerical solution of partial differential equations in science and engineering*. John Wiley, New York, 1982.

E. Larsson and B. Fornberg. Theoretical and computational aspects of multivariate interpolation with increasingly flat radial basis functions. *Computers and Mathematics with Applications*, 49:103–130, 2005.

P. D. Lax and R. D. Richtmyer. Survey of the stability of linear finite difference equations. *Comm. Pure Appl. Math.*, 9:267–293, 1956.

F. Legoll. *Partial Differential Equations and the Finite Element Method*. Ecole des Ponts Paristech, 2019.

H. Lemonnier. Application de la méthode des éléments de frontière aux écoulements diphasiques et aux techniques de mesure correspondantes. Technical report, CEA, 1996. Cours de l'école CEA/EDF/INRIA.

A. Lepourhiet. *Résolution Numérique des Equations aux Dérivées Partielles*. Cepadues, Toulouse, 1988.

R. J. LeVeque. *Finite Difference Methods for Ordinary and Partial Differential Equations*. SIAM, Philadelphia, 2007.

R. J. LeVeque and J. Randall. *Numerical Methods for Conservation Laws*. Springer, Basel, 1990.

R. J. LeVeque and J. Randall. *Finite Volume Methods for Hyperbolic Problems*. Cambridge University Press, Cambridge, 2002.

M. T. Lewandowski. Implementation of the boundary element method to two-dimensional heat transfer with thermal bridge effects. *Task Quaterly*, 17(1–2): 109–117, 2013.

R. W. Lewis, P. Nithiarasu, and K. N. Seetharamu. *Fundamentals of the Finite Element Method for Heat and Fluid Flow*. Chichester, England, Wiley, 2004.

S. Li, L. R. Petzold, and Y. Ren. Stability of moving mesh systems of partial differential equations. *SIAM J. Sci. Comp.*, 20(2):719–738, 1998.

Z. Li, Z. Qiao, and T. Tang. *Numerical Solution of Differential Equations - Introduction to Finite Difference and Finite Element Methods*. Cambridge University Press, Cambridge, 2018.

S. Liszka and J. Orkisz. The finite difference method at arbitrary irregular grids and its application in applied mechanics. *Computers & Structures*, 11:83–95, 1980.

Y. Liu. *Fast Multipole Boundary Element Method: Theory and Applications in Engineering*. Cambridge University Press, Cambridge, 2009.

D. L. Logan. *A First Course in the Finite Element Method*. Thomson, 4th edition, 2016.

J. Loustau. *Numerical Differential Equations - Theory and Technique, ODE Methods, Finite Differences, Finite Elements and Collocation*. World Scientific, Hackensack, New Jersey, 2016.

M. Malinen and P. Råback. *Multiscale Modelling Methods for Applications in Material Science*, volume 19, chapter Elmer finite element solver for multiphysics and multiscale problems, pages 101–113. Forschungszentrum Jülich, 2013.

C. Manni, A. Reali, and H. Speleers. Isogeometric collocation methods with generalized b-splines. *Computers and Mathematics with Applications*, 70:1659–1675, 2015.

S. De Marchi and G. Santin. A new stable basis for radial basis function interpolation. *Journal of Computational and Applied Mathematics*, 2013.

J. J. Martinez and P. T. T. Esperanca. A Chebyshev collocation spectral method for numerical simulation of incompressible flow problems. *J. of the Braz. Soc. of Mech. Sci. & Eng.*, 29(3):317–328, 2007.

J. R. R. A. Martins, P. Sturdza, and J. J. Alonso. The complex-step derivative approximation. *ACM Transactions on mathematical Software*, 29(3):245–262, 2003.

P. L. Mills and P. A. Ramachandran. Mathematical modelling of chemical engineering systems by finite element analysis using PDE/PROTRAN. *Mathl Comput. Modelling*, 11:375–379, 1988.

M. Mongillo. Choosing basis functions and shape parameters for radial basis function methods. *SIAM SIURO*, 4:190–209, 2011.

K. W. Morton and D. F. Mayers. *Numerical Solution of Partial Differential Equations*. Cambridge University Press, Cambridge, 2nd edition, 2012.

H. V. Mott and Z. A. Green. On Danckwerts' boundary conditions for the plug-flow model with dispersion/reaction model. *Chem. Eng. Comm.*, 202(6):739–745, 2015.

F. Moukalled, L. Mangani, and M. Darwish. *The Finite Volume Method in Computational Fluid Dynamics - An Advanced Introduction with OpenFOAM and Matlab*. Springer, Cham, 2016.

R. M. Nicoud. *Chromatographic processes: modeling, simulation and design*. Academic Press, Cambridge, 2015.

J. T. Oden, I. Babuska, F. Nobile, Y. Feng, and R. Tempone. Theory and methodology for estimation and control of errors due to modeling, approximation and uncertainty. *Computer Methods in Applied Mechanics and Engineering*, 194(2–5): 195–204, 2005.

A. Otto. Methods of numerical simulation in fluids and plasmas. Technical report, University of Alaska, Fairbanks, Alaska, 2011.

S. V. Patankar. *Numerical heat transfer and fluid flow*. Taylor & Francis, New York, 1980.

S. V. Patankar, K. C. Karki, and K. M. Kelkar. *The handbook of fluid dynamics*, chapter Finite volume method, pages 27.1–27.26. CRC Press, 1998.

D. W. Pepper, A. Kassab, and E. Divo. *An introduction to finite element, boundary element, and meshless methods with applications to heat transfer and fluid flow*. ASME Press, New York, 2014.

R. Peyret. *Spectral Methods for Incompressible Viscous Flow*. Springer-Verlag, New York, 2002.

R. L. Pigford, B. Baker, and D. E. Blum. Equilibrium theory of the parametric pump. *Ind. Eng. Chem. Fundamentals*, 8(1):144–149, 1969.

R. H. Pletcher, J. C. Tannehill, and D. A. Anderson. *Computational Fluid Mechanics and Heat Transfer*. CRC Press, Boca Raton, 2013.

C. Pozrikidis. *Introduction to Finite and Spectral Element Methods Using Matlab*. CRC Press, Boca Raton, 2nd edition, 2014.

P. Råback, M. Malinen, J. Ruokolainen, A. Pursula, and T. Zwinger. *Elmer Models Manual*. CSC – IT Center for Science, Finland, 2020.

S. S. Rao. *The Finite Element Method in Engineering*. Butterworth-Heinemann, Oxford, 6th edition, 2018.

P. A. Raviart and J. M. Thomas. *Introduction à l'analyse numérique des équations aux dérivées partielles*. Masson, Paris, 1983.

J. N. Reddy. *An Introduction to Nonlinear Finite Element Analysis*. Oxford University Press, Oxford, 2nd edition, 2015.

J. N. Reddy and D. K. Gartling. *The Finite Element Method in Heat Transfer and Fluid Dynamics*. CRC Press, Boca Raton, 3rd edition, 2010.

J. Restrepo. Boundary value problems - the method of weighted residuals. Technical report, University of Arizona, Tucson, Arizona, 2003. http://www.physics.arizona.edu/ restrepo/475B/Notes/sourcehtml/node25.html.

H. Rocha. On the selection of the most adequate radial basis function. *Applied Mathematical Modelling*, 33:1573–1583, 2009.

J. Ruokolainen, M. Malinen, P. Råback, T. Zwinger, A. Pursula, and M. Byckling. *Elmer Solver Manual*. CSC – IT Center for Science, Finland, 2020.

R. D. Russell and L. F. Shampine. A collocation method for boundary value problems. *Numer. Math.*, 19:1–28, 1972.

P. Saucez, W. E. Schiesser, and A. Vande Wouwer. Upwinding in the method of lines. *Mathematics and Computers in Simulation*, 56:171–185, 2001.

S. A. Sauter and C. Schwab. *Boundary Element Methods*. Springer, Berlin, 2011.

W. Schiesser, A. Yücel, R. Carcagno, and J. Demko. A numerical investigation of spatial approximations for strongly convective flows. Technical Report SSCL-N-881, Superconducting Super Collider Laboratory, Dallas, 1994.

W. E. Schiesser. *The numerical method of lines: integration of partial differential equations*. Academic Press, San Diego, 1991.

P. Schlatter. Spectral methods. Technical report, KTH Royal Institute of Technology, Stockholm, Sweden, 2010.

C. Sereno, A. Rodrigues, and J. Villadsen. The moving finite element method with polynomial approximation of any degree. *Comp. Chem. Engng.*, 15(1):25–33, 1991.

C. Sereno, A. Rodrigues, and J. Villadsen. Solution of partial differential equations by the moving finite element method. *Comp. Chem. Engng.*, 16(1):583–592, 1992.

S. S. Shastry and R. M. Allen. Method of lines and enthalpy method for solving moving boundary problems. *Int. Comm. Heat Mass Transfer*, 25(4):531–540, 1998.

Y. Shi and P. K. Banerjee. Boundary element methods for convective heat transfer. *Computer Methods in Applied Mechanics and Engineering*, 105:261–284, 1993.

W. Smigaj, T. Becke, S. Arridge, J. Phillips, and M. Schweiger. Solving boundary integral problems with BEM++. *ACM Transactions on Mathematical Software,*, 41 (2), 2015.

N. Spillane. *Introduction à la méthode des éléments finis*. Ecole des Mines Paristech, Paris, 2017.

I. Stakgold and M. Holst. *Green's functions and boundary value problems*. Wiley, Hoboken, third edition, 2011.

A. Sutradhar and G. H. Paulino. The simple boundary element method for transient heat conduction in functionally graded materials. *Comput. Methods Appl. Mech. Engrg.*, 193:4511–4539, 2004.

T. Tang. *Recent Advances in Adaptive Computation*, volume 383, chapter Moving mesh methods for computational fluid dynamics, pages 141–173. American Mathematical Society, Contemporary Mathematics, 2005.

L. G. Tham and Y. K. Cheung. Numerical solution of heat conduction problems by parabolic time-space element. *International Journal for Numerical Methods*, 18: 467–474, 1982.

J. W. Thomas. *Numerical Partial Differential Equations*. Springer, New York, 1998.

L. N. Trefethen. Finite difference methods and spectral methods for ordinary and partial differential equations. Cornell University, 1996.

L. N. Trefethen. *Spectral Methods in MATLAB*. SIAM, 2000.

A. Vande Wouwer, P. Saucez, and W. E. Schiesser. Simulation of distributed parameter systems using a Matlab-based method of lines toolbox: Chemical engineering applications. *Ind. Eng. Chem. Res.*, 43:3469–3477, 2004.

A. Vande Wouwer, P. Saucez, W. E. Schiesser, and S. Thompson. A MATLAB implementation of upwind finite differences and adaptive grids in the method of lines. *Journal of Computational and Applied Mathematics*, 183:245–258, 2005.

A. Varma and M. Morbidelli. *Mathematical Methods in Chemical Engineering*. Oxford University Press, New York, 1997.

H. K. Versteeg and W. Malalasekera. *An Introduction to Computational Fluid Dynamics*. Logman, Harlow, 1995.

J. Gwinner and E. P. Stephan. *Advanced Boundary Element Methods*. Springer, Cham, 2018.

J. Villadsen and M. L. Michelsen. *Solution of differential equation models by polynomial approximation*. Englewood Cliffs, New Jersey, 1978.

J. V. Villadsen and W. E. Stewart. Solution of boundary-value problems by orthogonal collocation. *Chem. Eng. Sci.*, 22(11):1483–1501, 1967.

J. A. C. Weideman and S. C. Reddy. A MATLAB differentiation matrix suite. *ACM Transactions of Mathematical Software*, 26:465–519, 2000.

P. Wesseling. *Principles of Computational Fluid Dynamics*. Springer-Verlag, Heidelberg, 2001.

F. White. *Fluid Mechanics*. Mc Graw Hill, New York, 8th edition, 2016.

K. A. Woodbury. Method of weighted residuals. Technical report, University of Alabama, Tuscaloosa, Alabama, USA, 2002.

K. Schittkowski. The nonlinear programming method of Wilson, Han and Powell with an augmented Lagrange type line search function, Part 1: convergence analysis. *Numer. Math.*, 38:83–114, 1981.

L. C. Wrobel and A. J. Kassab. *Advances in Numerical Heat Transfer*, chapter Boundary Element Methods for Heat Conduction. CRC Press, 2000.

B. Wu and R. E. White. One implementation variant of the finite difference method for solving ODEs/DAEs. *Computers and Chemical Engineering*, 28:303–309, 2004.

J. R. Yu and T. R. Hsu. Analysis of heat conduction in solids by space-time finite element method. *International Journal for Numerical Methods in Engineering*, 21: 2001–2012, 1985.

Q. Yu and N. H. L. Wang. Computer simulations of the dynamics of multicomponent ion exchange and adsorption in fixed beds – gradient-directed moving finite element method. *Comp. Chem. Engng.*, 13(8):915–926, 1989.

Q. Zang, J. Liu, W. Ye, and G. Lin. Isogeometric boundary element method for steady-state heat transfer with concentrated/surface heat sources. *Engineering Analysis with Boundary Elements*, 122:202–213, 2021.

O. C. Zienkiewicz and J. Z. Zhu. A simple error estimator for practical engineering analysis. *Int. J. Num. Meth. Engng*, 24:337–357, 1987.

O. C. Zienkiewicz, R. L. Taylor, and P. Nithiarasu. *The Finite Element Method for Fluid Dynamics*. Butterworth–Heinemann, Oxford, 7th edition, 2013.

O. C. Zienkiewicz, R. L. Taylor, and J. Z. Zhu. *The Finite Element Method: Its Basis and Fundamentals*. Butterworth–Heinemann, Oxford, 7th edition, 2014.

Chapter 8
Analytical Methods for Optimization

8.1 Introduction

After some mathematical reminder, the analytical methods of optimization are exposed in the case of equality constraints with the Lagrangian and inequality constraints with Karush–Kuhn–Tucker parameters. The sensitivity analysis concludes the chapter. These fundamentals are essential for numerical solution of optimization problems. Many mathematical and numerical examples illustrate the different cases encountered.

8.2 Mathematical Reminder

We consider a function $f(x_1, \ldots, x_n)$ of several variables that is noted $f(\boldsymbol{x})$ where $\boldsymbol{x} \in \mathbb{R}^n$.

The *gradient* of function f is the vector

$$\text{grad } f = \begin{bmatrix} \frac{\partial f}{\partial x_1} \\ \ldots \\ \frac{\partial f}{\partial x_n} \end{bmatrix} \tag{8.2.1}$$

The *Hessian* matrix is defined by

$$\begin{bmatrix} \frac{\partial^2 f}{\partial x_1^2} & \frac{\partial^2 f}{\partial x_1 \partial x_2} & \cdots & \frac{\partial^2 f}{\partial x_1 \partial x_n} \\ \frac{\partial^2 f}{\partial x_2 \partial x_1} & \frac{\partial^2 f}{\partial x_2^2} & \cdots & \frac{\partial^2 f}{\partial x_2 \partial x_n} \\ & \vdots & & \\ \frac{\partial^2 f}{\partial x_n \partial x_1} & \frac{\partial^2 f}{\partial x_n \partial x_2} & \cdots & \frac{\partial^2 f}{\partial x_n^2} \end{bmatrix} \tag{8.2.2}$$

J.-P. Corriou, *Numerical Methods and Optimization*, Springer Optimization and Its Applications 187, https://doi.org/10.1007/978-3-030-89366-8_8

Let $\boldsymbol{f}(\boldsymbol{x})$ be a vector of p functions $f_i(\boldsymbol{x})$ where $\boldsymbol{x} \in \mathbb{R}^n$

$$\boldsymbol{f}(\boldsymbol{x}) = \begin{bmatrix} f_1(\boldsymbol{x}) \\ \vdots \\ f_p(\boldsymbol{x}) \end{bmatrix} \tag{8.2.3}$$

The *Jacobian* matrix of $\boldsymbol{f}$ (also called the Jacobian) is defined by

$$\begin{bmatrix} \dfrac{\partial f_1}{\partial x_1} & \cdots & \dfrac{\partial f_1}{\partial x_n} \\ \vdots & & \\ \dfrac{\partial f_p}{\partial x_1} & \cdots & \dfrac{\partial f_p}{\partial x_n} \end{bmatrix} \tag{8.2.4}$$

Thus, it will be noticed that the Jacobian matrix of the gradient vector grad $f(\boldsymbol{x})$ is equal to the Hessian matrix of function $f(\boldsymbol{x})$.

8.3 Introduction

Let us start by some generalities (Allaire 2007; Bonnans et al. 2003; Chong et al. 2014; Dennis and Schnabel 1983; Gill et al. 1981; Fletcher 1991; Polak 1997; Nocedal and Wright 2006; Rao 2009; Ray and Szekely 1973).

Optimization consists is seeking the minimum or the maximum of a function of one or several variables. This function is called *objective function*. We are interested, either by the value of the function or frequently by the values of the variables or parameters for which the function reaches its optimum, or by both of them. The applications are found in all the domains, physical sciences, economy, ... The methods of optimum search are either purely analytical or numerical.

In optimization, we are concerned about *continuous* functions at least piecewise (analytical functions for example), as well as *discrete* functions (for example, the price of goods, number of given equipment).

A function is called *unimodal* when it presents only one maximum or one minimum. If it presents several maxima or minima, it is called *multimodal*.

A function is *convex* if it satisfies the following definition:

$$f[(1-\theta)\boldsymbol{x}_1 + \theta\boldsymbol{x}_2] \leq (1-\theta)f(\boldsymbol{x}_1) + \theta f(\boldsymbol{x}_2) \quad \forall 0 \leq \theta \leq 1$$

that is: Any point on the straight line segment joining two points of the function is above the point of the function having the same "abscissa" $\boldsymbol{x}$ (indeed, the "abscissa" is a vector of $\mathbb{R}^n$) (Figure 8.1a). The domain of Figure 8.1b is convex: Any point belonging to the segment defined by any two points of the domain also belongs to the domain.

If a function satisfies the opposite inequality, it is *concave*.

A function f of $\mathbb{R}^n$ is defined by the knowledge of a vector $\boldsymbol{x}$ of $\mathbb{R}^n$. Its representation is impossible as soon as n becomes important. It is possible to use projections on planes

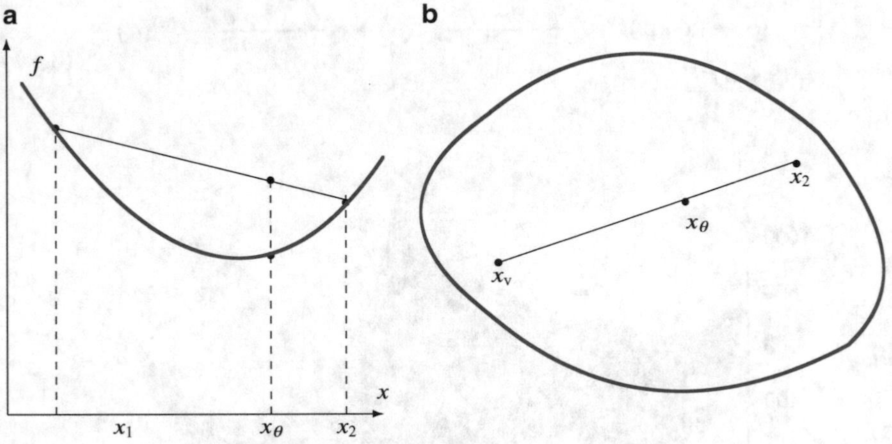

Fig. 8.1 Convex function (**a**), convex set (**b**)

formed by two axes. For example, let a function be defined in R^2. We can plot *iso-response* curves or *contours* of that function in the plane Ox_1x_2, they are sets of points for which the values of the function are constant. By choosing different values for the function, and visualizing the level curves thus obtained, we can imagine the form of the surface, the existence of local minima or maxima, of saddle points, refer to Figures 8.2, 8.3, 8.4, and 8.5 containing, for the same function $z = f(x, y)$, a three dimensional representation of the surface and a two-dimensional projection on two dimensions of the contours. The equation of the first function is $z = 3\,y^2 + x^4 + 2\,y^4$ and the second function $z = 2\,y\,x^2 - 2\,y^2 + x^3$.

Schwefel function (Bicking et al. 1994; Schwefel 1977) of equation

$$z = -x\sin(\sqrt{(|x|)}) - y\sin(\sqrt{(|y|)})$$

very well illustrates the multimodal character of a function either by the three dimensional representation of the surface (Figure 8.6) or by its contours (Figure 8.7).

8.4 Functions of One Variable

8.4.1 Infinite Interval

Suppose that the objective function f is continuous on the studied interval $]-\infty, +\infty[$ and that the related derivatives are also continuous. Moreover, suppose that the function f has a minimum at $x = \alpha$. By using a nth degree Taylor polynomial for the function f near $x = \alpha$

$$f(x) = f(\alpha) + (x-\alpha)f'(\alpha) + \frac{(x-\alpha)^2}{2!}f''(\alpha) + \cdots + \frac{(x-\alpha)^n}{n!}f^{(n)}(\alpha) + \frac{(x-\alpha)^{n+1}}{(n+1)!}f^{(n+1)}(\zeta) \tag{8.4.1}$$

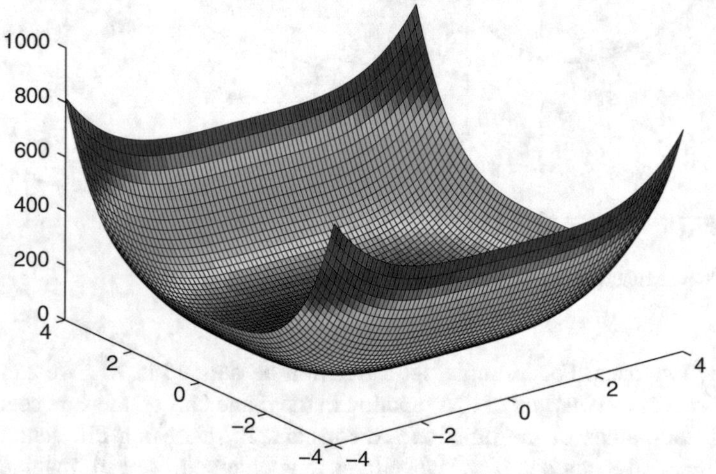

Fig. 8.2 Three dimensional surface of function $z = 3\,y^2 + x^4 + 2\,y^4$

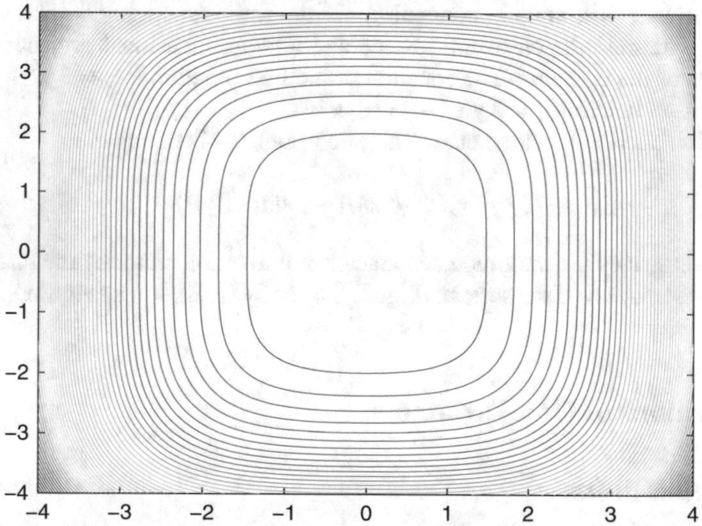

Fig. 8.3 Contours of function $z = 3\,y^2 + x^4 + 2\,y^4$

with $\zeta \in [x, \alpha]$, it can be shown that the necessary condition of minimum is that the point $(\alpha, f(\alpha))$ is such that

$$f'(\alpha) = 0 \tag{8.4.2}$$

This point is called a *stationary point*.

The condition of maximum is identical. Thus, in a general way, it is a necessary condition of existence of an optimum.

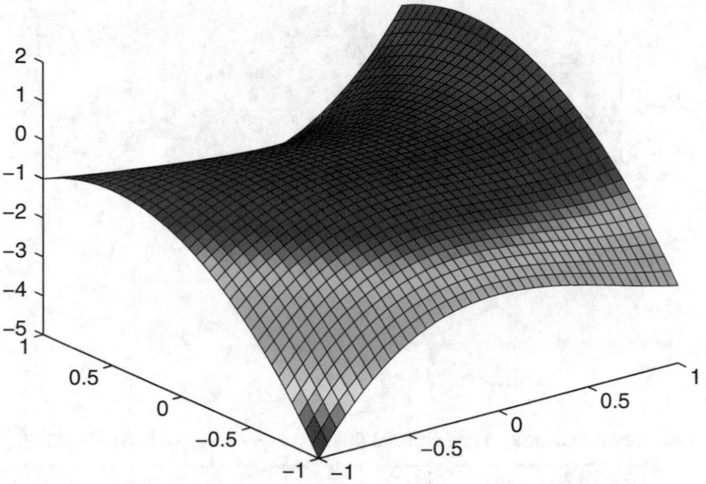

Fig. 8.4 Three dimensional surface of function $z = 2\,y\,x^2 - 2\,y^2 + x^3$

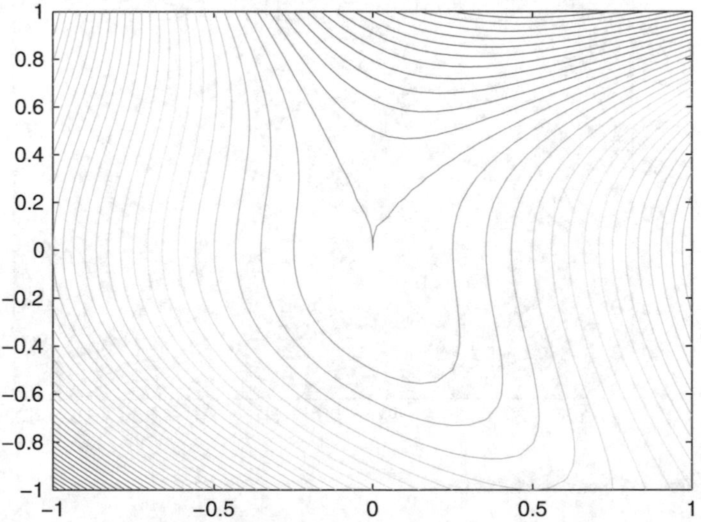

Fig. 8.5 Contours of function $z = 2\,y\,x^2 - 2\,y^2 + x^3$

This condition is not sufficient. Continuing on the example of minimum and considering the second degree Taylor polynomial, pose $h = x - \alpha$. Near α, we get

$$f(\alpha + h) - f(\alpha) = \frac{h^2}{2!} f''(\alpha) + 0(h^3) \tag{8.4.3}$$

The point $(\alpha, f(\alpha))$ is a minimum on the condition that

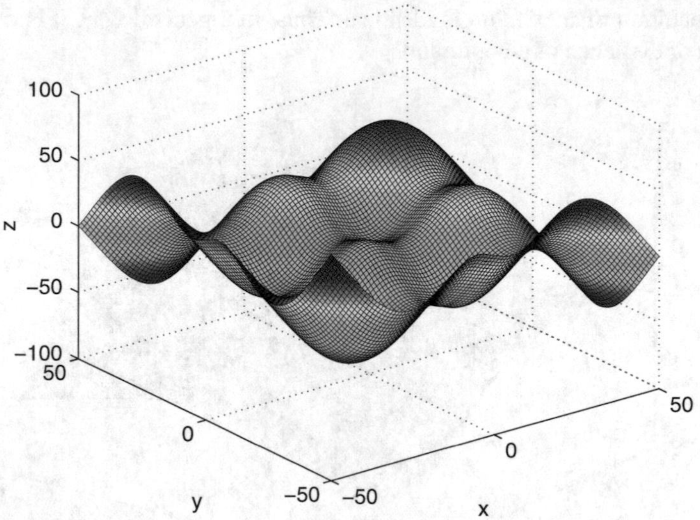

Fig. 8.6 Three dimensional surface of Schwefel function $z = -x\sin(\sqrt(|x|)) - y\sin(\sqrt(|y|))$

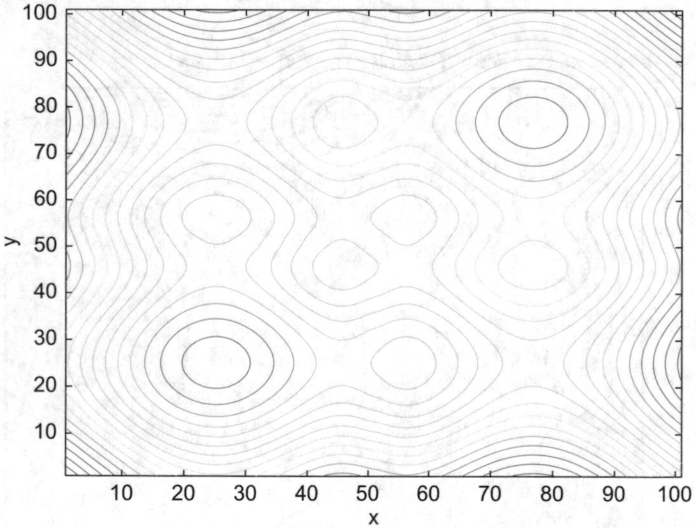

Fig. 8.7 Contours of Schwefel function

$$f''(\alpha) > 0 \tag{8.4.4}$$

Similarly, it will be a maximum if

$$f''(\alpha) < 0 \tag{8.4.5}$$

On the contrary, it will be an *inflexion point* or *saddle point* if

$$f''(\alpha) = 0 \quad \text{and} \quad f^{(3)}(\alpha) \neq 0 \tag{8.4.6}$$

In the particular case where the derivatives of order 3 or larger order would be zero, the reasoning could be pursued by considering the Taylor polynomial of appropriate degree.

As a general rule, we can show that

(a) If

$$f'(\alpha) = f''(\alpha) = \cdots = f^{(2k-1)}(\alpha) = 0 \quad \text{and} \quad f^{(2k)}(\alpha) \neq 0 \tag{8.4.7}$$

then the point $(\alpha, f(\alpha))$ is a minimum or a maximum depending on whether the derivative $f^{(2k)}(\alpha)$ is positive or negative.

(b) If

$$f'(\alpha) = f''(\alpha) = \cdots = f^{(2k)}(\alpha) = 0 \quad \text{and} \quad f^{(2k+1)}(\alpha) \neq 0 \tag{8.4.8}$$

then the point $(\alpha, f(\alpha))$ is a stationary point, but not an optimum.

8.4.2 Finite Interval

When the search is done in an unlimited interval $]-\infty, +\infty[$, the optimum, if it exists, of a continuous function with continuous derivatives is indeed a stationary point. If the interval is restricted, the optimum can be either a stationary point (Figure 8.8, left), or a limit point of the interval (optimum located at the constraints) (Figure 8.8, right).

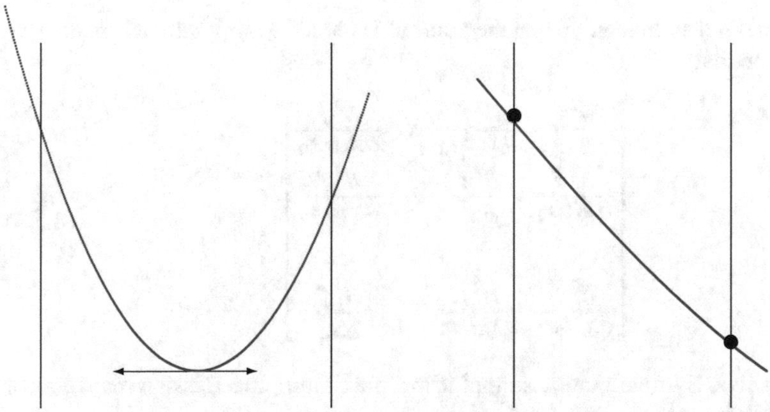

Fig. 8.8 Left: stationary point in the finite interval. Right: the optimum is a boundary point of the interval

8.4.3 Presence of Discontinuities

The previous reasoning is no more valid in the case of discontinuous functions or of functions having discontinuous derivatives. On the contrary, it is applicable in each of the subintervals where the function and its derivatives are continuous.

8.5 Functions of Several Variables

8.5.1 Infinite Interval

Consider the function $f(\boldsymbol{x})$ where $\boldsymbol{x}$ is a vector in $\mathbb{R}^n$, $\boldsymbol{x} = (x_1, x_2, \dots, x_n)$. The necessary condition of optimum is that the gradient of the function f is zero at a point $\boldsymbol{x}^*$

$$\text{grad } f = \begin{bmatrix} \frac{\partial f}{\partial x_1} \\ \dots \\ \frac{\partial f}{\partial x_n} \end{bmatrix} \tag{8.5.1}$$

hence

$$\frac{\partial f}{\partial x_1} = \dots = \frac{\partial f}{\partial x_n} = 0 \tag{8.5.2}$$

This condition is not sufficient. So that the point $\boldsymbol{x}^*$ is indeed a minimum; it is necessary that the Hessian matrix

$$\begin{bmatrix} \frac{\partial^2 f}{\partial x_1^2} & \frac{\partial^2 f}{\partial x_1 \partial x_2} & \cdots & \frac{\partial^2 f}{\partial x_1 \partial x_n} \\ \frac{\partial^2 f}{\partial x_2 \partial x_1} & \frac{\partial^2 f}{\partial x_2^2} & \cdots & \frac{\partial^2 f}{\partial x_2 \partial x_n} \\ & \vdots & & \\ \frac{\partial^2 f}{\partial x_n \partial x_1} & \frac{\partial^2 f}{\partial x_n \partial x_2} & \cdots & \frac{\partial^2 f}{\partial x_n^2} \end{bmatrix} \tag{8.5.3}$$

is definite positive. Symmetrically, so that it is a maximum, the Hessian matrix must be definite negative.

Conditions of positivity for a matrix:

A matrix $\boldsymbol{A}$ is definite positive if all its eigenvalues have their real part strictly positive.

8.5.2 Finite Interval

In the same manner as for a function of one variable, when we study the optimum of a function of several variables on a finite interval, the optimum is either inside the concerned interval and we are brought back to the case of an infinite interval, or at the limits of the domain.

8.5.3 *Presence of Discontinuities*

When discontinuities are present in the function or its derivatives, it is necessary to split the domain into subdomains where the function and its related derivatives will be continuous.

8.6 Function Subject to Equality Constraints

8.6.1 *Jacobi's Method*

Jacobi's method deals with the following problem:

$$\begin{aligned} &\min_{\boldsymbol{x}} f(\boldsymbol{x}), \quad x \in \mathbb{R}^n \\ &\text{subject to} \qquad g_j(\boldsymbol{x}) = 0, \quad j = 1, \ldots, m \end{aligned} \tag{8.6.1}$$

Supposing $n > m$, the state variables w_i, $i = 1, \ldots, m$, and the decision variables y_i, $i = 1, \ldots, n-m$. With respect to the new variables, the optimization problem becomes

$$\begin{aligned} &\sum_{i=1}^{m} \frac{\partial f(\boldsymbol{x})}{\partial w_i} dw_i + \sum_{i=1}^{n-m} \frac{\partial f(\boldsymbol{x})}{\partial y_i} dy_i = df(\boldsymbol{x}) \\ &\sum_{i=1}^{m} \frac{\partial g_j}{\partial w_i} dw_i + \sum_{i=1}^{n-m} \frac{\partial g_j}{\partial y_i} dy_i = 0, \quad j = 1, \ldots, m \end{aligned} \tag{8.6.2}$$

The Jacobian matrix $\boldsymbol{J}$ of the equality constraints is equal to

$$\boldsymbol{J} = \left[\frac{\partial g_j}{\partial w_k}\right], \quad j = 1, \ldots, m, \quad k = 1, \ldots, m \tag{8.6.3}$$

and the control matrix $\boldsymbol{C}$ equal to

$$\boldsymbol{C} = \left[\frac{\partial g_j}{\partial y_k}\right], \quad j = 1, \ldots, m, \quad k = 1, \ldots, n-m \tag{8.6.4}$$

The state and decision variables must be independent. $\boldsymbol{J}$ must be nonsingular. Equation (8.6.2) can be written under a condensed form

$$\begin{aligned} &\nabla_w f^T \, d\boldsymbol{w} + \nabla_y f^T \, d\boldsymbol{y} = df(\boldsymbol{w}, \boldsymbol{y}) \\ &\boldsymbol{J} \, d\boldsymbol{w} + \boldsymbol{C} \, d\boldsymbol{y} = 0 \end{aligned} \tag{8.6.5}$$

From Equation (8.6.5), it results

$$d\boldsymbol{w} = -\boldsymbol{J}^{-1}\,\boldsymbol{C}\,d\boldsymbol{y} \tag{8.6.6}$$

Using that result in Equation (8.6.5), we deduce

$$df(\boldsymbol{w},\boldsymbol{y}) = (\nabla_y f^T - \nabla_w f^T\,\boldsymbol{J}^{-1}\,\boldsymbol{C})\,d\boldsymbol{y} \tag{8.6.7}$$

Then, the constrained gradient of f with respect to $\boldsymbol{y}$ is formed

$$\nabla_y^c f = \frac{\partial^c f(\boldsymbol{w},\boldsymbol{y})}{\partial^c \boldsymbol{y}} = \nabla_y f^T - \nabla_w f^T\,\boldsymbol{J}^{-1}\,\boldsymbol{C} \tag{8.6.8}$$

It can be shown that if $\boldsymbol{x}^*$ is an extremum, then

$$\nabla_y^c f(\boldsymbol{x}^*) = 0 \tag{8.6.9}$$

In this manner, all the stationary points can be identified. Then, it remains to study the second variations to find the global extremum, which can be done by only considering the matrix of the constrained second derivatives, that is

$$\frac{\partial^{c2} f(\boldsymbol{w},\boldsymbol{y})}{\partial^c \boldsymbol{y}^2} \tag{8.6.10}$$

which must be definite positive so that it is indeed a minimum.

To study the second variations, we can consider that the constrained gradient is a row vector function of the state and decision variables as

$$\nabla_y^c f = \boldsymbol{h}^T(\boldsymbol{w},\boldsymbol{y}) \tag{8.6.11}$$

where $\boldsymbol{h}$ is a column vector. The matrix of the second constrained derivatives is

$$\frac{\partial^{c2} f(\boldsymbol{w},\boldsymbol{y})}{\partial^c \boldsymbol{y}^2} = \frac{\partial \boldsymbol{h}}{\partial \boldsymbol{y}} + \frac{\partial \boldsymbol{h}}{\partial \boldsymbol{w}}\frac{\partial \boldsymbol{w}}{\partial \boldsymbol{y}} = \frac{\partial \boldsymbol{h}}{\partial \boldsymbol{y}} - \frac{\partial \boldsymbol{h}}{\partial \boldsymbol{w}}\boldsymbol{J}^{-1}\,\boldsymbol{C} \tag{8.6.12}$$

This method is much less used than the method of Lagrange multipliers which follows in Section 8.6.2.

Example 8.1 :
Jacobi's method

Consider the following problem to be solved by Jacobi's method:

$$\max_{\boldsymbol{x}} f(\boldsymbol{x}) = -2\,x_1^2 - x_2^2 - 3\,x_3^2 \tag{8.6.13}$$

subject to two equality constraints

$$\begin{aligned} g_1(\boldsymbol{x}) &= x_1 + 2\,x_2 + x_3 - 1 = 0 \\ g_2(\boldsymbol{x}) &= 4\,x_1 + 3\,x_2 + 2\,x_3 - 2 = 0 \end{aligned} \tag{8.6.14}$$

We deduce $n = 3$ and $m = 2$. Several choices of variables are possible. Let us take $\boldsymbol{w} = \{x_1, x_2\}$ and $\boldsymbol{y} = \{x_3\}$. The Jacobian and control matrices are

$$\boldsymbol{J} = \begin{bmatrix} 1 & 2 \\ 4 & 3 \end{bmatrix}, \quad \boldsymbol{C} = \begin{bmatrix} 1 \\ 2 \end{bmatrix} \tag{8.6.15}$$

On another side

$$\nabla_y f = -6x_3 = -6y_1 , \quad \text{and} \quad \nabla_w f = \begin{bmatrix} -4x_1 \\ -2x_2 \end{bmatrix} = \begin{bmatrix} -4w_1 \\ -2w_2 \end{bmatrix} \tag{8.6.16}$$

The constrained gradient is equal to

$$\nabla_y^c f = \nabla_y f^T - \nabla_w f^T \boldsymbol{J}^{-1} \boldsymbol{C} = -6y_1 + 0.8w_1 + 0.8w_2 \tag{8.6.17}$$

Eq. $\nabla_y^c f = 0$ at the optimum, joined to the two equations of constraints $g_1 = 0$ and $g_2 = 0$, yields the following linear system:

$$\begin{aligned} &-6y_1 + 0.8w_1 + 0.8w_2 = 0 \\ &w_1 + 2w_2 + y_1 - 1 = 0 \\ &4w_1 + 3w_2 + 2y_1 - 2 = 0 \end{aligned} \tag{8.6.18}$$

from which we get $w_1 = 0.185$, $w_2 = 0.3704$, $y_1 = 0.074$, and x_i^*.

Now, let us study the second variations by using Equation (8.6.12) and setting $\boldsymbol{h} = \nabla_y^c f$

$$\frac{\partial^{c2} f(\boldsymbol{w}, \boldsymbol{y})}{\partial^c \boldsymbol{y}^2} = -6 + \begin{bmatrix} 0.8 & 0 \\ 0 & 0.8 \end{bmatrix} \left[- \begin{bmatrix} 0.2 \\ 0.4 \end{bmatrix} \right] = -6 - 0.16 - 0.32 < 0 \tag{8.6.19}$$

The point of coordinates $x^* = 0.185\ ,\ 0.3704\ ,\ 0.074$ is thus the sought maximum.

8.6.2 *Lagrange Multipliers*

A common case of optimization in physical and economical problems is the search of the minimum of a function f of $\mathbb{R}^n$

$$\text{minimum of } f(\boldsymbol{x}), \quad \boldsymbol{x} \in \mathbb{R}^n \tag{8.6.20}$$

subject to m equality constraints

$$g_i(\boldsymbol{x}) = 0, \quad i = 1, \ldots, m \tag{8.6.21}$$

Remark: The problem of maximum search of f is equivalent to the problem of minimum search of $-f$

$$\text{minimum of } f(\boldsymbol{x}) \Longleftrightarrow \text{maximum of } - f(\boldsymbol{x}) \tag{8.6.22}$$

An elegant method of solving this type of problem consists in using the *Lagrange multipliers* (also noted Euler–Lagrange). The function called *Lagrangian* is formed

$$\mathcal{L}(\boldsymbol{x}, \boldsymbol{\lambda}) = f(\boldsymbol{x}) + \sum_{i=1}^{m} \lambda_i \, g_i(\boldsymbol{x}) \tag{8.6.23}$$

for which we search the stationary points in the considered domain. The unknowns are, in addition to the vector $\boldsymbol{x}$ of $\mathbb{R}^n$, the Lagrange multipliers λ_i, $i = 1, \ldots, m$, that must be determined.

The gradient of $\mathcal{L}$ must be zero so that it comes

$$\begin{cases} \dfrac{\partial \mathcal{L}}{\partial x_j} = 0\,, & j = 1, \ldots, n \\ \dfrac{\partial \mathcal{L}}{\partial \lambda_i} = 0\,, & i = 1, \ldots, m \end{cases} \tag{8.6.24}$$

We observe that each of the conditions

$$\frac{\partial \mathcal{L}}{\partial \lambda_i} = 0\,, \quad i = 1, \ldots, m \tag{8.6.25}$$

indeed corresponds to impose the nullity of the corresponding constraint

$$g_i(\boldsymbol{x}^*) = 0 \tag{8.6.26}$$

at the optimum $\boldsymbol{x}^*$.

Example 8.2 :
Lagrange multipliers

Find the maximum of the surface area of a rectangle while imposing that the perimeter is constant.

Let x_1 be the width and x_2 the length. We impose $P = 2\,x_1 + 2\,x_2$, that we write under the form $2\,x_1 + 2\,x_2 - P = 0$. The surface area is equal to $S = x_1\,x_2$. Thus, we search the maximum of S, P being constant.

We form the Lagrangian function

$$\mathcal{L} = x_1\,x_2 + \lambda_1\,(2\,x_1 + 2\,x_2 - P) \tag{8.6.27}$$

The gradient is zero at the optimum, thus

$$\begin{cases} \dfrac{\partial \mathcal{L}}{\partial x_1} = x_2 + 2\,\lambda_1 = 0 \\ \dfrac{\partial \mathcal{L}}{\partial x_2} = x_1 + 2\,\lambda_1 = 0 \\ \dfrac{\partial \mathcal{L}}{\partial \lambda_1} = 2\,x_1 + 2\,x_2 - P = 0 \end{cases} \tag{8.6.28}$$

From the two first equalities, we draw

$$\lambda_1 = -x_2/2 = -x_1/2 \tag{8.6.29}$$

which imposes at the optimum

$$x_1^* = x_2^* \tag{8.6.30}$$

The rectangle of optimal surface area is in fact the square.

From the third equality, we deduce

$$-\,8\,\lambda_1 - P = 0 \quad \text{that is} \quad \lambda_1 = -P/8 \tag{8.6.31}$$

hence the values of x_1^* and x_2^*

$$x_1^* = x_2^* = P/4 \tag{8.6.32}$$

It remains to verify that it is a maximum. Consider the function f and calculate an expansion of f near the optimum. We have the general second degree Taylor polynomial for a function of two variables at the optimum

$$
\begin{aligned}
f(x_1, x_2) = f(x_1^*, x_2^*) + \Sigma_i \left(\frac{\partial f}{\partial x_i}\right)_{\boldsymbol{x}_*} (x_i - x_i^*) + \\
\frac{1}{2!} \Sigma_{i,j} \left(\frac{\partial^2 f}{\partial x_i\, x_j}\right)_{\boldsymbol{x}_*} (x_i - x_i^*)(x_j - x_j^*) + 0(\boldsymbol{x} - \boldsymbol{x}_*)^3
\end{aligned}
\tag{8.6.33}
$$

The partial derivatives of order 1 and of order 2 at the optimum are equal to

$$
\left(\frac{\partial f}{\partial x_1}\right)_{\boldsymbol{x}^*} = x_2^* = \frac{P}{4}, \qquad \left(\frac{\partial f}{\partial x_2}\right)_{\boldsymbol{x}^*} = x_1^* = \frac{P}{4} \tag{8.6.34}
$$

$$
\left(\frac{\partial^2 f}{\partial x_1^2}\right)_{\boldsymbol{x}^*} = 0, \qquad \left(\frac{\partial^2 f}{\partial x_2^2}\right)_{\boldsymbol{x}^*} = 0, \qquad \left(\frac{\partial^2 f}{\partial x_1 \partial x_2}\right)_{\boldsymbol{x}^*} = 1 \tag{8.6.35}
$$

Near the optimum, the function is thus equal to

$$
f(x_1, x_2) = P^2/16 + P/4\,(x_1 - P/4) + P/4\,(x_2 - P/4) + (x_1 - P/4)(x_2 - P/4) + 0(\boldsymbol{x} - \boldsymbol{x}^*)^3 \tag{8.6.36}
$$

The variation of f is equal to

$$
\begin{aligned}
df &= f(x_1, x_2) - f^*(x_1, x_2) \\
&= P/4\,dx_1 + P/4\,dx_2 + dx_1 dx_2 + 0(d\boldsymbol{x})^3
\end{aligned}
\tag{8.6.37}
$$

We use the constraint that relates dx_1 and dx_2 near the optimum, that is

$$
2x_1 + 2x_2 - P = 0 \Longrightarrow 2\,dx_1 + 2\,dx_2 = 0 \tag{8.6.38}
$$

hence

$$
df(x_1, x_2) = -dx_1^2 + 0(d\boldsymbol{x})^3 \quad \leq 0 \tag{8.6.39}
$$

The optimum indeed corresponds to a maximum.

It must be noticed that this variational study is possible because of the low dimension of the present problem but it can be hardly generalized and thus more general techniques will be recommended.

8.6.3 Signification of Lagrange Multipliers

According to the definition (8.6.23) of the Lagrangian, it is clear that the Lagrange multipliers represent the derivative of the Lagrangian at the optimum with respect to the considered constraint

$$
\lambda_i = \frac{\partial \mathcal{L}(\boldsymbol{x}^*)}{\partial g_i} \tag{8.6.40}
$$

The multipliers are also called *sensitivity* coefficients or *influence* functions as they indicate the rate at which the value of the studied function varies near the optimum when the corresponding constraint is perturbed. This remark can be very useful in some problems, for example in economy.

8.6.4 Conditions of Minimum

The Lagrangian being equal to

$$\mathcal{L} = f(\boldsymbol{x}) + \sum_{i=1}^{m} \lambda_i \, g_i(\boldsymbol{x}) \tag{8.6.41}$$

the stationary points are first determined by the equations

$$\begin{cases} \dfrac{\partial \mathcal{L}}{\partial x_j} = 0\,, & j = 1, \ldots, n \\ \dfrac{\partial \mathcal{L}}{\partial \lambda_i} = 0\,, & i = 1, \ldots, m \end{cases} \tag{8.6.42}$$

8.6.5 Conditions of Minimum by the Projected Gradient in the Case of Equality Constraints

Gill et al. (1981) described the method of the *projected gradient* (*reduced gradient* according to Fletcher 1991) first in the case of linear equality constraints, then nonlinear. It will be presented here directly in the nonlinear case. The idea of the projected gradient comes from the method of gradient projection (Rosen 1960, 1961). The projected gradient uses the exact satisfaction of a subset of constraints while decreasing the value of the function to be minimized to reduce the dimension of the optimization problem. Remaining exactly on the constraints constitutes a difficult problem when the constraints are strongly nonlinear, which requires a form of correction with respect to the linear case. The method of generalized reduced gradient (Abadie and Carpentier 1969) is an efficient form of reduced gradient consisting in exactly satisfying at each iteration a set of constraints that are supposed to be active for the optimal solution. Different algorithms exist for the generalized reduced gradient (Sargent 1974).

The optimum of a function f of $\mathbb{R}^n$ subject to equality constraints $g_i = 0$, $i = 1, \ldots, m$ is searched. In our case, only the method of the projected gradient is presented.

To determine the nature of the optimum, we do a series expansion near any stationary point as

$$df = f - f^* = \sum_i \left(\frac{\partial f}{\partial x_i}\right)^* dx_i + \frac{1}{2} \sum_{ij} \left(\frac{\partial^2 f}{\partial x_i \partial x_j}\right)^* dx_i \, dx_j \tag{8.6.43}$$

From the expression of the Lagrangian, we deduce

$$d\mathcal{L} = df(\boldsymbol{x}) + \sum_{i=1}^{m} \lambda_i \, dg_i(\boldsymbol{x}) \tag{8.6.44}$$

However, when the displacement is near a stationary point (Figure 8.9), we remain on the constraints, as they are equalities. Consider a given constraint g.

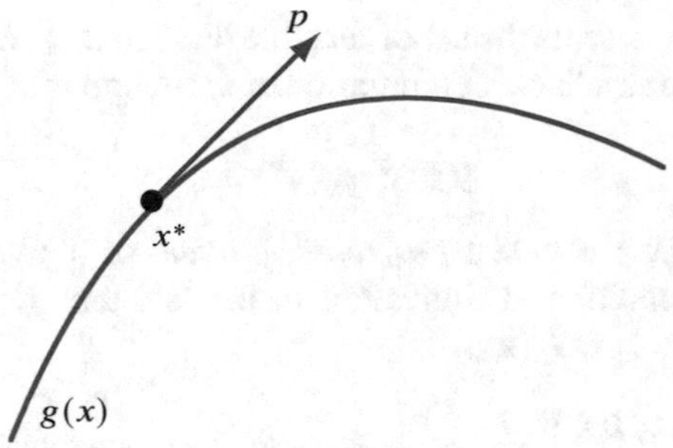

Fig. 8.9 Authorized displacement along a constraint from a stationary point

A displacement from the given stationary point along that constraint takes place on an arc and it can be represented by means of only one parameter θ that can be assumed to be zero at the stationary point $\boldsymbol{x}^*$. Let $\alpha(\theta)$ be the length of the displacement on this arc. Consider a vector p tangent to the arc and do a second degree Taylor polynomial near the stationary point $\boldsymbol{x}^*$ in the direction of $\boldsymbol{p}$

$$g(\boldsymbol{x}^* + \epsilon \boldsymbol{p}) = g(\boldsymbol{x}^*) + \epsilon \left(g_x^T\right)_{\boldsymbol{x}^*} \boldsymbol{p} + \frac{1}{2}\epsilon \boldsymbol{p}^T (g_{xx})_{\boldsymbol{x}^*} \epsilon \boldsymbol{p} + \dots \tag{8.6.45}$$

where g_x is the gradient of g and g_{xx} the Hessian matrix of g. The function $g(\alpha(\theta))$ representing the constraint remains zero at any point of the arc, hence

$$\left(\frac{d}{d\theta} g(\alpha(\theta))\right)_{\theta=0} = g_x(\alpha(0))^T \boldsymbol{p} = g_x(\boldsymbol{x}^*)^T \boldsymbol{p} = 0 \tag{8.6.46}$$

As this equality is valid whatever the considered constraint g_i, along a possible arc for all the equality constraints, we deduce

$$G_x(\boldsymbol{x}^*)\boldsymbol{p} = 0 \tag{8.6.47}$$

where G_x is the Jacobian matrix of the constraints, thus each row is equal to $g_{x,i}(\boldsymbol{x})^T$. The matrix G_x must be of full rank with respect to the rows, i.e. the gradients of the constraints must be linearly independent at $\boldsymbol{x}^*$.

On another side, the function f must be stationary at $\boldsymbol{x}^*$ along any possible arc (with respect to θ), that is

$$\left(\frac{d}{d\theta} f(\alpha(\theta))\right)_{\theta=0} = 0 \tag{8.6.48}$$

The condition results

$$f_x(\boldsymbol{x}^*)^T \boldsymbol{p} = 0 \tag{8.6.49}$$

where the vectors $\boldsymbol{p}$ verify Equation (8.6.47). Let us note $Z(\boldsymbol{x}^*)$ be a matrix whose columns form a basis for the set of vectors orthogonal to the rows of $G_x(\boldsymbol{x}^*)$ (the columns of $Z(\boldsymbol{x}^*)$ form a basis for the kernel of $G_x(\boldsymbol{x}^*)$), hence

$$G_x(\boldsymbol{x}^*)Z(\boldsymbol{x}^*) = 0 \tag{8.6.50}$$

Thus, $Z(\boldsymbol{x}^*)$ indeed represents the set of feasible directions $\boldsymbol{p}$. Any vector $\boldsymbol{p}$ verifying (8.6.47) can be written as a linear combination of the columns of $Z(\boldsymbol{x}^*)$. Then, we must have

$$Z(\boldsymbol{x}^*)^T f_x(\boldsymbol{x}^*) = 0 \tag{8.6.51}$$

the vector $Z(\boldsymbol{x}^*)^T f_x(\boldsymbol{x}^*)$ is called *projected gradient* of f at $\boldsymbol{x}^*$ (Gill et al. 1981). As previously, this condition is equivalent to the fact that $f_x(\boldsymbol{x}^*)$ must be a linear combination of the rows of $G_x(\boldsymbol{x}^*)$

$$\left(\frac{\partial \mathcal{L}}{\partial \boldsymbol{x}}\right)_{\boldsymbol{x}^*} = 0 \Rightarrow f_x(\boldsymbol{x}^*) = -G_x(\boldsymbol{x}^*)^T \boldsymbol{\lambda}^* \tag{8.6.52}$$

for the vector λ^* of the Lagrange multipliers of dimension m.

The condition of minimum is that the curvature of f must be nonnegative at $\boldsymbol{x}^*$ along any possible arc

$$\left(\frac{d^2}{d\theta^2} f(\alpha(\theta))\right)_{\theta=0} \geq 0 \tag{8.6.53}$$

It results

$$\boldsymbol{p}^T \left[f_{xx}(\boldsymbol{x}^*) + \sum_{i=1}^{m} \lambda_i^* g_{i,xx}(\boldsymbol{x}^*) \right] \boldsymbol{p} \geq 0 \tag{8.6.54}$$

which thus makes use of the Hessian matrix of the Lagrangian. This condition must be fulfilled for any p satisfying Equation (8.6.47) $G_x(\boldsymbol{x}^*)p = 0$. The condition (8.6.54) is equivalent to the condition that the matrix called *projected Hessian* (Gill et al. 1981) (*reduced Hessian* according to Fletcher 1991) of the Lagrangian

$$Z(\boldsymbol{x}^*)^T \left[f_{xx}(\boldsymbol{x}^*) + \sum_{i=1}^{m} \lambda_i^* g_{i,xx}(\boldsymbol{x}^*) \right] Z(\boldsymbol{x}^*) \tag{8.6.55}$$

is semi-definite positive.

Summary of problem solving:

$$\min f(\boldsymbol{x}), \quad \boldsymbol{x} \in \mathbb{R}^n \tag{8.6.56}$$

subject to m equality constraints

$$g_i(\boldsymbol{x}) = 0, \quad i = 1, \dots, m \tag{8.6.57}$$

The necessary conditions so that $\boldsymbol{x}^*$ minimizes f subject to equality constraints are successively:

(a) The nullity of the constraints (condition of realizability)

$$\mathcal{L}_\lambda = 0 \quad \Leftrightarrow \quad g_i(\boldsymbol{x}^*) = 0 \tag{8.6.58}$$

(b) The determination of the possible optima (condition of stationarity)

$$\mathcal{L}_x = 0 \quad \Leftrightarrow \quad f_x(\boldsymbol{x}^*) = -G_x(\boldsymbol{x}^*)^T \boldsymbol{\lambda}^* \tag{8.6.59}$$

Note that the union of both equalities (8.6.58) and (8.6.59) is equivalent to write that the gradient of $\mathcal{L}$ with respect to $\boldsymbol{\lambda}$ and $\boldsymbol{x}$ respectively is zero, hence Equations (8.6.42).

(c) The determination of the directions $\boldsymbol{p}$ along the constraints

$$G_x(\boldsymbol{x}^*)\, \boldsymbol{p} = 0 \tag{8.6.60}$$

hence $Z(\boldsymbol{x}^*)$.

(d) The condition about the nature of the optimum, so that the optimum is a minimum, it is necessary that the projected Hessian matrix

$$Z(\boldsymbol{x}^*)^T \left[f_{xx}(\boldsymbol{x}^*) + \sum_{i=1}^{m} \lambda_i^* g_{i,xx}(\boldsymbol{x}^*) \right] Z(\boldsymbol{x}^*) \tag{8.6.61}$$

is semi-definite positive. On another side, it is sufficient that this matrix is definite positive so that it is a minimum. The other conditions (a), (b), and (c) are necessary and sufficient.

These conditions are sufficient conditions so that $\boldsymbol{x}^*$ is a *local minimum* . Note that if several local minima are found, it suffices to compare the value of the function at these different points to find the *global minimum.*

It is possible to present the problem under a simplified form. Let us do a series expansion of f near the optimum

$$\begin{aligned} f(\boldsymbol{x}^* + \delta) &= \mathcal{L}(\boldsymbol{x}^* + \delta, \lambda^*) \\ &= \mathcal{L}(\boldsymbol{x}^*, \lambda^*) + \delta^T\, \mathcal{L}_x(\boldsymbol{x}^*, \lambda^*) + \frac{1}{2}\, \delta^T\, \mathcal{L}_{xx}(\boldsymbol{x}^*, \lambda^*)\, \delta + 0(\delta^T \delta) \\ &= f(\boldsymbol{x}^*) + \frac{1}{2}\, \delta^T\, \mathcal{L}_{xx}(\boldsymbol{x}^*, \lambda^*)\, \delta + 0(\delta^T \delta) \end{aligned} \tag{8.6.62}$$

To obtain this equality, we successively used the fact that the constraints are equality constraints, then that the gradient of the Lagrangian is zero at the optimum.

Like in the previous demonstration, we know that the possible directions p must satisfy

$$G_x^* \, p = 0 \Longleftrightarrow A^{*T}\, p = 0 \tag{8.6.63}$$

where A is the transpose matrix of the Jacobian matrix of the gradient, equal to

$$A = \begin{bmatrix} g_{1,1} & \cdots & g_{m,1} \\ \vdots & & \vdots \\ g_{1,n} & \cdots & g_{m,n} \end{bmatrix} \tag{8.6.64}$$

where $g_{i,j}$ is the jth component of the gradient $g_{x,i}$ of the constraint g_i (the number of constraints equality is m).

The second order condition of local minimum is then

$$p^T \mathcal{L}_{xx}(\boldsymbol{x}^*, \lambda^*) p \geq 0 \tag{8.6.65}$$

and for a strict local minimum

$$p^T \mathcal{L}_{xx}(\boldsymbol{x}^*, \lambda^*) p > 0 \tag{8.6.66}$$

It is then possible to verify the above condition, or it is equivalent to verify that the roots w of the equation

$$\det \begin{bmatrix} w\boldsymbol{I} - \mathcal{L}_{xx} & G_x^T \\ G_x & O \end{bmatrix} \Longleftrightarrow \det \begin{bmatrix} \mathcal{L}_{xx} - w\boldsymbol{I} & -G_x^T \\ -G_x & O \end{bmatrix} = 0 \tag{8.6.67}$$

are positive or zero. Note that the calculation of a determinant or of eigenvalues is not always easy, whereas the calculation according to equation (8.6.65) is a simple matrix calculation.

Example 8.3 :

Optimization with equality constraints and projected gradient

Application to Example 8.2 of search of the maximum of the surface area of a rectangle by imposing that the perimeter is constant.

The problem is transformed into the search of the minimum of the opposite of the surface area

$$f(\boldsymbol{x}) = -x_1\, x_2\,, \quad g_1(\boldsymbol{x}) = 2x_1 + 2x_2 - P = 0 \tag{8.6.68}$$

- The condition (8.6.58) induces

$$2x_1^* + 2x_2^* - P = 0 \tag{8.6.69}$$

- Apply the condition (8.6.59), $f_x(\boldsymbol{x}^*) = -G_x(\boldsymbol{x}^*)^T \lambda^*$. We calculate

$$f_x(\boldsymbol{x}^*) = \begin{bmatrix} -x_2 \\ -x_1 \end{bmatrix}, \quad G_x(\boldsymbol{x}) = \begin{bmatrix} \frac{\partial g_1}{\partial x_1} & \frac{\partial g_1}{\partial x_2} \end{bmatrix} = \begin{bmatrix} 2 & 2 \end{bmatrix} \tag{8.6.70}$$

hence

$$\begin{bmatrix} -x_2 \\ -x_1 \end{bmatrix} = - \begin{bmatrix} 2 \\ 2 \end{bmatrix} \lambda \Longrightarrow x_1^* = x_2^* \tag{8.6.71}$$

This equation coupled to Equation (8.6.69) gives

$$x_1^* = x_2^* = \frac{P}{4}\,, \quad \lambda^* = \frac{P}{8} \tag{8.6.72}$$

- The condition (8.6.60)

$$G_x(\boldsymbol{x}^*)\boldsymbol{p} = 0 \tag{8.6.73}$$

gives

$$\begin{bmatrix} 2 & 2 \end{bmatrix} \begin{bmatrix} p_1 \\ p_2 \end{bmatrix} = 0 \Longrightarrow p_1 = -p_2 \tag{8.6.74}$$

hence the matrix $Z(\boldsymbol{x}^*)$ representing the set of feasible directions $\boldsymbol{p}$

$$Z(\boldsymbol{x}^*) = \begin{bmatrix} 1 \\ -1 \end{bmatrix} p_1 \tag{8.6.75}$$

- The condition (8.6.61) gives

$$p_1 \begin{bmatrix} 1 & -1 \end{bmatrix} \left(\begin{bmatrix} 0 & -1 \\ -1 & 0 \end{bmatrix} + \lambda \begin{bmatrix} 0 & 0 \\ 0 & 0 \end{bmatrix} \right) \begin{bmatrix} 1 \\ -1 \end{bmatrix} p_1 = 2p_1^2 \geq 0 \tag{8.6.76}$$

The function f reaches a minimum, thus it corresponds to a maximum for the surface area of the rectangle subject to the constraints.

8.7 Function Subject to Inequality Constraints

We search the optimum of a function f of $\mathbb{R}^n$ subject to inequality constraints $h_i \leq 0,\ i = 1, \ldots, p$. Note that the inequality constraints are expressed as "lower than or equal to." If constraints of type "larger than or equal to" are present, it is sufficient to express them by their opposite to be brought back to the studied case.

8.7.1 Use of Slack Variables

The main idea consists in introducing positive *slack variables* $E(x_i),\ i = n+1, \ldots, n+p$, so that the inequality constraints $h_i \leq 0,\ i = 1, \ldots, p$ are transformed into equality constraints $g_i = 0,\ i = 1, \ldots, p$.

Thus, the problem with inequality constraints has been replaced by an augmented problem with equality constraints in which there are $n+p$ unknowns x_i and p Lagrange multipliers λ_i.

Several choices of slack variables are possible, for example the positive variables $x_i,\ i = n+1, \ldots, n+p$ such that $g_i = h_i + x_{n+i} = 0,\ i = 1, \ldots, p$

Example 8.4 :
Optimization with inequality constraints

Search the minimum of $f = 4\,x_1^2 + 5\,x_2^2$ subject to $x_1 \leq 1$.

The inequality constraint is written as

$$h = x_1 - 1 \leq 0 \tag{8.7.1}$$

which is replaced by an equality constraint

$$g = x_1 - 1 + x_3 = 0 \tag{8.7.2}$$

with the condition $x_3 \geq 0$. The Lagrangian function is thus equal to

$$\mathcal{L} = f + \lambda\, g = 4\,x_1^2 + 5\,x_2^2 + \lambda\,(x_1 - 1 + x_3) \tag{8.7.3}$$

At the optimum, the gradient of $\mathcal{L}$ is zero or the total derivative of $\mathcal{L}$ is zero. x_1, x_2, and x_3 are considered as independent variables. Thus, we get

$$8x_1\, dx_1 + 10x_2\, dx_2 + \lambda(dx_1 + dx_3) + (x_1 - 1 + x_3)\, d\lambda = 0 \tag{8.7.4}$$

that is

$$\begin{cases} 8x_1 + \lambda = 0 \\ 10x_2 = 0 \\ \lambda = 0 \\ x_1 - 1 + x_3 = 0 \end{cases} \tag{8.7.5}$$

which gives $\lambda = 0,\ x_1 = 0,\ x_2 = 0$, and $x_3 = 1$. The condition of positivity of x_3 is verified. The solution is inner to the domain.

The choice of the slack variable x_{n+i} with the condition $x_{n+i} \geq 0$ can be usefully replaced by the following slack variable x_{n+i}^2 which avoids to verify the previous condition.

Example 8.5 :
Optimization with inequality constraints

Consider again Example 8.4,
Search the minimum of $f = 4\,x_1^2 + 5\,x_2^2$ subject to $x_1 \leq 1$.

The constraint is replaced by an equality constraint

$$g = x_1 - 1 + x_3^2 = 0 \tag{8.7.6}$$

The Lagrangian function is thus equal to

$$\mathcal{L} = f + \lambda\, g = 4\,x_1^2 + 5\,x_2^2 + \lambda\,(x_1 - 1 + x_3^2) \tag{8.7.7}$$

The condition of gradient equal to zero at the optimum gives

$$\begin{cases} 8x_1 + \lambda = 0 \\ 10x_2 = 0 \\ \lambda\, 2\, x_3 = 0 \\ x_1 - 1 + x_3^2 = 0 \end{cases} \tag{8.7.8}$$

which gives either $\lambda = 0$, or $x_3 = 0$.
If $x_3 = 0$, we get an optimum at the limit of the domain $x_1 = 1,\ x_2 = 0$ with $\lambda = -8$. In this case, the function $f = 4$.
If $\lambda = 0$, we get $x_1 = 0,\ x_2 = 0$, and $x_3 = \pm 1$. The solution is inside the domain. In this case, the function f is equal to $f = 0$.
Conclusion: The found minimum is thus $f = 0$ at point (0,0).

8.7.2 Karush–Kuhn–Tucker (KKT) Parameters

Karush–Kuhn–Tucker conditions (also known as Kuhn–Tucker conditions) are a generalization of Lagrange multipliers in the case of inequality constraints.
Remark (Polak 1997): Kuhn-Tucker parameters are frequently called Karush–Kuhn–Tucker parameters as (Karush 1939) had introduced them but without any other publication as his Master thesis. Later, Kuhn and Tucker (1951) discovered them again and published them. Karush–Kuhn–Tucker conditions are related to John John (1948) or Fritz John conditions.

In the same manner as the previously studied example, the optimum of the function f of $\mathbb{R}^n$ subject to p inequality constraints $h_i \leq 0$ can be treated in a general manner by choosing the slack variables so that we are brought back to a problem with p equality constraints of the form

$$g_i = h_i + x_{n+i}^2 \,, \; i = 1, \ldots, p \tag{8.7.9}$$

The Lagrangian is then equal to

$$\mathcal{L} = f + \sum_i \lambda_i \,(h_i + x_{n+i}^2) \,, \quad i = 1, \ldots, p \tag{8.7.10}$$

and the stationary points satisfy the conditions

$$\begin{cases} \dfrac{\partial \mathcal{L}}{\partial x_j} = 0 \,, & j = 1, \ldots, n+p \\ \dfrac{\partial \mathcal{L}}{\partial \lambda_i} = 0 \,, & i = 1, \ldots, p \end{cases} \tag{8.7.11}$$

equivalent to the following system of equations:

$$\begin{cases} \dfrac{\partial f}{\partial x_j} + \sum_i \lambda_i \dfrac{\partial h_i}{\partial x_j} = 0 \,, & j = 1, \ldots, n \\ 2\,\lambda_j \, x_{n+j} = 0 \,, & j = 1, \ldots, p \\ h_i + x_{n+i}^2 = 0 \,, & i = 1, \ldots, p \end{cases} \tag{8.7.12}$$

Among the $n+2p$ equations related to the condition of zero gradient at the optimum, we find p equations of the form

$$x_{n+j} \lambda_j = 0 \,, \quad j = 1, \ldots, p \tag{8.7.13}$$

- If the Lagrange multiplier is zero, $\lambda_j = 0$, hence in general $x_{n+j} \neq 0 \Longrightarrow h_j \neq 0$; in general, the stationary point is not at the boundary.
- If the Lagrange multiplier λ_j is different from zero, the stationary point is at the corresponding boundary corresponding to the constraint $h_j = 0$.

A typical situation is illustrated in Figure 8.10.

Consider the case where only one inequality constraint h_i exists. Note the inequality constraint h and the associated equality constraint g, the Lagrange parameter λ. Two cases are possible:

- $\lambda > 0$: At the boundary, the corresponding constraint h is zero

$$h(x^*) = 0 \Longrightarrow g(x^*) = 0 \tag{8.7.14}$$

and the only displacements near the optimum are such that

$$dh \leq 0 \Longrightarrow dg \leq 0 \tag{8.7.15}$$

because of the sign of the inequality $h \leq 0$. The total derivative dg is equal to

$$dg = \sum_{i=1}^{n} \frac{\partial g}{\partial x_i} dx_i \tag{8.7.16}$$

From the form of the Lagrangian function $\mathcal{L} = f + \lambda g$ such that at the optimum

$$\frac{\partial \mathcal{L}}{\partial x_i} = 0 \tag{8.7.17}$$

we deduce

$$\frac{\partial f}{\partial x_i} = -\lambda \frac{\partial g}{\partial x_i} \quad \forall\, i \quad \Longrightarrow \quad df = -\lambda\, dg \tag{8.7.18}$$

As, on one side, the Lagrange multiplier is positive and, on another side, we have the inequality $dg \leq 0$, it results $df \geq 0$, the function f can only increase. Thus, the optimum located at the boundary cannot be a maximum.

- $\lambda < 0$: This is the opposite case. The optimum is located at the boundary and cannot be a minimum.

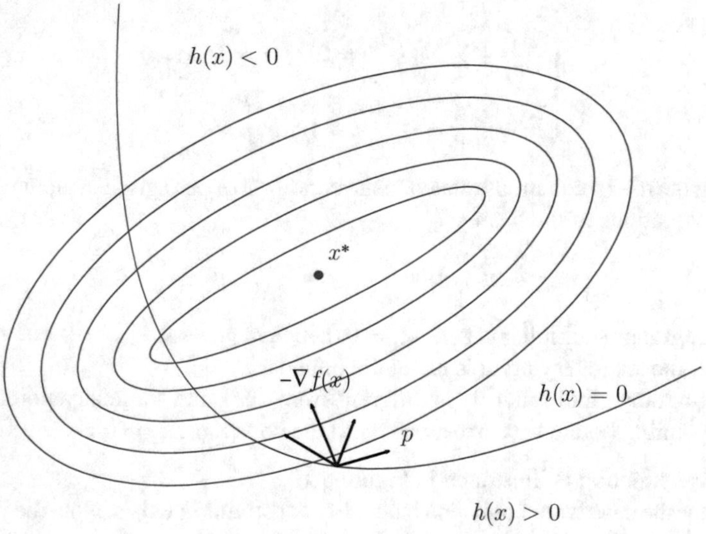

Fig. 8.10 Illustration of the contours of function $f(\boldsymbol{x})$, a constraint $h(\boldsymbol{x})$, possible displacements $\boldsymbol{p}$, and the opposite of gradient $\nabla f(\boldsymbol{x})$

Conclusion:

According to Karush, Kuhn, and Tucker, the KKT *necessary* conditions (non-sufficient) at the optimum at some point of a function f subject to m inequality constraints h_i such that $h_i(x) \leq 0$ are as follows:

1. The modified objective function

$$\mathcal{L} = f + \sum_{i=1}^{m} \lambda_i\, h_i \tag{8.7.19}$$

must be stationary at this point (its gradient is zero)

$$\left(\frac{\partial \mathcal{L}}{\partial x_j}\right)_{x^*} = 0\,, \quad j = 1, \ldots, n \tag{8.7.20}$$

2. The complementary slackness condition

$$\lambda_i^* \, h_i(x^*) = 0 \qquad \forall i = 1, \ldots, m \tag{8.7.21}$$

is verified in the following way, either a point is inside the domain (then all the Lagrange multipliers λ_i are zero), or it is at a boundary (then the constraint associated to the boundary is zero, its Lagrange parameter nonzero, all the other constraints being non zero and their Lagrange parameters zero), or it is at the intersection of constraints (this induces a combinatory aspect).

3. **If the function f must be a maximum, then necessarily**

$$\lambda_i^* \leq 0\,, \quad i = 1, \ldots, m \tag{8.7.22}$$

If the function f must be a minimum, then necessarily

$$\lambda_i^* \geq 0\,, \quad i = 1, \ldots, m \tag{8.7.23}$$

Example 8.6 :
Optimization with inequality constraints and KKT parameters

Determine the minimum of the function

$$f = x_1^3 + 2x_2^2 - 10x_1 + 6 + 2x_2^3 \tag{8.7.24}$$

subject to the constraints

$$x_1 x_2 \leq 10\,, \quad x_1 \geq 0\,, \quad x_2 \geq 0 \tag{8.7.25}$$

The constraints are again written under the form

$$h_1 = x_1 x_2 - 10 \leq 0\,, \quad h_2 = -x_1 \leq 0\,, \quad h_3 = -x_2 \leq 0 \tag{8.7.26}$$

The Lagrangian function is then equal to (Stage ♯ 1)

$$\mathcal{L} = x_1^3 + 2x_2^2 - 10x_1 + 6 + 2x_2^3 + \lambda_1(x_1 x_2 - 10) + \lambda_2(-x_1) + \lambda_3(-x_2) \tag{8.7.27}$$

The stationary points are obtained by setting to zero the gradient of the Lagrangian (Stage ♯ 2)

$$\begin{cases} \dfrac{\partial \mathcal{L}}{\partial x_1} = 3x_1^2 - 10 + \lambda_1 x_2 - \lambda_2 = 0 \\ \dfrac{\partial \mathcal{L}}{\partial x_2} = 4x_2 + 6x_2^2 + \lambda_1 x_1 - \lambda_3 = 0 \end{cases} \tag{8.7.28}$$

We review the different possibilities of position of the optimum (Stage ♯ 3)

- Suppose that a stationary point exists inside the domain, then all the Lagrange multipliers are zero, hence

$$3x_1^2 - 10 = 0\,, \quad 4x_2 + 6x_2^2 = 0 \tag{8.7.29}$$

The solutions are $x_1 = \pm 1.8257,\ x_2 = 0$ or $x_1 = \pm 1.8257,\ x_2 = -2/3$. As according to the third constraint, x_2 cannot be negative, thus there remains the possibility $x_2 = 0$. Thus, we find point A(1.8257, 0) and point B(−1.8257, 0) associated to the multipliers $\lambda_i = 0,\ i = 1, 2, 3$. The point B is not feasible, as it violates the second constraint dealing with x_1.

Thus, we must calculate the values of the constraints for each stationary point that was found, verify if the point is feasible, and then calculate the value of the function f.
In the final Table 8.1, it can be verified that point A satisfies all the constraints, thus belongs to the feasible domain, whereas B violates the second constraint and is not feasible.

- Now assume that a stationary point is on the boundary corresponding to the zero of the first constraint. In this case, the Lagrange multiplier λ_1 is different from 0, whereas the two other multipliers are zero, $\lambda_i = 0, \ i = 2, 3$. The stationary point verifies the system

$$\begin{cases} 3x_1^2 - 10 + \lambda_1 x_2 = 0 \\ 4x_2 + 6x_2^2 + \lambda_1 x_1 = 0 \\ g_1 = x_1 x_2 - 10 = 0 \end{cases} \tag{8.7.30}$$

After solving, either we obtain the point C(3.847, 2.6) and $\lambda_1 = -13.238$, or the point D(−3.568, −2.802) and $\lambda'_1 = 10.063$. As we search a minimum of f and as λ_1 is negative, point C is not feasible. It can be a maximum. With regard to D, after calculation, we notice that both constraints h_2 and h_3 are not verified, thus D is not feasible.

- Similarly, we assume that the stationary point is on the boundary corresponding to the zero of the second constraint. Then $\lambda_i = 0, \ i = 1, 3$ and the stationary point verifies the system

$$\begin{cases} 3x_1^2 - 10 - \lambda_2 = 0 \\ 4x_2 + 6x_2^2 = 0 \\ -x_1 = 0 \end{cases} \tag{8.7.31}$$

giving point E(0,0) and $\lambda_2 = -10$. As λ_2 is negative, it cannot be a minimum.

- Finally, we assume that the stationary point is on the boundary corresponding to the zero of the third constraint. Then, $\lambda_i = 0, \ i = 1, 2$ and the stationary point verifies the system

$$\begin{cases} 3x_1^2 - 10 = 0 \\ 4x_2 + 6x_2^2 - \lambda_3 = 0 \\ -x_2 = 0 \end{cases} \tag{8.7.32}$$

giving point F(1.8257, 0) and point G(−1.8257, 0) with $\lambda_3 = 0$. Indeed, F and G are confused with A and B respectively and are not represented in the final Table, as all the characteristics of points F and G are identical to A and B respectively.

- Assume that a point belongs to the two first constraints $x_1 x_2 = 10$ and $x_1 = 0$. Impossible.
- Assume that a point belongs to the first constraint $x_1 x_2 = 10$ and to the third constraint $x_2 = 0$. Impossible.
- Assume that a point belongs to the second constraint $x_1 = 0$ and to the third constraint $x_2 = 0$. The first constraint is verified. We find point E of Table 8.1.

The final Table 8.1 results.

The function has only one minimum, but also two maxima. This technique will be improved in the following by the use of the projected gradient.

8.7.3 Conditions of Minimum by the Projected Gradient in the Case of Inequality Constraints

The problem is of the form

$$\text{minimum of } f(\boldsymbol{x}), \quad \boldsymbol{x} \in \mathbb{R}^n \tag{8.7.33}$$

subject to m inequality constraints

Table 8.1 Classification of the extrema

	A	B	C	D	E
x_1	1.8257	−1.8257	3.847	−3.568	0
x_2	0	0	2.6	−2.802	0
λ_1	0	0	−13.238	10.063	0
λ_2	0	0	0	0	−10
λ_3	0	0	0	0	0
h_1	−10	−10	0	0	−10
h_2	−1.8257	1.8257	− 3.847	3.568	0
h_3	0	0	−2.6	2.802	0
f	−6.1716	18.1716	73.1353	−32.0386	6
KKT conclusion	No conclusion	No conclusion	Not a minimum	Not a maximum	Not a minimum
Final conclusion	Minimum	Nonfeasible		Nonfeasible	

$$g_i(\boldsymbol{x}) \leq 0, \; i = 1, \ldots, m \tag{8.7.34}$$

This problem is brought back to the problem of equality constraints.[1] We search, in the set of the m constraints, the constraints that are zero at the stationary point $\boldsymbol{x}^*$, these constraints are called *active constraints* $g^a = 0$, while the other constraints are called inactive constraints $g^{ina} \neq 0$. Near the stationary point, the inactive constraints are not concerned and only the active constraints are concerned. Then, the problem is the same as previously by considering only the active constraints in the set of constraints, when we calculate $g_x^a(\boldsymbol{x}^*)$ and $g_{xx}^a(\boldsymbol{x}^*)$.

Summary of problem solving:

$$\min f(\boldsymbol{x}), \quad \boldsymbol{x} \in \mathbb{R}^n \tag{8.7.35}$$

subject to m inequality constraints

$$g_i(\boldsymbol{x}) \leq 0, \; i = 1, \ldots, m \tag{8.7.36}$$

The necessary conditions so that $\boldsymbol{x}^*$ minimizes f subject to inequality constraints are:

(a) The consideration of active or inactive constraints (combinatorial aspect: We can have 1 active constraint, 2 active constraints, ..., m active constraints) (condition of realizability)

$$g_i^a(\boldsymbol{x}^*) = 0 \quad i \in [1, m], \quad g_i^{ina}(\boldsymbol{x}^*) < 0 \quad i \in [1, m] \tag{8.7.37}$$

[1] For reasons that appear in the following (in the calculation, we especially consider the active constraints, the other ones must be simply verified), now we note the constraints g and no more h like in the previous demonstration.

(b) The determination of the possible optima (condition of stationarity)

$$\mathcal{L}_x = 0 \quad \Leftrightarrow \quad f_x(\boldsymbol{x}^*) = -G_x(\boldsymbol{x}^*)^T \boldsymbol{\lambda}^* \tag{8.7.38}$$

where G_x takes into account only the active constraints. For that reason, the different cases must be taken one by one in the same manner as in (8.7.21).

(c) Karush–Kuhn–Tucker conditions (nonnegativity)

$$\lambda_i^* \geq 0 \quad \forall i = 1, \ldots, m \tag{8.7.39}$$

Note that the set of three conditions (a), (b), (c), is equivalent to the conditions developed about Equations (8.7.19)–(8.7.23).

(d) The determination of the feasible directions $\boldsymbol{p}$

$$G_x(\boldsymbol{x}^*)\, \boldsymbol{p} = 0 \tag{8.7.40}$$

hence $Z(\boldsymbol{x}^*)$

(e) The condition on the nature of the optimum: So that the optimum is a minimum, it is necessary that the projected Hessian matrix

$$Z(\boldsymbol{x}^*)^T \left[f_{xx}(\boldsymbol{x}^*) + \sum_{i=1}^{m} \lambda_i^* g_{i,xx}^a(\boldsymbol{x}^*) \right] Z(\boldsymbol{x}^*) \tag{8.7.41}$$

is semi-definite positive. Furthermore, it is sufficient that this matrix is definite positive so that it is a minimum.

The condition (c) about KKT parameters is a necessary condition. When the KKT parameters are strictly positive, the condition is sufficient. The other conditions (a), (b), and (e) are necessary and sufficient.

When the Lagrange multipliers are strictly equal to zero, the discussion is a little more difficult (Gill et al. 1981).

Remark on the implementation:

By exploiting Equation (8.7.38) and reviewing all the possible cases related to the combinatorial aspect, we find the different possible extrema $\boldsymbol{x}^*$. We start by eliminating those which are not feasible as they do not respect the constraints or as they do not respect the KKT conditions. We consider the remaining possible extrema $\boldsymbol{x}^*$ and, for each of them, we calculate the value of the objective function $f(\boldsymbol{x}^*)$. In order to spare time, in the case of the search of a minimum (resp. maximum), we start by checking the point $\boldsymbol{x}^*$ corresponding to the lowest (resp. largest) value of $f(\boldsymbol{x}^*)$ by studying Equations (8.7.40) and (8.7.41). If this point is feasible, it is the global optimum. If not, we consider the point corresponding to the second lowest (resp. largest) value of f and so on until finding a feasible point.

Example 8.7 :

Optimization with inequality constraints, KKT parameters, and projected gradient

Application to Example 8.6:

Search of the minimum of the function

$$f(\boldsymbol{x}) = x_1^3 + 2x_2^2 - 10x_1 + 6 + 2x_2^3 \tag{8.7.42}$$

subject to the constraints

$$x_1 x_2 \leq 10\,, \quad x_1 \geq 0\,, \quad x_2 \geq 0 \tag{8.7.43}$$

written as lower inequalities

$$\begin{aligned} g_1(\boldsymbol{x}) &= x_1 x_2 - 10 \leq 0 \\ g_2(\boldsymbol{x}) &= -x_1 \leq 0 \\ g_3(\boldsymbol{x}) &= -x_2 \leq 0 \end{aligned} \tag{8.7.44}$$

From the study already done (Table 8.1), the point A($x_1 = 1.8257; x_2 = 0$) on the constraint g_3 and satisfying $g_1 \leq 0$ and $g_2 \leq 0$ is a possible minimum.

- The conditions of the projected gradient are applied to this point. The active constraint is $g_3(\boldsymbol{x}) = 0$ at the optimum, thus the condition (8.7.37) $g_i^a(\boldsymbol{x}^*) = 0$ is satisfied for $i = 3$ (third constraint), that is

$$x_2^* = 0 \tag{8.7.45}$$

- Use the condition (8.7.38), $f_x(\boldsymbol{x}^*) = -G_x(\boldsymbol{x}^*)^T \lambda^*$. Only the active constraints are considered, hence

$$f\!x = \begin{bmatrix} 3x_1^2 - 10 \\ 4x_2 + 6x_2^2 \end{bmatrix} \qquad G_x = [0 - 1] \tag{8.7.46}$$

thus condition (8.7.38) implies

$$\begin{bmatrix} 3x_1^2 - 10 \\ 4x_2 + 6x_2^2 \end{bmatrix}_* = - \begin{bmatrix} 0 \\ -1 \end{bmatrix} \lambda^* \tag{8.7.47}$$

hence by taking into account $x_2^* = 0$

$$x_1 = \pm\sqrt{\frac{10}{3}}\,, \quad \lambda^* = 0 \tag{8.7.48}$$

Only the positive root is feasible (constraint g_2) thus

$$x_1^* = \sqrt{\frac{10}{3}} \tag{8.7.49}$$

- The condition (8.7.39), $\lambda^* \geq 0$, is respected as $\lambda^* = 0$.
- Calculate the feasible directions $\boldsymbol{p}$ with the relation (8.7.40)

$$G_x{}^* \boldsymbol{p} = 0 \Longrightarrow [0 - 1] \begin{bmatrix} p_1 \\ p_2 \end{bmatrix} = 0 \tag{8.7.50}$$

hence

$$\boldsymbol{p} = \begin{bmatrix} 1 \\ 0 \end{bmatrix} p_1\,, \quad Z(\boldsymbol{x}) = \begin{bmatrix} p_1 \\ 0 \end{bmatrix} \tag{8.7.51}$$

- For the condition (8.7.41), let us calculate

$$f_{xx} = \begin{bmatrix} 6x_1 & 0 \\ 0 & 4 + 12x_2 \end{bmatrix}\,, \quad g_{xx} = \begin{bmatrix} 0 & 0 \\ 0 & 0 \end{bmatrix} \tag{8.7.52}$$

hence

$$\begin{aligned} & Z(\boldsymbol{x}^*)^T \left[f_{xx}(\boldsymbol{x}^*) + \textstyle\sum_{i=1}^m \lambda^* g_{xx}(\boldsymbol{x}^*) \right] Z(\boldsymbol{x}^*) \\ &= [p_1\ 0] \begin{bmatrix} 6x_1^* & 0 \\ 0 & 4 + 12x_2^* \end{bmatrix} \begin{bmatrix} p_1 \\ 0 \end{bmatrix} = p_1^2\, 6\, x_1^* \end{aligned} \tag{8.7.53}$$

As $x_1^* \geq 0$, the condition (8.7.41) so that this matrix is semi-definite positive is verified.

Point A is a minimum for f subject to the constraints.

Example 8.8 :
Optimization with inequality constraints, KKT parameters, and projected gradient

Consider a second example:
Minimize the function

$$f(\boldsymbol{x}) = 3\,x_1^2 + x_2 + 2\,x_3^2 \tag{8.7.54}$$

subject to the following constraints:

$$\begin{aligned} x_1 + 2\,x_2 + x_3 &\geq 2 \\ -x_1 + 3\,x_2 - x_3 &\geq 1 \end{aligned} \tag{8.7.55}$$

We must not forget that the constraints must be presented under the form $g_i(\boldsymbol{x}) \leq 0$, hence

$$\begin{aligned} g_1(\boldsymbol{x}) &= 2 - x_1 - 2\,x_2 - x_3 \\ g_2(\boldsymbol{x}) &= 1 + x_1 - 3\,x_2 + x_3 \end{aligned} \tag{8.7.56}$$

Calculate the Lagrangian

$$\mathcal{L} = 3\,x_1^2 + x_2 + 2\,x_3^2 + \lambda_1\,(2 - x_1 - 2\,x_2 - x_3) + \lambda_2\,(1 + x_1 - 3\,x_2 + x_3) \tag{8.7.57}$$

The gradient is zero

$$\frac{\partial \mathcal{L}}{\partial x} = 0 \quad \Rightarrow \begin{cases} 6\,x_1 - \lambda_1 + \lambda_2 = 0 \\ 1 - 2\,\lambda_1 - 3\,\lambda_2 = 0 \\ 4\,x_3 - \lambda_1 + \lambda_2 = 0 \end{cases} \tag{8.7.58}$$

- Points inside the domain: $\lambda_1 = \lambda_2 = 0$. Impossible.
- Point on the first constraint $g_1(x) = 0$: $\lambda_1 \neq 0$, $\lambda_2 = 0$. We find $\lambda_1 = \frac{1}{2} \geq 0$, the point can be a minimum.
  Possible minimum point

$$\boldsymbol{x}^* = \left|\begin{matrix} \frac{1}{12} \\ \frac{43}{48} \\ \frac{1}{8} \end{matrix}\right.$$

  Calculation of the constraint: $g_2(\boldsymbol{x}^*) \leq 0$ thus it is verified.
- Point on the second constraint $g_2(x) = 0$: $\lambda_1 = 0$, $\lambda_2 \neq 0$. We find $\lambda_2 = \frac{1}{3} \geq 0$, the point can be a minimum.
  Possible minimum point

$$\boldsymbol{x}^* = \left|\begin{matrix} -\frac{1}{18} \\ \frac{31}{108} \\ -\frac{1}{12} \end{matrix}\right.$$

  Calculation of the constraint: $g_1(\boldsymbol{x}^*) \geq 0$ thus it is not verified. The point is not feasible.
- Point at the intersection of constraints $g_1(x) = 0$ and $g_2(x) = 0$: $\lambda_1 \neq 0$, $\lambda_2 \neq 0$.
  We get the following system of equations:

$$\begin{cases} 6\,x_1 - \lambda_1 + \lambda_2 = 0 \\ 1 - 2\,\lambda_1 - 3\,\lambda_2 = 0 \\ 4\,x_3 - \lambda_1 + \lambda_2 = 0 \\ 2 - x_1 - 2\,x_2 - x_3 = 0 \\ 1 + x_1 - 3\,x_2 + x_3 = 0 \end{cases} \tag{8.7.59}$$

  for which the solution is

$$\begin{bmatrix} x_1^* \\ x_2^* \\ x_3^* \\ \lambda_1^* \\ \lambda_2^* \end{bmatrix} = \begin{bmatrix} 0.32 \\ 0.6 \\ 0.48 \\ 1.352 \\ -0.568 \end{bmatrix} \tag{8.7.60}$$

As $\lambda_2 < 0$, the point cannot be a minimum.

- Partial conclusion: The only possible minimum is thus on the first constraint which is considered as active in the study by the projected gradient as follows:

$$g_1^a(\boldsymbol{x}) = 2 - x_1 - 2\,x_2 - x_3 \tag{8.7.61}$$

The transpose gradient of the constraint is

$$G_{\boldsymbol{x}} = [-1 \ -2 \ -1] \tag{8.7.62}$$

The gradient of the function is

$$f_{\boldsymbol{x}} = \begin{vmatrix} 6\,x_1 \\ 1 \\ 4\,x_3 \end{vmatrix} \quad \Rightarrow \quad f_{\boldsymbol{x}^*} = \begin{vmatrix} \frac{1}{2} \\ 1 \\ \frac{1}{2} \end{vmatrix} \tag{8.7.63}$$

The condition (8.7.38), that is $f_x(\boldsymbol{x}^*) = -G_x(\boldsymbol{x}^*)^T\, \boldsymbol{\lambda}^*$ implies $\lambda = \frac{1}{2}$, thus as $\lambda \geq 0$, the condition (8.7.39) is verified.

Calculation of the feasible directions according to condition (8.7.40), that is $G_x(\boldsymbol{x}^*)\boldsymbol{p} = 0$ which gives

$$-p_1 - 2\,p_2 - p_3 = 0 \quad \Rightarrow \quad \boldsymbol{p} = \begin{vmatrix} p_1 \\ p_2 \\ -p_1 - 2\,p_2 \end{vmatrix} \quad \Rightarrow \quad Z(\boldsymbol{x}) = \begin{bmatrix} p_1 & 0 \\ 0 & p_2 \\ -p_1 & -2\,p_2 \end{bmatrix} \tag{8.7.64}$$

We check the condition (8.7.41)

$$Z(\boldsymbol{x}^*)^T \left[f_{xx}(\boldsymbol{x}^*) + \Sigma_{i=1}^m \lambda_i^* g_{i,xx}^a(\boldsymbol{x}^*) \right] Z(\boldsymbol{x}^*) =$$

$$\begin{bmatrix} p_1 & 0 & -p_1 \\ 0 & p_2 & -2p_2 \end{bmatrix} \begin{bmatrix} 6 & 0 & 0 \\ 0 & 0 & 0 \\ 0 & 0 & 4 \end{bmatrix} \begin{bmatrix} p_1 & 0 \\ 0 & p_2 \\ -p_1 & -2\,p_2 \end{bmatrix} = \begin{bmatrix} 10\,p_1^2 & 8p_1p_2 \\ 8p_1p_2 & 16\,p_2^2 \end{bmatrix} = \boldsymbol{A} \tag{8.7.65}$$

We search the eigenvalues of $\boldsymbol{A}$

$$\boldsymbol{A} - \lambda \boldsymbol{I} = \begin{bmatrix} 10\,p_1^2 - \lambda & 8p_1p_2 \\ 8p_1p_2 & 16\,p_2^2 - \lambda \end{bmatrix} \tag{8.7.66}$$

by setting that the determinant of $(\boldsymbol{A} - \lambda \boldsymbol{I})$ is zero, hence the second degree equation with respect to λ

$$\lambda^2 - \lambda(10\,p_1^2 + 16\,p_2^2) + 96p_1^2 p_2^2 = 0 \tag{8.7.67}$$

We notice that the sum and the product of the roots are positive. If the roots are real, both are positive. If the roots are complex conjugate, their real part is positive. Indeed, the discriminant is always positive or zero, thus the roots are real positive. The matrix $\boldsymbol{A}$ is thus semi-definite positive. It is confirmed that this point is a minimum.

8.8 Function Subject to Equality and Inequality Constraints

8.8.1 Setting of the Problem

In the case of simultaneous equality and inequality constraints, the problem is set as

$$\min_{\boldsymbol{x}} f(\boldsymbol{x}), \quad \boldsymbol{x} \in \mathbb{R}^n \tag{8.8.1}$$

subject to m equality constraints and p inequality constraints

$$\begin{aligned} g_i(\boldsymbol{x}) &\leq 0, \quad i = 1, \ldots, m \\ h_i(\boldsymbol{x}) &\leq 0 \quad, i = 1, \ldots, p \end{aligned} \tag{8.8.2}$$

In the Lagrangian function, simultaneously Lagrange multipliers λ_i and KKT parameters μ_i are present

$$\mathcal{L}(\boldsymbol{x}, \boldsymbol{\lambda}, \boldsymbol{\mu}) = f(\boldsymbol{x}) + \sum_{i=1}^{m} \lambda_i g_i + \sum_{i=1}^{p} \mu_i h_i \tag{8.8.3}$$

Then, the characteristics of the case of Lagrange multipliers and those of the case of KKT parameters are combined.

The treatment will be highlighted by Example 8.9.

Example 8.9 :
Optimization with equality and inequality constraints

Search the minimum of the function

$$f(\boldsymbol{x}) = x_2 + x_3 \tag{8.8.4}$$

subject to

$$\begin{aligned} x_1 + x_2 + x_3 &= 1 \\ x_1^2 + x_2^2 + x_3^2 &\leq 1 \end{aligned} \tag{8.8.5}$$

The constraints are written as

$$\begin{aligned} g_1(\boldsymbol{x}) &= x_1 + x_2 + x_3 - 1 = 0 \\ h_1(\boldsymbol{x}) &= x_1^2 + x_2^2 + x_3^2 - 1 \leq 0 \end{aligned} \tag{8.8.6}$$

To facilitate the understanding, we will adopt the following notations, λ_i for Lagrange multipliers and μ_i for KKT parameters. As we are searching a minimum, according to Karush–Kuhn–Tucker, we know that $\mu_i \geq 0$.

The Lagrangian function is thus equal to

$$\mathcal{L} = x_2 + x_3 + \lambda_1 (x_1 + x_2 + x_3 - 1) + \mu_1 (x_1^2 + x_2^2 + x_3^2 - 1) \tag{8.8.7}$$

It results

$$\begin{aligned}
\frac{\partial \mathcal{L}}{\partial x_1} &= \lambda_1 + 2\,\mu_1\,x_1 = 0\\
\frac{\partial \mathcal{L}}{\partial x_2} &= 1 + \lambda_1 + 2\,\mu_1\,x_2 = 0\\
\frac{\partial \mathcal{L}}{\partial x_3} &= 1 + \lambda_1 + 2\,\mu_1\,x_3 = 0\\
\frac{\partial \mathcal{L}}{\partial \lambda_1} &= x_1 + x_2 + x_3 - 1 = 0
\end{aligned} \tag{8.8.8}$$

Then, two possibilities deal with the inequality constraint. The point can be inside the domain or on the constraint.

- Point inside the domain with respect to the inequality constraint.
  We deduce $\mu_1 = 0$, hence

$$\begin{aligned}
&\lambda_1 = 0\\
&x_2 = x_3\\
&x_1 + 2\,x_2 - 1 = 0
\end{aligned} \tag{8.8.9}$$

  This is impossible.
- Point on the inequality constraint.

We deduce $\mu_1 \neq 0$ and moreover $\mu_1 \geq 0$ as it is a KKT parameter and we search a minimum of f, hence

$$\begin{aligned}
\frac{\partial \mathcal{L}}{\partial x_1} &= \lambda_1 + 2\,\mu_1\,x_1 = 0\\
\frac{\partial \mathcal{L}}{\partial x_2} &= 1 + \lambda_1 + 2\,\mu_1\,x_2 = 0\\
\frac{\partial \mathcal{L}}{\partial x_3} &= 1 + \lambda_1 + 2\,\mu_1\,x_3 = 0\\
\frac{\partial \mathcal{L}}{\partial \lambda_1} &= x_1 + x_2 + x_3 - 1 = 0\\
h_1 &= x_1^2 + x_2^2 + x_3^2 - 1 = 0
\end{aligned} \tag{8.8.10}$$

The solution of the system imposes $x_2 = x_3$, and $x_2 = 0$ or $x_2 = \frac{2}{3}$. Several solutions result

$$\boldsymbol{x} = \begin{bmatrix} 1\\ 0\\ 0 \end{bmatrix}, \quad \lambda_1 = -1\,, \quad \mu_1 = 0.5 \tag{8.8.11}$$

is feasible. For this point, $f = 0$.

$$\boldsymbol{x} = \begin{bmatrix} -\frac{1}{3}\\ \frac{2}{3}\\ \frac{2}{3} \end{bmatrix}, \quad \lambda_1 = -\frac{1}{3}\,, \quad \mu_1 = -\frac{1}{2} \tag{8.8.12}$$

is not feasible.

Verify that the first point indeed corresponds to a minimum. Two constraints are active, as $g_1(\boldsymbol{x}) = 0$ and $h_1(\boldsymbol{x}) = 0$.

$$G_x(\boldsymbol{x}) = \begin{bmatrix} 1 & 1 & 1\\ 2\,x_1 & 2\,x_2 & 2\,x_3 \end{bmatrix} \quad \text{hence} \quad G_x(\boldsymbol{x}^*) = \begin{bmatrix} 1 & 1 & 1\\ 2 & 0 & 0 \end{bmatrix} \tag{8.8.13}$$

$$G_x(\boldsymbol{x}^*)\boldsymbol{p} = 0 \Longrightarrow \begin{bmatrix} 1 & 1 & 1\\ 2 & 0 & 0 \end{bmatrix} \begin{bmatrix} p_1\\ p_2\\ p_3 \end{bmatrix} \Longrightarrow \begin{bmatrix} p_1 + p_2 + p_3 = 0\\ 2\,p_1 + 0 + 0 = 0 \end{bmatrix} \tag{8.8.14}$$

We draw of it

$$\boldsymbol{p} = \begin{bmatrix} 0\\ p_2\\ -p_2 \end{bmatrix} \Longrightarrow Z(\boldsymbol{x}^*) = \begin{bmatrix} 0\\ p_2\\ -p_2 \end{bmatrix} \tag{8.8.15}$$

We check the condition (8.7.41)

$$Z(\boldsymbol{x}^*)^T \left[f_{xx}(\boldsymbol{x}^*) + \lambda_1^* g^a_{1,xx}(\boldsymbol{x}^*) + \mu_1^* h^a_{1,xx}(\boldsymbol{x}^*) \right] Z(\boldsymbol{x}^*) =$$

$$[0 \; p_2 \; -p_2] \times 1 \times \begin{bmatrix} 2 & 0 & 0 \\ 0 & 2 & 0 \\ 0 & 0 & 2 \end{bmatrix} \times \begin{bmatrix} 0 \\ p_2 \\ -p_2 \end{bmatrix} = 4p_2^2 > 0 \tag{8.8.16}$$

by noting that f_{xx} and $g_{1,xx}$ are zero matrices. Thus, it is a minimum.

8.8.2 Lagrange Duality

For ease of notations, the symbols used by Boyd and Vandenberghe (2009) are adopted in this section. The problem (8.8.1) is reset as

$$\min_{\boldsymbol{x}} f_0(\boldsymbol{x}), \quad \boldsymbol{x} \in \mathbb{R}^n \tag{8.8.17}$$

subject to m equality constraints and p inequality constraints

$$\begin{aligned} f_i(\boldsymbol{x}) = 0, \quad i = 1, \ldots, m \\ h_i(\boldsymbol{x}) \leq 0, \; i = 1, \ldots, p \end{aligned} \tag{8.8.18}$$

Thus, the Lagrangian is

$$\mathcal{L}(\boldsymbol{x}, \boldsymbol{\lambda}, \boldsymbol{\mu}) = f_0(\boldsymbol{x}) + \sum_{i=1}^{m} \lambda_i f_i + \sum_{i=1}^{p} \mu_i h_i \tag{8.8.19}$$

According to the strong Lagrangian principle (Allaire 2007; Bertsekas 2009; Bonnans et al. 2003; Boyd and Vandenberghe 2009; Luenberger and Ye 2016), the Lagrange dual function for problem (8.8.17) is

$$g(\boldsymbol{\lambda}, \boldsymbol{\mu}) = \inf_{\boldsymbol{x}} \mathcal{L}(\boldsymbol{x}, \boldsymbol{\lambda}, \boldsymbol{\mu}) \tag{8.8.20}$$

where the Lagrange multipliers $\boldsymbol{\lambda}$ and KKT parameters $\boldsymbol{\mu}$ are the dual variables. According to KKT conditions, given the searched minimum of (8.8.17), the parameters $\boldsymbol{\mu}$ verify $\boldsymbol{\mu} \geq 0$. Assume that f_0^* is the optimal value for problem (8.8.17), then the function g verifies the theorem of weak duality

$$g(\boldsymbol{\lambda}, \boldsymbol{\mu}) \leq f_0^* \quad \forall \boldsymbol{\lambda}, \quad \forall \boldsymbol{\mu} \geq 0 \tag{8.8.21}$$

thus this property yields a lower bound for the optimal solution f_0^*. Moreover, if the problem (8.8.17) is convex ($f_i(x)$ are convex $i = 0, \ldots, m$), in general the strong duality is verified (Boyd and Vandenberghe 2009). The theorem of strong duality stipulates that

$$\inf_{\boldsymbol{x}} f_0(\boldsymbol{x}) = \max_{\boldsymbol{\lambda}, \boldsymbol{\mu} \geq 0} g(\boldsymbol{\lambda}, \boldsymbol{\mu}) \Rightarrow f_0^* = g^* \tag{8.8.22}$$

If the functions $f_i(x)$ are convex and moreover Slater's condition about the existence of a vector $\boldsymbol{x}$ respecting the constraints is valid, the strong duality is verified. This will be used in linear programming (Section 10.6.3).

8.9 Sensitivity Analysis

In many problems, in particular in simple optimization and in dynamic optimization, the equations that describe the system depend on parameters $\boldsymbol{p}$, for example according to the following very general system of differential algebraic equations:

$$\begin{aligned} & \boldsymbol{f}(t, \boldsymbol{x}, \dot{\boldsymbol{x}}, \boldsymbol{p}) = 0\,, \quad \boldsymbol{x}, \boldsymbol{f} \in \Re^n\,,\ \boldsymbol{p} \in \Re^m \\ & \boldsymbol{x}(0) = \boldsymbol{x}_0(\boldsymbol{p}) \end{aligned} \tag{8.9.1}$$

Consider an objective function $g(\boldsymbol{x}, \boldsymbol{p})$. The *sensitivity analysis* of the objective function with respect to the parameters (Petzold et al. 2006) consists in determining the vector of sensitivities defined by

$$\boldsymbol{s} = \frac{dg}{d\boldsymbol{p}} \tag{8.9.2}$$

If it were a vector of objective functions, we would get a matrix of sensitivities (Caracotsios and Stewart 1985). When the number of parameters is important like for example in dynamic optimization or in the case of partial differential equations, the direct methods are considered as non-realizable and the methods based on the adjoint are preferable (Petzold et al. 2006).

Petzold et al. (2006) considers the algebraic system

$$\boldsymbol{f}(t, \boldsymbol{x}, \boldsymbol{p}) = 0 \tag{8.9.3}$$

and a vector of objective functions $\boldsymbol{g}(\boldsymbol{x}, \boldsymbol{p})$, depending on the states and the parameters, for which the sensitivities $d\boldsymbol{g}/d\boldsymbol{p}$ are obtained by the total derivative

$$\frac{d\boldsymbol{g}}{d\boldsymbol{p}} = \frac{\partial \boldsymbol{g}}{\partial \boldsymbol{x}} \frac{\partial \boldsymbol{x}}{\partial \boldsymbol{p}} + \frac{\partial \boldsymbol{g}}{\partial \boldsymbol{p}}\,, \quad g \in \Re^q \tag{8.9.4}$$

In the direct method, the system (8.9.3) is linearized under the form

$$\frac{\partial \boldsymbol{f}}{\partial \boldsymbol{x}} \frac{\partial \boldsymbol{x}}{\partial \boldsymbol{p}} + \frac{\partial \boldsymbol{f}}{\partial \boldsymbol{p}} = 0 \tag{8.9.5}$$

$\partial \boldsymbol{x}/\partial \boldsymbol{p}$ is a matrix of dimension $(n \times m)$ that is obtained by solving m linear systems (8.9.5), which then allows to obtain $d\boldsymbol{g}/d\boldsymbol{p}$ by Equation (8.9.4). As far as m is not too large, this method works well.

When the number m of parameters $\boldsymbol{p}$ becomes large, we obtain $\partial \boldsymbol{x}/\partial \boldsymbol{p}$ by the adjoint method which consists in multiplying Equation (8.9.5) by the adjoint variable $\boldsymbol{\lambda}$ to obtain

$$\boldsymbol{\lambda}^T \frac{\partial \boldsymbol{f}}{\partial \boldsymbol{x}} \frac{\partial \boldsymbol{x}}{\partial \boldsymbol{p}} + \boldsymbol{\lambda}^T \frac{\partial \boldsymbol{f}}{\partial \boldsymbol{p}} = 0\,, \quad \boldsymbol{\lambda} \in \Re^{n \times q} \tag{8.9.6}$$

In the general case, $\boldsymbol{\lambda}$ is a matrix of dimension $(n \times q)$. Then, we impose that $\boldsymbol{\lambda}$ is the solution of the adjoint system

$$\boldsymbol{\lambda}^T \frac{\partial \boldsymbol{f}}{\partial \boldsymbol{x}} = \frac{\partial \boldsymbol{g}}{\partial \boldsymbol{x}} \tag{8.9.7}$$

which transforms Equation (8.9.6) into

$$\frac{\partial \boldsymbol{g}}{\partial \boldsymbol{x}} \frac{\partial \boldsymbol{x}}{\partial \boldsymbol{p}} + \boldsymbol{\lambda}^T \frac{\partial \boldsymbol{f}}{\partial \boldsymbol{p}} = 0 \tag{8.9.8}$$

from which we draw using Equation (8.9.4)

$$\frac{d\boldsymbol{g}}{d\boldsymbol{p}} = -\boldsymbol{\lambda}^T \frac{\partial \boldsymbol{f}}{\partial \boldsymbol{p}} + \frac{\partial \boldsymbol{g}}{\partial \boldsymbol{p}} \tag{8.9.9}$$

Whatever the number m of parameters, Equation (8.9.7) needs to be solved only once. Thus, the system of large dimension is not solved to obtain $\partial \boldsymbol{x}/\partial \boldsymbol{p}$, but only the adjoint system (8.9.7) (Petzold et al. 2006). This method has been implemented in codes solving differential algebraic systems (Li and Petzold 2002).

Feehery et al. (1997) presents the sensitivity analysis under a different angle, by studying the sensitivity of the states with respect to the parameters. We consider a system of differential algebraic equations described by

$$\begin{aligned} &\boldsymbol{f}(t, \boldsymbol{x}, \dot{\boldsymbol{x}}, \boldsymbol{p}) = 0\,, \quad \boldsymbol{x}, \boldsymbol{f} \in \Re^n\,,\ \boldsymbol{p} \in \Re^m \\ &\boldsymbol{h}(t_0, \boldsymbol{x}(t_0), \dot{\boldsymbol{x}}(t_0), \boldsymbol{p}) = 0 \end{aligned} \tag{8.9.10}$$

The sensitivities of the states with respect to the parameters are the vectors $s_i = \partial \boldsymbol{x}/\partial p_i$. The sensitivities are solutions of the following differential algebraic system of $(n \times m)$ equations:

$$\begin{aligned} &\frac{\partial \boldsymbol{f}}{\partial \boldsymbol{x}} s_i + \frac{\partial \boldsymbol{f}}{\partial \dot{\boldsymbol{x}}} \dot{s}_i + \frac{\partial \boldsymbol{f}}{\partial p_i} = 0\,, \quad i = 1, \ldots, m \\ &\frac{\partial \boldsymbol{h}}{\partial \boldsymbol{x}} s_i(t_0) + \frac{\partial \boldsymbol{h}}{\partial \dot{\boldsymbol{x}}} \dot{s}_i(t_0) + \frac{\partial \boldsymbol{h}}{\partial p_i} = 0 \end{aligned} \tag{8.9.11}$$

which can be presented under the form

$$\boldsymbol{A} \begin{bmatrix} s_i \\ \dot{s}_i \end{bmatrix} = -\frac{\partial \boldsymbol{f}}{\partial p_i}\,, \quad \text{with } \boldsymbol{A} = \begin{bmatrix} \frac{\partial f_k}{\partial \boldsymbol{x}} & \frac{\partial f_k}{\partial \dot{\boldsymbol{x}}} \end{bmatrix} \tag{8.9.12}$$

During the solving of this type of system by multi-step integration methods, Feehery et al. (1997) propose to solve (8.9.12) by a quasi-Newton method.

8.10 Discussion and Conclusion

The use of the Lagrangian is not reserved to problems for which an analytical solution is sought. In the numerical methods of linear and nonlinear optimization, the Lagrange and Karush–Kuhn–Tucker parameters are used. Thus, this chapter is important to settle the theoretical bases. The sensitivity analysis presented at the end of the present chapter is very often useful in problems dealing with optimization, parameter estimation, design of experiments, and optimal control.

8.11 Exercise Set

Preamble: In this section, the analytical methods are favored.

Exercise 8.11.1 (Easy)
Find the maximum of the function

$$f(\boldsymbol{x}) = -2x_1^2 - x_2^2 - 3x_3^2 \tag{8.11.1}$$

subject to the constraints

$$\begin{aligned} x_1 + 2x_2 + x_3 - 1 = 0 \\ 4x_1 + 3x_2 + 2x_3 - 2 = 0 \end{aligned} \tag{8.11.2}$$

Solution: $x_1 = 5/27$, $x_2 = 10/27$, $x_3 = 2/27$, $f = -0.222$.

Exercise 8.11.2 (Easy)
Find the maximum of the function

$$f(\boldsymbol{x}) = -2x_1^2 - x_2^2 - 3x_3^2 \tag{8.11.3}$$

subject to the constraint

$$x_1 + 2x_2 + x_3 - 1 = 0 \tag{8.11.4}$$

Solution: $x_1 = 3/29$, $x_2 = 12/29$, $x_3 = 2/29$, $f = -0.207$.

Exercise 8.11.3 (Easy)
Find the minimum of the function

$$f(\boldsymbol{x}) = -x_1 x_2 \tag{8.11.5}$$

subject to the constraints

$$\begin{aligned} 2 \geq x_1 + x_2 \geq 0 \\ 2 \geq x_1 - x_2 \geq -2 \end{aligned} \tag{8.11.6}$$

Solution: $x_1 = 1$, $x_2 = 1$, $f = -1$.

Exercise 8.11.4 (Easy)
Find the minimum of the function

$$f(\boldsymbol{x}) = -x_1 - x_2 + x_3 \tag{8.11.7}$$

subject to the constraints

$$\begin{aligned} 0 \leq x_3 \leq 1 \\ x_1^3 + x_3 \leq 1 \\ x_1^2 + x_2^2 + x_3^2 \leq 1 \end{aligned} \tag{8.11.8}$$

Solution: $x_1 = 1/\sqrt{2}$, $x_2 = 1/\sqrt{2}$, $x_3 = 0$, $f = -\sqrt{2}$.

Exercise 8.11.5 (Medium)
Find the minimum of the function

$$f(\boldsymbol{x}) = 2x_1^2 - x_2^2 - 3x_3^2 \tag{8.11.9}$$

subject to the constraint

$$4x_1 + 3x_2 + 2x_3 - 2 \leq 0 \tag{8.11.10}$$

Solution: no minimum (think: why?).

Exercise 8.11.6 (Medium)
Find the maximum of the function

$$f(\boldsymbol{x}) = 2x_1^2 - x_2^2 - 3x_3^2 \tag{8.11.11}$$

subject to the constraints

$$\begin{aligned} x_1 + 2x_2 + x_3 - 1 &\leq 0 \\ 4x_1 + 3x_2 + 2x_3 - 2 &\leq 0 \end{aligned} \tag{8.11.12}$$

Solution: no maximum (think: why?).

Exercise 8.11.7 (Medium)
Find the minimum of the function

$$f(\boldsymbol{x}) = \frac{1}{2}[(x_1 - 1)^2 + x_2^2] \tag{8.11.13}$$

subject to the constraint

$$x_1 + \beta x_2^2 = 0 \tag{8.11.14}$$

β is a fixed parameter. For which values of β, is the extremum $\boldsymbol{x}^* = (0, 0)$ a local minimum?
Solution: $\beta < 0.5$.

Exercise 8.11.8 (Easy)
Find the minimum of the function

$$f(\boldsymbol{x}) = 2x_1^2 - 8x_1 + x_2^2 - 2x_2 \tag{8.11.15}$$

subject to the constraints

$$\begin{aligned} x_1 - x_2 &\geq 0 \\ x_1 + x_2 &\leq 4 \\ 0 \leq x_1 &\leq 3 \\ x_2 &\geq 0 \end{aligned} \tag{8.11.16}$$

Solution: $x_1 = 2$, $x_2 = 1$, $f = -9$.

Exercise 8.11.9 (Medium)
Prove that the minimum and maximum distances for the origin to the curve defined by the intersection of

$$\frac{x_1^2}{a_1^2} + \frac{x_2^2}{a_2^2} + \frac{x_3^2}{a_3^2} = 1 \tag{8.11.17}$$

and

$$c_1 x_1 + c_2 x_2 + c_3 x_3 = 0 \tag{8.11.18}$$

are obtained by d solution of the following equation:

$$\frac{c_1^2 a_1^2}{a_1^2 - d^2} + \frac{c_2^2 a_2^2}{a_2^2 - d^2} + \frac{c_3^2 a_3^2}{a_3^2 - d^2} = 0 \tag{8.11.19}$$

Exercise 8.11.10 (Easy)
Find the maximum of the function

$$f(\boldsymbol{x}) = 2x_1 - x_1^2 + x_2 \tag{8.11.20}$$

subject to the constraints

$$\begin{aligned} &2x_1 + 3x_2 \le 6 \\ &2x_1 + x_2 \le 4 \\ &x_1 \ge 0\,, \quad x_2 \ge 0 \end{aligned} \tag{8.11.21}$$

Solution: $x_1 = 2/3\,,\ x_2 = 14/9\,,\ f = 22/9$.

Exercise 8.11.11 (Medium)
Find the minimum of the function

$$f(\boldsymbol{x}) = 2x_1^2 + x_2^2 + 3x_1 + 4x_2 + 9 \tag{8.11.22}$$

subject to the constraints

$$\begin{aligned} x_1^2 + x_2 + 3x_1 x_2 &= 11 \\ x_1 + x_2^2 + 4x_1 x_2 &= 12 \end{aligned} \tag{8.11.23}$$

During this problem, it may be useful to use Newton's method to find the roots of a polynomial of degree larger than 2.
Note: This exercise is mainly analytical, but it also requires numerical methods.
Solution: $x_1 = -1.9374\,,\ x_2 = -1.5059\,,\ f = 6.939$.

Exercise 8.11.12 (Easy)
Find the extrema of the function

$$f(\boldsymbol{x}) = x_1\, x_2^2\, x_3^3 \tag{8.11.24}$$

subject to the constraints

$$\begin{aligned} &x_1 + x_2 + x_3 = 6 \\ &x_i > 0 \end{aligned} \tag{8.11.25}$$

Solution: $x_1 = 1\,,\ x_2 = 2\,,\ x_3 = 3\,,\ f = 108$.

Exercise 8.11.13 (Easy)
Find the minimum of the function

$$f(\boldsymbol{x}) = 3\,x_1^2 + x_2 + 2\,x_3^2 \tag{8.11.26}$$

subject to the constraints

$$\begin{aligned} &x_1 + 2\,x_2 + x_3 \geq 2 \\ &-x_1 + 3\,x_2 - x_3 \geq 1 \end{aligned} \tag{8.11.27}$$

Exercise 8.11.14 (Easy)
Find the minimum of the function

$$f(\boldsymbol{x}) = x_1^3 - 3\,x_1\,x_2 \tag{8.11.28}$$

subject to the constraints

$$\begin{aligned} &5\,x_1 + 2\,x_2 \geq 20 \\ &-2\,x_1 + x_2 = 5 \\ &x_1 \geq 0 \\ &x_2 \geq 0 \end{aligned} \tag{8.11.29}$$

Solution: $x_1 = 5$, $x_2 = 15$, $f = -100$.

Exercise 8.11.15 (Easy)
Find the minimum of the function

$$f(\boldsymbol{x}) = -x_1 - x_2 + x_3 \tag{8.11.30}$$

subject to the constraints

$$\begin{aligned} &0 \leq x_3 \leq 1 \\ &x_1 + x_2 + x_3 \leq 1 \\ &x_1^2 + x_2^2 + x_3^2 \leq 2 \end{aligned} \tag{8.11.31}$$

Exercise 8.11.16 (Easy)
Find the minimum of the function

$$f(\boldsymbol{x}) = x_1^2 + 2\,x_2^2 - x_3^2 \tag{8.11.32}$$

subject to the constraints

$$\begin{aligned} &2\,x_1 + x_2 + x_3 = 1 \\ &|x_1 + x_2| \leq 0.4 \end{aligned} \tag{8.11.33}$$

Exercise 8.11.17 (Medium)
Find the global maximum of the function

$$f(\boldsymbol{x}) = 3\,x_1^2 - 2\,x_1 - 5\,x_2^2 + 30\,x_2 + x_3^2 \tag{8.11.34}$$

subject to the constraints

$$\begin{aligned} &8 \leq 2\,x_1 + 3\,x_2 \\ &2\,x_2 - 10 \leq 0 \\ &3\,x_1 + 2\,x_2 + x_3 \leq 15 \\ &x_3^2 - 1 \leq 0 \end{aligned} \tag{8.11.35}$$

Exercise 8.11.18 (Easy)
Find the nearest point on the surface of equation

$$z = x^2 + y^2 \tag{8.11.36}$$

to the point $(3, -6, 4)$.
Solution: $x_1 = 1$, $x_2 = -2$, $x_3 = 5$.

Exercise 8.11.19 (Medium)
Find the minimum of the function

$$f(\boldsymbol{x}) = \frac{1}{2}\,\boldsymbol{x}^T \begin{bmatrix} 3 & -1 & 0 \\ -1 & 2 & -1 \\ 0 & -1 & 1 \end{bmatrix} \boldsymbol{x} + \begin{bmatrix} 1 \\ 1 \\ 1 \end{bmatrix}^T \boldsymbol{x} \tag{8.11.37}$$

subject to the constraint $x_1 + 2\,x_2 + x_3 = 4$ by the two following methods:

1. Apply the method of Lagrange multipliers.
2. Confirm the previous result by first substituting x_1 with respect to x_2 and x_3 in $f(\boldsymbol{x})$, then by searching the solution by the conjugate gradient method (start from point $(x_2, x_3) = (0, 0)$).
   Solution: $x_1 = 0.38889$, $x_2 = 1.2222$, $x_3 = 1.16666$.

Exercise 8.11.20 (Medium)
Consider the function

$$f(\boldsymbol{x}) = 3\,x_1^2 - 2\,x_1 - 5\,x_2^2 + 30\,x_2 \tag{8.11.38}$$

subject to the constraints

$$\begin{aligned} 2\,x_1 + 3\,x_2 &\geq 8 \\ x_2 &\leq 5 \\ 3\,x_1 + 2\,x_2 &\leq 15 \end{aligned} \tag{8.11.39}$$

Determine the nature of all points of interest. Use Figure 8.11 by representing those points and noting their nature.
Solution: ($x_1 = 0.333$, $x_2 = 5$, $f = 24.666$) global minimum. ($x_1 = 3.9697$, $x_2 = 1.5455$, $f = 73.7576$) global maximum. ($x_1 = 2.7143$, $x_2 = 0.8571$, $f = 38.7143$) local minimum. ($x_1 = -3.5$, $x_2 = 5$, $f = 68.75$) local maximum. ($x_1 = 0.333$, $x_2 = 3$, $f = 44.666$) saddle point. ($x_1 = 5.8$, $x_2 = -1.2$, $f = 46.12$) saddle point. ($x_1 = 1.6667$, $x_2 = 5$, $f = 30.000$) saddle point.

Exercise 8.11.21 (Medium)
We study the function $f(\boldsymbol{x})$

$$f(\boldsymbol{x}) = (x_1 - 3)^2 + (x_2 - 3)^2 \tag{8.11.40}$$

subject to the constraints

$$\begin{aligned} 4\,x_1^2 + 9\,x_2^2 &\leq 36 \\ x_1^2 + 3\,x_2 &= 3 \end{aligned} \tag{8.11.41}$$

We desire to characterize all the extrema of the function.

1. Represent the constraints and explain the consequences.
2. We use the equality constraint to operate a substitution of x_2 with respect to x_1. We then solve the onevariable problem with respect to x_1 by the general method of the Lagrangian function. Characterize the different extrema without using the projected gradient. It may be interesting to do a graphical representation of the function.
3. We do not want to make the substitution of question 2 and we solve the problem by the general method. Explain how you proceed. Characterize completely the local extrema using if possible the projected gradient. Comment the results obtained by this question with respect to the results of the previous question.

Exercise 8.11.22 (Medium)

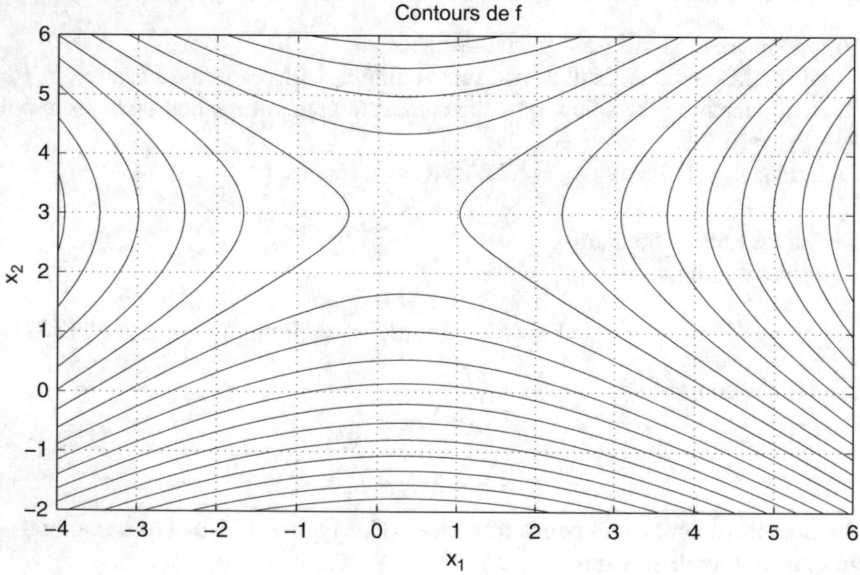

Fig. 8.11 Contours of function (8.11.38)

1. Find the global minimum of the function

$$f(\boldsymbol{x}) = x_1^3 + x_2^3 \tag{8.11.42}$$

subject to the constraint

$$x_1 + x_2 - 1 = 0 \tag{8.11.43}$$

2. Find the global minimum of the function

$$f(\boldsymbol{x}) = x_1^3 + x_2^3 + \mu\,(x_1 + x_2 - 1)^2 \tag{8.11.44}$$

with $\mu > 0$. With regard to function (8.11.42), a penalty function has been added. Determine the locus (= set of points) of the optimal solution, μ being a variable parameter. Make a graphical representation.
3. With respect to question 2, the following constraints are added: $|x_1| \leq 1$ and $|x_2| \leq 1$. Make a geometric representation of the optimal solution without taking into account the points which are on both constraints, which amounts to consider these constraints as strict inequalities.
4. Continuing the problem of question 3, find the solution when $\mu \longrightarrow +\infty$. Comment the obtained result. Why did we obtain it, what was the benefit of the constraints in question 3?

Exercise 8.11.23 (Difficult)
Find the minimum of

$$f(\boldsymbol{x}) = x_1 + x_2 \tag{8.11.45}$$

subject to the constraints

$$\begin{aligned} x_1^2 + x_2^2 &= 4 \\ -2\,x_1 - x_2 &\leq 4 \end{aligned} \tag{8.11.46}$$

Remark: During the discussion of the nature of the extremum by means of the projected gradient method, it is necessary to make a figure to illustrate the problem. When the projected gradient method is used, it will lead to a false conclusion about the nature of the extremum. Explain why. On the contrary, it will be possible to study the variation df by using a part of the work performed for the projected gradient and, then, the right conclusion will be obtained.

Exercise 8.11.24 (Difficult)
We determine the minimum of the function

$$f(\boldsymbol{x}) = \frac{1}{2}\,\boldsymbol{x}^T \begin{bmatrix} 3 & -1 & 0 \\ -1 & 2 & -1 \\ 0 & -1 & 1 \end{bmatrix} \boldsymbol{x} + \begin{bmatrix} 1 \\ 1 \\ 1 \end{bmatrix}^T \boldsymbol{x} \tag{8.11.47}$$

subject to the constraint

$$x_1 + 2\,x_2 + x_3 = 4 \tag{8.11.48}$$

This problem can be written under the general form

$$\min_{\boldsymbol{x}} f(\boldsymbol{x}) = \frac{1}{2}\,\boldsymbol{x}^T\,\boldsymbol{A}\,\boldsymbol{x} + \boldsymbol{b}^T\,\boldsymbol{x} \tag{8.11.49}$$

subject to the constraint

$$\boldsymbol{C}\,\boldsymbol{x} = \boldsymbol{d} \tag{8.11.50}$$

This problem is called quadratic programming with equality.
Note that the questions of the present exercise can be independently treated.

1. Exceptionally, we proceed by eliminating the variable x_1. Find the solution.
2. The question 1 is generalized. In this question, when necessary, define whether you consider a scalar, a vector, or a matrix and give their dimensions. It is supposed that there still exists only one constraint written as Equation (8.11.50).

(a) We pose the vector

$$\boldsymbol{y} = \begin{bmatrix} x_2 \\ x_3 \end{bmatrix} \tag{8.11.51}$$

and we partition the vector $\boldsymbol{x}$ under the form

$$\boldsymbol{x} = \begin{bmatrix} \tilde{\boldsymbol{x}} \\ \boldsymbol{y} \end{bmatrix} \tag{8.11.52}$$

Determine $\tilde{\boldsymbol{x}}$. We would like to transform the constraint (8.11.50) with respect to the new variables $\tilde{x}$ and $\boldsymbol{y}$. For that, we also partition the matrix $\boldsymbol{C}$ under the form

$$\boldsymbol{C} = [\boldsymbol{C}_1 \quad \boldsymbol{C}_2] \tag{8.11.53}$$

Explain $\boldsymbol{C}_1$ and $\boldsymbol{C}_2$ in the studied particular case. Give the new form of the constraint.

(b) Express the vector $\boldsymbol{x}$ with respect to vector $\boldsymbol{y}$ under a form

$$\boldsymbol{x} = \boldsymbol{E}\,\boldsymbol{y} + \boldsymbol{F} \tag{8.11.54}$$

by defining $\boldsymbol{E}$ and $\boldsymbol{F}$ in the particular case studied in 1 and in a general manner.

(c) We desire to maintain the form of the problem, i.e. we must express the function $f(\boldsymbol{x})$ under a new form similar to Equation (8.11.49):

$$h(\boldsymbol{y}) = \frac{1}{2}\,\boldsymbol{y}^T\,\boldsymbol{A}'\,\boldsymbol{y} + \boldsymbol{b}'^T\,\boldsymbol{y} + c' \tag{8.11.55}$$

where $\boldsymbol{A}'$ is a symmetrical matrix, $\boldsymbol{b}'$ is a column vector, and c' a constant. Of course, $h(\boldsymbol{y})$ will be numerically equal to $f(\boldsymbol{x})$. Find the form of $h(\boldsymbol{y})$ by expressing the matrix $\boldsymbol{A}'$, the vector $\boldsymbol{b}'$, the constant c' with respect to $\boldsymbol{A}$, $\boldsymbol{b}$, $\boldsymbol{C}$, $\boldsymbol{d}$, $\boldsymbol{E}$, $\boldsymbol{F}$,

(d) Calculate the general solution of this type of problem. Deduce the solution of the proposed numerical case.

3. Find the solution of the problem of Equations (8.11.47)–(8.11.48) by using Lagrange multipliers in this particular case.
4. Treat the general problem of Equations (8.11.49)–(8.11.50) in a case with several constraints by using Lagrange multipliers. Verify that, in the particular case (8.11.47)–(8.11.48), the general solution is the same as that found in 3.

Exercise 8.11.25 (Difficult)

Consider the function

$$h(\boldsymbol{x}) = \frac{1}{2}\boldsymbol{x}^T\boldsymbol{A}\boldsymbol{x} + \boldsymbol{b}^T\boldsymbol{x}\,, \quad \boldsymbol{x} \in \mathbb{R}^n \tag{8.11.56}$$

subject to the constraints

$$\begin{aligned} \boldsymbol{C}\boldsymbol{x} &\geq \boldsymbol{d} \\ \boldsymbol{E}\boldsymbol{x} &= \boldsymbol{f} \end{aligned} \tag{8.11.57}$$

It is assumed that $\boldsymbol{A}$ is a symmetrical, positive definite matrix. $\boldsymbol{C}$ is an $n_c \times n$ matrix. $\boldsymbol{E}$ is an $n_e \times n$ matrix.

1. Define the characteristics (dimension, type) of the vectors used in the problem.
2. Write the necessary first-order conditions of minimum for the problem (be cautious about the consistency with respect to the dimensions). Try to obtain a form as simple as possible.
3. Application:

$$\boldsymbol{A} = \begin{bmatrix} 6 & -2 & 0 \\ -2 & 10 & -1 \\ 0 & -1 & 2 \end{bmatrix}, \quad \boldsymbol{b} = \begin{bmatrix} -2 \\ 30 \\ 0 \end{bmatrix}$$
$$\boldsymbol{C} = \begin{bmatrix} 2 & 3 & 1 \end{bmatrix}, \quad \boldsymbol{d} = \begin{bmatrix} 6 \end{bmatrix}$$
$$\boldsymbol{E} = \begin{bmatrix} 1 & 2 & 1 \end{bmatrix}, \quad \boldsymbol{f} = \begin{bmatrix} 4 \end{bmatrix} \tag{8.11.58}$$

Exercise 8.11.26 (Difficult)

Geometric programming

The practical problem is as follows:

$$\min_{t} y(\boldsymbol{t}) = \frac{2}{t_1\, t_2} + \frac{3}{t_2} + 4\, t_1\, t_2^2 + 6\, t_2^2 \tag{8.11.59}$$

with the constraint $t_j \geq 0\,,\ \forall\, j$.

1. We decompose y under the form

$$y = \sum_{i=1}^{n} y_i \qquad \text{with} \quad y_i = c_i\, t_1^{a_{i,1}}\, t_2^{a_{i,2}} \ldots t_k^{a_{i,k}} \quad \text{and} \quad 0 \leq t_j \leq +\infty \tag{8.11.60}$$

   Do that decomposition on the proposed example by giving all the coefficients.
2. We introduce the functions x_i defined by

$$x_i = \sum_{j=1}^{k} a_{i,j} \ln(t_j) \tag{8.11.61}$$

   Explain the functions x_i for your example.
3. We deduce the expression of y with respect to x_i

$$y(\boldsymbol{x}) = \sum_{i=1}^{n} c_i\, e^{x_i} \tag{8.11.62}$$

   Give $y(\boldsymbol{x})$ for your example.
4. We introduce the partial derivatives d_i of $y(\boldsymbol{x})$

$$d_i = \frac{\partial y}{\partial x_i} = c_i\, e^{x_i} \qquad \text{thus} \quad x_i = \ln\left(\frac{d_i}{c_i}\right) \tag{8.11.63}$$

We introduce the Legendre transformation of $y(\boldsymbol{x})$ as the function

$$\psi(\boldsymbol{d}) = y(\boldsymbol{x}) - \sum_{i=1}^{n} d_i\, x_i = \sum_{i=1}^{n} d_i - \sum_{i=1}^{n} d_i \ln\left(\frac{d_i}{c_i}\right) \tag{8.11.64}$$

$\psi(\boldsymbol{d})$ is a dual, provided that $y(\boldsymbol{x})$ is convex and has a Legendre transformation. The d_i satisfy the equation

$$\sum_{i=1}^{n} x_i^*\, d_i = 0 \tag{8.11.65}$$

It remains to determine the d_i such that

$$\sum_{i=1}^{n} x_i\, d_i = \sum_{i=1}^{n} \left[\sum_{j=1}^{k} a_{i,j} \ln(t_j)\right] d_i = 0 \tag{8.11.66}$$

Even if ignoring the t_j, Equation (8.11.66) can be satisfied by determining the d_i such that

$$\sum_{i=1}^{n} a_{i,j}\, d_i = 0 \qquad \forall\ j = 1, \ldots, k \tag{8.11.67}$$

Justify Equation (8.11.67). Give Equation (8.11.67) for your example.

5. We introduce a normalization of the d_i by means of a normalization factor

$$\theta = \sum_{i=1}^{n} d_i \tag{8.11.68}$$

hence the normalized variables

$$\delta_i = \frac{d_i}{\theta} \tag{8.11.69}$$

so that

$$\sum_{i=1}^{n} \delta_i = 1 \tag{8.11.70}$$

The function ψ becomes

$$\psi(\boldsymbol{\delta}, \theta) = \theta - \theta \ln(\theta) - \theta \sum_{i=1}^{n} \ln\left(\left[\frac{\delta_i}{c_i}\right]^{\delta_i}\right) \tag{8.11.71}$$

Minimizing y amounts to maximize its dual ψ function of the parameter θ, hence the derivative

$$\begin{aligned} &\frac{d\psi}{d\theta} = 0 \Longrightarrow \\ &-\ln(\theta) - \sum_{i=1}^{n} \ln\left(\left[\frac{\delta_i}{c_i}\right]^{\delta_i}\right) = 0 \end{aligned} \tag{8.11.72}$$

and the optimal θ

$$\theta = \left(\frac{c_1}{\delta_1}\right)^{\delta_1} \left(\frac{c_2}{\delta_2}\right)^{\delta_2} \dots \left(\frac{c_n}{\delta_n}\right)^{\delta_n} \tag{8.11.73}$$

By replacing θ by its optimal value in Equation (8.11.71), a new expression of ψ results

$$\psi(\boldsymbol{\delta}) = \left(\frac{c_1}{\delta_1}\right)^{\delta_1} \left(\frac{c_2}{\delta_2}\right)^{\delta_2} \dots \left(\frac{c_n}{\delta_n}\right)^{\delta_n} \tag{8.11.74}$$

where the δ_i (positive) verify Equation (8.11.70) and the following equation:

$$\sum_{i=1}^{n} a_{i,j}\, \delta_i = 0 \qquad \forall\ j = 1, \dots, k \tag{8.11.75}$$

Find the expression of $\psi(\boldsymbol{\delta})$ in the case of the present example as well as the constraints verified by the variables δ_i.

6. If we suppose that only one of the variables δ_i is known, we can deduce the other variables δ_i. Of course, the function $\psi(\boldsymbol{\delta})$ has very little chance to be maximal for that set $\boldsymbol{\delta}$. Let ψ_c be the numerical value thus taken by $\psi(\boldsymbol{\delta})$. We thus have a lower bound of ψ^* (dual problem)

$$\psi_c \le \psi^* \tag{8.11.76}$$

where ψ^* is the maximal value of ψ.

Choose a value of δ_1. Deduce a consistent set $\boldsymbol{\delta}$. Calculate that lower bound of ψ.

7. For the primal problem, if we take a set $\boldsymbol{t}$ satisfying the constraints, we obtain a calculated value y_c that constitutes in a similar way an upper bound of the minimal value y^*

$$y_c \ge y^* \tag{8.11.77}$$

with moreover the following property:

$$y^* = \min\ y = \max\ \psi = \psi^* \tag{8.11.78}$$

We thus get an interval for y^*. Find an interval of y^* in this way.

8. From the constraints that exist about the variables δ_i, we can deduce $n-1$ of the variables δ_i with respect to one of these variables.

Express $\delta_1, \delta_2, \delta_3$ with respect to δ_4. Deduce the possible variation interval of δ_4.

9. We can express ψ as a function of a single variable δ_i. Express ψ as a function of the single variable δ_4.

10. We can easily deduce the maximum of ψ as now ψ is a function of a single variable. We deduce the optimal set δ^*. The dual problem is then solved.

Find the maximum ψ^* of ψ and the corresponding set δ^*. It may be helpful to use $\ln(\psi)$ and make the derivatives $(d\delta_i/d\delta_4)$ appear. Deduce the optimal value y^*.

11. We can come back to the primal problem.

Deduce the optimal set t^* for $y(\boldsymbol{t})$. Explain clearly how you proceed to get that result.

Exercise 8.11.27 (Medium)

Demonstrate that for a quadratic function

$$f(\boldsymbol{x}) = \frac{1}{2}\boldsymbol{x}^T \boldsymbol{A}\,\boldsymbol{x} - \boldsymbol{h}^T \boldsymbol{x} + c \tag{8.11.79}$$

the scalar α that minimizes $f(\boldsymbol{x} + \alpha\,\boldsymbol{r})$, where $\boldsymbol{r}$ is the residual, is equal to

$$\alpha = \frac{|\boldsymbol{r}^2|}{\boldsymbol{r}^T \boldsymbol{A}\,\boldsymbol{r}} \tag{8.11.80}$$

Exercise 8.11.28 (Difficult)
Minimize the rational fraction

$$f(\boldsymbol{x}) = \frac{\boldsymbol{p}^T \boldsymbol{x} + \alpha}{\boldsymbol{q}^T \boldsymbol{x} + \beta} \tag{8.11.81}$$

subject to the constraints

$$\boldsymbol{A}\boldsymbol{x} \le \boldsymbol{b}\,, \quad \boldsymbol{x} \ge 0 \tag{8.11.82}$$

Method:
We pose

$$z = \frac{1}{\boldsymbol{q}^T \boldsymbol{x} + \beta} \tag{8.11.83}$$

by supposing $\boldsymbol{q}^T \boldsymbol{x} + \beta > 0$ and $\boldsymbol{y} = z\,\boldsymbol{x}$.
It amounts to the problem

$$\min \quad \boldsymbol{p}^T \boldsymbol{y} + \alpha z \tag{8.11.84}$$

subject to the constraints

$$\boldsymbol{A}\boldsymbol{y} - \boldsymbol{b}z \le 0 \tag{8.11.85}$$

$$\boldsymbol{q}^T \boldsymbol{y} + \beta z = 1 \tag{8.11.86}$$

$$\boldsymbol{y}, z \ge 0 \tag{8.11.87}$$

From the solution $(\boldsymbol{y}^*, z^*)$, we deduce $\boldsymbol{x}^* = \boldsymbol{y}^*/z^*$.
To solve that problem, we use Equation (8.11.86) defining z by operating a substitution in the function and the other inequalities.
Application:
Find the minimum of the function

$$f(\boldsymbol{x}) = \frac{-2x_1 + x_2 + 2}{x_1 + 3x_2 + 4} \tag{8.11.88}$$

subject to the constraints

$$\begin{aligned} &-x_1 + x_2 \le 4 \\ &2\,x_1 + x_2 \le 14 \\ &x_2 \le 6 \\ &x_1, x_2 \ge 0 \end{aligned} \tag{8.11.89}$$

Exercise 8.11.29 (Difficult)
Minimize the sum

$$f(\boldsymbol{x}) = x_1^2 + x_2^2 + x_3^2 + x_4^2 + x_5^2 \tag{8.11.90}$$

subject to the constraints

$$\begin{aligned} x_1\, x_2\, x_3\, x_4\, x_5 &= 11 \\ x_i \geq 0 \quad &\forall i \end{aligned} \tag{8.11.91}$$

by dynamic programming explained in the following. The value 11 is purely indicative as well as the dimension of $\boldsymbol{x}$ equal to 5. Indeed, the proposed method is able to solve this type of problem whatever the product and the dimension of $\boldsymbol{x}$.

Dynamic programming method to use:

The dynamic programming method was originally proposed by Bellman (1957), Bellman and Dreyfus (1962).

We pose

$$p_n = \prod_{i=1}^{n} x_i \tag{8.11.92}$$

and we use the fundamental relation of dynamic programming

$$f_n(p_n) = \min_{x_n}\{f_{n-1}(p_{n-1}) + x_n^2\} \tag{8.11.93}$$

We will work by induction.

1. Justify the relation (8.11.93). Relate p_{n-1} and p_n.
2. Perform the study for $n = 1$. Deduce x_1.
3. Perform the study for $n = 2$. Deduce x_2.
4. Perform the study for $n = 3$. Deduce x_3.
5. (a) If you guess the recurrence relation at that level: Pose it, suppose it true until $n - 1$, and deduce the relation for n.
   (b) It is quite possible that the iteration 3 is not sufficient to guess the recurrence. In this case, make the fourth iteration. At that moment, the recurrence should be obvious. Pose the recurrence relation, suppose it true until $n - 1$, and deduce the relation for n.
6. Deduce the general relation.

References

J. Abadie and J. Carpentier. Generalization of the Wolfe reduced-gradient method to the case of nonlinear constraints. In R. Fletcher, editor, *Optimization*, pages 37–49. Academic Press, London, 1969.

G. Allaire. Numerical Analysis and optimization. Oxford University Press, Oxford, 2007.

M. Avriel. *Nonlinear programming: Analysis and methods*. Prentice-Hall, Englewood Cliffs, New Jersey, 1976.

R. Bellman. *Dynamic Programming*. Princeton University Press, Princeton, New Jersey, 1957.

R. Bellman and S. Dreyfus. *Applied Dynamic Programming*. Princeton University Press, Princeton, New Jersey, 1962.

D. P. Bertsekas. *Convex Optimization Theory*. Athena Scientific, Belmont, 2009.

G. S. G. Beveridge and R. S. Schechter. *Optimization: theory and practice*. Mc Graw Hill, New York, 1970.

F. Bicking, C. Fonteix, J. P. Corriou, and I. Marc. Global optimization by artificial life: a new technique using genetic population evolution. *Recherche Opérationnelle/Operations Research*, 28(1):23–36, 1994.

J. F. Bonnans, J. C. Gilbert, C. Lemaréchal, and C. A. Sagastizabal. *Numerical optimization*. Springer, Berlin, 2003.

S. Boyd and L. Vandenberghe. *Convex optimization*. Cambridge University Press, New York, 2009.

M. Caracotsios and W. E. Stewart. Sensitivity analysis of initial value problems with mixed odes and algebraic equations. *Comp. Chem. Engn.*, 9(4):359–365, 1985.

E. K. P. Chong, , and S. H. Zak. *An introduction to optimization*. Wiley, New York, 4th edition, 2014.

J. C. Culioli. *Introduction à l'optimisation*. Ellipses, Paris, 1994.

J. E. Dennis and R. Schnabel. *Numerical Methods for Unconstrained Optimization and Nonlinear Equations*. Prentice-Hall, Englewood Cliffs, 1983.

W. F. Feehery, J. E. Tolsma, and P. I. Barton. Efficient sensitivity analysis of large-scale differential-algebraic systems. *Applied Numerical Mathematics*, 25:41–54, 1997.

R. Fletcher. *Practical Methods of Optimization*. Wiley, Chichester, 1991.

L. R. Foulds. *Optimization Techniques: an Introduction*. Springer-Verlag, New York, 1981.

P. E. Gill, W. Murray, and M. H. Wright. *Practical optimization*. Academic Press, London, 1981.

F. John. Extremum problems with inequalities as side conditions. In K. O. Friedrichs, O. W. Neugebauer, and J. J. Stoker, editors, *Studies and Essays: Courant Anniversary Volume*, pages 187–204. Interscience Publishers, New York, 1948.

W. Karush. Minima of functions of several variables with inequalities as side conditions. Master's thesis, University of Chicago, Chicago, Ill, 1939.

H. W. Kuhn and A. W. Tucker. Nonlinear programming. In *Proc. Second Berkeley Symp. Mathematics, Statisticas and Probability*, pages 481–492, Berkeley, CA, 1951. University of California Press.

S. Li and L. Petzold. Description of DASPKADJOINT: an adjoint sensitivity solver for differential-algebraic equations. Technical report, UCSB, www.engineering.ucsb.edu.cse, 2002.

D. G. Luenberger and Y. Ye. *Linear and Nonlinear Programming*. Springer, Heidelberg, 4th edition, 2016.

J. Nocedal and S. J. Wright. *Numerical optimization*. Springer, New York, 2nd edition, 2006.

L. Petzold, S. T. Li, Y. Cao, and R. Serban. Sensitivity analysis of differential-algebraic equations and partial differential equations. *Comp. Chem. Engng.*, 30:1553–1559, 2006.

E. Polak. *Optimization - Algorithms and consistent approximations*. Springer, New York, 1997.

S. S. Rao. *Engineering Optimization - Theory and Practice*. John Wiley, New York, 4th edition, 2009.

W. H. Ray and J. Szekely. *Process Optimization with applications in metallurgy and chemical engineering*. John Wiley, New York, 1973.

J. B. Rosen. The gradient projection method for nonlinear programming, part I - Linear constraints. *SIAM J. Appl. Math.*, 8:181–217, 1960.

J. B. Rosen. The gradient projection method for nonlinear programming, part II - Nonlinear constraints. *SIAM J. Appl. Math.*, 9:514–532, 1961.

R. W. H. Sargent. Reduced-gradient and projection methods for nonlinear programming. In P. E. Gill and W. Murray, editors, *Numerical methods for constrained optimization*, pages 149–174. Academic Press, London, 1974.

H. P. Schwefel. *Numerische Optimierung von Computer-Modellen mittels der Evolutionsstrategie*. 26, Interdisciplinary System Research, Birkhäuser, Basel, 1977.

J. A. Snyman and D. N. Wilke. *Practical Mathematical optimization - Basic Optimization Theory and Gradient-based Algorithms*. Springer, Cham, 2nd edition, 2018.

Chapter 9
Numerical Methods of Optimization

9.1 Introduction

Analytical methods represent an important theoretical base for optimization, but, in general, optimization is performed by means of iterative numerical methods (Kelley 1999b), especially when it deals with large scale problems such as those encountered in engineering (Edgar and Himmelblau 2001; Rao 2009; Ray and Szekely 1973) and economics. Numerical optimization can take on different forms such as *linear programming* (Chapter 10) and *quadratic programming* (Chapter 11), but the most general problem of optimization of a nonlinear function of several variables subject to nonlinear constraints is called *nonlinear programming* (Chapter 11). Programming here means optimization.

9.2 Functions of One Variable

In the case of an optimization problem without constraints, for a function of one variable, the problem is to find the zero of the derivative

$$f'(x) = 0 \tag{9.2.1}$$

In Chapter 3, many methods have been described to find the zero of a function, for example, bisection (Section 3.6.1), secant, and Newton methods (Section 3.8). Some of these methods, bisection and regula falsi, require only the values of the function, while Newton's method imposes to know its derivative.

With respect to Chapter 3, the present problem consists in replacing the evaluation of the function by that of its derivative.

J.-P. Corriou, *Numerical Methods and Optimization*, Springer Optimization and Its Applications 187, https://doi.org/10.1007/978-3-030-89366-8_9

9.2.1 Bisection Method

In the case of root-finding for a function, the bisection method required the knowledge of three values of the function at each step (Section 3.6.1). In the case of the search of the optimum of a unimodal function, thus having only one minimum or maximum in an initial interval $[a, b]$, the derivative $f'(x)$ replaces the function $f(x)$. If a numerical approximation of the derivative is used instead of the analytical expression of the derivative, three numerical evaluations of the derivative will be necessary which imply the calculation of the function for each of them at two different points. Thus, this type of method requires at least six values of the function at each step, two of them being new ones.

Consider the search of a minimum. The bisection method is thus achievable according to the following algorithm:

1. Start from an interval $[a, b]$ to which this minimum belongs. Call $[x_{L0}, x_{U0}]$ this interval.
2. At iteration k, calculate x_m middle of $[x_{Lk}, x_{Uk}]$. The value of the derivative $f'(x)$ is estimated at each of the 3 points by a central divided difference formula

$$f'(x) \approx \frac{f(x(1+\epsilon)) - f(x(1-\epsilon))}{2\, x\, \epsilon} \tag{9.2.2}$$

3. In the case of a minimum

   (a) If $f'(x_{Lk}) f'(x_m) < 0$, then the new interval $[x_{L,k+1}, x_{U,k+1}]$ is $[x_{Lk}, x_m]$.
   (b) If $f'(x_m) f'(x_U) < 0$, then the new interval $[x_{L,k+1}, x_{U,k+1}]$ is $[x_m, x_U]$.

4. Verification of the tolerance on x and on f and the number of iterations. In the case of nonsatisfaction, return to step 2. Otherwise, go to step 5.
5. Estimation of the second derivative according to the central divided difference formula

$$f''(x) \approx \frac{f(x(1+\epsilon)) - 2 f(x) + f(x(1-\epsilon))}{(x\, \epsilon)^2} \tag{9.2.3}$$

   to determine the nature of the optimum.

Note that, in the case of a unimodal function, it would be possible to operate a bisection method which would only use the values of the function f at five points regularly spaced on the studied interval, in a manner similar to Fibonacci's method (Section 9.2.3).

Example 9.1 :
Optimization by the bisection method

The bisection method is used to search the minimum of the following function of one variable:

$$f(x) = \sqrt{\exp((x - 0.5)^2)} - x^2 \tag{9.2.4}$$

represented in Figure 9.1.

Using the search algorithm of an optimum by the bisection method, described in Section 9.2.1, the results of Table 9.1 are obtained with an acceptable precision in about 10 iterations

---

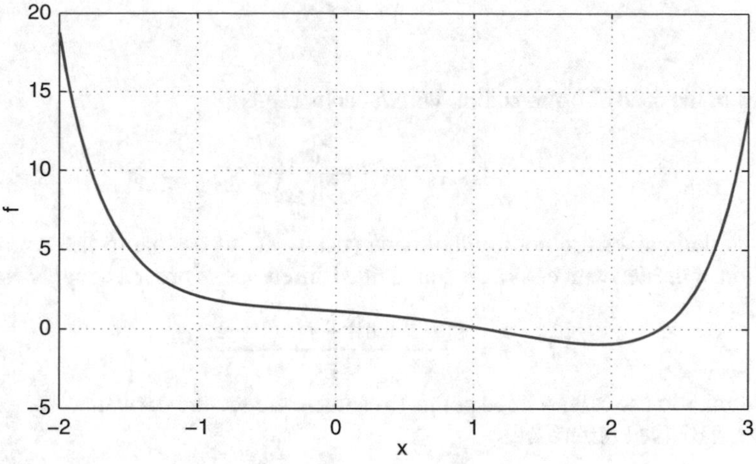

Fig. 9.1 Function of one variable used for the search of a minimum

Table 9.1 Bisection method: search of the minimum of a function of one variable

k	x_L	x_U	$f_L(x)$	$f_U(x)$
0	−2.0000	3.0000	18.7599	13.7599
1	0.5000	3.0000	0.7500	13.7599
2	1.7500	3.0000	−0.8783	13.7599
3	1.7500	2.3750	−0.8783	0.1591
4	1.7500	2.0625	−0.8783	−0.8643
5	1.9063	2.0625	−0.9459	−0.8643
6	1.9063	1.9844	−0.9459	−0.9285
7	1.9063	1.9453	−0.9459	−0.9424
8	1.9063	1.9258	−0.9459	−0.9453
9	1.9063	1.9160	−0.9459	−0.9459
10	1.9111	1.9160	−0.9460	−0.9459
11	1.9111	1.9136	−0.9460	−0.9459
12	1.9111	1.9124	−0.9460	−0.9460
13	1.9111	1.9117	−0.9460	−0.9460
14	1.9114	1.9117	−0.9460	−0.9460
15	1.9116	1.9117	−0.9460	−0.9460

9.2.2 Newton's Method

In the case of Newton's method, the calculation of the function is replaced by the calculation of the first derivative and that of the first derivative by the calculation of the second derivative so that Newton's algorithm which was

$$x_{k+1} = x_k - \frac{f(x_k)}{f'(x_k)} \tag{9.2.5}$$

becomes in the case of optimization without constraints

$$x_{k+1} = x_k - \frac{f'(x_k)}{f''(x_k)} \tag{9.2.6}$$

When the derivatives are not available analytically, it is possible to use a numerical estimation of the derivatives as a central divided difference of order 2 to apply Newton's method

$$f'(x) \approx \frac{f(x(1+\epsilon)) - f(x(1-\epsilon))}{2\,x\,\epsilon} \tag{9.2.7}$$

which implies to use at least three points to estimate the second derivative, for example, as a central divided difference

$$f''(x) \approx \frac{f(x(1+\epsilon)) - 2f(x) + f(x(1-\epsilon))}{(x\,\epsilon)^2} \tag{9.2.8}$$

where ϵ is small with respect to 1 and the increase of x is calculated in a relative way.

Example 9.2 :
Optimization by Newton's method

Using Newton's method, we search the minimum of the following function of one variable:

$$f(x) = \sqrt{\exp((x-0.5)^2) - x^2} \tag{9.2.9}$$

This is the same example as 9.1 for comparison purpose.

The initial search point is $x_0 = -1$ and the iterative formula (9.2.6) is used with the analytical first and second derivatives. The results are gathered in Table 9.2. It can be noticed that the particular characteristics (Figure 9.1) of the function with a very flat minimum located in a flat domain but with strongly increasing limits, rendered the search relatively long with respect to the number of iterations. In this very peculiar case, Newton's method does not present a neat advantage with respect to the bisection method. In a very large majority of cases, Newton's method would converge more rapidly.

9.2.3 Fibonacci's Method

Consider the search on the interval $[a, b]$ of the optimum of a unimodal function f, thus having only one local minimum or maximum in this interval. Instead of proceeding like in bisection to the division of the interval $[a, b]$ into two equal disjoint subintervals, $[a, b]$ is divided into two equal subintervals with a common part and the part without interest is eliminated; then, the search goes on in the same manner.

To start, the first interval is noted $[a_1, b_1]$. Both points located inside this interval, x_1 and x_2, are such that $x_2 > x_1$. Thus, two subintervals $[a_1, x_2]$ and $[x_1, b_1]$ result with a recovering on $[x_1, x_2]$. The convenient subintervals (containing either the sought minimum or the maximum) will be used for the following iteration.

Table 9.2 Newton's method: search of the minimum of a function of one variable

k	x	$f(x)$
0	−1	2.0802
1	−0.6729	1.5366
2	−0.3106	1.2925
3	1.3619	−0.4050
4	4.1595	791.70
5	3.9058	315.01
6	3.6373	123.94
7	3.3524	47.21
8	3.0517	16.62
9	2.7401	4.78
10	2.4336	0.5624
11	2.1667	−0.6842
12	1.9879	−0.9267
13	1.9199	−0.9457
14	1.9117	−0.9460
15	1.9116	−0.9460

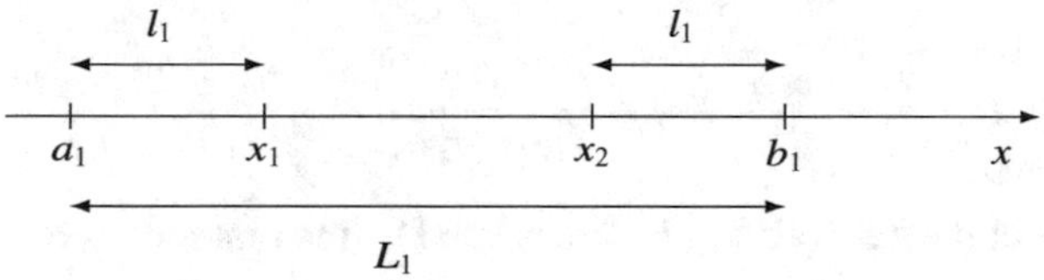

Fig. 9.2 Fibonacci's method

Define the length l_1 equal to

$$l_1 = x_1 - a_1 = b_1 - x_2 \tag{9.2.10}$$

while the initial interval $[a_1, b_1]$ has its length $L_1 = b_1 - a_1$.

The peculiarity of Fibonacci's method resides in the appropriate choice of this length l_1. The number of iterations to do must be given beforehand. Let N be that number. Fibonacci's numbers are defined by the following recurrence relation:

$$\begin{aligned} &F_i = F_{i-1} + F_{i-2}\,, \quad i = 1, \ldots, N \qquad \text{with } F_0 = 0,\ F_1 = 1, \\ &\text{thus } F_2 = 1,\ F_3 = 2,\ F_4 = 3,\ F_5 = 5,\ F_6 = 8, \ldots \end{aligned} \tag{9.2.11}$$

or according to Binet's formula

$$F_i = \frac{1}{\sqrt{5}}(\alpha^i - \beta^i) \qquad \text{with } \alpha = \frac{1+\sqrt{5}}{2} \quad \text{and } \beta = \frac{1-\sqrt{5}}{2} \tag{9.2.12}$$

where α and β are the roots of the characteristic equation of Fibonacci's sequence

$$x^2 - x - 1 = 0 \tag{9.2.13}$$

It can be noted that $\beta = -1/\alpha$ and α is called golden number, often noted by Φ. Numerically, $\alpha = 1.6180339887$. From the relation: $\alpha + \beta = 1$, it follows that $\beta = -0.6180339887$ presents the same decimal part. Equation (9.2.13) corresponds to the condition of division of a segment AB into two parts AC and CB such that AC is the geometric mean of AB and CB

$$\text{AC} = \sqrt{\text{AB}\,.\text{CB}} \tag{9.2.14}$$

It is possible to show that

$$\lim_{n\to\infty} \frac{F_{n+1}}{F_n} = \alpha \tag{9.2.15}$$

The total number N of iterations to do being fixed, the lengths L_1 and l_1 are such that

$$l_1 = L_1 - L_2 = (1 - \frac{F_{N-1}}{F_N})L_1 = \frac{F_{N-2}}{F_N}\,L_1 \tag{9.2.16}$$

from the recurrence relation (9.2.11), where L_2 is the length of interval at the end of the first iteration.

At the ith iteration, the subregion will be such that

$$l_i = \frac{F_{N-(i+1)}}{F_{N-(i-1)}}\,L_i \tag{9.2.17}$$

The lengths L_i of the intervals $[a_i, b_i]$ are related by the relation

$$L_i = \frac{F_{N-i+1}}{F_N} L_1\,, \quad i = 2, \ldots, N \tag{9.2.18}$$

and thus, the smallest possible length is

$$L_N = \frac{F_1}{F_N} L_1 = \frac{1}{F_N} L_1 \qquad \text{or} \quad \frac{L_N}{L_1} = \frac{1}{F_N} \tag{9.2.19}$$

which allows us to fix the number of iterations to achieve from the desired tolerance.

Suppose that the desired relative tolerance is $\epsilon = 10^{-5}$, and the initial bounds of the interval are $a = -2$ and $b = 3$ like in Example 9.3. Then, the required minimum length is $L_N = (b - a)\epsilon = 5\ 10^{-5}$. The number of iterations N is i such that

$$\frac{F_1}{F_i} > \epsilon > \frac{F_1}{F_{i+1}} \tag{9.2.20}$$

with F_i calculated by Equation (9.2.12), $F_1 = 1$, giving here $N = 25$ corresponding to $1/F_{25} = 1.33\ 10^{-5}$ and $1/F_{26} = 8.24\ 10^{-6}$. Thus, it ensures that

$$0 < x_{N,U} - x_{N,L} < L_N \tag{9.2.21}$$

Fibonacci's method is a method of sequential search considered theoretically as the most effective among search methods only requiring the values of the functions. This method converges much more rapidly than the bisection method. It is applicable to discrete functions as well as to continuous functions.

In the case of optimization, Fibonacci's method can work either by using the fact that the function is unimodal or by using the values of the derivative or of its estimation and by searching the zero of that derivative.

Consider the search of a minimum based on the hypothesis of a unimodal function. At iteration k, four points are available, successively $x_{k,L}$, $x_{k,1}$, $x_{k,2}$, and $x_{k,U}$, for which the value of the function is calculated. Then, the algorithm is as follows:

1. Search the minimum in the set $\{f(x_{k,L}), f(x_{k,1}), f(x_{k,2}), f(x_{k,U})\}$.
2. If this minimum is for $x = x_{k,L}$ or $x = x_{k,1}$, then $[x_{k+1,L}, x_{k+1,U}] = [x_{k,L}, x_{k,2}]$.
3. If this minimum is for $x = x_{k,2}$ or $x = x_{k,U}$, then $[x_{k+1,L}, x_{k+1,U}] = [x_{k,1}, x_{k,U}]$.

Example 9.3:
Optimization by Fibonacci's method

Using Fibonacci's method, we search the minimum of the following function of one variable:

$$f(x) = \sqrt{\exp((x-0.5)^2)} - x^2 \tag{9.2.22}$$

like in bisection Example 9.1 and Newton Example 9.2.

Using the search algorithm of an optimum described in Section 9.2.3, the results of Table 9.3 are obtained. The search ended when the tolerance on $f(x)$ was reached, although the tolerance on x was not yet reached because of the flat behavior of the function near its minimum.

Table 9.3 Fibonacci's method: search of the minimum of a function of several variables

k	x_L	x_U	$f_L(x)$	$f_U(x)$
0	−2.0000	3.0000	18.7599	13.7599
1	−0.0902	3.0000	18.7599	13.7599
2	1.0902	3.0000	1.1821	13.7599
3	1.0902	2.2705	0.0018	13.7599
4	1.5410	2.2705	0.0018	−0.3613
5	1.8197	2.2705	−0.6555	−0.3613
6	1.8197	2.0983	−0.9225	−0.3613
7	1.8197	1.9919	−0.9225	−0.8160
8	1.8854	1.9919	−0.9225	−0.9246
9	1.8854	1.9512	−0.9439	−0.9246
10	1.8854	1.9261	−0.9439	−0.9410
11	1.9010	1.9261	−0.9439	−0.9453
12	1.9010	1.9165	−0.9456	−0.9453
13	1.9069	1.9165	−0.9456	−0.9459
14	1.9069	1.9128	−0.9459	−0.9459
15	1.9092	1.9128	−0.9459	−0.9459
16	1.9106	1.9128	−0.9459	−0.9459
17	1.9106	1.9120	−0.9459	−0.9459

9.3 Functions of Several Variables

All methods lie on several common principles:

- Selection of a basis point: a set of feasible values of the variables must be found, that is belonging to the domain and satisfying the constraints.
- Calculation of the objective function at this point.
- Choice of a second feasible point according to a given method.
- Calculation of the objective function at this point.
- Comparison of the value of the objective function at the second point to that at the basis point.
- If the second point is better, it constitutes the new basis. If the initial point is better, modify the search direction or of the strategy or stop.

Two main classes of methods are distinguished, the methods of direct search and the gradient methods.

9.4 Methods of Direct Search

The methods of *direct search* do not necessitate the knowledge of the gradient of the objective function (Fletcher 1965; Lewis et al. 2000). It is sufficient to know the value of the function at some points depending on the chosen method. Most of them are progressively abandoned for the benefit of methods with a stronger theoretical base and more effective (conjugate gradients and quasi-Newton for example).

9.4.1 Simple One Variable Search

The simple one variable search consists in alternately modifying the value of only one variable, for example a displacement parallel to Ox_1 axis, then parallel to Ox_2 axis, and so on.

Let u_i be the unit vectors of the Cartesian plane. Note ξ the search direction. A new point $\boldsymbol{x}_{j+1}$ is obtained from the previous point $\boldsymbol{x}_j$ by the relation

$$\boldsymbol{x}_{j+1} = \boldsymbol{x}_j + a_j \, \xi_j \tag{9.4.1}$$

If initially $\xi_1 = u_1$, there remains the choice of the scalar a_1 for which no information exists. The method will continue according to the general indications expressed at the beginning: comparison of the performance of the new point with respect to the old one, change of direction, …

9.4.2 Simplex Method

This basic simplex differs slightly from the simplex developed for linear programming in Chapter 10. Generally speaking, the simplex is an n-dimensional convex polytope bounded by $n + 1$ hyperplanes (Matousek and Gärtner 2007). Furthermore, in Chapter 10, the simplex is more or less identified with the simplex method.

Two-dimensional simplex

The two-dimensional case is perfect to explain how this simplex works.

Consider a regular geometric figure, called *simplex* (Walters et al. 1991), as the base. In the two-dimensional case, the simplex is an equilateral triangle. Then, we proceed according to a set of rules (Figure 9.3):

(a) Calculate the objective function at the vertices of the triangle.
(b) Rule 1: Reject the worst point (from the point of view of the minimum or the maximum of the objective function). Replace it by its symmetrical with respect to the centroid of the other two points.
(c) Calculate the objective function at this new point.

If this new point gives a better result than the rejected point, this new point is adopted and forms the third point of the simplex. Restart in b/. Thus, it goes away from the bad direction.

Rule 2: If this new point gives a result worse than the rejected point, keep it and choose the second worse point as the rejected point, and then act like in b/ by replacing the rejected point by its symmetrical with respect to the centroid of the other two points.

Rule 3: It may also happen that the new point is not feasible, as it crosses one or several constraints. In this case, it cannot be accepted. The second worse point will be chosen as the worst point.

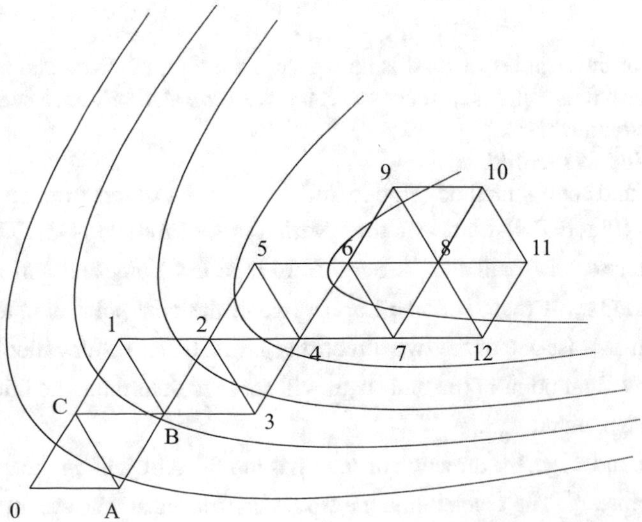

Fig. 9.3 Simplex method

It can be noticed that, at the end, the figure, an equilateral triangle, turns around the optimum and that there is no more progression toward the optimum, which is a drawback of this method. In this case, the size of the simplex must be reduced, for example, by decreasing by one half the length of a side of the triangle (Rule 4).

The search is stopped when it is estimated that the vertex is localized with enough precision.

Multidimensional simplex

In 2 dimensions, the figure is simply an equilateral triangle. In 3 dimensions, it will be a regular tetrahedron. In n dimensions, it will be a regular polyhedron with $n+1$ vertices. The rules are of the same type as in 2 dimensions.

An initial base figure must be formed by $n+1$ points $\boldsymbol{x}_j$ of coordinates $x_{i,j}$. Then, the worst point is detected and noted $\boldsymbol{x}_R$. This point is replaced by its new point $\boldsymbol{x}_N$ symmetrical with respect to the centroid noted $\boldsymbol{x}_G$ of the n other points.

$$\boldsymbol{x}_{i,G} = \frac{1}{n}\left(\sum_{j=1}^{n+1} x_{i,j} - x_{i,R}\right) \tag{9.4.2}$$

hence the coordinates of the new point

$$\boldsymbol{x}_{i,N} = \frac{2}{n}\left(\sum_{j=1}^{n+1} x_{i,j} - x_{i,R}\right) - x_{i,R} \tag{9.4.3}$$

It must be verified that this point gives a better result than the rejected point and that it is in the feasible domain in the same way as for the two-dimensional simplex.

9.4.3 Acceleration Methods

A drawback of the simplex method is its rigidity: the figure is fixed and no account is taken of the previous failures or successes. *Acceleration methods* have been developed to rectify these drawbacks.

Hooke and Jeeves method

In Hooke and Jeeves method (Hooke and Jeeves 1961), start from an initial point $\boldsymbol{x}_1$ noted $\boldsymbol{x}_0^{(1)}$ (Figure 9.4). Start by a fixed variation $\pm\delta$ parallel to Ox_1. The first point giving a better result is called $\boldsymbol{x}_1^{(1)}$. Start from this new point and make a variation $\pm\delta$ parallel to Ox_2. If there is some improvement, the new point is noted $\boldsymbol{x}_2^{(1)}$. It is supposed that at least one of the two directions gave a better result called $\boldsymbol{x}_2^{(1)}$ distinct from $\boldsymbol{x}_0^{(1)}$. This final point of the first stage will serve to determine the initial point for the second stage, noted $\boldsymbol{x}_0^{(2)}$.

For the second stage, the direction of search is modified by joining the first point $\boldsymbol{x}_0^{(1)}$ to the new point $\boldsymbol{x}_2^{(1)}$. The acceleration method is introduced at this level by multiplying the distance from $\boldsymbol{x}_0^{(1)}$ to $\boldsymbol{x}_2^{(1)}$ by a factor α yielding the point $\boldsymbol{x}_0^{(2)}$

$$x_0^{(2)} = x_0^{(1)} + \alpha(x_2^{(1)} - x_0^{(1)}) = \alpha x_2^{(1)} + (1-\alpha)x_0^{(1)} \tag{9.4.4}$$

The recommended factor is $\alpha = 2$. From this point $x_0^{(2)}$, a search is made completely identical to the first stage (parallel to the axes with $\pm\delta$). A new better point $x_2^{(2)}$ will be obtained, which will serve by acceleration between $x_2^{(2)}$ and $x_0^{(2)}$ to find the initial point $x_0^{(3)}$ of the third stage, and so on.

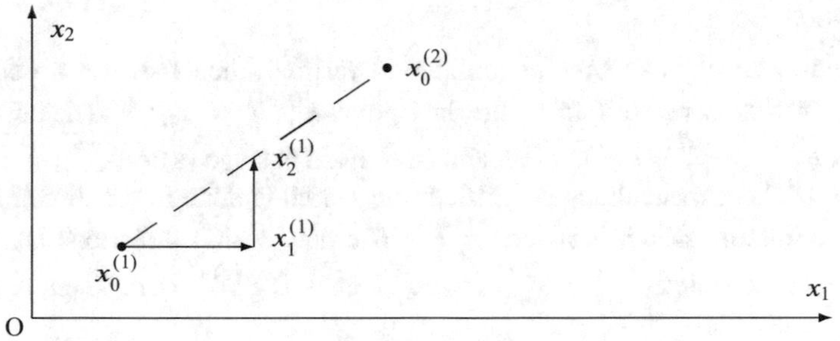

Fig. 9.4 Hooke and Jeeves method

In the case of a failure, that is when the final point of a stage is not better than the initial point, the search is started again at the very beginning, but in a cautious way, with a small step.

Hooke and Jeeves method has been modified by many authors. Furthermore, it can be easily generalized to a space of dimension n.

Powell's method

When a function is examined in the neighborhood of its optimum, frequently it is represented by an approximate quadratic function. Even if it is not always justified, Powell's method (Powell 1964) is based on that observation.

In n dimensions, the search follows the next stages:

(a) Suppose that the $(k+1)$th iteration starts from the point $x_0^{(k)}$ and that n search directions $\xi_1^{(k)}, \xi_2^{(k)}, \ldots, \xi_n^{(k)}$ are known and they are linearly independent (thus forming a base of the space). The search starts by seeking the optimum $x_1^{(k)}$ along a straight line passing by $x_0^{(k)}$ and parallel to $\xi_1^{(k)}$. Going from the optimum thus found, a new optimum $x_2^{(k)}$ is searched in the direction $\xi_2^{(k)}$ and so on for the n directions.

(b) Note $x_m^{(k)}$ the point for which the best improvement of the objective function f with respect to the previous point $x_{m-1}^{(k)}$ has been obtained. Let $\Delta = |f(x_m^{(k)}) - f(x_{m-1}^{(k)})|$ and $\mu = x_n^{(k)} - x_0^{(k)}$

(c) Calculate $f(2x_n^{(k)} - x_0^{(k)}) = f_t^{(k)}$

(d) Check if

$$f_t^{(k)} \geq f_0^{(k)} \tag{9.4.5}$$

(where $f_0^{(k)}$ is the value of the function f at point $x_0^{(k)}$) and/or if

$$(f_0^{(k)} - 2f_n^{(k)} + f_t^{(k)})(f_0^{(k)} - f_n^{(k)} - \Delta)^2 \geq \frac{\Delta(f_0^{(k)} - f_t^{(k)})^2}{2} \tag{9.4.6}$$

The first inequality is used to examine if the point $\boldsymbol{x}_t = 2\boldsymbol{x}_n^{(k)} - \boldsymbol{x}_0^{(k)}$ tested in the direction μ and placed at the same distance of $\boldsymbol{x}_n^{(k)}$ as the point $\boldsymbol{x}_0^{(k)}$ gives a better value of the objective function (here a minimum is sought).

The second inequality is used to examine if the function does not suddenly increase from the point $\boldsymbol{x}_n^{(k)}$ to the tested point $\boldsymbol{x}_t = 2\boldsymbol{x}_n^{(k)} - \boldsymbol{x}_0^{(k)}$ after having decreased from $\boldsymbol{x}_0^{(k)}$ to $\boldsymbol{x}_n^{(k)}$.

If at least one of these two inequalities is verified, then μ is not a good search direction. In this case, start from the last point $\boldsymbol{x}_0^{(k+1)} = \boldsymbol{x}_n^{(k)}$ and keep the same directions $\boldsymbol{\xi}_i^{(k+1)} = \boldsymbol{\xi}_i^{(k)} \forall i = 1, \ldots, n$ and start again at stage (a).

If none of these inequalities is verified, the search is done in the direction μ until finding a minimum which is noted $\boldsymbol{x}_0^{(k+1)}$. The new search directions are $\boldsymbol{\xi}_i^{(k+1)} = \boldsymbol{\xi}_i^{(k)}, i = 1, \ldots, m-1$; $\boldsymbol{\xi}_i^{(k+1)} = \boldsymbol{\xi}_{i+1}^{(k)}, i = m, \ldots, n-1$; $\boldsymbol{\xi}_n^{(k+1)} = \mu$. Then, repeat from stage (a).

Rosenbrock's method

Rosenbrock's method (Rosenbrock 1960) makes an acceleration both in direction and in distance.

To start the explanation of the method, first consider a two-dimensional example (Figure 9.5):

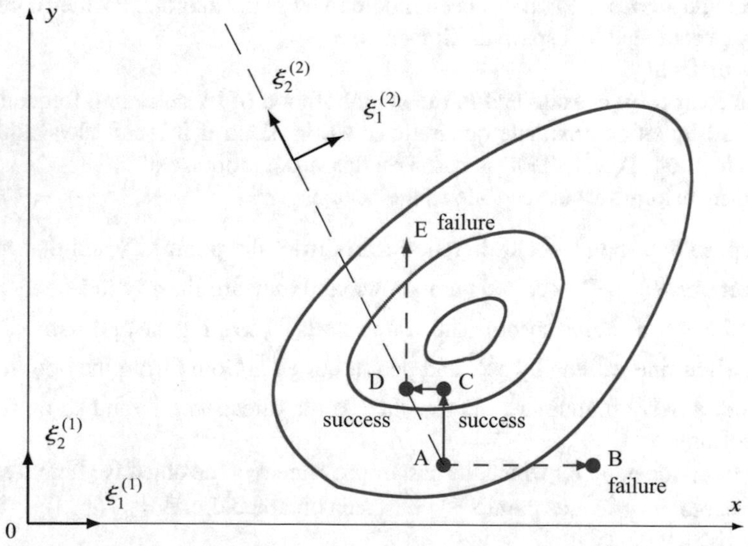

Fig. 9.5 Rosenbrock's method. Plain vectors correspond to a success and dotted vectors to a failure

- A stage is divided into cycles, each cycle corresponding to the successive sweeping parallel to the n unit vectors $\boldsymbol{\xi}$.

- Choose two step sizes s_1 and s_2, each corresponding to a direction.
- Consider the first stage. For the cycle noted $^{(1)}$, let a start point $\boldsymbol{x}_0^{(1)}$ be noted $A^{(1)}$. In this cycle $^{(1)}$, 2 tests are made around points successively noted B and C. First, a search is done in the direction of the unit vector $\xi_1^{(1)} = (1, 0)$ at a distance s_1 of $\boldsymbol{x}_0^{(1)}$; the point $\boldsymbol{x}_1^{(1)}$ or $B^{(1)}$ thus obtained is checked. Suppose that it gives a worse result (for the objective function) than $\boldsymbol{x}_0^{(1)}$, then at the following cycle for the point $D^{(1)}$, a test will be made in the direction $\xi_1^{(1)}$, opposite way $(-1, 0)$, at a lower distance βs_1 ($0 < \beta < 1$: deceleration, $\beta = 0.5$ recommended). As $B^{(1)}$ is worse than $A^{(1)}$, remain at $A^{(1)}$. Still in the cycle $^{(1)}$, make a test in the second direction $\xi_2^{(1)} = (0, 1)$ at a distance s_2 from point $A^{(1)}$. Suppose that the result of the new point $\boldsymbol{x}_2^{(1)}$ or $C^{(1)}$ is better than $\boldsymbol{x}_0^{(1)}$, then at the next cycle, a test is made for the point $E^{(1)}$ in the same direction $\xi_2^{(1)}$, same way $= (0, 1)$, at a larger distance αs_2 ($1 < \alpha$: acceleration, $\alpha = 3$ recommended). Start now from $C^{(1)}$, which is the best point of first cycle. The second cycle is done by testing the points $D^{(1)}$ and $E^{(1)}$. Suppose that $D^{(1)}$ is a success and $E^{(1)}$ a failure. The best point will be retained, that is, $D^{(1)}$. At least one success is necessary in each direction, which is done by doing a series of tests in one direction and the other one and alternating the way and decreasing the distance in the case of a failure. The first stage ends with at least one success and one failure in each direction $\xi_i^{(1)}$. The point $D^{(1)}$ will be the start point $A^{(2)}$ of the second stage in which a change of direction will be done.

  Example: To illustrate a series of tests with failures:

A test noted s_1 will mean a test in the direction $\xi_1^{(1)}$, $-s_1$ in the opposite way. A series can be

$$s_1,\ s_2,\ -\beta s_1,\ -\beta s_2,\ \beta^2 s_1,\ \beta^2 s_2,\ -\beta^3 s_1 \ldots \tag{9.4.7}$$

- For the second stage, a change of direction is done so that the new vector $\xi_1^{(2)}$ is in the direction and the way of the largest progression, the second vector $\xi_2^{(2)}$ being perpendicular. Show how $\xi_1^{(2)}$ is obtained.

  Let the vector

$$V_1^{(1)} = d_1\, \xi_1^{(1)} + d_2\, \xi_2^{(1)} \tag{9.4.8}$$

where d_i is the algebraic sum of all the successful moves in the direction $\xi_i^{(1)}$. This vector will be parallel to the new search direction $\xi_1^{(2)}$. The directional vectors $\xi_i^{(1)}$ are expressed with respect to their components

$$\xi_i^{(1)} = \{\xi_{1i}^{(1)}\,,\ \xi_{2i}^{(1)}\} \tag{9.4.9}$$

so that the components of $V_1^{(1)}$ are equal to

$$V_{i,1}^{(1)} = d_1\, \xi_{i1}^{(1)} + d_2\, \xi_{i2}^{(1)} \tag{9.4.10}$$

Coming back to the previous example, the search directions are $\xi_1^{(1)} = \{1, 0\}$ and $\xi_2^{(1)} = \{0, 1\}$. There was a success from A to C with $d_2 = s_2$ (direction 2) and success from C to D with $d_1 = -\beta s_1$

(direction 1). It results $V_1^{(1)} = \{-\beta s_1, s_2\}$. The new unit vector (direction 1) of the second stage will be the unit vector $\xi_1^{(2)}$ parallel to $V_1^{(1)}$.

As the new search direction is defined by the unit vector

$$\xi_1^{(2)} = \frac{V_1^{(1)}}{\|V_1^{(1)}\|} \tag{9.4.11}$$

the components of the search direction for the second stage can be expressed as

$$\xi_{i1}^{(2)} = \frac{V_{i,1}^{(1)}}{\sqrt{(V_{1,1}^{(1)})^2 + (V_{2,1}^{(1)})^2}} \tag{9.4.12}$$

and thus the new direction is

$$\xi_1^{(2)} = \{\xi_{11}^{(2)}, \xi_{21}^{(2)}\} \tag{9.4.13}$$

By means of Gram–Schmidt orthogonalization method, the search direction orthogonal to $\xi_1^{(2)}$ is associated

$$\xi_2^{(2)} = \frac{V_2^{(1)} - (\xi_1^{(2)} \cdot V_2^{(1)})\xi_1^{(2)}}{\|V_2^{(1)} - (\xi_1^{(2)} \cdot V_2^{(1)})\xi_1^{(2)}\|} \tag{9.4.14}$$

where $V_2^{(1)}$ is defined by

$$V_2^{(1)} = d_2 \xi_2^{(1)} \tag{9.4.15}$$

Coming back to the previous example, we have $d_2 = s_2$, $\xi_2^{(1)} = \{0, 1\}$ and consequently $V_2^{(1)} = \{0, s_2\}$.

The method is generalized to the case of an n-dimensional space.

At stage k:

- Choose n directions $\xi_i^{(k)}$ $(1 \leq i \leq n)$ and n step sizes. In general, for $k = 1$, the directions are parallel to the axes.
- Do a search by sweeping the n directions one after the other, keeping the found point if it is better and keeping the start point if the found point is worse. After a first sweeping, do a second one by multiplying the step by α if it is a success in the corresponding direction, by multiplying by $-\beta$ in the case of a failure. Continue the sweepings until having at least one success and one failure in each direction.
- For the new stage $k + 1$, do a change of direction parallel to the vectors $V_i^{(k)}$, which will define the unit vectors $\xi_i^{(k+1)}$ with the relation

$$V_i^{(k)} = \sum_{j=i}^{n} d_j \, \xi_i^{(k)} \tag{9.4.16}$$

where the d_j is the algebraic sum of all the successful moves in the direction $\xi_i^{(k)}$. The directing vectors $\xi_i^{(k+1)}$ are equal to

$$\xi_i^{(k+1)} = \frac{B_i^{(k)}}{\|B_i^{(k)}\|} \tag{9.4.17}$$

where the vectors $B_i^{(k)}$ are defined by the relations

$$\begin{cases} B_1^{(k)} = V_1^{(k)} \\ B_i^{(k)} = V_i^{(k)} - \sum_{j=1}^{i-1} (V_i^{(k)} . \xi_j^{(k+1)}) \, \xi_j^{(k+1)} \end{cases} \tag{9.4.18}$$

The procedure is continued until convergence to an optimum.

9.4.4 Nelder–Mead Simplex

Opposite to the simplex of Section 9.4.2, the Nelder–Mead simplex (Nelder and Mead 1965) is a deformable figure, called a polytope, which thus can adapt itself to the shape of the function for which an optimum is sought. It presents many similar points with the method of the regular simplex. This method of direct search has known a very large success (Gao and Han 2010; Kelley 1999a; Lagarias et al. 1998). It corresponds to the method "fminsearch" of Matlab. The method works relatively correctly when the dimension of the space is not too large.

Supposing that the search takes place in a space of dimension n and that the function is unimodal, the simplex is a geometric figure with n dimensions which is a convex polytope with $(n + 1)$ vertices, corresponding to a tetrahedron in three dimensions and a triangle in two dimensions. Starting from an initial point, Nelder–Mead algorithm generates a series of vertices to come closer to the sought optimum of the function f, assumed to be a minimum. It is supposed that at each iteration, the vertices of the simplex are ordered with respect to the value of the objective function as

$$f(\boldsymbol{x}_1) \leq f(\boldsymbol{x}_2) \leq \cdots \leq f(\boldsymbol{x}_{n+1}) \tag{9.4.19}$$

$\boldsymbol{x}_1$ is thus the best vertex and $\boldsymbol{x}_{n+1}$ the worst one. The algorithm uses four operations: reflection, expansion, contraction, and shrink, each of them being associated with a parameter, ρ for reflection, χ for expansion, γ for contraction, and σ for shrink (Figures 9.6 and 9.7).

Initially, the user must provide a nondegenerated simplex, with its $(n + 1)$ vertices. The centroid $\overline{\boldsymbol{x}}$ of the n best points is defined as

$$\overline{\boldsymbol{x}} = \frac{1}{n} \sum_{i=1}^{n} \boldsymbol{x}_i \tag{9.4.20}$$

An iteration of Nelder–Mead algorithm comprises the following stages:

1. Order
   Order the $(n + 1)$ vertices so that

$$f(\boldsymbol{x}_1) \le f(\boldsymbol{x}_2) \le \cdots \le f(\boldsymbol{x}_{n+1}) \tag{9.4.21}$$

and note $f_i = f(\boldsymbol{x}_i)$.

2. Reflection
   Calculate the reflected point with respect to the worst point as

$$\boldsymbol{x}_r = \overline{\boldsymbol{x}} + \rho(\overline{\boldsymbol{x}} - \boldsymbol{x}_{n+1}) \tag{9.4.22}$$

   Evaluate $f_r = f(\boldsymbol{x}_r)$. Then, apply the rule: if $f_1 \le f_r < f_n$, then replace $\boldsymbol{x}_{n+1}$ by $\boldsymbol{x}_r$ and terminate the iteration.

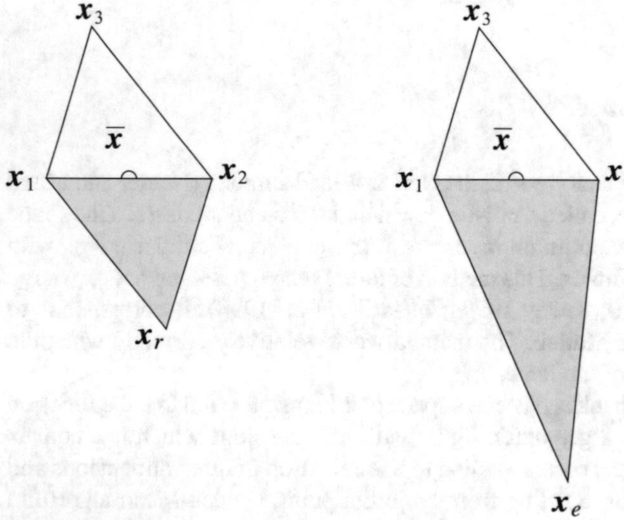

Fig. 9.6 Nelder–Mead algorithm: from left to right, from the simplex $\{\boldsymbol{x}_1, \boldsymbol{x}_2, \boldsymbol{x}_3\}$, the centroid $\overline{\boldsymbol{x}}$ of the best points, operations of reflection and expansion

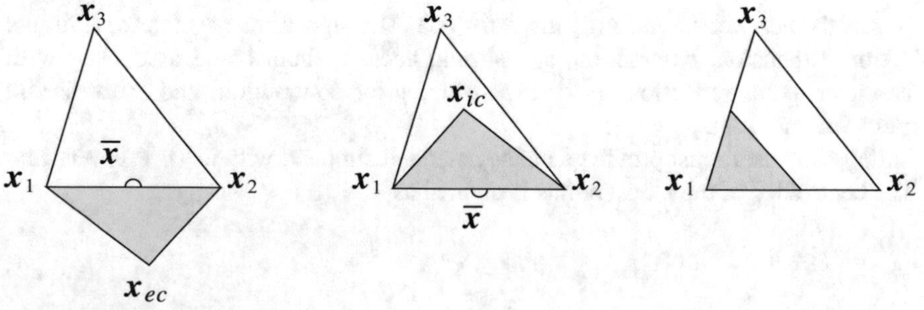

Fig. 9.7 Nelder–Mead algorithm: from left to right, from the simplex $\{\boldsymbol{x}_1, \boldsymbol{x}_2, \boldsymbol{x}_3\}$ and the centroid $\overline{\boldsymbol{x}}$ of the best points, operations of external contraction, internal contraction, shrink

3. Expansion
   If $f_r < f_1$, calculate the expansion point $\boldsymbol{x}_e$ such that

$$\boldsymbol{x}_e = \overline{\boldsymbol{x}} + \chi(\boldsymbol{x}_r - \overline{\boldsymbol{x}}) \tag{9.4.23}$$

   Evaluate $f_e = f(\boldsymbol{x}_e)$. If $f_e < f_r$, replace $\boldsymbol{x}_{n+1}$ by $\boldsymbol{x}_e$ and terminate the iteration, otherwise (if $f_e \geq f_r$), replace $\boldsymbol{x}_{n+1}$ by $\boldsymbol{x}_r$ and terminate the iteration.
4. Contraction
   If $f_r \geq f_{n+1}$, do a contraction between $\overline{\boldsymbol{x}}$ and the best of points $\boldsymbol{x}_{n+1}$ and $\boldsymbol{x}_r$.

   (a) External contraction
   If $f_n \leq f_r < f_{n+1}$, calculate the external contraction as

$$\boldsymbol{x}_{ec} = \overline{\boldsymbol{x}} + \gamma(\boldsymbol{x}_r - \overline{\boldsymbol{x}}) \tag{9.4.24}$$

   Evaluate $f_{ec} = f(\boldsymbol{x}_{ec})$. If $f_{ec} \leq f_r$, replace $\boldsymbol{x}_{n+1}$by $\boldsymbol{x}_{ec}$ and terminate the iteration. Otherwise, proceed to the shrink stage.

   (b) Internal contraction
   If $f_r \geq f_{n+1}$, calculate the internal contraction point as

$$\boldsymbol{x}_{ic} = \overline{\boldsymbol{x}} - \gamma(\overline{\boldsymbol{x}} - \boldsymbol{x}_{n+1}) \tag{9.4.25}$$

   Evaluate $f_{ic} = f(\boldsymbol{x}_{ic})$. If $f_{ic} < f_{n+1}$, replace $\boldsymbol{x}_{n+1}$ by $\boldsymbol{x}_{ic}$ and terminate the iteration. Otherwise, proceed to the shrink stage.

5. Shrink
   For $2 \leq i \leq n+1$, calculate the new points $\boldsymbol{v}_i$

$$\boldsymbol{v}_i = \boldsymbol{x}_1 + \sigma(\boldsymbol{x}_i - \boldsymbol{x}_1) \tag{9.4.26}$$

   The simplex is then formed by the unordered points $\boldsymbol{x}_1, \boldsymbol{v}_2, \ldots, \boldsymbol{v}_{n+1}$.

The recommended values for the parameters are $\rho = 1$, $\chi = 2$, $\chi > \rho$, $0 < \gamma < 1$, $0 < \sigma < 1$. In a standard way, $\gamma = 1/2$, $\sigma = 1/2$ can be taken.

Lagarias et al. (1998) give additional rules in the case of equality of the functions, which were not present in the initial algorithm of Nelder and Mead (1965).

Example 9.4 :
Optimization by Nelder–Mead method

Search the minimum of the function f

$$f = (x_1 - 0.5)^2 + (x_2 - 1)^2 \tag{9.4.27}$$

This is the same as Example 9.5, but without constraints. The domain of variation of Example 9.5 has been kept to generate the points of the initial simplex. The initial point chosen is (0, 0) and the two other points of the first simplex have been generated randomly like in Box's method. Then, Nelder–Mead rules have been applied until satisfying a convergence criterion. Figure 9.8 clearly shows the evolution of the simplexes and their deformation toward the sought minimum (0.5, 1). As random numbers are used to generate the points during the search, if two different searches of the extremum of a given function are compared, the followed path is never the same.

9.4.5 Box Complex

The simplex method described in Section 9.4.2 presents drawbacks due to the rigidity of the geometric figure of the simplex. Nelder–Mead method (Nelder and Mead 1965) allowed to alleviate these drawbacks, but it is an algorithm without constraints. To improve that latter method, Box has developed the method of complex which is based on a flexible figure of k vertices ($k \geq n + 1$), n being dimension of the space (Box 1965).

Suppose that there are m constraints of the form $g_i(\boldsymbol{x}) \leq 0$.

Each vertex must satisfy all the constraints.

Moreover, all variables x_i must be lower and upper bounded

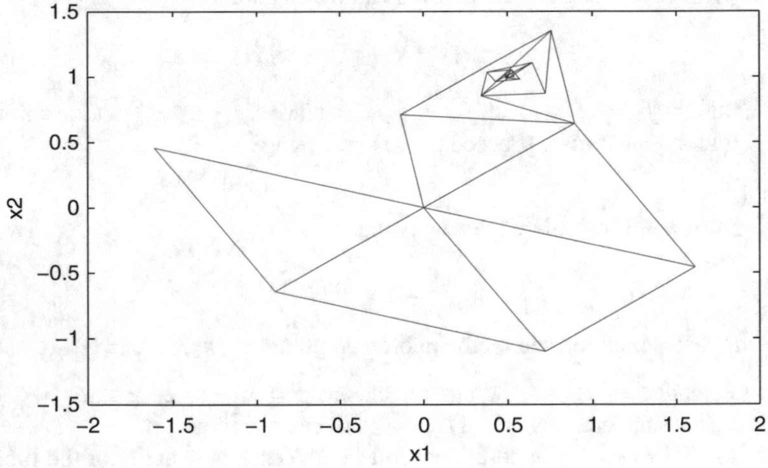

Fig. 9.8 Nelder–Mead method: evolution of the simplexes during the search

$$x_{i,L} \leq x_i \leq x_{i,U} \tag{9.4.28}$$

Start from a first point $\boldsymbol{x}_1$ satisfying all constraints. Generate the $(k - 1)$ missing points of the complex by the relations

$$x_{i,j} = x_{i,L} + r_{i,j}\,(x_{i,U} - x_{i,L})\,, \quad j = 2, \ldots, k\,, \quad i = 1, \ldots, n \tag{9.4.29}$$

where $r_{i,j}$ are random numbers between 0 and 1.

The points thus generated may not satisfy all the constraints.

Rule 1: If a constraint is violated, the point tried is moved half-way in direction of the centroid of the vertices already determined.

If the domain is convex, necessarily by this procedure (repeated if necessary), a convenient point will be obtained.

Repeat the generation until obtaining the k vertices (Box recommends $k = 2n$ especially if n is lower than 5). Evaluate the objective function at each vertex.
Rule 2: the worst point is rejected and replaced by a point located at a distance α ($\alpha \geq 1$) times far from the centroid of the remaining points with respect to the distance of centroid of the rejected vertex and in a direction pointing from the rejected vertex toward the centroid (Figure 9.9).

Let R be the rejected point and G the centroid of the remaining points (all except R). R is replaced by N so that

$$x_{i,N} = \alpha\,(x_{i,G} - x_{i,R}) + x_{i,G} \tag{9.4.30}$$

with the centroid defined by

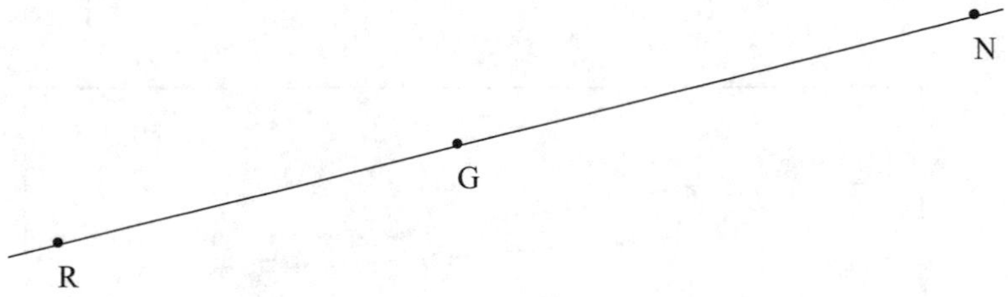

Fig. 9.9 Box complex: replacement of the rejected point R by the new point N

$$x_{i,G} = \frac{1}{k-1}\left(\sum_{j=1}^{k} x_{i,j} - x_{i,R}\right) \tag{9.4.31}$$

Box recommends $\alpha = 1.3$.

If N is feasible, evaluate the objective function at N and continue the process of rejection and regeneration of vertex.
Rule 3: If the new point N has a worse value than the other vertices of the new figure, N is rejected and moved mid-way in direction of the previous centroid, hence the new point N'

$$x_{i,N'} = \frac{1}{2}\,(x_{i,N} + x_{i,G}) \tag{9.4.32}$$

Rule 4: If a constraint in the possible domain is violated at any time, the vertex is again moved mid-way in direction of the centroid. If the constraint is still violated, again do this process until a success.

Still again, for a convex region, the process necessarily succeeds. For nonconvex regions, a problem occurs in particular when the vertex is outside the acceptable region.
Rule 5: If a generated point does not satisfy one of the explicit constraints on x_i (lower and upper bounds), this variable is modified to a low distance inside the domain to give a feasible point.

Many optimization methods by direct search are variants of Box complex method.

Example 9.5 :
Optimization by Box's complex

Search the minimum of function f

$$f = (x_1 - 0.5)^2 + (x_2 - 1)^2 \tag{9.4.33}$$

subject to the implicit constraint

$$g_1 = x_1^2 + 2\,x_2^2 - 4 \le 0 \tag{9.4.34}$$

and to explicit constraints (domain on x_i):

$$-2 \le x_1 \le 2\,, \quad -\sqrt{2} \le x_2 \le \sqrt{2} \tag{9.4.35}$$

To generate the first 4 points, random numbers noted r_1 and r_2 must be used.

Figure 9.10 represents all the generated points during the successive stages starting from (0, 0). The "random" character of the generation of the successive points during the search can be noted. The function converges to the value 0 at (0.5; 1) (Figure 9.10 and Table 9.4). Due to the use of random numbers, if two different searches of the extremum of a given function are compared, the followed path is never the same.

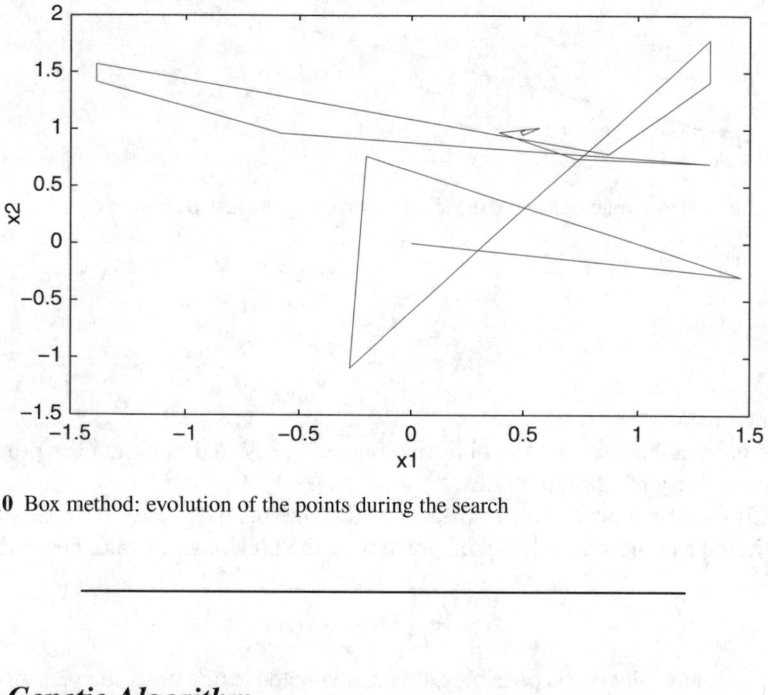

Fig. 9.10 Box method: evolution of the points during the search

9.4.6 Genetic Algorithm

Genetic or evolutionary algorithms (Brownlee 2011; De Jong 2006; Deb 2001; Eiben and Smith 2007; Goldberg 1989, 2002; Holland 1992; Poli et al. 2008) are frequently used to estimate the optimum of a function according to an approach which is qualified as quasi-global, that is the space of variables is sufficiently explored so that we can hope to have obtained the global optimum and not only a local optimum. When this optimum is correctly estimated, it is then possible to continue with a gradient type method to

Table 9.4 Box complex method: search of the minimum of a function subject to constraints

n	r_1	r_2	x_1	x_2	f	g_1
1	0.000	0.000	0.0000	0.0000	1.250	−4.000
2	0.865	0.396	1.460	−.2941	2.596	−1.695
3	0.449	0.770	−.2040	0.7636	0.5515	−2.792
4	0.431	0.113	−.2760	−1.094	4.989	−1.528
n			x_1	x_2	f	g_1
5			1.322	1.783	1.288	4.103
6			1.322	1.413	0.8455	1.738
7			0.8702	0.7845	0.1835	−2.012
8			−1.387	1.569	3.886	2.849
9			−1.387	1.413	3.732	1.915
10			−.5826	0.9643	1.173	−1.801
20			1.319	0.6982	0.7620	−1.285
30			0.7298	0.7422	0.1193	−2.366
40			0.3926	0.9738	0.1223×10^{-1}	−1.949
50			0.5686	1.014	0.4886×10^{-2}	−1.622
60			0.4960	0.9514	0.2380×10^{-2}	−1.944
70			0.4879	0.9794	0.5691×10^{-3}	−1.843
80			0.4860	0.9824	0.5045×10^{-3}	−1.833
90			0.5041	0.9936	0.5693×10^{-4}	−1.771
100			0.4972	1.004	0.2353×10^{-4}	−1.737

improve the search. The inspiration of genetic algorithms comes from biology and they belong to the class of meta-heuristic optimization methods. Many variants of genetic or evolutionary algorithms have been developed, for example, swarm algorithms.

Such a method is described in the following procedure corresponding to the minimization of a function in the absence of constraints. Population will be successively created with evolution mechanisms.

Algorithm:

1. Creation of the initial population: The counter indicating the number of the population is initialized at $i_p = 1$ (first population). Suppose that the space of variables x is of dimension n. An initial population of size m is created so that each individual x^j (j subscript for the individual) belongs to the prescribed initial domain:

$$x_{i,\min} \le x_i^j \le x_{i,\max} \tag{9.4.36}$$

Each individual is generated by using a random number r_i^j belonging to the interval [0, 1] such that each coordinate is obtained as

$$x_i^j = x_{i,\min} + (x_{i,\max} - x_{i,\min})\, r_i^j \tag{9.4.37}$$

2. For each individual x^j, the function f to be minimized is evaluated, and its value is equal to f^j. The set of coordinates x_i^j and values of the functions is gathered in an array *tab* of dimension $(m, n+1)$, where each row corresponds to an individual j, the n first columns being occupied by the coordinates x_i^j, the column $n+1$ being occupied by the value of function f^j.

3. An ordering is done in the array *tab* so that the values of the function f are ordered increasingly. While reordering the functions, the associated individuals are also reordered. After reorder, it gives

$$f^1 \le f^j \le f^m \qquad \forall\ 1 < j < m \tag{9.4.38}$$

4. This stage corresponds to the stop tests. Two stop tests are integrated:

   (a) Test if the number of population i_p is larger than the maximum number of population chosen by the user:

   $$i_p > i_{p,\max} \quad ? \tag{9.4.39}$$

   If this is the case, stop the algorithm.

   (b) Test if the difference between the maximum value and the minimum value of the function for all the population is lower than some threshold ϵ fixed by the user

   $$(f_{\max} - f_{\min}) < \epsilon \quad ? \tag{9.4.40}$$

   If it is the case, stop the algorithm.

5. Evolution of the population by mutation and growth.
A threshold of mortality s $(0 < s < 1)$ is fixed so that a fraction s of the population giving the lower values of the function f is kept, and thus the fraction $(1 - s)$ of the population giving higher values of f is eliminated. The total number of births compensating the mortality will thus be equal to $m(1 - s)$. Among these births, a low fraction β (for example 10%) is taken as a mutation.

   (a) Mutation
   The mutants are created by Equation (9.4.37), but only the mutants for which the function f is lower than the largest value f of the surviving individuals, that is $tab(sm, n + 1)$, are kept. In this way, the accepted mutants are generally a number m_{mutant} lower than the possible maximum equal to $m(1 - s)\beta$.

   (b) Births
   A number of children equal to

   $$m(1 - s) - m_{mutant} \tag{9.4.41}$$

   are created according to

   $$child_j^i = \alpha\, tab(j_{p1}, j) + (1 - \alpha)\, tab(j_{p2}, j) \tag{9.4.42}$$

   where α is calculated as

   $$\alpha = \alpha_{\min} + (\alpha_{\max} - \alpha_{\min})\, r_j^i \tag{9.4.43}$$

   r_j^i being a random number in the interval [0, 1], whereas $\alpha_{\min}$ and $\alpha_{\max}$ surround the values 0 and 1, for example, $\alpha_{\min} = -0.2$ and $\alpha_{\max} = 1.2$. Moreover, j_{p1}^i and j_{p2}^i are the indexes of two parents belonging to the surviving population calculated as

$$j_{p1}^i = 1 + int(r_1^i\, s\, m)\,, \quad j_{p2}^i = 1 + int(r_2^i\, s\, m) \tag{9.4.44}$$

where r_1^i and r_2 are two random numbers in [0, 1] and *int* designates the integer part of a number. Only the children such that the corresponding function is lower than the larger value f of the surviving individuals, that is $tab(s * m, n + 1)$, are kept.

When all the births have been performed in such a way that the total number of individuals is again m, an order is done according to stage 3 and the new population is then obtained. Return to stage 4.

A genetic algorithm can also be used under a different form when a function is minimized in the presence of constraints.

Example 9.6 :
Optimization of Rosenbrock's function by a genetic algorithm

The genetic algorithm has been tested in the case of the minimization of Rosenbrock's function

$$f(\boldsymbol{x}) = \alpha(x_2 - x_1^2)^2 + (1 - x_1)^2 \tag{9.4.45}$$

with $\alpha = 100$ in the studied case. The optimum is at (1, 1). The contours of this function are represented in Figure 9.11. The domain has the banana shape and is not convex.

The genetic algorithm was applied with an initial population equal to 500 individuals. The dimension of the variable space is 2. The threshold of mortality was chosen equal to 60%. It is interesting to compare the contours of Figure 9.11 to the population obtained after 5 generations (Figure 9.12). In Figure 9.12, the progressive concentration of points around the theoretical solution located at (1, 1) can be noticed; however, the "banana" dispersion remains important even after 20 generations.

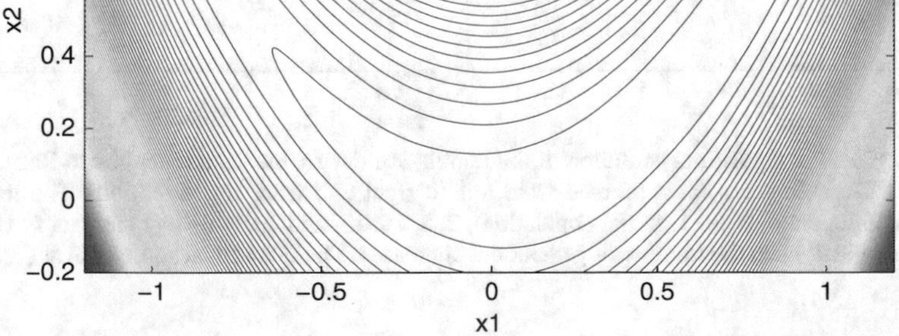

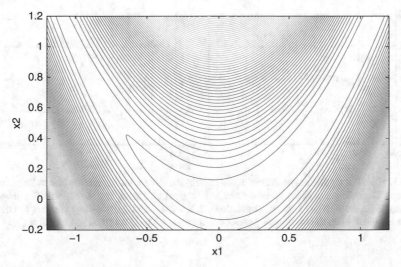

Fig. 9.11 Rosenbrock's function: contours

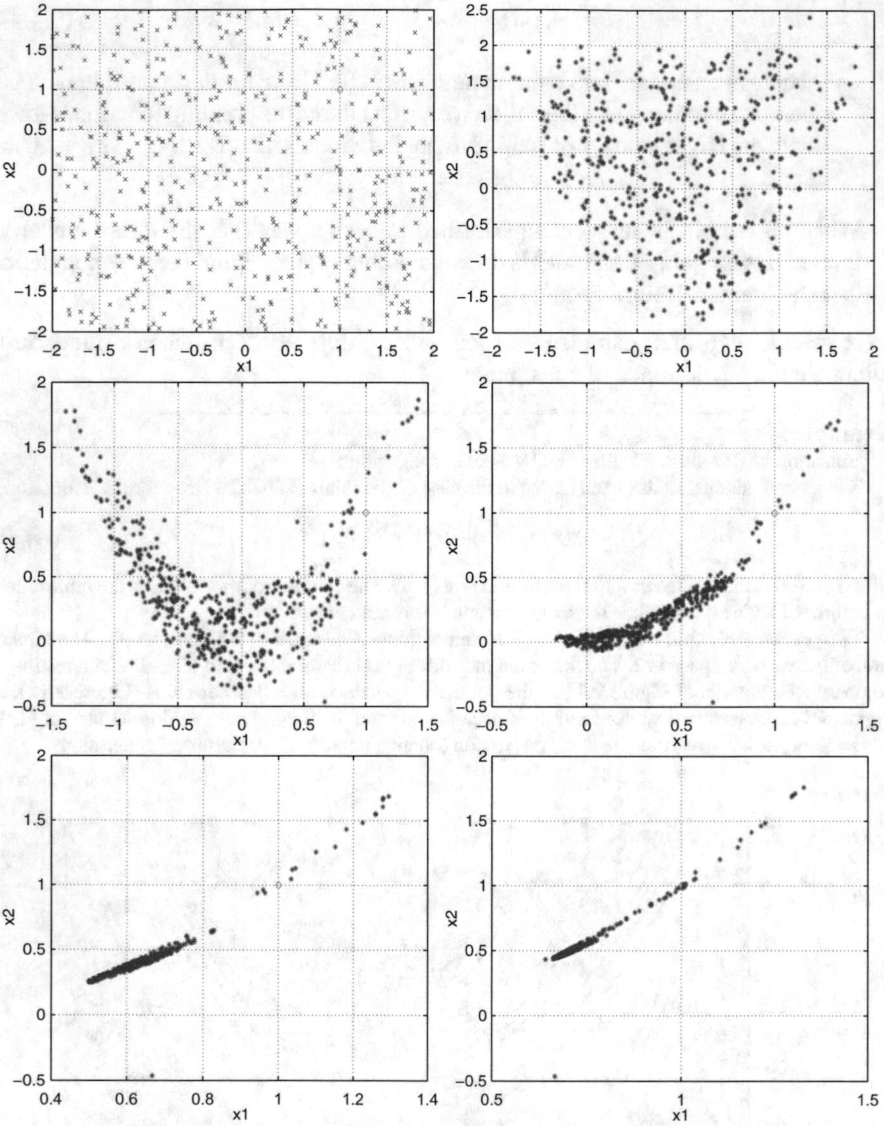

Fig. 9.12 Genetic algorithm: evolution of the population during the search applied to Rosenbrock's function. The subfigures must be read from left to right and from top to bottom. The respective numbers of generations are 1 (initial population), 2, 5, 10, 15, and 20. Note that the scales of the axes are not identical when the number of generations increases. The theoretical solution is at (1, 1)

9.5 Gradient Methods

All the *gradient methods* can be considered as stemming from the following model algorithm:
Let $\boldsymbol{x}_k$ be the estimation of $\boldsymbol{x}^*$, sought optimum.

1. Test of convergence. If the conditions of convergence are satisfied, the algorithm is stopped providing a solution $\boldsymbol{x}_k$.
2. Calculation of a search direction. Calculate a nonzero vector $\boldsymbol{p}_k$ that is the search direction.
3. Calculation of the displacement length. Calculate a scalar α_k that is the step length for which
$$f(\boldsymbol{x}_k + \alpha_k \, \boldsymbol{p}_k) < f(\boldsymbol{x}_k) \tag{9.5.1}$$
supposing that a minimum of f is sought.

In gradient methods, the displacement is done parallel to the *gradient* of the objective function (method of steepest descent), which gives the most pronounced variation of the function (Figure 9.13).

A usual condition ensuring that f decreases at the kth iteration is that $\boldsymbol{p}_k$ be a descent direction at k, that is a vector satisfying the condition

$$\nabla f(\boldsymbol{x}_k)^T \, \boldsymbol{p}_k < 0 \tag{9.5.2}$$

where $\nabla f(\boldsymbol{x}_k)$ is the gradient of f at $\boldsymbol{x}_k$. Often, in the following, the gradient $\nabla f(\boldsymbol{x}_k)$ will be simply noted g_k.

9.5.1 Case of a Quadratic Function

In the neighborhood of the optimum, the behavior of a gradient method for any function is in general similar to the one for a quadratic function (second degree Taylor polynomial for any function). For that reason, the case of a quadratic function for which the minimum is sought is first studied, thus of the form

$$f(\boldsymbol{x}) = \frac{1}{2} \boldsymbol{x}^T \, \boldsymbol{A} \, \boldsymbol{x} - \boldsymbol{h}^T \, \boldsymbol{x} + c \tag{9.5.3}$$

where $\boldsymbol{A}$ (Hessian matrix of f) is symmetrical definite positive.

The minimum $\boldsymbol{x}_0$ is solution of the equation

$$f_x(\boldsymbol{x}) = \boldsymbol{A} \, \boldsymbol{x} - \boldsymbol{h} = 0 \tag{9.5.4}$$

and thus equal to $\boldsymbol{x}_0 = \boldsymbol{A}^{-1} \, \boldsymbol{h}$, but the inverse of matrix $\boldsymbol{A}$ is supposed to be unknown, as it can be ill-conditioned.

The *residual* vector of f at $\boldsymbol{x}$ is defined by

$$\boldsymbol{r} = -f_x(\boldsymbol{x}) = \boldsymbol{A}(\boldsymbol{x}_0 - \boldsymbol{x}) \tag{9.5.5}$$

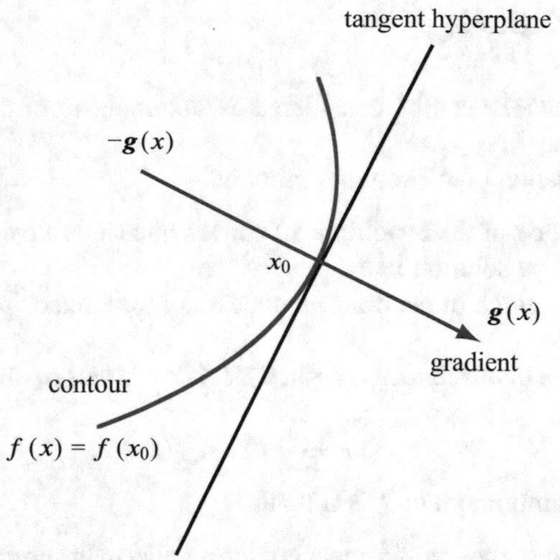

Fig. 9.13 Tangent hyperplane and gradient

Note that $\boldsymbol{r}$ is the *opposite of the gradient* and is thus in the direction of the *steepest descent*. Moreover, the residual is useful to quantify the variations of the function f in the neighborhood on a point. Indeed,

$$f(\boldsymbol{x} + \Delta\boldsymbol{x}) = f(\boldsymbol{x}) - \boldsymbol{r}^T \Delta\boldsymbol{x} + \frac{1}{2} \Delta\boldsymbol{x}^T A \Delta\boldsymbol{x} \tag{9.5.6}$$

Suppose that $\Delta\boldsymbol{x} = \boldsymbol{x}_0 - \boldsymbol{x}$ is chosen, it results

$$f(\boldsymbol{x}_0) = f(\boldsymbol{x}) - (\boldsymbol{x}_0 - \boldsymbol{x})^T \boldsymbol{A}^T (\boldsymbol{x}_0 - \boldsymbol{x}) + \frac{1}{2} (\boldsymbol{x}_0 - \boldsymbol{x})^T \boldsymbol{A} (\boldsymbol{x}_0 - \boldsymbol{x}) \tag{9.5.7}$$

that is

$$f(\boldsymbol{x}) - f(\boldsymbol{x}_0) = \frac{1}{2} (\boldsymbol{x} - \boldsymbol{x}_0)^T \boldsymbol{A} (\boldsymbol{x} - \boldsymbol{x}_0) = \frac{1}{2} \boldsymbol{r}^T \boldsymbol{A}^{-1} \boldsymbol{r} \tag{9.5.8}$$

Posing m and M the smallest and largest eigenvalues of $\boldsymbol{A}$, these latter verify the inequality

$$m|\boldsymbol{v}|^2 \leq \boldsymbol{v}^T \boldsymbol{A} \boldsymbol{v} \leq M|\boldsymbol{v}|^2 \quad \text{or} \quad \frac{|\boldsymbol{v}|^2}{M} \leq \boldsymbol{v}^T \boldsymbol{A}^{-1} \boldsymbol{v} \leq \frac{|\boldsymbol{v}|^2}{m} \quad \forall \boldsymbol{v} \tag{9.5.9}$$

and the inequality

$$\frac{4Mm}{(M+m)^2} \leq \frac{|\boldsymbol{v}|^4}{(\boldsymbol{v}^T \boldsymbol{A} \boldsymbol{v})(\boldsymbol{v}^T \boldsymbol{A}^{-1} \boldsymbol{v})} \leq 1 \tag{9.5.10}$$

By posing $\boldsymbol{v} = (\boldsymbol{x} - \boldsymbol{x}_0)$, using the left-hand side of (9.5.8) and the left-hand side of (9.5.9), the following inequality results

$$\frac{m}{2} |\boldsymbol{x} - \boldsymbol{x}_0|^2 \leq f(\boldsymbol{x}) - f(\boldsymbol{x}_0) \leq \frac{M}{2} |\boldsymbol{x} - \boldsymbol{x}_0|^2 \tag{9.5.11}$$

or still, by using the right-hand side of (9.5.8) and the right-hand side of (9.5.9), with respect to the residual

$$\frac{|\boldsymbol{r}|^2}{2M} \leq f(\boldsymbol{x}) - f(\boldsymbol{x}_0) \leq \frac{|\boldsymbol{r}|^2}{2m} \tag{9.5.12}$$

Let $\boldsymbol{x}$ be a vector of R^n, and the contour of f defined by $f(\boldsymbol{x}) =$ is an $(n-1)$-ellipsoid of center $\boldsymbol{x}_0$ (minimum of the surface). Pose

$$a = \frac{|\boldsymbol{r}|^2}{\boldsymbol{r}^T \boldsymbol{A}\boldsymbol{r}} = \arg\{\min_a f(\boldsymbol{x} + a\,\boldsymbol{r})\} \tag{9.5.13}$$

corresponding to the minimum of $f(\boldsymbol{x} + a\boldsymbol{r})$ with respect to a, intensity of the displacement in the direction $\boldsymbol{r}$. A geometric property of that ellipsoid is that if $\boldsymbol{x}$ is a point of the ellipsoid, $(\boldsymbol{x} + 2a\boldsymbol{r})$ belongs to the same contour of the ellipsoid. It results that

$$f(\boldsymbol{x}) = f(\boldsymbol{x} + 2a\boldsymbol{r}) > f(\boldsymbol{x} + \beta a\boldsymbol{r}) \geq f(\boldsymbol{x} + a\boldsymbol{r}) \qquad \text{with } 0 < \beta < 2 \tag{9.5.14}$$

Moreover, by taking δ such that

$$0 < \delta < 1 \qquad \text{and} \qquad \delta \leq \beta \leq 2 - \delta \tag{9.5.15}$$

it gives

$$f(\boldsymbol{x} + \beta a\boldsymbol{r}) - f(\boldsymbol{x}_0) \leq L[f(\boldsymbol{x}) - f(\boldsymbol{x}_0)] \quad \text{with} \quad L = 1 - \frac{\delta(2-\delta)4Mm}{(M+m)^2} \tag{9.5.16}$$

From this property, we deduce

$$-f(\boldsymbol{x} + a\boldsymbol{r}) + f(\boldsymbol{x}) \leq \left(\frac{M-m}{M+m}\right)^2 (f(\boldsymbol{x}) - f(\boldsymbol{x}_0)) \quad \forall \boldsymbol{x} \neq \boldsymbol{x}_0 \tag{9.5.17}$$

M/m, the ratio of the largest eigenvalue of $\boldsymbol{A}$ to its smallest eigenvalue, is the condition number of $\boldsymbol{A}$. If this ratio is close to 1, the matrix is well-conditioned, and if it is large, it is ill-conditioned.

In the case where the matrix **A** *is well-conditioned*, the *optimal gradient method* is efficient to find the minimum of f in the direction of the steepest descent. It is defined by the following recurrence:

$$\boldsymbol{x}_{k+1} = \boldsymbol{x}_k + a_k \boldsymbol{r}_k \tag{9.5.18}$$

with

$$\begin{cases} a_k &= \dfrac{|\boldsymbol{r}_k|^2}{\boldsymbol{r}_k^T \boldsymbol{A}\boldsymbol{r}_k} \\ \boldsymbol{r}_{k+1} &= -f_x(\boldsymbol{x}_{k+1}) = \boldsymbol{r}_k - a_k \boldsymbol{A}\boldsymbol{r}_k \end{cases} \tag{9.5.19}$$

starting from any $\boldsymbol{x}_1$, the first residual being defined by $\boldsymbol{r}_1 = \boldsymbol{h} - \boldsymbol{A}\boldsymbol{x}_1$.

The calculation is stopped when the modulus of the residual vector becomes lower to a given ϵ. a_k coefficient of the residual in the recurrence formula was determined by minimizing

$$\min_{a_k} f(\boldsymbol{x}_k + a_k \boldsymbol{r}_k) \tag{9.5.20}$$

Thus, it is an optimal procedure. Indeed, the residual $\boldsymbol{r}_{k+1}$ is orthogonal to $\boldsymbol{r}_k$ (as $\boldsymbol{r}_{k+1}^T \boldsymbol{r}_k = 0$).

When the matrix $\boldsymbol{A}$ *is ill-conditioned*, it may be interesting to use an algorithm of the *relaxed gradient* where, with respect to the algorithm of the optimal gradient, the only change comes from the choice of a_k

$$\boldsymbol{x}_{k+1} = \boldsymbol{x}_k + a_k\, \boldsymbol{r}_k \tag{9.5.21}$$

$$a_k = \beta_k\, a_{ko} \qquad \text{with } a_{ko} = \frac{|\boldsymbol{r}_k|^2}{\boldsymbol{r}_k^T \boldsymbol{A} \boldsymbol{r}_k}, \qquad 0 < \beta_k < 2 \tag{9.5.22}$$

When $\beta_k = 1$, it corresponds to the algorithm of the optimal gradient. By posing δ $(0 < \delta \le 1)$ such that $\delta \le \beta_k \le 2 - \delta$, and

$$L = 1 - \delta(2 - \delta)\frac{4Mm}{(M+m)^2} \tag{9.5.23}$$

it is shown that the convergence of f to the minimum x_0 is ruled by

$$f(\boldsymbol{x}_{k+1}) - f(\boldsymbol{x}_0) \le L^k\, [f(\boldsymbol{x}_1) - f(\boldsymbol{x}_0)] \tag{9.5.24}$$

so that $f(\boldsymbol{x}_k)$ tends linearly toward $f(\boldsymbol{x}_0)$ with a constant L and $\boldsymbol{x}_k$ tends linearly toward $\boldsymbol{x}_0$ with a constant $L^{0,5}$.

9.5.2 Case of a Non-quadratic Function

Suppose that the function f possesses a unique minimum $\boldsymbol{x}_0$ in a given search domain. The Hessian matrix $f_{xx}(\boldsymbol{x})$ is definite positive. It is possible to define, in a similar manner to the algorithm of optimal gradient for a quadratic function, the following algorithm of generalized gradient:

$$\boldsymbol{x}_{k+1} = \boldsymbol{x}_k + a_k\, \boldsymbol{H}_k\, \boldsymbol{r}_k \tag{9.5.25}$$

where $\boldsymbol{H}_k$ is a symmetric definite positive matrix and $\boldsymbol{r}_k$ is the residual vector equal to $\boldsymbol{r}_k = -f_x(\boldsymbol{x}_k)$. Indeed, this amounts to choose a quadratic model for the function f, which can be approximated by a second degree Taylor polynomial. This method is often effective in practice.

Whatever the vector $\boldsymbol{v}$, then, there exist positive real numbers K_1 and K_2 such that

$$K_1|\boldsymbol{v}|^2 \le \boldsymbol{v}^T\, \boldsymbol{H}_k\, \boldsymbol{v} \le K_2|\boldsymbol{v}|^2\,, \quad \frac{|\boldsymbol{v}|^2}{K_2} \le \boldsymbol{v}^T\, \boldsymbol{H}_k^{-1}\, \boldsymbol{v} \le \frac{|\boldsymbol{v}|^2}{K_1} \tag{9.5.26}$$

The scalar a_k must satisfy the following constraint:

$$f(\boldsymbol{x}_k + (a_k + a)\, \boldsymbol{H}_k\, \boldsymbol{r}_k) \le f(\boldsymbol{x}_k)\,, \quad \text{with } 0 < a \le a_k \tag{9.5.27}$$

where a is a small scalar.

Indeed, the previous constraint can be replaced by one of the four following constraints:

- $f(\boldsymbol{x}_k + 2\, a_k\, \boldsymbol{H}_k\, \boldsymbol{r}_k) \leq f(\boldsymbol{x}_k)\,, \quad \text{with } 0 < a \leq a_k$
- $f(\boldsymbol{x}_k + 2\, a_k\, \boldsymbol{H}_k\, \boldsymbol{r}_k) \leq f(\boldsymbol{x}_k) \leq f(\boldsymbol{x}_k + 4\, a_k\, \boldsymbol{H}_k\, \boldsymbol{r}_k)$
- $\alpha = a_k$ minimizes the function $\phi_k(\alpha) = f(\boldsymbol{x}_k + \alpha\, \boldsymbol{H}_k\, \boldsymbol{r}_k)$
- $a_k = -\phi'_k(0)/\phi''_k(0)$

The previous generalized algorithm can be presented under a form that does not use the matrix $\boldsymbol{H}_k$; it is then called *generalized descent algorithm* and is written as

$$\boldsymbol{x}_{k+1} = \boldsymbol{x}_k + a_k\, \boldsymbol{v}_k \tag{9.5.28}$$

where the vector $\boldsymbol{v}_k$ must satisfy the conditions

$$\boldsymbol{v}_k^T\, \boldsymbol{r}_k \geq \delta |\boldsymbol{v}_k|\, |\boldsymbol{r}_k|\,, \quad K_1 |\boldsymbol{r}_k| \leq |\boldsymbol{v}_k| \leq K_2 |\boldsymbol{r}_k| \tag{9.5.29}$$

with $0 < \delta \leq 1$ and $0 < K_1 \leq K_2$. The scalar a_k is always subject to the same constraint

$$f(\boldsymbol{x}_k + (a_k + a)\, \boldsymbol{H}_k\, \boldsymbol{r}_k) \leq f(\boldsymbol{x}_k)\,, \quad \text{with } 0 < a \leq a_k \tag{9.5.30}$$

where a is a small scalar.

This algorithm can be modified by replacing the previous constraint by one of the four constraints previously expressed about the algorithm of generalized gradient.

When the Hessian matrix is undefinite, we could think about replacing it by a close definite positive matrix, for example, by means of Cholesky's factorization method ($\boldsymbol{H} \rightarrow \boldsymbol{L}\boldsymbol{D}\boldsymbol{L}^T$, where $\boldsymbol{L}$ is lower triangular and $\boldsymbol{D}$ is positive diagonal). $\boldsymbol{D}$ and $\boldsymbol{L}$ are defined by

$$\begin{aligned} d_j &= h_{jj} - \sum_{s=1}^{j-1} d_s\, l_{js}^2 \\ l_{ij} &= \frac{1}{d_j} (h_{ij} - \sum_{s=1}^{j-1} d_s\, l_{js}\, l_{is}) \end{aligned} \tag{9.5.31}$$

However, two drawbacks are present; on one side, Cholesky's factorization of an undefinite matrix may not exist, and on another side, its calculus is numerically unstable. Under some conditions, modified Cholesky factorization may be necessary (Gill et al. 1981, pages 108–109).

9.5.3 Method of Steepest Descent

The *method of steepest descent* will be a descent for a minimization, an ascent for a maximization. Let f be the function to minimize (respectively, maximize). The problem is to find a small step ds such that the function decreases (respectively, increases) as much as possible. Noting dx_i the projections of ds on the axes, it verifies

$$ds^2 = \sum_{i=1}^{n} (dx_i)^2 \tag{9.5.32}$$

The corresponding variation of the function f is

$$df = \sum_{i=1}^{n} \left(\frac{\partial f}{\partial x_i}\right) dx_i \quad \Longrightarrow \left(\frac{df}{ds}\right) = \sum_{i=1}^{n} \frac{\partial f}{\partial x_i}\frac{dx_i}{ds} \tag{9.5.33}$$

df/ds must be minimized (respectively, maximized) with the constraint on $(ds)^2$, that is,

$$\begin{aligned} &\min_{\frac{dx_i}{ds}} \text{ (or max) } \frac{df}{ds} = \sum_{i=1}^{n} \frac{\partial f}{\partial x_i}\frac{dx_i}{ds} \\ &\text{subject to the constraint} \qquad ds^2 = \sum_{i=1}^{n} (dx_i)^2 \end{aligned} \tag{9.5.34}$$

This optimization yields the vector of components dx_i/ds. Then, the optimum of the Lagrangian function must be found

$$L = \sum_{i=1}^{n} \left(\frac{\partial f}{\partial x_i}\right) \frac{dx_i}{ds} + \lambda \left[1 - \sum_{i=1}^{n} \left(\frac{dx_i}{ds}\right)^2\right] \tag{9.5.35}$$

By differentiating with respect to the variable (dx_i/ds), it gives

$$\frac{\partial f}{\partial x_i} - 2\lambda \frac{dx_i}{ds} = 0 \Longrightarrow \frac{dx_i}{ds} = \frac{1}{2\lambda}\frac{\partial f}{\partial x_i} \quad \forall i \tag{9.5.36}$$

Consider a point $\boldsymbol{x}_0$. According to the previous equation, a movement from $\boldsymbol{x}_0$ in the gradient direction leads to $\boldsymbol{x}$ such that

$$x_i - x_{i,0} = \frac{dx_i}{ds}\, ds = \frac{1}{2\,\lambda}\left(\frac{\partial f}{\partial x_i}\right)_0 ds \tag{9.5.37}$$

From the previous equality, note that the vector $[dx_1 \; \dots \; dx_n]^T$ is parallel to the gradient vector $[\frac{\partial f}{\partial x_1} \; \dots \; \frac{\partial f}{\partial x_n}]^T$. Using the constraint on ds, it yields

$$1 = \sum_{i=1}^{n} \left(\frac{dx_i}{ds}\right)^2 = \frac{1}{4\,\lambda^2} \sum_{i=1}^{n} \left(\frac{\partial f}{\partial x_i}\right)^2 \tag{9.5.38}$$

hence λ

$$\lambda = \pm\frac{1}{2} \sqrt{\sum_{i=1}^{n} \left(\frac{\partial f}{\partial x_i}\right)^2} \tag{9.5.39}$$

and

$$\frac{df}{ds} = \pm\sqrt{\sum_{i=1}^{n} \left(\frac{\partial f}{\partial x_i}\right)^2} \tag{9.5.40}$$

the negative sign corresponding to the largest decrease of f and the positive sign to the largest increase.

It must be remembered that the gradient method is local, and the found direction changing during the search is given by the vector ξ of component

$$\xi_i = \pm \frac{\dfrac{\partial f}{\partial x_i}}{\sqrt{\displaystyle\sum_{i=1}^{n} \frac{\partial f}{\partial x_i}^2}} \tag{9.5.41}$$

The points according to the recurrence

$$\boldsymbol{x}_{i+1} = \boldsymbol{x}_i + \xi_i \, \boldsymbol{s} \tag{9.5.42}$$

are successively obtained.

9.5.4 Search in a Given Direction s

Frequently, the problem arises to find the displacement length in a given direction, that is, α, which minimizes $f(\boldsymbol{x} + \alpha\, \boldsymbol{s})$. Several algorithms are possible. Set $g(\alpha) = f(\boldsymbol{x} + \alpha\, \boldsymbol{s})$.

9.5.4.1 Algorithm of Bracketing–Shrinkage

Fletcher (1991) describes an algorithm containing a section of *bracketing* of the value α in an interval $[a, b]$ and then a section of *shrinkage* of that interval.

α responds to the condition of zero of the derivative

$$\frac{dg(\alpha)}{d\alpha} = 0 \tag{9.5.43}$$

There exist different types of practical conditions:

- Goldstein conditions

$$g(\alpha) \leq g(0) + \alpha\, \rho\, g'(0)\,, \quad g(\alpha) \geq g(0) + \alpha\,(1 - \rho)\, g'(0) \tag{9.5.44}$$

  with ρ fixed parameter $\in]0, 0.5[$.
- Wolfe condition

$$g'(\alpha) \geq \sigma g'(0) \tag{9.5.45}$$

  with $\sigma \in [\rho, 1[$.

(a) **Bracketing algorithm** (search of an interval [a,b])
Initialization of the section:
$\alpha = 0$
Give a lower bound of f, called $f_{lim,inf}$.
$\tau_1 = 9$
Fletcher advises to choose $\sigma = 0.1$ for a precise search and $\sigma = 0.9$ for a low search.
σ must verify the condition $\sigma \geq \rho$ (with $\rho = 0.01$).

Initialization: $i = 0$
Calculations with $\alpha_0 = 0$.
Calculate $g(\alpha_0) = g(0)$ and $g'(\alpha_0) = g'(0)$.
The search of α can be limited to the interval $[0, \mu]$.

$$\mu = \frac{f_{lim,inf} - g(0)}{\rho\, g'(0)}$$

```
Loop:
do i := 1, 2, ...
    if i = 1 then
        the first value α_1 must be in [0, μ]
        Choose α_1
    endif
    Calculate g(α_i)
    if g(α_i) ≤ f_lim,inf then
        the value α_i is convenient
        go to 20
    endif
    if g(α_i) > (g(0) + α_i g'(0)) or g(α_i) ≥ g(α_{i-1}) then
        the interval [a, b] is convenient
        a = α_{i-1} , b = α_i
        go to 10
    endif
    Calculate g'(α_i)
    if |g'(α_i)| ≤ −σ g'(0) then
        the value α_i is convenient
        go to 20
    endif
    if g'(α_i) ≥ 0 then
        the interval [a, b] is convenient
        a = α_i , b = α_{i-1}
        go to 10
    endif
    if μ ≤ (2 α_i − α_{i-1}) then
        α_{i+1} = μ
    else
        choose α_{i+1} ∈ [2 α_i − α_{i-1}, min(μ, α_i + τ_1 (α_i − α_{i-1}))]
    endif
enddo
10 continue
```

(b) **Shrinkage algorithm**:

```
    τ_2 = 0.1
    τ_3 = 0.5
Loop:
do j := i, i + 1, ...
    Choose α_j ∈ [a_j + τ_2 (b_j − a_j), b_j − τ_3 (b_j − a_j)]
    Calculate g(α_j)
```

```
    Calculate g(a_j)
    if g(α_j) > (g(0) + ρ α g'(0)) or g(α_j) ≥ g(a_j) then
        a_{j+1} = a_j , b_{j+1} = α_j
    else
        Calculate g'(α_j)
        if |g'( α_j)| ≤ −σ g'(0) then
            the value α_j is convenient
            goto 20
        endif
        a_{j+1} = α_j
        if (b_j − a_j) g'(α_j) ≥ 0 then
            b_{j+1} = a_j
        else
            b_{j+1} = b_j
        endif
    endif
enddo
20 continue
end
```

Given a point $\boldsymbol{x}$ and a direction $\boldsymbol{s}$, the calculation of the function $g(\alpha)$ is done by calculating the new point $\boldsymbol{x} + \alpha\,\boldsymbol{s}$ and by evaluating the function $f(\boldsymbol{x} + \alpha\,\boldsymbol{s})$ at this point.

The derivative with respect to α of the function $g(\alpha)$ can be done numerically by doing a small variation of $f(\boldsymbol{x} + \alpha\,\boldsymbol{s})$ on both sides of a given value α.

9.5.4.2 Wolfe Algorithm

Wolfe (Wolfe 1959) gives a rule for the search method following a given direction. Two coefficients ρ and σ are chosen such that $0 < \rho < \sigma < 1$. It is supposed that α belongs to an interval $[a, b]$. Three cases exist:

$$\begin{array}{lll} \text{If } g(\alpha) \leq g(0) + \rho\alpha g'(0) \quad \text{and} \quad g'(\alpha) \geq \sigma g'(0), & & \text{then end} \\ \text{If } g(\alpha) > g(0) + \rho\alpha g'(0), & & \text{then } b = \alpha \\ \text{If } g(\alpha) \leq g(0) + \rho\alpha g'(0) \quad \text{and} \quad g'(\alpha) < \sigma g'(0), & & \text{then } a = \alpha \end{array} \tag{9.5.46}$$

Wolfe's rule can be partly found in the previous algorithm by Fletcher and Wolfe's rule can be combined with any descent algorithm (Bonnans et al. 2003; Gill et al. 1981). In practice, at iteration k, the angle between the descent direction p_k and the gradient $f_{\boldsymbol{x},k}$ must not be too pronounced, which means that the gradient must not be too orthogonal to the descent direction. The cosinus

$$\cos(\theta_k) = -\frac{f_{\boldsymbol{x},k}\, p_k}{|f_{\boldsymbol{x},k}|.|p_k|} \tag{9.5.47}$$

must be sufficiently large, close to 1, which will ensure a global convergence.

In general, the Wolfe conditions are presented as follows. The displacement intensity α_k in the direction p_k must verify Wolfe conditions (Nocedal and Wright 2006)

$$\begin{aligned} f(\boldsymbol{x}_k + \alpha_k \boldsymbol{p}_k) &\leq f(\boldsymbol{x}_k) + \mu_1 \alpha_k \nabla f(\boldsymbol{x}_k)^T \boldsymbol{p}_k \\ \mu_2 \nabla f(\boldsymbol{x}_k)^T \boldsymbol{p}_k &\leq \nabla f(\boldsymbol{x}_k + \alpha_k \boldsymbol{p}_k)^T \boldsymbol{p}_k \end{aligned} \tag{9.5.48}$$

The first inequality is known as Armijo condition or Goldstein–Armijo condition and the second inequality is a curvature condition. μ_1 and μ_2 are scalars such that $0 < \mu_1 \leq \mu_2 < 1$, typically $\mu_1 \approx 0.01$ and μ_2 close to 1, thus giving a control of α_k. The purpose of the first condition is that f decreases sufficiently with a value of α_k different from the optimal value of α_k in the direction $\boldsymbol{p}_k$. The purpose of the second condition, the curvature condition, is to avoid too small steps so that the algorithm progresses reasonably toward the solution.

9.5.4.3 Algorithm of Cubic Approximation

Let $g(\alpha) = f(\boldsymbol{x} + \alpha \boldsymbol{s})$. This method (Polak 1997) is based on an approximation according to a cubic interpolation polynomial P of $g(\alpha)$, defined at iteration i by

$$P_i(\alpha) = g(\alpha_{i-1}) + g'(\alpha_{i-1})[\alpha - \alpha_{i-1}] + c_i\,[\alpha - \alpha_{i-1}]^2 + d_i\,[\alpha - \alpha_{i-1}]^2\,[\alpha - \alpha_i] \tag{9.5.49}$$

Let two parameters $a \in [0, 0.5]$ and $b \in [0.5, 1]$ and two parameters σ_{min} and σ_{max}.

1. Let α_0 and α_1 be two initial values.
2. Calculate

$$P''(\alpha_i) = 2\,c_i + 4\,d_i(\alpha_i - \alpha_{i-1}) \tag{9.5.50}$$

with

$$\begin{aligned} c_i &= \frac{1}{\alpha_i - \alpha_{i-1}} \left[\frac{g(\alpha_i) - g(\alpha_{i-1})}{\alpha_i - \alpha_{i-1}} - g'(\alpha_{i-1}) \right] \\ d_i &= \frac{1}{(\alpha_i - \alpha_{i-1})^2} \left[g'(\alpha_i) - 2\,\frac{g(\alpha_i) - g(\alpha_{i-1})}{\alpha_i - \alpha_{i-1}} + g'(\alpha_{i-1}) \right] \end{aligned} \tag{9.5.51}$$

3.

$$\eta_i = \begin{cases} -\dfrac{g'(\alpha_i)}{P''(\alpha_i)} & \text{if} \quad \sigma_{min} \leq P''(\alpha_i) \leq \sigma_{max} \\ -g'(\alpha_i) & \text{otherwise} \end{cases} \tag{9.5.52}$$

4. Calculate the step length according to Armijo

$$l_i = \max_{k \in \mathbb{N}} \left\{ b^k \mid g(\alpha_i + b^k\,\eta_i) - g(\alpha_i) \leq a\,b^k\,\eta_i\,g'(\alpha_i) \right\} \tag{9.5.53}$$

5. Set $\alpha_{i+1} = \alpha_i + l_i\,\eta_i$. Set $i = i + 1$. Return to 2.

9.5.4.4 Algorithm of the Golden Section

This algorithm (Polak 1997) lies on Fibonacci sequence (Section 9.2.3). Setting $F = 0.618033$, $l_i = F^i\,l_0$.

1. Initialize with an interval $[a, b]$ containing α which minimizes $f(\boldsymbol{x} + \alpha \boldsymbol{s})$. $l_0 = b - a$.
2. At iteration i, calculate $a'_i = a_i + l_i(1 - F)$ and $b'_i = b_i - l_i(1 - F)$.

3. If $f(\boldsymbol{x} + b'_i \boldsymbol{s}) \leq \min\{f(\boldsymbol{x} + a'_i \boldsymbol{s}), f(\boldsymbol{x} + b_i \boldsymbol{s})\}$, then $a_{i+1} = a'_i, b_{i+1} = b_i$. Otherwise, $a_{i+1} = a_i, b_{i+1} = b'_i$.
4. $i = i + 1$, return to 2.

9.5.5 Conjugate Gradient Method

The reference book about the conjugate gradient method is Hestenes (1980) and more recent literature (Andrei 2020; Lucambio Perez and Prudente 2018; Zhang et al. 2009).

Notion of k-planes

Consider the quadratic function

$$f(\boldsymbol{x}) = \frac{1}{2}\boldsymbol{x}^T \boldsymbol{A}\boldsymbol{x} - \boldsymbol{h}^T \boldsymbol{x} + c \tag{9.5.54}$$

whose Hessian matrix $\boldsymbol{A}$ is symmetric definite positive.

An important problem deals with the minimization of a quadratic function subject to a set of linear constraints

$$\boldsymbol{q}_i^T \boldsymbol{x} = \rho_i \,, \quad i = 1, \ldots, n-k \,, \quad 1 \leq k < n \tag{9.5.55}$$

where $\boldsymbol{q}_i$ are linearly independent vectors. The set of points $\boldsymbol{x}$ satisfying this system of equations is called a ***k**-plane* noted π_k. A 1-plane π_1 is a straight line. In dimension $n = 3$, the intersection of two 2 hyperplanes (2-planes) is a straight line 1-plane. An $(n-1)$-plane is a hyperplane defined by only one equation $\boldsymbol{q}^T \boldsymbol{x} = \rho$. Thus, a k-plane can be considered as the intersection of $n - k$ hyperplanes.

To express that a point $\boldsymbol{x}_1$ belongs to π_k (Figure 9.14), it is sufficient to write

$$\boldsymbol{q}_i^T (\boldsymbol{x} - \boldsymbol{x}_1) = 0 \,, \quad i = 1, \ldots, n-k \tag{9.5.56}$$

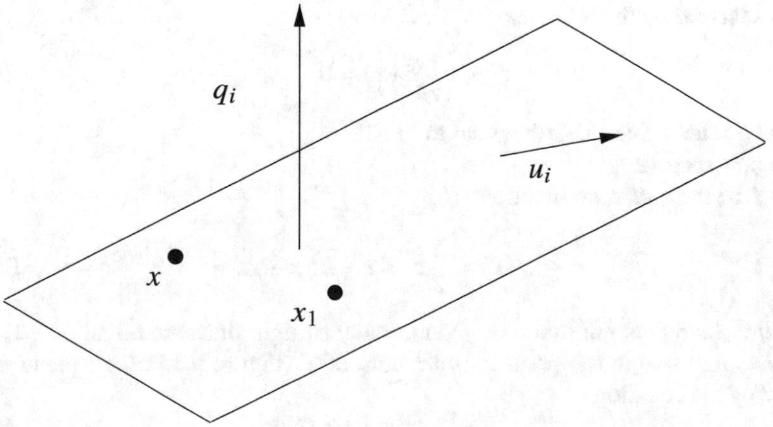

Fig. 9.14 k-Plane π_k

The vector $\boldsymbol{x} - \boldsymbol{x}_1$ is thus orthogonal to the vectors $\boldsymbol{q}_i$. A condition for a vector $\boldsymbol{u}$ to be in π_k is that $\boldsymbol{u}$ is orthogonal to the vectors $\boldsymbol{q}_i$, which are normal to π_k

$$\boldsymbol{q}_i^T \boldsymbol{u} = 0\,, \quad i = 1, \ldots, n-k \tag{9.5.57}$$

Any vector normal to π_k is a linear combination of $(n-k)$ vectors $\boldsymbol{q}_i$. The previous system of equations has k linearly independent solutions, so that there exist k vectors $\boldsymbol{u}_1, \ldots, \boldsymbol{u}_k$ linearly independent in π_k and any vector $\boldsymbol{u}$ of π_k is expressed as a linear combination of the vectors $\boldsymbol{u}_i$

$$\boldsymbol{u} = \sum_{i=1}^{k} \alpha_i \, \boldsymbol{u}_i \tag{9.5.58}$$

Given the point $\boldsymbol{x}_1$ of π_k, any point x of π_k can be written as

$$\boldsymbol{x} = \boldsymbol{x}_1 + \boldsymbol{U}\boldsymbol{y} \tag{9.5.59}$$

where $\boldsymbol{U}$ is the matrix of the column vectors $\boldsymbol{u}_i$ and $\boldsymbol{y}$ the column vector of coefficients α_i.

For any point x of π_k, the function f is thus

$$f(\boldsymbol{x}_1 + \boldsymbol{U}\boldsymbol{y}) = \frac{1}{2}\boldsymbol{y}^T \boldsymbol{U}^T \boldsymbol{A}\boldsymbol{U}\boldsymbol{y} + \boldsymbol{U}^T f_x(\boldsymbol{x}_1)\boldsymbol{y} + f(\boldsymbol{x}_1) = g(\boldsymbol{y}) \tag{9.5.60}$$

The function g has the minimum

$$\boldsymbol{y}_0 = (\boldsymbol{U}^T \boldsymbol{A}\boldsymbol{U})^{-1}(-\boldsymbol{U}^T f_x(\boldsymbol{x}_1)) \tag{9.5.61}$$

and f admits as a minimum on π_k

$$\boldsymbol{x}_0 = \boldsymbol{x}_1 + \boldsymbol{U}\,\boldsymbol{y}_0 \tag{9.5.62}$$

As $g_y(\boldsymbol{y}_0) = 0$ induces $\boldsymbol{U}^T f_x(\boldsymbol{x}_0) = 0$, it results that the point $\boldsymbol{x}_0$ minimum of f on π_k is characterized by the relations

$$\boldsymbol{u}_i^T f_x(\boldsymbol{x}) = 0 \tag{9.5.63}$$

thus the gradient $f_x(\boldsymbol{x})$ is orthogonal to π_k.

Conjugate vectors

Let f be the quadratic function

$$f(\boldsymbol{x}) = \frac{1}{2}\boldsymbol{x}^T \boldsymbol{A}\boldsymbol{x} - \boldsymbol{h}^T \boldsymbol{x} + c \tag{9.5.64}$$

Theorem: the points minimum of f on parallel straight lines are on an $(n-1)$-plane $\hat{\pi}_{n-1}$ passing through the point $\boldsymbol{x}_0$ minimum of f (Figure 9.15). This plane $\hat{\pi}_{n-1}$ is defined by the equation

$$\boldsymbol{u}^T(\boldsymbol{A}\boldsymbol{x} - \boldsymbol{h}) = 0 \tag{9.5.65}$$

where $\boldsymbol{u}$ is a vector parallel to these straight lines. The vector $\boldsymbol{A}\boldsymbol{u}$ is normal to $\hat{\pi}_{n-1}$.

The fact that the vector $\boldsymbol{A}\boldsymbol{u}$ is normal to $\hat{\pi}_{n-1}$ is expressed as saying that the $\boldsymbol{u}$ is *conjugate* (or $\boldsymbol{A}$-*orthogonal*) to $\hat{\pi}_{n-1}$ and reciprocally that $\hat{\pi}_{n-1}$ is conjugate to $\boldsymbol{u}$.

The directions defined by the vectors $\boldsymbol{u}$ and $\boldsymbol{v}$ are conjugate when the relation

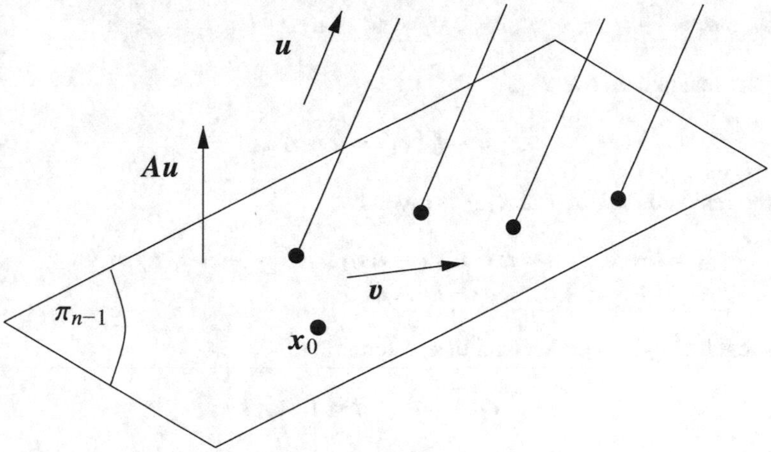

Fig. 9.15 Points minimum of f on parallel straight lines belonging to an $(n-1)$-plane π_{n-1}

$$\boldsymbol{u}^T \boldsymbol{A}\, \boldsymbol{v} = 0 \tag{9.5.66}$$

is true whatever the vector $\boldsymbol{v}$ of $\hat{\pi}_{n-1}$.

This relation implies that the vector joining the points minimum of f on two parallel straight lines L_1 and L_2 of direction $\boldsymbol{u}$ is conjugate to $\boldsymbol{u}$ (as the vector is in $\hat{\pi}_{n-1}$).

The dual theorem is as follows:

The points minimum of f on parallel $(n-1)$-planes are found on a straight line L conjugate to these hyperplanes and passing through the minimum $\boldsymbol{x}_0$.

Let $\boldsymbol{v}$ be a nonzero vector, then whatever the number d, the point $\boldsymbol{x}_1$ minimum of f on the $(n-1)$-plane $\hat{\pi}_{n-1}$ of equation

$$\boldsymbol{v}^T \boldsymbol{x} = d \tag{9.5.67}$$

is on the straight line L of equation

$$\boldsymbol{x} = \boldsymbol{x}_0 + \alpha \boldsymbol{A}^{-1} v \tag{9.5.68}$$

which passes through the minimum $\boldsymbol{x}_0$ of f in the direction $\boldsymbol{u} = \boldsymbol{A}^{-1}\boldsymbol{v}$. The vector $\boldsymbol{u}$ (or the straight line L) is conjugate to $\hat{\pi}_{n-1}$.

Minimization of a quadratic function on k-planes

The equation of the k-plane π_k passing through $\boldsymbol{x}_1$ and determined by a conjugate system $\boldsymbol{u}_1, \ldots, \boldsymbol{u}_k$ (a set of vectors of π_k mutually conjugate: $\boldsymbol{u}_i^T \boldsymbol{A}\, \boldsymbol{u}_j = 0$ if $i \neq j$) is

$$\boldsymbol{x} = \boldsymbol{x}_1 + \alpha_1 \boldsymbol{u}_1 + \cdots + \alpha_k \boldsymbol{u}_k \tag{9.5.69}$$

The point $\boldsymbol{x}_{k+1}$ minimum of f on π_k is given by the formula

$$\boldsymbol{x}_{k+1} = \boldsymbol{x}_1 + a_1\,\boldsymbol{u}_1 + \cdots + a_k\,\boldsymbol{u}_k \tag{9.5.70}$$

with

$$a_i = \frac{c_i}{d_i}\,, \quad c_i = \boldsymbol{u}_i^T\,\boldsymbol{r}_i\,, \quad d_i = \boldsymbol{u}_i^T\,\boldsymbol{A}\,\boldsymbol{u}_i\,, \quad i = 1, \ldots, k \tag{9.5.71}$$

and $\boldsymbol{r}_1$ the residual of f at $\boldsymbol{x}_1$

$$\boldsymbol{r}_1 = -f_x(\boldsymbol{x}_1) = \boldsymbol{h} - \boldsymbol{A}\,\boldsymbol{x}_1 \tag{9.5.72}$$

while the residual $\boldsymbol{r}_{k+1}$ of f at $\boldsymbol{x}_{k+1}$ is given by

$$\begin{aligned}\boldsymbol{r}_{k+1} &= \boldsymbol{h} - \boldsymbol{A}\,\boldsymbol{x}_{k+1} = \boldsymbol{h} - \boldsymbol{A}\,\boldsymbol{x}_1 - \boldsymbol{A}\,a_1\,\boldsymbol{u}_1 - \cdots - \boldsymbol{A}\,a_k\,\boldsymbol{u}_k = \\ &= \boldsymbol{r}_1 - a_1\,\boldsymbol{A}\,\boldsymbol{u}_1 - \cdots - a_k\,\boldsymbol{A}\,\boldsymbol{u}_k\end{aligned} \tag{9.5.73}$$

The residual $\boldsymbol{r}_{k+1}$ is orthogonal to π_k, hence

$$\boldsymbol{u}_i^T\,\boldsymbol{r}_{k+1} = 0\,, \quad i = 1, \ldots, k \tag{9.5.74}$$

The minimum value of the function f on π_k is equal to

$$f(\boldsymbol{x}_{k+1}) = f(\boldsymbol{x}_1) - \frac{1}{2}\,(a_1\,c_1 + \cdots + a_k\,c_k) \tag{9.5.75}$$

Algorithm of conjugate directions
First stage:
Choose a point $\boldsymbol{x}_1$, the initial estimation of the minimum $\boldsymbol{x}_0$ of f. Calculate the residual $\boldsymbol{r}_1 = -f_x(\boldsymbol{x}_1) = \boldsymbol{h} - \boldsymbol{A}\,\boldsymbol{x}_1$ and choose an initial direction $\boldsymbol{u}_1$.
Iterative stages:
Knowing $\boldsymbol{x}_k$, estimation of $\boldsymbol{x}_0$, its residual $\boldsymbol{r}_k$, and the direction $\boldsymbol{u}_k$, calculate a new estimation $\boldsymbol{x}_{k+1}$ and its residual from the formulas

$$a_k = \frac{c_k}{d_k} \quad \text{with} \quad c_k = \boldsymbol{u}_k^T\,r_k\,, \quad d_k = \boldsymbol{u}_k^T\,A\,\boldsymbol{u}_k \tag{9.5.76}$$

$$\boldsymbol{x}_{k+1} = \boldsymbol{x}_k + a_k\,\boldsymbol{u}_k\,, \quad \boldsymbol{r}_{k+1} = \boldsymbol{r}_k - a_k\,\boldsymbol{A}\,\boldsymbol{u}_k \tag{9.5.77}$$

Then choose a vector $\boldsymbol{u}_{k+1}$ conjugate to the k previous vectors, thus such that

$$\boldsymbol{u}_j^T\,\boldsymbol{A}\,\boldsymbol{u}_{k+1} = 0 \qquad \forall j = 1, \ldots, k \tag{9.5.78}$$

Stop at the mth stage when $\boldsymbol{r}_{m+1} = 0$. Then $m \leq n$ and $\boldsymbol{x}_{m+1}$ is the sought minimum $\boldsymbol{x}_0$ of f.

Note that nothing is said in the previous algorithm about the manner to obtain the conjugate vectors $\boldsymbol{u}_k$. It will be the purpose of the conjugate gradient algorithm.
Conjugate gradient algorithm
This algorithm can be considered as an improved method of steepest descent.

First stage:
Choose a point $\boldsymbol{x}_1$, the initial estimation of the minimum $\boldsymbol{x}_0$ of f. Calculate the residual $\boldsymbol{r}_1 = -f_x(\boldsymbol{x}_1) = \boldsymbol{h} - \boldsymbol{A}\,\boldsymbol{x}_1$; the first vector is equal to $\boldsymbol{u}_1 = -f_x(\boldsymbol{x}_1)$.
Iterative stages:
Knowing $\boldsymbol{x}_k$ and $\boldsymbol{u}_k$, determine $\boldsymbol{x}_{k+1}$ and $\boldsymbol{u}_{k+1}$ by the rules:

(a) Find the point $\boldsymbol{x}_{k+1} = \boldsymbol{x}_k + a_k\,\boldsymbol{u}_k$ minimum of f on the straight line of equation

$$\boldsymbol{x} = \boldsymbol{x}_k + \alpha\,\boldsymbol{u}_k \tag{9.5.79}$$

(b) Determine the new direction $\boldsymbol{u}_{k+1}$ by the formulas

$$\boldsymbol{u}_{k+1} = \boldsymbol{r}_{k+1} + b_k\,\boldsymbol{u}_k\,, \quad \boldsymbol{r}_{k+1} = -f_x(\boldsymbol{x}_{k+1})\,, \quad b_k = \frac{|\boldsymbol{r}_{k+1}|^2}{|\boldsymbol{r}_k|^2} \tag{9.5.80}$$

Stop at the mth stage when $\boldsymbol{r}_{m+1} = 0$. Necessarily, $m \leq n$, and $\boldsymbol{x}_{m+1}$ is the sought minimum $\boldsymbol{x}_0$ of f. Thus, *the number of iterations necessary to find the minimum is lower than or equal to the dimension of the space.*

To find the point minimum of f on the straight line, it is sufficient to write that the gradient at this point, thus the residual $\boldsymbol{r}_{k+1}$ is orthogonal to the direction of the straight line, that is $\boldsymbol{u}_k$

$$\boldsymbol{u}_k^T\,\boldsymbol{r}_{k+1} = 0 \tag{9.5.81}$$

Already cited formulas are given

$$a_k = \frac{c_k}{d_k} \quad \text{with} \quad c_k = \boldsymbol{u}_k^T\,\boldsymbol{r}_k\,, \quad d_k = \boldsymbol{u}_k^T\,\boldsymbol{A}\,\boldsymbol{u}_k \quad \text{hence } b_k = -\frac{\boldsymbol{u}_k^T\,\boldsymbol{A}\,\boldsymbol{r}_{k+1}}{d_k} \tag{9.5.82}$$

The following formulas can also be used:

$$c_k = |\boldsymbol{r}_k|^2\,, \quad b_k = \frac{|\boldsymbol{r}_{k+1}|^2}{c_k} \tag{9.5.83}$$

then the recurrence is used

$$\boldsymbol{u}_{k+1} = \boldsymbol{r}_{k+1} + b_k\,\boldsymbol{u}_k \tag{9.5.84}$$

Fletcher (1991) notes that the conjugate gradient methods are less effective and less robust than methods of quasi-Newton type, but they can be adapted for large scale problems (Yuan and Zhang 2013).

Example 9.7 :
Conjugate gradient method: minimization of a quadratic function

The conjugate gradient method is used to minimize the following quadratic function:

$$f(\boldsymbol{x}) = 1/2\, \boldsymbol{x}^T \boldsymbol{A}\, \boldsymbol{x} - \boldsymbol{h}^T\, \boldsymbol{x} + c \tag{9.5.85}$$

with

$$\boldsymbol{A} = \begin{bmatrix} 1 & 2 & -1 & 1 \\ 2 & 5 & 0 & 2 \\ -1 & 0 & 6 & 0 \\ 1 & 2 & 0 & 3 \end{bmatrix}, \quad \boldsymbol{h} = \begin{bmatrix} 0 \\ 2 \\ -1 \\ 1 \end{bmatrix} \tag{9.5.86}$$

The value is c does not influence the final vector $\boldsymbol{x}$. Here, $c = 1$. The initial vector is as simple as possible

$$\boldsymbol{x}^T = [1\ 0\ 0\ 0] \tag{9.5.87}$$

Thus, the results of Table 9.5 are obtained. It is verified that the residual vector becomes zero after 4 iterations at the most, in accordance with the dimension of vector $\boldsymbol{x}$. It is also verified that the function f decreases at each iteration.

Table 9.5 Conjugate gradient method: search of the minimum of a quadratic function

Iteration k	$\boldsymbol{x}$	$\boldsymbol{r}$	$\boldsymbol{u}$	a_k	b_{k-1}	c_k	d_k	f
1	$\begin{bmatrix} 1 \\ 0 \\ 0 \\ 0 \end{bmatrix}$	$\begin{bmatrix} -1 \\ 0 \\ 0 \\ 0 \end{bmatrix}$	$\begin{bmatrix} -1 \\ 0 \\ 0 \\ 0 \end{bmatrix}$	1		1	1	1.5
2	$\begin{bmatrix} 0 \\ 0 \\ 0 \\ 0 \end{bmatrix}$	$\begin{bmatrix} 0 \\ 2 \\ -1 \\ 1 \end{bmatrix}$	$\begin{bmatrix} -6 \\ 1 \\ -1 \\ 1 \end{bmatrix}$	6	6	6	1	1
3	$\begin{bmatrix} -36 \\ 12 \\ -6 \\ 6 \end{bmatrix}$	$\begin{bmatrix} 0 \\ 2 \\ -1 \\ -5 \end{bmatrix}$	$\begin{bmatrix} -30 \\ 12 \\ -6 \\ 0 \end{bmatrix}$	0.833	5	30	36	−17
4	$\begin{bmatrix} -61 \\ 22 \\ -11 \\ 6 \end{bmatrix}$	$\begin{bmatrix} 0 \\ 2 \\ 4 \\ 0 \end{bmatrix}$	$\begin{bmatrix} -20 \\ 10 \\ 0 \\ 0 \end{bmatrix}$	0.20	0.667	20	100	−29.5
5	$\begin{bmatrix} -65 \\ 24 \\ -11 \\ 6 \end{bmatrix}$	$\begin{bmatrix} 0 \\ 0 \\ 0 \\ 0 \end{bmatrix}$						−31.5

Application to a nonquadratic function

The conjugate gradient algorithm can be used in the case of a nonquadratic function f (Hestenes 1980) by building the approximation $F(z)$ at second order of the function f in the neighborhood of a point $\boldsymbol{x}$

$$f(\boldsymbol{x}+z) \approx F(z) = f(\boldsymbol{x}) + f_x(\boldsymbol{x})^T z + \frac{1}{2} z^T f_{xx}(\boldsymbol{x}) z \tag{9.5.88}$$

such that

$$\begin{aligned} &F(z) = \tfrac{1}{2} z^T \boldsymbol{A} z - \boldsymbol{h}^T z + c \\ &\text{with} \quad \boldsymbol{A} = f_{xx}(\boldsymbol{x}), \quad \boldsymbol{h} = \boldsymbol{A}\,\boldsymbol{x} - f_x(\boldsymbol{x}), \quad c = f(\boldsymbol{x}) \end{aligned} \tag{9.5.89}$$

The algorithm works in the following way:

(a) Choose a point $\boldsymbol{x}_1$, the initial estimation of the minimum $\boldsymbol{x}_0$ of f.
(b) Build the approximation $F(z)$ at second order such that

$$\begin{aligned} &F(z) = \tfrac{1}{2} z^T \boldsymbol{A} z - \boldsymbol{h}^T z + c \\ &\text{with} \quad \boldsymbol{A} = f_{xx}(\boldsymbol{x}_1), \quad \boldsymbol{h} = \boldsymbol{A}\,\boldsymbol{x}_1 - f_x(\boldsymbol{x}_1), \quad c = f(\boldsymbol{x}_1) \end{aligned} \tag{9.5.90}$$

(c) Take the initial value $z_1 = 0$. Use a conjugate gradient algorithm (n stages at the most) according to the following equations:

(c1) Calculate the residual $\boldsymbol{r}_1 = -f_x(\boldsymbol{x}_1)$. $\boldsymbol{u}_1 = \boldsymbol{r}_1$,

(c2) For $k = 1, \ldots, n$,

$$\begin{aligned} &c_k = \boldsymbol{u}_k^T \boldsymbol{r}_k\,, \quad d_k = \boldsymbol{u}_k^T \boldsymbol{A}\,\boldsymbol{u}_k\,, \quad a_k = \frac{c_k}{d_k} \\ &z_{k+1} = z_k + a_k\,\boldsymbol{u}_k\,, \quad \boldsymbol{r}_{k+1} = \boldsymbol{r}_k - a_k\,\boldsymbol{A}\,\boldsymbol{u}_k \\ &b_k = \frac{|\boldsymbol{r}_{k+1}|^2}{|\boldsymbol{r}_k|^2}\,, \quad \boldsymbol{u}_{k+1} = \boldsymbol{r}_{k+1} + b_k\,\boldsymbol{u}_k \end{aligned} \tag{9.5.91}$$

until obtaining the minimum point z_{n+1} of F.

(d) Repeat stages b/ and c/ by replacing $\boldsymbol{x}_1$ at each time by its new value $\boldsymbol{x}'_1 = \boldsymbol{x}_1 + z_{n+1}$ as new estimation of the minimum point $\boldsymbol{x}_0$ of f. Stop when the norm of the gradient $|f_x(\boldsymbol{x}'_1)|$ is sufficiently small so that $\boldsymbol{x}'_1$ can be considered as a good estimation of $\boldsymbol{x}_0$. Otherwise, restart at (a)

Hestenes (1980) notes that the choice of a_k does not minimize $f(z_k + a\,\boldsymbol{u}_k)$. In stage c, Hestenes (1980) proposes eventually to use $c_k = \boldsymbol{u}_k^T \boldsymbol{r}_1$ and $\boldsymbol{r}_{k+1} = -f_x(\boldsymbol{x}_{k+1})$.

By following the previous algorithm, the minimization of a nonquadratic function is very close to that of a quadratic function. This is the case with the relation

$$b_k = |\boldsymbol{r}_{k+1}|^2 / |\boldsymbol{r}_k|^2 \tag{9.5.92}$$

used in Fletcher–Reeves. However, in practice, difficulties may occur and other methods that seem less rigorous theoretically are more effective (Bonnans et al. 2003). Polak–Ribière formula (Polak 1997) which thus proposes

$$b_k = \frac{< \boldsymbol{r}_{k+1} - \boldsymbol{r}_k, \boldsymbol{r}_{k+1} >}{|\boldsymbol{r}_k|^2} \tag{9.5.93}$$

converges more rapidly so that Fletcher–Reeves formula is no more used. The notation $< \boldsymbol{V_1}, \boldsymbol{V_2} >$ represents the scalar product of vectors. Moreover, Polak–Ribière algorithm uses a minimization in the descent direction and can be summarized as follows: Choose a point $\boldsymbol{x}_0$, the initial estimation of the minimum $\boldsymbol{x}^*$ of f.

(a) $i = 0$, $\boldsymbol{h}_0 = -f_x(\boldsymbol{x}_0)$.
(b) Calculate $a > 0$ minimizing $f(x_i + a\,\boldsymbol{h}_i)$ (cf. algorithm of search in Section 9.5.4).
(c) Calculate

$$\begin{aligned} \boldsymbol{x}_{i+1} &= \boldsymbol{x}_i + a\,\boldsymbol{h}_i \\ \boldsymbol{g}_{i+1} &= f_x(\boldsymbol{x}_{i+1}) \\ b_i &= \frac{< \boldsymbol{g}_{i+1} - \boldsymbol{g}_i, \boldsymbol{g}_{i+1} >}{|\boldsymbol{g}_i|^2} \\ \boldsymbol{h}_{i+1} &= -\boldsymbol{g}_{i+1} + b_i\,\boldsymbol{h}_i \end{aligned} \tag{9.5.94}$$

(d) $i = i + 1$, return to (b).

Fletcher–Reeves algorithm corresponds to the replacement in the previous algorithm of b_i by Equation (9.5.92).

Example 9.8 :
Conjugate gradient method: minimization of a nonquadratic function
The conjugate gradient method was applied to minimize the nonquadratic function

$$f(\boldsymbol{x}) = (x_1 + 2\,x_2)^2 + (2\,(x_1 + 1)^2 + (x_2 + 1)^2 - 1)^2 \tag{9.5.95}$$

The previously described algorithm based on the conjugate gradient method but using the quadratic approximation of $f(\boldsymbol{x})$ was executed and provided the results of Table 9.6. It is noteworthy that these results are strictly identical to those obtained by the use of Newton–Raphson.

Table 9.6 Conjugate gradient method: minimization of a nonquadratic function (9.5.95)

Iteration	x_1	x_2	f
1	0.5000	0.5000	35.3125
2	0.0542	0.0374	5.3014
3	−0.2420	−0.1055	1.1063
4	−0.4662	−0.0096	0.5389
5	−0.6417	0.0982	0.4125
6	−0.7110	0.1219	0.3995
7	−0.7190	0.1235	0.3993
8	−0.7191	0.1235	0.3993

9.5.6 Newton–Raphson Method

In the case of the search of the optimum of a function $f(\boldsymbol{x})$, the gradient must be zero, hence

$$\frac{\partial f}{\partial x_i} = 0\,, \quad i = 1, \ldots, n \tag{9.5.96}$$

This condition is necessary so that it is a minimum or a maximum, but not sufficient. It is called a critical point. Thus, the nullity of the n components f_i of the gradient amounts to a system of n functions that must be simultaneously zero, which is written by posing

$$g_i(\boldsymbol{x}) = \frac{\partial f}{\partial x_i}\,, \quad i = 1, \ldots, n \tag{9.5.97}$$

where the g_i are the components of a vector $\boldsymbol{g}(\boldsymbol{x})$. The system to solve is

$$\boldsymbol{g}(\boldsymbol{x}) = 0 \tag{9.5.98}$$

The gradient of $\boldsymbol{g}$ at point $\boldsymbol{x}$ is indeed the Jacobian matrix of $\boldsymbol{g}$ at $\boldsymbol{x}$ which is noted

$$\boldsymbol{G}(\boldsymbol{x}) = \frac{\partial g_i}{\partial x_j}\,, \quad i, j = 1, \ldots, n \tag{9.5.99}$$

Note that the Jacobian matrix $\boldsymbol{G}$ of $\boldsymbol{g} = f_x$ is the Hessian matrix f_{xx} of f. A critical point of f is called nondegenerated if $f_{xx}(\boldsymbol{x}_0)$ is nonsingular ($\boldsymbol{x}_0$ is a minimum if f_{xx} is definite positive, $\boldsymbol{x}_0$ is a maximum if f_{xx} is definite negative, and $\boldsymbol{x}_0$ is a saddle point if f_{xx} is indefinite).

The Newton–Raphson algorithm is

$$\boldsymbol{x}_{k+1} = \boldsymbol{x}_k - G(\boldsymbol{x}_k)^{-1}\, g(\boldsymbol{x}_k) \tag{9.5.100}$$

Apparently, Equation (9.5.100) imposes the explicit inversion of the Jacobian matrix $\boldsymbol{G}$. Indeed, it is possible to avoid it by solving the linear system at each iteration k

$$G(\boldsymbol{x}_k)\,\boldsymbol{\eta}_k = -g(\boldsymbol{x}_k) \tag{9.5.101}$$

which gives both the direction $\boldsymbol{p}_k$ and the intensity of the displacement α_k under the form $\boldsymbol{\eta}_k = \alpha_k \boldsymbol{p}_k$, and then by using the gradient formula

$$\boldsymbol{x}_{k+1} = \boldsymbol{x}_k + \boldsymbol{\eta}_k \tag{9.5.102}$$

Indeed, if the Jacobian matrix $\boldsymbol{G}$ is ill-conditioned, solving the linear system (9.5.101) does not remove the difficulty, and the other methods of quasi-Newton type are then necessary.

Example 9.9 :
Newton–Raphson method: minimization of Rosenbrock's function

Newton–Raphson method is used to find the minimum of Rosenbrock's function

$$f(\boldsymbol{x}) = 100\,(x_2 - x_1^2)^2 + (1 - x_1)^2 \tag{9.5.103}$$

Rosenbrock's function is particularly interesting to assess minimization methods as it has a banana shape and thus is nonconvex.

First, the nullity of the gradient of f is written as

$$\begin{aligned} \frac{\partial f}{\partial x_1} &= g_1 = -400\, x_1\, (x_2 - x_1^2) - 2\,(1 - x_1) = 0 \\ \frac{\partial f}{\partial x_2} &= g_2 = 200\,(x_2 - x_1^2) = 0 \end{aligned} \tag{9.5.104}$$

The Jacobian matrix of $\boldsymbol{g}$ is the following symmetrical matrix:

$$\boldsymbol{G} = \begin{bmatrix} G_{11} & G_{12} \\ G_{21} & G_{22} \end{bmatrix} \tag{9.5.105}$$

with

$$\begin{aligned} G_{11} &= -400\,(x_2 - x_1^2) + 800\, x_1^2 + 2 \\ G_{12} &= g_{21} = -400\, x_1 \\ G_{22} &= 200 \end{aligned} \tag{9.5.106}$$

1. First method based on the inversion of the matrix $\boldsymbol{G}$:
The determinant of $\boldsymbol{G}$ is equal to

$$\det(\boldsymbol{G}) = G_{11}\, G_{22} - G_{12}\, G_{21} \tag{9.5.107}$$

The inverse of a two-dimensional matrix is easily obtained. Let $\boldsymbol{A}$ be the matrix and the inverse of which is

$$\boldsymbol{A}^{-1} = \frac{1}{\det(\boldsymbol{A})} \begin{bmatrix} a_{22} & -a_{12} \\ -a_{21} & a_{11} \end{bmatrix} \tag{9.5.108}$$

hence the recurrence

$$\begin{aligned} x_{k+1,1} &= x_{k,1} - G_{11}^{-1}\, g_1 - G_{12}^{-1}\, g_2 = x_{k,1} - (G_{22}\, g_1 - G_{12}\, g_2)/\det(\boldsymbol{G}) \\ x_{k+1,2} &= x_{k,2} - G_{21}^{-1}\, g_1 - G_{22}^{-1}\, g_2 = x_{k,2} - (-G_{21}\, g_1 + G_{11}\, g_2)/\det(\boldsymbol{G}) \end{aligned} \tag{9.5.109}$$

where G_{ij}^{-1} represents the element (i, j) of matrix $\boldsymbol{G}^{-1}$.

Starting from point (2, 2), the recurrence leads to Table 9.7.

Table 9.7 Newton–Raphson method: minimization of Rosenbrock's function

Iteration	x_1	x_2	g_1	g_2	f
1	2	2	1602	−400	401
2	1.998	3.99	2.00	-1.244×10^{-3}	0.995
3	1.000	4.981×10^{-3}	398	−199	99.01
4	1.000	1.000	-2.46×10^{-3}	1.23×10^{-3}	3.781×10^{-9}
5	1.000	1.000	3.13×10^{-6}	-1.56×10^{-6}	6.131×10^{-15}

Figure 9.16 shows both the contours of Rosenbrock's function and the trajectory followed during the iterations by Newton–Raphson method. Note the fast convergence following a complex trajectory, first toward the valley, then following the valley and finally directly to the optimum.

The calculation ended when each component of the gradient of f became lower than 10^{-6}. The convergence to the final solution is very rapid in spite of a "difficult" function.

2. Second method avoiding the inversion of matrix $\boldsymbol{G}$:
At each iteration, the linear system (9.5.101) is solved with respect to the unknowns $\boldsymbol{\eta}_k$

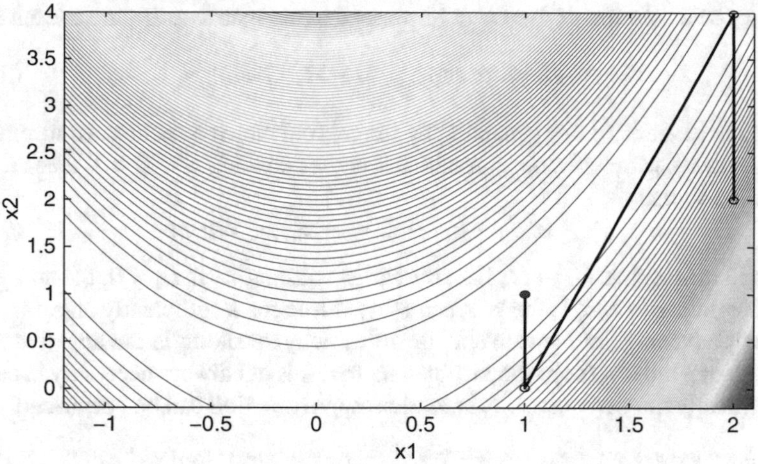

Fig. 9.16 Newton–Raphson method: contours of Rosenbrock's function and trajectory followed during the iterations

$$\boldsymbol{G}(\boldsymbol{x}_k)\,\boldsymbol{\eta}_k = -\boldsymbol{g}(\boldsymbol{x}_k) \tag{9.5.110}$$

or under developed form

$$\begin{array}{ll} \left[-400\,(x_2 - x_1^2) + 800\,x_1^2 + 2\right]^{(k)} \boldsymbol{\eta}_1^{(k)} & +[-400\,x_1]^{(k)}\,\boldsymbol{\eta}_2^{(k)} = \\ & -\left[-400\,x_1\,(x_2 - x_1^2) - 2\,(1 - x_1)\right]^{(k)} \\ [-400\,x_1]^{(k)}\,\boldsymbol{\eta}_1^{(k)} & +200\,\boldsymbol{\eta}_2^{(k)} = \left[200\,(x_2 - x_1^2)\right]^{(k)} \end{array} \tag{9.5.111}$$

Of course, the same result is obtained.

Remarks:

- Hestenes (1980) notes that in some difficult cases, it is better, not to say necessary, to use a *relaxed* Newton algorithm. Instead of using the recurrence

$$\boldsymbol{x}_{k+1} = \boldsymbol{x}_k - \boldsymbol{G}(\boldsymbol{x}_k)^{-1}\,\boldsymbol{g}(\boldsymbol{x}_k) \tag{9.5.112}$$

  the following recurrence will be used:

$$\begin{aligned} \boldsymbol{x}_{k+1} &= \boldsymbol{x}_k - a_k\,\boldsymbol{G}(\boldsymbol{x}_k)^{-1}\,\boldsymbol{g}(\boldsymbol{x}_k) \\ &= \boldsymbol{x}_k - a_k\,\boldsymbol{H}_k\,\boldsymbol{g}(\boldsymbol{x}_k) \end{aligned} \tag{9.5.113}$$

  with the constraints

$$\lim_{k\to\infty} a_k = 1 \qquad \text{and} \qquad \lim_{k\to\infty} [\boldsymbol{H}_k - \boldsymbol{G}(\boldsymbol{x}_k)^{-1}] = 0 \tag{9.5.114}$$

- The convergence of Newton sequence is superlinear and even quadratic when $\boldsymbol{g}$ is twice differentiable.
  Note on convergence:

Let $\boldsymbol{x}_0$ be a solution of $\boldsymbol{g}(\boldsymbol{x}) = 0$. Suppose that the matrices $\boldsymbol{H}_k$ are chosen so that

$$L_0 = \lim\sup_{k\to\infty} \|I - \boldsymbol{H}_k\, \boldsymbol{G}(\boldsymbol{x}_0)\| < 1 \tag{9.5.115}$$

then the sequence $\{\boldsymbol{x}_k\}$ generated by the relaxed Newton recurrence formula converges to $\boldsymbol{x}_0$. Moreover, if a constant L is chosen so that $L_0 < L < 1$, then an integer m exists so that

$$|\boldsymbol{x}_{k+1} - \boldsymbol{x}_0| \leq L|\boldsymbol{x}_k - \boldsymbol{x}_0|\,, \quad k \geq m \tag{9.5.116}$$

thus, it means that $\boldsymbol{x}_k \to \boldsymbol{x}_0$ linearly with a constant L. If $L_0 = 0$, the convergence is superlinear ($\forall\, L < 1$, the last inequality is true for k sufficiently large).

- The disadvantage of Newton's algorithm strictly speaking is the apparent need of calculating the inverse of the Jacobian matrix. It is not always necessary to calculate analytically the derivatives. A numerical approximation can be performed

$$\frac{g_i(\boldsymbol{x} + \epsilon\, \boldsymbol{e}_j) - g_i(\boldsymbol{x})}{\epsilon} \approx \frac{\partial g_i(\boldsymbol{x})}{\partial x_j} \quad \text{or:} \quad \frac{g_i(\boldsymbol{x} + \epsilon\, \boldsymbol{e}_j) - g_i(\boldsymbol{x} - \epsilon\, \boldsymbol{e}_j)}{2\epsilon} \approx \frac{\partial g_i(\boldsymbol{x})}{\partial x_j} \tag{9.5.117}$$

where $\boldsymbol{e}_j$ is the unit vector in direction j.

- By posing $\boldsymbol{H}_k = \boldsymbol{G}(x_k)^T$ and using the following algorithm:

$$\boldsymbol{x}_{k+1} = \boldsymbol{x}_k - a_k\, \boldsymbol{G}(\boldsymbol{x}_k)^T\, \boldsymbol{g}(\boldsymbol{x}_k) \tag{9.5.118}$$

the inversion of the matrix can be avoided as well as solving the linear system on condition of properly choosing the sequence a_k. This algorithm will converge linearly and its convergence rate depends on the condition number of matrix $\boldsymbol{G}(\boldsymbol{x})$ at point $\boldsymbol{x}_0$, solution of $\boldsymbol{g}(\boldsymbol{x}) = 0$. Recall that the condition number of a matrix is equal to the ratio of its largest singular value to its lowest singular value (the singular values of $\boldsymbol{G}$ are the square roots of the eigenvalues of the product matrix $\boldsymbol{G}^T\boldsymbol{G}$). A possible choice for the sequence $\{a_k\}$ is

$$a_k = \frac{|\boldsymbol{p}_k|^2}{|\boldsymbol{G}(\boldsymbol{x}_k)\, \boldsymbol{p}_k|^2} \tag{9.5.119}$$

where $\boldsymbol{p}_k$ is any nonzero vector. However, we can take $\boldsymbol{p}_k = \boldsymbol{g}(\boldsymbol{x}_k)$ or $\boldsymbol{p}_k = \boldsymbol{G}(\boldsymbol{x}_k)^T\, \boldsymbol{g}(\boldsymbol{x}_k)$.

It must be noticed that the gradient of the function

$$f(\boldsymbol{x}) = 1/2\, \|\boldsymbol{g}(\boldsymbol{x})\|^2 = 1/2\, \boldsymbol{g}(\boldsymbol{x})^T\, \boldsymbol{g}(\boldsymbol{x}) \tag{9.5.120}$$

is equal to

$$\boldsymbol{f}_x(\boldsymbol{x}) = \boldsymbol{G}(\boldsymbol{x})^T\, \boldsymbol{g}(\boldsymbol{x}) \tag{9.5.121}$$

and that this algorithm written under the form

$$\boldsymbol{x}_{k+1} = \boldsymbol{x}_k - a_k\, \boldsymbol{f}_x(\boldsymbol{x}_k) \tag{9.5.122}$$

corresponds exactly to Equation (9.5.118). This is a gradient algorithm minimizing f.

- When $\boldsymbol{G}(\boldsymbol{x})$ is a symmetrical definite positive matrix, $\boldsymbol{H}_k = \boldsymbol{I}$ can be chosen, hence the simple algorithm

$$\boldsymbol{x}_{k+1} = \boldsymbol{x}_k - a_k\, \boldsymbol{g}(\boldsymbol{x}_k) \tag{9.5.123}$$

and a_k is frequently taken equal to

$$a_k = \frac{|\boldsymbol{v}_k|^2}{\boldsymbol{v}_k^T\, \boldsymbol{G}(\boldsymbol{x}_k)\, \boldsymbol{v}_k} \tag{9.5.124}$$

with, for example, $\boldsymbol{v}_k = \boldsymbol{g}(\boldsymbol{x}_k)$.
Again, it must be noticed that, in this case, $\boldsymbol{g}$ is the gradient of the function

$$f(\boldsymbol{x}) = \int_{\boldsymbol{x}_1}^{\boldsymbol{x}} \boldsymbol{g}(t)\, dt \tag{9.5.125}$$

so that this is a gradient algorithm minimizing f.
- A simple but important case occurs when the equation $\boldsymbol{g}(\boldsymbol{x}) = 0$ is linear. It can be set

$$\boldsymbol{g}(\boldsymbol{x}) = \boldsymbol{A}\,\boldsymbol{x} - \boldsymbol{h} \tag{9.5.126}$$

and solved by an algorithm such as

$$\boldsymbol{x}_{k+1} = \boldsymbol{x}_k - \boldsymbol{H}_k\,(\boldsymbol{A}\,\boldsymbol{x}_k - \boldsymbol{h}) \tag{9.5.127}$$

The solution is found in n iterations at most.

9.5.7 Quasi-Newton Method

An important drawback of Newton–Raphson method is the need to provide the Hessian matrix (matrix of second partial derivatives) or the Jacobian matrix.

The most obvious method to avoid that difficulty is to calculate the elements of the Hessian matrix of f by a finite difference method, for example:
the ith column of $\boldsymbol{G}$ is estimated by $[\boldsymbol{g}(\boldsymbol{x}^{(k)} + h_i\,\boldsymbol{e}_i) - \boldsymbol{g}(\boldsymbol{x}^{(k)})]/h_i$ or even better $[\boldsymbol{g}(\boldsymbol{x}^{(k)} + h_i\,\boldsymbol{e}_i) - \boldsymbol{g}(\boldsymbol{x}^{(k)} - h_i\,\boldsymbol{e}_i)]/(2h_i)$, where h_i represents an increment in direction $\boldsymbol{e}_i$. Then, an estimation $\hat{\boldsymbol{G}}$ of the matrix $\boldsymbol{G}$ is estimated. Knowing that the Hessian matrix is symmetrical, we take the matrix $\frac{1}{2}\,(\hat{\boldsymbol{G}} + \hat{\boldsymbol{G}}^T)$. However, the matrix thus obtained may not be definite positive.

The drawbacks previously cited are avoided by the quasi-Newton method. In this method, the inverse matrix $\boldsymbol{G}^{-1}$ is approximated by a definite positive matrix $\boldsymbol{H}^{(k)}$ and updated at each iteration. The algorithm has the following general form at iteration k:

(a) Pose $\boldsymbol{s}^{(k)} = -\boldsymbol{H}^{(k)}\,\boldsymbol{g}^{(k)}$.
(b) A search is done in the direction $\boldsymbol{s}^{(k)}$, which gives $\boldsymbol{x}^{(k+1)} = \boldsymbol{x}^{(k)} + \alpha^{(k)}\,\boldsymbol{s}^{(k)}$.
(c) Update $\boldsymbol{H}^{(k)}$, hence $\boldsymbol{H}^{(k+1)}$.

The initial matrix $\boldsymbol{H}^{(1)}$ can be any definite positive matrix. In the absence of choice criterion, take $\boldsymbol{H}^{(1)} = \boldsymbol{I}$. For the search of α, refer to the algorithms described in Section 9.5.4.

A major part of the algorithm resides in the update of the matrix $\boldsymbol{H}^{(k)}$. It is desired that $\boldsymbol{H}^{(k)}$ is similar to $[\boldsymbol{G}^{(k)}]^{-1}$. Set

$$\begin{aligned} \boldsymbol{\delta}^{(k)} &= \alpha^{(k)} \boldsymbol{s}^{(k)} = \boldsymbol{x}^{(k+1)} - \boldsymbol{x}^{(k)} \\ \boldsymbol{\gamma}^{(k)} &= \boldsymbol{g}^{(k+1)} - \boldsymbol{g}^{(k)} \end{aligned} \tag{9.5.128}$$

A first degree Taylor polynomial gives

$$\boldsymbol{\gamma}^{(k)} = \boldsymbol{G}^{(k)} \boldsymbol{\delta}^{(k)} + 0(\|\boldsymbol{\delta}^{(k)}\|) \tag{9.5.129}$$

So that the condition $\boldsymbol{H} \approx \boldsymbol{G}^{-1}$, equivalent to $\boldsymbol{H}^{(k)} \boldsymbol{\gamma}^{(k)} \approx \boldsymbol{\delta}^{(k)}$, is fulfilled, and we impose

$$\boldsymbol{H}^{(k+1)} \boldsymbol{\gamma}^{(k)} = \boldsymbol{\delta}^{(k)} \Longleftrightarrow \boldsymbol{H}^{(k+1)} (\boldsymbol{g}^{(k+1)} - \boldsymbol{g}^{(k)}) = \boldsymbol{x}^{(k+1)} - \boldsymbol{x}^{(k)} \tag{9.5.130}$$

This condition is sometimes called *quasi-Newton condition.*

Different manners of achieving this condition exist. One of them is simply to add a symmetrical matrix to $\boldsymbol{H}^{(k)}$

$$\boldsymbol{H}^{(k+1)} = \boldsymbol{H}^{(k)} + a\, \boldsymbol{u}\, \boldsymbol{u}^T \tag{9.5.131}$$

By relating this equation to quasi-Newton condition, it gives

$$\boldsymbol{H}^{(k)} \boldsymbol{\gamma}^{(k)} + a\, \boldsymbol{u}\, \boldsymbol{u}^T \boldsymbol{\gamma}^{(k)} = \boldsymbol{\delta}^{(k)} \tag{9.5.132}$$

thus $\boldsymbol{u}$ is proportional to $\boldsymbol{\delta}^{(k)} - \boldsymbol{H}^{(k)} \boldsymbol{\gamma}^{(k)}$. We can pose $\boldsymbol{u} = \boldsymbol{\delta}^{(k)} - \boldsymbol{H}^{(k)} \boldsymbol{\gamma}^{(k)}$, hence a. The formula of rank 1 results

$$\boldsymbol{H}^{(k+1)} = \boldsymbol{H} + \frac{(\boldsymbol{\delta} - \boldsymbol{H}\, \boldsymbol{\gamma})(\boldsymbol{\delta} - \boldsymbol{H}\, \boldsymbol{\gamma})^T}{(\boldsymbol{\delta} - \boldsymbol{H}\, \boldsymbol{\gamma})^T \boldsymbol{\gamma}} \tag{9.5.133}$$

Note that, in the right-hand member, the exponents (k) were omitted to avoid a too cumbersome notation.

It may happen that the denominator is zero. On another side, the matrix $\boldsymbol{H}^{(k)}$ does not remain definite positive. For those reasons, formulas of rank 2 have been developed. They take the form

$$\boldsymbol{H}^{(k+1)} = \boldsymbol{H}^{(k)} + a\, \boldsymbol{u}\, \boldsymbol{u}^T + b\, \boldsymbol{v}\, \boldsymbol{v}^T \tag{9.5.134}$$

Thus, Davidon–Fletcher–Powell method (Davidon 1959, 1991) gives

$$\boldsymbol{H}^{(k+1)} = \boldsymbol{H} + \frac{\boldsymbol{\delta}\, \boldsymbol{\delta}^T}{\boldsymbol{\delta}^T \boldsymbol{\gamma}} - \frac{\boldsymbol{H}\, \boldsymbol{\gamma}\, \boldsymbol{\gamma}^T \boldsymbol{H}}{\boldsymbol{\gamma}^T \boldsymbol{H}\, \boldsymbol{\gamma}} \tag{9.5.135}$$

Provided that $\boldsymbol{\delta}^T \boldsymbol{\gamma} > 0$, the matrix $\boldsymbol{H}^{(k)}$ remains definite positive (descent property).

When applied to quadratic functions, this method generates conjugate directions, and when $\boldsymbol{H}^{(1)} = \boldsymbol{I}$, it generates conjugate gradients.

When applied to any function, this method presents a superlinear order of convergence.

More recently, a method due to Broyden, Fletcher, Goldfarb, and Shanno (BFGS) (Broyden 1965, 1967, 1970) would give the “best” quasi-Newton method

$$\boldsymbol{H}^{(k+1)} = \boldsymbol{H} + \left(1 + \frac{\boldsymbol{\gamma}^T \boldsymbol{H} \boldsymbol{\gamma}}{\boldsymbol{\delta}^T \boldsymbol{\gamma}}\right) \frac{\boldsymbol{\delta} \boldsymbol{\delta}^T}{\boldsymbol{\delta}^T \boldsymbol{\gamma}} - \left(\frac{\boldsymbol{\delta} \boldsymbol{\gamma}^T \boldsymbol{H} + \boldsymbol{H} \boldsymbol{\gamma} \boldsymbol{\delta}^T}{\boldsymbol{\delta}^T \boldsymbol{\gamma}}\right) \tag{9.5.136}$$

This formula is said to be dual of the previous formula as, to pass from one to the other one, it suffices to exchange on one side $\boldsymbol{B}$ and $\boldsymbol{H}$ and on another side $\boldsymbol{\gamma}$ and $\boldsymbol{\delta}$ (by noting $\boldsymbol{B} = \boldsymbol{H}^{-1}$). In quadratic optimization and in successive quadratic optimization (Chapter 11), the BFGS formula is often used to approximate the Hessian matrix. In this same type of optimization, Powell (1978) proposes a modified version of the BFGS formula.

To find the value of α along the direction $\boldsymbol{s}$, one of the algorithms described in Section 9.5.4 can be used.

Example 9.10 :

Quasi-Newton method (Davidon, Fletcher, and Powell): minimization of a quadratic function

The following quadratic function

$$f(\boldsymbol{x}) = 10\, x_1^2 + x_2^2 \tag{9.5.137}$$

is minimized by a quasi-Newton method, presently Davidon, Fletcher, and Powell method.

The initial point is $\boldsymbol{x} = (0.1, 1)$. The initial matrix $\boldsymbol{H}$ is the identity matrix.

The search algorithm of α minimizing $f(\boldsymbol{x} + \alpha\, \boldsymbol{s})$ by bracketing–shrinkage is used (Section 9.5.4).

Table 9.8 summarizes the values found by the different variables in the order of the calculations.

Table 9.8 Quasi-Newton method (Davidon, Fletcher, and Powell): minimization of function f

n	$\boldsymbol{x}^{(k)}$	$\boldsymbol{g}^{(k)}$	$\boldsymbol{\delta}^{(k-1)}$	$\boldsymbol{\gamma}^{(k-1)}$	$\boldsymbol{H}^{(k)}$	$\boldsymbol{s}^{(k)}$	$\alpha^{(k)}$
1	$\begin{bmatrix}0.10\\1.0\end{bmatrix}$	$\begin{bmatrix}2\\2\end{bmatrix}$			$\begin{bmatrix}1&0\\0&1\end{bmatrix}$	$\begin{bmatrix}-2.0\\-2.0\end{bmatrix}$	0.09116
2	$\begin{bmatrix}-0.082321\\0.81768\end{bmatrix}$	$\begin{bmatrix}-1.6464\\1.6464\end{bmatrix}$	$\begin{bmatrix}-0.18232\\-0.18232\end{bmatrix}$	$\begin{bmatrix}-3.6464\\-0.36464\end{bmatrix}$	$\begin{bmatrix}0.055356&-0.053555\\-0.053555&1.0356\end{bmatrix}$	$\begin{bmatrix}0.17872\\-1.7817\end{bmatrix}$	0.46173
3	$\begin{bmatrix}0.00020\\-0.00497\end{bmatrix}$	$\begin{bmatrix}0.00400\\-0.00995\end{bmatrix}$	$\begin{bmatrix}0.082521\\-0.82266\end{bmatrix}$	$\begin{bmatrix}1.6504\\-1.6453\end{bmatrix}$	$\begin{bmatrix}0.05000&0.196\ 10^{-7}\\0.196\ 10^{-7}&0.5000\end{bmatrix}$	$\begin{bmatrix}-0.00020\\0.00498\end{bmatrix}$	1.0057

The calculations were executed with $\sigma = 0.01$. The solution is thus

$$\boldsymbol{x} = ().00020, -0.00497) \tag{9.5.138}$$

close to the origin (within tolerances).

Example 9.11 :

Quasi-Newton method (BFGS: Broyden, Fletcher, Goldfarb, and Shanno): minimization of Rosenbrock’s function

The BFGS method (Broyden, Fletcher, Goldfarb, and Shanno) has been assessed on the example of Rosenbrock’s function

$$f(\boldsymbol{x}) = 100\,(x_2 - x_1^2)^2 + (1 - x_1)^2 \tag{9.5.139}$$

The initial point is (2, 2). Table 9.9 shows the excellent convergence of this method which can be compared to Levenberg–Marquardt method (Table 9.12). The algorithms of bracketing and shrinkage described in section 9.5.4 were used. The gradient and the Jacobian matrix were provided analytically. For a more complex function, they could be calculated numerically. It will be noted that, compared to Newton–Raphson method (Table 9.7), the convergence is slower, but it was not necessary to invert a matrix. In Figure 9.17, it can be noticed that, except the first iteration where the matrix $\boldsymbol{H}$ is taken as identity, the points converge rapidly to the sought optimum as, at the third iteration, the point is already very close to the optimum. With respect to Newton–Raphson method (Figure 9.16), the trajectory to convergence is quite different.

Table 9.9 Quasi-Newton method (BFGS method): minimization of Rosenbrock's function

k	x_1	x_2	$f(x)$
0	2.0000	2.0000	0.4010×10^{3}
1	1.2119	2.1968	5.3054×10^{1}
2	0.8447	0.3702	1.1818×10^{1}
3	0.9680	0.8883	2.3831×10^{-1}
4	0.8618	0.9716	2.8053×10^{-4}
5	0.9866	0.9734	1.8053×10^{-4}
6	0.9993	0.9988	4.4638×10^{-6}
7	1.0003	1.0006	1.1797×10^{-7}
8	1.0000	1.0000	2.9513×10^{-10}

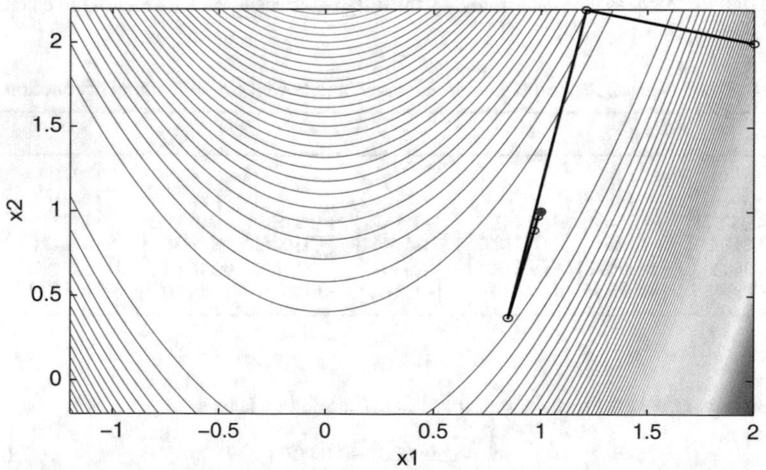

Fig. 9.17 Quasi-Newton method (BFGS method): contours of Rosenbrock's function and trajectory followed during the iterations

Example 9.12 :

Quasi-Newton method (BFGS): parameter estimation for saturated vapor equation

An experimenter measured saturated vapor pressures of ethane at moderate pressures and obtained the results of Table 9.10. He desires to use Antoine's law to represent these results

$$\ln(P) = A - \frac{B}{C+T} \tag{9.5.140}$$

Table 9.10 Experimental saturated vapor pressure of ethane versus temperature

T (K)	150	160	170	180	190	200
P (atm)	0.095	0.21	0.42	0.78	1.34	2.15

The model parameters are determined by using Broyden, Fletcher, Goldfarb, and Shanno method (BFGS). The criterion to minimize with respect to the vector of parameters is

$$J = \sum_{i=1}^{n} (y_i^{exp} - y_i^{mod})^2 \tag{9.5.141}$$

where n is the number of experiments, y^{exp} the experimental pressure, and y^{mod} the pressure given by the model (9.5.140) depending on the parameter vector.

The method of parameter search α by bracketing and shrink (Section 9.5.4) was used and σ was chosen equal to 0.1 for a precise search of α. In the absence of preliminary information, the following initial point chosen for the parameters is $[A = 1, B = 1, C = 0]$. The BFGS method yields the following parameters:
$[A = 9.1894, B = 1.5550\,10^3, C = -15.3166]$ with a relative accuracy of 10^{-6} in 26 iterations (Table 9.11).

The agreement between the experimental points and the model appears clearly in Figure 9.18.

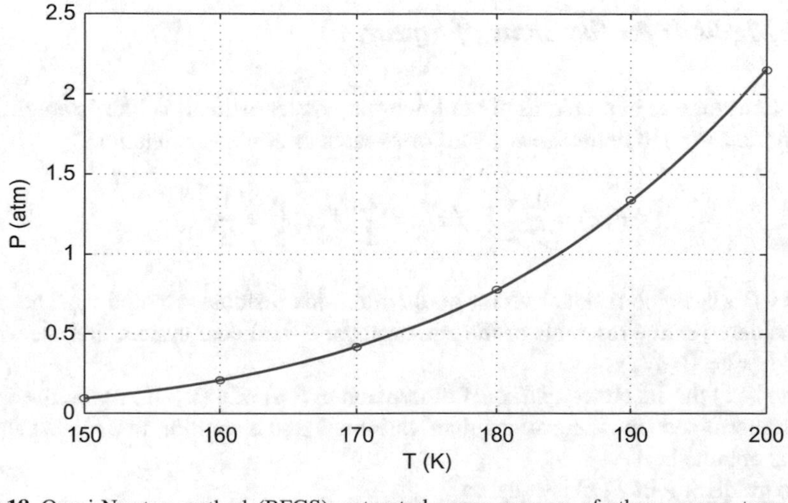

Fig. 9.18 Quasi-Newton method (BFGS): saturated vapor pressure of ethane versus temperature (model and experimental points) after optimization

Table 9.11 Quasi-Newton method (BFGS): search of parameters of Antoine's law during the first 20 iterations

i	A	B	C	J
0	1.0000	1.0000	0.0000	0.2328×10^{2}
1	−0.5815	1.0097	−0.0001	0.6858×10^{1}
2	−0.6575	1.0176	−0.0002	0.6825×10^{1}
3	−0.6526	1.0443	−0.0005	0.6825×10^{1}
4	11.3140	1766.6587	−21.1094	0.1090×10^{1}
5	11.8406	1858.9720	−22.2873	0.8392
6	8.8411	1473.6583	−18.3121	0.3142×10^{-2}
7	8.9811	1490.8260	−18.4831	0.1576×10^{-3}
8	8.9846	1491.6234	−18.4943	0.1169×10^{-3}
9	8.9880	1492.0946	−18.4996	0.1161×10^{-3}
10	8.9878	1492.0650	−18.4992	0.1160×10^{-3}
11	8.9884	1492.1682	−18.4972	0.1160×10^{-3}
12	9.0188	1502.6588	−17.9127	0.1110×10^{-3}
13	9.0916	1524.7525	−16.8039	0.9980×10^{-4}
14	9.0875	1523.0352	−16.9216	0.9735×10^{-4}
15	9.1197	1532.5638	−16.4628	0.9546×10^{-4}
16	9.1657	1547.4662	−15.6897	0.9206×10^{-4}
17	9.1596	1545.6542	−15.7802	0.9158×10^{-4}
18	9.1864	1554.0163	−15.3636	0.9106×10^{-4}
19	9.1892	1554.9021	−15.3200	0.9105×10^{-4}
20	9.1894	1554.9598	−15.3171	0.9105×10^{-4}

9.5.8 Methods for the Sums of Squares

Very often, such as in problems of nonlinear parameter estimation like Example 9.12, the function $F(\boldsymbol{x})$ to minimize is a sum of squares of nonlinear functions

$$F(\boldsymbol{x}) = \frac{1}{2} \sum_{i=1}^{m} f_i(\boldsymbol{x})^2 = \frac{1}{2} \|\boldsymbol{f}(\boldsymbol{x})\|_2^2 = \frac{1}{2} \boldsymbol{f}^T \boldsymbol{f} \tag{9.5.142}$$

where $\|\boldsymbol{f}(\boldsymbol{x})\|$ is the residual vector at $\boldsymbol{x}$ (often this residual is noted $\boldsymbol{r}$). The f_i can be considered as the residuals of the equations of a nonlinear system that we want to solve ($f_i(\boldsymbol{x}) = 0 \quad \forall i$).

Note $J(\boldsymbol{x})$ the Jacobian matrix (of dimension $m \times n$) of $\boldsymbol{f}(\boldsymbol{x})$. If $m > n$, the system is overdetermined (more equations than variables) and a solution in the least squares sense is obtained.

The gradient $\boldsymbol{g}$ of $F(\boldsymbol{x})$ is equal to

$$\boldsymbol{g}(\boldsymbol{x}) = \boldsymbol{J}(\boldsymbol{x})^T \, \boldsymbol{f}(\boldsymbol{x}) \tag{9.5.143}$$

and the Hessian $\boldsymbol{H}$ of $F(\boldsymbol{x})$

$$\boldsymbol{H}(\boldsymbol{x}) = \boldsymbol{J}(\boldsymbol{x})^T \boldsymbol{J}(\boldsymbol{x}) + \boldsymbol{Q}(\boldsymbol{x}) \tag{9.5.144}$$

with

$$\boldsymbol{Q}(\boldsymbol{x}) = \sum_{i=1}^{m} f_i(\boldsymbol{x})\, \boldsymbol{G}_i(\boldsymbol{x}) \tag{9.5.145}$$

where $\boldsymbol{G}_i(\boldsymbol{x})$ is the Hessian matrix of $f_i(\boldsymbol{x})$:

$$\boldsymbol{G}_i(\boldsymbol{x}) = \begin{bmatrix} \dfrac{\partial^2 f_i}{\partial x_1^2} & \cdots & \dfrac{\partial^2 f_i}{\partial x_1 \partial x_n} \\ \vdots & & \vdots \\ \dfrac{\partial^2 f_i}{\partial x_n \partial x_1} & \cdots & \dfrac{\partial^2 f_i}{\partial x_n^2} \end{bmatrix} \tag{9.5.146}$$

Frequently, an approximation of $\boldsymbol{H}(\boldsymbol{x})$ is $\boldsymbol{J}(\boldsymbol{x})^T \boldsymbol{J}(\boldsymbol{x})$, which amounts to take a linear approximation for the residuals f_i.

9.5.9 Gauss–Newton Method

For a quadratic function of the form

$$\frac{1}{2}\boldsymbol{p}_k^T \boldsymbol{H}_k \boldsymbol{p}_k + \boldsymbol{g}_k^T \boldsymbol{p}_k + c_k \tag{9.5.147}$$

the condition that a point $\boldsymbol{p}_k$ is stationary is that it verifies the following linear system:

$$\boldsymbol{H}_k\, \boldsymbol{p}_k = -\boldsymbol{g}_k \tag{9.5.148}$$

which gives

$$(\boldsymbol{J}_k^T \boldsymbol{J}_k + \boldsymbol{Q}_k)\, \boldsymbol{p}_k = -\boldsymbol{J}_k^T \boldsymbol{f}_k \tag{9.5.149}$$

Suppose that $\boldsymbol{p}_N$ is the solution of the system, called Newton's direction. If $\|\boldsymbol{f}_k\| \to 0$, when $\boldsymbol{x}_k$ gets closer to the solution, then $\boldsymbol{Q}_k \to 0$, thus Newton's direction can be approximated by the solution of the system

$$\boldsymbol{J}_k^T \boldsymbol{J}_k \boldsymbol{p}_k = -\boldsymbol{J}_k^T \boldsymbol{f}_k \tag{9.5.150}$$

and then

$$\boldsymbol{x}_{k+1} = \boldsymbol{x}_k + \boldsymbol{p}_k \tag{9.5.151}$$

The solution of problem (9.5.149) is a solution of the problem of linear least squares

$$\min_{p} \frac{1}{2} \|\boldsymbol{J}_k\, \boldsymbol{p} + \boldsymbol{f}_k\|_2^2 \tag{9.5.152}$$

and is unique when $\boldsymbol{J}_k$ is of full rank. The vector solution of this problem is called Gauss–Newton direction and noted $\boldsymbol{p}_{GN}$. If $\boldsymbol{J}_k$ is of full column rank, Gauss–Newton direction approaches Newton's direction when $\|\boldsymbol{Q}_k\| \to 0$.

A drawback of this method, applied in this manner, is that the condition number of $(\boldsymbol{J}_k^T \boldsymbol{J}_k)$ is the square of that of $\boldsymbol{J}_k$; thus, $(\boldsymbol{J}_k^T \boldsymbol{J}_k)$ is less well-conditioned than $\boldsymbol{J}_k$ and it produces errors. Note that a corrected Gauss–Newton method exists.

9.5.10 Levenberg–Marquardt Method

The Levenberg–Marquardt method is an efficient alternative to Gauss–Newton method. The search direction of Levenberg–Marquardt is the solution of equation

$$(\boldsymbol{G}_k + \lambda_k \boldsymbol{I})\, \boldsymbol{p}_k = -\boldsymbol{g}_k \tag{9.5.153}$$

where λ_k is a nonnegative scalar. A unit displacement along $\boldsymbol{p}_k$ is always chosen so that

$$\boldsymbol{x}_{k+1} = \boldsymbol{x}_k + \boldsymbol{p}_k \tag{9.5.154}$$

λ_k must be chosen adequately:
if $\lambda_k = 0$, $\boldsymbol{p}_k$ is Gauss–Newton direction, and
if $\lambda_k \to \infty$, $\boldsymbol{p}_k$ becomes parallel to the direction of the steepest descent.

9.5.10.1 Case of Any Function

First, consider the case of any function $f(\boldsymbol{x})$ to be minimized.

Supposing a displacement $\boldsymbol{p}_k$ from $\boldsymbol{x}_k$, the function $f(\boldsymbol{x}_k + \boldsymbol{p}_k)$ is approximated as

$$f(\boldsymbol{x}_{k+1}) = f(\boldsymbol{x}_k + \boldsymbol{p}_k) \approx q_k(\boldsymbol{p}_k) = \frac{1}{2}\boldsymbol{p}_k^T \boldsymbol{G}_k \boldsymbol{p}_k + \boldsymbol{g}_k^T p_k + f(\boldsymbol{x}_k) \tag{9.5.155}$$

by noting $\boldsymbol{g}_k$ the gradient of f and $\boldsymbol{G}_k$ the Hessian matrix of f.

Levenberg–Marquardt search direction is the solution of equation

$$(\boldsymbol{G}_k + \lambda_k \boldsymbol{I})\, \boldsymbol{p}_k = -\boldsymbol{g}_k \tag{9.5.156}$$

which consists to add a bias in the direction of the steepest descent with respect to Newton's method which would have a direction $\boldsymbol{p}_k$ such that

$$\boldsymbol{G}_k\, \boldsymbol{p}_k = -\boldsymbol{g}_k \tag{9.5.157}$$

A usual Levenberg–Marquardt algorithm is as follows:

1. Given $\boldsymbol{x}_k$ and λ_k, calculate $\boldsymbol{g}_k$ and $\boldsymbol{G}_k$.
2. Factorize $(\boldsymbol{G}_k + \lambda_k \boldsymbol{I})$. If this matrix is not definite positive, pose $\lambda_k = 4\lambda_k$ and restart.

3. Calculate $\boldsymbol{p}_k$ by Equation (9.5.156).
4. Evaluate $f(\boldsymbol{x}_k + \boldsymbol{p}_k)$ and the ratio

$$\rho_k = \frac{\Delta f_k}{\Delta q_k} = \frac{f_k - f_{k+1}}{f_k - q_k} = \frac{f(\boldsymbol{x}_k) - f(\boldsymbol{x}_k + \boldsymbol{p}_k)}{f(\boldsymbol{x}_k) - q(\boldsymbol{p}_k)}$$

5. If $\rho_k < 0.25$, pose $\lambda_{k+1} = 4\lambda_k$,
   if $\rho_k > 0.75$, pose $\lambda_{k+1} = \lambda_k/2$, and
   if $0.25 \leq \rho_k \leq 0.75$, pose $\lambda_{k+1} = \lambda_k$.
6. If $\rho_k \leq 0$, pose $\boldsymbol{x}_{k+1} = \boldsymbol{x}_k$. If $\rho_k > 0$, calculate $\boldsymbol{x}_{k+1} = \boldsymbol{x}_k + \boldsymbol{p}_k$.

The ratio ρ_k comes from the use of a trust region (Moré and Wright 1994; Nocedal and Wright 2006), which is an alternative to the use of a search in a given direction. In some cases, the approximation q_k by Equation (9.5.155) is valid only in the neighborhood of $\boldsymbol{x}_k$. It may even happen that $\boldsymbol{G}_k$ is not definite, and thus q_k would not be bounded. In this case, it is possible to find the step $\boldsymbol{p}_k$ by solving

$$\min_{\boldsymbol{p}} \; q_k(\boldsymbol{p}) \quad \text{such that} \quad ||\boldsymbol{D}_k \boldsymbol{p}||_2 \leq \Delta_k \tag{9.5.158}$$

where $\boldsymbol{D}_k$ is a transformation matrix which acts on $\boldsymbol{p}$ and Δ_k the radius of the trust region. The ratio ρ_k gives an indication of the precision of the approximation of $f(\boldsymbol{x}_k + \boldsymbol{p}_k)$ by q_k, which depends on vector $\boldsymbol{p}_k$. The "dogleg" method (Nocedal and Wright 2006; Wang 2006) is a particular case of trust region methods.

9.5.10.2 Case of a Sum of Squares

Consider now the minimization problem of the function $F(\boldsymbol{x})$ defined by Equation (9.5.142).

Pose

$$\boldsymbol{G}_k \approx \boldsymbol{J}_k^T \boldsymbol{J}_k \quad \text{and} \quad \boldsymbol{g}_k = \boldsymbol{J}_k^T \boldsymbol{f}_k \tag{9.5.159}$$

In this case, $\boldsymbol{G}_k$ is not equal to the Hessian matrix of F but constitutes an approximation of it.

The remaining of Levenberg–Marquardt method is identical on the condition of replacing f of the previous part by F.

Supposing a displacement $\boldsymbol{p}_k$ from $\boldsymbol{x}_k$, the function $F(\boldsymbol{x}_k + \boldsymbol{p}_k)$ is approximated as

$$F(\boldsymbol{x}_k + \boldsymbol{p}_k) \approx q_k(\boldsymbol{p}_k) = \frac{1}{2}\boldsymbol{p}_k^T \boldsymbol{G}_k \boldsymbol{p}_k + \boldsymbol{g}_k^T p_k + F(\boldsymbol{x}_k)$$

Levenberg–Marquardt algorithm becomes:

1. Given $\boldsymbol{x}_k$ and λ_k, calculate $\boldsymbol{g}_k$ and $\boldsymbol{G}_k$.
2. Factorize $(\boldsymbol{G}_k + \lambda_k \boldsymbol{I})$. If this matrix is not definite positive, pose $\lambda_k = 4\lambda_k$ and restart.

3. Calculate $\boldsymbol{p}_k$ by Equation (9.5.153).
4. Evaluate $F(\boldsymbol{x}_k + \boldsymbol{p}_k)$ and the ratio

$$\rho_k = \frac{\Delta F_k}{\Delta q_k} = \frac{F_k - F_{k+1}}{F_k - q_k} = \frac{F(\boldsymbol{x}_k) - F(\boldsymbol{x}_k + \boldsymbol{p}_k)}{F(\boldsymbol{x}_k) - q(\boldsymbol{p}_k)}$$

5. If $\rho_k < 0.25$, pose $\lambda_{k+1} = 4\lambda_k$,
   if $\rho_k > 0.75$, pose $\lambda_{k+1} = \lambda_k/2$, and
   if $0.25 \leq \rho_k \leq 0.75$, pose $\lambda_{k+1} = \lambda_k$.
6. If $\rho_k \leq 0$, pose $\boldsymbol{x}_{k+1} = \boldsymbol{x}_k$. If $\rho_k > 0$, calculate $\boldsymbol{x}_{k+1} = \boldsymbol{x}_k + \boldsymbol{p}_k$.
   The ratio ρ_k gives an indication of the precision of the approximation of $F(\boldsymbol{x}_k + \boldsymbol{p}_k)$ by q_k, which depends on vector $\boldsymbol{p}_k$.

Example 9.13 :
Levenberg–Marquardt method: minimization of Rosenbrock's function

Levenberg–Marquardt method was tested on the example of Rosenbrock's function

$$f(\boldsymbol{x}) = 100\,(x_2 - x_1^2)^2 + (1 - x_1)^2 \tag{9.5.160}$$

The initial point is (2, 2). Table 9.12 shows the excellent convergence of this method. λ was initially chosen equal to 1. The gradient and the Jacobian matrix were given analytically. Figure 9.19 shows that, except the first iteration, the points remarkably follow the valley of Rosenbrock's function. With respect to BFGS method (Figure 9.17), the behavior is relatively different and shows a little slower convergence. The point at first iteration is nearly the same as for Newton–Raphson method (Figure 9.16), and then the trajectory is totally different.

Table 9.12 Levenberg–Marquardt method: minimization of Rosenbrock's function

k	x_1	x_2	$f(x)$
0	2.0000	2.0000	0.4010×10^3
1	1.9878	3.9416	0.9855
2	1.8486	3.3991	0.7526
3	1.7155	2.9258	0.5412
4	1.5777	2.4706	0.3687
5	1.4637	2.1294	0.2316
6	1.3392	1.7781	0.1389
7	1.2574	1.5744	0.7072×10^{-1}
8	1.1485	1.3072	0.3609×10^{-1}
9	1.1046	1.2182	0.1131×10^{-1}
10	1.0294	1.0540	0.4056×10^{-2}
11	1.0156	1.0313	0.2475×10^{-3}
12	1.0006	1.0010	0.5444×10^{-5}
13	1.0000	1.0001	0.6750×10^{-9}

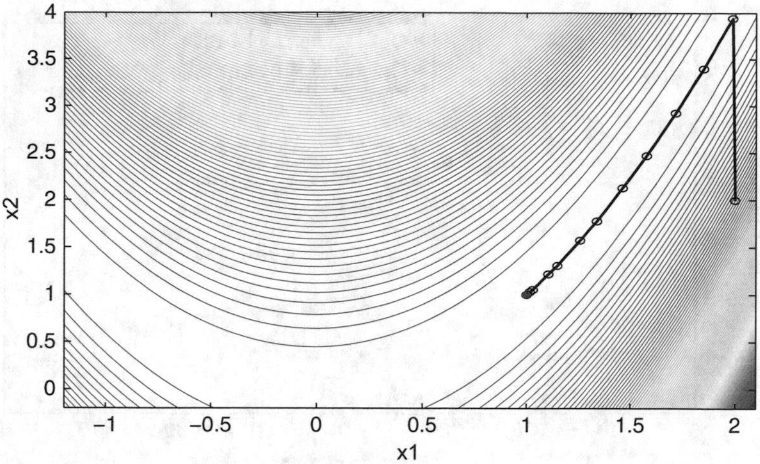

Fig. 9.19 Levenberg–Marquardt method: contours of Rosenbrock's function and trajectory followed during the iterations

Example 9.14 :
Levenberg–Marquardt method: estimation of Antoine's law parameters

Consider again Example 9.12 of estimation of Antoine's law parameters. The parameters of this model are determined by using Levenberg–Marquardt method. The same initial point (1, 1, 0) is chosen.

Levenberg–Marquardt method gives the following parameters:
[$A = 9.1894$, $B = 1.5550\ 10^3$, and $C = -15.3165$] after 76 iterations.

Figure 9.20 compares the convergence of Broyden, Fletcher, Goldfarb, and Shanno (BFGS) and Levenberg–Marquard methods by plotting the optimized criterion along the iterations. The BFGS method converges much faster than Levenberg–Marquard method, and however this is done at the price of a heavy search by the bracketing and shrinkage algorithm renewed at each iteration. Another search algorithm could be used (Section 9.5.4). The convergence seems to be executed by levels. The iterations of Levenberg–Marquard method require less calculations and are faster. In this case, the convergence is more regular and slower.

9.5.11 Quasi-Newton Approximation

For the problems where the residual $\|f(x)\|$ is large, Gauss–Newton and Levenberg–Marquardt methods are not well adapted. It is considered that a problem is at large

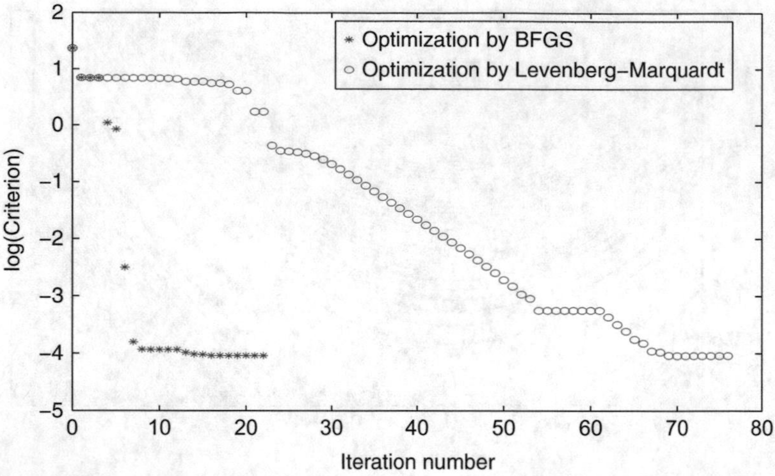

Fig. 9.20 Evolution along iterations of the optimized criterion for Antoine's law by BFGS and Levenberg–Marquard methods

residual $\|\boldsymbol{f}(\boldsymbol{x})\|$ when $\|\boldsymbol{f}(\boldsymbol{x}^*)\|$ is larger than the eigenvalues of $[\boldsymbol{J}(\boldsymbol{x}^*)^T\, \boldsymbol{J}(\boldsymbol{x}^*)]$. Then, we can include a quasi-Newton approximation $\boldsymbol{M}_k$ of the unknown term $\boldsymbol{Q}(x)$. The search direction $\boldsymbol{p}_k$ in this case is given by

$$(\boldsymbol{J}_k^T\, \boldsymbol{J}_k + \boldsymbol{M}_k)\boldsymbol{p}_k = -\boldsymbol{J}_k^T \boldsymbol{f}_k \tag{9.5.161}$$

and the condition on $\boldsymbol{M}_{k+1}$ is

$$(\boldsymbol{J}_{k+1}^T\, \boldsymbol{J}_{k+1} + \boldsymbol{M}_{k+1})\boldsymbol{s}_k = \boldsymbol{y}_k \tag{9.5.162}$$

with

$$\boldsymbol{s}_k = \boldsymbol{x}_{k+1} - \boldsymbol{x}_k\,, \quad \boldsymbol{y}_k = \boldsymbol{J}_{k+1}^T\, \boldsymbol{f}_{k+1} - \boldsymbol{J}_k^T \boldsymbol{f}_k \tag{9.5.163}$$

To calculate the new value $\boldsymbol{M}_k$, the BFGS formula can be used

$$\boldsymbol{M}_{k+1} = \boldsymbol{M}_k - \frac{1}{\boldsymbol{s}_k^T\, \boldsymbol{W}_k\, \boldsymbol{s}_k} \boldsymbol{W}_k\, \boldsymbol{s}_k\, \boldsymbol{s}_k^T\, \boldsymbol{W}_k + \frac{1}{\boldsymbol{y}_k^T\, \boldsymbol{s}_k} \boldsymbol{y}_k\, \boldsymbol{y}_k^T \tag{9.5.164}$$

with

$$\boldsymbol{W}_k = \boldsymbol{J}_{k+1}^T\, \boldsymbol{J}_{k+1} + \boldsymbol{M}_k \tag{9.5.165}$$

9.5.12 *Systems of Nonlinear Equations*

When the following system of nonlinear equations is to be solved:

$$\boldsymbol{f}(\boldsymbol{x}) = 0 \tag{9.5.166}$$

where $\boldsymbol{f}$ is a vector, this problem can be transformed into solving the following problem:

$$\min \boldsymbol{f}^T(\boldsymbol{x})\,\boldsymbol{f}(\boldsymbol{x}) \qquad \text{or} \qquad \min \sum \frac{1}{2} f_i^2(\boldsymbol{x}) \tag{9.5.167}$$

Most often, it is necessary to weight the functions to guarantee similar orders of magnitude, which gives

$$\min \sum \frac{1}{2} w_i\, f_i^2(\boldsymbol{x}), \quad w_i \geq 0 \tag{9.5.168}$$

9.6 Discussion and Conclusion

The present chapter proposes a large number of methods of direct search or of gradient type. The great advantage of the methods of direct search is that they only need the value of the function, without any gradient. The great disadvantage of all these methods is the impossible prediction of the rate of convergence and their restriction to problems of small dimension. The methods of evolutionary type such as the genetic algorithms are very popular, as they are apparently simple to understand, they possess properties of quasi-global optimization, but they would earn a lot to be coupled to other more rigorous methods. The gradient methods offer qualities of robustness and guaranty of convergence, and they are well adapted to large scale problems. Among the methods of quasi-Newton type inspired by Newton–Raphson method, the BFGS method of Broyden, Fletcher, Goldfarb, and Shanno, and the Levenberg–Marquardt method are recommended. The drawback of gradient methods is the risk of a local solution. The systems of nonlinear equations, even of large dimension, are in general solved by these methods by transforming the problem under the form of the minimization of a sum of squares. The use of specialized libraries, open (MINPACK) or commercial (GAMS, NAG, IMSL), is recommended.

9.7 Exercise Set

Preamble: In this section, the students can use existing codes. However, if they develop their own codes written in order to solve general problems, this will constitute an excellent project for each proposed method.

Exercise 9.7.1 (Medium)
Find the minimum of the function

$$f(\boldsymbol{x}) = x_1^2 - x_1 x_2 + x_2^2 - 3x_1 \tag{9.7.1}$$

starting from point (0,0) by different methods:

1. Newton–Raphson method.
2. Conjugate gradient method.
3. Rosenbrock's method.

Solution: $x_1 = 2$, $x_2 = 1$, $f = -3$.

Exercise 9.7.2 (Easy)
Find the minimum of the function

$$f(\boldsymbol{x}) = 9x_1^2 + 4x_1 x_2 + 4x_2^2 - 18x_1 - 4x_2 + 9 \tag{9.7.2}$$

by the conjugate gradient method, starting from point (2,2).
Solution: $x_1 = 1$, $x_2 = 0$, $f = 0$.

Exercise 9.7.3 (Easy)
Find the minimum of the function

$$f(\boldsymbol{x}) = 5x_1^2 + x_2^2 + x_3^2 - 4x_1 x_2 - 2x_1 - 6x_3 \tag{9.7.3}$$

by the conjugate gradient method, starting from point (0,0,0).
Solution: $x_1 = 1$, $x_2 = 2$, $x_3 = 3$, $f = -10$.

Exercise 9.7.4 (Easy)
Find the minimum of the function

$$f(\boldsymbol{x}) = 5x_1^2 + 6x_1 x_2 + 5x_2^2 + 8x_1 + 24x_2 + 32 \tag{9.7.4}$$

by the conjugate gradient method, starting from point (0,0).
Solution: $x_1 = 1$, $x_2 = -3$, $f = 0$.

Exercise 9.7.5 (Easy)
Determine the nature of the extremum of the function

$$f(\boldsymbol{x}) = (x_1 - 2)^2 + (x_2 - 5)^2 - 4x_1^2 x_2 \tag{9.7.5}$$

calculated by Newton–Raphson method, starting from point (0,0).
Solution: $x_1 = -0.104774$, $x_2 = 5.02194$, $f = 4.2105$, saddle point.

Exercise 9.7.6 (Medium)
Find the minimum of the function

$$f(\boldsymbol{x}) = 2x_1^4 - x_1^2 x_2 + x_2^2 + 3x_1 - 7 \tag{9.7.6}$$

starting from point $x^{(1)} = (2, 1)$ by two different methods:

1. Newton–Raphson method.
2. Conjugate gradient method.
   Write a program for each method.

Solution: $x_1 = -0.753947$, $x_2 = -0.28428$, $f = -8.6964$.

Exercise 9.7.7 (Medium)
Find the minimum of the function

$$f(\boldsymbol{x}) = x_1^3 - 2x_1 + 7 + 5x_2^2 + 3x_1x_2^2 \tag{9.7.7}$$

starting from point (1, 1) by two different methods:

1. Newton–Raphson method.
2. Conjugate gradient method.

Write a program for each method.
Solution: $x_1 = 0.816496$, $x_2 = 0$, $f = 5.911$.

Exercise 9.7.8 (Medium)
Find the minimum of the function

$$f(\boldsymbol{x}) = 3x_1^2 - 2x_1 + 4x_1x_2 - x_2 + 2x_2^2 + 1 \tag{9.7.8}$$

starting from point $x^{(1)} = (1, 1)$ by two different methods:

1. Newton–Raphson method
2. Conjugate gradient method.

Solution: $x_1 = 0.5$, $x_2 = -0.25$.

Exercise 9.7.9 (Medium)
Find the minimum of the function

$$f(\boldsymbol{x}) = (x_1 + 2\,x_2)^2 + (2\,(x_1 + 1)^2 + (x_2 + 1)^2 - 1)^2 \tag{9.7.9}$$

by Newton–Raphson method starting from point (0.5, 0.5). Calculate the value of f at the extremum. Verify that it is a minimum.
Do the similar calculations using the conjugate gradient method.

Exercise 9.7.10 (Medium)
Find the minimum of the function

$$f(\boldsymbol{x}) = x_1 + 2\,x_2^2 + \exp(x_1^2 + x_2^2) \tag{9.7.10}$$

starting from point (1, 1) by two different methods:

1. Newton–Raphson method
2. Conjugate gradient method.

Write a program for each method.

Exercise 9.7.11 (Medium)
Find the maximum of the function

$$f(\boldsymbol{x}) = (x_1 - 2)^2 + (x_2 - 5)^2 - 4\,x_1^2\,x_2 \tag{9.7.11}$$

1. Use Newton–Raphson method starting from initial point (0, 0).
   Present the results in a table giving at each iteration the new values of $\boldsymbol{x}$ and the value of the function f. Note $\boldsymbol{x}^*$ the result obtained after convergence. Comment these results.
2. Study the Hessian matrix and discuss.
3. Setting $x_2^* = \alpha\, x_1^*$, make a cross section by the vertical plane of equation $x_2 = \alpha\, x_1$ and discuss. Make a graphical representation.

Exercise 9.7.12 (Easy)
Find the minimum of the function $f(\boldsymbol{x})$ found in a design of experiment

$$\begin{aligned} f(\boldsymbol{x}) = 0.4323 + 0.3996\, x_1 + 0.1289\, x_2 + 0.06340\, x_3 + \\ 1.1681\, x_1^2 + 0.3632\, x_2^2 + 0.2106\, x_3^2 + \\ 1.1353\, x_1\, x_2 + 0.7415\, x_1\, x_3 + 0.5349\, x_2\, x_3 \end{aligned} \tag{9.7.12}$$

by the conjugate gradient method starting from initial point (1 , 1 , 1).

Exercise 9.7.13 (Easy)
Find the minimum of the function

$$y(\boldsymbol{t}) = \frac{2}{t_1\, t_2} + \frac{3}{t_2} + 4\, t_1\, t_2^2 + 6\, t_2^2 \tag{9.7.13}$$

by Newton–Raphson method. Verify a posteriori the constraint $t_i \geq 0 \quad \forall i$. If this constraint was not respected, solve the problem by the usual methods.

Remark: as always, an optimization method depends on the initial point. In the present case, according to the initialization, Newton–Raphson method may converge rather slowly (25 to 30 iterations) even passing through negative values of t_j (which is an infeasible path) before giving the converged result. Do not hesitate to take another initialization in the case of doubt.
Solution: $t_1 = 1.7863$, $t_2 = 0.5391$, $y = 11.4621$.

Exercise 9.7.14 (Difficult)

Conjugate gradient method according to Fletcher and Reeves

We desire to find the minimum of the function $f(\boldsymbol{x})$

$$f(\boldsymbol{x}) = (x_1 - 2)^4 + (x_1 - 2\, x_2)^2 \tag{9.7.14}$$

We propose to use the conjugate gradient method according to Fletcher and Reeves (1964) that can be applied to a nonquadratic function in $\mathcal{R}^n$. Below, find the description of the method:

- Initialization stage:
  Choose a scalar $\epsilon > 0$ and an initial point $\boldsymbol{x}_1$. Let $\boldsymbol{y}_1 = \boldsymbol{x}_1$ and $\boldsymbol{d}_1 = -\boldsymbol{f}_x(\boldsymbol{y}_1)$, $k = j = 1$. Proceed to the main steps.
- Main stages:

  1. If $\|\boldsymbol{f}_x(\boldsymbol{y}_1)\| < \epsilon$, stop. Otherwise, let λ_j be an optimal solution to the minimization problem of the function $f(\boldsymbol{y}_j + \lambda\, \boldsymbol{d}_j)$ with $\lambda > 0$. We pose $\boldsymbol{y}_{j+1} = \boldsymbol{y}_j + \lambda\, \boldsymbol{d}_j$. If $j < n$, go to stage 2, otherwise go to stage 3.

We can choose

$$\lambda = \frac{\boldsymbol{g}^T \boldsymbol{g}}{\boldsymbol{g}^T \boldsymbol{H} \boldsymbol{g}} \tag{9.7.15}$$

where $\boldsymbol{g}$ is the gradient $\boldsymbol{f}_x$ of f and $\boldsymbol{H}$ the Hessian matrix of f at the considered point.

2. Let $\boldsymbol{d}_{j+1} = -\boldsymbol{f}_x(\boldsymbol{y}_{j+1}) + \alpha_j \boldsymbol{d}_j$ with

$$\alpha_j = \frac{\|\boldsymbol{f}_x(\boldsymbol{y}_{j+1})\|^2}{\|\boldsymbol{f}_x(\boldsymbol{y}_j)\|^2} \tag{9.7.16}$$

Replace j by $j + 1$, and go to stage 1.

3. Let $\boldsymbol{y}_1 = \boldsymbol{x}_{k+1} = \boldsymbol{y}_{n+1}$ and $\boldsymbol{d}_1 = -\boldsymbol{f}_x(\boldsymbol{y}_1)$. Let $j = 1$, replace k by $k + 1$ and go to stage 1.

Questions:

1. Justify the method at the theoretical level.
2. Verify the technique starting from point $(0, 0)$.

Exercise 9.7.15 (Difficult)

Study of function with logarithmic barrier

(A) Introduction.
Consider the following optimization problem:

$$\min_{\boldsymbol{x}} f(\boldsymbol{x}), \quad \boldsymbol{x} \in \mathbb{R}^n \tag{9.7.17}$$

subject to the constraints

$$g_i(\boldsymbol{x}) \geq 0 \tag{9.7.18}$$

We consider the associated function

$$P(\boldsymbol{x}, \mu) = f(\boldsymbol{x}) - \mu \sum_i \log(g_i(\boldsymbol{x})) \tag{9.7.19}$$

where log is the Naperian logarithm and μ the barrier parameter. The function $P(\boldsymbol{x}, \mu)$ is called a logarithmic barrier function.

(B) First example.
We consider

$$f(x) = x, \quad x \in \mathbb{R} \tag{9.7.20}$$

subject to the constraints

$$\begin{aligned} g_1(x) &= x \geq 0 \\ g_2(x) &= 1 - x \geq 0 \end{aligned} \tag{9.7.21}$$

Pose the function $P(\boldsymbol{x}, \mu)$. Plot the function $P(\boldsymbol{x}, \mu)$ in the considered domain of x for the following values of μ: $\mu = 1$, $\mu = 0.1$, and $\mu = 0.01$. Find the solution of this optimization problem proposed as a first example and deduce a remark on this solving technique.

(C) Second example.
We consider

$$f(\boldsymbol{x}) = (x_1 + 0.5)^2 + (x_2 - 0.5)^2 , \quad x \in \mathbb{R}^2 \tag{9.7.22}$$

subject to the constraints

$$\begin{aligned} g_1(x) &= x_1 \geq 0 \\ g_2(x) &= x_1 \leq 1 \\ g_3(x) &= x_2 \geq 0 \\ g_4(x) &= x_2 \leq 1 \end{aligned} \tag{9.7.23}$$

Pose the function $P(\boldsymbol{x}, \mu)$. The contours of the function $P(\boldsymbol{x}, \mu)$ have been plotted in the considered domain of $\boldsymbol{x}$ for the following values of μ: $\mu = 1$, $\mu = 0.1$, and $\mu = 0.01$. Comment on Figure 9.21. Find the solution of this optimization problem proposed as a second example and deduce a remark on this solving technique with an additional and more general comment taking into account both questions (B) and (C).

(D) Verification by a classic approach:
Verify the result of question (C) by a conventional approach. Comment the interest of the approach (C).

Exercise 9.7.16 (Medium)
We study Rosenbrock's function

$$f(\boldsymbol{x}) = \alpha(x_2 - x_1^2)^2 + (1 - x_1)^2 \tag{9.7.24}$$

where α is an adjustable parameter. In the present case, we chose $\alpha = 100$.

1. Find the minimum of this function for $\boldsymbol{x} \in \mathbb{R}^2$. Clearly place this optimum in Figures 9.22 and 9.23 plotted with different scales.
2. Use Newton–Raphson method from the initial point $(-1, 1)$. Explain the algorithm in the case of function $f(\boldsymbol{x})$. Make three iterations at the most. Note that this function can present very large variations related to the choice of α. Represent the results of the three iterations in Figure 9.23 with a symbol different from the conjugate gradients.
3. Use the conjugate gradients method to find the optimum. Start from the same initial point $(-1, 1)$. Explain how you execute the conjugate gradient method with function $f(\boldsymbol{x})$ and make two iterations. Represent the initial point and the results of the iterative calculations in Figure 9.23.

Exercise 9.7.17 (Medium)
Given the physical model

$$f(x, \boldsymbol{a}) = x\, a_1 \exp(a_2\, x) \tag{9.7.25}$$

clearly explain how, by a general optimization method, you determine the parameters a_1 and a_2 with Table 9.13 of experimental results. Find these parameters.
Hint: pose the complete problem at the mathematical level, and then, explain how to proceed at the numerical level to obtain an estimation of parameters a_1 and a_2.

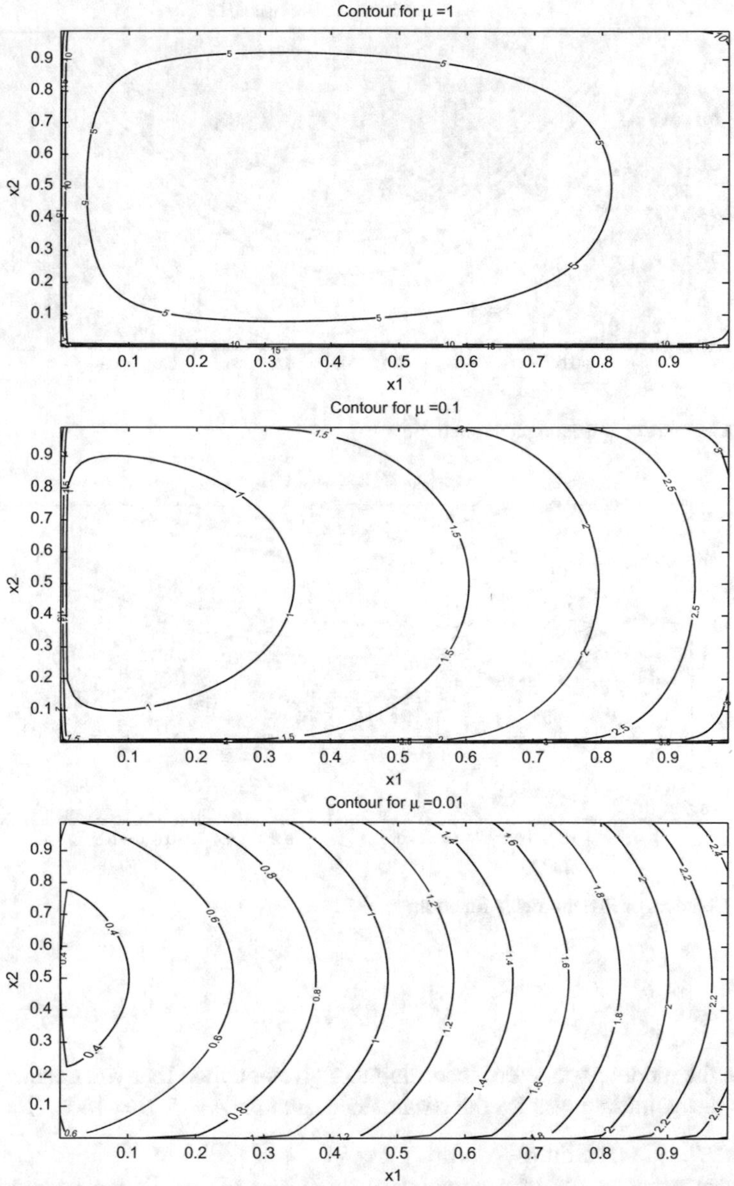

Fig. 9.21 Contours of function $P(\boldsymbol{x}, \mu)$ for successive values of μ (from top to bottom $\mu = 1, \mu = 0.1$, and $\mu = 0.01$)

Exercise 9.7.18 (Medium)

An experimenter measured saturated vapor pressures of ethane at moderate pressures (Yaws and Yang 1989) and obtained the results of Table 9.14. He/she uses Antoine's law to represent the results

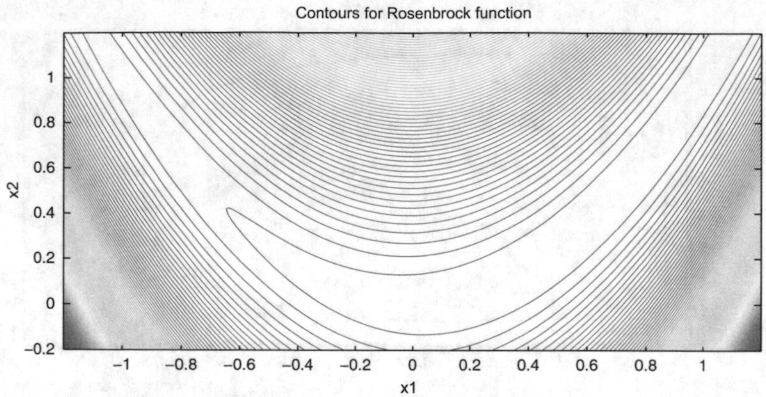

Fig. 9.22 Contours of Rosenbrock's function

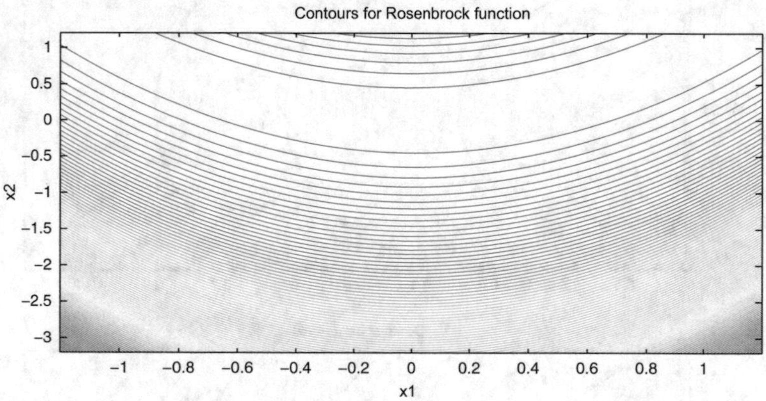

Fig. 9.23 Contours of Rosenbrock's function

$$\ln(P) = A - \frac{B}{C + T} \tag{9.7.26}$$

Calculate the model parameters according to a given method that you detail.

A suggested initial point for the parameter search is $[A = 5, B = 1000, C = -10]$.

Exercise 9.7.19 (Medium)

The heat capacity C_p of a component (Poling et al. 2001) can be written in a restricted domain under the form

Table 9.13 Experimental results

x	-1	1	2	3	4
f	-0.043099	3.17639	54.5378	702.299	8038.856

Table 9.14 Saturated vapor pressure with respect to temperature for ethane

T (K)	150	160	170	180	190	200
P (atm)	0.095	0.21	0.42	0.78	1.34	2.15

Table 9.15 Heat capacity of ethane with respect to temperature

Temperature T (K)	300	325	350	375	400	425	450	475	500
C_p (J.mol^{-1}.K^{-1})	52.8	56.2	59.6	62.9	66.1	69.2	72.3	75.3	78.2

$$C_p = a_0 + a_1\,T + a_2\,T^2 \qquad (9.7.27)$$

We make n experiments noted as pairs $\{T_i, C_{pi}\}$. The vector of parameters to determine is noted $\boldsymbol{\theta}$. n is sufficiently large to determine the parameters as least squares.

1. Set the least squares criterion to minimize to determine $\boldsymbol{\theta}$.
2. Notice that the parameters intervene in a linear way in the model and set the criterion under a matrix form. To do that, it suffices to think that each experiment corresponds to a linear expression, write all these expressions for the experiments and deduce a matrix representation making use of the vector of parameters $\boldsymbol{\theta}$, the vector $\boldsymbol{Y}$ of experimental results and some matrix that will be noted $\boldsymbol{A}$ which is nonsquare.
3. Due to the minimization of the criterion, deduce the optimal vector of parameters $\boldsymbol{\theta}^*$ under a matrix form (explain some intermediate calculations that seem necessary). Application: in the case of ethane, the experimental results are given in Table 9.15. Calculate the values of the coefficients a_i.

Exercise 9.7.20 (Medium)

A correlation allowing us to calculate the saturated vapor pressure of a liquid with respect to temperature is

$$\ln P = a_1 + \frac{a_2}{a_3 + T} + a_4\,T + a_5\,T^2 + a_6\,\ln T \qquad (9.7.28)$$

For propane, when the pressure P is expressed in Pa and the temperature T in K, measurements at different temperatures gave the results of Table 9.16.

Explain how you determine the model parameters a_i. Write a program for their estimation.

Table 9.16 Saturated temperature and vapor pressure for propane

Temperature (K)	$10^{-7} P$ (Pa)
300	0.1491
310	0.1907
320	0.2403
330	0.2989
340	0.3678
350	0.4479
360	0.5412
370	0.6497
380	0.7756
390	0.9219
400	1.0921
410	1.2909

Exercise 9.7.21 (Medium)

A physical model is given by the following equation giving f with respect to time t:

$$f(\boldsymbol{t},\boldsymbol{\theta}) = \left(1 - \frac{\theta_1 t}{\theta_2}\right)^{\left(\frac{1}{c\,\theta_1}-1\right)} \tag{9.7.29}$$

with $c = 96.05$. The objective is to find how both parameters θ_1 and θ_2 are determined. The experimental results are given in Table 9.17. Explain how you determine the model parameters. Make a program and solve the problem.

Table 9.17 Experimental results

t_i	2000	5000	10000	20000	30000	40000
f_i	0.9427	0.8616	0.7384	0.5362	0.3739	0.3096

References

N. Andrei. *Nonlinear Conjugate Gradient Methods for Unconstrained Optimization*. Springer, Cham, Switzerland, 2020.

J. F. Bonnans, J. C. Gilbert, C. Lemaréchal, and C. A. Sagastizabal. *Numerical optimization*. Springer, Berlin, 2003.

M. J. Box. A new method of constrained optimization and a comparison with other methods. *The Computer Journal*, 8(1):42–52, 1965.

J. Brownlee. *Clever Algorithms - Nature-Inspired Programming Recipes*. Creative Commons, San Francisco, 2011.

C. G. Broyden. A class of methods for solving nonlinear simultaneous equation. *Math. Comp.*, 19:577–593, 1965.

C. G. Broyden. Quasi-Newton methods and their application to function minimisation. *Math. Comp.*, 21:368–381, 1967.

C. G. Broyden. The convergence of a class of double-rank minimization algorithms. *Journal of the Institute of Mathematics and Its Applications*, 6:76–90, 1970.

W. C. Davidon. Variable metric method for minimization. Technical report, A.E.C. Research and Development Rept. ANL-5990, 1959.

W. C. Davidon. Variable metric method for minimization. *SIAM Journal on Optimization*, 1:1–17, 1991.

K. A. De Jong. Evolutionary Algorithms - A Unified Approach. MIT Press, Cambridge, 2006.

K. Deb. *Multi-objective optimization using evolutionary algorithms*. Wiley, Chichester, UK, 2001.

T. F. Edgar and D. M. Himmelblau. *Optimization of Chemical Processes*. Mc Graw-Hill, 2nd edition, 2001.

A. E. Eiben and J. E. Smith. *Introduction to Evolutionary Computing*. Springer, New York, 2nd edition, 2007.

R. Fletcher. Function minimization without evaluating derivatives – a review. *The Computer Journal*, 8(1):33–41, 1965.

R. Fletcher. *Practical Methods of Optimization*. Wiley, Chichester, 1991.

R. Fletcher and C. M. Reeves. Function minimization by conjugate gradients. *The Computer Journal*, 7(2):149–154, 1964.

F. Gao and L. Han. Implementing the Nelder-Mead simplex algorithm with adaptive parameters. *Comput. Optim. Appl.*, DOI 10.1007/s10589-010-9329-3, 2010.

P. E. Gill, W. Murray, and M. H. Wright. *Practical optimization*. Academic Press, London, 1981.

D. E. Goldberg. *Genetic algorithms in search, optimization and machine learning*. Addison-Wesley, Reading, MA, 1989.

D. E. Goldberg. *The Design of Innovation: Lessons from and for Competent Genetic Algorithms*. Springer-Verlag, New York, 2002.

M. Hestenes. *Conjugate Direction Methods in Optimization*. Springer-Verlag, New-York, 1980.

J. Holland. *Adaptation in Natural and Artificial Systems*. MIT Press, Cambridge, 1992.

R. Hooke and T. A. Jeeves. Direct search solution of numerical and statistical problems. *J. Assoc. Computer Machines*, 8:212–229, 1961.

C. T. Kelley. Detection and remediation of stagnation in the Nelder-Mead algorithm using a sufficient decrease condition. *SIAM Journal on Optimization*, 10:43–55, 1999a.

C. T. Kelley. *Iterative Methods for Optimization*. SIAM, Philadelphia, 1999b.

J. C. Lagarias, J. A. Reeds, M. H. Wright, and P. E. Wright. Convergence properties of the Nelder-Mead simplex method in low dimensions. *SIAM Journal on Optimization*, 9(1):112–147, 1998.

R. M. Lewis, V. Torczon, and M. W. Trosset. Direct search methods: then and now. *Journal of Computational and Applied Mathematics*, 124:191–207, 2000.

L. R. Lucambio Perez and L. F. Prudente. Nonlinear conjugate gradient methods for vector optimization. *SIAM Journal on Optimization*, 28(3):2690–2720, 2018.

J. Matousek and B. Gärtner. *Understanding and using linear programming*. Springer, Berlin, 2007.

J. J. Moré and S. J. Wright. *Optimization software guide*. SIAM, Philadelphia, 1994.

J. A. Nelder and R. Mead. A simplex method for function minimization. *The Computer Journal*, 7:308–313, 1965.

J. Nocedal and S. J. Wright. *Numerical optimization*. Springer, New York, 2nd edition, 2006.

E. Polak. *Optimization - Algorithms and consistent approximations*. Springer, New York, 1997.

R. Poli, W. B. Langdon, and N. F. McPhee. *A Field Guide to Genetic Programming*. Creative Commons, San Francisco, 2008.

B. E. Poling, J. M. Prausnitz, and J. P. O'Connel. *The properties of gases and liquids*. McGraw-Hill, New York, 5th edition, 2001.

M. J. D. Powell. An efficient method for finding the minimum of a function of several variables without calculating derivatives. *Computer J.*, 7:155–162, 1964.

M. J. D. Powell. A fast algorithm for nonlinearly constrained optimization calculations. In G. A. Watson, editor, *Proceedings of 1977 Dundee Biennial Conference on Numerical Analysis*, pages 144–157. Springer, 1978.

S. S. Rao. *Engineering Optimization - Theory and Practice*. John Wiley, New York, 4th edition, 2009.

W. H. Ray and J. Szekely. *Process Optimization with applications in metallurgy and chemical engineering*. John Wiley, New York, 1973.

H. H. Rosenbrock. An automatic method for finding the greatest or least value of a function. *The Computer Journal*, 3:175–184, 1960.

F. H. Walters, S. L. Morgan, L. R. Parker, and S. N. Deming. *Sequential simplex optimization*. CRC Press, Boca Raton, 1991.

C. J. Wang. Dogleg paths and trust region methods with back tracking technique for unconstrained optimization. *Applied Mathematics and Computation*, 177(1): 159–169, 2006.

P. Wolfe. The secant method for solving nonlinear equations. *Comm. Assoc. Comput. Mach.*, 2:12–13, 1959.

G. Yuan and M. Zhang. A modified Hestenes-Stiefel conjugate gradient algorithm for large-scale optimization. *Numerical Functional Analysis and Optimization*, 34(8): 914–937, 2013.

C. L. Yaws and H. C. Yang. To estimate vapor pressure easily. Antoine coefficients relate vapor pressure to temperature for almost 700 major organic compounds. *Hydrocarbon Processing*, 68(10):65–68, 1989.

J. Zhang, Y. Xiao, and Z. Wei. Nonlinear conjugate gradient methods with sufficient descent condition for large-scale unconstrained optimization. *Mathematical Problems in Engineering*, 2009.

Chapter 10
Linear Programming

10.1 Introduction

In the case where the function $f(\boldsymbol{x})$ to minimize is linear and all the constraints $g_i(\boldsymbol{x})$ are linear, the appropriate solving method is linear programming (Chretienne et al. 1980; Dantzig 1987; Dantzig and Thapa 1997, 2003; Maurin 1967; Luenberger and Ye 2016; Panik 1996; Vanderbei 2010). The books by Dantzig and Thapa (1997, 2003) remarkably illustrate linear programming and must be recommended at all points of view, as well for their practical approach rich in various examples as for their theoretical developments. The method was imagined by Leonid Kantorovich, a Russian mathematician and only Russian Nobel Prize in economic sciences. In USA, Wassily Leontief had already proposed in 1932 a matrix method called "Interindustry Input-Output Model of the American Economy" (Dantzig and Thapa 1997), and he won Nobel Prize in 1976 for that development. Dantzig, after his PhD, worked since 1946 for US Air Force, with the objective to generalize and transform the previous steady-state method into a dynamic model adaptable to computer calculation. Dantzig published the simplex method in 1947 (Dantzig 1955, 1956, 1957, 1963; Dantzig et al. 1955). During the initial development period, he particularly cites the fructuous discussions with the economist T.C. Koopmans and the mathematician John Von Neumann who had just achieved a book on game theory. Von Neumann himself made him discover the concept of duality. Linear programming often occurs in economics, management, optimal allocation of resources, optimization of production, game theory, problems of mixing, storage, transport, and networks (Matousek and Gärtner 2007; Thie and Keough 2008).

The problem is posed under the following form: minimize the linear function f of n variables x_j

$$\min f(\boldsymbol{x}) = \min \sum_{j=1}^{n} c_j \, x_j \tag{10.1.1}$$

subject to linear positive or negative inequality constraints

J.-P. Corriou, *Numerical Methods and Optimization*, Springer Optimization and Its Applications 187, https://doi.org/10.1007/978-3-030-89366-8_10

$$g_i(\boldsymbol{x}) = \sum_{j=1}^{n} a_{ij} x_j - b_i \leq 0 \,, \; 1 \leq i \leq m \tag{10.1.2}$$

and possibly linear equality constraints

$$g_i(\boldsymbol{x}) = \sum_{j=1}^{n} a_{ij} x_j - b_i = 0 \,, \; 1 \leq i \leq p \tag{10.1.3}$$

thus a total of $m + p$ constraints to which must be added the condition of positivity of the variables

$$x_i \geq 0 \qquad \forall i \tag{10.1.4}$$

Note that the search is restricted to the first quadrant ($x_i \geq 0$). The simplex method that will be developed lies on this hypothesis, which, in reality, poses no real problem.

If a function is to be maximized, it is sufficient to come back to the minimization of its opposite. If a constraint is expressed under the form of a positive inequality, by multiplying it by (-1), it is transformed into a negative inequality.

When linear equality constraints (10.1.3) are present, to take them into account in the simplex method, it is possible to transform each equality constraint under the equivalent form of two inequality constraints

$$\begin{aligned} \sum_{j=1}^{n} a_{ij} x_j - b_i \leq 0 \\ \sum_{j=1}^{n} a_{ij} x_j - b_i \geq 0 \end{aligned} \tag{10.1.5}$$

It is also possible to directly treat the equality constraints as such (Section 10.5).

10.2 Formulation of the Problem Based on Examples

10.2.1 Use of Slack Variables

The use of slack variables is illustrated by Example 10.1.

Example 10.1 :
Linear programming: use of slack variables

We desire to minimize the linear function

$$f(\boldsymbol{x}) = 2\, x_1 - x_2 \tag{10.2.1}$$

subject to linear inequality constraints

$$\begin{array}{l} 3\,x_1 - 2\,x_2 \geq -2 \\ 2\,x_1 - 4\,x_2 \leq 3 \\ x_1 + x_2 \leq 6 \\ x_1 \geq 0 \\ x_2 \geq 0 \end{array} \tag{10.2.2}$$

The previous problem is transformed as

$$\min f = 2\,x_1 - x_2 \tag{10.2.3}$$

The constraint of positive inequality is transformed into a negative constraint

$$-\,3\,x_1 + 2\,x_2 \leq 2 \tag{10.2.4}$$

Finally, the inequalities are transformed into equalities by adding positive *slack variables*

$$\begin{array}{llllll} -3\,x_1 & +2\,x_2 & +x_3 & & & = 2 \\ 2\,x_1 & -4\,x_2 & & +x_4 & & = 3 \\ x_1 & +x_2 & & & +x_5 & = 6 \end{array} \tag{10.2.5}$$

The slack variables are also called complementary.

The problem is definitively posed under the form:
Minimize f such that

$$f = \sum_{j=1}^{n} c_j\, x_j \tag{10.2.6}$$

subject to m equality constraints

$$\sum_{j=1}^{n} a_{ij} x_j + x_{n+i} - b_i = 0\,, \quad 1 \leq i \leq m\,,\ 1 \leq j \leq n \tag{10.2.7}$$

$$x_j \geq 0\,, \quad x_{n+i} \geq 0 \tag{10.2.8}$$

- The n variables x_j are called *main variables* or *structural*. They must be positive.
- The m variables x_{n+i} are called *slack variables*. They must also be positive.
- This is a problem with $(n + m)$ variables and m equations.
- Any set of variables $\{x_j, x_{n+i}\}$ verifying the system (10.2.7) is a *solution of the problem*. When, moreover, this set verifies Equation (10.2.8), it is a *realizable* or *feasible* solution.
- A *basis* is constituted by a set of m variables whose coefficients in the m equations form a square nonsingular (of nonzero determinant) matrix. These m variables are called *basic variables*. The n other variables are *nonbasic variables*. Frequently, the nonbasic variables will be chosen equal to 0. Thus, the basic variables are perfectly determined using the m equations. If the basic variables are all positive or zero, they are called *feasible basic solution*.

When the basic variables are strictly positive, they are *nondegenerate*. The case where a basic variable is zero constitutes a degeneracy of second kind (Faure et al. 2009). It may be the cause of cycling for which anticycling procedures are implemented in computing codes.

Example 10.2 :
Linear programming: determination of a feasible solution
Consider again Example 10.1. The nonbasic variables are chosen as x_1 and x_2 that are set equal to zero. Then, the basic variables x_3, x_4, and x_5 are determined. The set

$$x_1 = 0\,,\ x_2 = 0\,,\ x_3 = 2\,,\ x_4 = 3\,,\ x_5 = 6 \tag{10.2.9}$$

constitutes a feasible solution.

10.2.2 *Use of Slack and Artificial Variables*

In some cases, artificial variables must be introduced; this is illustrated by Example 10.3.

Example 10.3 :
Linear programming: slack and artificial variables
We desire to maximize the following objective function:

$$f = 4\,x_1 + 5\,x_2 \tag{10.2.10}$$

subject to the constraints

$$\begin{aligned} &2\,x_1 + x_2 \le 6 \\ &x_1 + 2\,x_2 \le 5 \\ &x_1 + x_2 \ge 1 \\ &x_1 + 4\,x_2 \ge 2 \end{aligned} \tag{10.2.11}$$

The quantities x_1 and x_2 must be positive

$$x_1 \ge 0\,, \quad x_2 \ge 0 \tag{10.2.12}$$

The inequality constraints are transformed by adding slack variables

$$\begin{array}{llllll} 2\,x_1 & +x_2 & +x_3 & & & = 6 \\ x_1 & +2\,x_2 & & +x_4 & & = 5 \\ x_1 & +x_2 & & & -x_5 & = 1 \\ x_1 & +4\,x_2 & & & \quad -x_6 & = 2 \end{array} \tag{10.2.13}$$

$$x_i \ge 0\,, \quad i = 1, 6 \tag{10.2.14}$$

The system contains 4 equations and 6 unknowns. First, a feasible solution must be found. The first trial consists in taking the nonbasic variables $x_1 = 0$ and $x_2 = 0$ and calculating the basic variables from x_3 to x_6. We notice that the obtained solution cannot be accepted as two variables, x_5 and x_6, would take negative values.

This first trial having failed, we use the method of *artificial variables*. On both equations that gave negative basic variables, we add an artificial variable which must be positive or zero

$$\begin{array}{llllll} x_1 & +x_2 & -x_5 & & +x_7 & & = 1 \\ x_1 & +4\,x_2 & & -x_6 & & +x_8 & = 2 \end{array} \tag{10.2.15}$$

while the two other equations of constraints remain the same. The system now contains 4 equations and 8 unknowns. We still take x_1 and x_2 as nonbasic variables, to which we add x_5 and x_6, which posed a difficulty. All the nonbasic variables are set equal to zero.

$$x_1 = x_2 = x_5 = x_6 = 0 \tag{10.2.16}$$

The value of the basic variables results

$$x_3 = 6\,,\ x_4 = 5\,,\ x_7 = 1\,,\ x_8 = 2 \tag{10.2.17}$$

The initial problem called main having been transformed, the obtained solution is not feasible for the main problem. Indeed, it is sufficient to solve the *auxiliary linear problem*

$$\min f' = x_7 + x_8 \tag{10.2.18}$$

If we find a minimum of f' equal to zero, as the artificial variables are positive, each of them will be zero. After this phase of solving the auxiliary problem, we know a solution of the main problem that we then solve.

If we do not find a minimum of f' equal to zero, the artificial variables cannot be zero, and the solution is not feasible for the main problem which thus admits no solution.

10.2.3 Conditions of Optimality

Whatever the need or not of artificial variables, the problem constituted by Equations (10.2.6), (10.2.7), and (10.2.8) can be written in the general form
Minimize f such that

$$f(\boldsymbol{x}) = \boldsymbol{c}^T\,\boldsymbol{x} \tag{10.2.19}$$

subject to

$$\begin{aligned} \boldsymbol{A}\,\boldsymbol{x} &= \boldsymbol{b} \\ \boldsymbol{x} &\geq 0 \end{aligned} \tag{10.2.20}$$

Pose the Lagrangian equal to

$$\mathcal{L}(\boldsymbol{x}, \boldsymbol{\lambda}, \boldsymbol{\mu}) = \boldsymbol{c}^T\,\boldsymbol{x} + \boldsymbol{\lambda}^T(\boldsymbol{A}\,\boldsymbol{x} - \boldsymbol{b}) - \boldsymbol{\mu}^T\,\boldsymbol{x} \tag{10.2.21}$$

where $\boldsymbol{\lambda}$ are Lagrange multipliers and $\boldsymbol{\mu}$ Karush–Kuhn–Tucker parameters. The conditions of optimality at $(\boldsymbol{x}^*, \boldsymbol{\lambda}^*, \boldsymbol{\mu}^*)$ are

$$\begin{aligned} &\frac{\partial \mathcal{L}}{\partial \boldsymbol{x}} = 0 \\ &\frac{\partial \mathcal{L}}{\partial \boldsymbol{\lambda}} = 0 \\ &\mu_i\,x_i = 0\,, \qquad \forall\ i = 1, \ldots, n \\ &\boldsymbol{\mu} \geq 0 \\ &\boldsymbol{x} \geq 0 \end{aligned} \tag{10.2.22}$$

that is

$$\begin{aligned} &\boldsymbol{c} + \boldsymbol{A}^T\,\boldsymbol{\lambda}^* - \boldsymbol{\mu}^* = 0 \\ &\boldsymbol{A}\,\boldsymbol{x}^* = \boldsymbol{b} \\ &\mu_i^*\,x_i^* = 0\,, \qquad \forall\ i = 1, \ldots, n \\ &\boldsymbol{\mu}^* \geq 0 \\ &\boldsymbol{x}^* \geq 0 \end{aligned} \tag{10.2.23}$$

Using the previous equalities, we get

$$(\boldsymbol{c} + \boldsymbol{A}^T \boldsymbol{\lambda} - \boldsymbol{\mu})^T \boldsymbol{x}^* = 0 \Rightarrow \boldsymbol{c}^T \boldsymbol{x}^* + \boldsymbol{\lambda}^{*T} \boldsymbol{A} \boldsymbol{x}^* = 0 \Rightarrow \boldsymbol{c}^T \boldsymbol{x}^* - \boldsymbol{\lambda}^{*T} \boldsymbol{b} = 0 \quad (10.2.24)$$

hence the remarkable equation of the *dual problem* (Section 10.6)

$$\boldsymbol{c}^T \boldsymbol{x}^* = \boldsymbol{\lambda}^{*T} \boldsymbol{b} \quad (10.2.25)$$

During solving, several issues can arise:

1. It is not possible to satisfy all the constraints.
2. The domain is not bounded, which means that there exist solutions satisfying the constraints and such that the objective function is infinite.
3. There exist several different solutions giving the same extremum value of the objective function.

10.3 Solving the Problem: Simplex Tableau

10.3.1 Geometric Interpretation on Example 10.1

To solve the following problem, as it is a two-dimensional case, it is useful to make a graphical representation of the feasible domain. Names are given to the vertices of the polygon where the solutions must be located.

According to a *basic theorem of linear programming*,
a feasible basic solution of the linear programming problem is a vertex of the convex polytope of the feasible solutions. This convex domain is called simplex.

In dimension 2, a polytope is called a polygon. In dimension 3, it is a polyhedron. The polytope is thus the generalization in dimension n of the polyhedron. In general, the polytope is supposed to be convex and bounded.

A feasible solution means the points satisfying the set of constraints $\boldsymbol{A}\,\boldsymbol{x} \le \boldsymbol{b}$. Any vertex of the polytope is the intersection of n hyperplanes. If the polytope is bounded, the problem admits at least one solution. Finally, a vertex is optimum if and only if the neighboring vertices give a worse value of the objective function.

10.3.1.1 Minimization of f

Search the minimization of f

$$f = 2\,x_1 - x_2 \quad (10.3.1)$$

subject to the constraints with slack variables

$$\begin{array}{llllll} -3\,x_1 & +2\,x_2 & +x_3 & & & = 2 \\ 2\,x_1 & -4\,x_2 & & +x_4 & & = 3 \\ x_1 & +x_2 & & & +x_5 & = 6 \end{array} \quad (10.3.2)$$

To start solving this problem (Figure 10.1), the nonbasic solutions are taken equal to zero, and the basic solutions are calculated. We verify that the solution is feasible

$$x_1 = 0\,,\ x_2 = 0 \text{ (point A)}\,,\ x_3 = 2\,,\ x_4 = 3\,,\ x_5 = 6 \tag{10.3.3}$$

The function $f = 0$ results.
First stage:
The basic variables can be expressed with respect to the nonbasic variables so that we obtain

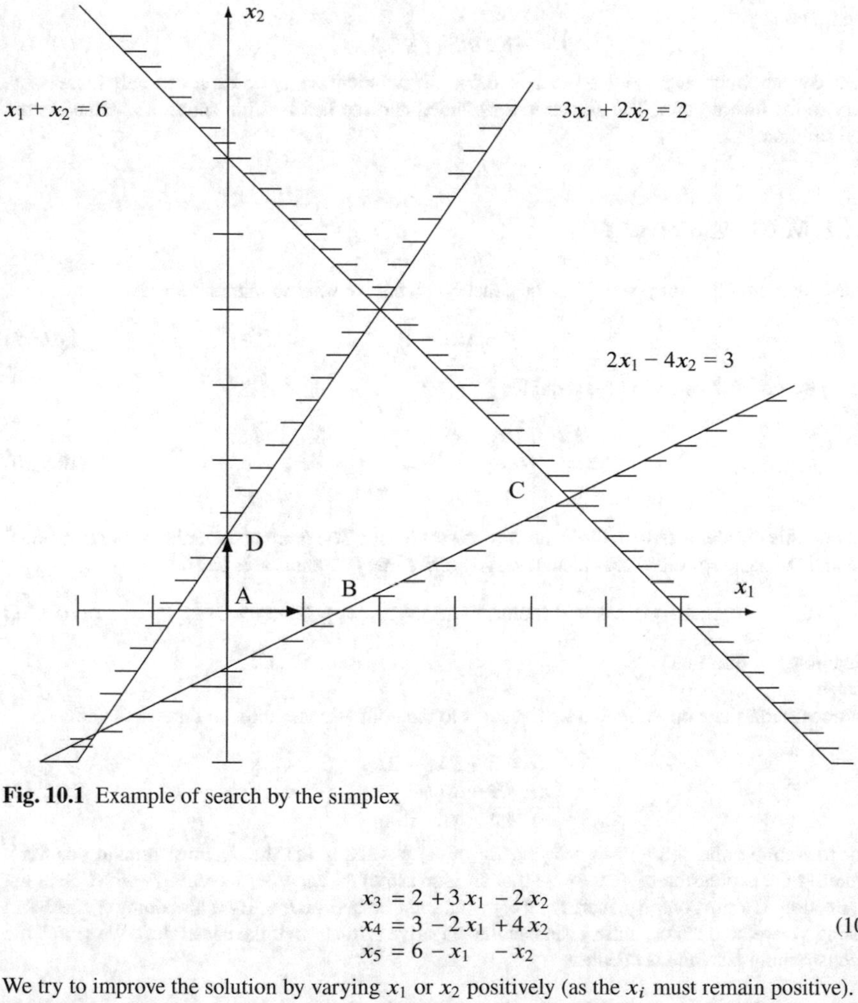

Fig. 10.1 Example of search by the simplex

$$\begin{aligned} x_3 &= 2 + 3\,x_1 - 2\,x_2 \\ x_4 &= 3 - 2\,x_1 + 4\,x_2 \\ x_5 &= 6 - x_1 \quad - x_2 \end{aligned} \tag{10.3.4}$$

We try to improve the solution by varying x_1 or x_2 positively (as the x_i must remain positive). If we consider the expression of f, we note that an increase of x_1 can only increase f, which is opposite to the sought minimization. On the contrary, we can increase x_2, but not in any proportion. By keeping x_1 zero and increasing x_2, we note that x_3 and x_5 may become negative (violation of the constraint). We search the most constraining fact, and we deduce such that

$$x_3 = 2 - 2\,x_2 = 0 \Longrightarrow x_2 = 1 \tag{10.3.5}$$

then the set of values

$$x_1 = 0\,,\; x_2 = 1 \text{ (point D)}\,,\; x_3 = 0\,,\; x_4 = 7\,,\; x_5 = 5 \tag{10.3.6}$$

The new nonbasic variables are thus x_1 and x_3, and the basic variables are x_2, x_4, and x_5. This operation corresponds to a Gauss–Jordan elimination. The function is now $f = -1$.

Now, express the basic variables with respect to the nonbasic variables

$$\begin{aligned} x_2 &= 1 + 1.5\,x_1 - 0.5\,x_3 \\ x_4 &= 7 + 4\,x_1 \quad - 2\,x_3 \\ x_5 &= 5 - 2.5\,x_1 + 0.5\,x_3 \end{aligned} \tag{10.3.7}$$

and the function f

$$f = -1 + 0.5\,x_1 + 0.5\,x_3 \tag{10.3.8}$$

Consider the function $f = -1 + 0.5\,x_1 + 0.5\,x_3$. If we increase x_1 or x_3, it can only increase f, contrary to the desired aim. The search is thus ended and the last feasible solution obtained is the optimal solution.

10.3.1.2 Maximization of f

Consider the same function f and the same constraints, but we want to maximize f

$$f = 2\,x_1 - x_2 \tag{10.3.9}$$

subject to the constraints with slack variables

$$\begin{aligned} -3\,x_1 + 2\,x_2 + x_3 \qquad\qquad &= 2 \\ 2\,x_1 - 4\,x_2 \qquad + x_4 \qquad &= 3 \\ x_1 + x_2 \qquad\qquad + x_5 &= 6 \end{aligned} \tag{10.3.10}$$

Solving this problem starts in the same way as previously. The nonbasic solutions are taken equal to zero and the basic solutions are calculated. We verify that the solution is feasible

$$x_1 = 0\,,\; x_2 = 0 \text{ (point A)}\,,\; x_3 = 2\,,\; x_4 = 3\,,\; x_5 = 6 \tag{10.3.11}$$

The function $f = 0$ results.
First stage:
The basic variables can be expressed with respect to the nonbasic variables so that we obtain

$$\begin{aligned} x_3 &= 2 + 3\,x_1 - 2\,x_2 \\ x_4 &= 3 - 2\,x_1 + 4\,x_2 \\ x_5 &= 6 - x_1 \quad - x_2 \end{aligned} \tag{10.3.12}$$

We try to improve the solution by varying x_1 or x_2 positively (as the x_i must remain positive). Considering the expression of f, we note that an increase of x_1 can only increase f, which is in the right direction. On the contrary, increasing x_2 would not be acceptable. By maintaining x_2 zero and increasing x_1, we note that x_4 and x_5 can become negative (violation of the constraint). We search the most constraining fact, and we deduce

$$x_4 = 3 - 2\,x_1 = 0 \Longrightarrow x_1 = 1.5 \tag{10.3.13}$$

then all the variables

$$x_1 = 1.5\,,\; x_2 = 0 \text{ (point B)}\,,\; x_3 = 6.5\,,\; x_4 = 0\,,\; x_5 = 4.5 \tag{10.3.14}$$

The new nonbasic variables are thus x_2 and x_4, and the basic variables are x_1, x_3, and x_5. The function is now $f = 3$ (note the increase).

Second stage:
The basic variables are expressed with respect to the nonbasic variables

$$\begin{aligned} x_1 &= 1.5 + 2\,x_2 - 0.5\,x_4 \\ x_3 &= 6.5 + 4\,x_2 - 1.5\,x_4 \\ x_5 &= 4.5 - 3\,x_2 + 0.5\,x_4 \end{aligned} \tag{10.3.15}$$

and the function f becomes

$$f = 3 + 3\,x_2 - x_4 \tag{10.3.16}$$

Considering the expression of f, we note that an increase of x_2 can only increase f, which is in the right direction. Maintaining x_4 zero and increasing x_2, we note that x_5 can become negative (violation of the constraint). It results

$$x_5 = 4.5 - 3\,x_2 = 0 \Longrightarrow x_2 = 1.5 \tag{10.3.17}$$

then all the variables

$$x_1 = 4.5\,,\ x_2 = 1.5 \text{ (point C)}\,,\ x_3 = 12.5\,,\ x_4 = 0\,,\ x_5 = 0 \tag{10.3.18}$$

The new nonbasic variables thus are x_4 and x_5, and the basic variables are x_1, x_2, and x_3. The function is now $f = 7.5$ (still increase).
Third stage:
The basic variables are expressed with respect to the nonbasic variables

$$\begin{aligned} x_1 &= 4.5 - 0.167\,x_4 - 0.667\,x_5 \\ x_2 &= 1.5 + 0.167\,x_4 - 0.333\,x_5 \\ x_3 &= 12.5 - 0.833\,x_4 - 1.333\,x_5 \end{aligned} \tag{10.3.19}$$

and the function f

$$f = 7.5 - 0.5\,x_4 - x_5 \tag{10.3.20}$$

We note that, if we increase x_4 or x_5, f can only decrease. The last feasible solution is thus optimal, and we keep $x_1 = 4.5$, $x_2 = 1$, and $f = 7.5$.

10.3.1.3 Simplex Tableau

It is possible to represent all the stages under the form of an array called *simplex tableau*.

We consider again Example 10.1, but in the case of *maximization* of f

$$f = 2\,x_1 - x_2 \tag{10.3.21}$$

Presentation of the problem under *canonical form*:

$$\max_x f = 2\,x_1 - x_2 \tag{10.3.22}$$

subject to equality constraints with slack variables

$$\begin{array}{lllll} -3\,x_1 & +2\,x_2 & +x_3 & & = 2 \\ 2\,x_1 & -4\,x_2 & & +x_4 & = 3 \\ x_1 & +x_2 & & & +x_5 = 6 \\ x_i \geq 0\,, & \forall i = 1, \ldots, 5 & & & \end{array} \tag{10.3.23}$$

- In the first column, note the iteration number.
- In the following columns, the vectors of the canonical form noted $\boldsymbol{V}_i$ represent the coefficients of all the variables x_i in the equality constraints.
- In the second-to-last column $\boldsymbol{B}$, the values of the basic variables.
- In the last column $\boldsymbol{B}$, the basic variables.

- The row ending an iteration gives the information about the function under the form $f = \sum_{j=1}^{n} c_j x_j$. We first read the coefficients of f in the columns of the vectors of the canonical form and then the value of $-f$ in the column $\boldsymbol{B}$.

Remark: The columns containing only one nonzero value equal to 1 and all the other values equal to 0 are those of vectors corresponding to the vectors of the basis, and the other columns (except the column $\boldsymbol{B}$) correspond to the nonbasic vectors.

We thus get Table 10.1.

Table 10.1 Simplex tableau: Example 10.1 with slack variables

n	V_1	V_2	V_3	V_4	V_5	B	Basis		
I	−3	2	1	0	0	2	x_3		
	[2]	−4	0	1	0	3	x_4		
	1	1	0	0	1	6	x_5		
	2	−1	0	0	0	0	$-f$	$\leftrightarrow$	$2x_1 - x_2 - f = 0$
II	0	−4	1	1.5	0	6.5	x_3		
	1	−2	0	0.5	0	1.5	x_1		
	0	[3]	0	−0.5	1	4.5	x_5		
	0	3	0	−1	0	−3	$-f$	$\leftrightarrow$	$3x_2 - x_4 - f = -3$
III	0	0	1	0.833	1.333	12.5	x_3		
	1	0	0	0.167	0.667	4.5	x_1		
	0	1	0	−0.167	0.333	1.5	x_2		
	0	0	0	−0.5	−1	−7.5	$-f$	$\leftrightarrow$	$-0.5x_4 - x_5 - f = -7.5$

At each iteration, the following sequence of search stages must be executed (the first iteration is taken as an example):

1. Examine the coefficients of the row of the cost function. In the case of a minimization, if all are positive, the present solution is optimal. In the case of a maximization, if all are negative, the solution is optimal (in the present case, we consider the coefficients 2 and -1 at the first iteration).
2. If this is not the case, in the case of a minimization, choose the most negative coefficient. In the case of a maximization, choose the most positive coefficient (here 2) (application of Dantzig's first criterion). The corresponding vector is here $\boldsymbol{V}_1$. The corresponding variable x_1 becomes a basic variable.
3. Divide each element of vector $\boldsymbol{B}$ (except the one of the row of f) by the corresponding elements of the chosen vector (here $\boldsymbol{V}_1$). At this level, intervene the limitation by the constraint: among all the strictly positive ratios, choose the smallest one, thus limiting (here 3/2=1.5) (application of Dantzig's second criterion). The corresponding basic variable (here x_4) becomes a nonbasic variable.
4. If all the ratios are strictly negative, it means that no constraint is limiting. It would then be possible to increase as much as wanted the variable which is thus nonlimited, the domain is unbounded. Stop the problem.
5. For the new iteration, swap the basic variable and the nonbasic variable previously noted (here x_1 and x_4). The element intersection of the column of this old nonbasic variable (x_1) and of the row of this old basic variable (x_4) is called *pivot element* (noted e_p, here 2).
6. Create the part of the tableau corresponding to the new iteration (here II), for which the basic and nonbasic variables are now known, from the elements of the previous tableau (iteration I). The only columns that are really interesting for the transformation are those which are complete (columns $\boldsymbol{V}_1$ and $\boldsymbol{V}_2$ of iteration I). It is possible to do a provisory tableau (only by thought) by keeping all the elements identical except that we swap the column ($\boldsymbol{V}_1$) of the pivot (elements of the old nonbasic variable (x_1)) and the column ($\boldsymbol{V}_4$) of the new nonbasic variable. Thus, in the new tableau (iteration II), the columns $\boldsymbol{V}_4$ and $\boldsymbol{V}_2$ will be full.

7. Start by the columns corresponding to the basic variables which contain only one 1 (for example, x_3 and $\boldsymbol{V}_3$).
8. Replace the pivot e_p by its inverse $1/e_p$.
9. Replace each element e_r of the pivot row by e_r/e_p and each element e_c of the pivot column by $(-e_c/e_p)$. For all the other elements e, replace them by $(e - e_c\, e_r/e_p)$, where e_c is the value in the pivot column and the same row as e, and e_r is the value in the pivot row and the same column as e. Do not forget to swap the columns as described at stage 6.

Remark a: the operations 8 and 9 can be done on the entire tableau, which allows an easier programming. The exchange of columns described at stage 6 must not be forgotten.
Remark b: the operations 8 and 9 correspond to the pivot method of Gauss–Jordan algorithm to invert a matrix when solving a linear system (Equation (10.4.9)).
Remark c: it may occur that stage 2 fails and that the iterations never end. In that case, other strategies are possible, including a random draw of the column among the selectable columns (Faure et al. 2009). The degeneracy occurs because the coefficients of the right column are only related to the coefficients of the basic variables. Dantzig and Thapa (2003) propose different possibilities to solve that difficulty, including perturbing the right column. This type of degeneracy is very common (Fletcher 1991, page 178), for example, in problems of linear networks.
Remark d: it may occur that the optimal solution is not restricted to a single point but is constituted by a whole hyperplane when the function to minimize is parallel to that hyperplane. The degeneracy is called of first kind (Faure et al. 2009).
Remark e: theoretically, the complexity of simplex method is exponential (in $\exp n$, $\boldsymbol{x} \in \mathcal{R}^n$) (Karmarkar 1984). In practice, this difficulty is rarely encountered and the complexity is in $0(nmL)$ (m constraints, data coded in strings of L bits).
Remark f: a basic solution is called degenerate if one or several basic variables are equal to 0. The degeneracy is not an exception, but it is the rule (Dantzig and Thapa 1997), i.e. it is common.
Remark g: if several variables are likely to be selected for the pivot rule, choose the entering variable which has the smallest index (similarly for the exiting variable). This Bland rule avoids cycling (Matousek and Gärtner 2007); however, it considerably slows the simplex. Cycling rarely occurs in practice.

10.3.2 Simplex Tableau with Slack and Artificial Variables

Consider Example 10.2. Recall that, in addition to the slack variables x_3 to x_6, two artificial variables x_7 and x_8 had to be introduced. Hence, in the first stage of solving this auxiliary problem, the function to optimize is no more f

$$f = 4\,x_1 + 5\,x_2 \tag{10.3.24}$$

for which the maximum is sought, but f' whose minimum must be zero due to the nullity of the artificial variables at the optimum

$$\min f' = x_7 + x_8 \tag{10.3.25}$$

The equations related to the constraints are

$$\begin{array}{lllllllll} 2\,x_1 & +x_2 & +x_3 & & & & & & = 6 \\ x_1 & +2\,x_2 & & +x_4 & & & & & = 5 \\ x_1 & +x_2 & & & -x_5 & & +x_7 & & = 1 \\ x_1 & +4\,x_2 & & & & -x_6 & & +x_8 & = 2 \end{array} \tag{10.3.26}$$

$$x_i \geq 0\,, \quad i = 1, \ldots, 8 \tag{10.3.27}$$

This problem can be solved by the simplex method.
Express f' by replacing the artificial variables with respect to the nonbasic variables

$$f' = (1 - x_1 - x_2 + x_5) + (2 - x_1 - 4\,x_2 + x_6) = 3 - 2\,x_1 - 5\,x_2 + x_5 + x_6 \tag{10.3.28}$$

At the first iteration, the nonbasic variables that are zero are

$$x_1 = x_2 = x_5 = x_6 = 0 \tag{10.3.29}$$

from which the value of the basic variables results

$$x_3 = 6\,,\ x_4 = 5\,,\ x_7 = 1\,,\ x_8 = 2 \tag{10.3.30}$$

First iteration:
The only columns completely filled are those of the vectors corresponding to the basic vectors.

- the smallest negative coefficient of the row of f' is -5. In this column (V_2), the smallest positive ratio is 2/4 (row x_8), and thus the pivot is 4. x_2 becomes a basic variable and x_8 a nonbasic variable.
Second iteration:
- the smallest negative coefficient of the row of f' is -0.75 . In this column (V_1), the smallest positive ratio is 0.5/0.75 (row x_7), and thus the pivot is 0.75. x_1 becomes a basic variable and x_7 a nonbasic variable.
Third iteration:
- all the coefficients of the row of f' are positive or zero. No more progress is possible. The minimum of f' is reached as indicated by $f' = 0$, corresponding to the nullity of the artificial variables. The value of the function f results at this minimum $f = 4.33$ and the values of the quantities $x_1 = 0.67$, $x_2 = 0.33$, $x_3 = 4.33$, $x_4 = 3.67$, $x_5 = 0$, and $x_6 = 0$, which will constitute the start point of the second simplex tableau without artificial variables.
By using the simplex method previously described, we get the first simplex tableau (Table 10.2) with artificial variables.

Table 10.2 First simplex tableau for Example 10.2 with slack and artificial variables

n	$\boldsymbol{V}_1$	$\boldsymbol{V}_2$	$\boldsymbol{V}_3$	$\boldsymbol{V}_4$	$\boldsymbol{V}_5$	$\boldsymbol{V}_6$	$\boldsymbol{V}_7$	$\boldsymbol{V}_8$	$\boldsymbol{B}$	Basis
I	2	1	1	0	0	0	0	0	6	x_3
	1	2	0	1	0	0	0	0	5	x_4
	1	1	0	0	−1	0	1	0	1	x_7
	1	4	0	0	0	−1	0	1	2	x_8
	−2	−5	0	0	1	1	0	0	−3	$-f'$
II	1.75	0	1	0	0	0.25	0	−0.25	5.5	x_3
	0.5	0	0	1	0	0.5	0	−0.5	4	x_4
	0.75	0	0	0	−1	0.25	1	−0.25	0.5	x_7
	0.25	1	0	0	0	−0.25	0	0.25	0.5	x_2
	−0.75	0	0	0	1	−0.25	0	1.25	−0.5	$-f'$
III	0	0	1	0	2.33	−0.33	−2.33	0.33	4.33	x_3
	0	0	0	1	0.67	0.33	−0.67	−0.33	3.67	x_4
	1	0	0	0	−1.33	0.33	1.33	−0.33	0.67	x_1
	0	1	0	0	0.33	−0.33	−0.33	0.33	0.33	x_2
	0	0	0	0	0	0	1	1	0	$-f'$

Now that a solution is available for the main problem, the second simplex tableau (Table 10.3) is built by taking the part of the first tableau (Table 10.2) corresponding to the last iteration and by suppressing the columns of the artificial variables (here columns V_7 and V_8 for x_7 and x_8). Now, the function f is used in the way it was originally defined.
First iteration:
The only columns totally filled are those of the vectors corresponding to the basic vectors.
We take the original definition of f

$$\max\ f = 4\,x_1 + 5\,x_2 \tag{10.3.31}$$

and the constraints with only the slack variables

$$\begin{array}{lllllll} 2\,x_1 & +x_2 & +x_3 & & & & = 6 \\ x_1 & +2\,x_2 & & +x_4 & & & = 5 \\ x_1 & +x_2 & & & -x_5 & & = 1 \\ x_1 & +4\,x_2 & & & & -x_6 & = 2 \end{array} \tag{10.3.32}$$

The basic variables are x_1, x_2, x_3, and x_4 and the nonbasic variables are x_5 and x_6. The cost function f must be expressed with respect to the nonbasic variables, hence

$$f = 4\,(\frac{2}{3} + \frac{4}{3}x_5 - \frac{1}{3}x_6) + 5\,(\frac{1}{3} - \frac{1}{3}x_5 + \frac{1}{3}x_6) = \frac{13}{3} + \frac{11}{3}\,x_5 + \frac{1}{3}\,x_6 \tag{10.3.33}$$

Indeed, this problem consists in considering among the constraints (10.3.32) those where the artificial variables intervened and expressing each of the basic variables of f with respect to the nonbasic variables appearing in these constraints. In general, a matrix inversion is necessary to find the solution. Then, we simply do the replacement in f.

Table 10.3 Second simplex tableau for Example 10.2 with slack variables and suppression of artificial variables

n	V_1	V_2	V_3	V_4	V_5	V_6	B	Basis
I	0	0	1	0	2.33	−0.33	4.33	x_3
	0	0	0	1	0.67	0.33	3.67	x_4
	1	0	0	0	−1.33	0.33	0.67	x_1
	0	1	0	0	[0.33]	−0.33	0.33	x_2
	0	0	0	0	3.67	0.33	−4.33	$-f$
II	0	−7	1	0	0	[2]	2	x_3
	0	−2	0	1	0	1	3	x_4
	1	4	0	0	0	−1	2	x_1
	0	3	0	0	1	−1	1	x_5
	0	−11	0	0	0	4	−8	$-f$
III	0	−3.5	0.5	0	0	1	1	x_6
	0	[1.5]	−0.5	1	0	0	2	x_4
	1	0.5	0.5	0	0	0	3	x_1
	0	−0.5	0.5	0	1	0	2	x_5
	0	3	−2	0	0	0	−12	$-f$
IV	0	0	−0.67	2.33	0	1	5.67	x_6
	0	1	−0.33	0.67	0	0	1.33	x_2
	1	0	0.67	−0.33	0	0	2.33	x_1
	0	0	0.33	0.33	1	0	2.67	x_5
	0	0	−1	−2	0	0	−16	$-f$

– We desire to maximize f. The largest positive coefficient of the row of f is 3.67. In this column (V_5), the smallest positive ratio is 0.33/0.33 (row x_2), and thus the pivot is 0.33. x_5 becomes a basic variable and x_2 a nonbasic variable.

Second iteration:

- The largest positive coefficient of the row of f is 4. In this column (V_6), the smallest positive ratio is 2/2 (row x_3), and thus the pivot is 2. x_6 becomes a basic variable and x_3 a nonbasic variable.
  Third iteration:

- The largest positive coefficient of the row of f is 3. In this column (V_2), the smallest positive ratio is 2/1.5 (row x_4), and thus the pivot is 1.5. x_2 becomes a basic variable and x_4 a nonbasic variable. Fourth iteration:
- All the coefficients of the row of f are negative. Thus, the maximum of f has been reached and it is equal to 16, the quantities x_1 and x_2 being equal to 2.33 and 1.33, respectively.

10.4 Theoretical Solution

Let the general problem of linear programming

$$\max f(\boldsymbol{x}) = \max \sum_{j=1}^{n} c_j\, x_j \tag{10.4.1}$$

subject to linear inequality constraints

$$g_i(\boldsymbol{x}) = \sum_{j=1}^{n} a_{ij} x_j - b_i \leq 0 \tag{10.4.2}$$

$$x_i \geq 0 \Longrightarrow -x_i \leq 0 \tag{10.4.3}$$

that is written under the most general matrix form including slack artificial variables

$$\max f = \boldsymbol{c}^T\, \boldsymbol{x} \tag{10.4.4}$$

subject to constraints

$$\boldsymbol{A}\, \boldsymbol{x} = \boldsymbol{b} \tag{10.4.5}$$

$$\boldsymbol{x} \geq 0 \tag{10.4.6}$$

To express the solution, it is necessary to distinguish the *basic variables* which are noted $\boldsymbol{x}_B$ and the *nonbasic variables* $\boldsymbol{x}_N$. The notation B is introduced to emphasize the fact that a basis exists, which means that a nonsingular square matrix $\boldsymbol{A}_B$ exists.

We again write f by separating the basic and nonbasic variables

$$f = \boldsymbol{c}_B^T\, \boldsymbol{x}_B + \boldsymbol{c}_N^T\, \boldsymbol{x}_N \tag{10.4.7}$$

subject to constraints

$$\boldsymbol{A}_B\, \boldsymbol{x}_B + \boldsymbol{A}_N\, \boldsymbol{x}_N = \boldsymbol{b} \tag{10.4.8}$$

The basic solution is equal to

$$\boldsymbol{x}_B = \boldsymbol{A}_B^{-1}\, (\boldsymbol{b} - \boldsymbol{A}_N\, \boldsymbol{x}_N) \tag{10.4.9}$$

hence the value of function f at the optimum

$$f = \boldsymbol{c}_B^T \boldsymbol{x}_B + \boldsymbol{c}_N^T \boldsymbol{x}_N = \boldsymbol{c}_B^T\, \boldsymbol{A}_B^{-1}\, \boldsymbol{b} + (\boldsymbol{c}_N^T - \boldsymbol{c}_B^T\, \boldsymbol{A}_B^{-1}\, \boldsymbol{A}_N)\, \boldsymbol{x}_N \tag{10.4.10}$$

Note that the value of the basic variables by means of the previous relation depends on the value of the nonbasic variables which is arbitrary. However, this relation is very general and always valid.

When all the nonbasic variables are zero, and when the coefficients $\boldsymbol{c}_N$ are positive or zero, then the solution is called *extremal basic solution* and is equal to

$$\boldsymbol{x}_B = \boldsymbol{A}_B^{-1}\,\boldsymbol{b} \tag{10.4.11}$$

Moreover, if the basis B is chosen so that the product $[\boldsymbol{A}_B^{-1}\,\boldsymbol{b}]$ is positive or zero (condition on x_i), then the extremal basic solution is called *realizable*. The maximum of the cost function is equal to

$$f^* = \boldsymbol{c}_B^T\,\boldsymbol{A}_B^{-1}\,\boldsymbol{b} \tag{10.4.12}$$

Example 10.4:

Linear programming: solution using the basic and nonbasic matrices

Consider again Example 10.1 of maximization of f

$$\max\ f = 2\,x_1 - x_2 \tag{10.4.13}$$

subject to constraints with slack variables

$$\begin{array}{llllll} -3\,x_1 & +2\,x_2 & +x_3 & & & = 2 \\ 2\,x_1 & -4\,x_2 & & +x_4 & & = 3 \\ x_1 & +x_2 & & & +x_5 & = 6 \end{array} \tag{10.4.14}$$

with

$$\boldsymbol{A} = \left[\begin{array}{rr|rrr} -3 & 2 & 1 & 0 & 0 \\ 2 & -4 & 0 & 1 & 0 \\ 1 & 1 & 0 & 0 & 1 \end{array}\right] = [\boldsymbol{A}_N \mid \boldsymbol{A}_B]$$
$$\boldsymbol{b}^T = [2\ 3\ 6] \quad , \quad \boldsymbol{c}^T = [2\ -1 \mid 0\ 0\ 0] = [\boldsymbol{c}_N^T \mid \boldsymbol{c}_B^T] \tag{10.4.15}$$

The coefficients c of function f are equal to those of the row of the simplex tableau (Table 10.3).

At iteration 1, we have

$$\boldsymbol{x}_B = \begin{bmatrix} x_3 \\ x_4 \\ x_5 \end{bmatrix}, \quad \boldsymbol{x}_N = \begin{bmatrix} x_1 \\ x_2 \end{bmatrix}, \quad \text{such that } \boldsymbol{A}_B\,\boldsymbol{x}_B + \boldsymbol{A}_N\,\boldsymbol{x}_N = \boldsymbol{b} \tag{10.4.16}$$

$$\boldsymbol{x}_B = \begin{bmatrix} 2 \\ 3 \\ 6 \end{bmatrix}, \quad \boldsymbol{x}_N = \begin{bmatrix} 0 \\ 0 \end{bmatrix}, \quad f = \boldsymbol{c}_B^T\,\boldsymbol{A}_B^{-1}\,\boldsymbol{b} = 0 \tag{10.4.17}$$

All the coefficients $\boldsymbol{c}_N^T$ are nonnegative or zero, and thus the solution can be improved.

At iteration 2, by applying the rules of the simplex tableau (Table 10.3), the new basic and nonbasic variables are determined, corresponding to the swap of x_1 and x_4

$$x_B = \begin{bmatrix} x_3 \\ x_1 \\ x_5 \end{bmatrix}, \quad x_N = \begin{bmatrix} x_4 \\ x_2 \end{bmatrix} \tag{10.4.18}$$

With respect to these new basic and nonbasic variables, the matrix $\boldsymbol{A}$ must be reordered by swapping the columns 1 and 4 as

$$A = \begin{bmatrix} 0 & 2 & 1 & -3 & 0 \\ 1 & -4 & 0 & 2 & 0 \\ 0 & 1 & 0 & 1 & 1 \end{bmatrix} = \begin{bmatrix} A_N \mid A_B \end{bmatrix} \tag{10.4.19}$$

which gives

$$x_B + A_B^{-1} A_N x_N = A_B^{-1} b \tag{10.4.20}$$

hence

$$\begin{bmatrix} x_3 \\ x_1 \\ x_5 \end{bmatrix} + \begin{bmatrix} 1.5 & -4 \\ 0.5 & -2 \\ -0.5 & 3 \end{bmatrix} \begin{bmatrix} x_4 \\ x_2 \end{bmatrix} = \begin{bmatrix} 6.5 \\ 1.5 \\ 4.5 \end{bmatrix} \tag{10.4.21}$$

or still

$$\begin{aligned} x_3 + 1.5\, x_4 - 4\, x_2 &= 6.5 \\ x_1 + 0.5\, x_4 - 2\, x_2 &= 1.5 \\ x_5 - 0.5\, x_4 + 3\, x_2 &= 4.5 \end{aligned} \tag{10.4.22}$$

which allows to express the basic variables with respect to the nonbasic variables to transform the function f as

$$\begin{aligned} f &= c_B^T x_B + c_N^T x_N = c_B^T (A_B^{-1} b - A_B^{-1} A_N x_N) + c_N^T x_N \\ &= 2\, x_1 - x_2 \\ &= 2(1.5 - 0.5\, x_4 + 2\, x_2) - x_2 \\ &= 3 + 3\, x_2 - x_4 \end{aligned} \tag{10.4.23}$$

from the previous calculations, the result of the second iteration of simplex tableau (Table 10.3) is deduced, and the new matrices, that is,

$$\begin{aligned} A &= \begin{bmatrix} 1.5 & -4 & 1 & 0 & 0 \\ 0.5 & -2 & 0 & 1 & 0 \\ -0.5 & 3 & 0 & 0 & 1 \end{bmatrix} = \begin{bmatrix} A_N \mid A_B \end{bmatrix} \\ b^T &= \begin{bmatrix} 6.5 \; 1.5 \; 4.5 \end{bmatrix} \quad , \quad c^T = \begin{bmatrix} -1 \; 3 \mid 0 \; 0 \; 0 \end{bmatrix} = \begin{bmatrix} c_N^T \mid c_B^T \end{bmatrix} \end{aligned} \tag{10.4.24}$$

All the coefficients c_N^T are nonnegative or zero, and thus the solution can still be improved. The nonzero coefficients c of function f are those that are obtained when f is expressed with respect to the nonbasic variables.

At iteration 3, by applying the rules of the simplex tableau (Table 10.3), the new basic and nonbasic variables are determined, corresponding to the swap of x_2 and x_5

$$x_B = \begin{bmatrix} x_3 \\ x_1 \\ x_2 \end{bmatrix} , \quad x_N = \begin{bmatrix} x_4 \\ x_5 \end{bmatrix} \tag{10.4.25}$$

Taking into account these new basic and nonbasic variables, the matrix A is reordered as

$$A = \begin{bmatrix} 1.5 & 0 & 1 & 0 & -4 \\ 0.5 & 0 & 0 & 1 & -2 \\ -0.5 & 1 & 0 & 0 & 3 \end{bmatrix} = \begin{bmatrix} A_N \mid A_B \end{bmatrix} \tag{10.4.26}$$

which gives

$$x_B + A_B^{-1} A_N x_N = A_B^{-1} b \tag{10.4.27}$$

hence

$$\begin{bmatrix} x_3 \\ x_1 \\ x_2 \end{bmatrix} + \begin{bmatrix} 0.833 & 1.333 \\ 0.167 & 0.667 \\ -0.167 & 0.333 \end{bmatrix} \begin{bmatrix} x_4 \\ x_5 \end{bmatrix} = \begin{bmatrix} 12.5 \\ 4.5 \\ 1.5 \end{bmatrix} \tag{10.4.28}$$

or still

$$\begin{aligned} x_3 + 0.833\, x_4 + 1.333\, x_5 &= 12.5 \\ x_1 + 0.167\, x_4 + 0.667\, x_5 &= 4.5 \\ x_2 - 0.167\, x_4 + 0.333\, x_5 &= 1.5 \end{aligned} \tag{10.4.29}$$

which allows to express the basic variables with respect to the nonbasic variables to transform the function f as

$$\begin{aligned} f &= \boldsymbol{c}_B^T \boldsymbol{x}_B + \boldsymbol{c}_N^T \boldsymbol{x}_N = \boldsymbol{c}_B^T (\boldsymbol{A}_B^{-1} \boldsymbol{b} - \boldsymbol{A}_B^{-1} \boldsymbol{A}_N \boldsymbol{x}_N) + \boldsymbol{c}_N^T \boldsymbol{x}_N \\ &= 3 + 3\,x_2 - x_4 \\ &= 3 + 3\,(1.5 + 0.167\,x_4 - 0.333\,x_5) - x_4 \\ &= 7.5 - 0.5\,x_4 - x_5 \end{aligned} \tag{10.4.30}$$

All the coefficients $\boldsymbol{c}_N^T$ are negative or zero, and thus the solution cannot be improved any more. By imposing the nonbasic variables equal to zero, we obtain the maximum of f equal to 7.5 for $x_1 = 4.5$ and $x_2 = 1.5$.

Finally, we deduce the result of the third and last iteration of the tableau (Table 10.3), and the new matrices, that is,

$$\begin{aligned} \boldsymbol{A} &= \left[\begin{array}{cc|ccc} 0.833 & 1.333 & 1 & 0 & 0 \\ 0.167 & 0.667 & 0 & 1 & 0 \\ -0.167 & 0.333 & 0 & 0 & 1 \end{array}\right] = \left[\boldsymbol{A}_N \mid \boldsymbol{A}_B\right] \\ \boldsymbol{b}^T &= \left[12.5\ 4.5\ 1.5\right] \quad , \quad \boldsymbol{c}^T = \left[-0.5\ -1 \mid 0\ 0\ 0\right] = \left[\boldsymbol{c}_N^T \mid \boldsymbol{c}_B^T\right] \end{aligned} \tag{10.4.31}$$

Example 10.5 :

Linear programming: solution for the basic and nonbasic matrices and general expression of the cost function

To explain the general calculation of the cost function, consider again Example 10.3 with artificial variables.

$$\max\ f = 4\,x_1 + 5\,x_2 \tag{10.4.32}$$

and the constraints with only slack variables

$$\begin{array}{llllll} 2\,x_1 & +x_2 & +x_3 & & & = 6 \\ x_1 & +2\,x_2 & & +x_4 & & = 5 \\ x_1 & +x_2 & & & -x_5 & = 1 \\ x_1 & +4\,x_2 & & & \quad -x_6 & = 2 \end{array} \tag{10.4.33}$$

It was already explained that the first iteration of the second simplcx tablcau (Table 10.3) is obtained by taking all the rows (except the row of the cost function) of the last iteration of the first tableau (Table 10.2) with artificial variables and suppressing the columns of the artificial variables.

To obtain the row of the cost function of the first iteration of the second simplex tableau, we calculate the cost function according to the relation

$$f = \boldsymbol{c}_B^T \boldsymbol{A}_B^{-1} \boldsymbol{b} + [\boldsymbol{c}_N^T - \boldsymbol{c}_B^T \boldsymbol{A}_B^{-1} \boldsymbol{A}_N] \boldsymbol{x}_N \tag{10.4.34}$$

This is particularly important in that case. The first part of the expression of f gives the constant value of f and the second part the coefficients of f with respect to the nonbasic variables considered in the order of vector $\boldsymbol{x}_N$.

Knowing that the vector of basic variables is $\boldsymbol{x}_B = [x_3\ x_4\ x_1\ x_2]^T$, the vector $\boldsymbol{c}_B$ is equal to $\boldsymbol{c}_B = [0\ 0\ 4\ 5]^T$ taking into account the definition (10.4.32) of f. As f does not contain any variable of the nonbasic vector $\boldsymbol{x}_N = [x_5\ x_6]$, it results $\boldsymbol{c}_N = [0\ 0]^T$. Using on another side

$$\boldsymbol{A}_N = \begin{bmatrix} 1 & 0 & 0 & 0 \\ 0 & 1 & 0 & 0 \\ 0 & 0 & 1 & 0 \\ 0 & 0 & 0 & 1 \end{bmatrix} \quad , \quad \boldsymbol{A}_B = \begin{bmatrix} 2.33 & -0.33 \\ 0.67 & 0.33 \\ -1.33 & 0.33 \\ 0.33 & -0.33 \end{bmatrix} \quad , \quad \boldsymbol{b} = \begin{bmatrix} 4.33 \\ 3.67 \\ 0.67 \\ 0.33 \end{bmatrix} \tag{10.4.35}$$

hence f with respect to the nonbasic variables

$$f = 4.33 + 3.67\,x_5 + 0.33\,x_6 \tag{10.4.36}$$

In the tableau (Table 10.3), the last row of the first iteration makes the value of f appear as well as the coefficients of f with respect to the nonbasic variables according to the second expression of (10.4.36) as required by the tableau.

10.5 Case of Simultaneous Inequality and Equality Constraints

In the case where inequality and equality constraints are simultaneously present, the problem is set under the following form:

$$\begin{aligned}
&\min\ f(\boldsymbol{x}) = \sum_{j=1}^{n} c_j\, x_j \\
&\text{subject to} \\
&g_i(\boldsymbol{x}) = \sum_{j=1}^{n} a_{ij} x_j - b_i \leq 0\,, \quad 1 \leq i \leq m \\
&g_i(\boldsymbol{x}) = \sum_{j=1}^{n} a_{ij} x_j - b_i = 0\,, \quad 1 \leq i \leq p
\end{aligned} \tag{10.5.1}$$

To start the method of simplex tableau, a feasible point is needed, that is, a point that in particular satisfies the equality constraints. To find such a point, several methods are possible (Ferris et al. 2007). A method similar to that used for the alone inequality constraints is to introduce an artificial variable for each equality constraint. Moreover, it can be necessary to introduce artificial variables for some inequality constraints. Then, we must minimize the sum of all the artificial variables introduced for these inequality and equality constraints. This will constitute the first simplex tableau which will lead to obtain or not a feasible point. If a feasible point exists, a second simplex tableau will be implemented like in the previous cases.

Suppose that the linear programming problem is written as

$$\begin{aligned}
&\min\ f(\boldsymbol{x}) = \boldsymbol{c}^T\, \boldsymbol{x} \\
&\text{subject to} \\
&\boldsymbol{A}\, \boldsymbol{x} \geq \boldsymbol{b} \\
&\boldsymbol{E}\, \boldsymbol{x} = \boldsymbol{F}
\end{aligned} \tag{10.5.2}$$

another method cited by Ferris et al. (2007) consists in decomposing $\boldsymbol{x}$

$$\boldsymbol{x} = \boldsymbol{x}^+ - \boldsymbol{x}^- \quad \text{with } \boldsymbol{x}^+\,,\ \boldsymbol{x}^- \geq 0 \tag{10.5.3}$$

which transforms the problem into

$$
\begin{aligned}
&\min \boldsymbol{c}^T \boldsymbol{x}^+ - \boldsymbol{c}^T \boldsymbol{x}^- \\
&\text{subject to} \\
&\boldsymbol{A}\,\boldsymbol{x}^+ - \boldsymbol{A}\,\boldsymbol{x}^- \geq \boldsymbol{b} \\
&\boldsymbol{E}\,\boldsymbol{x}^+ - \boldsymbol{E}\,\boldsymbol{x}^- \geq \boldsymbol{F} \\
&-\boldsymbol{E}\,\boldsymbol{x}^+ + \boldsymbol{E}\,\boldsymbol{x}^- \geq -\boldsymbol{F}
\end{aligned}
\tag{10.5.4}
$$

by transforming each equality constraint as lower and upper inequalities. Posing z the sought vector equal to

$$
z = \begin{pmatrix} \boldsymbol{x}^+ \\ \boldsymbol{x}^- \end{pmatrix}
\tag{10.5.5}
$$

the problem (10.5.4) can be easily formulated under the standard form.

To illustrate the approach followed by the introduction of an artificial variable for each equality constraint, two examples are solved.

Example 10.6 :

Linear programming: simultaneous inequality and equality constraints

Consider the following problem:

$$
\begin{aligned}
&\max f(\boldsymbol{x}) = 2\,x_1 - x_2 + x_3 \\
&\text{subject to} \\
&-x_1 + x_2 - 4\,x_3 \leq 1 \\
&-x_1 + x_2 + x_3 \leq -2 \\
&x_1 + 3\,x_2 + 2\,x_3 = 3
\end{aligned}
\tag{10.5.6}
$$

By introducing the slack variables x_4 and x_5 and the artificial variables x_6 and x_7, the constraints are transformed as

$$
\begin{aligned}
&-x_1 + x_2 - 4\,x_3 + x_4 = 1 \\
&x_1 - x_2 - x_3 - x_5 + x_6 = 2 \\
&x_1 + 3\,x_2 + 2\,x_3 + x_7 = 3
\end{aligned}
\tag{10.5.7}
$$

and the first part of the problem is

$$
\min f'(\boldsymbol{x}) = x_6 + x_7
\tag{10.5.8}
$$

Initially, the basic variables are $x_4 = 1$, $x_6 = 2$, and $x_7 = 3$, and the function f' must be expressed with respect to the nonbasic variables, that is,

$$
f'(\boldsymbol{x}) = 2 - (x_1 - x_2 - x_3 - x_5) + 3 - (x_1 + 3\,x_2 + 2\,x_3) = 5 - 2\,x_1 - 2\,x_2 - x_3 + x_5
\tag{10.5.9}
$$

The tableau (Table 10.4) thus results.

At the end of the first simplex tableau (Table 10.4), f' is zero; thus, the solution $x_1 = 2.25$, $x_2 = 0.25$, $x_3 = 0$ is feasible and the second simplex tableau (Table 10.5) can be achieved. The cost function f must be expressed with respect to the nonbasic variables x_3 and x_5, hence $f = 4.25 + 2.25\,x_3 + 1.75\,x_5$. By achieving the second simplex tableau (Table 10.5), the maximum of f is obtained at $x_1 = 3$, $x_2 = 0$, and $x_3 = 0$.

If we had searched the minimum of f with the same constraints, the first simplex tableau (Table 10.4) would have been identical, and the second tableau would have started in the same way as Table 10.5. However, as the coefficients of the row of f are positive in the first iteration of the second tableau (Table 10.5), the minimum of f would have been found at $x_1 = 2.25$, $x_2 = 0.25$, $x_3 = 0$ at the first iteration.

Example 10.7 :

Linear programming: simultaneous inequality and equality constraints

Let the following problem:

Table 10.4 First simplex tableau for Example 10.6 with inequality and equality constraints

n	V_1	V_2	V_3	V_4	V_5	V_6	V_7	B	Basis
I	−1	1	−4	1	0	0	0	1	x_4
	[1]	−1	−1	0	−1	1	0	2	x_6
	1	3	2	0	0	0	1	3	x_7
	−2	−2	−1	0	1	0	0	−5	$-f'$
II	0	0	−5	1	−1	1	0	3	x_4
	1	−1	−1	0	−1	1	0	2	x_1
	0	[4]	3	0	1	−1	1	1	x_7
	0	−4	−3	0	−1	2	0	−1	$-f'$
III	0	0	−5	1	−1	1	0	3	x_4
	1	0	−0.25	0	−0.75	0.75	0.25	2.25	x_1
	0	1	0.75	0	0.25	−0.25	0.25	0.25	x_2
	0	0	0	0	0	1	1	0	$-f'$

Table 10.5 Second simplex tableau for Example 10.6 with inequality and equality constraints

n	V_1	V_2	V_3	V_4	V_5	B	Basis
I	0	0	−5	1	−1	3	x_4
	1	0	−0.25	0	−0.75	2.25	x_1
	0	1	[0.75]	0	0.25	0.25	x_2
	0	0	2.25	0	1.75	−4.25	$-f$
II	0	6.667	0	1	0.667	4.667	x_4
	1	0.333	0	0	−0.667	2.333	x_1
	0	1.333	1	0	[0.333]	0.333	x_3
	0	−3	0	0	1	−5	$-f$
III	0	4	−2	1	0	4	x_4
	1	3	2	0	0	3	x_1
	0	4	3	0	1	1	x_5
	0	−7	−3	0	0	−6	$-f$

$$\begin{aligned} &\min f(\boldsymbol{x}) = 2x_1 - x_2 + x_3 \\ &\text{subject to} \\ &-x_1 + x_2 - 4x_3 \leq 1 \\ &-x_1 + x_2 + x_3 \leq -2 \\ &2x_1 + 4x_2 + 5x_3 \leq 3 \\ &-x_1 - 2x_2 - x_3 \leq -3 \\ &x_1 + 3x_2 + 2x_3 = 3 \end{aligned} \tag{10.5.10}$$

By the introduction of slack and artificial variables, the constraints become

$$\begin{aligned} &-x_1 + x_2 - 4x_3 + x_4 = 1 \\ &x_1 - x_2 - x_3 - x_5 + x_8 = 2 \\ &2x_1 + 4x_2 + 5x_3 + x_6 = 3 \\ &x_1 + 2x_2 + x_3 - x_7 + x_9 = 3 \\ &x_1 + 3x_2 + 2x_3 + x_{10} = 3 \end{aligned} \tag{10.5.11}$$

Table 10.6 First simplex tableau for Example 10.7 with inequality and equality constraints

n	V_1	V_2	V_3	V_4	V_5	V_6	V_7	V_8	V_9	V_{10}	B	Basis
I	−1	1	−4	1	0	0	0	0	0	0	1	x_4
	1	−1	−1	0	−1	0	0	1	0	0	2	x_8
	2	4	5	0	0	1	0	0	0	0	3	x_6
	1	2	1	0	0	0	−1	0	1	0	3	x_9
	1	3	2	0	0	0	0	0	0	1	3	x_{10}
	−3	−4	−2	0	1	0	1	0	0	0	−8	$-f'$
II	−1.5	0	−5.25	1	0	−0.25	0	0	0	0	0.25	x_4
	1.5	0	0.25	0	−1	0.25	0	1	0	0	2.75	x_8
	0.5	1	1.25	0	0	0.25	0	0	0	0	0.75	x_2
	0	0	−1.5	0	0	−0.5	−1	0	1	0	1.5	x_9
	−0.5	0	−1.75	0	0	−0.75	0	0	0	1	0.75	x_{10}
	−1	0	3	0	1	1	1	0	0	0	−5	$-f'$
III	0	3	−1.5	1	0	0.5	0	0	0	0	2.5	x_4
	0	−3	−3.5	0	−1	−0.5	0	1	0	0	0.5	x_8
	1	2	2.5	0	0	0.5	0	0	0	0	1.5	x_1
	0	0	−1.5	0	0	−0.5	−1	0	1	0	1.5	x_9
	0	1	−0.5	0	0	−0.5	0	0	0	1	1.5	x_{10}
	0	2	5.5	0	1	1.5	1	0	0	0	−3.5	$-f'$

The function f' sum of the artificial variables is equal to

$$\begin{aligned} f' &= x_8 + x_9 + x_{10} \\ &= 8 - 3\,x_1 - 4\,x_2 - 2\,x_3 + x_5 + x_7 \end{aligned} \tag{10.5.12}$$

It is then expressed with respect to the nonbasic variables x_1, x_2, x_3, x_5, and x_7.

At the end of the first simplex tableau (Table 10.6), all coefficients of the row of f' are positive, meaning that the solution cannot be anymore improved. On another side, the function f' sum of the artificial variables is $f' = 3.5$ and thus is strictly positive, which implies that there exists no feasible solution.

10.6 Duality

When the scale of the linear programming problem is large, i.e. when, on one side, the number of variables is important and, on another side, the number of constraints is also important, the computing time can become a non-negligible factor. It seems that the computing time is more sensitive to the number of constraints than to the number of variables. For that reason, the treatment of a dual problem that would possibly have less constraints can turn out to be interesting.

The duality theorem is due to Von Neumann and the term of primal to George Dantzig's father, Tobias Dantzig, mathematician of Russian origin (Dantzig and Thapa 1997). The demonstration of the duality theorem is assigned to Gale et al. (1951).

10.6.1 Example of Duality

Consider again Example 10.1 already treated, search of the maximum of f

$$\max\ f = 2\,x_1 - x_2 \tag{10.6.1}$$

subject to the following constraints:

$$\begin{array}{lll} -3\,x_1 & +2\,x_2 & \leq 2 \\ 2\,x_1 & -4\,x_2 & \leq 3 \\ x_1 & +x_2 & \leq 6 \end{array} \tag{10.6.2}$$

$$x_i \geq 0 \tag{10.6.3}$$

This problem is qualified as *primal*.
The vectors and matrices of the example are as follows:

$$\boldsymbol{c} = \begin{bmatrix} 2 \\ -1 \end{bmatrix} \quad , \quad \boldsymbol{x} = \begin{bmatrix} x_1 \\ x_2 \end{bmatrix} \quad , \quad \boldsymbol{A} = \begin{bmatrix} -3 & 2 \\ 2 & -4 \\ 1 & 1 \end{bmatrix} \quad , \quad \boldsymbol{b} = \begin{bmatrix} 2 \\ 3 \\ 6 \end{bmatrix} \tag{10.6.4}$$

The following *dual problem* is associated, search of the minimum of h

$$\min\ h = 2\,w_1 + 3\,w_2 + 6\,w_3 \tag{10.6.5}$$

subject to the following constraints:

$$\begin{array}{llll} -3\,w_1 & +2\,w_2 & +w_3 & \geq 2 \\ 2\,w_1 & -4\,w_2 & +w_3 & \geq -1 \end{array} \tag{10.6.6}$$

$$w_i \geq 0 \tag{10.6.7}$$

with

$$\boldsymbol{w} = \begin{bmatrix} w_1 \\ w_2 \\ w_3 \end{bmatrix} \tag{10.6.8}$$

The *duality theorem* demonstrates that both functions f and h have the same value at the optimum and that the *maximum of f is equal to the minimum of h.*

10.6.2 Demonstration of the Duality Theorem

In the optimality conditions (Section 10.2.3), an important relation was shown.
Let us pose the primal problem P_1 under a theoretical form

$$\max\ f(\boldsymbol{x}) = \max \sum_{j=1}^{n} c_j\,x_j \tag{10.6.9}$$

subject to m linear inequality constraints g_i

$$\begin{aligned} & g_i(\boldsymbol{x}) = \sum_{j=1}^{n} a_{ij} x_j - b_i \leq 0\,, \quad i = 1,\ldots,m \\ & x_j \geq 0\,, \quad j = 1,\ldots,n \end{aligned} \tag{10.6.10}$$

or under matrix form

$$\max\ f = \boldsymbol{c}^T \boldsymbol{x} \tag{10.6.11}$$

subject to constraints

$$\boldsymbol{A}\,\boldsymbol{x} \le \boldsymbol{b}\,, \quad \boldsymbol{x} \ge 0 \tag{10.6.12}$$

The dual problem P_2 is as follows:

$$\min\ h = \boldsymbol{b}^T \boldsymbol{w} \tag{10.6.13}$$

subject to constraints

$$\boldsymbol{A}^T \boldsymbol{w} \ge \boldsymbol{c}\,, \quad \boldsymbol{w} \ge 0 \tag{10.6.14}$$

Note that $\boldsymbol{b}$, $\boldsymbol{c}$, $\boldsymbol{x}$, and $\boldsymbol{w}$ are column vectors. Note also the first matrix characteristic of the primal problem and the second matrix characteristic of the dual problem which is the transpose of the primal matrix

$$\begin{bmatrix} \boldsymbol{A} & \boldsymbol{b} \\ \boldsymbol{c}^T & \end{bmatrix}, \quad \begin{bmatrix} \boldsymbol{A}^T & \boldsymbol{c} \\ \boldsymbol{b}^T & \end{bmatrix} \tag{10.6.15}$$

Theorem 1:
The dual of the dual is the primal.
Demonstration:
Pose P_2 under a slightly different form, equivalent to

$$\max\ -h = -\boldsymbol{b}^T \boldsymbol{w} \tag{10.6.16}$$

subject to constraints

$$-\boldsymbol{A}^T \boldsymbol{w} \le -\boldsymbol{c} \tag{10.6.17}$$

$$\boldsymbol{w} \ge 0 \tag{10.6.18}$$

and take its dual

$$\min\ -\boldsymbol{c}^T \boldsymbol{x} \tag{10.6.19}$$

subject to constraints

$$-\boldsymbol{A}\,\boldsymbol{x} \ge -\boldsymbol{b} \tag{10.6.20}$$

$$\boldsymbol{x} \ge 0 \tag{10.6.21}$$

equivalent to the problem

$$\max\ \boldsymbol{c}^T \boldsymbol{x} \tag{10.6.22}$$

subject to constraints

$$\boldsymbol{A}\,\boldsymbol{x} \le \boldsymbol{b} \tag{10.6.23}$$

$$\boldsymbol{x} \ge 0 \tag{10.6.24}$$

that is the problem P_1.
Theorem 2:
If both primal and dual problems admit a feasible solution, each of them has a finite optimum and the optimal values of the cost functions are equal.

Demonstration of the relation between the cost functions (under a different form with respect to Section 10.2.3):
Suppose that the primal problem is written under canonical form (with slack variables and possibly artificial variables) (Matousek and Gärtner 2007). In the following, the notations consider only m slack variables, but it could take into account additional artificial variables. Thus, the primal problem is

$$\max\ f = \bar{\boldsymbol{c}}^T\, \bar{\boldsymbol{x}}\,, \quad \bar{\boldsymbol{x}} \in \mathbb{R}^{n+m} \tag{10.6.25}$$

subject to

$$\begin{aligned} \bar{\boldsymbol{A}}\, \bar{\boldsymbol{x}} &= \boldsymbol{b} \\ \bar{\boldsymbol{x}} &\geq 0 \end{aligned} \tag{10.6.26}$$

with $\bar{\boldsymbol{c}} = (c_1, \ldots, c_n, 0, \ldots, 0)$ and $\bar{\boldsymbol{A}} = [\boldsymbol{A}|\boldsymbol{I}_m]$.

The main constraints are now written under the form of an equality

$$\bar{\boldsymbol{A}}\, \bar{\boldsymbol{x}} = \boldsymbol{b} \tag{10.6.27}$$

This equality is transposed, and then both sides are multiplied by a vector $\boldsymbol{w}$, hence

$$\bar{\boldsymbol{x}}^T\, \bar{\boldsymbol{A}}^T\, \boldsymbol{w} = \boldsymbol{b}^T\, \boldsymbol{w} \tag{10.6.28}$$

Notice that the right-hand side is equal to $h(\boldsymbol{w})$. On another side, by using the relation of constraints of problem P_2, we get

$$\bar{\boldsymbol{x}}^T\, \bar{\boldsymbol{A}}^T\, \boldsymbol{w} \geq \bar{\boldsymbol{c}}^T\, \boldsymbol{x} \tag{10.6.29}$$

Again, notice that the right-hand side is equal to $f(\bar{\boldsymbol{x}})$. We deduce

$$h(\boldsymbol{w}) \geq f(\bar{\boldsymbol{x}}) \tag{10.6.30}$$

As this is true whatever $\bar{\boldsymbol{x}}$ and $\boldsymbol{w}$, this is in particular true at the optimum, and thus the minimum of h (dual problem P_2) is larger than or equal to the maximum of f (primal problem P_1).
Demonstration that the optimum is finite:
Let $\bar{\boldsymbol{x}}^*$ be the optimum for f and $\boldsymbol{w}^*$ that for h, the respective feasible solutions. According to the previous theorem, we can write

$$h(\boldsymbol{w}) \geq \bar{\boldsymbol{c}}^T\, \bar{\boldsymbol{x}}^* \tag{10.6.31}$$

thus the minimum of h being lower bounded by a finite number is finite.

We also have

$$f(\bar{\boldsymbol{x}}) \leq \boldsymbol{b}^T\, \boldsymbol{w}^* \tag{10.6.32}$$

thus the maximum of f being upper bounded by a finite number is finite.
Demonstration that the optimum is the same for the primal and the dual:
It was already shown that the solution of the primal problem is

$$\bar{\boldsymbol{x}}_B = \bar{\boldsymbol{A}}_B^{-1}\, (\boldsymbol{b} - \bar{\boldsymbol{A}}_N\, \bar{\boldsymbol{x}}_N) \tag{10.6.33}$$

and the value of the cost function at the optimum

$$f = \bar{c}_B^T \bar{A}_B^{-1} b + (\bar{c}_N^T - \bar{c}_B^T \bar{A}_B^{-1} \bar{A}_N) x_N \tag{10.6.34}$$

Pose

$$w_0^T = \bar{c}_B^T \bar{A}_B^{-1} \tag{10.6.35}$$

hence

$$w_0^T \bar{A}_B = \bar{c}_B^T \tag{10.6.36}$$

and

$$w_0^T \bar{A}_N = \bar{c}_B^T \bar{A}_B^{-1} \bar{A}_N \tag{10.6.37}$$

The matrix $\bar{A}$ is formed by two submatrices

$$\bar{A} = \left[\bar{A}_B \,|\, \bar{A}_N\right] \tag{10.6.38}$$

so that we can write

$$w_0^T \bar{A} = w_0^T \left[\bar{A}_B \,|\, \bar{A}_N\right] = \left[\bar{c}_B^T \,|\, \bar{c}_B^T \bar{A}_B^{-1} \bar{A}_N\right] \tag{10.6.39}$$

If the condition

$$\left[\bar{c}_B^T \,|\, \bar{c}_B^T \bar{A}_B^{-1} \bar{A}_N\right] \geq \left[\bar{c}_B^T \,|\, \bar{c}_N^T\right] \tag{10.6.40}$$

is fulfilled, it induces the sought inequality for the constraints of the dual problem

$$w_0^T \bar{A} \geq \bar{c}^T \tag{10.6.41}$$

Yet, the previous condition is equivalent to

$$\bar{c}_B^T \bar{A}_B^{-1} \bar{A}_N \geq \bar{c}_N^T \tag{10.6.42}$$

a condition that can be written under a form that is recognized in the term which is a factor of the nonbasic variables in the general expression of the cost function of the primal problem

$$\bar{c}_N^T - \bar{c}_B^T \bar{A}_B^{-1} \bar{A}_N \leq 0 \tag{10.6.43}$$

Thus, it is the optimality condition of the primal problem. If the basis $\bar{A}_B$ is optimal, the vector w^* that was defined as the solution of the dual problem and the cost function of this problem is equal to

$$h^* = b^T w^* \tag{10.6.44}$$

Still supposing that the basis $\bar{A}_B$ is optimal, the nonbasic variables being zero, the maximum of the cost function of the primal problem f is equal to

$$\max f = f^* = \bar{c}_B^T \bar{A}_B^{-1} b \tag{10.6.45}$$

a relation in which we recognize the expression of w^* so that we have

$$f^* = w^{*T} b = b^T w^* = \bar{c}^T \bar{x}^* = c^T x^* \tag{10.6.46}$$

It results that

$$h^* = f^* \tag{10.6.47}$$

From the first part of the demonstration of the theorem, h was larger than or equal to f, and thus the smallest possible value of h is f^* maximum of f; thus, h^* is the minimum of h and $\boldsymbol{w}^*$ is the optimal solution of the dual problem.

Knowing $\boldsymbol{w}^*$, by using the equality

$$\bar{\boldsymbol{c}}^T\,\bar{\boldsymbol{x}}^* = \boldsymbol{c}^T\boldsymbol{x}^* = \boldsymbol{b}^T\,\boldsymbol{w}^* \tag{10.6.48}$$

and the active constraints, $\boldsymbol{x}^*$ results.

Properties relating the primal problem and the dual problem:

1. At the optimum, the value of the objective function of the primal problem is equal to that of the objective function of the dual problem.
2. When a constraint is saturated at the optimum, the associated slack variable is zero and the marginal cost (coefficient of the objective function of the variable in the column corresponding to this slack variable) is equal in absolute value to the corresponding dual variable. It must be noticed that, to the constraint i of the primal problem, corresponds the variable w_i of the dual and vice versa.
3. A zero dual variable corresponds to an unsaturated constraint at the optimum.
4. The sign of a dual variable indicates the direction in which it would be interesting to modify the constraint to improve the optimum of the function.

10.6.3 Demonstration of the Duality Theorem Based on the Lagrangian

The duality in linear programming (Matousek and Gärtner 2007) can be approached differently (Bertsekas 2009; Bonnans et al. 2003; Boyd and Vandenberghe 2009; Luenberger and Ye 2016) by means of the Lagrangian, as explained in Section 8.8.2, for the primal problem (10.6.11).

First, the primal problem (10.6.11) is transformed under the form (10.6.25)

$$\max\ f = \bar{\boldsymbol{c}}^T\,\bar{\boldsymbol{x}}\,, \quad \bar{\boldsymbol{x}} \in \mathbb{R}^{n+m} \tag{10.6.49}$$

subject to

$$\begin{aligned} \bar{\boldsymbol{A}}\,\bar{\boldsymbol{x}} &= \boldsymbol{b} \\ \bar{\boldsymbol{x}} &\geq 0 \end{aligned} \tag{10.6.50}$$

with $\bar{\boldsymbol{c}} = (c_1, \ldots, c_n, 0, \ldots, 0)$ and $\bar{\boldsymbol{A}} = [\boldsymbol{A}|\boldsymbol{I}_m]$.

The dual problem is

$$\min\ h = \boldsymbol{b}^T\,\boldsymbol{w} \tag{10.6.51}$$

subject to constraints

$$\boldsymbol{A}^T\,\boldsymbol{w} \geq \boldsymbol{c}\,, \quad \boldsymbol{w} \geq 0 \tag{10.6.52}$$

The Lagrangian is

$$\mathcal{L}(x,\mu) = \boldsymbol{c}^T \boldsymbol{x} + \mu^T (\boldsymbol{A}\boldsymbol{x} - \boldsymbol{b}) = -\boldsymbol{b}^T \mu + (\boldsymbol{A}^T \mu + \boldsymbol{c})^T x\,, \quad \mu \geq 0 \tag{10.6.53}$$

where μ are KKT parameters such that $\mu \geq 0$. The Lagrange dual function is

$$g(\boldsymbol{\mu}) = \inf_{\boldsymbol{x}} \mathcal{L}(\boldsymbol{x},\mu) = -\boldsymbol{b}^T \mu + \inf_{\boldsymbol{x}} (\boldsymbol{A}^T \mu + \boldsymbol{c})^T \boldsymbol{x} \tag{10.6.54}$$

As $g(\boldsymbol{\mu})$ is a linear function with respect to $\boldsymbol{x}$, and the linear term with respect to $\boldsymbol{x}$ is unbounded, it verifies

$$g(\boldsymbol{\mu}) = \begin{cases} -\boldsymbol{b}^T \mu & \text{if } \boldsymbol{A}^T \mu + \boldsymbol{c} = 0 \\ -\infty & \text{otherwise} \end{cases} \tag{10.6.55}$$

Thus, $\boldsymbol{\mu}$ is dual feasible if $\mu \geq 0$ and $\boldsymbol{A}^T \mu + \boldsymbol{c} = 0$. The optimal solution of the primal problem is noted p^* and d^* the optimal solution of the dual problem. From the theorem of weak duality, it results that

$$g(\boldsymbol{\mu}) \leq p^* \quad \forall \lambda\,, \quad \forall \mu \tag{10.6.56}$$

or

$$d^* \leq p^* \tag{10.6.57}$$

The difference $p^* - d^*$ is called the duality gap. If the primal problem is unbounded, then the dual problem is infeasible

$$p^* = -\infty \Rightarrow d^* = -\infty \tag{10.6.58}$$

and conversely, if the dual problem is unbounded, then the primal problem is infeasible

$$d^* = \infty \Rightarrow p^* = \infty \tag{10.6.59}$$

From the theorem of strong duality,

$$\inf_{\boldsymbol{x}} f_0(\boldsymbol{x}) = \max_{\boldsymbol{\mu} \geq 0} g(\boldsymbol{\mu}) \Rightarrow p^* = d^* \tag{10.6.60}$$

thus, the duality gap is zero. In the general case, the strong duality is verified when the functions are convex and Slater's constraint condition is valid (Section 8.8.2).

Example 10.8 :

Linear programming: primal and dual problems

Consider Example 10.1 in the case of maximization of the cost function. This is the primal problem

$$\max f = 2\,x_1 - x_2 \tag{10.6.61}$$

subject to

$$\begin{cases} -3\,x_1 & +2\,x_2 & \leq 2 \\ 2\,x_1 & -4\,x_2 & \leq 3 \\ x_1 & +x_2 & \leq 6 \end{cases} \tag{10.6.62}$$

which gave simplex tableau (Table 10.7).

Table 10.7 Simplex tableau of the primal problem. The values in bold characters are also referred in Table 10.9 and commented with respect to Tables 10.7 and 10.9 (primal and dual tableaux)

n	V_1	V_2	V_3	V_4	V_5	B	Basis
I	−3	2	1	0	0	2	x_3
	[2]	−4	0	1	0	3	x_4
	1	1	0	0	1	6	x_5
	2	−1	0	0	0	0	$-f$
II	0	−4	1	1.5	0	6.5	x_3
	1	−2	0	0.5	0	1.5	x_1
	0	[3]	0	−0.5	1	4.5	x_5
	0	3	0	−1	0	−3	$-f$
III	0	0	1	0.833	1.333	12.5	x_3
	1	0	0	0.167	0.667	**4.5**	x_1
	0	1	0	−0.167	0.333	**1.5**	x_2
	0	0	0	**−0.5**	**−1**	**−7.5**	$-f$

The dual problem of (10.6.61) is

$$\min h = 2\,w_1 + 3\,w_2 + 6\,w_3 \tag{10.6.63}$$

subject to

$$\begin{cases} -3\,w_1 + 2\,w_2 + w_3 \geq 2 \\ 2\,w_1 - 4\,w_2 + w_3 \geq -1 \end{cases} \iff \begin{cases} 3\,w_1 - 2\,w_2 - w_3 \leq -2 \\ -2\,w_1 + 4\,w_2 - w_3 \leq 1 \end{cases} \tag{10.6.64}$$

Equations (10.6.64) can be written again with slack w_4 and w_5 and artificial w_6 variables as

$$\begin{array}{l} -3\,w_1 + 2\,w_2 + w_3 - w_4 \qquad + w_6 = 2 \\ -2\,w_1 + 4\,w_2 - w_3 \qquad + w_5 \qquad = 1 \end{array} \tag{10.6.65}$$

The simplex for the dual problem requires the use of artificial variables, hence two tableaux. The first simplex tableau of the dual problem is Table 10.8, where $f' = w_6$ is minimized, thus with respect to the nonbasic variables

$$f' = 2 + 3\,w_1 - 2\,w_2 - w_3 + w_4 \tag{10.6.66}$$

The second simplex tableau of the dual problem is Table 10.9. By using the two rows of the constraints on w_2 and w_3 in the third part of Table 10.8, we obtain

$$\begin{array}{l} -0.833\,w_1 + w_2 \qquad -0.167\,w_4 + 0.167\,w_5 = 0.5 \\ -1.333\,w_1 \qquad + w_3 - 0.667\,w_4 - 0.333\,w_5 = 1 \end{array} \tag{10.6.67}$$

hence w_2 and w_3 with respect to the nonbasic variables. The function h to minimize is then expressed with respect to the nonbasic variables as

$$\begin{aligned} h &= 2\,w_1 + 3(0.833\,w_1 + 0.167\,w_4 - 0.167\,w_5 + 0.5) + \\ &\quad 6(1.333\,w_1 + 0.667\,w_4 + 0.333\,w_5 + 1) \\ &= 12.5\,w_1 + 4.5\,w_4 + 1.5\,w_5 + 7.5 \end{aligned} \tag{10.6.68}$$

The optimal solution is found since iteration I of the second simplex tableau.

Table 10.8 First simplex tableau of the dual problem

n	V_1	V_2	V_3	V_4	V_5	V_6	B	Basis
I	−2	[4]	−1	0	1	0	1	w_5
	−3	2	1	−1	0	1	2	w_6
	3	−2	−1	1	0	0	−2	$-h'$
II	−0.5	1	−0.25	0	0.25	0	0.25	w_2
	−2	0	[1.5]	−1	−0.4	1	1.5	w_6
	2	0	−1.5	1	0.5	0	−1.5	$-h'$
III	−0.833	1	0	−0.167	0.167	0.167	0.5	w_2
	−1.333	0	1	−0.667	−0.333	0.667	1	w_3
	0	0	0	0	0	1	0	$-h'$

Table 10.9 Second simplex tableau of the dual problem. The values in bold characters are also referred in Table 10.7 and commented with respect to Tables 10.7 and 10.9 (primal and dual tableaux)

n	V_1	V_2	V_3	V_4	V_5	B	Basis
I	−0.833	1	0	−0.167	0.167	**0.5**	w_2
	−1.333	0	1	−0.667	−0.333	**1**	w_3
	12.5	0	0	**4.5**	**1.5**	**−7.5**	$-h$

In Tables 10.7 and 10.9, the following points can be verified:

1. Note that the optimal value $f(\boldsymbol{x}^*) = 7.5$ of the primal is equal to the optimal value $h(\boldsymbol{w}^*) = 7.5$ of the dual.
2. In Table 10.7, the second and third constraints are saturated at the optimum (the corresponding slack variables x_4 and x_5 are zero). To the constraint i of the primal problem, corresponds the variable w_i of the dual.
   To the slack variable x_4 equal to zero, for the second constraint of the primal problem, correspond the basic variable $x_1 = 4.5$ and the column $\boldsymbol{V}_4$ of the primal problem. The marginal cost −0.5 is equal in absolute value to the corresponding variable w_2 of the dual problem.
   To the slack variable x_5 equal to zero, for the third constraint of the primal problem, correspond the basic variable $x_2 = 1.5$ and the column $\boldsymbol{V}_5$ of the primal problem. The marginal cost −1 is equal in absolute value to the corresponding variable w_3 of the dual problem.
3. In Table 10.7, the first constraint is not saturated at the optimum. The corresponding slack variable x_3 belongs to the basis and is nonzero. Its marginal cost is zero (column $\boldsymbol{V}_2$ of the primal). The dual variable associated with the constraint is w_1 which is zero.
4. In Table 10.9, the two constraints of the dual problem are saturated (the corresponding slack variables w_4 and w_5 are zero). To the constraint i of the dual problem, corresponds the variable x_i of the primal.
   To the slack variable w_4 equal to zero, for the first constraint of the dual problem, correspond the basic variable $w_2 = 0.5$ and the column V_4 of the dual problem. The marginal cost 4.5 is equal in absolute value to the corresponding variable x_1 of the primal problem (dual of the dual).
   To the slack variable w_5 equal to zero, for the second constraint of the dual problem, correspond the basic variable $w_3 = 1$ and the column V_5 of the dual problem. The marginal cost 1.5 is equal in absolute value to the corresponding variable x_2 of the primal problem.

10.7 Interior Point Methods

With respect to the traditional simplex method, Karmarkar's algorithm (Karmarkar 1984) represents a very different approach as it uses interior points to the simplex and consequently is considered as a revolution (Nazareth 2003). At the beginning, the notion of interior point was reserved to points belonging to the interior of the feasible polytope. Presently, the term of interior can be extended to the positive quadrant of $\boldsymbol{x}$ (Nazareth 2003). When the equality constraints of the primal or dual problem are not respected, they can be called nonfeasible interior points. With respect to the exponential complexity of the simplex method, the complexity of Karmarkar's algorithm is strongly reduced, which led to much research in this interior point direction, named as mathematical programming (Barnes 1986; Boyd and Vandenberghe 2009; Forsgren et al. 2002; Gill et al. 1986; Li and Xuan 1998; Nemirovski and Todd 2009; Luenberger and Ye 2016; Roos et al. 2005; Terlaky 1996; Wright 1997). It must be noted that the first algorithm of polynomial complexity is the method of ellipsoids by Khachian in 1979, thus before Karmarkar, but which was not fruitful at a practical level with respect to the simplex. Dantzig and Thapa (1997) mention several interior point approaches for the simplex as soon as 1952, including the nonlinear method by R. Frisch in 1954, the affine transformation method by Dikin (1967).

10.7.1 Karmarkar Projection Method

Karmarkar (1984) poses the general linear programming problem under the following form:

$$\begin{aligned}
&\min \quad \boldsymbol{c}'^T\,\boldsymbol{x}' \\
&\text{subject to} \\
&\boldsymbol{A}'\,\boldsymbol{x}' = 0 \\
&\boldsymbol{e}^T\,\boldsymbol{x}' = 1 \\
&\boldsymbol{x}' \geq 0
\end{aligned} \tag{10.7.1}$$

$\boldsymbol{x}' \in \mathcal{R}^{n+1}$, $\boldsymbol{A}'$ is of dimension $m \times (n+1)$. $\boldsymbol{e} = [1, 1, \ldots, 1]^T$ is a vector of dimension $n+1$. From Equation (10.7.1), it results that

$$\sum_{i=1}^{n+1} x_i' = 1 \tag{10.7.2}$$

10.7.1.1 Passage from the Standard Form to Karmarkar's Form

1. The original linear programming problem is constituted by Equations (10.2.19) and (10.2.20), thus under primal form

$$\begin{aligned}&\min \quad \boldsymbol{c}^T \boldsymbol{x}\,, && \boldsymbol{x} \in \mathcal{R}^n\\&\text{subject to}\\&\boldsymbol{A}\,\boldsymbol{x} \geq \boldsymbol{b}\,, && \boldsymbol{b} \in \mathcal{R}^m\,,\ \boldsymbol{A} \in \mathcal{R}^{m\times n}\\&\boldsymbol{x} \geq 0\end{aligned} \tag{10.7.3}$$

or under dual form

$$\begin{aligned}&\max \quad \boldsymbol{b}^T \boldsymbol{y}\,, && \boldsymbol{y} \in \mathcal{R}^m\\&\text{subject to}\\&\boldsymbol{A}^T \boldsymbol{y} \leq \boldsymbol{c}\\&\boldsymbol{y} \geq 0\end{aligned} \tag{10.7.4}$$

with, at the optimum, the equality of the respective objective functions

$$\boldsymbol{c}^T \boldsymbol{x} = \boldsymbol{b}^T \boldsymbol{y} \tag{10.7.5}$$

2. Then, slack variables $\boldsymbol{v}$ and $\boldsymbol{w}$ are introduced to transform the inequality constraints into equalities

$$\begin{aligned}&\boldsymbol{A}\,\boldsymbol{x} - \boldsymbol{v} = \boldsymbol{b}\\&\boldsymbol{A}^T \boldsymbol{y} + \boldsymbol{w} = \boldsymbol{c}\\&\boldsymbol{v} \geq 0\,,\ \boldsymbol{w} \geq 0\end{aligned} \tag{10.7.6}$$

3. To create an initial interior point, an artificial variable λ is introduced. Let $\boldsymbol{x}_0$, $\boldsymbol{y}_0$, $\boldsymbol{v}_0$, and $\boldsymbol{w}_0$ be points strictly interior to the first quadrant (positive quadrant $= \{\boldsymbol{x} \in \mathcal{R}^n \mid \boldsymbol{x} \geq 0\}$). The minimization problem of the artificial variable artificial is posed

$$\begin{aligned}&\min\ \lambda\\&\text{subject to}\\&\boldsymbol{A}\,\boldsymbol{x} - \boldsymbol{v} + (\boldsymbol{b} - \boldsymbol{A}\,\boldsymbol{x}_0 + \boldsymbol{v}_0)\lambda = \boldsymbol{b}\\&\boldsymbol{A}^T \boldsymbol{y} + \boldsymbol{w} + (\boldsymbol{c} - \boldsymbol{A}^T \boldsymbol{y}_0 - \boldsymbol{w}_0)\lambda = \boldsymbol{c}\\&\boldsymbol{c}^T \boldsymbol{x} - \boldsymbol{b}^T \boldsymbol{y} + (-\boldsymbol{c}^T \boldsymbol{x}_0 + \boldsymbol{b}^T \boldsymbol{y}_0)\lambda = 0\\&\boldsymbol{x} \geq 0\,,\ \boldsymbol{y} \geq 0\,,\ \boldsymbol{v} \geq 0\,,\ \boldsymbol{w} \geq 0\,,\ \lambda \geq 0\end{aligned} \tag{10.7.7}$$

The solution $\boldsymbol{x} = \boldsymbol{x}_0$, $\boldsymbol{y} = \boldsymbol{y}_0$, $\boldsymbol{v} = \boldsymbol{v}_0$, $\boldsymbol{w} = \boldsymbol{w}_0$, $\lambda = 1$ is a feasible strictly interior solution.

4. Following (Karmarkar 1984), the problem (10.7.7) can be posed under the form

$$\begin{aligned}&\min \quad \boldsymbol{c}^{eT} \boldsymbol{x}^e\\&\text{subject to}\\&\boldsymbol{A}^e\,\boldsymbol{x}^e = \boldsymbol{b}^e\\&\boldsymbol{x}^e > 0\end{aligned} \tag{10.7.8}$$

by posing $\boldsymbol{x}^e$, $\boldsymbol{A}^e$, $\boldsymbol{b}^e$, $\boldsymbol{c}^e$ extended vectors or matrices defined by

$$
\boldsymbol{x}^e = \begin{bmatrix} \boldsymbol{x} \\ \boldsymbol{y} \\ \boldsymbol{v} \\ \boldsymbol{w} \\ \lambda \end{bmatrix}, \quad \boldsymbol{b}^e = \begin{bmatrix} \boldsymbol{b} \\ \boldsymbol{c} \\ 0 \end{bmatrix}, \quad \boldsymbol{c}^e = \begin{bmatrix} \boldsymbol{c} \\ -\boldsymbol{b} \\ \boldsymbol{0}_m \\ \boldsymbol{0}_n \\ 1 \end{bmatrix}
$$
$$
\boldsymbol{A}^e = \begin{bmatrix} \boldsymbol{A} & \boldsymbol{0}_{m\times m} & -\boldsymbol{I}_m & \boldsymbol{0}_{m\times n} & (\boldsymbol{b} - \boldsymbol{A}\,\boldsymbol{x}_0 - \boldsymbol{v}_0) \\ \boldsymbol{0}_{n\times n} & \boldsymbol{A}^T & \boldsymbol{0}_{n\times m} & \boldsymbol{I}_n & (\boldsymbol{c} - \boldsymbol{A}^T\,\boldsymbol{y}_0 + \boldsymbol{w}_0) \\ \boldsymbol{c}^T & -\boldsymbol{b}^T & \boldsymbol{0}_{1\times m} & \boldsymbol{0}_{1\times n} & (-\boldsymbol{c}^T\,\boldsymbol{x}_0 + \boldsymbol{b}^T\,\boldsymbol{y}_0) \end{bmatrix} \tag{10.7.9}
$$

by integrating the cost functions of the primal and dual problems in the cost function of Equation (10.7.7). If only λ was taken into account in the cost function, only the last component in $\boldsymbol{c}^e$ would remain. The dimensions of zero and identity matrices have been given. To ensure in stage 3 strictly positive variables $\boldsymbol{x}_0$, $\boldsymbol{y}_0$, $\boldsymbol{v}_0$, $\boldsymbol{w}_0$, it may be necessary to introduce artificial variables in addition to the slack variables in stage 2.

Note that $\boldsymbol{x}^e$ must be strictly positive.

Karmarkar (1984) supposes that the initial point is $\boldsymbol{x}^e = \boldsymbol{a}$ and that the optimal solution of the objective function is 0.

To avoid clumsiness in the following of those explanations, the extended superscript “e” is omitted, but of course the operations deal with extended notations. Thus, the dimension n deals with the extended problem.

Karmarkar (1984) then performs a bijective projective transformation of the positive quadrant into the simplex by posing

$$
\begin{aligned} x'_i &= \frac{x_i/a_i}{\sum_{j=1}^{n}(x_j/a_j) + 1} \\ x'_{n+1} &= 1 - \sum_{i=1}^{n} x'_i \end{aligned} \tag{10.7.10}
$$

the simplex being defined by $\Delta = \left\{ \boldsymbol{x}' \in \mathcal{R}^{n+1} \,|\, \boldsymbol{x}' \geq 0\,, \sum_{i=1}^{n+1} x'_i = 1 \right\}$. The image of point $\boldsymbol{a}$ is the center of the simplex with $x'_i = 1/(n+1)$. The inverse transformation is defined by

$$
x_i = \frac{a_i x'_i}{x'_{n+1}} \tag{10.7.11}
$$

The columns of the matrix $\boldsymbol{A}$ are set as $\boldsymbol{A}_i$. From $\boldsymbol{A}\boldsymbol{x} = \boldsymbol{b}$, it results

$$
\begin{aligned} &\sum_{i=1}^{n} \boldsymbol{A}_i x_i = \boldsymbol{b} \qquad \text{hence} \qquad \sum_{i=1}^{n} \boldsymbol{A}_i \frac{a_i x'_i}{x'_{n+1}} = \boldsymbol{b} \\ &\sum_{i=1}^{n} \boldsymbol{A}_i a_i x'_i - \boldsymbol{b} x'_{n+1} = 0 \qquad \text{i.e.} \qquad \sum_{i=1}^{n} \boldsymbol{A}'_i x'_i = 0 \end{aligned} \tag{10.7.12}
$$

with the matrix $\boldsymbol{A}'$ formed in columns as

$$A' = \left[a_1 A_1 \mid a_2 A_2 \mid \dots \mid a_n A_n \mid -b\right] \tag{10.7.13}$$

The problem (10.7.3) is then transformed by Karmarkar under the form considered as canonical

$$\begin{array}{l} \min \quad c'^T x' \\ \text{subject to} \\ A' x' = 0 \\ e^T x' = 1 \\ x' \geq 0 \end{array} \tag{10.7.14}$$

$x' \in \mathcal{R}^{n+1}$, A' is of dimension $m \times (n+1)$. $e = [1, 1, \dots, 1]^T$ is a vector of dimension $n + 1$.

10.7.1.2 Karmarkar's Algorithm

Karmarkar's algorithm (Karmarkar 1984) takes place as

1. Initialization. $k = 0$. Determine an extended point $x^{(0)}$ strictly positive such that the image of a defined by

$$a_i = \frac{x_i^{(0)}}{\sum_j x_j^{(0)}} \tag{10.7.15}$$

by the transformation (10.7.10) is the center of the simplex, as

$$a^{(0)} = \frac{1}{n+1} e \tag{10.7.16}$$

The image x' of $x^{(0)}$ by the transformation (10.7.10) verifies the relations (10.7.14), in particular $A'x' = 0$. This point $x'^{(0)}$ constitutes the initial point for the iterations. Pose the cost function c' defined by

$$\begin{array}{l} c'_i = a_i \, c_i \,, \quad i = 1, \dots, n \\ c'_{n+1} = 0 \end{array} \tag{10.7.17}$$

2. Iteration k.
Pose the diagonal matrix: $D = \text{diag}(x_1'^{(k)}, \dots, x_{n+1}'^{(k)})$.
The transformation T_k is defined by

$$T_k(x) = \frac{D^{-1} x}{e^T D^{-1} x} \tag{10.7.18}$$

2.1 Pose the augmented matrix

$$G = \begin{bmatrix} A' D \\ e^T \end{bmatrix} \tag{10.7.19}$$

Thus, the kernel of G belongs to the hyperplane $\left\{y \mid \sum_{i=1}^{n+1} y_i = 1, \; y_i \geq 0\right\}$.

2.2 Project $\boldsymbol{D}\,\boldsymbol{c}'$ orthogonally on the kernel of $\boldsymbol{G}$, hence

$$\boldsymbol{c}'_p = [\boldsymbol{I} - \boldsymbol{G}^T\,(\boldsymbol{G}\,\boldsymbol{G}^T)^{-1}\,\boldsymbol{G}]\boldsymbol{D}\,\boldsymbol{c}' \tag{10.7.20}$$

2.3 Normalize $\boldsymbol{c}'_p$ as

$$\bar{\boldsymbol{c}}'_p = \frac{\boldsymbol{c}'_p}{\|\boldsymbol{c}'_p\|} \tag{10.7.21}$$

2.4 Calculate $\boldsymbol{y}$ as

$$\boldsymbol{y} = \frac{\boldsymbol{e}}{n+1} - \alpha\, r\, \bar{\boldsymbol{c}}'_p \tag{10.7.22}$$

$\alpha\, r$ is the displacement length in the direction $\bar{\boldsymbol{c}}'_p$. α is a parameter in $(0, 1)$. Karmarkar chooses $\alpha = 1/4$. r is the radius of the largest sphere inscribed in the simplex, that is, $r = 1/\sqrt{(n+1)n}$.

2.5 Apply the inverse transformation T_k^{-1} on $\boldsymbol{y}$ such that

$$\boldsymbol{x}'^{(k+1)} = \frac{\boldsymbol{D}\,\boldsymbol{y}}{\boldsymbol{e}^T\,\boldsymbol{D}\,\boldsymbol{y}} \tag{10.7.23}$$

3. Verify the convergence with

$$\boldsymbol{c}'^T\,\boldsymbol{x}'^{(k)} \le 2^{-q}\,\boldsymbol{c}'^T\,\boldsymbol{x}'^{(0)} \tag{10.7.24}$$

q being a constant parameter. In the case of non-convergence, return to 2.

When the convergence is reached, note that $\boldsymbol{x}'$ is the transformed vector and, to come back to the extended vector $\boldsymbol{x}$ according to Equation (10.7.9), the transformation (10.7.11) must be applied.

To this algorithm, a potential function is associated

$$p(\boldsymbol{x}') = \sum_{i=1}^{n+1} \ln\left(\frac{\boldsymbol{c}'^T\,\boldsymbol{x}'}{x'_i}\right) \tag{10.7.25}$$

for which an improvement δ can be hoped at each iteration. If $\beta = 1/4$, then $\delta \ge 1/8$, and we can take $\delta = 1/8$. If the improvement expected is not achieved (for example, if $p(\boldsymbol{x}'^{(k+1)}) > p(\boldsymbol{x}'^{(k)}) - \delta$), stop by concluding that the minimum of the objective function should be strictly positive (the case where the original problem has no solution or is not bounded). Starting from problems (10.7.3) and (10.7.4), the following potential function (Nemirovski and Todd 2009) is also defined by

$$p(\boldsymbol{x}, \boldsymbol{s}) = q \ln\left(\boldsymbol{x}^T\,\boldsymbol{s}\right) - \sum_{i=1}^{n+1} \ln x_i - \sum_{i=1}^{n+1} \ln s_i\,, \quad q > n \tag{10.7.26}$$

The first right-hand term measures the duality jump (between the primal problem and the dual problem), the second term the "distance" with respect to the boundary of the primal problem, and the third term the "distance" with respect to the boundary of the dual problem. The duality jump $\boldsymbol{s}^T\,\boldsymbol{x} = \boldsymbol{c}^T\,\boldsymbol{x} - \boldsymbol{b}^T\,\boldsymbol{y}$ is zero at the optimum. The potential p should tend to $-\infty$ as fast as possible. Then, there exist algorithms with a primal and a dual step and algorithms of potential reduction.

Example 10.9 :
Linear programming: Karmarkar's algorithm

Karmarkar's algorithm was implemented for the problem (10.2.11) already solved by the simplex and reformulated as

$$\begin{aligned} &\max\ f = 4\,x_1 + 5\,x_2 \\ &\text{subject to} \\ &2\,x_1 + x_2 \le 6 \\ &x_1 + 2\,x_2 \le 5 \\ &-x_1 - x_2 \le -1 \\ &-x_1 - 4\,x_2 \le -2 \\ &x_i \ge 0 \end{aligned} \tag{10.7.27}$$

This problem is considered as primal, while its dual is

$$\begin{aligned} &\min\ f = 6\,y_1 + 5\,y_2 - y_3 - 2\,y_4 \\ &\text{subject to} \\ &2\,y_1 + y_2 - y_3 - y_4 \ge 4 \\ &y_1 + 2\,y_2 - y_3 - 4\,y_4 \ge 5 \\ &y_i \ge 0 \end{aligned} \tag{10.7.28}$$

This problem was solved under the primal–dual form (10.7.7). The vector result of the iterations thus contains the original variables, the slack variables and possibly the artificial variables of the primal problem, those of the dual problem, and finally the artificial variable λ. The cost function contains the cost function of the primal problem, that of the dual problem, and that of the artificial variable λ.

Starting from problem (10.7.27), the initialization of the algorithm needs a specific preparation so that the problem is under the form (10.7.14), where all the variables are strictly positive and the initial point is strictly positive. The extended initial point chosen is $\boldsymbol{x}^{(0)} = [1\ 1\ 3\ 2\ 1\ 3\ 1\ 1\ 1\ 1\ 1\ 1\ 4\ 8\ 1]$, deduced from $(x_1, x_2) = (1, 1)$ for the primal problem and $\boldsymbol{y} = (1, 1, 1, 1)$ for the dual problem. Note the important number of variables due to all the variables of the primal and dual problems and to λ which is the last value of $\boldsymbol{x}^{(0)}$. *Stricto sensu*, the initialization chosen in our study does not perfectly respond to the phase of the algorithm using Equation (10.7.7), where $\boldsymbol{x}^{(0)}$ and $\boldsymbol{y}^{(0)}$ should be feasible. In our example, they simply were taken strictly positive. As such, in agreement with thc original article by Karmarkar (1984), the algorithm converged to the optimum of the simplex tableau (Table 10.2). Only this first stage is represented in Figure 10.2, the convergence occurring to the point $(x_1, x_2) = (2.3333, 1.3333)$, with $\lambda = 8.17 \times 10^{-6}$, after 95 iterations with an absolute precision of $|f^{k+1} - f^k| < 10^{-10}$. The solution of the dual problem is (1, 2, 0, 0). The difference between the cost functions of the primal problem and the dual problem was lower than 10^{-5}.

Table 10.10 Karmarkar's method: influence of initial points $\boldsymbol{x}^{(0)}$ used in Figure 10.3 and corresponding subfigures (from left to right and top to bottom)

	Left	Middle	Right
Top	(0.05,0.05) Nonfeasible	(0.25,0.25) Nonfeasible	(0.5,0.5) Feasible
Middle	(0.55,0.55) Feasible	(0.6,0.6) Feasible	(0.7,0.7) Feasible
Bottom	(0.9,0.9) Feasible	(1,1) Feasible	(2,2) Nonfeasible

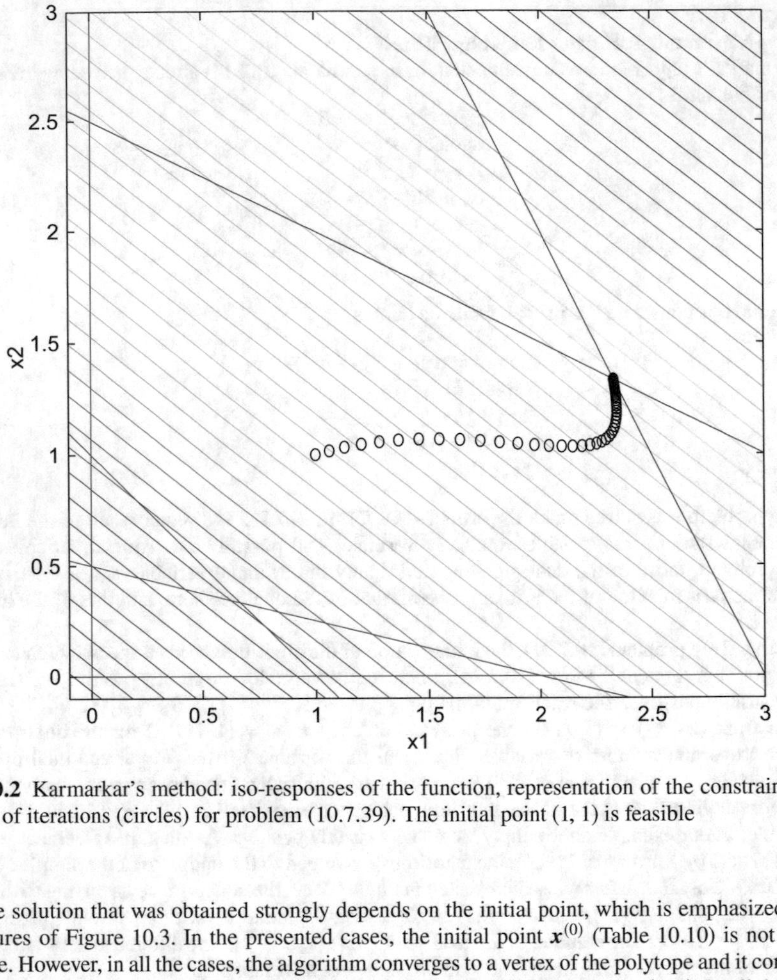

Fig. 10.2 Karmarkar's method: iso-responses of the function, representation of the constraints, and results of iterations (circles) for problem (10.7.39). The initial point (1, 1) is feasible

The solution that was obtained strongly depends on the initial point, which is emphasized in the subfigures of Figure 10.3. In the presented cases, the initial point $\boldsymbol{x}^{(0)}$ (Table 10.10) is not always feasible. However, in all the cases, the algorithm converges to a vertex of the polytope and it converged to the optimal solution in several cases, in particular when the initial point was (1, 1). In all the cases, λ converged to 0 and the difference between the costs of the primal problem and the dual problem was zero. Equations (10.7.7) was verified. To do all the stages of the simplex tableau, the problem should be reformulated according to the method by Karmarkar (1984) at the end of each stage, by restarting from a strictly positive point and formulating the matrices with respect to the basic variables, slack variables, and possibly artificial variables.

10.7.2 Affine Transformation

In Karmarkar's algorithm, the projection of a vector in the kernel of a matrix modified at each iteration is considered as penalizing from the computational point of view and solutions of approximation of the projection are proposed (Dennis et al. 1986). In view of the improvement of Karmarkar's algorithm, of its simplification, and of

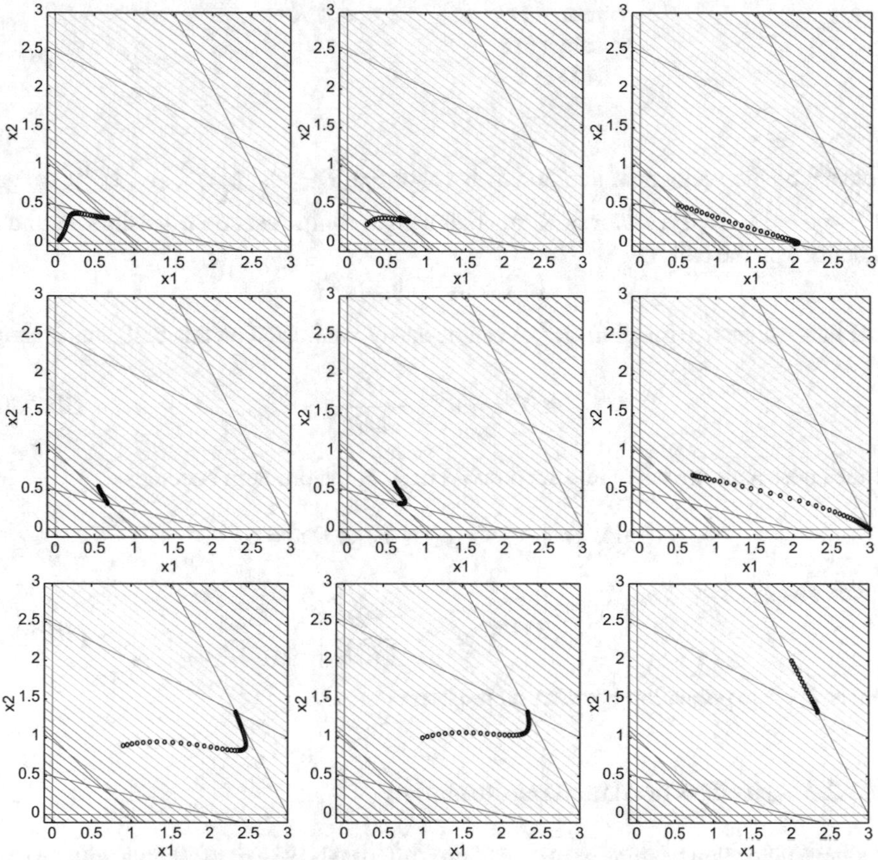

Fig. 10.3 Karmarkar's method: iso-responses of the function, representation of the constraints, and results of iterations (circles) from different initial points for problem (10.7.39). The different initial points are commented in Table 10.10

the improvement of its convergence rate, many variants have been developed among which simpler algorithms of affine transformation ("affine scaling") were proposed (Monteiro et al. 1990; Saigal 1993, 1997; Tsuchiya 1992) according to a method of predictor–corrector type such as Euler–Newton (Mizuno 1992).

10.7.2.1 Consequences of the Affine Transformation

The idea of affine transformation is related to the easy optimization of a linear function inside an ellipsoid compared to the optimization of the same function on the envelope of the polytope $\mathrm{A}\boldsymbol{x} = \boldsymbol{b}$. Thus, the optimization is carried out in an iterative manner on a sequence of ellipsoids.

Let $\boldsymbol{D} \in \mathcal{R}^{n \times n}$ be a diagonal matrix. The matrix $\boldsymbol{D}$ transforms $\boldsymbol{x}$ into $\boldsymbol{x} = \boldsymbol{D}\boldsymbol{u}$ or still $\boldsymbol{u} = \boldsymbol{D}^{-1}\boldsymbol{x}$ and leads to a linear problem equivalent to the original problem (10.7.3)

$$\begin{array}{ll} \min \quad \boldsymbol{D}\,\boldsymbol{c}^T\,\boldsymbol{u} \quad , & \boldsymbol{u} \in \mathcal{R}^n \\ \text{subject to} & \\ \boldsymbol{A}\,\boldsymbol{D}\,\boldsymbol{u} = \boldsymbol{b} & \\ \boldsymbol{u} \geq 0 & \end{array} \tag{10.7.29}$$

Let $\boldsymbol{x}^{(k)}$ be the vector $\boldsymbol{x}$ at iteration k. By choosing $\boldsymbol{D} = \text{diag}\left\{x_1^{(k)}, \ldots, x_n^{(k)}\right\}$, we get $\boldsymbol{u}^{(k)} = \boldsymbol{e} = [1, \ldots, 1]$ (vector of Karmarkar's algorithm). A set on $\boldsymbol{u}$ feasible around $\boldsymbol{e}$ is defined by the ball

$$\{\boldsymbol{u} \mid \quad \|\boldsymbol{u} - \boldsymbol{e}\| \leq 1\}$$

The linear objective function $\boldsymbol{D}\boldsymbol{c}^T\,\boldsymbol{u}$ can be easily minimized on this ball, which gives

$$\boldsymbol{u}^{(k+1)} = \boldsymbol{u}^{(k)} - \frac{\boldsymbol{D}\boldsymbol{c}}{\|\boldsymbol{D}\boldsymbol{c}\|} \tag{10.7.30}$$

If the constraints are taken into account $\boldsymbol{A}\boldsymbol{D}\boldsymbol{u} = \boldsymbol{b}$, the problem becomes

$$\text{Find: } \{\boldsymbol{u} \mid \quad \|\boldsymbol{u} - \boldsymbol{e}\| \leq 1\,,\ \boldsymbol{A}\,\boldsymbol{D}\,\boldsymbol{u} = \boldsymbol{b}\}$$

which gives

$$\boldsymbol{u}^{(k+1)} = \boldsymbol{u}^{(k)} - \frac{\boldsymbol{P}\boldsymbol{D}\boldsymbol{c}}{\|\boldsymbol{P}\boldsymbol{D}\boldsymbol{c}\|} \tag{10.7.31}$$

where $\boldsymbol{P}$ is the projection operator on the kernel of $\boldsymbol{A}\boldsymbol{D}$.

10.7.2.2 Algorithm of Affine Transformation

A simple algorithm of affine transformation (Adler et al. 1989) is presented with respect to the original problem of linear programming

$$\begin{array}{l} \max \quad \mathrm{c}^T\,\boldsymbol{x} \\ \text{subject to} \\ \mathrm{A}\,\boldsymbol{x} \leq \boldsymbol{b}\,,\ \boldsymbol{x} \geq 0 \end{array} \tag{10.7.32}$$

where $\boldsymbol{x} \in \mathcal{R}^n$, $\boldsymbol{b} \in \mathcal{R}^m$

The algorithm follows the stages:

1. Initialization. $k = 0$. Let the point $\boldsymbol{x}^{(0)}$. Let $0 < \gamma \leq 1$.
2. Iteration k.

2.1

$$\boldsymbol{v}^{(k)} = \boldsymbol{b} - \mathrm{A}\,\boldsymbol{x}^{(k)} \tag{10.7.33}$$

2.2 Calculation of the diagonal matrix

$$\boldsymbol{D} = \text{diag}\left\{v_1^{(k)}, \ldots, v_m^{(k)}\right\} \tag{10.7.34}$$

2.3

$$\boldsymbol{h}_x = [\mathrm{A}^T\,\boldsymbol{D}^{-2}\,\mathrm{A}]^{-1}\,\mathrm{c} \tag{10.7.35}$$

2.4

$$\boldsymbol{h}_v = -\mathrm{A}\,\boldsymbol{h}_x \tag{10.7.36}$$

2.5

$$\text{if } h_{v,i} \geq 0 \quad \forall i\text{, nonconvex domain}$$
$$\alpha = \gamma \min\left\{\frac{-v_i^{(k)}}{h_{v,i}} \mid h_{v,i} < 0\,, i = 1,\ldots,m\right\} \tag{10.7.37}$$

2.6

$$\boldsymbol{x}^{(k+1)} = \boldsymbol{x}^{(k)} + \alpha\,\boldsymbol{h}_x \tag{10.7.38}$$

Test if convergence. In the case of non-convergence, $k = k + 1$, return to 2.1.

Note that the stage (10.7.38) is a corrector of Newton type.

Example 10.10 :
Linear programming: Karmarkar's method and affine transformation

The algorithm of affine transformation was used to solve the following problem, same as Example 10.9,

$$\begin{aligned}
&\max\ f = 4\,x_1 + 5\,x_2\\
&\text{subject to}\\
&2\,x_1 + x_2 \leq 6\\
&x_1 + 2\,x_2 \leq 5\\
&x_1 + x_2 \geq 1\\
&x_1 + 4\,x_2 \geq 2\\
&x_i \geq 0
\end{aligned} \tag{10.7.39}$$

The results of iterations during solving the problem (10.7.39) by the method of affine transformation are given in Table 10.11. They show that the solution is obtained with a good precision after 15 iterations.

Table 10.11 Method of affine transformation: results of iterations during solving the problem (10.7.39)

k	x_1	x_2	f
0	0.1000	0.1000	0.9000
1	0.7959	0.9271	7.8188
2	1.2405	1.2923	11.4232
3	1.5479	1.4323	13.3531
4	1.8874	1.4094	14.5968
5	2.1196	1.3530	15.2433
6	2.2301	1.3358	15.5997
7	2.2823	1.3335	15.7965
8	2.3078	1.3333	15.8981
9	2.3206	1.3333	15.9490
10	2.3270	1.3333	15.9745
11	2.3301	1.3333	15.9873
12	2.3317	1.3333	15.9936
13	2.3325	1.3333	15.9968
14	2.3329	1.3333	15.9984
15	2.3331	1.3333	15.9992

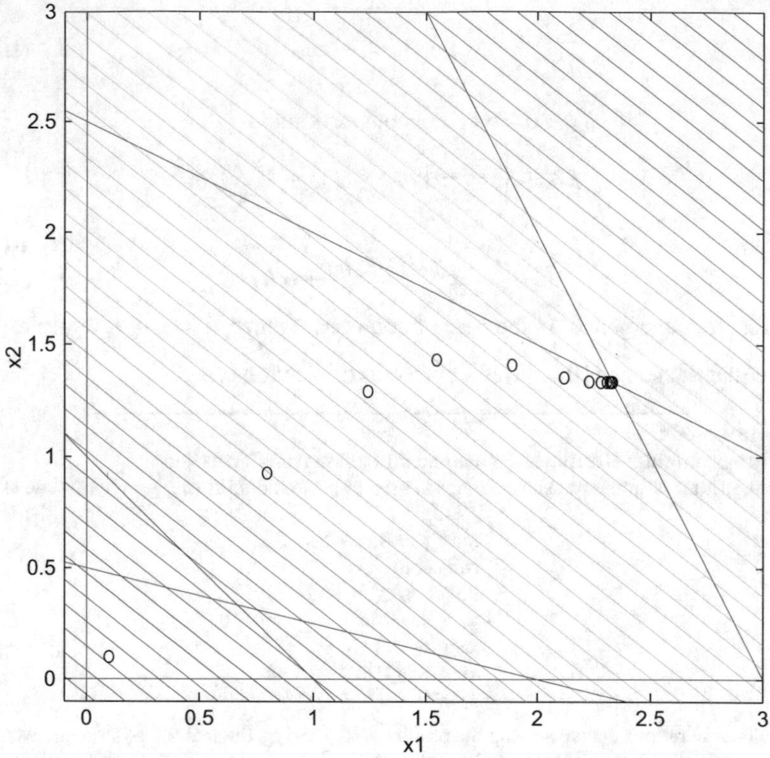

Fig. 10.4 Affine transformation: iso-responses of the function, representation of the constraints, results of iterations (circles), and optimal solution for problem (10.7.39). Note that the initial point (0.1, 0.1) is not feasible

Note that the initial point is not feasible as it does not respect the third and fourth constraints. Nevertheless, the second iterative point is feasible. Note the very rapid convergence to the point intersection of the first two constraints as well as the position of the optimum with respect to the parallel iso-responses (Figure 10.4).

10.8 Discussion and Conclusion

In this chapter, there apparently exists a limited choice of methods, as only the simplex method, Karmarkar's projection method, and the affine transformation are presented. The simplex method is easier to start than the interior point methods and is often more effective for moderate size problems (Ferris et al. 2007). With respect to the simplex method, the interior point methods such as the algorithm of affine transformation deduced from Karmarkar's method are theoretically more performing with respect to complexity in particular on large scale problems. The affine transformation is apparently simple to implement. Moreover, both types of methods can even be coupled, starting by an interior point method and then using the simplex when the final solution becomes close. It is important to know that, in practice, the simplex presents many difficulties, such as degeneracy or cycling, which were little mentioned in the present chapter and that only professional codes can solve. Presently, the codes based on the simplex or the interior point methods are very efficient. For actual problems beyond simple exercises, there is no doubt that this type of code must be used.

10.9 Exercise Set

Preamble: Most exercises can be solved by means of the usual simplex method and do not need a computer.

Exercise 10.9.1 (Easy)

1. Find the maximum of the function

$$f(\boldsymbol{x}) = 4x_1 + 3x_2 \tag{10.9.1}$$

   subject to the constraints

$$\begin{aligned} &3x_1 + 4x_2 \leq 12 \\ &3x_1 + 3x_2 \leq 10 \\ &4x_1 + 2x_2 \leq 8 \\ &\text{with} \quad x_i \geq 0 \quad \forall i \end{aligned} \tag{10.9.2}$$

2. Find the dual of this problem and solve it.

Solution: $x_1 = 0.8$, $x_2 = 2.4$, $f = 10.4$.

Exercise 10.9.2 (Easy)

1. Find the maximum of the function

$$f(\boldsymbol{x}) = 4x_1 + 3x_2 \tag{10.9.3}$$

   subject to the constraints

$$\begin{aligned} &3x_1 + 4x_2 \leq 12 \\ &3x_1 + 3x_2 \leq 10 \\ &2x_1 + x_2 \leq 4 \\ &x_1 + x_2 \geq 1 \\ &\text{with} \quad x_i \geq 0 \quad \forall i \end{aligned} \tag{10.9.4}$$

   Solution: $x_1 = 0.8$, $x_2 = 2.4$, $f = 10.4$.
2. Find the dual of this problem and solve it.

Exercise 10.9.3 (Medium)
Find the maximum of the function

$$f(\boldsymbol{x}) = 5x_1 - 3x_2 + 4x_3 \tag{10.9.5}$$

subject to the constraints

$$\begin{aligned} &x_1 - x_2 \leq 1 \\ &-3x_1 + 2x_2 + 2x_3 \leq 1 \\ &4x_1 - x_3 = 1 \\ &x_2 \geq 0\,, \quad x_3 \geq 0 \end{aligned} \tag{10.9.6}$$

The variable x_1 can take any value.

Hint: pose $x_1 = x_{11} - x_{22}\,, \qquad x_{11} \geq 0\,, \quad x_{22} \geq 0$

Solution: $x_1 = 0.6$, $x_2 = 0$, $x_3 = 1.4$, $f = 8.6$.

Exercise 10.9.4 (Easy)

Find the maximum of the function

$$f(\boldsymbol{x}) = x_1 + 4x_2 + x_3 \tag{10.9.7}$$

subject to the constraints

$$\begin{aligned} &x_1 + 2x_2 - x_3 \geq 3 \\ &-2x_1 - x_2 + 4x_3 \geq 1 \\ &x_1 + 2x_2 + x_3 \leq 11 \\ &\text{with} \quad x_i \geq 0 \quad \forall i \end{aligned} \tag{10.9.8}$$

Solution: $x_1 = 0$, $x_2 = 4.778$, $x_3 = 1.444$, $f = 20.556$.

Exercise 10.9.5 (Easy)

1. Find the maximum of the function

$$f(\boldsymbol{x}) = 4x_1 + 5x_2 \tag{10.9.9}$$

   subject to the constraints

$$\begin{aligned} &2x_1 + x_2 \leq 6 \\ &x_1 + 2x_2 \leq 5 \\ &x_1 + x_2 \geq 1 \\ &x_1 + 4x_2 \geq 2 \\ &\text{with} \quad x_i \geq 0 \quad \forall i \end{aligned} \tag{10.9.10}$$

2. Find the dual of this problem and solve it.

Solution: $x_1 = 2.333$, $x_2 = 1.333$, $f = 16$.

Exercise 10.9.6 (Medium)

A company proposes to build units of type A, B, C, with a cost and a return on investment given in Table 10.12.

Demonstrate that this problem can be set under the following form:
Find the maximum of the given function

$$f(\boldsymbol{x}) = 50x_1 + 150x_2 + 300x_3 \tag{10.9.11}$$

subject to the constraints

Table 10.12 Cost and return on investment according to the type of unit

Type	Cost	Monthly return on investment
A	20,000	50
B	50,000	150
C	80,000	300

$$\begin{aligned} &2x_1 + 5x_2 + 8x_3 \leq 4000 \\ &-x_1 + x_2 + x_3 \leq 0 \\ &-x_1 - x_2 + 4x_3 \leq 0 \\ &\text{with} \quad x_i \geq 0 \quad \forall i \end{aligned} \tag{10.9.12}$$

Find the solution of this problem. For some reasons that you will explain, it may happen that you need to replace the constant (right-hand side) of the two last constraints by a very small value (for example, 0.001).
Solution: $x_1 = 488$, $x_2 = 293$, $x_3 = 195$, $f = 1.27 \times 10^5$.

Exercise 10.9.7 (Easy)

1. Find the maximum of the function

$$f(\boldsymbol{x}) = 4\,x_1 + 5\,x_2 \tag{10.9.13}$$

subject to the constraints

$$\begin{aligned} &2\,x_1 + x_2 \leq 6 \\ &x_1 + 2\,x_2 \leq 5 \\ &x_1 + x_2 \geq 1 \\ &\text{with} \quad x_i \geq 0 \quad \forall i \end{aligned} \tag{10.9.14}$$

2. Find the dual of this problem and solve it.

Solution: $x_1 = 2.333$, $x_2 = 1.333$, $f = 16$.

Exercise 10.9.8 (Easy)

1. Find the maximum of the function

$$f(\boldsymbol{x}) = 2\,x_1 + x_2 + 4\,x_3 \tag{10.9.15}$$

subject to the constraints

$$\begin{aligned} &x_1 + 3\,x_2 + x_3 && \leq 4 \\ &x_1 + 2\,x_2 && \geq 2 \\ &3\,x_1 + x_2 + 2\,x_3 && \leq 3 \\ &\text{with} \qquad x_i \geq 0 \quad \forall\, i \end{aligned} \tag{10.9.16}$$

2. Find the dual of this problem and solve it.

Solution: $x_1 = 0$, $x_2 = 1$, $x_3 = 1$, $f = 5$.

Exercise 10.9.9 (Easy)

1. Find the maximum of the function

$$f(\boldsymbol{x}) = x_1 + x_2 + x_3 \tag{10.9.17}$$

subject to the constraints

$$\begin{aligned} &2\,x_1 + x_2 + 2\,x_3 \le 2 \\ &4\,x_1 + 2\,x_2 + x_3 \le 2 \\ &\text{with} \quad x_i \ge 0 \quad \forall i \end{aligned} \tag{10.9.18}$$

2. Find the dual of this problem and solve it.

Exercise 10.9.10 (Easy)

1. Find the maximum of

$$f(\boldsymbol{x}) = 2\,x_1 + x_2 + x_3 \tag{10.9.19}$$

subject to the constraints

$$\begin{aligned} &4\,x_1 + 6\,x_2 + 3\,x_3 \le 8 \\ &-x_1 + 9\,x_2 - x_3 \ge 3 \\ &3\,x_1 + 3\,x_2 - 5\,x_3 \ge 4 \\ &x_i \ge 0 \quad \forall i \end{aligned} \tag{10.9.20}$$

2. Pose the dual problem and solve it.

Exercise 10.9.11 (Easy)

Find the minimum of the function

$$f(\boldsymbol{x}) = 2\,x_1 + 3\,x_2 \tag{10.9.21}$$

subject to the constraints

$$\begin{aligned} &2\,x_1 + x_2 \le 3 \\ &x_1 + x_2 \le 10 \\ &x_1 + 3\,x_2 \ge 8 \\ &x_i \ge 0 \quad \forall i \end{aligned} \tag{10.9.22}$$

Exercise 10.9.12 (Easy)

Find the maximum of the function

$$f(\boldsymbol{x}) = 2\,x_1 + x_2 + x_3 + x_4 \tag{10.9.23}$$

subject to the constraints

$$\begin{aligned} &x_1 - 2\,x_2 + x_3 + x_4 \le 11 \\ &-4\,x_1 - x_2 + 2\,x_3 \le 4 \\ &2\,x_1 - 2\,x_3 + x_4 \le 1 \\ &-x_3 + x_4 \ge 2 \\ &x_i \ge 0 \qquad \forall\, i \end{aligned} \tag{10.9.24}$$

1. Solve this problem.
2. Find the dual of this problem and solve it.

Exercise 10.9.13 (Medium)

Find the minimum of the function

$$f(\boldsymbol{x}) = x_1 + 2\,x_2 + 3\,x_3 + 4\,x_4 \tag{10.9.25}$$

subject to the constraints

$$\begin{aligned} x_1 + x_2 + x_3 + x_4 &= 1 \\ x_1 + x_3 - 3\,x_4 &= 0.5 \\ x_i \geq 0 \quad &\forall i \end{aligned} \tag{10.9.26}$$

by the simplex method.

Hint: add artificial variables x_5 and x_6, which are positive or zero, to the constraints (10.9.26) and minimize the function $g(\boldsymbol{x}) = x_5 + x_6$ by the simplex method. Justify the method and use it. Deduce the solution of the initial problem.

Exercise 10.9.14 (Medium)

An oil company possesses 2 refineries R_1 and R_2 from which it must distribute gasoline to 4 clients C_1, C_2, C_3, and C_4. The transportation costs c_{ji} from a refinery i to a client j are given in Table 10.13. The costs are given per kton.

Table 10.13 Transportation costs

Refinery	R_1	R_1	R_1	R_1	R_2	R_2	R_2	R_2
Client	C_1	C_2	C_3	C_4	C_1	C_2	C_3	C_4
Cost	20	55	70	45	15	45	80	55

The capacities of production of the refineries and the capacities of receiving by the clients are fixed and known (Table 10.14). They are given in kton.

Table 10.14 Capacities of production (for a refinery) and receiving (for a client)

Refinery or Client	R_1	R_2	C_1	C_2	C_3	C_4
Capacity	1.3	0.5	0.7	0.5	0.2	0.4

We search the distribution policy allowing us to minimize the total bill. It is assumed that the refineries perfectly satisfy the clients, that is, the client receives the requested quantity. On another side, the refineries have no overflow.

1. Pose the complete problem, precisely set all the elements allowing you to get the first optimization tableau and this tableau. Note that there is not a single manner to obtain this first tableau. It is important to correctly pose the problem: a good method is to think about the degrees of freedom before making any numerical calculation. Solve this problem.
2. What would happen if the refineries had overflows? Simply indicate the modifications that would result without going into intricate calculations.

Solution: $x_{11} = 0.7$, $x_{21} = 0.5$, $x_{31} = 0.1$, $x_{32} = 0.1$, $x_{42} = 0.4$, $f = 78.5$.

Exercise 10.9.15 (Medium)

I/ A manufacturer possesses a warehouse in each of the N towns noted W_1 to W_N. Each warehouse i has a capacity C_i (expressed in a number of trucks).

The clients are spread in M objective towns O_1 to O_M, and each town O_j needs to receive a given number of trucks T_j.

Finally, the transportation cost from W_i to O_j is known and is equal to C_{ij} for each truck.

We note x_{ij} the flux (the number of trucks) between a warehouse W_i and an objective O_j.

The objective is to minimize the transportation cost of the goods, while ensuring a sufficient supply to each town, in the limit of the stocks of each warehouse.

Formulate the problem in terms of linear programming.

Exercise 10.9.16 (Medium)

A manufacturer of cosmetics produces three types of products noted P_i. In a first approximation, each type of product is a mixture of pure ingredients noted B_i according to Table 10.15.

Table 10.15 Mass composition of the cosmetics and profit per product

	Proportion of basis B_i			
Product P_i	B_1	B_2	B_3	Profit per product
P_1	0.6	0.2	0.2	12
P_2	0.3	0.2	0.5	15
P_3	0.4	0.5	0.1	9

The manufacturer desires to fabricate at least 30% of P_1, 20% of P_2, and 40% of P_3.

1. Reasoning on a total of 100 products, how many products of each type must be manufactured to maximize the profit? What quantity of ingredients must be planned?
2. Solve the same problem by Lagrange method.

Remark: reason as if the number of products was a real number and not an integer. Note that the total number of constraints will be equal to 4 when the problem will be correctly set.

Exercise 10.9.17 (Medium)

A chemical plant fabricates three different products, whereas the maximum possible sales per week are 2000 kg for product A, 1000 kg for product B, and 3000 kg for product C, respectively. The plant works 12 h per day, 6 days per week. The fabrication of product A requires 3h/100 kg, that of product B 5h/100 kg, and that of product C 4h/100 kg. The profits per products A, B, and C are 12 E/kg, 20 E/kg, and 5 E/kg, respectively. The manufacturer desires to maximize the profit.

1. Working on 1 week, set the problem.
2. Solve this problem.

References

I. Adler, M. G. C. Resende, G. Veiga, and N. Karmarkar. An implementation of Karmarkar's algorithm for linear programming. *Mathematical programming*, 44: 297–335, 1989.

E. R. Barnes. A variation on Karmarkar's algorithm for solving linear programming problems. *Mathematical Programming*, 36:174–182, 1986.

D. P. Bertsekas. *Convex Optimization Theory*. Athena Scientific, Belmont, 2009.

J. F. Bonnans, J. C. Gilbert, C. Lemaréchal, and C. A. Sagastizabal. *Numerical optimization*. Springer, Berlin, 2003.

S. Boyd and L. Vandenberghe. *Convex optimization*. Cambridge University Press, New York, 2009.

P. Chretienne, Y. Pesqueux, and J.C. Grandjean. *Algorithmes et pratique de programmation linéaire*. Technip, Paris, 1980.

G. B. Dantzig. Linear programming under uncertainty. *Management Science*, 4–7: 197–206, 1955.

G. B. Dantzig. Recent advances in linear programming. *Management Science*, 1: 131–144, 1956.

G. B. Dantzig. Thoughts on linear programming and automation. *Management Science*, 1:131–139, 1957.

G. B. Dantzig. *Linear programming and extensions*. Princeton University Press, Princeton, NJ, 1963.

G. B. Dantzig. Origins of the simplex method. Technical Report AD-A182 708, Stanford University, Systems Optimization Laboratory, 1987.

G. B. Dantzig and M. N. Thapa. *Linear programming 1: Introduction*. Springer, New York, 1997.

G. B. Dantzig and M. N. Thapa. *Linear programming 2: Theories and extensions*. Springer, New York, 2003.

G. B. Dantzig, A. Orden, and P. Wolfe. Generalized simplex method for minimizing a linear form under linear inequality restraints. *Pacific J. Math.*, 5:183–193, 1955.

J. E. Dennis, A. M. Morshedi, and K. Turner. A variable-metric variant of the Karmarkar algorithm for linear programming. Technical Report 86-13, Mathematical Sciences Department, Rice University, Houston, Texas, 1986.

I. I. Dikin. Iterative solution of problems of linear and quadratic programming. *Doklady Akademiia Nauk USSR*, 174:747–748, 1967. Translated in *Soviet Mathematics Doklady*, 8, 674–675.

R. Faure, B. Lemaire, and C. Picouleau. *Précis de recherche opérationnelle*. Dunod, Paris, 6ème edition, 2009.

M. C. Ferris, O. L. Mangasarian, and S. J. Wright. *Linear programming with MATLAB*. SIAM, Philadelphia, 2007.

R. Fletcher. *Practical Methods of Optimization*. Wiley, Chichester, 1991.

A. Forsgren, P. E. Gill, and M. H. Wright. Interior methods for nonlinear optimization. *SIAM Review*, 44(4):525–597, 2002.

D. H. Gale, H. W. Kuhn, and A. W. Tucker. Convex polyhedral cones and linear inequalities. In T. C. Koopmans, editor, *Activity analysis of production and allocation*,

Proceedings of Linear Programming Conference, June 20-24, 1949, pages 317–329. John Wiley, New York, 1951.

P. E. Gill, W. Murray, M. A. Saunders, J.A. Tomlin, and M. H. Wright. On projected Newton barrier methods for linear programming and an equivalence to Karmarkar's projective method. *Mathematical Programming*, 36:183–209, 1986.

N. J. Karmarkar. A new polynomial-time algorithm of linear programming. *Combinatorica*, 4:374–395, 1984.

X. Li and Z. Xuan. An interior-point QP algorithm for structural optimization. *Structural Optimization*, 15(3–4):172–179, 1998.

D. G. Luenberger and Y. Ye. *Linear and Nonlinear Programming*. Springer, Heidelberg, 4th edition, 2016.

J. Matousek and B. Gärtner. *Understanding and using linear programming*. Springer, Berlin, 2007.

H. Maurin. *Programmation linéaire appliquée*. Technip, Paris, 1967.

S. Mizuno. A new polynomial time method for a linear complementary problem. *Mathematical programming*, 56:31–43, 1992.

R. D. C. Monteiro, I. Adler, and M. G. C. Resende. A polynomial-time primal–dual affine scaling algorithm for linear and convex quadratic programming and its power series extension. *Mathematics of Operations Research*, 15(2):191–214, 1990.

J. L. Nazareth. *Differentiable optimization and equation solving. A treatise on algorithmic science and the Karmarkar revolution*. Springer, 2003.

A. S. Nemirovski and M. J. Todd. Interior-point methods for optimization. *Acta Numerica*, pages 1–44, 2009.

M. J. Panik. *Linear Programming: Mathematics, Theory and Algorithms*. Kluwer, Dordrecht, 1996.

C. Roos, T. Terlaky, and J. P. Vial. *Theory and Algorithms for Linear Optimization: An Interior Point Approach*. Wiley, New York, 2nd edition, 2005.

R. Saigal. Efficient variants of the affine scaling method. Technical Report 93-21, Department of Industrial and Operations Engineering, University of Michigan, Ann Arbor, USA, 1993.

R. Saigal. A three step quadratically convergent version of primal affine scaling method. *Journal of the Operations Research, Society of Japan*, 40(3):310–327, 1997.

T. Terlaky, editor. *Interior Point Methods of Mathematical Programming*. Kluwer, 1996.

P. R. Thie and G. E. Keough. *An Introduction to Linear Programming and Game Theory*. Wiley, Hoboken, N.J., 3rd edition, 2008.

T. Tsuchiya. Global convergence property of the affine scaling method for primal degenerate linear programming problems. *Mathematics of Operations Research*, 17:527–557, 1992.

R. J. Vanderbei. *Linear Programming: Foundations and Extensions*. Springer, New York, 3rd edition, 2010.

S. J. Wright. *Primal–dual interior-point methods*. SIAM, Philadelphia, 1997.

Chapter 11
Quadratic Programming and Nonlinear Optimization

11.1 Introduction

An optimization problem is called *quadratic optimization* or *quadratic programming* (QP) when the function to minimize is quadratic and the inequality constraints are linear

$$\begin{aligned} &\min \; f(\boldsymbol{x}) = \frac{1}{2}\boldsymbol{x}^T \boldsymbol{H}\, \boldsymbol{x} + \boldsymbol{g}^T \boldsymbol{x} \\ &\text{subject to} \quad \boldsymbol{A}\, \boldsymbol{x} \leq \boldsymbol{b} \qquad \text{and} \quad \boldsymbol{x} \geq 0 \end{aligned} \tag{11.1.1}$$

where $\boldsymbol{H}$ is a symmetric matrix of dimension $n \times n$ and m the number of linear constraints (dimension of $\boldsymbol{b}$). This kind of problem arises frequently in various domains such as economics, process optimization, parameter identification, and model predictive control (Biegler 2010; Corriou 2018).

Several cases can occur:

- The problem is strictly convex, and then $\boldsymbol{H}$ is definite positive ($\boldsymbol{x}^T \boldsymbol{H}\, \boldsymbol{x} > 0\,,\ \forall\, \boldsymbol{x}$). If a local minimum exists, this minimum is global.
- The problem is convex, and then $\boldsymbol{H}$ is semi-definite positive ($\boldsymbol{x}^T \boldsymbol{H}\, \boldsymbol{x} \geq 0\,,\ \forall\, \boldsymbol{x}$). The minimum, if it exists, is unique. Note that, when $\boldsymbol{H}$ is zero, it is a linear programming problem.
- The problem is not convex, and then $\boldsymbol{H}$ is any. Its eigenvalues have real parts either positive or negative. It is called indefinite (Lucia et al. 1996). Several local minima can exist.

Many methods exist to solve quadratic programming problems (Bertsekas 2009; Best 2016; Bonnans et al. 2003; Boyd and Vandenberghe 2009; Dostal 2009; Lin and Pang 1987; Pang 1983). These methods can be classified into finite methods, ending in a finite number of operations, or iterative methods, converging to the optimum.

When the function to be optimized is nonquadratic or when the constraints are nonlinear, the problem is referred as *nonlinear optimization* (Avriel 1976; Bartholomew-Biggs 2010; Bazaraa et al. 1993; Izmailov and Solodov 2014). The problems of nonlinear optimization in the presence of equality and inequality constraints (Section 11.4) can be solved in a very effective manner by successive quadratic programming (SQP) that makes use of quadratic optimization.

J.-P. Corriou, *Numerical Methods and Optimization*, Springer Optimization and Its Applications 187, https://doi.org/10.1007/978-3-030-89366-8_11

11.2 Quadratic Optimization, Karush–Kuhn–Tucker Conditions and Solution by the Simplex

The solution of a quadratic programming problem can be obtained by an algorithm close to the simplex according to Dantzig's method or Wolfe's method (Bouzitat 1979; Wolfe 1959). Two main presentations can be formulated to transform the problem.

11.2.1 First Presentation

To the problem (11.1.1), we can associate the Lagrangian

$$\mathcal{L}(\boldsymbol{x}, \boldsymbol{\mu}) = \frac{1}{2}\boldsymbol{x}^T\,\boldsymbol{H}\,\boldsymbol{x} + \boldsymbol{g}^T\,\boldsymbol{x} + \boldsymbol{\mu}^T\,(\boldsymbol{A}\,\boldsymbol{x} - \boldsymbol{b}) \tag{11.2.1}$$

where $\boldsymbol{\mu}$ is the vector of Karush–Kuhn–Tucker (KKT) parameters. A point $\boldsymbol{x}^*$ is a solution if it verifies at $\boldsymbol{x} = \boldsymbol{x}^*$

$$\begin{aligned}
&\frac{\partial \mathcal{L}}{\partial \boldsymbol{x}} = \boldsymbol{g} + \boldsymbol{H}\,\boldsymbol{x} + \boldsymbol{A}^T\,\boldsymbol{\mu} = 0\\
&\boldsymbol{\mu}^T\,(\boldsymbol{A}\,\boldsymbol{x} - \boldsymbol{b}) = 0\\
&\boldsymbol{A}\,\boldsymbol{x} - \boldsymbol{b} \leq 0\\
&\boldsymbol{x} \geq 0\\
&\boldsymbol{\mu} \geq 0
\end{aligned} \tag{11.2.2}$$

This point is called Karush–Kuhn–Tucker (KKT) point.

11.2.2 Second Presentation

In the second presentation, the positivity of $\boldsymbol{x}$ is taken into account in the expression of the Lagrangian, that is

$$\mathcal{L}(\boldsymbol{x}, \boldsymbol{\mu}) = \frac{1}{2}\boldsymbol{x}^T\,\boldsymbol{H}\,\boldsymbol{x} + \boldsymbol{g}^T\,\boldsymbol{x} + \boldsymbol{\mu}^T\,(\boldsymbol{A}\,\boldsymbol{x} - \boldsymbol{b}) - \boldsymbol{\nu}^T\,\boldsymbol{x} \tag{11.2.3}$$

where $\boldsymbol{\mu}$ and $\boldsymbol{\nu}$ are the two vectors of KKT parameters. Now, a point $\boldsymbol{x}^*$ is a solution if it verifies at $\boldsymbol{x} = \boldsymbol{x}^*$

$$\begin{aligned}
&\frac{\partial \mathcal{L}}{\partial \boldsymbol{x}} = \boldsymbol{g} + \boldsymbol{H}\,\boldsymbol{x} + \boldsymbol{A}^T\,\boldsymbol{\mu} - \boldsymbol{\nu} = 0\\
&\boldsymbol{\mu}^T\,(\boldsymbol{A}\,\boldsymbol{x} - \boldsymbol{b}) = 0\\
&\boldsymbol{\nu}^T\,\boldsymbol{x} = 0\\
&\boldsymbol{A}\,\boldsymbol{x} - \boldsymbol{b} \leq 0\\
&\boldsymbol{x} \geq 0\\
&\boldsymbol{\mu} \geq 0\\
&\boldsymbol{\nu} \geq 0
\end{aligned} \tag{11.2.4}$$

11.2.3 Solution Under the Form of a Simplex Problem

Consider the second presentation. Positive slack variables $\boldsymbol{y}$ are added so that the linear inequality of the original problem (11.1.1) becomes an equality as

$$\boldsymbol{A}\,\boldsymbol{x} - \boldsymbol{b} + \boldsymbol{y} = 0 \quad \text{with} \quad \boldsymbol{y} \geq 0 \tag{11.2.5}$$

which implies $\boldsymbol{\mu}^T\,\boldsymbol{y} = 0$. The problem (11.2.4) can then be written again under the form

$$\boldsymbol{g} + \boldsymbol{H}\,\boldsymbol{x} + \boldsymbol{A}^T\,\boldsymbol{\mu} - \boldsymbol{\nu} = 0 \tag{11.2.6}$$
$$\boldsymbol{A}\,\boldsymbol{x} - \boldsymbol{b} + \boldsymbol{y} = 0 \tag{11.2.7}$$
$$\boldsymbol{\nu}^T\,\boldsymbol{x} = 0 \tag{11.2.8}$$
$$\boldsymbol{\mu}^T\,\boldsymbol{y} = 0 \tag{11.2.9}$$
$$\boldsymbol{x} \geq 0\,, \quad \boldsymbol{y} \geq 0\,, \quad \boldsymbol{\mu} \geq 0\,, \quad \boldsymbol{\nu} \geq 0 \tag{11.2.10}$$

The constraints (11.2.6) and (11.2.7) are the structural constraints of this problem. The constraints (11.2.6) and (11.2.7) are again written by placing the constant term in the right-hand side, so that we get

$$\boldsymbol{H}\,\boldsymbol{x} + \boldsymbol{A}^T\,\boldsymbol{\mu} - \boldsymbol{\nu} = -\boldsymbol{g} \tag{11.2.11}$$
$$\boldsymbol{A}\,\boldsymbol{x} + \boldsymbol{y} = \boldsymbol{b} \tag{11.2.12}$$
$$\boldsymbol{\nu}^T\,\boldsymbol{x} = 0 \tag{11.2.13}$$
$$\boldsymbol{\mu}^T\,\boldsymbol{y} = 0 \tag{11.2.14}$$
$$\boldsymbol{x} \geq 0\,, \quad \boldsymbol{y} \geq 0\,, \quad \boldsymbol{\mu} \geq 0\,, \quad \boldsymbol{\nu} \geq 0 \tag{11.2.15}$$

The problem is then set under a convenient form to be solved by the simplex method:

- If the right-hand side of one of the constraints (11.2.11) and (11.2.12) is negative, multiply this constraint by -1.
- Add an artificial variable to each constraint (11.2.11) and (11.2.12).
- Set the objective function equal to the sum of the artificial variables.
- Solve the problem as a linear programming problem.

During the iterations, if an artificial variable in the basis becomes equal to 0, it means that it should be considered as a nonbasic variable. The artificial variable then determines the pivot row, and we determine the pivot column by the usual rule of the most negative reduced cost. Then it is possible to swap the artificial variable and the variable of the pivot column.

The classical simplex problem must be slightly modified as the complementary conditions (11.2.13) and (11.2.14) must be verified at each iteration. The variables x_i and ν_i are complementary for $i = 1, \ldots, n$. The variables y_i and μ_i are complementary for $j = 1, \ldots, m$. The rule is that a variable is entering the basis when this variable is that for which the reduced cost is the most negative on the condition that its complementary variable is not in the basis or does not leave the basis during the same iteration. The problem is then called linear complementarity problem.

When the solution of the problem is obtained, the vector $\boldsymbol{x}$ gives the optimal solution and the vector $\boldsymbol{\mu}$ the dual optimal variables.

This method works well when the matrix $\boldsymbol{H}$ is strictly definite positive. When the matrix is simply definite positive, it is recommended to add a small constant to each element of the diagonal of $\boldsymbol{H}$ so that the matrix becomes strictly definite positive. An often mentioned drawback of this type of method is the increase of the number of variables caused by the introduction of slack variables.

Example 11.1 :
Quadratic programming: solution by the simplex method

Let us solve the following problem:

$$\begin{array}{l} \min f(\boldsymbol{x}) = (x_1 - 1)^2 + 2(x_2 - 1.5)^2 \\ \text{subject to} \\ 2\,x_1 + 3\,x_2 \le 4 \\ 1.5\,x_1 + x_2 \le 2 \\ x_1 \ge 0\,, \quad x_2 \ge 0 \end{array} \tag{11.2.16}$$

hence

$$\boldsymbol{H} = \begin{bmatrix} 2 & 0 \\ 0 & 4 \end{bmatrix}, \quad \boldsymbol{g} = \begin{bmatrix} -2 \\ -6 \end{bmatrix} \tag{11.2.17}$$

For this problem, the matrix H is symmetric definite positive, and thus the problem can be solved by the simplex method. Using the method of the simplex tableau adapted to the quadratic optimization problem, thus the minimum of the sum of the artificial variables must be determined

$$f' = a_1 + a_2 + a_3 + a_4 \tag{11.2.18}$$

where a_i are the artificial variables such that

$$\begin{array}{l} 2\,x_1 + 2\,\mu_1 + 1.5\,\mu_2 - \nu_1 + a_1 = 2 \\ 4\,x_2 + 3\,\mu_1 + \mu_2 - \nu_2 + a_2 = 6 \\ 2\,x_1 + 3\,x_2 + y_1 + a_3 = 4 \\ 1.5\,x_1 + x_2 + y_2 + a_4 = 2 \end{array} \tag{11.2.19}$$

The simplex tableau (Table 11.1) results where the columns V_1 and V_2 correspond to x_i, V_3 and V_4 to ν_i, V_5 and V_6 to μ_i, V_7 and V_8 to y_i, the columns from V_9 to V_{12} to a_i. It must be noted that, at each iteration, the conditions (11.2.13) and (11.2.14) are verified. In the last column of each iteration, the value of $-f'$ is indicated. The optimal solution, obtained when $f' = 0$, is

$$x^* = \begin{bmatrix} 0.4118 \\ 1.0588 \end{bmatrix}, \quad \nu^* = \begin{bmatrix} 0 \\ 0 \end{bmatrix}, \quad y^* = \begin{bmatrix} 0 \\ 0.3235 \end{bmatrix}, \quad \mu^* = \begin{bmatrix} 0.5882 \\ 0 \end{bmatrix} \tag{11.2.20}$$

The optimal solution can be verified in Figure 11.1 where the contours are represented as well as the constraints. The initial point and the points during iterations are indicated in Figure 11.1, starting from (0, 0), going to (0, 1.333), then to the solution. Note that the number of iterations of the simplex tableau is not the number of generated points (x_1, x_2), some points remaining the same during the simplex operations.

Table 11.1 Quadratic programming by the simplex method: simplex tableau to solve problem (11.2.16).

n	V_1	V_2	V_3	V_4	V_5	V_6	V_7	V_8	V_9	V_{10}	V_{11}	V_{12}	B	Basis
1	2	0	−1	0	2	1.5	0	0	1	0	0	0	2	a_1
	0	4	0	−1	3	1	0	0	0	1	0	0	6	a_2
	2	[3]	0	0	0	0	1	0	0	0	1	0	4	a_3
	1.5	1	0	0	0	0	0	1	0	0	0	1	2	a_4
	−5.5	−8	1	1	−5	−2.5	−1	−1	0	0	0	0	−14	$-f'$
2	2	0	−1	0	2	1.5	0	0	1	0	0	0	2	a_1
	−2.67	0	0	−1	[3]	1	−1.33	0	0	1	−1.33	0	0.667	a_2
	0.67	1	0	0	0	0	0.33	0	0	0	0.33	0	1.333	x_2
	0.83	0	0	0	0	0	−0.33	1	0	0	−0.33	1	0.667	a_4
	−0.17	0	1	1	−5	−2.5	1.67	−1	0	0	2.67	0	−3.333	$-f'$
3	[3.78]	0	−1	0.67	0	0.83	0.89	0	1	−0.67	0.89	0	1.556	a_1
	−0.89	0	0	−0.33	1	0.33	−0.44	0	0	0.33	−0.44	0	0.222	μ_1
	0.67	1	0	0	0	0	0.33	0	0	0	0.33	0	1.333	x_2
	0.83	0	0	0	0	0	−0.33	1	0	0	−0.33	1	0.667	a_4
	−4.61	0	1	−0.67	0	−0.83	−0.56	1	0	1.67	0.44	0	−2.222	$-f'$
4	1	0	−0.26	0.18	0	0.22	0.23	0	0.26	−0.18	0.23	0	0.412	x_1
	0	0	−0.235	−0.18	1	0.53	−0.23	0	0.23	0.18	−0.23	0	0.588	μ_1
	0	1	0.18	−0.12	0	−0.15	0.18	0	−0.18	0.12	0.18	0	1.059	x_2
	0	0	0.22	−0.15	0	−0.18	−0.53	[1]	−0.22	0.15	−0.53	1	0.323	a_4
	0	0	−0.22	0.15	0	0.18	0.53	−1	1.22	0.85	1.53	0	−0.323	$-f'$
5	1	0	−0.26	0.18	0	0.22	0.23	0	0.26	−0.18	0.23	0	0.412	x_1
	0	0	−0.23	−0.18	1	0.53	−0.23	0	0.23	0.18	−0.23	0	0.588	μ_1
	0	1	0.18	−0.12	0	−0.15	0.18	0	−0.176	0.12	0.18	0	1.059	x_2
	0	0	0.22	−0.15	0	−0.18	−0.53	1	−0.22	0.15	−0.53	1	0.323	y_2
	0	0	0	0	0	0	0	0	1	1	1	1	0	$-f'$

11.3 Quadratic Optimization, Barrier Method

The barrier method takes place among the interior point methods (Forsgren et al. 2002; Jansen 2010; Nemirovski and Todd 2009; Ye 1997) that have known a large development since the article by Karmarkar (1984). A great advantage of interior point methods is the reduced complexity with respect to the theoretically exponential complexity of the simplex method of combinatory nature. Interior point methods allowed a better understanding of optimization under constraints and led to algorithm improvements.

Let the problem

$$\begin{aligned} &\min_{\boldsymbol{x}} f(\boldsymbol{x}) \\ &\text{subject to} \quad h_i(\boldsymbol{x}) \geq 0\,, \quad i = 1, \ldots, m \\ &\boldsymbol{x}_l \leq \boldsymbol{x} \leq \boldsymbol{x}_u \end{aligned} \tag{11.3.1}$$

let $\mathcal{R}_p$ be the inside of the feasible region such that

$$\mathcal{R}_p = \{\boldsymbol{x} \mid h_i(\boldsymbol{x}) > 0 , \quad i = 1, \dots, m\} \tag{11.3.2}$$

Among the classical barrier functions, let us cite the logarithmic barrier function such that the problem (11.3.1) is transformed at iteration k as

$$\min_{\boldsymbol{x}} B(\boldsymbol{x}, \mu^{(k)}) = f(\boldsymbol{x}) - \mu^{(k)} \sum_{i=1}^{m} \ln(h_i(\boldsymbol{x})) \tag{11.3.3}$$

Then, it becomes an optimization problem without constraints (Polyak 1992) that is solved in the following manner. Fix an initial value of parameter μ, and then progressively reduce the value of μ. Thus, the solution of the transformed problems converges asymptotically to the solution of the original problem. This definition of the logarithmic barrier function is well adapted to mathematical demonstrations in particular with respect to convergence, but it suffers from some drawbacks and even can fail in simple cases (Forsgren et al. 2002). For that reason, it is not proposed here under this strict form.

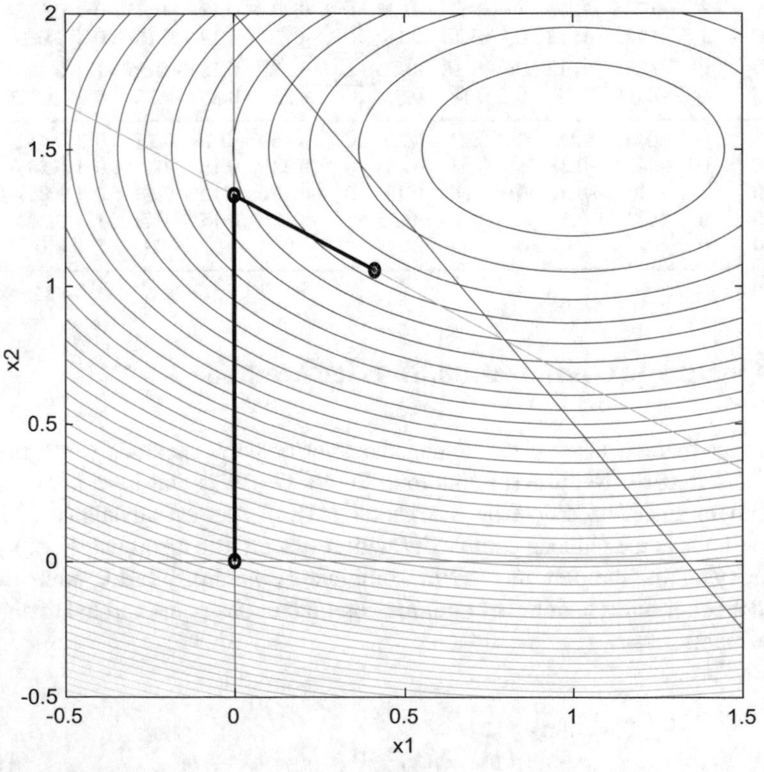

Fig. 11.1 Quadratic programming by the simplex method: contours of the function, representation of the constraints and iterative points (circles, solution by red disk) for problem (11.2.16)

The modified barrier method (Polyak 1992; Vassiliadis and Brooks 1998; Vassiliadis and Floudas 1997) consists of extending the previous formulation by taking into account Lagrange parameters as

$$\min_{\boldsymbol{x}} B_m(\boldsymbol{x}, \lambda^{(k)}, \mu^{(k)}) = f(\boldsymbol{x}) - \mu^{(k)} \sum_{i=1}^{m} \lambda_i^{(k)} \ln(r_i + \frac{h_i(\boldsymbol{x})}{\mu^{(k)}}) \tag{11.3.4}$$

$\lambda_i^{(k)}$ is a Lagrange parameter, and r_i a relaxation parameter, for each constraint. In this case, the constraints h_i can be zero and the convergence is finite instead of asymptotic like in the previous case.

Indeed, according to Vassiliadis and Floudas (1997), the modified barrier method is considered as

$$\begin{aligned} B_m(\boldsymbol{x}, \lambda^{(k)}, \mu^{(k)}) = f(\boldsymbol{x}) - \mu^{(k)} \Bigg[& \sum_{i=1}^{m} \lambda_i^{(k)} \ln(r_i + \frac{h_i(\boldsymbol{x})}{\mu^{(k)}}) + \\ & \sum_{i=1}^{n} \lambda_{i,l}^{(k)} \ln(r_{i,l} + \frac{x_i - x_{i,l}}{\mu^{(k)}}) + \sum_{i=1}^{n} \lambda_{i,u}^{(k)} \ln(r_{i,u} + \frac{x_{i,u} - x_i}{\mu^{(k)}}) \Bigg] \end{aligned} \tag{11.3.5}$$

to take into account lower and upper bounds of $\boldsymbol{x}$, $\boldsymbol{x}_l$, and $\boldsymbol{x}_u$, respectively. On another side, to avoid that some constraints are not verified during the iterative search, we define

$$F(\boldsymbol{x}, \lambda^{(k)}, \mu^{(k)}) = f(\boldsymbol{x}) - \mu^{(k)} \sum_{i=1}^{m} \lambda_i^{(k)} \Phi^{(k)}(h_i(\boldsymbol{x})) \tag{11.3.6}$$

with

$$\Phi^{(k)}(h_i(\boldsymbol{x})) = \begin{cases} \ln(r_i + \frac{h_i(\boldsymbol{x})}{\mu^{(k)}}) & \text{if } h_i(\boldsymbol{x}) \geq -\beta^{(k)} r_i \mu^{(k)} \quad \forall i = 1, \dots, m \\ Q^{(k)}(h_i(\boldsymbol{x})) & \text{if } h_i(\boldsymbol{x}) < -\beta^{(k)} r_i \mu^{(k)} \quad \forall i = 1, \dots, m \end{cases} \tag{11.3.7}$$

and

$$Q^{(k)}(h_i(\boldsymbol{x})) = \frac{1}{2} q_i^a \, h_i^2(\boldsymbol{x}) + q_i^b \, h_i(\boldsymbol{x}) + q_i^c \tag{11.3.8}$$

The coefficients of the quadratic function $Q^{(k)}$ are obtained by expressing the continuity of the function, of its first and second derivatives at point $h_i(\boldsymbol{x}) = -\beta^{(k)} r_i \mu^{(k)}$; hence

$$\begin{aligned} q_i^a &= \frac{1}{(r_i \mu^{(k)} (1 - \beta^{(k)}))^2} \\ q_i^b &= \frac{1 - 2\beta^{(k)}}{r_i \mu^{(k)} (1 - \beta^{(k)})^2} \\ q_i^c &= \frac{\beta^{(k)}(2 - 3\beta^{(k)})}{2(1 - \beta^{(k)})^2} + \ln(r_i + \frac{h_i(\boldsymbol{x})}{\mu^{(k)}}) \end{aligned} \tag{11.3.9}$$

The parameter $0 \leq \beta^{(k)} < 1$ controls the proximity of the extrapolation with respect to the asymptote of the logarithm.

The parameters are iterated in the following manner:

$$\begin{aligned}\lambda_i^{(k+1)} &= \mu^{(k)} \lambda_i^{(k)} \frac{d\Phi^{(k)}(h_i(\boldsymbol{x}))}{dh_i(\boldsymbol{x})} \\ \mu^{(k+1)} &= \frac{\mu^{(k)}}{\gamma} \\ \beta^{(k+1)} &= \min\left(1 - \frac{1-\beta^{(k)}}{\gamma}, \beta_{max}\right)\end{aligned} \tag{11.3.10}$$

with $\gamma > 1$ reduction parameter, $0 < \beta_{max} < 1$ chosen to avoid that the extrapolation is not too close to the asymptote of the logarithm. The parameter $\mu^{(0)}$ is in general chosen as 10^{-1} or 10^{-2}.

The algorithm is then:

1. $k = 0$. Initialize $\boldsymbol{x}$ and λ.
2. Iterate k. Minimize the function $F(\boldsymbol{x}, \lambda^{(k)}, \mu^{(k)})$ by a technique of optimization without constraints. For example, use the BFGS or Levenberg–Marquardt methods.
3. Calculate $\boldsymbol{\lambda}^{(k+1)}$.
4. Test the convergence. If convergence, exit. Otherwise, go to 5.
5. Calculate $\mu^{(k+1)}$ and $\beta^{(k+1)}$.
6. $k = k + 1$. Return to 2.

According to Vassiliadis and Brooks (1998) and Vassiliadis and Floudas (1997), the modified logarithmic barrier method can be used for large scale problems.

Example 11.2 :

Quadratic programming: solution by the modified logarithmic barrier method

The problem (11.2.16) previously treated in Example 11.2 is solved

$$\begin{aligned}&\min f(\boldsymbol{x}) = (x_1 - 1)^2 + 2(x_2 - 1.5)^2 \\ &\text{subject to} \\ &2\,x_1 + 3\,x_2 \le 4 \\ &1.5\,x_1 + x_2 \le 2 \\ &x_1 \ge 0\,, \quad x_2 \ge 0\end{aligned} \tag{11.3.11}$$

where the function is quadratic and the constraints linear. The initial point is $\boldsymbol{x} = (0, 0)$. The lower and upper bounds are $\boldsymbol{x}_l = (0, 0)$ and $\boldsymbol{x}_u = (10, 10)$, respectively. The initial Lagrange parameters are $\lambda_i = 1$. The other initial parameters are $\mu = 1, \beta = 0.5, \gamma = 2, r_i = 1$. Furthermore $\beta_{max} = 0.95$. The BFGS method was used for optimization, and the results along iterations are given in Table 11.2 with end of iterations when the relative tolerance equal to 10^{-6} on f is reached. The Lagrange parameter for the second constraint is zero, while it is nonzero for the first constraint; thus the solution is on this first constraint. The number of iterations largely depends on the value of γ. Thus, for $\gamma = 1.1$, we obtain an equivalent result in 19 iterations, and for $\gamma = 1.5$ in 11 iterations. Figure 11.2 shows the trajectory followed by the solution along the iterations. For the studied example, the convergence is fast.

Table 11.2 Quadratic programming by the modified logarithmic barrier method: results of iterations to solve problem (11.2.16)

k	x_1	x_2	f	λ_1	λ_2
0	0	0	0	1.0000	1.0000
1	0.0818	0.9757	−4.1072	0.5238	0.5259
2	0.3305	1.0875	−4.7115	0.4544	0.2868
3	0.3915	1.0822	−4.7807	0.5158	0.1235
4	0.4069	1.0658	−4.7713	0.5673	0.0344
5	0.4112	1.0598	−4.7658	0.5841	0.0056
6	0.4118	1.0589	−4.7649	0.5892	0.0005
7	0.4118	1.0588	−4.7647	0.5885	0.0000
8	0.4118	1.0588	−4.7647	0.5883	0.0000
9	0.4120	1.0587	−4.7647	0.5879	0.0000

11.4 Nonlinear Optimization by Successive Quadratic Programming

11.4.1 Introduction

Successive quadratic programming (SQP) consists of solving a general problem of nonlinear optimization noted P_1 of the form

$$\begin{array}{ll} \min_{\boldsymbol{x}} f(\boldsymbol{x}), & \boldsymbol{x} \in \mathbb{R}^n \\ \text{subject to} & \\ g_i(\boldsymbol{x}) = 0, & i = 1, \dots, m \\ h_i(\boldsymbol{x}) \leq 0, & i = 1, \dots, p \\ \boldsymbol{x}_l \leq \boldsymbol{x} \leq \boldsymbol{x}_u & \end{array} \tag{11.4.1}$$

as a succession of quadratic optimization problems, i.e. where the objective function is quadratic and the constraints are linear.

It must be noted that any equality constraint $g_i(\boldsymbol{x}) = 0$ can be replaced by two inequalities $g_i(\boldsymbol{x}) \leq 0$ and $g_i(\boldsymbol{x}) \geq 0$ that must be simultaneously verified.

Himmelblau (1972) gives a large number of test problems for nonlinear optimization that can be used to compare different algorithms (Avriel 1976; Bartholomew-Biggs 2010; Bazaraa et al. 1993; Borwein and Lewis 2006; Eiselt and Sandblom 2019; Jongen et al. 2000; Luenberger and Ye 2016).

11.4.2 Notion of Feasible Region and Tangent Cone

Given a point $\boldsymbol{x}$ and a set of active constraints at $\boldsymbol{x}$, the set of feasible linearized directions $\boldsymbol{d}$ such that

$$C(\boldsymbol{x}) = \left\{ \boldsymbol{d} | \begin{array}{ll} \boldsymbol{d}^T g_{i,x}(\boldsymbol{x}) = 0 & , i = 1, \ldots, m \\ \boldsymbol{d}^T h_{i,x}(\boldsymbol{x}) \leq 0 & , i = 1, \ldots, p \end{array} \right\} \tag{11.4.2}$$

forms a cone called *tangent cone*.

In the case of linear problems, the KKT conditions are necessary for optimality. This is no more the case for nonlinear constraints. Then, conditions called *regularity conditions* or *constraint qualification* must be considered (Forsgren et al. 2002). Thus, we distinguish:

- The qualification of linear independence: this condition is fulfilled at a feasible point $\boldsymbol{x}$ if the point $\boldsymbol{x}$ is strictly feasible (no constraint is active) or if the Jacobian of the constraints at $\boldsymbol{x}$ is of full rank, thus when the gradients of the constraints are linearly independent.

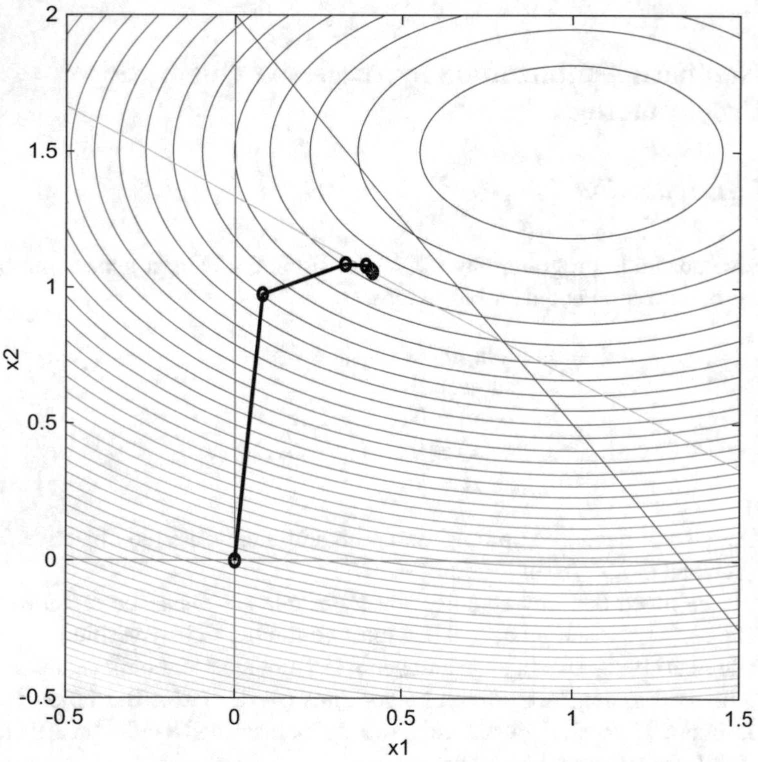

Fig. 11.2 Quadratic programming by the modified logarithmic barrier method: contours of the function, representation of the constraints and iterative points (circles) for problem (11.2.16)

- The qualification of Mangasarian–Fromovitz: this condition is fulfilled at a feasible point $\boldsymbol{x}$ if the point $\boldsymbol{x}$ is strictly feasible or if there exists a vector $\boldsymbol{p}$ such that $\boldsymbol{h}_{i,x}^T \boldsymbol{p} < 0$ for the set of the active constraints. This qualification is weaker than the qualification of linear independence. The satisfaction of this condition implies the existence of a path strictly inside the feasible region, but a path can exist without this qualification being verified.

A consequence is that if the qualification of Mangasarian–Fromovitz is fulfilled at a first order KKT point $\boldsymbol{x}$, then the set of all the multipliers μ is bounded and this point is a KKT point. This verification requires to solve a linear programming problem. The second order conditions are complex (Forsgren et al. 2002).

11.4.3 Successive Quadratic Programming (SQP)

Noting $\boldsymbol{\lambda}$ and $\boldsymbol{\mu}$ the Lagrange and KKT parameters, respectively, the Lagrangian of problem P_1 is written as

$$
\begin{aligned}
\mathcal{L}(\boldsymbol{x}, \boldsymbol{\lambda}, \boldsymbol{\mu}) &= f(\boldsymbol{x}) + \boldsymbol{\lambda}^T \boldsymbol{g}(\boldsymbol{x}) + \boldsymbol{\mu}^T \begin{bmatrix} \boldsymbol{h}(\boldsymbol{x}) \\ \boldsymbol{x}_l - \boldsymbol{x} \\ \boldsymbol{x} - \boldsymbol{x}_u \end{bmatrix} \\
&= f(\boldsymbol{x}) + \sum_{i=1}^{m} \lambda_i \, g_i(\boldsymbol{x}) + \sum_{i=1}^{p} \mu_i \, h_i(\boldsymbol{x}) + \\
&\sum_{i=1}^{n} \mu_{p+i} \, (x_{l,i} - x_i) + \sum_{i=1}^{n} \mu_{p+n+i} \, (x_i - x_{u,i})
\end{aligned}
\tag{11.4.3}
$$

hence the KKT relations

$$
\begin{aligned}
&\frac{\partial \mathcal{L}}{\partial \boldsymbol{x}} = 0 \\
&\frac{\partial \mathcal{L}}{\partial \boldsymbol{\lambda}} = 0 \Rightarrow g_i(\boldsymbol{x}) = 0 \,, \quad i = 1, \dots, m \\
&\boldsymbol{\mu}^T \begin{bmatrix} \boldsymbol{h}(\boldsymbol{x}) \\ \boldsymbol{x}_l - \boldsymbol{x} \\ \boldsymbol{x} - \boldsymbol{x}_u \end{bmatrix} = 0 \quad \text{with } \boldsymbol{\mu} \geq 0
\end{aligned}
\tag{11.4.4}
$$

that is

$$
\begin{aligned}
&\frac{\partial f}{\partial x_i} + \sum_{j=1}^{m} \lambda_j \frac{\partial g_j}{\partial x_i} + \sum_{j=1}^{p} \mu_j \frac{\partial h_j}{\partial x_i} - \mu_{p+i} + \mu_{p+n+i} = 0 \,, \quad i = 1, \dots, n \\
&g_i(\boldsymbol{x}) = 0 \,, \quad i = 1, \dots, m \\
&\mu_i \, h_i(\boldsymbol{x}) = 0 \,, \quad \mu_i \geq 0 \,, \quad i = 1, \dots, p \\
&\mu_{p+i} \, (x_{l,i} - x_i) = 0 \,, \quad \mu_{p+i} \geq 0 \,, \quad i = 1, \dots, n \\
&\mu_{p+n+i} \, (x_i - x_{u,i}) = 0 \,, \quad \mu_{p+n+i} \geq 0 \,, \quad i = 1, \dots, n
\end{aligned}
\tag{11.4.5}
$$

The SQP method comes from a second degree Taylor polynomial for the Lagrangian in the neighborhood of $(\boldsymbol{x}, \boldsymbol{\lambda}, \boldsymbol{\mu})$ and from the search of the solution $(d\boldsymbol{x}, d\boldsymbol{\lambda}, d\boldsymbol{\mu})$ by a Newton's iterative method.

The general SQP algorithm can be described (Lucia and Xu 1990) as:

1. Let $k = 1$. Initialize $\boldsymbol{x}^{(k)}$, $\boldsymbol{\lambda}^{(k)}$, $\boldsymbol{\mu}^{(k)}$. Initialize the Hessian matrix $\boldsymbol{H}$ of Lagrangian.
2. Calculate $f(\boldsymbol{x}^{(k)})$, $\boldsymbol{f}_x(\boldsymbol{x}^{(k)})$, $\boldsymbol{g}(\boldsymbol{x}^{(k)})$, $\boldsymbol{h}(\boldsymbol{x}^{(k)})$.
3. Verify KKT conditions
$$\begin{aligned} &\|\mathcal{L}_x\| \leq \epsilon \\ &\text{with } \boldsymbol{\mu} \geq 0 \end{aligned} \tag{11.4.6}$$
If these conditions are satisfied, stop. Otherwise, continue at stage 4.
4. Solve the corresponding QP subproblem to obtain a search direction $\boldsymbol{d}^{(k)}$
$$\begin{aligned} &\min_{\boldsymbol{d}^{(k)}} \frac{1}{2} \boldsymbol{d}^{(k)T}\, \boldsymbol{Q}^{(k)}\, \boldsymbol{d}^{(k)} + \boldsymbol{f}_x^T(\boldsymbol{x}^{(k)})\, \boldsymbol{d}^{(k)}\,, \quad \boldsymbol{d}^{(k)} \in \mathbb{R}^n \\ &\text{subject to} \\ &\begin{cases} \boldsymbol{g}(\boldsymbol{x}^{(k)}) + \boldsymbol{g}_x(\boldsymbol{x}^{(k)})\, \boldsymbol{d}^{(k)} = 0 \\ \boldsymbol{h}(\boldsymbol{x}^{(k)}) + \boldsymbol{h}_x(\boldsymbol{x}^{(k)}))\, \boldsymbol{d}^{(k)} \leq 0 \\ \boldsymbol{x}_l - \boldsymbol{x}^{(k)} \leq \boldsymbol{d}^{(k)} \leq \boldsymbol{x}^{(k)} - \boldsymbol{x}_u \end{cases} \end{aligned} \tag{11.4.7}$$
where $\boldsymbol{Q}$ is an approximation of the Hessian $\mathcal{L}_{xx}$ of the Lagrangian. $\boldsymbol{Q}$ can be obtained by a quasi-Newton type approximation of this Hessian matrix, in particular by the BFGS method (Wei et al. 2008) or BFGS modified by Powell (Schittkowski 1981). A first order approximation of the constraints has been achieved. $\boldsymbol{f}_x$ is the gradient of f, and $\boldsymbol{g}_x$ and $\boldsymbol{h}_x$ the Jacobian matrices of $\boldsymbol{g}$ and $\boldsymbol{h}$, respectively, of dimension $(m \times n)$ and $(p \times n)$.
5. Do a search in the direction $\boldsymbol{d}^{(k)}$ by a given method to obtain the displacement intensity $\alpha^{(k)}$.
6. Pose $\boldsymbol{x}^{(k+1)} = \boldsymbol{x}^{(k)} + \alpha^{(k)}\boldsymbol{d}^{(k)}$ and $k = k + 1$.
7. If the tolerance with respect to the stationarity of the solution is fulfilled, stop. Otherwise, return to stage 2.

Many variants of successive quadratic programming (SQP) exist. When the number of equations is large and the number of degrees of freedom is low, the SQP methods based on a reduced Hessian are effective. However, the number of inequality constraints may remain important, and the solution of the QP problem is essential for the improvement of these methods. According to Ternet and Biegler (1999), in the same way as interior point methods allow to solve efficiently and more rapidly linear optimization problems than the simplex, these methods can be applied to the QP subproblem.

Example 11.3 :
Quadratic programming: solution by successive quadratic programming

Using successive quadratic programming, we search the solution of the following problem:

$$\begin{aligned}
&\min\ f = (x_1 + 2\,x_2)^2 + (2\,(x_1+1)^2 + x_2)^2 \\
&\text{subject to} \\
&x_1^2 + x_2^2 - 0.25 = 0 \\
&x_1 + x_2 - 2 \leq 0 \\
&-2 \leq x_i \leq 2
\end{aligned} \tag{11.4.8}$$

Note that this function is not quadratic and the constraints are not linear. The minimum is found for $\boldsymbol{x}^* = (-0.4948, 0.0719)$, which belongs to the circle corresponding to the equality constraint and gives $f^* = 0.4623$.

Among the many existing SQP methods, two different methods have been tried: (a) corresponding to Ternet and Biegler (1999) and (b) corresponding to Albuquerque et al. (1999). Both are interior point methods (Forsgren et al. 2002) supposed to be able to handle large scale problems and close with respect to their concept.

In both cases, the QP problem (11.4.7) can be written under the simplified form (the iteration superscripts k are omitted)

$$\begin{aligned}
&\min_{\boldsymbol{d}}\ \frac{1}{2}\boldsymbol{d}^T\,\boldsymbol{Q}\,\boldsymbol{d} + \boldsymbol{f}_x^T\,\boldsymbol{d}\,, \quad \boldsymbol{d} \in \mathbb{R}^n \\
&\text{subject to} \\
&\begin{cases} \boldsymbol{g} + \boldsymbol{g}_x\,\boldsymbol{d} = 0 \\ \boldsymbol{h} + \boldsymbol{h}_x\,\boldsymbol{d} + \boldsymbol{s} = 0 \end{cases}
\end{aligned} \tag{11.4.9}$$

where $\boldsymbol{s}$ are positive slack variables for the set of inequality constraints $\boldsymbol{h}$ that include as well the inequalities on the bounds $\boldsymbol{x}_l$ and $\boldsymbol{x}_u$ of the domain as the inequalities in the strict sense. $\boldsymbol{Q}$ is a matrix that can be obtained as the Hessian matrix $\boldsymbol{H} = \mathcal{L}_{xx}$ of the Lagrangian $\mathcal{L}$ of the SQP problem equal to

$$\mathcal{L}(\boldsymbol{x}, \boldsymbol{\lambda}, \boldsymbol{\mu}) = f + \boldsymbol{\lambda}^T\,\boldsymbol{g} + \boldsymbol{\mu}^T\,\boldsymbol{h} \tag{11.4.10}$$

with the KKT parameters $\mu_i \geq 0$. The KKT conditions for the QP problem (11.4.9) are then written as

$$\begin{bmatrix} \boldsymbol{f}_x + \boldsymbol{B}\,\boldsymbol{d} + \boldsymbol{g}_x^T\,\boldsymbol{\lambda} + \boldsymbol{h}_x^T\,\boldsymbol{\mu} \\ \boldsymbol{g} + \boldsymbol{g}_x\,\boldsymbol{d} \\ \boldsymbol{h} + \boldsymbol{h}_x\,\boldsymbol{d} + \boldsymbol{s} \\ \boldsymbol{S}\,\boldsymbol{M}\,\boldsymbol{e} \end{bmatrix} = 0 \tag{11.4.11}$$

where $\boldsymbol{S}$ is the diagonal matrix formed by the positive slack variables s_i and $\boldsymbol{M}$ the diagonal matrix formed by KKT parameters μ_i. $\boldsymbol{e}$ is a vector $[1, \ldots 1]^T$ of dimension corresponding to the number of inequalities. The condition $\boldsymbol{S}\,\boldsymbol{M}\,\boldsymbol{e} = 0$ amounts to the KKT complementarity condition $s_i\mu_i = 0,\ \forall\, i$. The system (11.4.11) must be solved with respect to $\boldsymbol{d}, \boldsymbol{\lambda}, \boldsymbol{\mu}, \boldsymbol{s}$. For this solving, an iterative Newton type method can be used simultaneously as a search method to ensure $s_i, \mu_i \geq 0$. Indeed, the method may very little progress after a given number of iterations, and central-path methods are coupled to guarantee $s_i, \mu_i > 0$. A complementarity gap is defined as

$$\mu_{cp} = \frac{\boldsymbol{s}^T\,\boldsymbol{\mu}}{n_{in}} \tag{11.4.12}$$

where n_{in} is the number of inequalities, and $\boldsymbol{s}$ and $\boldsymbol{\mu}$ the vectors formed by s_i and μ_i, respectively.

(a) Interior point method according to Ternet and Biegler (1999).
   The QP subproblem (11.4.11) is solved according to the iterative procedure (at iteration k) of coupled prediction-centering

$$\begin{bmatrix} \boldsymbol{Q} & \boldsymbol{g}_x^T & \boldsymbol{h}_x^T & 0 \\ \boldsymbol{g}_x & 0 & 0 & 0 \\ \boldsymbol{h}_x & 0 & 0 & \boldsymbol{I} \\ 0 & 0 & \boldsymbol{S} & \boldsymbol{M} \end{bmatrix} \begin{bmatrix} \Delta \boldsymbol{d} \\ \Delta \boldsymbol{\lambda} \\ \Delta \boldsymbol{\mu} \\ \Delta \boldsymbol{s} \end{bmatrix} = - \begin{bmatrix} \boldsymbol{f}_x \\ \boldsymbol{g} \\ \boldsymbol{h} \\ \boldsymbol{S}_k \boldsymbol{M}_k \boldsymbol{e} - \sigma_k \mu_{cp,k} \boldsymbol{e} \end{bmatrix} - \begin{bmatrix} \boldsymbol{Q} & \boldsymbol{g}_x^T & \boldsymbol{h}_x^T & 0 \\ \boldsymbol{g}_x & 0 & 0 & 0 \\ \boldsymbol{h}_x & 0 & 0 & \boldsymbol{I} \\ 0 & 0 & \boldsymbol{S} & \boldsymbol{M} \end{bmatrix} \begin{bmatrix} \Delta \boldsymbol{d}_k \\ \Delta \boldsymbol{\lambda}_k \\ \Delta \boldsymbol{\mu}_k \\ \Delta \boldsymbol{s}_k \end{bmatrix} \quad (11.4.13)$$

with $\sigma_k \in [0, 1]$. $\sigma = 0$ corresponds to a Newton's method, and $\sigma = 1$ corresponds to a centering direction (Wright 1997). Having found the displacement direction d_{k+1}, the displacement intensity α is chosen so that the future variables $\mu_{i,k+1}$ and $s_{i,k+1}$ remain positive

$$\alpha = 0.995[\arg\ \max \left\{\alpha \in [0, 1] | \mu_{i,k} + \alpha \Delta \mu_{i,k} \geq 0\,,\ s_{i,k} + \alpha \Delta s_{i,k} \geq 0\right\}] \quad (11.4.14)$$

To test the stationarity of the solution, the following quantity issued from (11.4.11) was compared to a threshold ϵ

$$\|\boldsymbol{f}_x + \boldsymbol{B}\,\boldsymbol{d} + \boldsymbol{g}_x^T\,\boldsymbol{\lambda} + \boldsymbol{h}_x^T\,\boldsymbol{\mu}\| + \|\boldsymbol{g} + \boldsymbol{g}_x\,\boldsymbol{d}\| + \|\boldsymbol{h} + \boldsymbol{h}_x\,\boldsymbol{d} + \boldsymbol{s}\| \quad (11.4.15)$$

In the present case, the matrix $\boldsymbol{Q}$ was simply taken equal to $\mathcal{L}_{xx}$. The initial point was (0.4, 0.4). The parameters were $\sigma = 0.5$. Figure 11.3 shows that the optimal solution is found in 41 iterations for a threshold $\epsilon = 10^{-6}$ following a smooth trajectory that turns around the circle of the equality constraint.

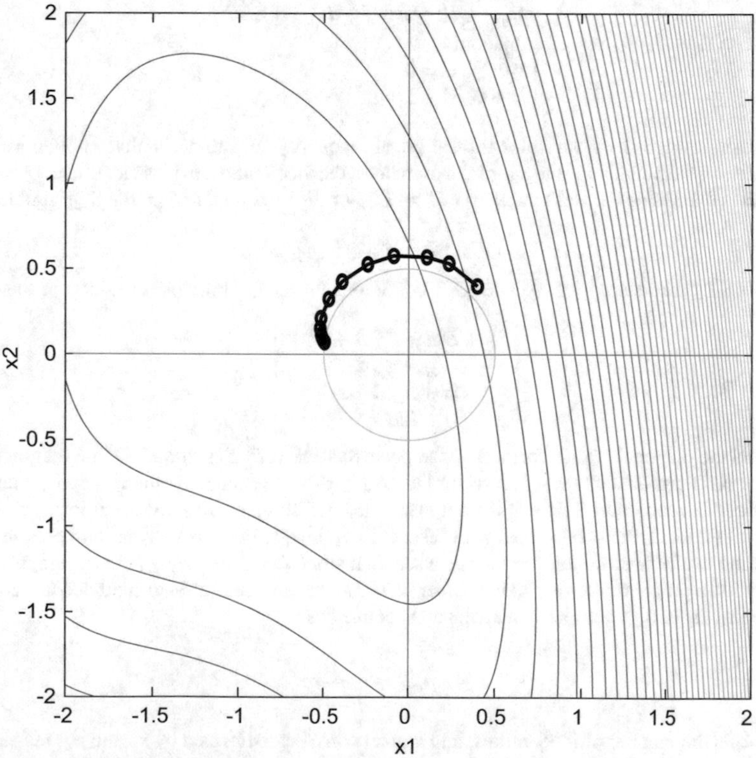

Fig. 11.3 Successive quadratic programming: contours of the function, representation of the equality constraint and iterative points (circles) for problem (11.4.8). Algorithm by Ternet and Biegler (1999)

(b) Interior point method according to Albuquerque et al. (1999).
With respect to Ternet and Biegler (1999), the solving of the QP subproblem is relatively similar except that the system (11.4.11) is solved according to an iterative procedure (at iteration k) containing a main prediction stage by the affine transformation method, then a centering stage, and finally a correction stage (Mehrotra 1992; Tapia et al. 1990). These stages are detailed successively:

- Prediction stage

$$\begin{bmatrix} \boldsymbol{Q} & \boldsymbol{g}_x^T & \boldsymbol{h}_x^T & 0 \\ \boldsymbol{g}_x & 0 & 0 & 0 \\ \boldsymbol{h}_x & 0 & 0 & \boldsymbol{I} \\ 0 & 0 & \boldsymbol{S} & \boldsymbol{M} \end{bmatrix} \begin{bmatrix} \Delta\boldsymbol{d}_p \\ \Delta\boldsymbol{\lambda}_p \\ \Delta\boldsymbol{\mu}_p \\ \Delta\boldsymbol{s}_p \end{bmatrix} = - \begin{bmatrix} \boldsymbol{r}_1 \\ \boldsymbol{r}_2 \\ \boldsymbol{r}_3 \\ \boldsymbol{S}\,\boldsymbol{M}\,\boldsymbol{e} \end{bmatrix} \tag{11.4.16}$$

- Centering stage

$$\begin{bmatrix} \boldsymbol{Q} & \boldsymbol{g}_x^T & \boldsymbol{h}_x^T & 0 \\ \boldsymbol{g}_x & 0 & 0 & 0 \\ \boldsymbol{h}_x & 0 & 0 & \boldsymbol{I} \\ 0 & 0 & \boldsymbol{S} & \boldsymbol{M} \end{bmatrix} \begin{bmatrix} \Delta\boldsymbol{d}_c \\ \Delta\boldsymbol{\lambda}_c \\ \Delta\boldsymbol{\mu}_c \\ \Delta\boldsymbol{s}_c \end{bmatrix} = \begin{bmatrix} 0 \\ 0 \\ 0 \\ \sigma\,\mu\,\boldsymbol{e} \end{bmatrix} \tag{11.4.17}$$

- Correction stage

$$\begin{bmatrix} \boldsymbol{Q} & \boldsymbol{g}_x^T & \boldsymbol{h}_x^T & 0 \\ \boldsymbol{g}_x & 0 & 0 & 0 \\ \boldsymbol{h}_x & 0 & 0 & \boldsymbol{I} \\ 0 & 0 & \boldsymbol{S} & \boldsymbol{M} \end{bmatrix} \begin{bmatrix} \Delta\boldsymbol{d}_{cor} \\ \Delta\boldsymbol{\lambda}_{cor} \\ \Delta\boldsymbol{\mu}_{cor} \\ \Delta\boldsymbol{s}_{cor} \end{bmatrix} = - \begin{bmatrix} 0 \\ 0 \\ 0 \\ \Delta\boldsymbol{S}_p\,\Delta\boldsymbol{\mu}_p\,\boldsymbol{e} \end{bmatrix} \tag{11.4.18}$$

with the residuals $\boldsymbol{r}_i$ equal to

$$\begin{aligned} \boldsymbol{r}_1 &= \boldsymbol{f}_x + \boldsymbol{B}\,\boldsymbol{d} + \boldsymbol{g}_x^T\,\boldsymbol{\lambda} + \boldsymbol{h}_x^T\,\boldsymbol{\mu} \\ \boldsymbol{r}_2 &= \boldsymbol{g} + \boldsymbol{g}_x\,\boldsymbol{d} \\ \boldsymbol{r}_3 &= \boldsymbol{h} + \boldsymbol{h}_x\,\boldsymbol{d} + \boldsymbol{s} \end{aligned} \tag{11.4.19}$$

Two algorithms are possible (Albuquerque et al. 1999) to calculate the displacement direction. Here, only one algorithm is cited, by posing $\boldsymbol{p} = [\boldsymbol{d}, \boldsymbol{\lambda}, \boldsymbol{\mu}, \boldsymbol{s}]$, at iteration i

$$\boldsymbol{p}_{i+1}(\alpha) = \boldsymbol{p}_i + \alpha\Delta\boldsymbol{p}_p + \alpha^2(\Delta\boldsymbol{p}_c + \Delta\boldsymbol{p}_{cor}) \tag{11.4.20}$$

The displacement intensity $\alpha \in [0, 1]$ must verify

$$\begin{aligned} g_1 &= \min_j(\boldsymbol{S}(\alpha)\boldsymbol{V}(\alpha)\boldsymbol{e})_j - \gamma\boldsymbol{s}(\alpha)^T\,\boldsymbol{\mu}(\alpha)/n_{in} \\ g_2 &= \boldsymbol{s}(\alpha)^T\,\boldsymbol{\mu}(\alpha) - (1-\alpha)\boldsymbol{s}_i^T\,\boldsymbol{\mu}_i \end{aligned} \tag{11.4.21}$$

where $\boldsymbol{s}$ and $\boldsymbol{\mu}$ are the vectors of s_j and μ_j, respectively. γ is a positive constant such that

$$\min_j[(\boldsymbol{S}^0\boldsymbol{V}^0\boldsymbol{e}]_j/(\boldsymbol{s}^{0,T}\,\boldsymbol{\mu}^0/n_{in}) \geq \gamma \tag{11.4.22}$$

Figure 11.4 shows that the trajectory followed by the points from the same initial point (0.4, 0.4) is very different from that of Figure 11.3, first deviating from the equality constraint toward the absolute minimum of the function, then coming back very rapidly to this constraint. The parameters were $\sigma = 0.5$. The optimum was found in only 12 iterations with the same threshold $\epsilon = 10^{-6}$, thus more rapidly than in the case a.

11.4.4 Specificities and Difficulties of the SQP Problem

Solving according to a SQP algorithm indeed poses several issues (Nocedal and Wright 2006) that many methods attempt to solve, among which gradient projection, trust region, interior points, and successive quadratic programming. In the physical problems, it is usual that the points generated during the iterations must verify all or partly the constraints, hence the feasible direction methods.

The simplex method to solve the quadratic optimization problems presents the theoretical drawback of the combinatory aspect related to the simplex and its theoretical exponential complexity, even if the simplex can effectively solve large scale problems. For that reason, the interior point methods have known a huge development following the article by Karmarkar (1984). They led to many variants (Forsgren et al. 2002) that had important consequences on nonlinear optimization.

11.4.4.1 Merit Functions

A serious limitation of SQP optimization is that the linearization of the nonlinear constraints can make the QP subproblem impossible, which has generated many propositions. To overcome this dead end, Xue et al. (2009) propose to solve a relaxed auxiliary problem with a penalty function instead of the QP problem (11.4.7).

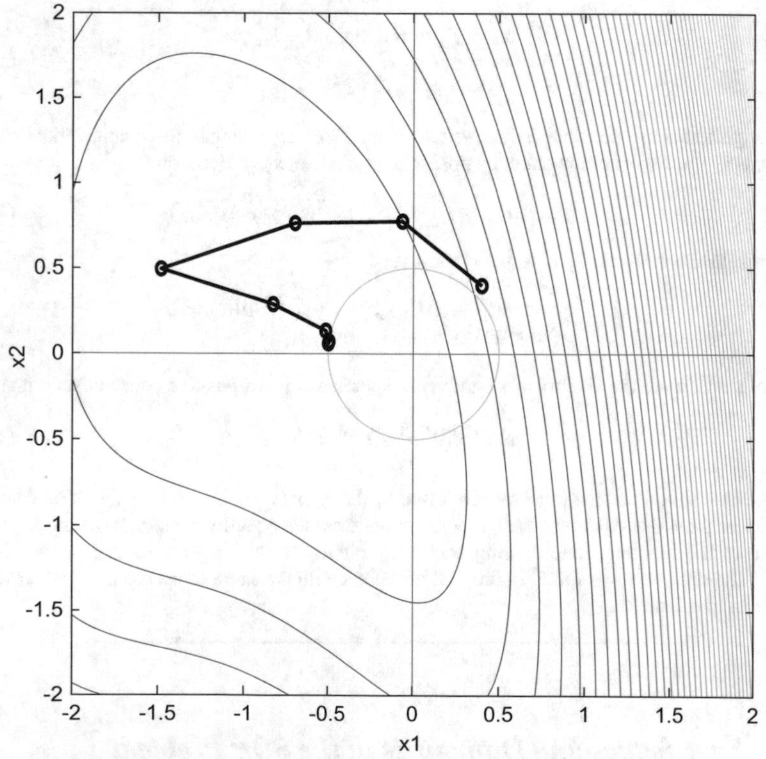

Fig. 11.4 Successive quadratic programming: contours of the function, representation of the equality constraint and iterative points (circles) for problem (11.4.8). Algorithm by Albuquerque et al. (1999)

Supposing that there are only inequality constraints, Powell (1978) proposes to solve the following problem at each iteration:

$$\begin{aligned} &\min_{d_k} \boldsymbol{f}_x^T \boldsymbol{d}_k + \frac{1}{2}\boldsymbol{d}_k^T \boldsymbol{Q}\boldsymbol{d}_k + \frac{1}{2}\delta_k(1-\mu)^2 \\ &\text{subject to} \quad \mu_i h_{i,x}^T(\boldsymbol{x}_k)\boldsymbol{d}_k \leq 0 \qquad \forall i = 1, \ldots, p \\ &\text{with} \quad \mu_i = \begin{cases} 1, & h_i(\boldsymbol{x}_k) < 0 \\ \mu, & h_i(\boldsymbol{x}_k) \geq 0 \end{cases} \quad \text{and} \quad 0 \leq \mu \leq 1, \quad \delta_k > 0 \end{aligned} \tag{11.4.23}$$

the μ_i being penalty parameters. This method, in general efficient, can nevertheless lead to a cycling.

The difficulty of the constrained optimization problems is that the function must be minimized while satisfying the constraints and that these two requirements are often contradictory. After the solution of the QP subproblem that gives a direction $\boldsymbol{d}_k$, we must seek α so that $\boldsymbol{x}_{k+1} = \boldsymbol{x}_k + \alpha \boldsymbol{d}_k$ produces a sufficient decrease of a function. For that reason, merit functions or penalty functions are often used (Schittkowski 1981, 1985) during the search following a given direction like in the penalty function l_1 (Fletcher 1991; Powell 1978)

$$\mathcal{P}_1(\boldsymbol{x}, \boldsymbol{\nu}) = f(\boldsymbol{x}) + \sum_{i \in \mathcal{E}} \nu_i |g_i(\boldsymbol{x})| + \sum_{i \in \mathcal{I}} \nu_i \max\{0, h_i(\boldsymbol{x})\} \tag{11.4.24}$$

where the ν_i are penalty parameters, and $\mathcal{E}$, $\mathcal{I}$, the respective sets of equality and inequality constraints. For sufficiently large values of ν, the local solutions of the nonlinear optimization problem are equivalent to the solutions minimizing the penalty function l_1. To ensure the convergence, the penalty parameters ν_i should verify $\nu_i \geq |\lambda_i|$ with respect to the Lagrange multipliers and $\nu_i \geq \mu_i \geq 0$ with respect to KKT parameters.

The augmented Lagrangian consists of adjoining a penalty function to the Lagrangian. It is also used under many variants as a merit function for example by Schittkowski (1985)

$$\begin{aligned} &\mathcal{L}_A(\boldsymbol{x}, \boldsymbol{\lambda}, \boldsymbol{\mu}) = f(\boldsymbol{x}) - \sum_{i \in \mathcal{E}} \lambda_i g_i(\boldsymbol{x}) + \frac{1}{2} \sum_{i \in \mathcal{E}} \nu_i g_i^2(\boldsymbol{x}) + \sum_{i \in \mathcal{I}} \psi_i(\boldsymbol{x}, \boldsymbol{\mu}) \\ &\text{with } \psi_i(\boldsymbol{x}, \boldsymbol{\mu}) = \begin{cases} \mu_i h_i(\boldsymbol{x}) + \frac{1}{2}\nu_i h_i^2(\boldsymbol{x}) & \text{if } |h_i(\boldsymbol{x})| \leq \frac{\mu_i}{\nu_i} \\ -\frac{1}{2}\frac{\mu_i^2}{\nu_i} & \text{otherwise} \end{cases} \end{aligned} \tag{11.4.25}$$

Note that, in the NLPQL code, Schittkowski (1985) uses the merit function (11.4.24) and the augmented Lagrangian (11.4.25). This augmented Lagrangian method can avoid the Maratos effect (Nocedal and Wright 2006) often cited as an issue. That effect occurs when a step that would be beneficial with respect to the sought solution is refused, either because the objective function increases, or because a constraint is violated. This effect slows down or blocks the convergence. Nonmonotonous searches possibly accepting steps that sometimes increase the merit function may also be suitable (Su and Yu 2009; Xue et al. 2009). A second order correction of the step may also prevent that issue. To obtain a set of points satisfying the constraints, Zhu and Jian

(2009), Zhu and Zhang (2004) and Zhu et al. (2010) propose different techniques including solving three QP subproblems or two subproblems and a linear least squares problem. According to Zhu et al. (2010), given an initial point $\boldsymbol{x}_k$, a descent direction $\boldsymbol{d}_k$ is found by solving a first QP problem, Maratos correction $\tilde{\boldsymbol{d}}_k$ is done by solving a second QP problem, and then a search along the arc $\boldsymbol{x}_k + \alpha \boldsymbol{d}_k + \alpha^2 \tilde{\boldsymbol{d}}_k$ ends the problem for iteration k, thus ensuring the global and superlinear convergence.

It is noticeable that a penalty function such as l_1 may fail (Fletcher 1991). For that reason, other methods can be coupled inside a given code such as the augmented Lagrangian (Schittkowski 1985) or the trust region used in Levenberg–Marquardt method.

11.4.4.2 Filter Methods

The drawback of merit function methods can be the difficulty to estimate the penalty parameter, for which filter methods propose solutions (Fletcher and Leyffer 2002; Shen et al. 2010; Wächter and Biegler 2005) based on a trust region. According to Fletcher and Leyffer (2002), a filter admits a step when this latter decreases either the objective function, or the constraint violation function. The construction of the filter consists of the following procedure:

- Initially, the 0th order filter is empty.

- A violation function is defined, for example (Shen et al. 2010)

$$V(\boldsymbol{x}) = \sum_{i \in \mathcal{E}} |g_i(\boldsymbol{x})| + \sum_{i \in \mathcal{I}} \max\{0, h_i(\boldsymbol{x})\} \tag{11.4.26}$$

  The filter is formed by pairs $\{f(\boldsymbol{x}), V(\boldsymbol{x})\}$.
- Fletcher and Leyffer (2002) use the domination concept coming from multi-objective optimization: a pair $\{f^{(k)}, V^{(k)}\}$ dominates another pair $\{f^{(l)}, V^{(l)}\}$ if and only if, at the same time, we verify $f^{(k)} < f^{(l)}$ and $V^{(k)} < V^{(l)}$.
- A filter is a list of pairs such that no pair dominates the other ones.
- A point $\{f^{(k)}, V^{(k)}\}$ is acceptable to the filter if it is not dominated by any point of the filter.
- For Fletcher and Leyffer (2002), a point $\hat{\boldsymbol{x}} = \boldsymbol{x}^{(k)} + \boldsymbol{d}$ results in the solution of the usual QP subproblem to which $\|d\|_\infty \le \rho$ is added. Two possibilities exist:

(a) If the point $\boldsymbol{x}^{(k+1)} = \hat{\boldsymbol{x}}$ is accepted, the pair $\{f^{(k+1)}, V^{(k+1)}\}$ is added to the filter, the other points dominated by $\{f^{(k+1)}, V^{(k+1)}\}$ are removed from the filter, and it is possible to increase the radius ρ of the trust region.
(b) If the point $\hat{\boldsymbol{x}}$ is not accepted in the filter, the step is rejected and the radius ρ of the trust region is decreased.

- If the point $\boldsymbol{x}^{(k)}$ is not feasible, the step reduction leads to an infeasible QP subproblem. To avoid that case, a restoration phase is necessary to come back into a feasible region of the nonlinear problem.

- For Shen et al. (2010), a point $\hat{\boldsymbol{x}} = \boldsymbol{x}^{(k)} + \boldsymbol{d}$ is called feasible for $\boldsymbol{x}_l$ of the filter if

$$\begin{aligned} & V(\hat{\boldsymbol{x}}) - V(\boldsymbol{x}_l) \leq -\gamma_1 V(\hat{\boldsymbol{x}}) \\ \text{or} \quad & f(\hat{\boldsymbol{x}}) - f(\boldsymbol{x}_l) < -\gamma_2 V(\hat{\boldsymbol{x}}) \end{aligned} \tag{11.4.27}$$

with the constants $\gamma_i \in [0, \frac{1}{2}]$. It follows that when a pair $\{f, V\}$ is accepted in the filter, then the point $\hat{\boldsymbol{x}}$ is acceptable for all the other pairs of the filter.

The main idea of the filter is to conceive a criterion of step acceptation or rejection for a SQP method. Fletcher and Leyffer (2002) note that the basic algorithm is convenient in most cases, but it can be improved by means of heuristics and it is possible to incorporate a second order step correction. The filter allows some degree of non-monotonicity. Other conditions are added to ensure the convergence to the optimum.

11.4.4.3 Reduced Hessian Methods

When the problems are of large scale, solving equations such as (11.4.11) is made easier by employing reduced space techniques.

Supposing that the nonlinear optimization problem is written as

$$\begin{aligned} & \min_{\boldsymbol{x}} f(\boldsymbol{x}) \\ & \text{subject to} \quad g_i(\boldsymbol{x}) = 0\,, \quad i = 1, \dots, m \end{aligned} \tag{11.4.28}$$

which can always be done by adding slack variables to the inequality constraints, at iteration k, the SQP method amounts to the following QP subproblem:

$$\begin{aligned} & \min_{\boldsymbol{d}} \frac{1}{2} \boldsymbol{d}^{(k)T} \boldsymbol{Q}^{(k)} \boldsymbol{d}^{(k)} + \boldsymbol{f}_x^T(\boldsymbol{x}^{(k)}) \boldsymbol{d}^{(k)} \\ & \text{subject to} \quad \boldsymbol{g}(\boldsymbol{x}^{(k)}) + \boldsymbol{g}_x(\boldsymbol{x}^{(k)}) \boldsymbol{d}^{(k)} = 0 \end{aligned} \tag{11.4.29}$$

where the constraints are consistent, and thus $\boldsymbol{g}_x$ is of full rank. The KKT conditions yield

$$\begin{bmatrix} \boldsymbol{Q} & \boldsymbol{g}_x^T \\ \boldsymbol{g}_x & 0 \end{bmatrix} \begin{bmatrix} \boldsymbol{d} \\ \boldsymbol{\lambda} \end{bmatrix} = - \begin{bmatrix} \boldsymbol{f}_x \\ \boldsymbol{g} \end{bmatrix} \tag{11.4.30}$$

and the left-hand side matrix of (11.4.30) is called KKT matrix. The search space is partitioned into two subspaces such that

$$\begin{aligned} & \boldsymbol{Z} \in \mathbb{R}^{n \times (n-m)} \quad \text{such that} \quad g_x(\boldsymbol{d}^{(k)}) \boldsymbol{Z}^{(k)} = 0 \\ & \boldsymbol{Y} \in \mathbb{R}^{n \times m} \qquad \text{such that} \quad [YZ] \text{ is nonsingular} \end{aligned} \tag{11.4.31}$$

the columns of $\boldsymbol{Z}$ form a basis for the kernel of g_x, and $\boldsymbol{Z}$ is of full rank (Nocedal and Wright 2006). $\boldsymbol{Y}$ and $\boldsymbol{Z}$ thus generate all the space, and the search direction can be decomposed according to two subspaces

$$\boldsymbol{d}^{(k)} = \boldsymbol{Y} \boldsymbol{d}_Y + \boldsymbol{Z} \boldsymbol{d}_Z \tag{11.4.32}$$

If the approximated reduced Hessian matrix $\boldsymbol{Z}^T\boldsymbol{Q}\boldsymbol{Z}$ is definite positive, the KKT matrix is nonsingular; hence the unique solution $(\boldsymbol{d}, \boldsymbol{\lambda})$. Moreover, if $\boldsymbol{g}_x$ is of full rank, $\boldsymbol{d}$ is a unique global solution. Several choices are possible for $\boldsymbol{Y}$ and $\boldsymbol{Z}$ (Gill and Leonard 2001; Schmid and Biegler 1994). From Equations (11.4.30) and (11.4.31), it comes

$$\boldsymbol{g}_x\boldsymbol{d} = -\boldsymbol{g} \Rightarrow \boldsymbol{g}_x[\boldsymbol{Y}\boldsymbol{d}_Y + \boldsymbol{Z}\boldsymbol{d}_Z] = \boldsymbol{g}_x\boldsymbol{Y}\boldsymbol{d}_Y = -\boldsymbol{g} \Rightarrow \boldsymbol{d}_Y = -(\boldsymbol{g}_x\boldsymbol{Y})^{-1}\boldsymbol{g} \tag{11.4.33}$$

Now that $\boldsymbol{d}_Y$ is supposed to be known as a linear problem, the QP subproblem can be rewritten under a reduced form only with respect to $\boldsymbol{d}_Z$

$$\min_{\boldsymbol{d}_Z} \frac{1}{2}\boldsymbol{d}_Z^T\boldsymbol{Z}^T\boldsymbol{Q}\boldsymbol{Z}\boldsymbol{d}_Z + (\boldsymbol{Z}^T\boldsymbol{f}_x + \boldsymbol{Z}^T\boldsymbol{Q}\boldsymbol{Y}\boldsymbol{d}_Y)^T\boldsymbol{d}_Z \tag{11.4.34}$$

which in general is a problem of smaller scale than (11.4.29). The reduced Hessian $\boldsymbol{Z}^T\boldsymbol{Q}\boldsymbol{Z}$ is in general definite positive and can be approximated by a quasi-Newton method such as BFGS.

Different techniques of system reduction exist. Some of them use the fact that the approximated Hessian or its inverse can be written as the sum of a diagonal matrix and a given number of matrices of rank equal to 1 (Albuquerque et al. 1999; Gill and Leonard 2001). In the following, two different techniques are shown:

- Equation (11.4.11) under the form of an iterative algorithm (Albuquerque et al. 1999) is rewritten under the form

$$\left[\begin{array}{cccc|c} \boldsymbol{Q} & \boldsymbol{g}_x^T & \boldsymbol{h}_x^T & 0 & 0 \\ \boldsymbol{g}_x & 0 & 0 & 0 & 0 \\ \boldsymbol{h}_x & 0 & 0 & \boldsymbol{I} & 0 \\ 0 & 0 & \boldsymbol{S} & \boldsymbol{M} & -\boldsymbol{e} \end{array}\right] \begin{bmatrix} \Delta\boldsymbol{d} \\ \Delta\boldsymbol{\lambda} \\ \Delta\boldsymbol{\mu} \\ \Delta\boldsymbol{s} \\ \Delta q \end{bmatrix} = -\begin{bmatrix} \boldsymbol{r}_1 \\ \boldsymbol{r}_2 \\ \boldsymbol{r}_3 \\ \boldsymbol{S}\boldsymbol{M}\boldsymbol{e} - q_k\boldsymbol{e} \end{bmatrix} \tag{11.4.35}$$

  by addition of the independent variable q with respect to the $(n-1)$ other dependent variables composing the vector $\boldsymbol{w} = (\boldsymbol{d}, \boldsymbol{\lambda}, \boldsymbol{\mu}, \boldsymbol{s}, q)$. It is then possible to do a partition of $\boldsymbol{A}$

$$\boldsymbol{A} = \left[\begin{array}{cccc|c} \boldsymbol{Q} & \boldsymbol{g}_x^T & \boldsymbol{h}_x^T & 0 & 0 \\ \boldsymbol{g}_x & 0 & 0 & 0 & 0 \\ \boldsymbol{h}_x & 0 & 0 & \boldsymbol{I} & 0 \\ 0 & 0 & \boldsymbol{S} & \boldsymbol{M} & -\boldsymbol{e} \end{array}\right] = [\boldsymbol{A}_1 \mid \boldsymbol{A}_2] \tag{11.4.36}$$

  where $\boldsymbol{A}_2$ is simply the last column vector of $\boldsymbol{A}$. When the QP problem has a unique solution, $\boldsymbol{A}_1$ is nonsingular. The basis matrices of the kernel of $\boldsymbol{A}$ and of the rest of the space are $\boldsymbol{Z}$ and $\boldsymbol{Y}$, such that

$$\boldsymbol{Z} = \begin{bmatrix} -\boldsymbol{A}_1^{-1}\boldsymbol{A}_2 \\ 1 \end{bmatrix}, \quad \boldsymbol{Y} = \begin{bmatrix} \boldsymbol{I}_{n-1} \\ 0 \end{bmatrix} \tag{11.4.37}$$

  It can be verified that: $\boldsymbol{A}\boldsymbol{Z} = 0$. The search direction is expressed under the form

$$\Delta\boldsymbol{w} = \boldsymbol{Y}\boldsymbol{p}_Y + \boldsymbol{Z}\boldsymbol{p}_Z \tag{11.4.38}$$

with $\boldsymbol{p}_Y \in \mathbb{R}^{n-1}$ and $\boldsymbol{p}_Z \in \mathbb{R}$. The step $\boldsymbol{p}_Y$ is found by solving the linear system

$$\boldsymbol{A}\boldsymbol{Y}\boldsymbol{p}_Y = -\boldsymbol{r} \tag{11.4.39}$$

where $\boldsymbol{r}$ if the right-hand vector of Equation (11.4.35).
To calculate $\boldsymbol{p}_Z$, Albuquerque et al. (1999) choose an objective function

$$f = q^{(2-\sigma)/(1-\sigma)} \tag{11.4.40}$$

hence

$$\boldsymbol{p}_Z = -(\boldsymbol{Z}^T\boldsymbol{W}\boldsymbol{Z})^{-1}[\boldsymbol{Z}^T\boldsymbol{f}_w + \boldsymbol{Z}^T\boldsymbol{W}\boldsymbol{Y}\boldsymbol{p}_Y] \tag{11.4.41}$$

where $\boldsymbol{W}$ is the Hessian of the Lagrangian

$$\boldsymbol{W} = \boldsymbol{f}_{ww} + \sum_j \lambda_j \boldsymbol{r}_{2jww} + \sum_j \mu_j \boldsymbol{r}_{3jww} \tag{11.4.42}$$

- Ternet and Biegler (1999) propose to solve the linear system (11.4.13) written under the form

$$\begin{bmatrix} \boldsymbol{Q} & \boldsymbol{g}_x^T & \boldsymbol{h}_x^T & 0 \\ \boldsymbol{g}_x & 0 & 0 & 0 \\ \boldsymbol{h}_x & 0 & 0 & \boldsymbol{I} \\ 0 & 0 & \boldsymbol{S} & \boldsymbol{M} \end{bmatrix} \begin{bmatrix} \Delta\boldsymbol{d} \\ \Delta\boldsymbol{\lambda} \\ \Delta\boldsymbol{\mu} \\ \Delta\boldsymbol{s} \end{bmatrix} = - \begin{bmatrix} \boldsymbol{r}_1 \\ \boldsymbol{r}_2 \\ \boldsymbol{r}_3 \\ \boldsymbol{r}_4 \end{bmatrix} \tag{11.4.43}$$

- Noticing that $\boldsymbol{M}$ is diagonal and supposing $\mu_i > 0$, by eliminating the slack variables $\Delta\boldsymbol{s}$ with respect to $\Delta\boldsymbol{\mu}$ by means of the last row, we get

$$\Delta\boldsymbol{s} = -\boldsymbol{M}^{-1}[\boldsymbol{S}\Delta\boldsymbol{\mu} + \boldsymbol{r}_4] \tag{11.4.44}$$

and the new system

$$\begin{bmatrix} \bar{\boldsymbol{Q}} & \boldsymbol{g}_x^T & \boldsymbol{h}_x^T \\ \boldsymbol{g}_x & 0 & 0 \\ \boldsymbol{h}_x & 0 & -\boldsymbol{M}^{-1}\boldsymbol{S} \end{bmatrix} \begin{bmatrix} \Delta\boldsymbol{d} \\ \Delta\boldsymbol{\lambda} \\ \Delta\boldsymbol{\mu} \end{bmatrix} = - \begin{bmatrix} \boldsymbol{r}_1 \\ \boldsymbol{r}_2 \\ \boldsymbol{r}_3 - \boldsymbol{M}^{-1}\boldsymbol{r}_4 \end{bmatrix} \tag{11.4.45}$$

- By noticing that $\boldsymbol{M}^{-1}\boldsymbol{S}$ is diagonal and supposing $s_i > 0$, the KKT parameters $\Delta\boldsymbol{\mu}$ are eliminated with respect to $\Delta\boldsymbol{d}$ by means of the last row

$$\Delta\boldsymbol{\mu} = \boldsymbol{S}^{-1}\boldsymbol{M}[\boldsymbol{h}_x\Delta\boldsymbol{d} + (\boldsymbol{r}_3 - \boldsymbol{M}^{-1}\boldsymbol{r}_4)] \tag{11.4.46}$$

and the new system results

$$\begin{bmatrix} \bar{\boldsymbol{Q}} + \boldsymbol{h}_x^T(\boldsymbol{S}^{-1}\boldsymbol{M})\boldsymbol{h}_x & \boldsymbol{g}_x^T \\ \boldsymbol{g}_x & 0 \end{bmatrix} \begin{bmatrix} \Delta\boldsymbol{d} \\ \Delta\boldsymbol{\lambda} \end{bmatrix} = - \begin{bmatrix} \boldsymbol{r}_1 + \boldsymbol{h}_x^T(\boldsymbol{S}^{-1}\boldsymbol{M})(\boldsymbol{r}_3 - \boldsymbol{M}^{-1}\boldsymbol{r}_4) \\ \boldsymbol{r}_2 \end{bmatrix} \tag{11.4.47}$$

The reduced space form is then used (Ternet and Biegler 1999) to partition the Jacobian matrix $\boldsymbol{g}_x$ as in Equation (11.4.31) and reduce the system (11.4.47).

11.5 Discussion and Conclusion

Through this chapter, three main types of methods have been used:

- Solving a quadratic optimization problem by a simplex type method
- Solving a quadratic optimization problem by a logarithmic barrier method
- Solving a nonlinear optimization problem by successive quadratic programming

In some types of problems, such as linear model predictive control, where a quadratic criterion subject to linear constraints is minimized, quadratic programming is necessary and sufficient to get the control vector to apply at each sampling period.

Successive quadratic programming (SQP) is the most effective optimization method as it allows to solve large scale nonlinear optimization problems in the presence of nonlinear equality and inequality constraints. However, successive quadratic programming is a general principle and the available codes make use of different submethods or couple these latter to improve the global efficiency.

Moré and Wright (1994) summarize the available optimization software and the methods used. Even if this article is old, much information is still actual and interesting for the user. As for all complex large scale problems, the use of software developed by specialists is practically indispensable (Andrei 2017; Wikipedia).

Optimization is not limited to constrained nonlinear optimization problems with respect to real variables, as specialized methods exist for problems making use of integer variables, and often at the same time mixed integer variables (MILP and MINLP, that is Mixed Integer Linear Programming and Mixed Integer Nonlinear Programming, respectively). For these problems beyond the scope of the present book, it is recommended to refer to specialized books and articles (Floudas 2000; Grossmann 2002; Kronqvist et al. 2019; Lee and Leyffer 2012).

11.6 Exercise Set

Preamble: Many exercises can be solved using the typical technique of quadratic programming. However, students can use existing codes, but in general a mathematical preparation before using a code is necessary.

Exercise 11.6.1 (Medium)

Find the minimum of the function

$$f(\boldsymbol{x}) = x_1^2 + 0.25\,x_2^2 - 2\,x_1\,x_2 + 2\,x_1 + 3\,x_2 + 5 \tag{11.6.1}$$

subject to the constraints

$$\begin{aligned} &2\,x_1 + 3\,x_2 \leq 3 \\ &3\,x_1 + 4\,x_2 \geq 2 \\ &x_i \geq 0 \quad \forall\, i \end{aligned} \tag{11.6.2}$$

Draw the acceptable domain. Solve this problem by quadratic programming.

Exercise 11.6.2 (Medium)

Find the minimum of the function

$$f(\boldsymbol{x}) = 3\,x_1^2 - 2\,x_1 + 4\,x_1\,x_2 - x_2 + 2\,x_2^2 + 1 \tag{11.6.3}$$

subject to the constraints

$$\begin{aligned} x_1 + 3\,x_2 &\geq 0 \\ -10 \leq x_i &\leq 10 \quad \forall\, i \end{aligned} \tag{11.6.4}$$

Solve this problem by quadratic programming.

Exercise 11.6.3 (Medium)
Find the maximum of the function

$$f(\boldsymbol{x}) = -2x_1^2 - x_2^2 - 3x_3^2 \tag{11.6.5}$$

subject to the constraints

$$\begin{aligned} x_1 + 2x_2 + x_3 - 1 &= 0 \\ 4x_1 + 3x_2 + 2x_3 - 2 &= 0 \\ -5 \leq x_i &\leq 5 \quad \forall\, i \end{aligned} \tag{11.6.6}$$

Solve this problem by quadratic programming.

Exercise 11.6.4 (Medium)
Find the maximum of the function

$$f(\boldsymbol{x}) = -2x_1^2 - x_2^2 - 3x_3^2 \tag{11.6.7}$$

subject to the constraints

$$\begin{aligned} x_1 + 2x_2 + x_3 - 1 &= 0 \\ -5 \leq x_i &\leq 5 \quad \forall\, i \end{aligned} \tag{11.6.8}$$

Solve this problem by quadratic programming.

Exercise 11.6.5 (Medium)
Find the minimum of the function

$$f(\boldsymbol{x}) = -x_1 x_2 \tag{11.6.9}$$

subject to the constraints

$$\begin{aligned} 2 \geq x_1 + x_2 &\geq 0 \\ 2 \geq x_1 - x_2 &\geq -2 \\ -4 \leq x_i &\leq 4 \quad \forall\, i \end{aligned} \tag{11.6.10}$$

Solve this problem by quadratic programming.

Exercise 11.6.6 (Medium)
Find the minimum of the function

$$f(\boldsymbol{x}) = x_1^2 + 2\,x_2^2 - x_3^2 \tag{11.6.11}$$

subject to the constraints

$$\begin{aligned} 2\,x_1 + x_2 + x_3 &= 1 \\ |x_1 + x_2| &\leq 0.4 \end{aligned} \tag{11.6.12}$$

Solve this problem by quadratic programming.

Exercise 11.6.7 (Difficult)
Find the minimum of the function

$$f(\boldsymbol{x}) = x_1^3 - 3\,x_1\,x_2 \tag{11.6.13}$$

subject to the constraints

$$\begin{aligned} &5\,x_1 + 2\,x_2 \geq 20 \\ &-2\,x_1 + x_2 \leq 5 \\ &0 \leq x_i \leq 50 \quad \forall\, i \end{aligned} \tag{11.6.14}$$

Solve this problem by successive quadratic programming.

Exercise 11.6.8 (Difficult)
Find the maximum of the function

$$f(\boldsymbol{x}) = (x_1 - 2)^2 + (x_2 - 5)^2 - 4x_1^2 x_2 \tag{11.6.15}$$

subject to the constraints

$$\begin{aligned} &x_1^2 + 2\,x_2^2 \leq 6 \\ &-10 \leq x_i \leq 10 \quad \forall\, i \end{aligned} \tag{11.6.16}$$

Solve this problem by successive quadratic programming.

Exercise 11.6.9 (Difficult)
Find the minimum of the function

$$f(\boldsymbol{x}) = 2\,x_1^4 - x_1^2\,x_2 + x_2^2 + 3\,x_1 - 7 \tag{11.6.17}$$

subject to the constraints

$$\begin{aligned} &x_1^3 + 2\,x_2^3 \geq 10 \\ &3\,x_1^2 + 2\,x_2^2 \leq 10 \end{aligned} \tag{11.6.18}$$

Solve this problem by successive quadratic programming.

Exercise 11.6.10 (Difficult)
Find the minimum of the function

$$f(\boldsymbol{x}) = x_1^3 - 2x_1 + 7 + 5x_2^2 + 3x_1x_2^2 \tag{11.6.19}$$

subject to the constraints

$$\begin{aligned} &x_1 + x_2 \geq 0 \\ &-10 \leq x_i \leq 10 \quad \forall\, i \end{aligned} \tag{11.6.20}$$

Solve this problem by successive quadratic programming.

Exercise 11.6.11 (Difficult)
Find the minimum of the function

$$f(\boldsymbol{x}) = \frac{2}{x_1\,x_2} + \frac{3}{x_2} + 4\,x_1\,x_2^2 + 6\,x_2^2 \tag{11.6.21}$$

subject to the constraints

$$0 \leq x_i \leq 3 \quad \forall\, i \tag{11.6.22}$$

Solve this problem by successive quadratic programming.

Exercise 11.6.12 (Difficult)
Find the minimum of the function

$$f(\boldsymbol{x}) = (x_1 - 2)^4 + (x_1 - 2\,x_2)^2 \tag{11.6.23}$$

subject to the constraints

$$\begin{aligned} 2\,x_1^2 + x_2^4 &\geq 1 \\ -10 \leq x_i &\leq 10 \quad \forall\, i \end{aligned} \tag{11.6.24}$$

Solve this problem by successive quadratic programming.

Exercise 11.6.13 (Difficult)
Find the minimum of the function

$$f(\boldsymbol{x}) = x_1^2 + 2\,x_2^2 + \exp(x_1^2 + x_2^2) \tag{11.6.25}$$

subject to the constraints

$$\begin{aligned} x_1 + x_2 &\geq 10 \\ -10 \leq x_i &\leq 10 \quad \forall\, i \end{aligned} \tag{11.6.26}$$

Solve this problem by successive quadratic programming.

Exercise 11.6.14 (Difficult)
Find the maximum of the function

$$f(\boldsymbol{x}) = (x_1 - 2)^2 + (x_2 - 5)^2 - 4\,x_1^2\,x_2 \tag{11.6.27}$$

subject to the constraints

$$-10 \leq x_i \leq 10 \quad \forall\, i \tag{11.6.28}$$

Solve this problem by successive quadratic programming.

Exercise 11.6.15 (Difficult)
Find the global minimum of the function

$$f(\boldsymbol{x}) = 2\,x_1^2 + x_2^2 + 2\,x_3^2 - x_1\,x_2 + x_1\,x_3 + x_1 + 2\,x_3 \tag{11.6.29}$$

subject to the constraints

$$\begin{aligned} & x_1^2 + x_2^2 - x_3 \leq 0 \\ & x_1 + x_2 + 2\,x_3 \leq 16 \\ & x_1 + x_2 \geq 3 \\ & x_i \geq 0 \qquad \forall\, i \end{aligned} \tag{11.6.30}$$

Solve this problem by successive quadratic programming.

Exercise 11.6.16 (Difficult)
The simplex method can be applied to linear programming. It has been extended to quadratic programming that consists of solving the Karush–Kuhn–Tucker conditions

for the following problem:

$$\min \ f(\boldsymbol{x}) = \frac{1}{2}\boldsymbol{x}^T G \boldsymbol{x} + \boldsymbol{g}^T \boldsymbol{x}\,, \quad \boldsymbol{x} \in \mathcal{R}^n \tag{11.6.31}$$

subject to the constraints

$$\begin{aligned} &\boldsymbol{A}^T \boldsymbol{x} \geq \boldsymbol{b} \\ &\boldsymbol{x} \geq 0 \end{aligned} \tag{11.6.32}$$

Karush–Kuhn–Tucker parameters $\boldsymbol{\lambda}$ and $\boldsymbol{\mu}$ are introduced for inequality and equality constraints, respectively.

1. Demonstrate that it amounts to equations of the form

$$\begin{aligned} &\boldsymbol{w} - \boldsymbol{M}\,\boldsymbol{z} = \boldsymbol{q} \\ &\boldsymbol{w}^T\,\boldsymbol{z} = 0 \\ &\boldsymbol{w} \geq 0\,, \quad \boldsymbol{z} \geq 0\,, \end{aligned} \tag{11.6.33}$$

   by introducing these new variables that you will define with respect to the previous variables.
2. Verify the question 1 on the following example:

$$\min \ f(\boldsymbol{x}) = x_1^2 + x_1\,x_2 + 6x_2^2 - 2\,x_1 + 8\,x_2 \tag{11.6.34}$$

   subject to the constraints

$$\begin{aligned} &x_1 + 2\,x_2 \leq 4 \\ &2\,x_1 + x_2 \leq 5 \\ &x_1,\, x_2 \geq 0 \end{aligned} \tag{11.6.35}$$

Exercise 11.6.17 (Medium)

The following problem was solved in Exercise 8.11.24 by means of the fundamental techniques of Lagrange multipliers.

Using the technique of quadratic programming, determine the minimum of the function

$$f(\boldsymbol{x}) = \frac{1}{2}\,\boldsymbol{x}^T \begin{bmatrix} 3 & -1 & 0 \\ -1 & 2 & -1 \\ 0 & -1 & 1 \end{bmatrix} \boldsymbol{x} + \begin{bmatrix} 1 \\ 1 \\ 1 \end{bmatrix}^T \boldsymbol{x} \tag{11.6.36}$$

subject to the constraint

$$x_1 + 2\,x_2 + x_3 = 4 \tag{11.6.37}$$

Exercise 11.6.18 (Difficult)

The following problem was examined by means of the fundamental techniques of Lagrange multipliers in Exercise 8.11.25.

It is proposed to solve the same problem by quadratic programming.

Minimize the function

$$h(\boldsymbol{x}) = \frac{1}{2}\boldsymbol{x}^T \boldsymbol{A}\boldsymbol{x} + \boldsymbol{b}^T \boldsymbol{x}\,, \quad \boldsymbol{x} \in \mathbb{R}^n \tag{11.6.38}$$

subject to the constraints

$$\begin{aligned} \boldsymbol{C}\boldsymbol{x} &\geq \boldsymbol{d} \\ \boldsymbol{E}\boldsymbol{x} &= \boldsymbol{f} \end{aligned} \tag{11.6.39}$$

Application:

$$\begin{aligned} \boldsymbol{A} &= \begin{bmatrix} 6 & -2 & 0 \\ -2 & 10 & -1 \\ 0 & -1 & 2 \end{bmatrix}, \quad \boldsymbol{b} = \begin{bmatrix} -2 \\ 30 \\ 0 \end{bmatrix} \\ \boldsymbol{C} &= \begin{bmatrix} 2 & 3 & 1 \end{bmatrix}, \quad \boldsymbol{d} = \begin{bmatrix} 6 \end{bmatrix} \\ \boldsymbol{E} &= \begin{bmatrix} 1 & 2 & 1 \end{bmatrix}, \quad \boldsymbol{f} = \begin{bmatrix} 4 \end{bmatrix} \end{aligned} \tag{11.6.40}$$

References

J. Albuquerque, V. Gopal, G. Staus, L. T. Biegler, and B. E. Ydstie. Interior point SQP strategies for large-scale, structured process optimization problems. *Comp. Chem. Engng*, 23:543–554, 1999.

N. Andrei. *Continuous Nonlinear Optimization for Engineering Applications in GAMS Technology*. Springer, Cham, Switzerland, 2017.

M. Avriel. *Nonlinear programming: Analysis and methods*. Prentice-Hall, Englewood Cliffs, New Jersey, 1976.

M. Bartholomew-Biggs. *Nonlinear Optimization with Engineering Applications*. Springer, New York, 2010.

M. A. Bazaraa, H. D. Sherali, and C. M. Shetty. *Nonlinear programming - Theory and algorithms*. Wiley, Hoboken, 2nd edition, 1993.

D. P. Bertsekas. *Convex Optimization Theory*. Athena Scientific, Belmont, 2009.

M. J. Best. *Quadratic Programming with Computer Programs*. CRC Press, Boca Raton, 2016.

L. T. Biegler. *Nonlinear Programming - Concepts, Algorithms and Applications to Chemical Processes*. SIAM, Philadelphia, 2010.

J. F. Bonnans, J. C. Gilbert, C. Lemaréchal, and C. A. Sagastizabal. *Numerical optimization*. Springer, Berlin, 2003.

J. Borwein and A. S. Lewis. *Convex Analysis and Nonlinear Optimization: Theory and Examples*. Springer, New York, 2nd edition, 2006.

J. Bouzitat. On Wolfe's method and Dantzig's method for convex quadratic programming. *RAIRO Recherche Opérationnelle*, 13(2):151–184, 1979.

S. Boyd and L. Vandenberghe. *Convex optimization*. Cambridge University Press, New York, 2009.

J. P. Corriou. *Process control - Theory and applications*. Springer, London, 2nd edition, 2018.

Z. Dostal. *Optimal Quadratic Programming Algorithms: with Applications to Variational Inequalities*. Springer, New York, 2009.

H. A. Eiselt and C. L. Sandblom. *Nonlinear Optimization*. Springer, Cham, 2019.

R. Fletcher. *Practical Methods of Optimization*. Wiley, Chichester, 1991.

R. Fletcher and S. Leyffer. Nonlinear programming without a penalty function. *Mathematical Programming*, 91(2):239–269, 2002.

C. A. Floudas. *Deterministic Global Optimization - Theory, Methods and Applications*. Springer, Dordrecht, 2000.

A. Forsgren, P. E. Gill, and M. H. Wright. Interior methods for nonlinear optimization. *SIAM Review*, 44(4):525–597, 2002.

P. E. Gill and M. W. Leonard. Reduced-Hessian quasi-Newton methods for unconstrained optimization. *SIAM Journal on Optimization*, 12:209–237, 2001.

I. E. Grossmann. Review of nonlinear mixed-integer and disjunctive programming techniques. *Optimization and Engineering*, 3:227–252, 2002.

D. M. Himmelblau. *Applied nonlinear programming*. McGraw-Hill, 1972.

A. F. Izmailov and M. V. Solodov. *Newton-Type Methods for Optimization and Variational Problems*. Springer, Cham, Switzerland, 2014.

B. Jansen. *Interior Point Techniques in Optimization: Complementarity, Sensitivity and Algorithms*. Kluwer, Dordrecht, 2010.

H. T. Jongen, P. Jonker, and F. Twilt. *Nonlinear Optimization in Finite Dimensions*. Springer, Berlin, 2000.

N. J. Karmarkar. A new polynomial-time algorithm of linear programming. *Combinatorica*, 4:374–495, 1984.

J. Kronqvist, D. E. Bernal, A. Lundell, and I. E. Grossmann. A review and comparison of solvers for convex MINLP. *Optimization and Engineering*, 20:397–455, 2019.

J. Lee and S. Leyffer, editors. *Mixed Integer Nonlinear Programming*. Springer, New York, 2012.

Y. Lin and J. Pang. Iterative methods for large convex quadratic programs: a survey. *SIAM Journal on Control and Optimization*, 25(2):383–411, 1987.

A. Lucia and J. Xu. Chemical process optimization using Newton-like methods. *Comp. Chem. Engng*, 14(2):119–138, 1990.

A. Lucia, J. Xu, and K. M Layn. Nonconvex process optimization. *Comp. Chem. Engng*, 20(12):1375–1398, 1996.

D. G. Luenberger and Y. Ye. *Linear and Nonlinear Programming*. Springer, Heidelberg, 4th edition, 2016.

S. Mehrotra. On the implementation of a primal-dual interior point method. *SIAM Journal on Optimization*, 2(4):575–601, 1992.

J. J. Moré and S. J. Wright. *Optimization software guide*. SIAM, Philadelphia, 1994.

A. S. Nemirovski and M. J. Todd. Interior-point methods for optimization. *Acta Numerica*, pages 1–44, 2009.

J. Nocedal and S. J. Wright. *Numerical optimization*. Springer, New York, 2nd edition, 2006.

J. S. Pang. Methods for quadratic programming: a survey. *Comp. Chem. Engng*, 7(5): 583–594, 1983.

R. Polyak. Modified barrier functions (theory and methods). *Math. Programming*, 54: 177–222, 1992.

M. J. D. Powell. A fast algorithm for nonlinearly constrained optimization calculations. In G. A. Watson, editor, *Proceedings of 1977 Dundee Biennial Conference on Numerical Analysis*, pages 144–157. Springer, 1978.

K. Schittkowski. The nonlinear programming method of Wilson, Han and Powell with an augmented Lagrange type line search function, Part 1: convergence analysis. *Numer. Math.*, 38:83–114, 1981.

K. Schittkowski. NLPQL: a Fortran subroutine solving constrained nonlinear programming problems. *Annals of Operations Research*, 5:485–500, 1985.

C. Schmid and L. T. Biegler. Quadratic programming methods for reduced Hessian SQP. *Comp. Chem. Engng*, 18(9):817–832, 1994.

C. Shen, W. Xue, and X. Chen. Global convergence of a robust filter SQP algorithm. *European Journal of Operational Research*, 206:34–45, 2010.

K. Su and Z. Yu. A modified SQP method with nonmonotone technique and its global convergence. *Computers and Mathematics with Applications*, 57:240–247, 2009.

R. Tapia, Y. Zhang, M. Saltzmann, and A. Weiser. The Mehrotra predictor-corrector interior-point method as a perturbed composite Newton method. Tr90-17, Center for Research on Parallel Computation, Rice University, Houston, Texas, 1990.

D.J. Ternet and L. T. Biegler. Interior-point methods for reduced Hessian successive quadratic programming. *Comp. Chem. Engng.*, 23:859–873, 1999.

V. S. Vassiliadis and S. A. Brooks. Application of the modified barrier method in large-scale quadratic programming problems. *Comp. Chem. Engng*, 22(9):1197–1205, 1998.

V. S. Vassiliadis and C. A. Floudas. The modified barrier function approach for large-scale optimization. *Comp. Chem. Engng*, 21(8):855–874, 1997.

A. Wächter and L. T. Biegler. Line search filter methods for nonlinear programming: motivation and global convergence. *SIAM Journal on Optimization*, 16:1–31, 2005.

Z. Wei, L. Liu, and S. Yao. The superlinear convergence of a new quasi-Newton-SQP method for constrained optimization. *Applied Mathematics and Computation*, 196: 791–801, 2008.

Wikipedia. List of optimization software. https://en.wikipedia.org/wiki/List_of_optimization_software.

P. Wolfe. The simplex method for quadratic programming. *Econometrica*, 27(3): 382–398, 1959.

S. J. Wright. *Primal-dual interior-point methods*. SIAM, Philadelphia, 1997.

W. Xue, C. Shen, and D. Pu. A penalty-function-free line search SQP method for nonlinear programming. *Journal of Computational and Applied Mathematics*, 228: 313–325, 2009.

Y. Ye. *Interior Point Algorithms: Theory and Analysis*. Wiley, New York, 1997.

Z. Zhu and J. Jian. An efficient feasible SQP algorithm for inequality constrained optimization. *Nonlinear analysis: real world applications*, 10:1220–1228, 2009.

Z. Zhu and K. Zhang. A new SQP method of feasible directions for nonlinear programming. *Applied Mathematics and Computation*, 148:121–134, 2004.

Z. Zhu, W. Zhang, and Z. Geng. A feasible SQP method for nonlinear programming. *Applied mathematics and computation*, 215:3956–3969, 2010.

Chapter 12
Dynamic Optimization

12.1 Introduction

Frequently, the engineer in charge of a process is faced with optimization problems. In fact, this may cover relatively different ideas.

For example, consider a chemical reaction studied in a laboratory; the chemist seeks the best kinetic parameters, stoichiometries, reaction orders, rate constants, activation energies, i.e. the parameters that will help to optimally represent the reacting system. Then, the chemist performs a *static* optimization, which most often consists of minimizing the distance between a set of experimental data and the corresponding prediction given by the model.

Now, consider the engineer in charge of a process, thus responsible for the real plant production. The laboratory chemist will have given to the engineer the information that seemed necessary with respect to the reaction and has proposed a recipe for the experimental operation guide. The engineer knows that the reactor, being continuous, batch, or fed-batch, and more or less similar to a perfectly stirred reactor or a plug-flow reactor, will behave, in reality, relatively different from the lab reactor and will be closer to the pilot reactor, if the latter was previously operated. For example, he or she knows that the reactive feed flow rate profile for a fed-batch reactor and the temperature or pressure profile to be followed for a batch reactor will have an influence on the yield, selectivity, or product quality. The engineer wishing to optimize production must then seek a time profile and realize a *dynamic* optimization with respect to the variables that he or she can manipulate while respecting the constraints of the system, such as the bounds on temperature and temperature rise rate, the constraints related to the possible runaway of the reactor. Similarly, an engineer realizing a reaction in a tubular reactor can seek the optimal temperature profile along the reactor. In the latter case, it is a spatial optimization very close to the dynamic optimization where the time is replaced by the abscissa along the reactor. The profile thus determined is calculated in open loop and will be applied as the set point in closed loop, which may lead to deviations between the effective result and the desired result. Thus, dynamic optimization occupies an important place in many human optimization issues (Bertsekas 2017; Bryson 1999; Chiang 1992; Kamien and Schwartz 1991). The direct closed-loop calculation of the

J.-P. Corriou, *Numerical Methods and Optimization*, Springer Optimization and Its Applications 187, https://doi.org/10.1007/978-3-030-89366-8_12

profile in the nonlinear case is not studied here. On the contrary, in the framework of optimal control, the linear case is treated in linear quadratic control and Gaussian linear quadratic control by Corriou (2018) using the theory of dynamic optimization.

In a continuous process, problems of dynamic optimization can also be considered with respect to the process changes from the nominal regime. For example, the quality of raw petroleum feeding the refineries changes very often. The economic optimization realized off-line imposes set point variations on the distillation columns. An objective for the engineer can be to find the optimal profile to be followed during the change from one set of set points to another set.

Consider a satellite around earth that must move from one orbit to another orbit. The energy available in the satellite is very limited, and the objective of the engineers is to calculate the best trajectory that saves the energy (Harada and Miyazawa 2013; Hur et al. 2021). Also, consider a rocket launching astronauts to the International Space Station. This latter is a moving object, and the engineers must calculate the best trajectory to reach this final destination. Similar issues apply for airplanes (Hagelauer and Mora-Camino 1998).

In business and economics, dynamic optimization is also used (Cao and Illing 2019; Chiang 1992; Kamien and Schwartz 1991; Mitra 2000; Sethi and Thompson 2000; Todorova 2011) in all activity sectors. In the economic domain, early applications of dynamic optimization are related to the pioneering works of Hotelling (1931), Ramsey (1928) using the calculus of variations. The economic planning problem can deal with the neo-classical growth model (Barro and Sala-I-Martin 2004), the determination of a price profile leading to a maximum profit, the optimal consumption planning, an inventory problem to satisfy the demand (Minner 2003), nonrenewable resources (Blanco Fonseca and Flichman 2011; Kennedy 1988), investment planning, a bank loan (Cretegny and Rutherford 2004; Kamien and Schwartz 1991), treatment, and health-care resource in pharmacoeconomics (Vernon and Hughen 2006).

Dynamic optimization is also used in the communications industry (Hampshire and Massey 2010) to solve dynamic rate queues, leading to the field of queueing theory (Chan 2014; Kleinrock 1975; Puterman 1994; Sennott 1999; Sundarapandian 2009).

In all cases, the engineer or the economist needs a dynamic model that is sufficiently representative of the behavior of the process, and that is also of a reasonable complexity with respect to the difficulty of the mathematician and numerical task of solving.

Thus, the criteria that the engineers wish to optimize can be any technical or economic criterion that simultaneously takes into account technical objectives and production or investment costs.

Dynamic optimization is divided into two main parts: continuous-time and discrete-time problems.

Optimal control is the formulation of the dynamic optimization methods in the framework of a control problem that can be either continuous or discrete.

12.2 Problem Statement

The dynamic optimization problem is first set in *continuous time*. The studied system is assumed to be nonlinear. Applications that will follow in control will be developed only for linear systems.

The fixed aim in this problem is the determination of the control $\boldsymbol{u}(t)$ minimizing a criterion $J(\boldsymbol{u})$ while verifying initial and final conditions and respecting constraints. The optimal control thus denoted by $\boldsymbol{u}^*(t)$ makes the state $\boldsymbol{x}(t)$ follow a trajectory $\boldsymbol{x}^*(t)$ that must belong to the set of admissible trajectories.

The formulation of the dynamic optimization problem is the following:
Consider a system described in state space by the set of differential equations

$$\dot{\boldsymbol{x}}(t) = \boldsymbol{f}(\boldsymbol{x}(t), \boldsymbol{u}(t)), \quad t_0 \le t \le t_f \tag{12.2.1}$$

with $\boldsymbol{x}$ being a state vector of dimension n and $\boldsymbol{u}$ a control vector of dimension m. The system is subject to initial and final conditions, called terminal (or at the boundaries)

$$\boldsymbol{k}(\boldsymbol{x}(t_0), t_0) = 0\,, \quad \boldsymbol{l}(\boldsymbol{x}(t_f), t_f) = 0 \tag{12.2.2}$$

Moreover, the system can be subject to instantaneous inequality constraints

$$\boldsymbol{p}(\boldsymbol{x}(t), \boldsymbol{u}(t), t) \le 0 \quad \forall t \tag{12.2.3}$$

or integral constraints (depending only on t_0 and t_f)

$$\int_{t_0}^{t_f} q(\boldsymbol{x}(t), \boldsymbol{u}(t), t) dt \le 0 \tag{12.2.4}$$

The question is to find the set of the admissible controls $\boldsymbol{u}(t)$ that minimize a technical or economic performance criterion $J(\boldsymbol{u})$

$$J(\boldsymbol{u}) = G(\boldsymbol{x}(t_0), t_0, \boldsymbol{x}(t_f), t_f) + \int_{t_0}^{t_f} F(\boldsymbol{x}(t), \boldsymbol{u}(t), t)\, dt \tag{12.2.5}$$

where G is called the algebraic part of the criterion. F is a functional.[1] Frequently, the initial instant is taken as $t_0 = 0$.

The performance index J depends on the type of problem. Some examples are:

1. For a minimum-time problem (minimize t_f)

$$J = \int_0^{t_f} dt = t_f \tag{12.2.6}$$

[1] A functional is a function of functions; the function $F(\boldsymbol{x}(t), \boldsymbol{u}(t), t)$ depends on functions $\boldsymbol{x}(t)$ and $\boldsymbol{u}(t)$.

2. Minimization of the state error variance

$$J = \int_0^{t_f} (\boldsymbol{x}(t) - \boldsymbol{x}^{\text{ref}}(t))^2 dt \tag{12.2.7}$$

where $\boldsymbol{x}^{\text{ref}}$ is the reference state.
3. Combination of performance indices

$$J = \int_0^{t_f} dt + \mu \int_0^{t_f} (\boldsymbol{x}(t) - \boldsymbol{x}^{\text{ref}}(t))^2 dt \tag{12.2.8}$$

where μ is a weighting factor.
4. Weighted sum of the state error variance and the control error variance

$$J = \int_0^{t_f} \{\mu_1 [\boldsymbol{x}(t) - \boldsymbol{x}^{\text{ref}}(t)]^2 + \mu_2 [\boldsymbol{u}(t) - \boldsymbol{u}^{\text{ref}}(t)]^2\} \, dt \tag{12.2.9}$$

5. Use of a nonlinear functional

$$J = \int_0^{t_f} F(\boldsymbol{x}(t), \boldsymbol{u}(t), t) \, dt \tag{12.2.10}$$

6. Combination with a functional

$$J = G(\boldsymbol{x}(0), \boldsymbol{u}(0), \boldsymbol{x}(t_f), \boldsymbol{u}(t_f)) + \int_0^{t_f} F(\boldsymbol{x}(t), \boldsymbol{u}(t), t) \, dt \tag{12.2.11}$$

Notice that an inequality constraint such as (12.2.3) can be replaced (Boudarel et al. 1969) by an equality constraint, by adding an auxiliary function according to Valentine's method, as

$$\boldsymbol{p}(\boldsymbol{x}(t), \boldsymbol{u}(t), t) + \boldsymbol{y}^2(t) = 0 \quad \forall t \tag{12.2.12}$$

Similarly, the inequality (12.2.4) becomes

$$\int_{t_0}^{t_f} [q(\boldsymbol{x}(t), \boldsymbol{u}(t), t) + z(t)^2] dt = 0 \tag{12.2.13}$$

In fact, by introducing the variable $w(t)$ such that

$$w(t) = \int_{t_0}^{t} [q(\boldsymbol{x}(\tau), \boldsymbol{u}(\tau), \tau) + z(\tau)^2] d\tau \tag{12.2.14}$$

the integral constraint is transformed into an instantaneous constraint

$$\dot{w}(t) - [q(\boldsymbol{x}(t), \boldsymbol{u}(t), t) + z(t)^2] = 0 \tag{12.2.15}$$

where w must verify the terminal conditions $w(t_0) = 0$ and $w(t_f) = 0$.

In this very general form, this problem makes use of the equality constraints corresponding to the state differential equations, terminal equality constraints, possibly instantaneous or integral inequality constraints, and m independent functions, which are the controls $\boldsymbol{u}(t)$. The term $G(\boldsymbol{x}(t_0), t_0, \boldsymbol{x}(t_f), t_f)$ represents a contribution of the terminal conditions to the criterion, whereas the integral term of Equation (12.2.5) represents a time-accumulation contribution.

Several methods allow us to solve this type of problem: variational methods (Kirk 1970), Pontryagin maximum principle (Pontryaguine et al. 1974), and Bellman dynamic programming (Bellman 1957). The books cited here Bryson and Ho (1975), Bryson (1999), Feldbaum (1973), Pun (1972); Ray and Szekely (1973) propose compared approaches.

Variational methods are first presented in a mathematical form that can be qualified as dynamic optimization, in order to distinguish the tools independently from the control problem. Then, the control problem is treated by a specification of some variables.

12.3 Variational Method in the Mathematical Framework

In the most general mathematical form, in the classical variational method, the performance index to be minimized with respect to the vector of functions $\boldsymbol{f}$ (of dimension n) of the variable x is the following:

$$J(\boldsymbol{f}) = G(x_0, \boldsymbol{f}(x_0), x_1, \boldsymbol{f}(x_1)) + \int_{x_0}^{x_1} F(x, \boldsymbol{f}(x), \dot{\boldsymbol{f}}(x))dx \tag{12.3.1}$$

where $\boldsymbol{f}$ and $\dot{\boldsymbol{f}}$ are considered as independent variables. The vector of functions $\boldsymbol{f}$ must satisfy the following equations:

$$\phi_i(x, \boldsymbol{f}(x), \dot{\boldsymbol{f}}(x)) = 0\,, \quad i = 1, \dots, n_\phi < n \tag{12.3.2}$$

$$k_j(x_0, \boldsymbol{f}(x_0)) = 0\,, \quad j = 1, \dots, n_0 \tag{12.3.3}$$

$$l_j(x_1, \boldsymbol{f}(x_1)) = 0\,, \quad j = n_0 + 1, \dots, n_0 + n_1 \le 2n + 2 \tag{12.3.4}$$

In this form, the problem is called a Bolza problem. If the functional F of Equation (12.3.1) is zero, this is a Mayer problem. If G is zero, it is a Lagrange problem.
General comments:

1. Notice that, in the performance index J, the term G depends only on the initial (or lower) limit x_0 and on the final (or upper) limit x_1, while the integral term depends on the whole history between the initial and final limits.
2. Equation (12.3.2) is a system of differential equations.
3. Equation (12.3.3) represents the constraints at the initial limit.
4. Equation (12.3.4) represents the constraints at the final limit.
5. If we define an additional variable

$$f_{n+1}(x) = G(x_0, \boldsymbol{f}(x_0), x, \boldsymbol{f}(x)) + \int_{x_0}^{x} F(\xi, \boldsymbol{f}(\xi), \dot{\boldsymbol{f}}(\xi))d\xi \tag{12.3.5}$$

equation equivalent to

$$\dot{f}_{n+1}(x) = G_{\boldsymbol{f}}^T \dot{\boldsymbol{f}} + G_x + F(x, \boldsymbol{f}(x), \dot{\boldsymbol{f}}(x)) \qquad \text{with } f_{n+1}(x_0) = 0 \tag{12.3.6}$$

and the performance index J equal to

$$J = f_{n+1}(x_1) = G(x_0, \boldsymbol{f}(x_0), x_1, \boldsymbol{f}(x_1)) + \int_{x_0}^{x_1} F(x, \boldsymbol{f}(x), \dot{\boldsymbol{f}}(x))dx \tag{12.3.7}$$

the problem is analogous to the Pontryagin method.

12.3.1 Variation of the Criterion

Assuming that $\boldsymbol{f}^*$ is the optimal solution belonging to the domain of admissible trajectories, we study the influence of the variations[2] of $\boldsymbol{f}$ in the neighborhood of $\boldsymbol{f}^*$ on the criterion J.

In the most general case, supposing that the boundaries x_0 and x_1 are not fixed, noting that $f_0 = f(x_0)$ and likewise for f_1, and $F_0 = F(x_0, \boldsymbol{f}(x_0), \dot{\boldsymbol{f}}(x_0))$ and similarly for F_1, the variation of criterion J of Equation (12.3.1) related to a variation of these boundaries is equal to

$$\begin{aligned}\delta J = &\int_{x_0}^{x_1} \left[\left(\frac{\partial F}{\partial \boldsymbol{f}}\right)^T \delta \boldsymbol{f} + \left(\frac{\partial F}{\partial \dot{\boldsymbol{f}}}\right)^T \delta \dot{\boldsymbol{f}} \right] dx + [F_1\, \delta x_1 - F_0\, \delta x_0] \\ &+ \left[\left(\frac{\partial G}{\partial x_0}\right) \delta x_0 + \left(\frac{\partial G}{\partial \boldsymbol{f}_0}\right)^T \delta \boldsymbol{f}_0 \right] + \left[\left(\frac{\partial G}{\partial x_1}\right) \delta x_1 + \left(\frac{\partial G}{\partial \boldsymbol{f}_1}\right)^T \delta \boldsymbol{f}_1 \right]\end{aligned} \tag{12.3.8}$$

The second part of the integral term can be expressed as

---

[2] Several mathematical relations are useful:

(a) We denote by y_z the partial derivative $\partial y / \partial z$, where z is a scalar. If y is scalar and $\boldsymbol{z}$ a vector, the notation $y_{\boldsymbol{z}}$ is the gradient vector of partial derivatives $\partial y / \partial z_i$. If $\boldsymbol{y}$ and $\boldsymbol{z}$ are vectors, the notation $\boldsymbol{y}_{\boldsymbol{z}}$ represents the Jacobian matrix of typical element $\partial y_i / \partial z_j$.

(b) The derivative with respect to $\boldsymbol{f}$ of the integral with fixed boundaries

$$I = \int_{x_0}^{x_1} F(x, \boldsymbol{f}(x), \dot{\boldsymbol{f}}(x))dx$$

is equal to

$$\frac{dI}{d\boldsymbol{f}} = \int_{x_0}^{x_1} \left[F_{\boldsymbol{f}} - \frac{d}{dx} F_{\dot{\boldsymbol{f}}} \right] dx$$

(c) According to the Euler–Lagrange lemma (Cartan 1967), if $\boldsymbol{C}(x)$ is a continuous function (vector) on $[a, b]$ verifying

$$\int_a^b \boldsymbol{C}^T(x) \boldsymbol{v}(x) dx = 0$$

for all function (vector) $\boldsymbol{v}(x)$ that is continuous and becomes zero at the boundaries, then $\boldsymbol{C}(x)$ is zero everywhere on $[a, b]$.

$$\int_{x_0}^{x_1}\left(\frac{\partial F}{\partial \dot{f}}\right)^T \delta\dot{f}\,dx = \int_{x_0}^{x_1}\left(\frac{\partial F}{\partial \dot{f}}\right)^T \frac{d}{dx}[\delta f]dx$$
$$= \left[\left(\frac{\partial F}{\partial \dot{f}}\right)^T \delta f\right]_{x_0}^{x_1} - \int_{x_0}^{x_1}\frac{d}{dx}\left(\frac{\partial F}{\partial \dot{f}}\right)^T \delta f\,dx$$
$$= \left(\frac{\partial F}{\partial \dot{f}}\right)_1^T \delta f(x_1) - \left(\frac{\partial F}{\partial \dot{f}}\right)_0^T \delta f(x_0) - \int_{x_0}^{x_1}\frac{d}{dx}\left(\frac{\partial F}{\partial \dot{f}}\right)^T \delta f\,dx \tag{12.3.9}$$

On the contrary, the following relations expressing the variation of the terminal functions as the sum of two contributions are used:

$$\delta f_0 = \delta f(x_0) + \dot{f}(x_0)\delta(x_0) \quad \text{and} \quad \delta f_1 = \delta f(x_1) + \dot{f}(x_1)\delta(x_1) \tag{12.3.10}$$

Using an integration by parts of the second term of the integral, the criterion variation equation can be transformed into

$$\begin{aligned}\delta J = &\int_{x_0}^{x_1}\left[\frac{\partial F}{\partial f} - \frac{d}{dx}\left(\frac{\partial F}{\partial \dot{f}}\right)\right]^T \delta f\,dx \\ &+ \left[\frac{\partial G}{\partial x_1} + \left(F - \left(\frac{\partial F}{\partial \dot{f}}\right)^T \dot{f}\right)_1\right]\delta x_1 + \left[\frac{\partial G}{\partial \boldsymbol{f_1}} + \left(\frac{\partial F}{\partial \dot{f}}\right)_1\right]^T \delta f_1 \\ &- \left[-\frac{\partial G}{\partial x_0} + \left(F - \left(\frac{\partial F}{\partial \dot{f}}\right)^T \dot{f}\right)_0\right]\delta x_0 - \left[-\frac{\partial G}{\partial \boldsymbol{f_0}} + \left(\frac{\partial F}{\partial \dot{f}}\right)_0\right]^T \delta f_0\end{aligned} \tag{12.3.11}$$

Moreover, the variations at the boundaries are dependent because of the constraints (12.3.3) and (12.3.4), giving the relations

$$\begin{aligned}\left(\frac{\partial \boldsymbol{k}}{\partial x}\right)_0 \delta x_0 + \left(\frac{\partial \boldsymbol{k}}{\partial f}\right)_0 \delta f_0 = 0 \\ \left(\frac{\partial \boldsymbol{l}}{\partial x}\right)_1 \delta x_1 + \left(\frac{\partial \boldsymbol{l}}{\partial f}\right)_1 \delta f_1 = 0\end{aligned} \tag{12.3.12}$$

A necessary, but not sufficient, condition of the minimum of the criterion is

$$\delta J = 0\,, \quad \forall\ \delta f,\ \delta f_0,\ \delta f_1,\ \delta x_0,\ \delta x_1 \tag{12.3.13}$$

12.3.2 Variational Problem Without Constraints: Fixed Boundaries

Consider the simple case where the criterion to be optimized is simply in the form:

$$J = \int_{x_0}^{x_1} F(x, f(x), \dot{f}(x))dx \tag{12.3.14}$$

and the boundaries x_0 and x_1 are fixed. We seek the admissible optimal trajectory f^*, minimizing the criterion J with respect to f

$$f^* = \arg\left\{\min_f J\right\} \tag{12.3.15}$$

From Equation (12.3.11), using the Euler–Lagrange lemma, we deduce the necessary Euler conditions so that J has a local extremum in f^* (necessary condition of stationarity), i.e. the following vector is zero:

$$\left(\frac{\partial F}{\partial f}\right)_* - \left(\frac{d}{dx}\frac{\partial F}{\partial \dot{f}}\right)_* = 0 \tag{12.3.16}$$

The system (12.3.16) can be written again in the following form:

$$\frac{\partial F}{\partial f} - \frac{\partial\left(\frac{\partial F}{\partial \dot{f}}\right)}{\partial x} - \frac{\partial\left(\frac{\partial F}{\partial \dot{f}}\right)}{\partial f}\dot{f} - \frac{\partial\left(\frac{\partial F}{\partial \dot{f}}\right)}{\partial \dot{f}}\ddot{f} = 0 \tag{12.3.17}$$

and thus constitutes n second order differential equations. The $2n$ degrees of freedom are filled by the n initial and n final conditions.

One of the simplest problems that can be solved by this method, using Euler conditions, is to find the function $y = f(x)$ that yields the minimum distance between two points in a two-dimensional Cartesian coordinate system (x, y).

12.3.3 Variational Problem with Constraints: General Case

Consider the simple case where the criterion to be optimized is

$$J = G(x_0, f(x_0), x_1, f(x_1)) + \int_{x_0}^{x_1} F(x, f(x), \dot{f}(x))dx \tag{12.3.18}$$

and the boundaries x_0 and x_1 are not fixed. We seek the admissible optimal trajectory f^*, minimizing the criterion J with respect to f

$$f^* = \arg\left\{\min_f J\right\} \tag{12.3.19}$$

When the problem includes equality constraints such as Equation (12.3.2)

$$\phi(x, f(x), \dot{f}(x)) = 0 \quad \forall\, x \in [x_0, x_1] \tag{12.3.20}$$

or inequality constraints transformed into equality constraints such as Equation (12.2.12)

$$p(x, f(x), \dot{f}(x)) + y^2(x) = 0 \quad \forall\, x \in [x_0, x_1] \tag{12.3.21}$$

and such as Equation (12.2.13)

$$\int_{x_0}^{x_1} [q(x, \boldsymbol{f}(x), \dot{\boldsymbol{f}}(x)) + z(x)^2] dx = 0 \tag{12.3.22}$$

we apply the *Euler conditions* (as well as all the terminal conditions, discontinuity conditions, and conditions relative to the second variations) to the augmented function F^a

$$\begin{aligned} F^a(x, \boldsymbol{f}, \dot{\boldsymbol{f}}, \boldsymbol{y}, \boldsymbol{z}, \boldsymbol{w}) = & \; F(x, \boldsymbol{f}, \dot{\boldsymbol{f}}) \\ & + \boldsymbol{\lambda}^T(x) \boldsymbol{\phi} + \boldsymbol{\mu}^T(x) [\boldsymbol{p} + \boldsymbol{y}^2] + \boldsymbol{\nu}^T(x) [\dot{\boldsymbol{w}} - \boldsymbol{q} - \boldsymbol{z}^2] \end{aligned} \tag{12.3.23}$$

obtained by introducing for each constraint a Lagrange or Kuhn–Tucker parameter. In this case, the variables concerned by the Euler conditions are $\boldsymbol{f}$, $\boldsymbol{y}$, $\boldsymbol{z}$, and $\boldsymbol{w}$. The Euler conditions applied to the augmented function F^a give the following equations:

- With respect to variable $\boldsymbol{f}$

$$\begin{aligned} & \left(\frac{\partial F^a}{\partial \boldsymbol{f}}\right)_* - \left(\frac{d}{dx} \frac{\partial F^a}{\partial \dot{\boldsymbol{f}}}\right)_* = 0 \quad \Longrightarrow \\ & F_{\boldsymbol{f}} + \boldsymbol{\phi}_{\boldsymbol{f}}^T \boldsymbol{\lambda} + \boldsymbol{p}_{\boldsymbol{f}}^T \boldsymbol{\mu} - \boldsymbol{q}_{\boldsymbol{f}}^T \boldsymbol{\nu} - \frac{d}{dx}(F_{\dot{\boldsymbol{f}}} + \boldsymbol{\phi}_{\dot{\boldsymbol{f}}}^T \boldsymbol{\lambda} + \boldsymbol{p}_{\dot{\boldsymbol{f}}}^T \boldsymbol{\mu} - \boldsymbol{q}_{\dot{\boldsymbol{f}}}^T \boldsymbol{\nu}) = 0 \end{aligned} \tag{12.3.24}$$

- With respect to variable $\boldsymbol{y}$,

$$2\boldsymbol{\mu}^T \boldsymbol{y} = 0 \tag{12.3.25}$$

- With respect to variable $\boldsymbol{z}$,

$$2\boldsymbol{\nu}^T \boldsymbol{z} = 0 \tag{12.3.26}$$

- With respect to variable $\boldsymbol{w}$,

$$\dot{\boldsymbol{\nu}} = 0 \quad \Longrightarrow \boldsymbol{\nu} = \text{constant} \tag{12.3.27}$$

After applying the Euler conditions, the boundary conditions and the trajectory discontinuities must be taken into account.

Terminal Conditions

The constraints (12.3.3) acting on the initial boundary x_0 and (12.3.4) acting on the final boundary x_1 are called the terminal conditions. They express that the terminal values of functions $\boldsymbol{f}$, x, belong to hypersurfaces or, more simply, that relations link the terminal values of functions $\boldsymbol{f}(x)$.

Transversality conditions

Taking into account the variation of criterion J described by Equation (12.3.11), the Euler condition of stationarity, and the constraints acting on the boundary variations $\delta \boldsymbol{f}_0$, δx_0, $\delta \boldsymbol{f}_1$, δx_1, we obtain the transversality conditions, which are written as follows.

At the initial boundary x_0,

$$\begin{aligned} & \left[-\frac{\partial G}{\partial x_0} + \left(F^a - \left(\frac{\partial F^a}{\partial \dot{\boldsymbol{f}}}\right)^T \dot{\boldsymbol{f}} \right)_0 \right] \delta x_0 + \left[-\frac{\partial G}{\partial \boldsymbol{f}_0} + \left(\frac{\partial F^a}{\partial \dot{\boldsymbol{f}}}\right)_0 \right]^T \delta \boldsymbol{f}_0 = 0 \\ & \text{with} \quad \left(\frac{\partial \boldsymbol{k}}{\partial x}\right)_0 \delta x_0 + \left(\frac{\partial \boldsymbol{k}}{\partial \boldsymbol{f}}\right)_0 \delta \boldsymbol{f}_0 = 0 \end{aligned} \tag{12.3.28}$$

At the final boundary x_1,

$$\left[\frac{\partial G}{\partial x_1} + \left(F^a - \left(\frac{\partial F^a}{\partial \dot{f}}\right)^T \dot{f}\right)_1\right] \delta x_1 + \left[\frac{\partial G}{\partial f_1} + \left(\frac{\partial F^a}{\partial \dot{f}}\right)_1\right]^T \delta f_1 = 0$$
$$\text{with} \quad \left(\frac{\partial l}{\partial x}\right)_1 \delta x_1 + \left(\frac{\partial l}{\partial f}\right)_1 \delta f_1 = 0 \tag{12.3.29}$$

This means that the extremizing trajectory must have, in the phase space of variables f_i, the same slope as the trajectories $\boldsymbol{k}$ and $\boldsymbol{l}$ at the initial and final boundaries x_0 and x_1, respectively.

If the initial boundary x_0 is fixed, this leads to $\delta x_0 = 0$ (similarly, $\delta x_1 = 0$ if the final boundary x_1 is fixed). If the initial extremity of the trajectory $\boldsymbol{f}_0$ is fixed, this leads to $\delta \boldsymbol{f}_0 = 0$ (similarly, $\delta \boldsymbol{f}_1 = 0$ if the final extremity $\boldsymbol{f}_1$ is fixed).

Discontinuity Condition

This condition is also called the Weierstrass–Erdmann condition. The discontinuous extremizing trajectories are composed of sub-arcs joined by discontinuities. At these points, the partial derivatives of two sub-arcs must be equal

$$\left(\frac{\partial F^a}{\partial \dot{f}_i}\right)_- = \left(\frac{\partial F^a}{\partial \dot{f}_i}\right)_+, \quad i = 1, \ldots, n \tag{12.3.30}$$

$$\left(-F^a + \sum_{i=1}^{n} \frac{\partial F^a}{\partial \dot{f}_i} \dot{f}_i\right)_- = \left(-F^a + \sum_{i=1}^{n} \frac{\partial F^a}{\partial \dot{f}_i} \dot{f}_i\right)_+, \quad i = 1, \ldots, n \tag{12.3.31}$$

In optimal control, an extremizing trajectory thus can be comprised of a first arc, where the control is saturated at its minimum value u_-, followed by an optimal arc and then followed by an arc where the control is saturated at its maximum value u_+. In practice, it is common that the control takes only the minimum and maximum values (bang–bang control), in particular for minimum-time problems.

Partial conclusion and first solution

After having applied the Euler equations and the terminal and discontinuity conditions, we have, in fact, realized a set of necessary conditions for obtaining a stationary solution, but not sufficient for the minimum solution that will be called a first solution.

Conditions relative to the second variations

Weierstrass–Erdmann and Legendre–Clebsch conditions are necessary conditions for the criterion J to be minimum. Weierstrass conditions are relative to large variations and Legendre conditions to small variations.

Weierstrass–Erdmann Condition

The necessary condition for the performance index J to be minimum is that the Weierstrass function W verifies

$$W = F^a(\boldsymbol{f}^*, \dot{\boldsymbol{f}}, x) - F^a(\boldsymbol{f}^*, \dot{\boldsymbol{f}}^*, x) - (\dot{\boldsymbol{f}} - \dot{\boldsymbol{f}}^*)^T \left(\frac{\partial F^a}{\partial \dot{\boldsymbol{f}}}\right)_* \geq 0 \tag{12.3.32}$$

at any point of the extremum arc, for all large variations $\Delta \dot{f}_i$ that are compatible with constraints ϕ_i.

Legendre–Clebsch Condition

This condition is relative to small variations $\delta \dot{f}$ in the neighborhood of $\dot{f}^*$

$$\sum_{i=1}^{n} \sum_{j=1}^{n} \frac{\partial^2 F^a}{\partial \dot{f}_i \partial \dot{f}_j} \delta \dot{f}_i \, \delta \dot{f}_j \geq 0 \tag{12.3.33}$$

If a maximum performance index was sought, the two previous inequalities would have an opposite sign.

By applying the conditions relative to the second variations, the optimal solution is completely determined.

12.3.4 Hamilton–Jacobi Equation

The problem formulation based on the use of the Hamiltonian allows us to express Euler equations in the canonical form of Hamilton equations. The Hamilton–Jacobi equation bearing on the criterion optimization leads us to obtain the optimal trajectory. The discrete Hamilton–Jacobi equation can be very well compared to the Bellman optimality principle in dynamic programming (Section 12.5).

If, on the optimal trajectory, we introduce the variable $\boldsymbol{\psi}(x)$, which is defined as

$$\boldsymbol{\psi}(x) = \frac{\partial F}{\partial \dot{\boldsymbol{f}}} \tag{12.3.34}$$

the Euler condition (12.3.16) becomes

$$\frac{d}{dx}\boldsymbol{\psi}(x) = \frac{\partial F}{\partial \boldsymbol{f}} \tag{12.3.35}$$

Provided that the matrix

$$\frac{\partial^2 F}{\partial \dot{\boldsymbol{f}}^2} \tag{12.3.36}$$

is nonsingular, the function $\dot{\boldsymbol{f}}$ can be expressed as a solution of implicit Equation (12.3.34) as

$$\dot{\boldsymbol{f}}(x) = \boldsymbol{p}(\boldsymbol{f}(x), \boldsymbol{\psi}(x), x) \tag{12.3.37}$$

The Hamiltonian function or Hamiltonian (similar to the Lagrangian function for classical optimization problems) is defined as

$$H(\boldsymbol{f}(x), \boldsymbol{\psi}(x), x) = -F(\boldsymbol{f}(x), \dot{\boldsymbol{f}}(x), x) + \boldsymbol{\psi}^T \boldsymbol{p}(\boldsymbol{f}(x), \boldsymbol{\psi}(x), x) \tag{12.3.38}$$

The partial derivatives of the Hamiltonian are equal to

$$\begin{aligned}
H_{\boldsymbol{f}} &= -F_{\boldsymbol{f}} - \dot{\boldsymbol{f}}^T_{\boldsymbol{f}} F_{\dot{\boldsymbol{f}}} + \boldsymbol{p}^T_{\boldsymbol{f}} \boldsymbol{\psi} = -F_{\boldsymbol{f}} - \boldsymbol{p}^T_{\boldsymbol{f}} F\boldsymbol{p} + \boldsymbol{p}^T_{\boldsymbol{f}} \boldsymbol{\psi} = -F_{\boldsymbol{f}} \\
H_{\boldsymbol{\psi}} &= -\dot{\boldsymbol{f}}^T_{\boldsymbol{\psi}} F_{\dot{\boldsymbol{f}}} + \boldsymbol{p} + \boldsymbol{p}^T_{\boldsymbol{\psi}} \boldsymbol{\psi} = -\boldsymbol{p}^T_{\boldsymbol{\psi}} F\boldsymbol{p} + \boldsymbol{p} + \boldsymbol{p}^T_{\boldsymbol{\psi}} \boldsymbol{\psi} = \boldsymbol{p} \\
H_x &= -F_x - F^T_{\boldsymbol{f}} \boldsymbol{f}_x - F^T_{\dot{\boldsymbol{f}}} \dot{\boldsymbol{f}}_x + \boldsymbol{\psi}^T_x \boldsymbol{p} + \boldsymbol{\psi}^T \boldsymbol{p}_x \\
&= -F_x - \dot{\boldsymbol{\psi}}^T \boldsymbol{p} - \boldsymbol{\psi}^T \boldsymbol{p}_x + \dot{\boldsymbol{\psi}}^T \boldsymbol{p} + \boldsymbol{\psi}^T \boldsymbol{p}_x = -F_x
\end{aligned} \tag{12.3.39}$$

On the other side, using the first two expressions of the partial derivatives, the derivative of the Hamiltonian gives

$$\frac{dH}{dx} = H^T_{\boldsymbol{f}} \dot{\boldsymbol{f}} + H^T_{\boldsymbol{\psi}} \dot{\boldsymbol{\psi}} + H_x = -F_x \tag{12.3.40}$$

from which we get the Hamilton canonical conditions for optimality

$$\begin{aligned}
\dot{\boldsymbol{f}} &= H_{\boldsymbol{\psi}} \\
\dot{\boldsymbol{\psi}} &= -H_{\boldsymbol{f}} \\
\frac{dH}{dx} &= -F_x
\end{aligned} \tag{12.3.41}$$

The first two equations of this system constitute a set of $2n$ equations equivalent to Euler conditions.

In the neighborhood of the optimal trajectory $\boldsymbol{f}^*$, let us realize a variation of the trajectory at the boundary x but with always verifying the final condition, and consider the criterion at the optimal trajectory $\boldsymbol{f}^*$

$$\mathcal{J}(\boldsymbol{f}^*, x) = G(\boldsymbol{f}^*_1, x_1) + \int_x^{x_1} F(\xi, \boldsymbol{f}^*(\xi), \dot{\boldsymbol{f}}^*(\xi))d\xi \tag{12.3.42}$$

According to Equation (12.3.11), the variation of criterion $\mathcal{J}$ in the neighborhood of the optimal solution $\boldsymbol{f}^*$ is given by

$$\delta\mathcal{J} = -\left(F - F^T_{\dot{\boldsymbol{f}}} \dot{\boldsymbol{f}}^*\right) \delta x - F^T_{\dot{\boldsymbol{f}}} \delta \boldsymbol{f} = H(\boldsymbol{f}^*(x), \boldsymbol{\psi}(x), x)\delta x - \boldsymbol{\psi}^T(x)\, \delta \boldsymbol{f}(x) \tag{12.3.43}$$

hence,

$$\begin{aligned}
\mathcal{J}_x &= H(\boldsymbol{f}^*(x), \boldsymbol{\psi}(x), x) \\
\mathcal{J}_{\boldsymbol{f}} &= -\boldsymbol{\psi}(x)
\end{aligned} \tag{12.3.44}$$

The system of equations (Equation (12.3.44)) provides the Hamilton–Jacobi equation

$$\mathcal{J}_x = H(\boldsymbol{f}^*(x), -\mathcal{J}_{\boldsymbol{f}}, x) \tag{12.3.45}$$

which is a first order partial derivative equation that admits as a solution the integral $\mathcal{J}(\boldsymbol{f}^*, x)$ defined by Equation (12.3.42). Along an optimal trajectory, the criterion solution of Equation (12.3.45) is optimal. The boundary condition of $\mathcal{J}$ takes into account the end part that is eventually present in criterion (12.3.42)

$$\mathcal{J}(\boldsymbol{f}^*, x_1) = G(\boldsymbol{f}^*_1, x_1) \tag{12.3.46}$$

The transversality conditions are deduced from Equations (12.3.28) and (12.3.29): at the initial boundary x_0, where the constraint is satisfied for all δx_0 and $\delta \boldsymbol{f}_0$

$$\left(\frac{\partial \boldsymbol{k}}{\partial x}\right)_0 \delta x_0 + \left(\frac{\partial \boldsymbol{k}}{\partial \boldsymbol{f}}\right)_0 \delta \boldsymbol{f}_0 = 0 \tag{12.3.47}$$

the transversality condition is imposed

$$\left[-\frac{\partial G}{\partial x_0} - H_0\right] \delta x_0 + \left[-\frac{\partial G}{\partial \boldsymbol{f_0}} + \boldsymbol{\psi}_0\right]^T \delta \boldsymbol{f}_0 = 0 \tag{12.3.48}$$

at the final boundary x_1, where the constraint is satisfied for all δx_1 and $\delta \boldsymbol{f}_1$

$$\left(\frac{\partial \boldsymbol{l}}{\partial x}\right)_1 \delta x_1 + \left(\frac{\partial \boldsymbol{l}}{\partial \boldsymbol{f}}\right)_1 \delta \boldsymbol{f}_1 = 0 \tag{12.3.49}$$

the transversality condition is imposed

$$\left[\frac{\partial G}{\partial x_1} - H_1\right] \delta x_1 + \left[\frac{\partial G}{\partial \boldsymbol{f_1}} + \boldsymbol{\psi}_1\right]^T \delta \boldsymbol{f}_1 = 0 \tag{12.3.50}$$

If x_1 is fixed, from Equation (12.3.50), the frequently encountered condition results in

$$\boldsymbol{\psi}(x_1) = -\frac{\partial G}{\partial \boldsymbol{f_1}} \tag{12.3.51}$$

12.4 Dynamic Optimization in Continuous Time

12.4.1 Variational Methods

As, henceforth, we place ourselves in the framework of optimal control, x plays the role of time and the variables f_k are divided into two types: state variables x_i $(1 \le i \le n)$ and control variables u_j $(1 \le j \le m)$, so that the new problem is formulated as follows. Given a criterion

$$J(\boldsymbol{u}) = G(\boldsymbol{x}(t_0), \boldsymbol{u}(t_0), \boldsymbol{x}(t_f), \boldsymbol{u}(t_f)) + \int_{t_0}^{t_f} F(\boldsymbol{x}(t), \boldsymbol{u}(t), t) dt \tag{12.4.1}$$

we seek the optimal control trajectory $\boldsymbol{u}^*(t)$ that minimizes $J(\boldsymbol{u})$

$$\boldsymbol{u}^*(t) = \arg \left\{\min_{\boldsymbol{u}} J(\boldsymbol{u})\right\} \tag{12.4.2}$$

the state and control variables being subject to the constraints

$$\phi_i = \dot{x}_i - f_i(\boldsymbol{x}, \boldsymbol{u}, t) = 0\,, \quad i = 1, \ldots, n \tag{12.4.3}$$

$$k_j(\boldsymbol{x}(t_0), \boldsymbol{u}(t_0), t_0) = 0\,, \quad j = 1, \dots, n_0 \tag{12.4.4}$$

$$l_j(\boldsymbol{x}(t_f), \boldsymbol{u}(t_f), t_f) = 0\,, \quad j = n_0 + 1, \dots, n_0 + n_1 \le 2n + 2 \tag{12.4.5}$$

In the criterion (12.4.1), the first term G is called the algebraic part and the second term the integral part. Note that Equations (12.4.3) represent the dynamic model of the process.

When the state can only vary in a given domain, it is preferable to introduce new variables. For example, in the case of a one-dimensional state x, such that $a \le x \le b$, we can set the variable z such that $(x - a)(b - x) = z^2$. It is also possible to do the same for the control u when the latter is bounded between two minimum and maximum values.

12.4.2 Variation of the Criterion

Three general, but different, ideas will be evoked to describe the criterion variation.

- In the most general case, the variation of the criterion (12.4.1) is equal to

$$\begin{aligned} \delta J = & \int_{t_0}^{t_f} \left[\left(\frac{\partial F}{\partial \boldsymbol{x}}\right)^T \delta \boldsymbol{x} + \left(\frac{\partial F}{\partial \boldsymbol{u}}\right)^T \delta \boldsymbol{u} \right] dt + F(\boldsymbol{x}_f, \boldsymbol{u}_f, t_f)\, \delta t_f - F(\boldsymbol{x}_0, \boldsymbol{u}_0, t_0)\, \delta t_0 \\ & + \left[\left(\frac{\partial G}{\partial t_0}\right) \delta t_0 + \left(\frac{\partial G}{\partial \boldsymbol{x}_0}\right) \delta \boldsymbol{x}_0 + \left(\frac{\partial G}{\partial \boldsymbol{u}_0}\right) \delta \boldsymbol{u}_0 \right] \\ & + \left[\left(\frac{\partial G}{\partial t_f}\right) \delta t_f + \left(\frac{\partial G}{\partial \boldsymbol{x}_f}\right) \delta \boldsymbol{x}_f + \left(\frac{\partial G}{\partial \boldsymbol{u}_f}\right) \delta \boldsymbol{u}_f \right] \end{aligned} \tag{12.4.6}$$

The derivative of variations $\delta \boldsymbol{x}$ can be expressed from the state equations as

$$\frac{d}{dt}\delta \boldsymbol{x} = \left(\frac{\partial \boldsymbol{f}}{\partial \boldsymbol{x}}\right) \delta \boldsymbol{x} + \left(\frac{\partial \boldsymbol{f}}{\partial \boldsymbol{u}}\right) \delta \boldsymbol{u} \quad \Longrightarrow \quad \frac{d}{dt}\delta \boldsymbol{x} - \left(\frac{\partial \boldsymbol{f}}{\partial \boldsymbol{x}}\right) \delta \boldsymbol{x} - \left(\frac{\partial \boldsymbol{f}}{\partial \boldsymbol{u}}\right) \delta \boldsymbol{u} = 0 \tag{12.4.7}$$

The latter equation can be multiplied by the Lagrange multipliers and integrated between t_0 and t_f, so that

$$\int_{t_0}^{t_f} \boldsymbol{\psi}(t)^T \left\{ \frac{d}{dt}\delta \boldsymbol{x} - \left(\frac{\partial \boldsymbol{f}}{\partial \boldsymbol{x}}\right) \delta \boldsymbol{x} - \left(\frac{\partial \boldsymbol{f}}{\partial \boldsymbol{u}}\right) \delta \boldsymbol{u} \right\} dt = 0 \tag{12.4.8}$$

By summing this equation and Equation (12.4.6), one obtains

$$
\begin{aligned}
\delta J = & \int_{t_0}^{t_f} \left\{ \left[\left(\frac{\partial F}{\partial \boldsymbol{x}} \right)^T - \boldsymbol{\psi}(t)^T \frac{\partial \boldsymbol{f}}{\partial \boldsymbol{x}} \right] \delta \boldsymbol{x} + \left[\left(\frac{\partial F}{\partial \boldsymbol{u}} \right)^T - \boldsymbol{\psi}(t)^T \frac{\partial \boldsymbol{f}}{\partial \boldsymbol{u}} \right] \delta \boldsymbol{u} \right\} dt \\
& + F(\boldsymbol{x}_f, \boldsymbol{u}_f, t_f)\, \delta t_f - F(\boldsymbol{x}_0, \boldsymbol{u}_0, t_0)\, \delta t_0 \\
& + \left[\left(\frac{\partial G}{\partial t_0} \right) \delta t_0 + \left(\frac{\partial G}{\partial \boldsymbol{x}_0} \right) \delta \boldsymbol{x}_0 + \left(\frac{\partial G}{\partial \boldsymbol{u}_0} \right) \delta \boldsymbol{u}_0 \right] \\
& + \left[\left(\frac{\partial G}{\partial t_f} \right) \delta t_f + \left(\frac{\partial G}{\partial \boldsymbol{x}_f} \right) \delta \boldsymbol{x}_f + \left(\frac{\partial G}{\partial \boldsymbol{u}_f} \right) \delta \boldsymbol{u}_f \right] \\
& + \int_{t_0}^{t_f} \boldsymbol{\psi}(t)^T \frac{d}{dt} \delta \boldsymbol{x}\, dt
\end{aligned}
\qquad (12.4.9)
$$

The last integral of (12.4.9) can be integrated into parts, thus

$$
\int_{t_0}^{t_f} \boldsymbol{\psi}(t)^T \frac{d}{dt} \delta \boldsymbol{x}\, dt = \boldsymbol{\psi}(t_f)^T \delta \boldsymbol{x}_f - \boldsymbol{\psi}(t_0)^T \delta \boldsymbol{x}_0 - \int_{t_0}^{t_f} \dot{\boldsymbol{\psi}}(t)^T \delta \boldsymbol{x}\, dt \qquad (12.4.10)
$$

- Note that it would have been possible to consider an augmented criterion of the form:

$$
\begin{aligned}
J^a(\boldsymbol{u}) = & \; G(\boldsymbol{x}(t_0), \boldsymbol{u}(t_0), \boldsymbol{x}(t_f), \boldsymbol{u}(t_f)) + \\
& \int_{t_0}^{t_f} \left\{ F(\boldsymbol{x}(t), \boldsymbol{u}(t), t) + \boldsymbol{\psi}^T \left[\dot{\boldsymbol{x}}(t) - \boldsymbol{f}(\boldsymbol{x}(t), \boldsymbol{u}(t), t) \right] \right\} dt
\end{aligned}
\qquad (12.4.11)
$$

where $\boldsymbol{\psi}(t)$ are Lagrange multipliers as the equation

$$
\dot{\boldsymbol{x}}(t) - \boldsymbol{f}(\boldsymbol{x}(t), \boldsymbol{u}(t), t) = 0 \qquad (12.4.12)
$$

corresponds to an equality constraint. Then, we consider a variation of J^a related to the variation $\delta \boldsymbol{u}(t)$ that induces a variation $\delta \boldsymbol{x}(t)$ to deduce Hamilton equations (12.4.37).

- If, according to the Hamilton–Jacobi formalism (Section 12.4.5), we furthermore introduce the Hamiltonian H equal to

$$
H(\boldsymbol{x}, \boldsymbol{u}, \boldsymbol{\psi}, t) = -F(\boldsymbol{x}, \boldsymbol{u}, t) + \boldsymbol{\psi}(t)^T \boldsymbol{f}(\boldsymbol{x}, \boldsymbol{u}, t) \qquad (12.4.13)
$$

the criterion (12.4.1) becomes

$$
J(\boldsymbol{u}) = G(\boldsymbol{x}(t_0), \boldsymbol{u}(t_0), \boldsymbol{x}(t_f), \boldsymbol{u}(t_f)) + \int_{t_0}^{t_f} \left[\boldsymbol{\psi}(t)^T \boldsymbol{f}(\boldsymbol{x}, \boldsymbol{u}, t) - H(\boldsymbol{x}, \boldsymbol{u}, \boldsymbol{\psi}, t) \right] dt \qquad (12.4.14)
$$

or

$$
J(\boldsymbol{u}) = G(\boldsymbol{x}(t_0), \boldsymbol{u}(t_0), \boldsymbol{x}(t_f), \boldsymbol{u}(t_f)) + \int_{t_0}^{t_f} \left[\boldsymbol{\psi}(t)^T \dot{\boldsymbol{x}}(t) - H(\boldsymbol{x}, \boldsymbol{u}, \boldsymbol{\psi}, t) \right] dt \qquad (12.4.15)
$$

Using the integration by parts, the variation of the criterion becomes

$$\delta J = \int_{t_0}^{t_f} \left\{ -\left[\left(\frac{\partial H}{\partial \boldsymbol{x}}\right)^T + \dot{\boldsymbol{\psi}}(t)^T \right] \delta \boldsymbol{x} - \left[\left(\frac{\partial H}{\partial \boldsymbol{u}}\right)^T \right] \delta \boldsymbol{u} \right\} dt$$
$$+ \left[\left(\frac{\partial G}{\partial t_0}\right) + H(\boldsymbol{x}_0, \boldsymbol{u}_0, \boldsymbol{\psi}_0, t_0) \right] \delta t_0 + \left[\left(\frac{\partial G}{\partial \boldsymbol{x}_0}\right) - \boldsymbol{\psi}(t_0)^T \right] \delta \boldsymbol{x}_0 + \left(\frac{\partial G}{\partial \boldsymbol{u}_0}\right) \delta \boldsymbol{u}_0$$
$$+ \left[\left(\frac{\partial G}{\partial t_f}\right) - H(\boldsymbol{x}_f, \boldsymbol{u}_f, \boldsymbol{\psi}_f, t_f) \right] \delta t_f + \left[\left(\frac{\partial G}{\partial \boldsymbol{x}_f}\right) + \boldsymbol{\psi}(t_f)^T \right] \delta \boldsymbol{x}_f + \left(\frac{\partial G}{\partial \boldsymbol{u}_f}\right) \delta \boldsymbol{u}_f \quad (12.4.16)$$

This equation, giving the variation of the criterion, is necessary for understanding the origin of Hamilton–Jacobi equations (Section 12.4.5).

12.4.3 Euler Conditions

According to the performance index, the augmented function F^a is defined as

$$F^a(\boldsymbol{x}, \dot{\boldsymbol{x}}, \boldsymbol{\lambda}, \boldsymbol{u}, t) = F(\boldsymbol{x}, \boldsymbol{u}, t) + \sum_{i=1}^{n} \lambda_i \, \phi_i \quad (12.4.17)$$

Notice that the function G does not intervene in this augmented function, as G depends only on the terminal conditions. G would only intervene in F^a if the terminal conditions were varying.

The variables are the control vector $u(t)$, the state vector $x(t)$, and the Euler–Lagrange multipliers λ. Euler conditions give

$$\begin{aligned} \frac{\partial F^a}{\partial u_j} - \frac{d}{dt} \frac{\partial F^a}{\partial \dot{u}_j} &= 0\,, \quad j = 1, \ldots, m \\ \frac{\partial F^a}{\partial x_i} - \frac{d}{dt} \frac{\partial F^a}{\partial \dot{x}_i} &= 0\,, \quad i = 1, \ldots, n \\ \frac{\partial F^a}{\partial \lambda_i} - \frac{d}{dt} \frac{\partial F^a}{\partial \dot{\lambda}_i} &= 0\,, \quad i = 1, \ldots, n \end{aligned} \quad (12.4.18)$$

The third group of this system of equations corresponds to the constraints $\phi_i = 0$, thus is a system of differential equations with respect to $\dot{x}$. The first group is a system of algebraic equations. The second group is a system of differential equations with respect to $\dot{\lambda}$.

If inequality constraints of the type (12.2.3) or (12.2.4) are present, the Valentine's method already discussed in Equations (12.2.12) and (12.2.13) should be used to modify F^a consequently.

On the contrary, the terminal conditions (12.4.4) and (12.4.5), which are transversality and discontinuity conditions, as well as the conditions relative to the second variations will have to be verified.

The transversality equations deduced from Equations (12.3.28) and (12.3.29) are as follows.

At the initial time t_0,

$$
\begin{aligned}
&\left[-\frac{\partial G}{\partial t_0} + \left(F^a - \boldsymbol{\lambda}^T \dot{\boldsymbol{x}}\right)_0\right] \delta t_0 + \left[-\frac{\partial G}{\partial \boldsymbol{x_0}} + \boldsymbol{\lambda}(t_0)\right]^T \delta \boldsymbol{x}_0 = 0 \\
&\text{with} \quad \left(\frac{\partial \boldsymbol{k}}{\partial t}\right)_0 \delta t_0 + \left(\frac{\partial \boldsymbol{k}}{\partial \boldsymbol{x}}\right)_0 \delta \boldsymbol{x}_0 = 0
\end{aligned}
\tag{12.4.19}
$$

At the final time t_f,

$$
\begin{aligned}
&\left[\frac{\partial G}{\partial t_f} + \left(F^a - \boldsymbol{\lambda}^T \dot{\boldsymbol{x}}\right)_f\right] \delta t_f + \left[\frac{\partial G}{\partial \boldsymbol{x_f}} + \boldsymbol{\lambda}(t_f)\right]^T \delta \boldsymbol{x}_f = 0 \\
&\text{with} \quad \left(\frac{\partial \boldsymbol{l}}{\partial t}\right)_f \delta t_f + \left(\frac{\partial \boldsymbol{l}}{\partial \boldsymbol{x}}\right)_f \delta \boldsymbol{x}_f = 0
\end{aligned}
\tag{12.4.20}
$$

At a fixed final time, from Equation (12.4.20), the following condition results:

$$
\boldsymbol{\lambda}(t_f) = -\frac{\partial G}{\partial \boldsymbol{x_f}} \tag{12.4.21}
$$

Example 12.1 :
Linear Quadratic Control using Euler Conditions
A one-dimensional first-order linear system is defined by the following differential equation:

$$
\dot{x} = a x + b u \tag{12.4.22}
$$

We wish to minimize the following quadratic criterion with respect to control $u(t)$:

$$
J = \frac{1}{2} \int_0^{t_f} (u^2 + x^2)\, dt \tag{12.4.23}
$$

This corresponds to a single-input single-output continuous linear quadratic control with a finite horizon t_f.

The augmented function is defined as

$$
F^a = \frac{1}{2}\left(u^2 + x^2\right) + \lambda(\dot{x} - a x - b u) \tag{12.4.24}
$$

The Euler conditions for optimality give

$$
\begin{aligned}
&\frac{\partial F^a}{\partial u} - \frac{d}{dt}\frac{\partial F^a}{\partial \dot{u}} = u - \lambda b = 0 \\
&\frac{\partial F^a}{\partial x} - \frac{d}{dt}\frac{\partial F^a}{\partial \dot{x}} = x - \lambda a - \dot{\lambda} = 0 \qquad \text{with} \quad \lambda(t_f) = -\frac{\partial G}{\partial x_f} = 0 \\
&\frac{\partial F^a}{\partial \lambda} - \frac{d}{dt}\frac{\partial F^a}{\partial \dot{\lambda}} = \dot{x} - a x - b u = 0 \qquad \text{with} \quad x(0) = x_0
\end{aligned}
\tag{12.4.25}
$$

that is, two differential equations and an algebraic equation. By replacing λ with respect to u from the first equation into the second one, and by eliminating u by differentiating the third equation, we obtain a unique differential equation with respect to the state

$$
\ddot{x} - (a^2 + b^2) x = 0 \tag{12.4.26}
$$

hence, the general solution

$$
x(t) = \alpha \exp(-\sqrt{a^2 + b^2}\, t) + \beta \exp(\sqrt{a^2 + b^2}\, t) \tag{12.4.27}
$$

where α and β are two constants to be determined from the terminal condition on the state $x(0)$ and the transversality condition on the adjoint variable $\lambda(t_f)$. However, this is a two-boundary problem that is difficult to solve. The control $u(t)$ is deduced from the model Equation (12.4.22).
Numerical Issue:

Indeed, a possibility for obtaining the numerical solution is the following. Given an initial value of the control vector, the differential Equation (12.4.22) describing the model is integrated forward in time, then the adjoint variable differential Equation (12.4.25) is integrated backward in time on the basis of the previous states. A new control profile is obtained, for example, by a gradient method, and the process is repeated iteratively until there is practically no change in the profiles or the criterion is no more improved.

12.4.4 Weierstrass Condition and Hamiltonian Maximization

The Weierstrass condition (12.3.32) relative to second variations is applied to the augmented function

$$F^a(\boldsymbol{x}, \dot{\boldsymbol{x}}, \boldsymbol{u}, t) = F(\boldsymbol{x}, \boldsymbol{u}, t) + \boldsymbol{\lambda}^T \, [\dot{\boldsymbol{x}} - \boldsymbol{f}(\boldsymbol{x}, \boldsymbol{u}, t)] \tag{12.4.28}$$

in the neighborhood of the optimum, thus

$$F^a(\boldsymbol{x}^*, \dot{\boldsymbol{x}}, \boldsymbol{u}, t) - F^a(\boldsymbol{x}^*, \dot{\boldsymbol{x}}^*, \boldsymbol{u}^*, t) - (\dot{\boldsymbol{x}} - \dot{\boldsymbol{x}}^*)^T \left(\frac{\partial F^a}{\partial \dot{\boldsymbol{x}}}\right)_* \geq 0 \tag{12.4.29}$$

By clarifying these terms and using the constraints

$$\begin{aligned} \dot{\boldsymbol{x}} &= \boldsymbol{f}(\boldsymbol{x}^*, \boldsymbol{u}, t) \\ \dot{\boldsymbol{x}}^* &= \boldsymbol{f}(\boldsymbol{x}^*, \boldsymbol{u}^*, t) \end{aligned} \tag{12.4.30}$$

the Weierstrass condition is simplified as

$$\begin{aligned} &F(\boldsymbol{x}^*, \boldsymbol{u}, t) - F(\boldsymbol{x}^*, \boldsymbol{u}^*, t) - \boldsymbol{\lambda}^T \, (\boldsymbol{f}(\boldsymbol{x}^*, \boldsymbol{u}, t) - \boldsymbol{f}(\boldsymbol{x}^*, \boldsymbol{u}^*, t)) \geq 0 \iff \\ &\left[\boldsymbol{\lambda}^T \boldsymbol{f}(\boldsymbol{x}^*, \boldsymbol{u}^*, t) - F(\boldsymbol{x}^*, \boldsymbol{u}^*, t)\right] - \left[\boldsymbol{\lambda}^T \boldsymbol{f}(\boldsymbol{x}^*, \boldsymbol{u}, t) - F(\boldsymbol{x}^*, \boldsymbol{u}, t)\right] \geq 0 \end{aligned} \tag{12.4.31}$$

in which we recognize the expression of the Hamiltonian (setting $\boldsymbol{\lambda} = \boldsymbol{\psi}$)

$$H(\boldsymbol{x}^*, \boldsymbol{u}, \boldsymbol{\lambda}, t) = -F(\boldsymbol{x}^*, \boldsymbol{u}, t) + \boldsymbol{\lambda}^T \, \boldsymbol{f}(\boldsymbol{x}^*, \boldsymbol{u}, t) \tag{12.4.32}$$

We then obtain the fundamental result that the optimal control maximizes the Hamiltonian while respecting the constraints

$$H(\boldsymbol{x}^*, \boldsymbol{u}^*, \boldsymbol{\lambda}, t) \geq H(\boldsymbol{x}^*, \boldsymbol{u}, \boldsymbol{\lambda}, t) \tag{12.4.33}$$

which will be generalized as Pontryagin's maximum principle.

Legendre–Clebsch condition for small variations would have allowed us to obtain the stationarity condition at the optimal trajectory, in the absence of constraints, as

$$\left(\frac{\partial H}{\partial u}\right)_* = 0 \tag{12.4.34}$$

and

$$\left(\frac{\partial^2 H}{\partial u^2}\right)_* \leq 0 \tag{12.4.35}$$

12.4.5 Hamilton–Jacobi Conditions and Equation

The Hamiltonian[3] is deduced from the criterion (12.4.1) and from constraints (12.4.3). It is equal to

$$H(\boldsymbol{x}(t), \boldsymbol{u}(t), \boldsymbol{\psi}(t), t) = -F(\boldsymbol{x}(t), \boldsymbol{u}(t), t) + \boldsymbol{\psi}^T(t)\, \boldsymbol{f}(\boldsymbol{x}(t), \boldsymbol{u}(t), t) \tag{12.4.36}$$

The variation of the criterion has been expressed with respect to the Hamiltonian through Equation (12.4.16). It results in the canonical system of Hamilton conditions

$$\begin{aligned} \dot{\boldsymbol{x}} &= H_{\boldsymbol{\psi}} \\ \dot{\boldsymbol{\psi}} &= -H_{\boldsymbol{x}} \end{aligned} \tag{12.4.37}$$

that are equivalent to Euler conditions, to which the following equation must be added:

$$H_t = -F_t \tag{12.4.38}$$

The second equation of (12.4.37) is, in fact, a system of equations called the *costate* equations, and $\boldsymbol{\psi}$ is called the costate or the vector of *adjoint* variables.

The derivative of the Hamiltonian is equal to

$$\frac{dH}{dt} = H_{\boldsymbol{x}}^T\, \dot{\boldsymbol{x}} + H_{\boldsymbol{u}}^T\, \dot{\boldsymbol{u}} + H_{\boldsymbol{\psi}}^T\, \dot{\boldsymbol{\psi}} + H_t = H_{\boldsymbol{u}}^T\, \dot{\boldsymbol{u}} + H_t \tag{12.4.39}$$

If $u(t)$ is an optimal control, one deduces

$$\dot{H} = H_t \tag{12.4.40}$$

Generally, the concerned physical system is time invariant, so that time does not intervene explicitly in $\boldsymbol{f}$ and also in the functional F and that Equation (12.4.40) becomes

$$\dot{H} = 0 \tag{12.4.41}$$

In this case, the Hamiltonian is constant along the optimal trajectory.

[3] Other authors use the definition of the Hamiltonian with an opposite sign before the functional, that is,

$$H(\boldsymbol{x}(t), \boldsymbol{u}(t), \boldsymbol{\psi}(t), t) = F(\boldsymbol{x}(t), \boldsymbol{u}(t), t) + \boldsymbol{\psi}^T(t)\boldsymbol{f}(\boldsymbol{x}(t), \boldsymbol{u}(t), t)$$

which changes nothing, as long as we remain at the level of first-order conditions. However, the sign changes in condition (12.4.21). See also the footnote in Section 12.4.6.

The transversality conditions are deduced from Equations (12.3.48) and (12.3.50) or from Equation (12.4.16) as follows.
At initial time t_0,

$$\begin{aligned} &\left[\frac{\partial G}{\partial t_0} + H(t_0)\right] \delta t_0 + \left[\frac{\partial G}{\partial \boldsymbol{x_0}} - \boldsymbol{\psi}(t_0)\right]^T \delta \boldsymbol{x}_0 + \frac{\partial G}{\partial u_0} \delta u_0 = 0 \\ &\text{with} \quad \left(\frac{\partial \boldsymbol{k}}{\partial t}\right)_0 \delta t_0 + \left(\frac{\partial \boldsymbol{k}}{\partial \boldsymbol{x}}\right)_0 \delta \boldsymbol{x}_0 = 0 \end{aligned} \tag{12.4.42}$$

At final time t_f,

$$\begin{aligned} &\left[\frac{\partial G}{\partial t_f} - H(t_f)\right] \delta t_f + \left[\frac{\partial G}{\partial \boldsymbol{x_f}} + \boldsymbol{\psi}(t_f)\right]^T \delta \boldsymbol{x}_f + \frac{\partial G}{\partial u_f} \delta u_f = 0 \\ &\text{with} \quad \left(\frac{\partial \boldsymbol{l}}{\partial t}\right)_f \delta t_f + \left(\frac{\partial \boldsymbol{l}}{\partial \boldsymbol{x}}\right)_f \delta \boldsymbol{x}_f = 0 \end{aligned} \tag{12.4.43}$$

It is possible to calculate the variation $\delta \mathcal{J}$ associated with the variation δt and with the trajectory change of $\delta \boldsymbol{x}$, the extremity $\boldsymbol{x}_f$ being fixed, for the criterion $\mathcal{J}$ defined in a similar way to Equation (12.3.42) as

$$\mathcal{J}(\boldsymbol{x}^*, t) = G(\boldsymbol{x}^*(t_f), t_f) + \int_t^{t_f} F(\boldsymbol{x}^*, \boldsymbol{u}^*, \tau) d\tau \tag{12.4.44}$$

The variation of the criterion can be expressed with respect to the Hamiltonian

$$\begin{aligned} \delta \mathcal{J}(\boldsymbol{x}^*, t) &= \mathcal{J}(\boldsymbol{x}^* + \delta \boldsymbol{x}(t), t + \delta t) - \mathcal{J}(\boldsymbol{x}^*, t) \\ &= -\left(F - F_{\dot{\boldsymbol{x}}}^T \dot{\boldsymbol{x}}^*\right) \delta t - F_{\dot{\boldsymbol{x}}}^T \delta \boldsymbol{x}(t) \\ &= -F(\boldsymbol{x}^*, \boldsymbol{u}^*, t)\, \delta t \\ &= [H(\boldsymbol{x}^*, \boldsymbol{u}^*, \boldsymbol{\psi}, t) - \boldsymbol{\psi}^T(t)\, \boldsymbol{f}(\boldsymbol{x}^*, \boldsymbol{u}^*, t)]\, \delta t \\ &= H(\boldsymbol{x}^*, \boldsymbol{u}^*, \boldsymbol{\psi}, t)\, \delta t - \boldsymbol{\psi}^T(t)\, \delta \boldsymbol{x}(t) \end{aligned} \tag{12.4.45}$$

The optimal control corresponds to a maximum of the Hamiltonian. Frequently, the control vector is bounded in a domain U defined by $\boldsymbol{u}_{\min}$ and $\boldsymbol{u}_{\max}$. In this case, the condition that the Hamiltonian is maximum can be expressed in two different ways:

- When a constraint u_i is reached, the function H defined by Equation (12.4.36) must be a maximum.
- When the control is located strictly inside the feasible domain U defined by $\boldsymbol{u}_{\min}$ and $\boldsymbol{u}_{\max}$, the derivative of function H defined by Equation (12.4.36) with respect to $\boldsymbol{u}$ is zero

$$\frac{\partial H}{\partial u} = 0 \tag{12.4.46}$$

This equation provides an implicit equation that allows us to express the optimal control with respect only to variables $\boldsymbol{x}, \boldsymbol{\psi}, t$

$$\boldsymbol{u}^* = \boldsymbol{u}^*(\boldsymbol{x}, \boldsymbol{\psi}, t) \tag{12.4.47}$$

hence, the new expression of the criterion

$$
\begin{aligned}
\delta\mathcal{J}(\boldsymbol{x}^*, t) &= H(\boldsymbol{x}^*, \boldsymbol{u}^*(\boldsymbol{x}, \boldsymbol{\psi}, t), \boldsymbol{\psi}, t)\delta t - \boldsymbol{\psi}^T(t)\,\delta\boldsymbol{x}(t) \\
&= \mathcal{J}_t \delta t + \mathcal{J}_{\boldsymbol{x}}^T \delta\boldsymbol{x}
\end{aligned}
\qquad (12.4.48)
$$

thus, by identification

$$
\begin{aligned}
\mathcal{J}_t &= H(\boldsymbol{x}^*, \boldsymbol{u}^*(\boldsymbol{x}, \boldsymbol{\psi}, t), \boldsymbol{\psi}, t) \\
\mathcal{J}_{\boldsymbol{x}} &= -\boldsymbol{\psi}(t)
\end{aligned}
\qquad (12.4.49)
$$

This equation shows that the optimal value of the Hamiltonian is equal to the derivative of criterion (12.4.44) with respect to time. The Hamilton–Jacobi equation results in

$$
\mathcal{J}_t - H(\boldsymbol{x}^*, \boldsymbol{u}^*(\boldsymbol{x}, -\mathcal{J}_{\boldsymbol{x}}, t), -\mathcal{J}_{\boldsymbol{x}}, t) = 0 \qquad (12.4.50)
$$

with boundary condition

$$
\mathcal{J}(\boldsymbol{x}_f^*, t_f) = G(\boldsymbol{x}^*(t_f), t_f) \qquad (12.4.51)
$$

The Hamilton–Jacobi equation is a first order partial derivative equation with respect to the sought function $\mathcal{J}$. Its solving is, in general, analytically impossible for a nonlinear system. In the case of a linear system encountered in Linear Quadratic control (LQ control) such as

$$
\begin{cases}
\dot{\boldsymbol{x}}(t) = \boldsymbol{A}\,\boldsymbol{x}(t) + \boldsymbol{B}\,\boldsymbol{u}(t) \\
\boldsymbol{y}(t) = \boldsymbol{C}\,\boldsymbol{x}(t)
\end{cases}
\qquad (12.4.52)
$$

its solving is possible and leads to a Riccati differential equation (Corriou 2018). Thus, it is possible to calculate the optimal control law with state feedback. Recall that the Hamilton–Jacobi Equation (12.4.50) in discrete form corresponds to the Bellman optimality principle in dynamic programming (Section 12.5).

12.4.5.1 Case with Constraints on Control and State Variables

Assume that general constraints of the form:

$$
g(x(t), u(t), t) = 0 \qquad (12.4.53)
$$

are to be respected in the considered problem. In that case, the augmented Hamiltonian is to be considered

$$
H(x(t), u(t), \psi(t), t) = -F(x(t), u(t), t) + \psi^T(t) f(x(t), u(t), t) + \mu^T g(x(t), u(t), t) \qquad (12.4.54)
$$

where μ is a vector of additional Lagrange multipliers. Equation (12.4.46) then yields

$$
\frac{\partial H}{\partial u} = -\frac{\partial F}{\partial u} + \psi^T(t)\frac{\partial f}{\partial u} + \mu^T \frac{\partial g}{\partial u} = 0 \qquad (12.4.55)
$$

together with Equation (12.4.37) as

$$
\dot{\psi} = -H_{\boldsymbol{x}} = F_{\boldsymbol{x}} - \psi^T(t)\, f_{\boldsymbol{x}} - \mu^T\, g_{\boldsymbol{x}} \qquad (12.4.56)
$$

Particular cases of (12.4.53) are those where the constraints g depend only on the states or where a constraint on the state is valid only for a specific time t_1, such as

$$g(x(t_1), t_1) = 0 \tag{12.4.57}$$

called interior point constraints (Bryson and Ho 1975). In that latter case, the state is continuous, but the Hamiltonian H and the adjoint variables ψ are no more continuous. Noting t_1^- and t_1^+ the times just before and after t_1 and given the criterion J, they must verify the following relations:

$$\psi^T(t_1^+) = \frac{\partial J}{\partial x(t_1)}, \quad H(t_1^+) = -\frac{\partial J}{\partial t_1} \tag{12.4.58}$$

and

$$\psi^T(t_1^+) = \psi^T(t_1^-) - \nu^T \frac{\partial g}{\partial x(t_1)}, \quad H(t_1^+) = H(t_1^-) + \nu^T \frac{\partial g}{\partial t_1} \tag{12.4.59}$$

where ν are Lagrange multipliers such that constraints (12.4.57) are satisfied.

12.4.5.2 Case with Terminal Constraints

A case frequently encountered in dynamic optimization is the one where terminal constraints are imposed

$$l_j(\boldsymbol{x}(t_f), \boldsymbol{u}(t_f), t_f) = 0 \tag{12.4.60}$$

The transversality equation (12.4.43) becomes

$$\left[\frac{\partial G}{\partial t_f} - H(t_f) + \frac{\partial \boldsymbol{l}^T}{\partial t_f}\boldsymbol{\nu}\right] \delta t_f + \left[\frac{\partial G}{\partial \boldsymbol{x}_f} + \boldsymbol{\psi}(t_f) + \frac{\partial \boldsymbol{l}^T}{\partial \boldsymbol{x}_f}\boldsymbol{\nu}\right]^T \delta \boldsymbol{x}_f + \frac{\partial G}{\partial u_f}\delta u_f = 0 \tag{12.4.61}$$

where $\boldsymbol{\nu}$ is a vector of Lagrange parameters. If the final time is fixed, the first term of Equation (12.4.61) disappears. If the component $\boldsymbol{x}_i(t_f)$ is fixed at final time, that component disappears in Equation (12.4.61).

12.4.6 *Maximum Principle*

Now, examine briefly the Maximum Principle[4] (Pontryaguine et al. 1974) about process optimal control. Pontryagin emphasizes several points:

---

[4] In many articles, authors refer to the Minimum Principle, which simply results from the definition of the Hamiltonian H with an opposite sign of the functional. Comparing to definition (12.4.36), they define their Hamiltonian as

$$H(\boldsymbol{x}(t), \boldsymbol{u}(t), \boldsymbol{\psi}(t), t) = F(\boldsymbol{x}(t), \boldsymbol{u}(t), t) + \boldsymbol{\psi}^T(t)\boldsymbol{f}(\boldsymbol{x}(t), \boldsymbol{u}(t), t) \tag{12.4.62}$$

With that definition, the optimal control $\boldsymbol{u}^*$ minimizes the Hamiltonian.

- An important difference with respect to variational methods is that it is not necessary to consider two close controls in the admissible control domain.
- The control variables u_i are physical; thus, they are constrained, e.g. $|u_1| \leq u_{\max}$, and we consider that they belong to a domain U. The admissible controls are piecewise continuous; that is, they are continuous nearly everywhere, except at some instants where they can undergo first order discontinuities (jump from one value to another).
- Very frequently, the optimal control is composed by piecewise continuous functions, i.e. the control jumps from one summit of the polyhedron defined by U to another. These cases of control occupying only extreme positions cannot be solved by classical methods.

The process is described by a system of differential equations

$$\dot{x}^i(t) = f^i(\boldsymbol{x}(t), \boldsymbol{u}(t)) \qquad i = 1, \ldots, n \tag{12.4.63}$$

We seek an admissible control $\boldsymbol{u}$ that transfers the system from point $\boldsymbol{x}_0$ in the phase space to point $\boldsymbol{x}_f$ and minimizes the criterion

$$J = G(\boldsymbol{x}_0, t_0, \boldsymbol{x}_f, t_f) + \int_{t_0}^{t_f} F(\boldsymbol{x}(t), \boldsymbol{u}(t))\, dt \tag{12.4.64}$$

To the n coordinates x^i in the phase space, we add the coordinate x^0 defined[5] as

$$x^0 = G(\boldsymbol{x}_0, t_0, \boldsymbol{x}(t), t) + \int_{t_0}^{t} F(\boldsymbol{x}(\tau), \boldsymbol{u}(\tau))\, d\tau \tag{12.4.65}$$

so that if $\boldsymbol{x} = \boldsymbol{x}_f$, then $x^0(t_f) = J$. Its derivative is equal to

$$\frac{dx^0}{dt} = G_{\boldsymbol{x}}^T \boldsymbol{f} + G_t + F(\boldsymbol{x}(t), \boldsymbol{u}(t)) \tag{12.4.66}$$

If time occurs explicitly in the terminal conditions (12.2.2), or if it is not first fixed, we add the coordinate x^{n+1} to the state (Boudarel et al. 1969), such that

$$\begin{aligned} x^{n+1} &= t \\ \dot{x}^{n+1} &= 1 \end{aligned} \tag{12.4.67}$$

The complete system of differential equations would then be of dimension $n + 2$. In the following, in order not to make the notations cumbersome, we will only consider stationary problems of dimension $n + 1$ under the form:

$$\dot{x}^i = f^i(\boldsymbol{x}(t), \boldsymbol{u}(t)), \quad i = 0, \ldots, n \tag{12.4.68}$$

by deducing f^0 from Equation (12.4.65) through derivation (extended notation $\boldsymbol{f}$).

---

[5] This notation is that of Pontryaguine et al. (1974). The superscript corresponds to the rank i of the coordinate, while the subscripts (0 and 1) or (0 and f), according to the authors, are reserved for the terminal conditions.

In the phase space of dimension $n+1$, we define the initial point $\boldsymbol{x}_0$ and a straight line π parallel to the axis x^0 (i.e. the criterion), passing through the final point $\boldsymbol{x}_f$. The optimal control is, among the admissible controls such that the solution $\boldsymbol{x}(t)$, having as the initial condition $\boldsymbol{x}_0$, intersects the line π, the one that minimizes the coordinate x^0 at the intersection point with π.

We introduce the costate variables ψ such that

$$\dot{\boldsymbol{\psi}} = -\boldsymbol{f}_{\boldsymbol{x}}^T \boldsymbol{\psi} \Longleftrightarrow \dot{\psi}_i = -\sum_{j=0}^{n} \frac{\partial f^j(\boldsymbol{x}(t), \boldsymbol{u}(t))}{\partial x^i} \psi_j \,, \quad i = 0, \ldots, n \tag{12.4.69}$$

This system admits a unique solution $\boldsymbol{\psi}$ comprised of piecewise continuous functions, corresponding to the control $\boldsymbol{u}$ and presenting the same discontinuity points.

Then, we consider the Hamiltonian to be equal to the scalar product of functions ψ and f

$$H(\boldsymbol{\psi}, \boldsymbol{x}, \boldsymbol{u}) = \boldsymbol{\psi}^T \boldsymbol{f} = \sum_{i=0}^{n} \psi_i f^i \,, \quad i = 0, \ldots, n \tag{12.4.70}$$

The systems can be written again in the Hamilton canonical form:

$$\begin{aligned} \frac{dx^i}{dt} &= \frac{\partial H}{\partial \psi_i} \,, \quad i = 0, \ldots, n \\ \frac{d\psi_i}{dt} &= -\frac{\partial H}{\partial x^i} \,, \quad i = 0, \ldots, n \end{aligned} \tag{12.4.71}$$

When the solutions $\boldsymbol{x}$ and $\boldsymbol{\psi}$ are fixed, the Hamiltonian depends only on the admissible control $\boldsymbol{u}$, hence the notation

$$\mathcal{M}(\boldsymbol{\psi}, \boldsymbol{x}) = \sup_{u \in U} H(\boldsymbol{\psi}, \boldsymbol{x}, \boldsymbol{u}) \tag{12.4.72}$$

in order to mean that $\mathcal{M}$ is the maximum of H at fixed $\boldsymbol{x}$ and $\boldsymbol{\psi}$, or further

$$H(\boldsymbol{\psi}^*, \boldsymbol{x}^*, \boldsymbol{u}^*) \geq H(\boldsymbol{\psi}^*, \boldsymbol{x}^*, \boldsymbol{u}^* + \delta \boldsymbol{u}) \qquad \forall \; \delta \boldsymbol{u} \tag{12.4.73}$$

We consider the admissible controls, defined on $[t_0, t_f]$, to be responding to the previous definition: the trajectory $\boldsymbol{x}(t)$ issued from $\boldsymbol{x}_0$ at t_0 intersects the straight line π at t_f. According to Pontryaguine et al. (1974), the first theorem of the Maximum Principle is expressed as:
So that the control $\boldsymbol{u}(t)$ and the trajectory $\boldsymbol{x}(t)$ are optimal, it is necessary that the continuous and nonzero vector, $\boldsymbol{\psi}(t) = [\psi_0(t), \psi_1(t), \ldots, \psi_n(t)]$ satisfying Hamilton canonical system (12.4.71), is such that:

1. The Hamiltonian $H[\boldsymbol{\psi}(t), \boldsymbol{x}(t), \boldsymbol{u}(t)]$ reaches its maximum at point $\boldsymbol{u} = \boldsymbol{u}(t) \; \forall \, t \in [t_0, t_f]$, thus
$$H[\boldsymbol{\psi}(t), \boldsymbol{x}(t), \boldsymbol{u}(t)] = \mathcal{M}[\boldsymbol{\psi}(t), \boldsymbol{x}(t)] \tag{12.4.74}$$
2. At the end-time t_f, the relations
$$\psi_0(t_f) \leq 0 \,, \quad \mathcal{M}[\boldsymbol{\psi}(t_f), \boldsymbol{x}(t_f)] = 0 \tag{12.4.75}$$
are satisfied.

With Equation (12.4.71) and condition (12.4.74) being verified, the time functions $\psi_0(t)$ and $\mathcal{M}[\boldsymbol{\psi}(t), \boldsymbol{x}(t)]$ are constant. In this case, the relation (12.4.75) is verified at any instant t included between t_0 and t_f.

12.4.7 Singular Arcs

In optimal control problems, it often occurs for some time intervals that the Maximum Principle does not give an explicit relation between the control and the state and costate variable; this is a singular optimal control problem that yields singular arcs.

Following Lamnabhi-Lagarrigue (1987), an extremal control has a singular arc $[a, b]$ in $[t_0, t_f]$ if and only if $H_u(\boldsymbol{\psi}^*, \boldsymbol{x}^*, \boldsymbol{u}^*) = 0$ and $H_{uu}(\boldsymbol{\psi}^*, \boldsymbol{x}^*, \boldsymbol{u}^*) = 0$, for all $t \in [a, b]$ and whatever $\boldsymbol{\psi}^*$ satisfies the Maximum Principle.

On the arcs corresponding to control constraints, it gives $H_u \neq 0$. Thus, a transversality condition must be verified at the junctions between the arcs. Stengel (1994) notes that if a smooth transition of u is possible for some problems, in some cases, it is necessary to perform a Dirac impulse on the control to link the arcs.

Among problems of singular arcs, a frequently encountered case is the one where the Hamiltonian is linear with respect to the control $\boldsymbol{u}$

$$H(\boldsymbol{x}(t), \boldsymbol{\psi}(t), \boldsymbol{u}(t)) = \alpha(\boldsymbol{x}(t), \boldsymbol{\psi}(t), t)\, \boldsymbol{u}(t) \tag{12.4.76}$$

In that case, the condition

$$\frac{\partial H}{\partial \boldsymbol{u}} = 0 \tag{12.4.77}$$

depends on the sign of α and does not allow us to determine the control with respect to the state and the adjoint vector. To maximize $H(\boldsymbol{u})$, it results in

$$\boldsymbol{u}(t) = \begin{cases} u_{min} & \text{if } \alpha < 0 \\ \text{non defined} & \text{if } \alpha = 0 \\ u_{max} & \text{if } \alpha > 0 \end{cases} \tag{12.4.78}$$

The case where $\alpha = 0$ on a given time interval $[t_1, t_2]$ corresponds to a singular arc. It must then be imposed that the time derivatives of $\partial H / \partial u$ be zero along the singular arc. For a unique control u, the generalized Legendre–Clebsch conditions, also called Kelley conditions that must be verified are

$$(-1)^i \frac{\partial}{\partial u}\left(\frac{d^{2i}}{dt^{2i}} \frac{\partial H}{\partial u}\right) \geq 0, \quad i = 0, 1, \ldots \tag{12.4.79}$$

so that the singular arc be optimal.

Example 12.2:
Linear Quadratic Control using Hamilton–Jacobi and Pontryagin Methods
The previous example is considered again and first treated in the framework of Hamilton–Jacobi equations and then in the framework of the Maximum Principle.

Minimize the performance index

$$J = \frac{1}{2} \int_0^{t_f} (u^2 + x_1^2)\, dt \tag{12.4.80}$$

given

$$\dot{x}_1 = f_1 = a x_1 + b u \tag{12.4.81}$$

Assume that no constraint exists on the control u.

(a) In the context of Hamilton–Jacobi equations, the Hamiltonian would be equal to

$$H = -F + \psi_1 f_1 = -\frac{1}{2} (u^2 + x_1^2) + \psi_1 (a x_1 + b u) \tag{12.4.82}$$

The Hamilton canonical conditions are

$$\begin{cases} \dot{\boldsymbol{x}} = H_{\boldsymbol{\psi}} \\ \dot{\boldsymbol{\psi}} = -H_{\boldsymbol{x}} \end{cases} \Longrightarrow \begin{cases} \dot{x}_1 = a x_1 + b u & \text{with} \quad x_1(0) = x_{1_0} \\ \dot{\psi}_1 = x_1 - a \psi_1 & \text{with} \quad \psi_1(t_f) = -\dfrac{\partial G}{\partial x_f} = 0 \end{cases} \tag{12.4.83}$$

H must be maximized with respect to the control u. As the control u is not constrained, it results that

$$\frac{\partial H}{\partial u} = -u + b \psi_1 = 0 \tag{12.4.84}$$

hence, the optimal control

$$u = b \psi_1 \tag{12.4.85}$$

From the previous equations, we draw

$$\ddot{x}_1 - (a^2 + b^2) x_1 = 0 \tag{12.4.86}$$

The initial state $x_1(0)$ (terminal condition) is given, and the final adjoint variable $\psi(t_f)$ (transversality condition) is also known. Thus, this is again a two-boundary problem with the same numerical issue previously stressed in the same example treated by Euler conditions.

(b) In the context of Pontryagin Maximum Principle, define the coordinate x_0 (corresponding to x^0 of Equation (12.4.65) and here denoted in subscript to avoid confusion with the powers in superscript), such that

$$\dot{x}_0 = f_0 = \frac{1}{2} (u^2 + x_1^2) \tag{12.4.87}$$

The Hamiltonian, defined according to the Maximum Principle, is equal to

$$H = \psi_0 f_0 + \psi_1 f_1 = \frac{1}{2} \psi_0 (u^2 + x_1^2) + \psi_1 (a x_1 + b u) \tag{12.4.88}$$

The Hamilton canonical equations are

$$\begin{cases} \dot{\boldsymbol{x}} = H_{\boldsymbol{\psi}} \\ \dot{\boldsymbol{\psi}} = -H_{\boldsymbol{x}} \end{cases} \Longrightarrow \begin{cases} \dot{x}_0 = \dfrac{\partial H}{\partial \psi_0} = \frac{1}{2} (u^2 + x_1^2) \\ \dot{x}_1 = \dfrac{\partial H}{\partial \psi_1} = a x_1 + b u \quad \text{with } x_1(0) = x_{1_0} \\ \dot{\psi}_0 = -\dfrac{\partial H}{\partial x_0} = 0 \\ \dot{\psi}_1 = -\dfrac{\partial H}{\partial x_1} = -\psi_0 x_1 - \psi_1 a \quad \text{with } \psi_1(t_f) = -\dfrac{\partial G}{\partial x_f} = 0 \end{cases} \tag{12.4.89}$$

The first two equations of this system are the state equations. By setting c_1 as a constant (equal to -1 according to the Hamiltonian given by Equation (12.4.82)), from the two following equations, we draw

$$\begin{aligned} \psi_0(t) &= c_1 \\ \dot{\psi}_1 &= -\psi_1 a - c_1 x_1 \end{aligned} \tag{12.4.90}$$

We notice that $\psi_0(t)$, which concerns the criterion, is constant as already mentioned in the general theory. H must be maximized with respect to the control u. As the control u is not constrained, it gives

$$\frac{\partial H}{\partial u} = \psi_0 \, u + \psi_1 \, b = 0 \tag{12.4.91}$$

hence, the optimal control

$$u = -\frac{\psi_1 \, b}{c_1} \tag{12.4.92}$$

By using the state Equations (12.4.81) and (12.4.90), the same differential equation as (12.4.26) results in

$$\ddot{x} - (a^2 + b^2)x = 0 \tag{12.4.93}$$

Indeed, in the framework of dynamic optimization, Equations (12.4.89) will be integrated with the terminal conditions, including the condition at the fixed initial state $x_1(0) = x_{1_0}$ and the transversality condition at the adjoint variable $\psi_1(t_f) = 0$ at the final time. Of course, the numerical difficulties previously evoked concerning Example 12.1 treated by the Euler equations are encountered here.

Example 12.3 :

Minimum-time Problem with Constraints on the Control

Consider the following classical example of mechanics: a system reduced to a point is described by its position y, its velocity v, and its acceleration u. The latter is thus the input that governs the system's position. Even outside the mechanics domain, this example presents many merits. Essentially, it is very simple so that the different cases studied provide analytical solutions that can be easily understood. The transposition to other engineering fields is not necessarily immediate, but, for example, the problems of minimum time are very similar. Also, the influence of constraints appears clearly and it illustrates perfectly how the Hamilton–Jacobi method is used.

The physical system is described by the following linear second order model:

$$y^{(2)} = u \tag{12.4.94}$$

By setting $x_1 = y$ and $x_2 = \dot{y}$, we get the equivalent state-space model

$$\begin{aligned} \dot{x}_1 &= x_2 \\ \dot{x}_2 &= u \end{aligned} \tag{12.4.95}$$

We wish to go from the state (0, 0) to the state $(A, 0)$ in a minimum time. This corresponds to a variation in amplitude A on the controlled output $y = x_1$ while imposing that its derivative remains zero when $t \to 0$ and $t \to t_f$. We impose that the control, i.e. the manipulated input, must stay in the interval $[-1, 1]$.

The performance index is thus

$$J = \int_0^{t_f} dt \tag{12.4.96}$$

We treat this problem with Hamilton–Jacobi equations, and the Hamiltonian is equal to

$$H = -1 + \psi_1 \, x_2 + \psi_2 \, u \tag{12.4.97}$$

We obtain

$$\dot{\boldsymbol{\psi}} = -H_{\mathbf{x}} \qquad \Longrightarrow \qquad \begin{cases} \dot{\psi}_1 = 0 \\ \dot{\psi}_2 = -\psi_1 \end{cases} \tag{12.4.98}$$

H can be represented with respect to the bounded control u, and we notice that, with this control being linear with respect to u, the maximum of H is reached when u is equal to -1 or $+1$ according to the sign of ψ_2 (Figure 12.1).

It results from Figure 12.1 that

$$u = \text{sign}(\psi_2) \tag{12.4.99}$$

hence,

$$H = -1 + \psi_1 \, x_2 + |\psi_2| \tag{12.4.100}$$

On the contrary, the Hamilton–Jacobi Equation (12.4.50) is

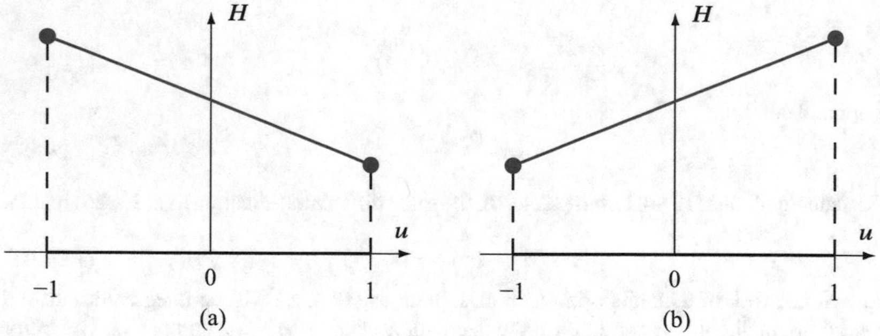

Fig. 12.1 Variation of the Hamiltonian with respect to the bounded control ($|u| \leq 1$) and the sign of ψ_2. **(a)** $\psi_2 < 0$. **(b)** $\psi_2 > 0$

$$\mathcal{J}_t - H(\boldsymbol{x}^*, \boldsymbol{u}^*(\boldsymbol{x}, -\mathcal{J}_{\boldsymbol{x}}, t), -\mathcal{J}_{\boldsymbol{x}}, t) = 0 \tag{12.4.101}$$

with the criterion $\mathcal{J}$ defined by Equation (12.4.44), which verifies

$$\mathcal{J}(\boldsymbol{x}^*, t) = \int_t^{t_f} d\tau \quad \Longrightarrow \quad \mathcal{J}_t = -1 \quad \text{and} \quad \mathcal{J}_{\boldsymbol{x}} = 0 \tag{12.4.102}$$

The Hamilton–Jacobi equation results in

$$H(\boldsymbol{x}^*, \boldsymbol{u}^*(\boldsymbol{x}, -\mathcal{J}_{\boldsymbol{x}}, t), -\mathcal{J}_{\boldsymbol{x}}, t) = -1 \quad \Longrightarrow \quad \psi_1 x_2^* + |\psi_2| = 0 \tag{12.4.103}$$

From the differential equations describing the variation of $\boldsymbol{\psi}$, we deduce

$$\begin{aligned} \psi_1 &= c_1 \\ \psi_2 &= -c_1\, t + c_2 \end{aligned} \tag{12.4.104}$$

where c_1 and c_2 are constant. As ψ_2 depends linearly on t, ψ_2 crosses the value 0 at the maximum only once at the commutation instant t_c, if the latter exists. The control u thus can take only two values at most, and each value only once

$$\text{if } t \in [0, t_c]\,,\ u = \pm 1\,, \quad \text{if } t \in]t_c, t_f]\,,\ u = \mp 1 \tag{12.4.105}$$

This type of control is called bang–bang. Let $u_0 = \pm 1$. We obtain the solutions of the trajectories

$$\begin{aligned} &\text{if } t \in [0, t_c], u = u_0, \quad \begin{cases} x_1 = \dfrac{u_0}{2}\, t^2 \\ x_2 = u_0\, t \end{cases} \\ &\text{if } t \in]t_c, t_f], u = -u_0, \quad \begin{cases} x_1 = -\dfrac{u_0}{2}\,(t - t_f)^2 + A \\ x_2 = -u_0\,(t - t_f) \end{cases} \end{aligned} \tag{12.4.106}$$

Note that $x_1(t)$ is continuous at the commutation time t_c, so that

$$x_1(t_c^-) = x_1(t_c^+) \qquad \begin{cases} x_1(t_c^-) = \dfrac{u_0}{2}\, t_c^2 \\ x_1(t_c^+) = -\dfrac{u_0}{2}\,(t_c - t_f)^2 + A \end{cases} \tag{12.4.107}$$

which gives

$$(t_c - t_f)^2 + t_c^2 = 2\frac{A}{u_0} \tag{12.4.108}$$

This equation implies that the right-hand term is positive, so that

$$u_0 = \text{sign}(A) \tag{12.4.109}$$

Equation (12.4.108) becomes

$$(t_c - t_f)^2 + t_c^2 = 2|A| \Longrightarrow t_f = t_c + \sqrt{2|A| - t_c^2} \tag{12.4.110}$$

As the criterion is $J = t_f$, it suffices to minimize the expression of t_f given by Equation (12.4.110) with respect to t_c to obtain the commutation time

$$\frac{d(t_c + \sqrt{2|A| - t_c^2})}{dt_c} = 0 \Longrightarrow t_c = \sqrt{|A|} \tag{12.4.111}$$

In summary, we obtain the three unknowns t_c, t_f, and u_0 as

$$\begin{aligned} u_0 &= \text{sign}(A) \\ t_c &= \frac{t_f}{2} \\ t_f &= 2\sqrt{|A|} \approx 2.828 \end{aligned} \tag{12.4.112}$$

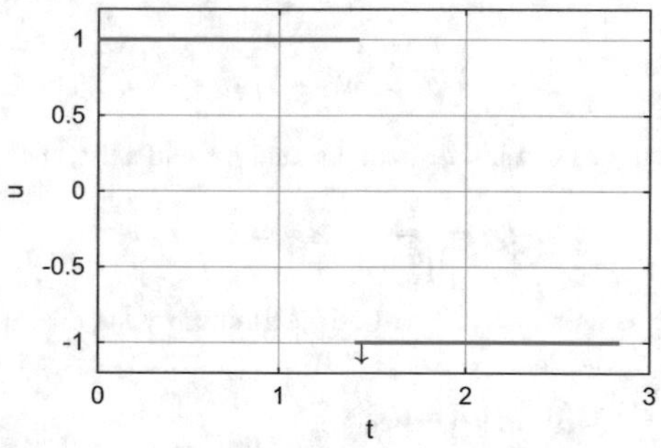

Fig. 12.2 Variation of the manipulated input

The manipulated input presents the bang–bang aspect of Figure 12.2. The commutation instant is indicated by an arrow. The phase portrait (Figure 12.3) shows the two arcs, and the output trajectory effectively presents a horizontal tangent at the beginning and the end.

This type of problem would correspond to a car driver wanting to go from one point to another in minimum time over a short distance. He first accelerates at the maximum, then brakes at the maximum. In chemical engineering, when the minimum time is searched for an operation, often the solution is that a valve is fully open, then fully closed.

Example 12.4 :
Minimum-time Problem with Constraints on the Control and on the State
Consider the same example as previous but further assume that the state x_2 can be bounded, which amounts to a limit for x_1 in its rate for reaching the final state, thus

$$|x_2| \le v \qquad \Longrightarrow \qquad x_2^2 - v^2 \le 0 \tag{12.4.113}$$

Taking into account the constraint, the Hamiltonian can be written as

$$H = -1 + \psi_1\, x_2 + \psi_2\, u + \mu_1(x_2^2 - v^2) \tag{12.4.114}$$

where μ_1 is a Kuhn–Tucker parameter. It results in

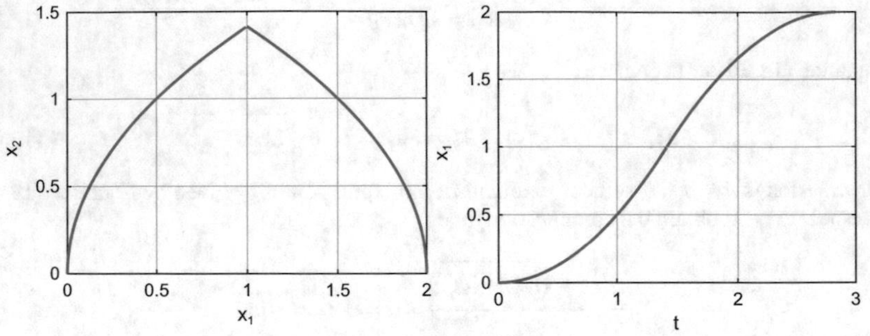

Fig. 12.3 Left: phase portrait, right: trajectory of the controlled output (chosen value of the final state x_1: $A = 2$)

$$\begin{array}{lll} \mu_1 = 0 & \text{if} & x_2^2 - v^2 < 0 \text{ (inside the domain, with respect to } |x_2| < v) \\ \mu_1 \le 0 & \text{if} & x_2^2 - v^2 = 0 \text{ (on the constraint } |x_2| = v) \end{array} \tag{12.4.115}$$

The condition $\mu_1 \le 0$ comes from the fact that H must be a maximum. We obtain

$$\begin{cases} \dot\psi_1 = 0 \\ \dot\psi_2 = -\psi_1 - 2\,\mu_1\,x_2 \end{cases} \tag{12.4.116}$$

At most, two commutation (or junction) instants t_{c1} and t_{c2} will exist, such that

$$t_{c1} = \frac{v}{|u_0|}\,, \quad t_{c2} = t_f - t_{c1} \tag{12.4.117}$$

Note that the final time t_f is unknown and will be determined by the final condition on the state x_1.

$$\begin{array}{ll} \text{if } t \in [0, t_{c1}],\, u = u_0, & \begin{cases} x_1 = \frac{u_0}{2}\,t^2 \\ x_2 = u_0\,t \end{cases} \\ \text{if } t \in]t_{c1}, t_{c2}],\, u = 0, & \begin{cases} x_1 = u_0\,\frac{t_{c1}}{2}(2\,t - t_{c1}) \\ x_2 = u_0\,t_{c1} \end{cases} \\ \text{if } t \in]t_{c2}, t_f],\, u = -u_0, & \begin{cases} x_1 = -\frac{u_0}{2}\,(t - t_f)^2 + A \\ x_2 = -u_0\,(t - t_f) \end{cases} \end{array} \tag{12.4.118}$$

We deduce

$$t_f = \frac{|u_0|}{u_0\,v}\,(A + \frac{v^2}{u_0}) = \frac{A}{v}\,\text{sign}(u_0) + \frac{v}{|u_0|} \tag{12.4.119}$$

This final time is equal to 3.12 in the case where the state x_2 is bounded by 0.9 and is larger than the final time without constraint on the state.

The time interval $[t_{c1}, t_{c2}]$ corresponds to a singular arc and is called a singular time interval.

The input (Figure 12.4) shows first its saturation, then the singular arc, and then again a saturation with an opposite value. Commutation instants are indicated by arrows. The phase portrait (Figure 12.5) displays the two commutation times and the saturation for $x_2 = 0.9$. We notice on the trajectory of the controlled output x_1 that during the saturation of x_2 (rate of x_1), while the manipulated input u being constant, x_1 increases linearly.

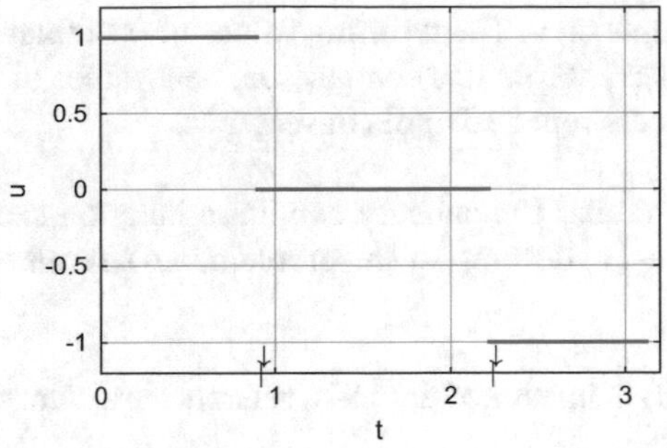

Fig. 12.4 Variation of the manipulated input

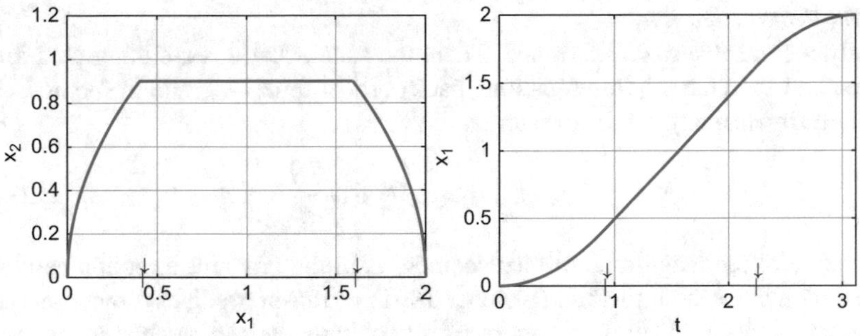

Fig. 12.5 Left: phase portrait, right: trajectory of the controlled output (chosen value for the final state x_1: $A = 2$ and the state x_2 is bounded: $|x_2| \leq 0.9$)

12.4.8 Numerical Issues

In general, the dynamic optimization problem results in a set of two systems of first order ordinary differential equations

$$\begin{aligned} \dot{\boldsymbol{x}} &= \dot{\boldsymbol{x}}(\boldsymbol{x}, \boldsymbol{\psi}, t) \quad \text{with} \quad \boldsymbol{x}(t_0) = \boldsymbol{x}_0 \\ \dot{\boldsymbol{\psi}} &= \dot{\boldsymbol{\psi}}(\boldsymbol{x}, \boldsymbol{\psi}, t) \quad \text{with} \quad \boldsymbol{\psi}(t_f) = \boldsymbol{\psi}_f \end{aligned} \tag{12.4.120}$$

where t_0 and t_f are initial and final time, respectively. Thus, it is a two-point boundary value problem. A criterion J is to be minimized with respect to a control vector. In general, in particular for nonlinear problems, there is no analytical solution.

The following general strategy is used to solve the two-point boundary value problem (12.4.120): an initial vector $\boldsymbol{x}(t)$ or $\boldsymbol{\psi}(t)$ or $\boldsymbol{\psi}(t_0)$ or $\boldsymbol{u}(t)$ is chosen, then by an iterative procedure, the vectors are updated until all equations are respected, including in particular the initial and final conditions.

Different numerical techniques (Bryson 1999; Latifi et al. 1998) can be used to find the optimal control, which are detailed as follows.

Boundary Condition Iteration:

First, the adjoint vector is initialized and the system (12.4.120) concerning both $\boldsymbol{x}$ and $\boldsymbol{\psi}$ is integrated. The control vector $u(t)$ results from the maximization of the Hamiltonian with possible constraints on u. The resulting values of $\boldsymbol{\psi}$ at final time t_f are compared to the required values $\boldsymbol{\psi}_f$. From this comparison, new values of the adjoint vector are deduced, and the process repeated until convergence.

Multiple Shooting:
This method is very similar to boundary condition iteration except that intermediate points are used in $[t_0, t_f]$ to decompose the problem into a series of boundary condition iteration problems.

Quasi-linearization:
The system (12.4.120) is linearized around a reference trajectory.

Invariant Embedding:
The initial two-point boundary value problem is transformed into an initial value problem that is now a system of partial differential equations.

Control Vector Iteration:
Initially, a control vector is assumed. Then, the state equations are integrated forward in time and then the adjoint equations backward in time. A gradient method can be used to estimate a new control vector as

$$u_{new}(t) = u_{old}(t) + \alpha \frac{\partial H}{\partial u} \tag{12.4.121}$$

The choice of the magnitude of displacement α in the gradient direction can be performed by a line search method (Fletcher 1991) or Rosen gradient projection method (Soeterboek 1992). In some minimum-time problems such as batch styrene polymerization (Farber and Laurence 1986), a method based on coordinate transformation can be used, which simply needs that one of the state variables be monotonic such as in general conversion (Kwon and Evans 1975).

Control Vector Parameterization:
In this method (Goh and Teo 1988; Teo et al. 1991), the control vector is approximated by basis functions. The control profile can be chosen in different ways, such as piecewise constant, piecewise linear, or piecewise polynomial, so that the control is expressed as

$$u(t) = \sum a_i \, \phi_i(t) \tag{12.4.122}$$

where $\phi(t)$ are adequate basis functions, which can be Lagrange polynomials. Thus, the optimization is realized with respect to parameters a_i that govern the control profile.

This method has been used by Fikar et al. (2000) to determine the optimal inputs during the transition from one couple of set points to another for a continuous industrial distillation column. A full model of the distillation column was used by the authors.

Corriou and Rohani (2008) searched optimal temperature profiles for a crystallizer by using an original method based on the concept of moving horizon, used in particular in model predictive control (Corriou 2018). Different criteria were minimized in the presence of various constraints. Figure 12.6 shows the optimal temperature profile and resulting concentration profiles obtained by that technique.

Collocation on Finite Elements or Control and State Parameterization:
Both state and control variables are approximated by basis functions such as Lagrange polynomials (Biegler 1984; Cuthrell and Biegler 1987). The final problem is solved by nonlinear programming.

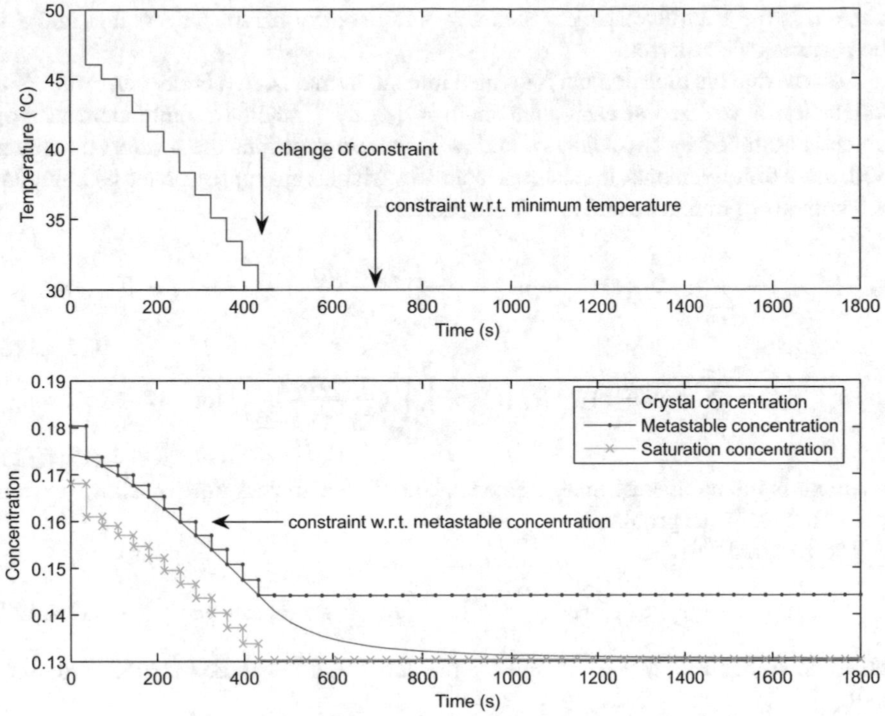

Fig. 12.6 Optimal temperature profile (top) and resulting concentration profiles (bottom) calculated by dynamic optimization with evidence of constraints, (objective function $J = \mu_3^n(t_f) - \mu_3^s(t_f)$) taken from Corriou and Rohani (2008)

Consider the following general optimization problem in the form:

$$(u, q)^* = \arg\{\min_{u(t),q} J\} \tag{12.4.123}$$

with the criterion to be minimized

$$J = \Psi[x(t_f), q, t_f] + \int_{t_o}^{tf} \Phi[x(t), u(t), q, t]dt \tag{12.4.124}$$

subject to the constraints

$$\begin{cases} g[x(t), u(t), q, t] \leq 0 & \text{instantaneous inequality constraints} \\ h[x(t), u(t), q, t] = 0 & \text{instantaneous equality constraints} \\ \dot{x}(t) = f[x(t), u(t), q, t] & \text{state model} \\ x(t_o) = x_o & \text{initial conditions} \\ x_{inf} \leq x(t) \leq x_{sup} & \text{state domain} \\ u_{inf} \leq u(t) \leq u_{sup} & \text{control domain} \\ q_{inf} \leq q \leq q_{sup} & \text{parameter domain} \end{cases} \tag{12.4.125}$$

where u is the control variable's vector, q is the control parameter's vector, and J is the performance criterion.

Assume that the time domain is divided into subdomains. In this case, an orthogonal collocation is realized at each subdomain $= [\alpha_i, \alpha_{i+1}]$, called a finite element $\Delta\alpha_i$, which is bounded by two knots α_i and α_{i+1}, whose positions are a priori unknown. With these finite elements, the state and control variables are approximated by Lagrange polynomials of orders nc and $nc - 1$, respectively,

$$x^i_{nc}(\tau) = \sum_{j=0}^{nc} a_{ij}\Phi_{ij}(\tau), \quad \Phi_{ij}(\tau) = \prod_{k=0,\neq j}^{nc} (\frac{\tau - \tau_{ik}}{\tau_{ij} - \tau_{ik}}) \qquad \text{for} \quad i = 1, \ldots, ne \tag{12.4.126}$$

$$u^i_{nc-1}(\tau) = \sum_{j=1}^{nc} b_{ij}\Psi_{ij}(\tau), \quad \Psi_{ij}(\tau) = \prod_{k=1,\neq j}^{nc} (\frac{\tau - \tau_{ik}}{\tau_{ij} - \tau_{ik}}) \qquad \text{for} \quad i = 1, \ldots, ne \tag{12.4.127}$$

where ne is the number of finite elements. The dimensionless time, τ, allows to treat easily free end-time problems.

The τ_{ij} are defined as

$$\tau_{ij} = \alpha_i + \gamma_j(\alpha_{i+1} - \alpha_i), \quad i = 1, \ldots, ne, \quad j = 0, \ldots, nc \tag{12.4.128}$$

where $\gamma_0 = 0$ and γ_j $(j = 1, \ldots, nc)$ are the zeros of a Legendre polynomial defined on $[0, 1]$.

Replacing the state and control variables with their polynomial approximations in the state system leads to the following algebraic residual equations:

$$r(\tau_{il}) = \sum_{j=0}^{nc} a_{ij}\dot{\Phi}_{ij}(\tau_{il}) - t_f.f[a_{il}, b_{il}, q, \tau_{il}] = 0, \quad l = 1, \ldots, nc, \; i = 1, \ldots, ne \tag{12.4.129}$$

where

$$\dot{\Phi}_{ij}(\tau_{il}) = \frac{\dot{\Phi}_j(\tau_l)}{\Delta\alpha_i} \tag{12.4.130}$$

Then, it is necessary to impose the continuity of the state variables between two successive finite elements and to bound the extrapolation of the control variables at both ends of the finite elements, as they are only defined inside each element.

Finally, the problem (12.4.123) can be approximated as follows:

$$(a_{il}, b_{il}, q, \alpha_i, t_f)^* = \arg\{\min_{a_{il}, b_{il}, q, \alpha_i, t_f} J[a_{il}, b_{il}, q, t_f]\} \tag{12.4.131}$$

subject to

$$\begin{cases} g[a_{il}, b_{il}, q, \tau_{il}] \le 0 \\ h[a_{il}, b_{il}, q, \tau_{il}] = 0 \\ r(\tau_{il}) = 0 \\ a_{10} = x_o \\ a_{i0} = \sum_{j=0}^{nc} a_{i-1j} \Phi_j(\tau = 1) \\ u_{inf} \le u^i_{nc-1}(\alpha_i) \le u_{sup} \\ u_{inf} \le u^i_{nc-1}(\alpha_{i+1}) \le u_{sup} \\ x_{inf} \le a_{il} \le x_{sup} \\ u_{inf} \le b_{il} \le u_{sup} \\ q_{inf} \le q \le q_{sup} \end{cases} \quad (12.4.132)$$

The optimization problem parameters are the state and control variables' values at the collocation points defined by Equation (12.4.128), the final time, and the position of the knots, which correspond at convergence to the control variable discontinuities. The state and control variables are then completely defined by Equations (12.4.126) and (12.4.127). This method allows us to easily handle all types of constraints.

The resulting nonlinear problem (12.4.131) can then be solved using a successive quadratic programming technique, such as that developed by Schittkowski (1985).

The drawback of this technique is that a large number of parameters to be optimized are generated. The advantage is that it does not require the use of Hamiltonian and adjoint equations. However, this method has been successfully used by Gentric et al. (1999) to calculate off-line the optimal temperature profile of a batch emulsion copolymerization reactor.

Iterative Dynamic Programming:

The system is simply described by the dynamic model

$$\dot{\boldsymbol{x}} = \boldsymbol{f}(\boldsymbol{x}, \boldsymbol{u}) \quad (12.4.133)$$

where the initial state vector $\boldsymbol{x}_0$ is given. A performance index $J(\boldsymbol{x}(t_f))$ is to be minimized, with the final time t_f being specified. The optimal control policy is searched by stage-to-stage optimization, where stages correspond to time subintervals. No gradients are necessary, and no auxiliary variables are introduced.

The general principle Luus (1996) is the following:

1. Divide the total time interval $[t_0, t_f]$ into P subintervals $[0, t_1], \ldots, [t_{i-1}, t_i]$, etc. of equal length L.
2. Choose an initial estimate of control u for each stage, giving the control policy vector $\boldsymbol{u}$ of dimension P. The control belongs to an initial domain denoted by $\boldsymbol{r}_{in}$. Choose the region contraction factor γ and the region restoration factor η.
3. Choose the number of iterations $j_{\max}$ for each pass and the number of passes $q_{\max}$.
4. Set pass index $q = 1$ and iteration index $j = 1$.
5. Set $\boldsymbol{r}^{(j)} = \eta^q \, \boldsymbol{r}_{in}$.
6. Integrate dynamic model (12.4.133) to get the state vector $\boldsymbol{x}(k-1)$ at each stage k.
7. From that stage, the principle of dynamic programming is really involved. Beginning at stage P, integrate model (12.4.133) from $t_f - L$ to t_f for each of the R allowable values of continuous control vector given by

$$\boldsymbol{u}(t) = \boldsymbol{u}(k) + [\boldsymbol{u}(k+1) - \boldsymbol{u}(k)](t - t_k)/L\,, \quad \forall\, k \tag{12.4.134}$$

with

$$\boldsymbol{u}(P-1) = \boldsymbol{u}^*(P-1)^{(j)} + \boldsymbol{D}_1\,\boldsymbol{r}^{(j)}\,, \quad \boldsymbol{u}(P) = \boldsymbol{u}^*(P)^{(j)} + \boldsymbol{D}_2\,\boldsymbol{r}^{(j)} \tag{12.4.135}$$

where the superscript * stands for the best value of the previous iteration and $\boldsymbol{D}_1$ and $\boldsymbol{D}_2$ are diagonal matrices of different random numbers between 0 and 1. The control values should satisfy the constraints $\boldsymbol{u}_{\min} \leq \boldsymbol{u} \leq \boldsymbol{u}_{\max}$. The control values that give the best performance index are retained as $\boldsymbol{u}(P-1)$ and $\boldsymbol{u}(P)$.

8. Proceeding backward, continue the procedure of step 7 until initial time $t = 0$.
9. Reduce the region for allowable control

$$\boldsymbol{r}^{(j+1)} = \gamma \boldsymbol{r}^{(j)} \tag{12.4.136}$$

using the best control policy coming from step 8 (denoted by *) as the mean value for the allowable control.

10. Set $j = j + 1$ and go to step 7. Continue until $j < j_{\max}$.
11. Set $q = 1$ and go to step 5. Continue until $q < q_{\max}$.

Variants of this method are given in the following articles, Banga and Carrasco (1998), Bojkov and Luus (1996), Carrasco and Banga (1997), Luus and Hennessy (1999), Mekarapiruk and Luus (1997). Examples of applications to batch chemical reactors are often cited in these articles.

12.5 Dynamic Programming (Discrete Time)

12.5.1 Classical Dynamic Programming

Dynamic programming (Bellman 1957; Bellman and Dreyfus 1962) has found many applications in all scientific and economic domains, among which is aeronautics (Hagelauer and Mora-Camino 1998; Harada and Miyazawa 2013; Hur et al. 2021). Frequently used in the 1960s and later, its principle remains valid, but, with computer progress, nonlinear optimization is more and more used. Staged systems can be optimized by dynamic programming. Among such typical processes in chemical engineering (Aris 1961; Roberts 1964) are the optimization of discontinuous reactors or reactors in series, catalyst replacement or regeneration, the optimization of the counter-current extraction process (Aris et al. 1960), the optimal temperature profile of a tubular chemical reactor (Aris 1960), and the optimization of a cracking reaction (Roberts and Laspe 1961).

In a different domain, a famous problem is the travelling salesman who, having to go from one point to another, must optimize his travel, which includes the possibility of going through different towns (Gutin and Punnen 2007; Hansknecht et al. 2021).

Optimality Principle (Bellman 1957):

> A policy is optimal if and only if, whatever the initial state and the initial decision, the decisions remaining to be taken constitute an optimal policy with respect to the state resulting from the first decision.

Because of the principle of continuity, the optimal final value of the criterion is entirely determined by the initial condition and the number of stages. In fact, it is possible to start from any stage, even from the last one. For this reason, Kaufmann and Cruon (1965) express the optimality principle in the following manner:

> A policy is optimal if, at a given time, whatever the previous decisions, the decisions remaining to be taken constitute an optimal policy with respect to the result of the previous decisions,

or further,

> Any sub-policy (from $\boldsymbol{x}_i$ to $\boldsymbol{x}_j$) extracted from an optimal policy (from $\boldsymbol{x}_0$ to $\boldsymbol{x}_N$) is itself optimal from $\boldsymbol{x}_i$ to $\boldsymbol{x}_j$.

At first, dynamic programming is discussed in the absence of constraints, which could be terminal constraints, constraints at any time (amplitude constraints) on the state $\boldsymbol{x}$ or on the control u, or inequality constraints. Moreover, we assume the absence of discontinuities.

In fact, as this is a numerical and not analytical solution, these particular cases previously mentioned would pose no problem and could be automatically considered.

In continuous form, the problem is the following:
Consider the state equation

$$\dot{\boldsymbol{x}} = f(\boldsymbol{x}, u) \quad \text{with } \boldsymbol{x}(0) = \boldsymbol{x}_0 \tag{12.5.1}$$

and the performance index to be minimized

$$J(u) = \int_0^{t_f} r(\boldsymbol{x}, u)dt \tag{12.5.2}$$

where r represents an income or revenue.

In discrete form, the problem becomes:
Consider the state equation

$$\boldsymbol{x}_{n+1} = \boldsymbol{x}_n + f(\boldsymbol{x}_n, u_n)\Delta t \tag{12.5.3}$$

with $\Delta t = t_{n+1} - t_n$. The control u_n brings the system from the state $\boldsymbol{x}_n$ to the state $\boldsymbol{x}_{n+1}$ and results in an elementary income $r(\boldsymbol{x}_n, u_n)$ (integrating, in fact, the control period Δt, which will be omitted in the following).

According to the performance index in the integral form, let us define the performance index or total income at instant N (depending on the initial state $\boldsymbol{x}_0$ and the policy $\mathcal{U}_0^{N-1}$ followed from 0 to $N-1$, bringing from the state $\boldsymbol{x}_0$ to the state $\boldsymbol{x}_N$) as the sum of the elementary incomes $r(\boldsymbol{x}_i, u_i)$

$$J_0 = \sum_{i=0}^{N-1} r(\boldsymbol{x}_i, u_i) \tag{12.5.4}$$

The values of the initial and final states are known

$$\boldsymbol{x}(t_0) = \boldsymbol{x}_0\,, \quad \boldsymbol{x}(t_N) = \boldsymbol{x}_N \tag{12.5.5}$$

If the initial instant is n, note the performance index J_n.

The problem is to find the optimal policy $\mathcal{U}_0^{*,N-1}$ constituted by the succession of controls u_k^* ($k = 0, \dots, N-1$) minimizing the performance index J_0. The optimal performance index $J^*(\boldsymbol{x}_0, 0)$ is defined as

$$J^*(\boldsymbol{x}_0, 0) = \min_{\{u_k\}} J_0 = \min_{\{u_k\}} \sum_{k=0}^{N-1} r(\boldsymbol{x}_k, u_k) \tag{12.5.6}$$

This performance index bears on the totality of the N stages and depends on the starting point $\boldsymbol{x}_0$. In fact, the optimality principle can be applied from any instant k, to which corresponds the optimal performance index $J^*(\boldsymbol{x}_k, k)$.

From the optimality principle, the following recursive algorithm of search of the optimal policy by *backward intuition* is derived:

$$J_b^*(\boldsymbol{x}_k, k) = \min_{u_k}[r(\boldsymbol{x}_k, u_k) + J_b^*(\boldsymbol{x}_k + f(\boldsymbol{x}_k, u_k), k+1)]\,, \quad k = N-1, \dots, 0 \tag{12.5.7}$$

or

$$J_{b,k}^*(\boldsymbol{x}_k, u_{k-1}) = \min_{u_k}[r(\boldsymbol{x}_k, u_k) + J_{b,k+1}^*(\boldsymbol{x}_{k+1}, u_k)] \tag{12.5.8}$$

which allows us to calculate the series $J_b^*(\boldsymbol{x}_{N-1}, N-1), \dots, J_b^*(\boldsymbol{x}_0, 0)$ from the final state $\boldsymbol{x}_N$ by determination of $u_{N-1}, \dots, u_0$ in this order (Figure 12.7). Thus, finally, the optimal performance index $J_b^*(\boldsymbol{x}_0, u_0) = J_0$ results.

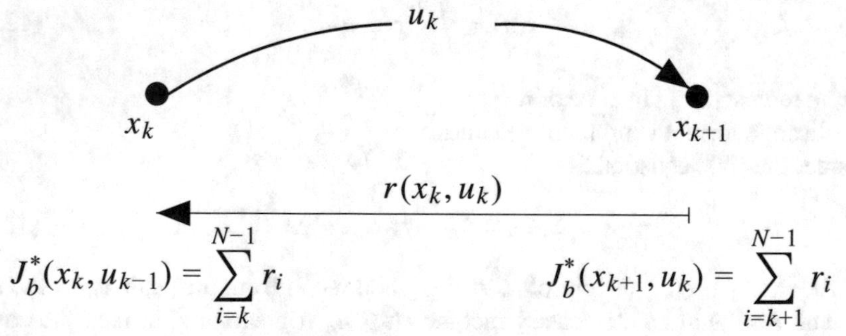

Fig. 12.7 Principle of dynamic programming by backward induction, $k = N-1, N-2, \dots, 0$

If the final state is free, choose $J^*(\boldsymbol{x}_N, N) = 0$. In the case where the final state is constrained, the last control u_{N-1}^* is calculated so as to satisfy the constraint.

The method resulting from Equation (12.5.7) is known as *backward intuition* because it starts at the final extremity and proceeds backward to the origin stage by stage.

Forward intuition also exists, progressing from the origin toward the final extremity. Obviously, it must be expressed differently as

$$J_f^*(\boldsymbol{x}_{k+1}, k+1) = \min_{u_k}[r(\boldsymbol{x}_k, u_k) + J_f^*(\boldsymbol{x}_k, k)], \quad k = 0, \dots, N-1 \tag{12.5.9}$$

or

$$J_{f,k+1}^*(\boldsymbol{x}_{k+1}, u_k) = \min_{u_k}[r(\boldsymbol{x}_k, u_k) + J_{f,k}^*(\boldsymbol{x}_k, u_{k-1})] \tag{12.5.10}$$

In this case, $u_0, u_1, \dots, u_{N-1}$ are successively determined (Figure 12.8). Then, finally, the optimal performance index is $J_f(x_N, u_{N-1}) = J_0$.

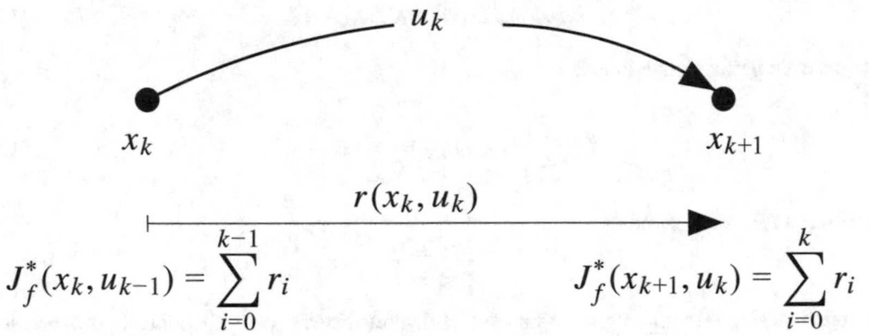

Fig. 12.8 Principle of dynamic programming by forward induction, $k = 0, 1, \dots, N-1$

It must be noticed that, according to the type of problems to be solved, dynamic programming by backward and forward inductions does not lead to the same solution (Hoppe 2007; Perea 2010, 2012). Dynamic programming by backward induction succeeds even if the model is stochastic, i.e. some uncertainty is present. Dynamic programming by forward induction succeeds only when the model is deterministic or perfect, i.e. in the absence of any uncertainty in the model. Indeed, dynamic programming by forward induction corresponds to an open loop control strategy where the user is fully confident with respect to the model of the system, whereas dynamic programming by backward induction corresponds to a closed loop control strategy where the user makes corrections at each new step by taking into account the model errors and the disturbances. Thus, in the forward induction approach, as the model is deterministic, the optimal policy depends only on the initial state x_0 and all actions $u_0, \dots, u_{N-1}$ are determined at initial time. Oppositely, in the backward induction approach, the optimal policy is recalculated at each new stage k corresponding to a state x_k and depends on the control law u_k that can be stated as some function of x_k; that is, some function $u_k = g(x_k)$. Thus, selecting the best approach for a particular issue must be seriously taken into account.

The algorithm (12.5.7) by backward induction could be written as

$$\begin{aligned} J_b^*(\boldsymbol{x}_k, k) &= \min_{u_k}[r(\boldsymbol{x}_k, u_k) + \min_{u_{n+1}}[r(\boldsymbol{x}_{k+1}, u_{k+1}) + J_b^*(\boldsymbol{x}_{k+2}, k+2)]] \\ &= \min_{u_k}[r(\boldsymbol{x}_k, u_k) + \min_{u_{k+1}}[r(\boldsymbol{x}_{k+1}, u_{k+1}) + \dots]] \end{aligned} \tag{12.5.11}$$

However, a difficulty resides frequently in the formulation of a given problem in an adequate form for the solving by means of dynamic programming and, with the actual progress of numerical calculation and nonlinear constrained optimization methods, the latter are nowadays more employed. A variant of dynamic programming (Luus 1990) called iterative dynamic programming can often provide good results with a lighter computational effort (Luus 1993, 1994; Luus and Bojkov 1994).

Example 12.5 :
Application of Dynamic Programming by backward induction
Consider again a particular form of example (12.4.22) in discrete form. The stable system of dimension 1 is then defined by the following discrete-time state equation:

$$x_{k+1} = f(x, u) = 0.5\, x_k + u_k \tag{12.5.12}$$

and the discrete performance index:

$$J = \sum_{k=0}^{3} r(x_k, u_k) = \sum_{k=0}^{3} (x_k^2 + u_k^2) \tag{12.5.13}$$

Terminal constraints are given

$$\begin{cases} x_0 = 0 \\ x_4 = 1 \end{cases} \tag{12.5.14}$$

In fact, this is a single-input single-output case of discrete linear quadratic control. This problem has been originally solved using Maple© by symbolic computation, and it can also be solved by Matlab again using symbolic computation.

We therefore wish to find the control u_k such that the sum of the costs r_k up to time k is minimum when the state goes from x_k to x_{k+1}. At each instant n, we will calculate the control u_n such that the performance index is minimized in backward recursive form as

$$J_b^*(x_n, n) = \min_{u_n} [r(x_n, u_n) + J_b^*(x_{n+1}, n+1)] \tag{12.5.15}$$

beginning from the last instant N.
The elementary income is equal to

$$r(x_k, u_k) = x_k^2 + u_k^2 \tag{12.5.16}$$

The process consists of four stages, thus $N = 4$. Set $a = 0.5$.
First stage:
The final state is fixed: $x_4 = 1$. It results that

$$x_4 = a x_3 + u_3 = 1 \tag{12.5.17}$$

which, when joined with the expression (12.5.12) of the state x_3,

$$x_3 = a x_2 + u_2 \tag{12.5.18}$$

gives the optimal control u_3^*

$$u_3^* = 1 - a(a x_2 + u_2) \tag{12.5.19}$$

which depends on the value of the state x_2 and the control u_2.
Second stage:
The performance index is transformed by using the expression of x_3 from Equation (12.5.12) and the expression (12.5.19) of the optimal control u_3^*

$$
\begin{aligned}
J(x_0, 0) &= (x_0^2 + u_0^2) + (x_1^2 + u_1^2) + (x_2^2 + u_2^2) + (x_3^2 + u_3^2) \\
&= (x_0^2 + u_0^2) + (x_1^2 + u_1^2) + (x_2^2 + u_2^2) + (a x_2 + u_2)^2 + [1 - a(a x_2 + u_2)]^2 \\
&= (x_0^2 + u_0^2) + (x_1^2 + u_1^2) + J_b(x_2, 2)
\end{aligned}
\tag{12.5.20}
$$

Note that at this instant, the performance index is comprised of two parts, one depending on instants from 0 to 1, the other one $J_b(x_2, 2)$, which is essential depending on the instant 2, thus on the state x_2 and the control u_2. The performance index $J_b(x_2, 2)$ must be minimum with respect to u_2, which gives the optimal control u_2^*

$$
\begin{aligned}
&\frac{dJ_b(x_2, 2)}{du_2} = 0 \Longrightarrow \\
&2u_2 + 2(a x_2 + u_2) - 2a[1 - a(a x_2 + u_2)] = 0 \Longrightarrow \\
&u_2^* = -\frac{a(x_2 - 1 + a^2 x_2)}{2 + a^2} = \frac{2}{9} x_4 - \frac{5}{9} x_2
\end{aligned}
\tag{12.5.21}
$$

Third stage:
The following is given numerically because of the complexity of analytical expressions. The performance index $J(x_0, 0)$ is transformed by using the expression of x_2 from Equation (12.5.12) and the expression (12.5.21) of the optimal control u_2^*

$$
\begin{aligned}
J(x_0, 0) &= (x_0^2 + u_0^2) + (x_1^2 + u_1^2) + (0.5 x_1 + u_1)^2 + (-0.139 x_1 - 0.278 u_1 + 0.222)^2 \\
&\quad + (0.111 x_1 + 0.222 u_1 + 0.222)^2 + (0.889 - 0.556 x_1 - 0.111 u_1)^2 \\
&= (x_0^2 + u_0^2) + J_b(x_1, 1)
\end{aligned}
\tag{12.5.22}
$$

The performance index $J_b(x_1, 1)$ must be minimum with respect to u_1, which gives the optimal control u_1^*

$$
\frac{dJ_b(x_1, 1)}{du_1} = 0 \Longrightarrow u_1^* = \frac{4}{77} x_4 - \frac{41}{154} x_1
\tag{12.5.23}
$$

Fourth stage:
Finally, the performance index is transformed by using the expression of x_1 from Equation (12.5.12) and the expression (12.5.23) of the optimal control u_1^*

$$
\begin{aligned}
J(x_0, 0) &= (x_0^2 + u_0^2) + (0.5 x_0 + u_0)^2 + (-0.133 x_0 - 0.266 u_0 + 0.0519)^2 \\
&\quad + (0.117 x_0 + 0.234 u_0 + 0.0519)^2 + (-0.0325 x_0 - 0.0649 u_0 + 0.208)^2 \\
&\quad + (0.0260 x_0 + 0.0519 u_0 + 0.234)^2 + (0.883 - 0.0130 x_0 - 0.0260 u_0)^2
\end{aligned}
\tag{12.5.24}
$$

This performance index must be minimum with respect to u_0, which gives the optimal control u_0^*

$$
\frac{dJ(x_0, 0)}{du_0} = 0 \Longrightarrow u_0^* = \frac{8}{657} x_4 - \frac{349}{1314} x_0
\tag{12.5.25}
$$

Table 12.1 Succession of the optimal states and controls obtained by dynamic programming

Instant k	State x_k	Control u_k^*
0	0	0.01218
1	0.01218	0.04871
2	0.05479	0.20700
3	0.23440	0.88280
4	1	

By knowing the value of the original state x_0, the succession of inputs and thus of the states is easily calculated in Table 12.1. The calculation was first performed by Maple using symbolic computation, and then the same calculation was performed by Matlab. Finally, the optimal discrete performance index is equal to 0.88280.

According to Table 12.1, the discrete inputs and associated states can be represented (Figure 12.9). The discrete state-space model does not allow us to obtain continuous variations of the states.

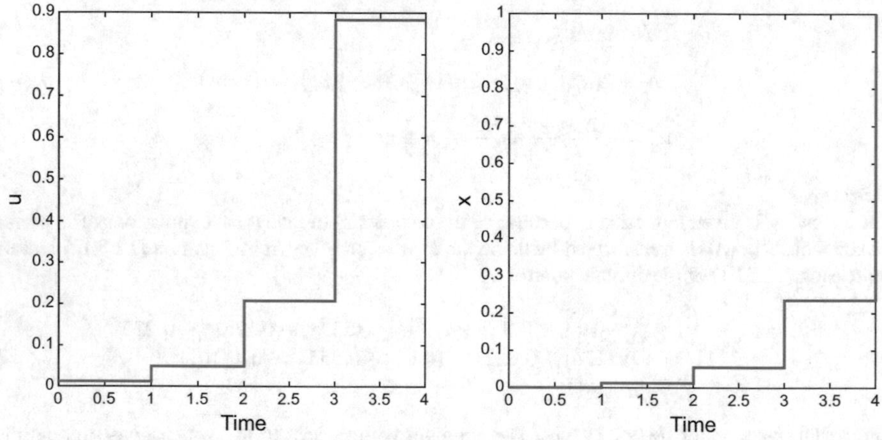

Fig. 12.9 Dynamic programming: Successive inputs (left) and states (right) determined by the application of backward induction

---

Example 12.6 :

Application of Dynamic Programming by forward induction

The example 12.5 used to demonstrate dynamic programming by backward induction is used here to decompose the method by forward induction.

Again, the problem is solved in four stages, but starting from x_0.

First stage:

In this stage, x_0 being fixed but assumed unknown and x_2 taking any value, the states from x_0 to x_2 are considered and the objective is to find the optimal value of u_0 that minimizes the part of the criterion J between x_0 and x_2, that is,

$$J_f(x_2, 2) = J_{f,2}(x_2, u_1) = x_0^2 + u_0^2 + x_1^2 + u_1^2 \tag{12.5.26}$$

In this expression, u_1 is first replaced by

$$u_1 = x_2 - a x_1 \tag{12.5.27}$$

with $a = 0.5$ from the discrete state-space model, and then x_1 is replaced by

$$x_1 = a x_0 + u_0 \tag{12.5.28}$$

This yields an expression of $J_f(x_2, 2)$ that now depends on x_0 and u_0, but also the state x_2 which is any, as

$$J_f(x_2, 2) = (u_0 + 0.5x_0)^2 + (0.5u_0 + 0.25x_0 - x_2)^2 + u_0^2 + x_0^2 \tag{12.5.29}$$

Differentiating $J_f(x_2, 2)$ with respect to u_0 and making the derivative equal to zero yield the optimal value of u_0 for any value of x_2

$$\frac{dJ_f(x_2, 2)}{du_0} = 0 \Longrightarrow u_0^* = \frac{2}{9}x_2 - \frac{5}{18}x_0 \tag{12.5.30}$$

hence, the optimal value of the state x_1

$$x_1^* = \frac{2}{9}x_0 + \frac{2}{9}x_2 \tag{12.5.31}$$

The optimal value of $J_f(x_2, 2)$ can be calculated as $J_f^*(x_2, 2)$ by replacing u_0 with u_0^*.

Second stage:

The states from x_0 to any value x_3 are considered, but the part from x_0 to x_1 is optimal from the previous calculation. The part of the criterion J that is considered is

$$\begin{aligned} J_f(x_3, 3) &= x_0^2 + u_0^2 + x_1^2 + u_1^2 + x_2^2 + u_2^2 \\ &= J_f^*(x_2, 2) + x_2^2 + u_2^2 \end{aligned} \tag{12.5.32}$$

Again, in this expression, u_2 is replaced by

$$u_2 = x_3 - a x_2 \tag{12.5.33}$$

then x_2 is replaced by

$$x_2 = a x_1^* + u_1 \tag{12.5.34}$$

This yields an expression of $J_f(x_3, 3)$ that now depends on x_0 and u_1, but also the state x_3 that is any. Differentiating $J_f(x_3, 3)$ with respect to u_1 and making this derivative equal to zero yield the optimal value of u_1

$$u_1^* = \frac{18}{77}x_3 - \frac{1}{9}x_2 - \frac{41}{693}x_0 \tag{12.5.35}$$

and the optimal value of the state x_2

$$x_2^* = a x_1^* + u_1^* = \frac{4}{77}x_0 + \frac{18}{77}x_3 \tag{12.5.36}$$

so that

$$u_1^* = \frac{16}{77}x_3 - \frac{5}{77}x_0 \tag{12.5.37}$$

Third stage:

The states from x_0 to any value x_4 are considered, but the part from x_0 to x_2 is optimal from the previous calculation. The part of the criterion J that is considered is

$$\begin{aligned} J_f(x_4, 4) &= x_0^2 + u_0^2 + x_1^2 + u_1^2 + x_2^2 + u_2^2 + x_3^2 + u_3^2 \\ &= J_f^*(x_3, 3) + x_3^2 + u_3^2 \end{aligned} \tag{12.5.38}$$

In this expression, u_3 is replaced by

$$u_3 = x_4 - a x_3 \tag{12.5.39}$$

then x_3 is replaced by

$$x_3 = a x_2^* + u_2 \tag{12.5.40}$$

This yields an expression of $J_f(x_4, 4)$ that depends on x_0 and u_2, but also the state x_4 considered as any. Differentiating $J_f(x_4, 4)$ with respect to u_2 and making this derivative equal to zero yield the optimal value of u_2

$$u_2^* = \frac{154}{657}x_4 - \frac{9}{77}x_3 - \frac{698}{50589}x_0 \tag{12.5.41}$$

and the optimal value of the state x_3

$$x_3^* = a x_2^* + u_2^* = \frac{8}{657}x_0 + \frac{154}{657}x_4 \tag{12.5.42}$$

so that

$$u_2^* = \frac{136}{657} x_4 - \frac{10}{657} x_0 \qquad (12.5.43)$$

Fourth stage:
As x_4 is the final imposed state equal to 1, the criterion $J_f(x_4, 4)$ is equal to J. The final value of x_4 is imposed, yielding x_3^*. The optimal value of u_3 can be calculated as

$$u_3^* = x_4 - a x_3^* = 1 - \frac{4}{657} x_0 - \frac{77}{657} x_4 \qquad (12.5.44)$$

The previous optimal control values u_i^* and the states x_i^* can be reconstructed by simple backward calculation. Finally, the same values as in Table 12.1 from backward induction are obtained by forward induction. Of course, the figures of the states and control values are identical as the model is perfect, i.e. without any uncertainty.

Example 12.7 :
Solution by Nonlinear Programming

The example of dynamic programming previously solved by backward induction (Example 12.5) or forward induction (Example 12.6) is now solved by Nonlinear Programming. Considering the states from $x_0 = 0$ to $x_4 = 1$, the control vector is $\{u_0, u_1, \ldots, u_3\}$.

The function "fmincon" of Matlab is used to minimize the criterion. In the criterion, it must be noted that the last value u_3 of the control vector is determined from the optimal state x_3 and the final value x_4, thus the control vector to be determined by optimization is $\{u_0, u_1, u_2\}$. The main program is written as follows:

```
global a xinit xfinal
a = 0.5;
xinit = 0;
xfinal = 1;
n = 4;
u0 = 0.25*ones(n-1,1);
uoptimal = fmincon(@dynamic_programming_criterion,u0);
```

The criterion is given by the function "dynamic_programming_criterion" as follows:

```
function J = dynamic_programming_criterion(u)
global a xinit xfinal
n = length(u);
for i=1:n+1
  if (i==1)
    x(i) = xinit;
  else
    x(i) = a*x(i-1) + u(i-1);
  end
end
J = 0;
for i=1:n
  J = J + x(i)^2 + u(i)^2;
end
% the final value of u is imposed by the final condition on x
x(n+2) = xfinal;
ulast = x(n+2) - a*x(n+1);
J = J + x(n+1)^2 + ulast^2;
```

Given the numerical tolerances, the results thus obtained by Nonlinear Programming are the same as those obtained by dynamic programming by backward induction (Example 12.5 and Table 12.1) or by forward induction (Example 12.6). The initialization before calling "fmincon" has little effect as far as it is reasonable.

12.5.2 Hamilton–Jacobi–Bellman Equation

Given the initial state $\boldsymbol{x}_0$ at time t_0, considering the state $\boldsymbol{x}$ and the control u, the optimal trajectory corresponds to the couple $(\boldsymbol{x}, u)$, such that

$$J^*(\boldsymbol{x}_0, t_0) = \min_{u(t)} J(\boldsymbol{x}_0, u, t_0) \tag{12.5.45}$$

thus the optimal criterion does not depend on the control u.

In an interval $[t, t + \Delta t]$, the Bellman optimality principle as given in the recursive Equation (12.5.7) can be formulated as

$$J^*(\boldsymbol{x}(t), t) = \min_{u(t)} \left\{ \int_t^{t+\Delta t} r(\boldsymbol{x}, u, \tau) d\tau + J^*(\boldsymbol{x}(t + \Delta t), t + \Delta t) \right\} \tag{12.5.46}$$

This can be expressed in continuous form as a Taylor series expansion in the neighborhood of the state $\boldsymbol{x}(t)$ and time t

$$J^*(\boldsymbol{x}(t), t) = \min_{u(t)} \left\{ \begin{array}{l} r(\boldsymbol{x}, u, t)\, \Delta t + J^*(\boldsymbol{x}(t), t) + \dfrac{\partial J^*}{\partial t}\, \Delta t + \\ \left(\dfrac{\partial J^*}{\partial \boldsymbol{x}}\right)^T f(\boldsymbol{x}, u, t)\, \Delta t + 0(\Delta t) \end{array} \right\} \tag{12.5.47}$$

Taking the limit when $\Delta t \to 0$ results in the Hamilton–Jacobi–Bellman equation

$$-\frac{\partial J^*}{\partial t} = \min_{u(t)} \left\{ r(\boldsymbol{x}, u, t) + \left(\frac{\partial J^*}{\partial \boldsymbol{x}}\right)^T f(\boldsymbol{x}, u, t) \right\} \tag{12.5.48}$$

As the optimal criterion does not depend on control u, it yields $J^*(\boldsymbol{x}(t_f), t_f) = W(\boldsymbol{x}(t_f))$, which gives the boundary condition for the Hamilton–Jacobi–Bellman Equation (12.5.48)

$$J^*(\boldsymbol{x}, t_f) = W(\boldsymbol{x}), \quad \forall\, \boldsymbol{x} \tag{12.5.49}$$

The solution of Equation (12.5.48) is the optimal control law

$$u^* = g(\frac{\partial J^*}{\partial \boldsymbol{x}}, \boldsymbol{x}, t) \tag{12.5.50}$$

which, when introduced into Equation (12.5.48), gives

$$-\frac{\partial J^*}{\partial t} = r(\boldsymbol{x}, g, t) + \left(\frac{\partial J^*}{\partial \boldsymbol{x}}\right)^T f(\boldsymbol{x}, g, t) \tag{12.5.51}$$

whose solution is $J^*(\boldsymbol{x}, t)$ subject to the boundary condition (12.5.49). Equation (12.5.51) should be compared to Hamilton–Jacobi Equation (12.3.45). Then, the gradient $\partial J^*/\partial \boldsymbol{x}$ should be calculated and returned in (12.5.50), which gives the optimal state-feedback control law

$$u^* = g(\frac{\partial J^*}{\partial \boldsymbol{x}}, \boldsymbol{x}, t) = h(\boldsymbol{x}, t) \tag{12.5.52}$$

This corresponds to a closed-loop optimal control law.

12.6 Conclusion

Dynamic optimization occupies a large place in advanced scientific domains such as aeronautics and aerospace but also economics or other domains. Some difficulty exists for many engineers to recognize a dynamic optimization issue in their problems. However, even if they have some idea about this issue, another gap is to be able to set it under an adequate mathematical form and the last gap is to solve the well-posed problem. Thus, this topic of dynamic optimization remains often ignored or not enough used even when it could offer large perspectives.

In this chapter, dynamic optimization has been split into continuous-time and discrete-time sections. In the first stage, continuous-time dynamic optimization is exposed in the mathematical framework with Euler and Jacobi's methods. Then, still in continuous-time, dynamic optimization is exposed by means of Euler, Jacobi, and Pontryagin's maximum principle methods by order of difficulty and efficiency. Finally, in discrete-time, dynamic optimization is developed by Bellman's dynamic programming. Different examples allow the reader to understand each of these methods.

12.7 Exercise Set

Preamble: Calogero (2020) has published a large and very well-presented set of exercises about dynamic optimization. They can easily be found on the Internet. They are ordered with respect to the different mathematical issues and their solutions are given.

Exercise 12.7.1 (Medium)

Using the concepts of dynamic optimization, find the curve of smallest length joining two points of $\mathbb{R}^2$. To simplify the representation, considering the points $M_1(x_1, y_1)$ and $M_2(x_2, y_2)$, assume that $x_2 > x_1$ and $y_2 > y_1$.

Exercise 12.7.2 (Medium)

Consider the set of all curves f joining two points (x_1, y_1) and (x_2, y_2) of $\mathbb{R}^2$. Suppose that the curve f turns around the horizontal axis Ox. The surface described by this rotation has the following area:

$$J = 2\pi \int_{x_1}^{x_2} f(x)\sqrt{1 + f'(x)^2}\, dx$$

Find the curves f that minimize the surface J.

Exercise 12.7.3 (Medium)
In $\mathbb{R}^2$, find the curve joining the points $(0,0)$ and $(4,4)$, such that the arc length is equal to 6 and the underlying surface area is maximum.

Exercise 12.7.4 (Difficult)
Using dynamic programming, determine

$$\max \sum_{i=1}^{N} \sqrt{x_i}$$

subject to

$$\sum_{i=1}^{N} x_i = d$$

with d a real positive constant and $x_i > 0 \quad \forall\, i$. This problem amounts to dividing a positive constant into N parts, such that the sum of the parts is maximum.
Indications:
Set $s_n = x_1 + \cdots + x_n$, $n = 1, \ldots, N$. Furthermore, set $f_1(s_1) = \sqrt{s_1}$ and show by induction that we can set

$$f_n(s_n) = \max_{\substack{x_n \\ 0<x_n\leq s_n}} \left\{ f_{n-1}(s_{n-1}) + \sqrt{x_n} \right\}$$

It is useful to notice that

$$s_{n-1} = s_n - x_n$$

and to introduce an additional function $F_n(x_n)$. Deduce the value of the maximum and the value of x_i.

Exercise 12.7.5 (Medium)
A linear first order system is represented by the following transfer function:

$$G(s) = \frac{10}{5s + 1}$$

equivalent to the linear state-space model

$$\begin{cases} \dot{x} = -0,2x + u \\ y = 2x \end{cases}$$

Find the control law $u(t)$ that minimizes the criterion

$$J = \int_{t_1=0}^{t_2=T} [4(x - x^{ref})^2 + (u - u^{ref})^2]\, dt$$

where u^{ref} is a fixed reference value and x^{ref} the corresponding reference state.

Exercise 12.7.6 (Difficult)

A particular linear dynamic system is defined under discrete form by the following state-space equation:

$$x_{n+1} = x_n + u_n \tag{12.7.1}$$

where x represents the state, u the control, and n the discrete time. The objective is to find the control u minimizing the following criterion:

$$J = \sum_{i=0}^{3} [x_i^2 + u_i^2] \tag{12.7.2}$$

and such that the following terminal constraints are satisfied:

$$\begin{cases} x_0 = 0 \\ x_4 = 1 \end{cases} \tag{12.7.3}$$

Moreover, it is assumed that $0 \leq x_i \leq 1$.
Indications: solve the problem by dynamic programming. To find the solution, even if x can take any value in the interval $[0, 1]$, in a first stage, it will be assumed that x_i can only take discrete values between 0 and 1, scaled by an increment equal to 0.1 and thus equal to 0 , 0.1 , 0.2 , ... , 0.9 , 1. The solution of the problem is obtained when providing the series of controls u_i and the followed states x_i.
Answer: $x_0^* = 0$, $u_0^* = 0.047619$, $x_1^* = 0.047619$, $u_1^* = 0.095238$, $x_2^* = 0.142857$, $u_2^* = 0.238095$, $x_3^* = 0.380952$, $u_3^* = 0.6190476$, $x_4^* = 1$.

Exercise 12.7.7 (Difficult)
Minimize the sum

$$f(\boldsymbol{x}) = x_1^2 + x_2^2 + x_3^2 + x_4^2 + x_5^2 \tag{12.7.4}$$

subject to

$$\begin{aligned} & x_1\, x_2\, x_3\, x_4\, x_5 = 11 \\ & x_i \geq 0 \quad \forall i \end{aligned} \tag{12.7.5}$$

by dynamic programming. The numerical value 11 is purely indicative as well as the dimension of $\boldsymbol{x}$ equal to 5. Indeed, the proposed method allows us to solve this type of problem whatever the product of x_i and the dimension of $\boldsymbol{x}$.
Method to use:
Set

$$p_n = \prod_{i=1}^{n} x_i \tag{12.7.6}$$

and use the dynamic programming relation

$$f_n(p_n) = \min_{x_n} \{f_{n-1}(p_{n-1}) + x_n^2\} \tag{12.7.7}$$

We will work by induction.

1. Justify the relation (12.7.7). Relate p_{n-1} and p_n.
2. Make the study for $n = 1$. Deduce x_1.
3. Make the study for $n = 2$. Deduce x_2.

4. Make the study for $n = 3$. Deduce x_3.
5. (a) If you have the intuition of the induction at this stage, set the induction relation, suppose it true until $n - 1$, and deduce the relation for n.
   (b) It is quite possible that the iteration for $n = 3$ is not sufficient to get the intuition of the induction. In this case, make the fourth iteration. At that stage, the induction should be obvious. Set the induction relation, suppose it true until $n - 1$, and deduce the relation for n.
6. Deduce the general relation.

Exercise 12.7.8 (Difficult)
Given the revenue r_i of stage i as

$$r_i = 7\,x_i - i\,x_i^2 \tag{12.7.8}$$

find the optimum policy that maximizes the total revenue for 3 stages ($i = 1, 2, 3$) while respecting the constraint

$$\sum_{i=1}^{3} x_i = 6 \qquad \text{with} \quad x_i \geq 0 \quad \forall\, i \tag{12.7.9}$$

To solve the exercise, the following indications are given. Set

$$s_i = \sum_{j=1}^{i} x_j \tag{12.7.10}$$

An intermediate variable can be used

$$f_n(s_n) = \max_{x_n}\{f_{n-1}(s_{n-1}) + r_n\} \tag{12.7.11}$$

while noticing that $s_n = s_{n-1} + x_n$.

Give the optimum policy and the corresponding total revenue.
Answer: $x_1^* = 36/11$, $x_2^* = 18/11$, $x_3^* = 12/11$, $J^* = 246/11$.

Exercise 12.7.9 (Difficult)
A plant manufactures a product whose stock is noted $x(t)$ and the production rate $\dot{x}(t)$. We desire to produce a quantity Q in an imposed time and with a minimum cost. The production cost follows the relation

$$J = \int_0^T \left[\frac{a}{2}\dot{x}^2(t) + b x(t)\right] dt \tag{12.7.12}$$

where the parameters a and b are generally positive. The objective is the optimum production rate. Assume that $\dot{x}(0) = 0$ (steady state regime at $t = 0$) and that $x(0) = x_0$ (initial stock).

1. Express the fact that Q is produced in the interval $[0, T]$.
2. Set the problem as an optimization problem with respect to the variable $x(t)$.

3. Solve this optimization problem. Give the production rate, stock, and cost.
4. Calculate the cost that would result from working at fixed production rate and compare both solutions.

Exercise 12.7.10 (Difficult)

Consider n perfectly stirred tank reactors in steady state (Figure 12.10). The inlet flow rate of the first reactor is constituted by the reactant A with a volume flow rate F and a concentration C_{A0}. All reactors work at the same temperature and are the place of a first order reaction

$$\mathrm{A} \rightarrow \mathrm{B} \qquad \text{with} \quad r_A = k\, C_A$$

where r_A is the reaction rate of A. Each reactor has a different volume V_i. However, the sum of the n volumes is constant and equal to V_t.

Fig. 12.10 Perfectly stirred tank reactors in series

The objective is to maximize the concentration of B at the outlet of the last reactor by acting on the volumes of the successive reactors.

1. Write the equations from the chemical engineering point of view (Fogler 2016; Levenspiel 1999), that is,

$$\begin{aligned}\text{Rate of component } i \text{ entering} = {} & \text{Rate of component } i \text{ exiting} \\ & + \text{Rate of component } i \text{ accumulated} \\ & + \text{Rate of component } i \text{ reacted}\end{aligned} \qquad (12.7.13)$$

   In steady state, the term of accumulation is zero. Do not forget that the stream entering the second reactor contains some product B.

   In the final result, to simplify the writing in view of the solution, the parameter $a = k/L$ should be used.
2. Find the solution for $n = 2$.
3. Indeed, the choice $n = 2$ is a very particular case. Set the problem for $n > 2$ in a mathematical manner and very precisely describe the solution method.
4. Solve the problem for $n = 3$ by using the general method. Describe precisely your reasoning.
5. Compare the final result to that obtained by assuming that all reactors have the same volume.

Numerical application: V_t =0.1 m^3, $F = 0.001$ m^3 s^{-1}, $k = 0.005$ s^{-1}, $C_{A0} =$ 1000 mol m^{-3}.

Exercise 12.7.11 (Difficult)

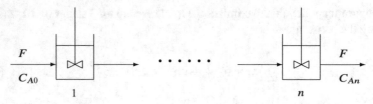

The reactions

$$A \rightarrow B \rightarrow C \rightarrow D \tag{12.7.14}$$

take place in a plug-flow reactor, i.e. a cylinder where the concentration depends on the space variable z, the distance along the reactor of total length L (Figure 12.11). A stream of flow rate F at pure A (mole fraction $y_A = 1$) enters the plug-flow reactor.

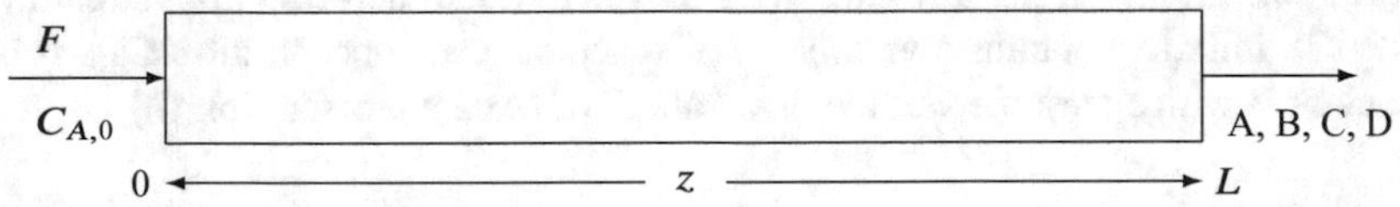

Fig. 12.11 Plug-flow reactor

The mole fractions noted y_i are such that

$$\begin{aligned} \frac{dy_A}{dz} &= -\alpha_1\, y_A + \beta_1\, T(z) \\ \frac{dy_B}{dz} &= \alpha_1\, y_A - \alpha_2\, y_B + \beta_2\, T(z)^2 \\ \frac{dy_C}{dz} &= \alpha_2\, y_B - \alpha_3\, y_C + \beta_3\, T(z)^3 \end{aligned} \tag{12.7.15}$$

with

$$\sum_{i=1}^{4} y_i = 1\,, \quad 0 \leq y_i \leq 1$$

where i represents A, B, C, or D. The kinetic parameters α_i are constant. The influence of temperature T is given by the parameters β_i.

The objective is to maximize the mole fraction of B at the reactor outlet ($z = L$) by using a feed of pure A, i.e. $y_A(0) = 1$.

1. Write the equations allowing us to obtain the optimum temperature profile $T(z)$.
2. Show that the optimum profile has the form:

$$T(z) = k\,[1 - \exp((\alpha_2 - \alpha_1)(z - L))] \tag{12.7.16}$$

   where k is a constant to be determined.
3. Compare the optimal solution to that obtained by working at a fixed temperature.

Remark: To find the solution of an ordinary differential equation of the form:

$$y' + f(x)\, y = g(x) \tag{12.7.17}$$

we first seek the general solution y_1 of the homogeneous ordinary differential equation

$$y' + f(x)\, y = 0 \tag{12.7.18}$$

then a particular solution y_2 of the general equation (12.7.17). Hence, the general solution of Equation (12.7.17) is written as

$$y(x) = a\, y_1(x) + y_2(x) \tag{12.7.19}$$

where a is a constant to be determined from the boundary condition of the equation. Numerical application: $F = 0.001$ m^3 s^{-1}, $L = 1$ m, $\alpha_1 = 10$, $\alpha_2 = 0.5$, $\alpha_3 = 0.1$, $T(0) = 400$, $\beta_1 = 3 \times 10^{-4}$, $\beta_1 = 10^{-6}$, $\beta_3 = 10^{-9}$.

Exercise 12.7.12 (Difficult)

A fed-batch reactor, i.e. a reactor with an inlet stream and no outlet stream (Figure 12.12), initially contains a volume V_0 of reactant A at concentration C_{A0}. It is fed in reactant B with a variable volume flow rate F of fixed concentration $C_{B,in}$.

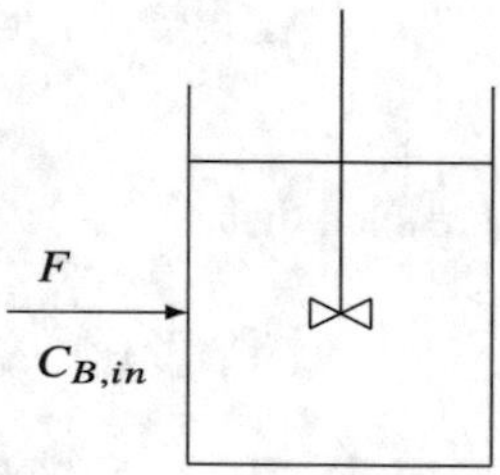

Fig. 12.12 Fed-batch reactor

The competitive reaction scheme takes place in the reactor

$$\begin{aligned} &\mathrm{A + B \rightarrow C} \qquad && r_1 = k_1\, C_A\, C_B \\ &\mathrm{2\,B \rightarrow D} && r_2 = k_2\, C_B^2 \end{aligned}$$

where the reaction rates r_1 and r_2 are expressed in mol s^{-1} m^{-3}.

The objective is to get the product C with a maximum concentration in a given time t_f.

1. Write the system of ordinary differential equations giving the evolutions of the volume V and of concentrations of A, B, C, and D in the reactor (refer to Equation (12.7.13)) (Fogler 2016; Levenspiel 1999). Show that the model can be finally reduced to the ordinary differential equations with respect to the volume V and the concentrations of A, B, and C.
   In the following, only the reduced model is considered.
2. Express the criterion under a canonical form. Express the Hamiltonian according to Hamilton–Jacobi theory.
3. Deduce the canonical system of Hamilton–Jacobi equations. Give the corresponding boundary conditions.
4. Let F^* be the flow rate profile that maximizes the final concentration of C. Give the expression of $\dfrac{\partial H}{\partial F}$.
5. The feed flow rate $F(t)$ is bounded: $0 \le F(t) \le F_{max}$. Give the optimum value of F at each time with respect to the sign of $\dfrac{\partial H}{\partial F}$.
6. Show that, at final time t_f, we have

$$\left(\frac{\partial H}{\partial F}\right)_{tf} = -\left(\frac{C_C}{V}\right)_{tf}$$

What must be the value of the flow rate F at final time t_f ?

7. Intuitively, find what must be the value of the flow rate F at initial time $t = 0$.
8. To completely solve the problem, the time interval $[t_0, t_f]$ is discretized in n equal subintervals. Supposing that, in each subinterval, the flow rate is constant, propose in a flow sheet a solving method allowing you to determine the n flow rates F_i that would maximize the final concentration of C.
9. Compare the optimum solution to that obtained by working at fixed flow rate.

Numerical application: $V_0 = 0.1\,\text{m}^3$, $F = 0.001\,\text{m}^3\,\text{s}^{-1}$,$C_{A0} = 2000\,\text{mol}\,\text{m}^{-3}$, $C_{B,in} = 1000\,\text{mol}\,\text{m}^{-3}$, $k_1 = 10^{-6}\ \text{mol}^{-1}\,\text{m}^3\,\text{s}^{-1}$, $k_2 = 2 \times 10^{-5}\,\text{mol}^{-1}\,\text{m}^3\,\text{s}^{-1}$, $t_f = 3600\,\text{s}$.

References

R. Aris. Studies in optimization. II. Optimal temperature gradients in tubular reactors. *Chem. Eng. Sci.*, 13(1):18–29, 1960.

R. Aris. *The Optimal Design of Chemical Reactors: a Study in Dynamic Programming*. Academic Press, New York, 1961.

R. Aris, D. F. Rudd, and N. R. Amundson. On optimum cross current extraction. *Chem. Eng. Sci.*, 12:88–97, 1960.

J. R. Banga and E. F. Carrasco. Rebuttal to the comments of Rein Luus on "Dynamic optimization of batch reactors using adaptive stochastic algorithms". *Ind. Eng. Chem. res.*, 37:306–307, 1998.

R. J. Barro and X. Sala-I-Martin. *Economic Growth*. McGraw-Hill, New York, 2004.

R. Bellman. *Dynamic Programming*. Princeton University Press, Princeton, New Jersey, 1957.

R. Bellman and S. Dreyfus. *Applied Dynamic Programming*. Princeton University Press, Princeton, New Jersey, 1962.

D. P. Bertsekas. *Dynamic Optimization and Optimal Control*. Athena Scientific, Belmont, 2017.

L. T. Biegler. Solution of dynamic optimization problems by successive quadratic programming and orthogonal collocation. *Comp. Chem. Eng.*, 8:243–248, 1984.

M. Blanco Fonseca and G. Flichman. *Bio-Economic Models applied to Agricultural Systems*, chapter Dynamic Optimisation Problems: Different Resolution Methods Regarding Agriculture and Natural Resource Economics, pages 29–57. Springer-Verlag, 2011.

B. Bojkov and R. Luus. Optimal control of nonlinear systems with unspecified final times. *Chem. Eng. Sci.*, 51(6):905–919, 1996.

P. Borne, G. Dauphin-Tanguy, J. P. Richard, F. Rotella, and I. Zambettakis. *Commande et optimisation des processus*. Technip, Paris, 1990.

R. Boudarel, J. Delmas, and P. Guichet. *Commande optimale des processus*. Dunod, Paris, 1969.

A. E. Bryson. *Dynamic Optimization*. Addison Wesley, Menlo Park, California, 1999.

A. E. Bryson and Y. C. Ho. *Applied Optimal Control*. Hemisphere, Washington, 1975.

A. Calogero. Notes on optimal control theory with economic models and exercises. Technical report, Università di Milano-Bicocca, 2020.

J. Cao and G. Illing. *Instructor's Manual for Money: Theory and Practice*. Springer Nature Switzerland, 2019.

E. F. Carrasco and J. R. Banga. Dynamic optimization of batch reactors using adaptive stochastic algorithms. *Ind. Eng. Chem. Res.*, 36:2252–2261, 1997.

H. Cartan. *Cours de calcul différentiel*. Hermann, Paris, 1967.

W. C. Chan. *An Elementary Introduction to Queueing Systems*. World Scientific, 2014.

A. C. Chiang. *Elements of Dynamic Optimization*. Waveland Press, Long Grove, IL, 1992.

J. P. Corriou. *Process control - Theory and applications*. Springer, London, 2nd edition, 2018.

J. P. Corriou and S. Rohani. A new look at optimal control of a batch crystallizer. *AIChE J.*, 54(12):3188–3206, 2008.

L. Cretegny and T. F. Rutherford. Worked examples in dynamic optimization: analytic and numeric methods. Technical report, Centre of Policy Studies, Monash University, Australia, 2004.

J. E. Cuthrell and L. T. Biegler. On the optimization of differential-algebraic process systems. *A.I.Ch.E. J.*, 33:1257–1270, 1987.

J. N. Farber and R. L. Laurence. The minimum time problem in batch radical polymerization: a comparison of two policies. *Chem. Eng. Commun.*, 46:347–364, 1986.

A. Feldbaum. *Principes théoriques des systèmes asservis optimaux*. Mir, Moscou, 1973. Edition Française.

M. Fikar, M. A. Latifi, J. P. Corriou, and Y. Creff. CVP-based optimal control of an industrial depropanizer column. *Comp. Chem. Engn.*, 24:909–915, 2000.

R. Fletcher. *Practical Methods of Optimization*. Wiley, Chichester, 1991.

H. S. Fogler. *Elements of Chemical Reaction Engineering*. Prentice-Hall, Boston, 5th edition, 2016.

C. Gentric, F. Pla, M. A. Latifi, and J. P. Corriou. Optimization and non-linear control of a batch emulsion polymerization reactor. *Chem. Eng. J.*, 75:31–46, 1999.

C. J. Goh and K. L. Teo. Control parametrization: a unified approach to optimal control problems with general constraints. *Automatica*, 24:3–18, 1988.

G. Gutin and A. P. Punnen, editors. *The Traveling Salesman Problem and Its Variations*. Springer, New York, 2007.

P. Hagelauer and F. Mora-Camino. A soft dynamic programming 4D-trajectory approach for on-line aircraft optimization. *European Journal of operational Research*, 107:87–95, 1998.

R. C. Hampshire and W. A. Massey. Dynamic optimization with applications to dynamic rate queues. *Tutorials in Operations Research*, pages 208–247, 2010.

C. Hansknecht, I. Joormann, and S. Stiller. Dynamic shortest paths methods for the time-dependent TSP. *Algorithms*, 14, 2021.

A Harada and Y. Miyazawa. Dynamic programming applications to flight trajectory optimization. In *19th IFAC Symposium on Automatic Control in Aerospace, Würzburg, Germany*, pages 441–446, 2013.

R. H. W. Hoppe. Optimization theory. Technical report, University of Houston, 2007.

H. Hotelling. The economics of exhaustible resources. *Journal of Political Economy*, 39:137–175, 1931.

S. W. Hur, S. H. Lee, Y. H. Nam, and C. J. Kim. Direct dynamic-simulation approach to trajectory optimization. *Chinese Journal of Aeronautics*, 2021.

M. L. Kamien and N. L. Schwartz. *Dynamic Optimization - The calculus of variations and optimal control in Economics and Management*. North-Holland, Amsterdam, 2nd edition, 1991.

A. Kaufmann and R. Cruon. *La programmation dynamique. Gestion Scientifique Séquentielle*. Dunod, Paris, 1965.

J. O. S. Kennedy. Principles of dynamic optimization in resource management. *Agricultural Economics*, 2:57–72, 1988.

D. E. Kirk. *Optimal control theory. An introduction*. Prentice-Hall, Englewood Cliffs, New Jersey, 1970.

L. Kleinrock. *Queueing Systems. Volume I: Theory*. Wiley, New York, 1975.

Y. D. Kwon and L. B. Evans. A coordinate transformation method for the numerical solution of non-linear minimum-time control problems. *AIChE J.*, 21:1158–, 1975.

F. Lamnabhi-Lagarrigue. Singular optimal control problems: on the order of a singular arc. *Systems & control letters*, 9:173–182, 1987.

M. A. Latifi, J. P. Corriou, and M. Fikar. Dynamic optimization of chemical processes. *Trends in Chem. Eng.*, 4:189–201, 1998.

E. B. Lee and L. Markus. *Foundations of Optimal Control Theory*. Krieger, Malabar, Florida, 1967.

O. Levenspiel. *Chemical Reaction Engineering*. Wiley, New York, 3rd edition, 1999.

R. Luus. Application of dynamic programming to high-dimensional nonlinear optimal control systems. *Int. J. Cont.*, 52(1):239–250, 1990.

R. Luus. Application of iterative dynamic programming to very high-dimensional systems. *Hung. J. Ind. Chem.*, 21:243–250, 1993.

R. Luus. Optimal control of bath reactors by iterative dynamic programming. *J. Proc. Cont.*, 4(4):218–226, 1994.

R. Luus. Numerical convergence properties of iterative dynamic programming when applied to high dimensional systems. *Trans. IChemE, part A*, 74:55–62, 1996.

R. Luus and B. Bojkov. Application of iterative dynamic programming to time-optimal control. *Chem. Eng. Res. Des.*, 72:72–80, 1994.

R. Luus and D. Hennessy. Optimization of fed-batch reactors by the Luus-Jaakola optimization procedure. *Ind. Eng. Chem. Res.*, 38:1948–1955, 1999.

W. Mekarapiruk and R. Luus. Optimal control of inequality state constrained systems. *Ind. Eng. Chem. Res.*, 36:1686–1694, 1997.

S. Minner. Multiple-supplier inventory models in supply chain management: A review. *Int. J. Production Economics*, 81–82:265–279, 2003.

T. Mitra. *Optimization and Chaos*, chapter Introduction to Dynamic Optimization Theory. Springer-Verlag, Heidelberg, 2000.

A. Perea. Backward induction versus forward induction reasoning. *Games*, 1:168–188, 2010.

A. Perea. *Epistemic Game Theory: Reasoning and Choice*. Cambridge University Press, Cambridge, 2012.

L. Pontryaguine, V. Boltianski, R. Gamkrelidze, and E. Michtchenko. *Théorie mathématique des processus optimaux*. Mir, Moscou, 1974. Edition Française.

L. Pun. *Introduction à la pratique de l'optimisation*. Dunod, Paris, 1972.

M. L. Puterman. *Markov Decision Processes: Discrete Stochastic Dynamic Programming*. Wiley, Chichester, 1994.

E. P. Ramsey. A mathematical theory of saving. *Economic Journal*, 38:543–559, 1928.

W. H. Ray and J. Szekely. *Process Optimization with applications in metallurgy and chemical engineering*. John Wiley, New York, 1973.

S. M. Roberts. *Dynamic Programming in Chemical Engineering and Process Control*. Academic Press, New York, 1964.

S. M. Roberts and C. G. Laspe. Computer control of a thermal cracking reaction. *Ind. Eng. Chem.*, 53(5):343–348, 1961.

K. Schittkowski. NLPQL: a Fortran subroutine solving constrained nonlinear programming problems. *Annals of Operations Research*, 5:485–500, 1985.

L. I. Sennott. *Stochastic Dynamic Programming and the Control of Queueing Systems*. Wiley, New York, 1999.

S. P. Sethi and G. Thompson. *Optimal Control Theory: Applications to Management Science and Economics*. Springer-Verlag, 2000.

R. Soeterboek. *Predictive Control - A Unified Approach*. Prentice Hall, Englewood Cliffs, New Jersey, 1992.

R. F. Stengel. *Optimal control and estimation*. Courier Dover Publications, 1994.

V. Sundarapandian. *Probability, Statistics and Queueing Theory*. PHI Learning, 2009.

K. L. Teo, C. J. Goh, and K. H. Wong. *A Unified Computational Approach to Optimal Control Problems*. Wiley and Sons, Inc., New York, 1991.

T. Todorova. *Mathematics for Economists: Problems Book*. Wiley-Blackwell, Hoboken, 2011.

J. A. Vernon and W. K. Hughen. A primer on dynamic optimization and optimal control in pharmacoeconomics. *Value in Health*, 9(2):106–113, 2006.

Index

J.-P. Corriou, *Numerical Methods and Optimization*, Springer Optimization and Its Applications 187, https://doi.org/10.1007/978-3-030-89366-8

N

O

Printed in the United States
by Baker & Taylor Publisher Services